hf	hyperfine
hfDC	hyperfine-decoupling
HMQC	heteronuclear multiple-quantum coherence
HPC	hydroxypropyl cellulose
HYPSO	hybrid polarizing solid
iCE	indirect cross effect
idHETCOR	indirectly detected heteronuclear correlation
INEPT	insensitive nuclei enhanced by polarization transfer
ISE	integrated solid effect
LAC	level anti-crossing
LAr	liquid argon
LCST	lowest critical solution temperature
LDH	lactate dehydrogenase
LEED	low-energy electron diffraction
lHe	liquid helium
LO	local oscillator
LPS	lipopolysaccharides
LT	low temperature
LT-MAS	low-temperature magic-angle spinning
LWHH	line width at half height
LZ	Landau–Zener
MAOSS	magic angle-oriented sample spinning
MAS	magic-angle spinning
MAS NMR	magic-angle spinning nuclear magnetic resonance
MAT	magic-angle turning
MC	multiple-contact
MFC	mass flow controller
MLF	Met-Leu-Phe-OH
MOF	metal-organic framework
MOP	mesoporous organic polymer
MP	mercaptopropyl
MQ	multiple-quantum
MR	magnetic resonance
MRS	magnetic resonance spectroscopy
MRSI	magnetic resonance spectroscopic imaging
MS	mass spectrometry
MSN	mesoporous silica nanoparticle
MW	microwave
NA	natural abundance
nc-AFM	noncontact atomic force microscopy
NMR	nuclear magnetic resonance
NMRD	nuclear magnetic relaxation dispersion
NOE	nuclear Overhauser effect
NOVEL	nuclear orientation via electron spin locking
NPS	nano-phase separation
NRF	nuclear rotating frame
Nups	nucleoporins
OE	Overhauser effect
OE-DNP	Overhauser enhancement DNP

OGT	
OMPs	proteins
OTP	ortho-terphenyl
PA	polarizing agent
PAGE	polyacrylamide gel electrophoresis
PARP	poly-ADP-ribose-polymerase
PBG	photonic bandgap
PDH	pyruvate dehydrogenase
PET	positron emission tomography
PFG	pulsed field gradient
PG	peptidoglycan
Ph	phenyl
PHIP	para-hydrogen induced polarization
PID	proportional-integral-derivative
PISEMA	polarization inversion spin exchange at the magic angle
PPP	pentose phosphate pathway
PR	proteorhodopsin
PRE	paramagnetic relaxation enhancement
PRESTO	phase-shifted recoupling effects a smooth transfer of polarization
PRESTO-QCPMG	phase-shifted recoupling effects a smooth transfer of order–quadrupolar Carr-Purcell-Meil-Gibson
PrIm	imidazolium-containing materials
PSA	pressure-swing adsorption
PVA	polyvinyl alcohol
QC	quality control
QCPMG	quadrupolar Carr-Purcell-Meiboom-Gill
QD	quantum dot
QI	quadrupolar interaction
qp	quasiperiodic
RA-NOVEL	ramped-amplitude nuclear orientation via electron spin locking
REDOR	rotational echo double-resonance
RESPDOR	rotational-echo saturation pulse double-resonance
RF	radio frequency
RFDR	radio frequency-driven recoupling
SE	solid effect
SEI	solid-electrolyte interphase
SENS	surface enhanced NMR spectroscopy
SMMW	submillimeter wave
SNR	signal-to-noise ratio
SOMO	singly occupied molecular orbital
SOSO	spinning-on/spinning-off
SP	square pyramidal geometry
SQ	single-quantum
SS	solid-state
SSE	stretched solid effect
ssNMR	solid-state nuclear magnetic resonance
STM	scanning tunneling microscopy
T4SScc	type 4 secretion system core complex
TAM	tetrathiatriarymethyl

TBI	traumatic brain injury	TWT	traveling wave tube
TBP	trigonal bipyramidal geometry	UDP-GlcNAc	uridine 5′-diphospho-N-acetylglucosamine
TC	thermally controlled		
TCA	tricarboxylic acid	VCO	voltage-controlled oscillator
TCE	1,1,2,2-tetrachloroethane	VDI	Virginia Diodes Inc
TE	transverse electric	VED	vacuum-electron device
TEMPO	2,2,6,6-tetramethylpiperidine-1-oxyl	VOCS	variable offset cumulative spectrum
		VT	variable temperature
TEMPONE	4-oxo-2,2,6,6-tetramethylpiperidine	VTI	variable temperature insert
		WC	whole cell
TEOS	tetraethylorthosilicate	wPDLF	windowed proton-detected local field
TM	thermal mixing	WURST	wideband, uniform-rate, and smooth-truncation
TMS	trimethylsilyl groups		
TOP-DNP	time-optimized pulsed dynamic nuclear polarization	YIG	yttrium iron garnet
		ZBD	zero-biased Schottky diode
TRIS	tris(hydroxymethyl) aminomethane	ZQ	zero-quantum

Handbook of High Field Dynamic Nuclear Polarization

eMagRes Books

eMagRes (formerly the *Encyclopedia of Magnetic Resonance*) publishes a wide range of online articles on all aspects of magnetic resonance in physics, chemistry, biology and medicine. The existence of this large number of articles, written by experts in various fields, is enabling the publication of a series of *eMagRes* Books – handbooks on specific areas of NMR and MRI. The chapters of each of these handbooks will comprise a carefully chosen selection of *eMagRes* articles.

Published *eMagRes* Books

NMR Crystallography
Edited by Robin K. Harris, Roderick E. Wasylishen,
   Melinda J. Duer
ISBN 978-0-470-69961-4

Multidimensional NMR Methods for the Solution State
Edited by Gareth A. Morris, James W. Emsley
ISBN 978-0-470-77075-7

Solid-State NMR Studies of Biopolymers
Edited by Ann E. McDermott, Tatyana Polenova
ISBN 978-0-470-72122-3

NMR of Quadrupolar Nuclei in Solid Materials
Edited by Roderick E. Wasylishen, Sharon E. Ashbrook,
   Stephen Wimperis
ISBN 978-0-470-97398-1

RF Coils for MRI
Edited by John T. Vaughan, John R. Griffiths
ISBN 978-0-470-77076-4

*MRI of Tissues with Short T_2s or T_2*s*
Edited by Graeme M. Bydder, Gary D. Fullerton,
   Ian R. Young
ISBN 978-0-470-68835-9

NMR Spectroscopy: A Versatile Tool for Environmental Research
Edited by Myrna J. Simpson, André J. Simpson
ISBN 978-1-118-61647-5

NMR in Pharmaceutical Sciences
Edited by Jeremy R. Everett, Robin K. Harris, John C. Lindon,
Ian D. Wilson
ISBN 978-1-118-66025-6

Handbook of Magnetic Resonance Spectroscopy In Vivo:
   MRS Theory, Practice and Applications
Edited by Paul A. Bottomley, John R. Griffiths
ISBN 978-1-118-99766-6

EPR Spectroscopy: Fundamentals and Methods
Edited by Daniella Goldfarb, Stefan Stoll
ISBN 978-1-119-16299-5

Handbook of High Field Dynamic Nuclear
   Polarization
Edited by Vladimir K. Michaelis, Robert G. Griffin,
   Björn Corzilius, Shimon Vega
ISBN 978-1-119-44164-9

Forthcoming *eMagRes* Books

Handbook of Safety and Biological Aspects in MRI
Edited by Devashish Shrivastava, John T. Vaughan
ISBN 978-1-118-82130-5

eMagRes

Edited by Sharon Ashbrook, Bella Bode, George A. Gray, John R. Griffiths, Tatyana Polenova, Roberta Pieratelli, Thomas Prisner, André J. Simpson, Myrna J. Simpson.

eMagRes (formerly the *Encyclopedia of Magnetic Resonance*) is based on the original publication of the *Encyclopedia of Nuclear Magnetic Resonance*, first published in 1996 with an updated volume added in 2000. The *Encyclopedia of Magnetic Resonance* was launched in 2007 online with all the existing published material, and was later relaunched as *eMagRes* in 2013. *eMagRes* captures every aspect of the interdisciplinary nature of magnetic resonance, providing all the essential information on the science, methodologies, engineering, technologies, applications, and the history of magnetic resonance, whilst encompassing a whole range of techniques, including MRI, MRS, NMR, and EPR/ESR.

For more information see: http://www.wileyonlinelibrary.com/ref/eMagRes.

Handbook of High Field Dynamic Nuclear Polarization

Editors

Vladimir K. Michaelis
University of Alberta, Edmonton, Alberta, Canada

Robert G. Griffin
Massachusetts Institute of Technology, Cambridge, MA, USA

Björn Corzilius
University of Rostock, Rostock, Germany

Shimon Vega
Weizmann Institute of Science, Rehovot, Israel

WILEY

Library of Congress Cataloging-in-Publication Data is available for this title.

978-1-119-44164-9 (hardback)

A catalogue record for this book is available from the British Library.

Cover Design: Wiley
Cover Images: Courtesy of Alessandra Lucini Paioni, Marie A.M. Renault and
Marc Baldus; Courtesy of Marina Bennati and Tomas Orlando; Courtesy of
Guy M. Bernard and Vladimir K. Michaelis; Background © Kjpargeter/Shutterstock

Set in 9.5/11.5 pt TimesLTStd by SPi Global, Chennai, India
Printed and bound in Singapore by Markono Print Media Pte Ltd

10 9 8 7 6 5 4 3 2 1

International Advisory Board

Gareth A. Morris
University of Manchester
Manchester
UK

C. Leon Partain
Vanderbilt University Medical
   Center
Nashville, TN
USA

Alexander Pines
University of California
   at Berkeley
Berkeley, CA
USA

George K. Radda
University of Oxford
Oxford
UK

Hans Wolfgang Spiess
Max-Planck Institute
   of Polymer Research
Mainz
Germany

**Charles P. Slichter
(deceased)**
University of Illinois
   at Urbana-Champaign
Urbana, IL
USA

John S. Waugh (deceased)
Massachusetts Institute
   of Technology (MIT)
Cambridge, MA
USA

**Bernd Wrackmeyer
(deceased)**
Universität Bayreuth
Bayreuth
Germany

Kurt Wüthrich
The Scripps Research
   Institute
La Jolla, CA
USA

Contents

Part B: Applications   289

Contributors

Jan H. Ardenkjaer-Larsen

Technical University of Denmark, Lyngby, Denmark; GE Healthcare, Brøndby, Denmark
Chapter 12: Introduction to Dissolution DNP: Overview, Instrumentation, and Human Applications

Marc Baldus

Utrecht University, Utrecht, The Netherlands
Chapter 15: DNP and Cellular Solid-state NMR

Burkhard Bechinger

Institut de Chimie, Université de Strasbourg/CNRS, UMR7177, Strasbourg, France
Chapter 17: DNP Solid-state NMR of Biological Membranes

Johanna Becker-Baldus

Biophysical Chemistry & Centre for Biomolecular Magnetic Resonance, Goethe University Frankfurt, Frankfurt, Germany
Chapter 16: Cryo-trapped Intermediates of Retinal Proteins Studied by DNP-enhanced MAS NMR Spectroscopy

Marina Bennati

Max Planck Institute for Biophysical Chemistry, Göttingen, Germany; University of Göttingen, Göttingen, Germany
Chapter 14: Overhauser DNP in Liquids on ^{13}C Nuclei

Guy M. Bernard

Gunning-Lemieux Chemistry Centre, University of Alberta, Edmonton, Alberta, Canada
Chapter 7: Instrumentation for High-field Dynamic Nuclear Polarization NMR Spectroscopy

Pierrick Berruyer

Institut des Sciences Analytiques, UMR 5280, University of Lyon, CNRS, Université Claude Bernard Lyon 1, ENS Lyon, Villeurbanne, France
Chapter 18: DNP in Materials Science: Touching the Surface

Monica Blank

Communications and Power Industries (CPI), Palo Alto, CA, USA
Chapter 8: Millimeter-wave Sources for DNP-NMR

Gilles Casano

CNRS, ICR, Aix Marseille University, Marseille, France
Chapter 5: Polarizing Agents: Evolution and Outlook in Free Radical Development for DNP

Arnaud Comment

General Electric Healthcare, Chalfont St Giles, UK
Chapter 21: In Vivo Hyperpolarized ^{13}C MRS and MRI Applications

Björn Corzilius   *University of Rostock, Rostock, Germany*
Chapter 6: Paramagnetic Metal Ions for Dynamic Nuclear
Polarization

Vasyl P. Denysenkov   *Goethe University, Frankfurt-am-Main, Germany*
Chapter 13: Liquid-state Overhauser DNP at High Magnetic Fields

Lyndon Emsley   *Institut des Sciences et Ingénierie Chimiques, Ecole Polytechnique
Fédérale de Lausanne (EPFL), Lausanne, Switzerland*
Chapter 18: DNP in Materials Science: Touching the Surface
Chapter 20: DNP-enhanced Solid-state NMR Spectroscopy of Active
Pharmaceutical Ingredients

Akiva Feintuch   *Chemical and Biological Physics Department, Weizmann Institute of
Science, Rehovot, Israel*
Chapter 2: DNP Mechanisms

Kevin L. Felch   *Communications and Power Industries (CPI), Palo Alto, CA, USA*
Chapter 8: Millimeter-wave Sources for DNP-NMR

Toshimichi Fujiwara   *Institute for Protein Research, Osaka University, Suita, Osaka, Japan*
Chapter 9: Cryogenic Platforms and Optimized DNP Sensitivity

Clemens Glaubitz   *Biophysical Chemistry & Centre for Biomolecular Magnetic Resonance, Goethe University Frankfurt, Frankfurt, Germany*
Chapter 16: Cryo-trapped Intermediates of Retinal Proteins Studied
by DNP-enhanced MAS NMR Spectroscopy

Robert G. Griffin   *Department of Chemistry, Francis Bitter Magnet Laboratory, Massachusetts Institute of Technology, Cambridge, MA, USA*
Chapter 3: Pulsed Dynamic Nuclear Polarization

Songi Han   *Department of Chemistry and Biochemistry, University of California
Santa Barbara, Santa Barbara, CA, USA; Department of Chemical
Engineering, University of California Santa Barbara, Santa Barbara,
CA, USA*
Chapter 10: Versatile Dynamic Nuclear Polarization Hardware with
Integrated Electron Paramagnetic Resonance Capabilities

Sabine Hediger   *Univ. Grenoble Alpes, CEA, CNRS, INAC-MEM, Grenoble, France*
Chapter 4: MAS-DNP Enhancements: Hyperpolarization,
Depolarization, and Absolute Sensitivity

Sami Jannin   *Université de Lyon, CNRS, Université Claude Bernard Lyon 1, ENS
de Lyon, Institut des Sciences Analytiques, UMR 5280, Villeurbanne,
France*
Chapter 11: Dissolution Dynamic Nuclear Polarization Methodology
and Instrumentation

Sudheer Jawla   *Plasma Science and Fusion Center, Massachusetts Institute of Technology, Cambridge, MA, USA*
Chapter 3: Pulsed Dynamic Nuclear Polarization

Ilia Kaminker

Department of Chemistry and Biochemistry, University of California Santa Barbara, Santa Barbara, CA, USA
Chapter 10: Versatile Dynamic Nuclear Polarization Hardware with Integrated Electron Paramagnetic Resonance Capabilities

Hakim Karoui

CNRS, ICR, Aix Marseille University, Marseille, France
Chapter 5: Polarizing Agents: Evolution and Outlook in Free Radical Development for DNP

Takeshi Kobayashi

U.S. DOE Ames Laboratory, Ames, IA, USA
Chapter 19: Growing Signals from the Noise: Challenging Nuclei in Materials DNP

Walter Köckenberger

University of Nottingham, Nottingham, UK
Chapter 22: Dissolution Dynamic Nuclear Polarization

Krishnendu Kundu

Chemical and Biological Physics Department, Weizmann Institute of Science, Rehovot, Israel
Chapter 2: DNP Mechanisms

Dennis Kurzbach

Laboratoire des biomolécules, LBM, Département de chimie, École normale supérieure, PSL University, Sorbonne Université, CNRS, Paris, France
Chapter 11: Dissolution Dynamic Nuclear Polarization Methodology and Instrumentation

Alisa Leavesley

Department of Chemistry and Biochemistry, University of California Santa Barbara, Santa Barbara, CA, USA
Chapter 10: Versatile Dynamic Nuclear Polarization Hardware with Integrated Electron Paramagnetic Resonance Capabilities

Daniel Lee

Univ. Grenoble Alpes, CEA, CNRS, INAC-MEM, Grenoble, France
Chapter 4: MAS-DNP Enhancements: Hyperpolarization, Depolarization, and Absolute Sensitivity

Anne Lesage

Institut des Sciences Analytiques, UMR 5280, University of Lyon, CNRS, Université Claude Bernard Lyon 1, ENS Lyon, Villeurbanne, France
Chapter 18: DNP in Materials Science: Touching the Surface

Alessandra Lucini Paioni

Utrecht University, Utrecht, The Netherlands
Chapter 15: DNP and Cellular Solid-state NMR

Irene Marco-Rius

Cancer Research UK Cambridge Institute, University of Cambridge, Cambridge, UK
Chapter 21: In Vivo Hyperpolarized ^{13}C MRS and MRI Applications

Yoh Matsuki

Institute for Protein Research, Osaka University, Suita, Osaka, Japan
Chapter 9: Cryogenic Platforms and Optimized DNP Sensitivity

Frédéric Mentink-Vigier
National High Magnetic Field Laboratory, Florida State University, Tallahassee, FL, USA
Chapter 2: DNP Mechanisms
Chapter 4: MAS-DNP Enhancements: Hyperpolarization, Depolarization, and Absolute Sensitivity

Vladimir K. Michaelis
Gunning-Lemieux Chemistry Centre, University of Alberta, Edmonton, Alberta, Canada
Chapter 7: Instrumentation for High-field Dynamic Nuclear Polarization NMR Spectroscopy

Tomas Orlando
Max Planck Institute for Biophysical Chemistry, Göttingen, Germany
Chapter 14: Overhauser DNP in Liquids on ^{13}C Nuclei

Olivier Ouari
CNRS, ICR, Aix Marseille University, Marseille, France
Chapter 5: Polarizing Agents: Evolution and Outlook in Free Radical Development for DNP

Gaël De Paëpe
Univ. Grenoble Alpes, CEA, CNRS, INAC-MEM, Grenoble, France
Chapter 4: MAS-DNP Enhancements: Hyperpolarization, Depolarization, and Absolute Sensitivity

Frédéric A. Perras
U.S. DOE Ames Laboratory, Ames, IA, USA
Chapter 19: Growing Signals from the Noise: Challenging Nuclei in Materials DNP

Arthur C. Pinon
Institut des Sciences et Ingénierie Chimiques, Ecole Polytechnique Fédérale de Lausanne (EPFL), Lausanne, Switzerland
Chapter 20: DNP-enhanced Solid-state NMR Spectroscopy of Active Pharmaceutical Ingredients

Thomas F. Prisner
Goethe University, Frankfurt-am-Main, Germany
Chapter 13: Liquid-state Overhauser DNP at High Magnetic Fields

Marek Pruski
U.S. DOE Ames Laboratory, Ames, IA, USA; Department of Chemistry, Iowa State University, Ames, IA, USA
Chapter 19: Growing Signals from the Noise: Challenging Nuclei in Materials DNP

Marie A.M. Renault
Utrecht University, Utrecht, The Netherlands
Chapter 15: DNP and Cellular Solid-state NMR

Aaron J. Rossini
Iowa State University, Ames, IA, USA; US DOE Ames Laboratory, Ames, IA, USA
Chapter 20: DNP-enhanced Solid-state NMR Spectroscopy of Active Pharmaceutical Ingredients

Charles P. Slichter
University of Illinois at Urbana-Champaign, Urbana, IL, USA
Chapter 1: The Discovery and Demonstration of Dynamic Nuclear Polarization–A Personal and Historical Account

Kong Ooi Tan
Department of Chemistry, Francis Bitter Magnet Laboratory, Massachusetts Institute of Technology, Cambridge, MA, USA
Chapter 3: Pulsed Dynamic Nuclear Polarization

Richard J. Temkin   *Plasma Science and Fusion Center, Massachusetts Institute of Technology, Cambridge, MA, USA*
Chapter 3: Pulsed Dynamic Nuclear Polarization

Shimon Vega   *Chemical and Biological Physics Department, Weizmann Institute of Science, Rehovot, Israel*
Chapter 2: DNP Mechanisms

Li Zhao   *Iowa State University, Ames, IA, USA; US DOE Ames Laboratory, Ames, IA, USA*
Chapter 20: DNP-enhanced Solid-state NMR Spectroscopy of Active Pharmaceutical Ingredients

Series Preface

The *Encyclopedia of Nuclear Magnetic Resonance* was originally published in eight volumes in 1996, in part to celebrate the fiftieth anniversary of the first publications describing the discovery of NMR (nuclear magnetic resonance) in January 1946. Volume 1 contained a historical overview and 200 articles by prominent NMR practitioners, while the remaining seven volumes consisted of 500 articles on a wide variety of topics in NMR, including MRI (magnetic resonance imaging). A ninth volume was brought out in 2000 and two "spin-off" volumes incorporating the articles on MRI and MRS (together with some new ones) were published in 2002. In 2006, the decision was made to publish all the articles electronically with the resulting Encyclopedia becoming available online in 2007. Since then, new articles have been published online every three months and many of the original articles have been updated. To recognize the fact that the *Encyclopedia of Magnetic Resonance* is a true online resource, the website was redesigned and new functionalities added, with a relaunch in January 2013 in a new volume and issue format, under the new name *eMagRes*. In December 2012, a new print edition of the *Encyclopedia of Nuclear Magnetic Resonance* was published in 10 volumes (6200 pages). This much needed update of the 1996 edition of the Encyclopedia encompassed the entire field of NMR.

As part of the development of *eMagRes*, a series of printed handbooks on specific areas of magnetic resonance has been introduced. The handbooks are planned in advance by specially selected editors and new articles written to give appropriate complete coverage of the subject area. The handbooks are intended to be of value and interest to research students, postdoctoral fellows, and other researchers learning about the topic in question and undertaking relevant experiments, whether in academia or industry.

This new handbook presents the basic principles and current practice of high-field DNP, a technique that is revolutionizing the field of NMR because of the sensitivity gains it affords. I wish to thank four experts in this research area: Björn Corzilius, Shimon Vega, Robert G. Griffin, and last but not least my colleague, Vladimir K. Michaelis for their efforts in putting together this authoritative handbook that I think will be essential reading for scientists wishing to understand the principles and practice of dynamic nuclear polarization. Finally, I wish to thank Jenny Cossham at Wiley for her assistance and patience in helping editors of this and previous handbooks associated with *eMagRes*.

Roderick E. Wasylishen
September 2019

Consult the *eMagRes* website at www. wileyonlinelibrary.com/ref/eMagRes for details of all our *eMagRes* handbooks.

Preface

Because it offers non-perturbing, site-specific resolution, and high-precision structural data, NMR is the method of choice to address a variety of chemical, physical, and biological problems. Nevertheless, the results that can be gleaned from NMR spectra are often limited by the low sensitivity of the experiment. In fact, limited signal-to-noise in spectroscopy, microscopy, and imaging experiments is the single factor that most often determines the success or failure of any application. The origin of the limited sensitivity in NMR is well known and originates in the small magnetic moment of nuclear spins, which leads to vanishing Boltzmann polarizations. In order to address the sensitivity problem, spectroscopists have invented many ingenious technological and methodological schemes that have shifted the boundaries of what is achievable and frequently opened new directions of research. The introduction of Fourier transform spectroscopy, which increased sensitivity $\sim 10^2$-fold and revolutionized NMR and other forms of spectroscopy, is an excellent example of these efforts.

A more direct approach to significantly impact the sensitivity of NMR is to address the fundamental problem of increasing the Boltzmann polarization of the nuclear spin reservoir. The straightforward, brute force solution is to increase the magnetic field, B_0, and this provides the rationale for the continuing development of higher field superconducting magnets, most recently with high-temperature superconductors. Another approach to enhancing polarization is to couple the nuclear spins of interest, say ^{1}H, to a reservoir of unpaired electrons that have a much higher polarization. This can then lead to an enhancement of the nuclear spin polarization by a factor $\gamma_e/\gamma_H \sim 660$, and dynamic nuclear polarization (DNP), which is the subject of this handbook, documents recent progress using this approach. Of course, this strategy is not new – it was originally proposed in 1953 by Overhauser and was first experimentally demonstrated by Carver and Slichter in lithium metal and in liquid

sodium-ammonia solutions, two systems that satisfy the criteria outlined by Overhauser, that the system has mobile electrons. The experiment involved irradiating EPR transitions, which then led to dynamic nuclear polarization of the ^{7}Li or ^{1}H spins. Thus, DNP is not a new area of scientific endeavor, but rather a latent idea that was explored at low fields during the 1950s and 1960s and is now undergoing a renaissance in transitioning from low to high fields and millimeter-wave frequencies.

Following the pioneering work of Overhauser and Carver and Slichter, the primary application of DNP was the production of polarized targets for nuclear scattering experiments. Although these experiments were performed primarily at low fields, they resulted in delineating and understanding the underlying physics of the polarization transfer mechanisms employed today. For example, in insulating solids with localized paramagnetic centers, which are the samples of interest in many contemporary experiments, the mechanisms that are operative are the solid effect, cross effect, or thermal mixing, which couple the nuclear spin(s) to one, two, or more electron spins, respectively. The dominant mechanism depends on the characteristics of the EPR spectrum of the polarizing agent. Furthermore, spin thermodynamic and quantum mechanical treatments were developed that describe these three mechanisms and predict transfer efficiencies that scale as ω_0^{-n}, where ω_0 is the Larmor frequency and $n = 1-2$. As a consequence, in the 1950s and 1960s there was little impetus to move DNP to higher magnetic fields or frequencies since the polarization enhancements would be attenuated. Thus, this field/frequency dependence, together with the dearth of high-frequency microwave sources operating above ~ 40 GHz, effectively relegated DNP to a position of an interesting scientific curiosity.

During the 1980s renewed interest in DNP materialized, stimulated by the development of magic angle spinning (MAS) experiments to detect ^{13}C, ^{15}N and other low abundance $I = 1/2$ nuclei. Accordingly, the labs of Wind, Schaefer and Yannoni reported DNP experiments at 40 GHz/60 MHz devoted primarily to studies of polymers. However, again because of the paucity of high-frequency microwave sources, these experiments were restricted to low fields and did not propagate beyond these laboratories. Subsequently, two developments changed the direction of DNP research and led us to the situation as it exists today and stimulated the preparation of this *Handbook of High Field Dynamic Nuclear Polarization.*

First, in 1993 and 1995 the two initial reports of 'high-frequency' DNP experiments (at the time 140 GHz/211 MHz for ^{1}H) utilizing gyrotrons appeared. Because gyrotrons are fast wave devices, they can produce $\sim$100 W of microwave power and their operating frequencies are determined by the cavity dimensions and the gyrotron magnetic field. Gyrotrons have operated up to $\sim$1 THz and therefore provide the possibility of performing DNP up to $\sim$1500 MHz for ^{1}H, with the highest frequency system currently functioning at 593 GHz/900 MHz. Second, in 2003 the first reports of 'dissolution DNP' (dDNP) appeared, which is an approach where the sample is polarized at low temperatures, $\sim$1.2 K, and then rapidly melted and transferred to another magnet for spectroscopy. The initial experiments were intended for applications of metabolic imaging, but, as can be seen in the reports below, a number of different application areas have emerged. Although progress in developing high-field DNP was slow at first, these initial reports provide platforms to test new experimental ideas. For example, biradical polarizing agents, where two TEMPO or similar molecules are tethered by an organic chain, were reported in 2004 and have become the polarizing agent of choice for cross-effect DNP. In addition, low-temperature MAS probes were developed first to perform experiments at $\sim$90 K and more recently in the range of $\sim$30 K. The stability and power output of the gyrotrons have improved substantially. Finally in the last few years, there have been reports of time-domain or pulsed DNP experiments. In the dDNP arena, sample handling techniques have improved – melting and transfer protocols are more efficient – and many new areas of application have emerged as a consequence. Two recent examples are studies of hyperpolarized water and its interactions with proteins and studies of cancerous tissue.

The handbook begins with a historical introduction written by the late C.P. Slichter that was published previously in a 2010 collection of DNP papers in Physical Chemistry Chemical Physics. It provides much insight into the origins of DNP and the people who were involved with the original experiments. Following Slichter are chapters devoted to a description of the fundamentals of DNP, instrumentation, dissolution DNP, DNP of liquids, and several chapters covering important applications of the techniques. We hope that these chapters, which cover the essential features of the field, are useful to both the novice and seasoned practitioner of the art.

Vladimir K. Michaelis
University of Alberta, Edmonton, Alberta, Canada

Robert G. Griffin
Massachusetts Institute of Technology, Cambridge, MA, USA

Björn Corzilius
University of Rostock, Rostock, Germany

Shimon Vega
Weizmann Institute of Science, Rehovot, Israel

August 2019

Acknowledgments

The editors would like to acknowledge the many authors who have contributed their time and energy assisting in bringing a unique handbook to the DNP community. The editors thank Professor Roderick E. Wasylishen who commissioned the content for this book as Editor-in-Chief of *eMagRes* and who is now co-chairman of the advisory board through Wiley. The editors dedicate this book to the memory of Professor Charles P. Slichter in acknowledgment of his pioneering efforts in NMR spectroscopy. Finally, the many past, present, and future colleagues and students who will continue to advance the DNP research field are thanked for their efforts.

PART A
Concepts, Theory, & Instrumentation

The discovery and demonstration of dynamic nuclear polarization—a personal and historical account

Charles P. Slichter

DOI: 10.1039/c003286g

The paper gives some background about prewar studies of paramagnetic relaxation, the author's postwar measurements of paramagnetic relaxation using 3 cm microwaves, the beginning studies of alkali metal NMR at Illinois, Overhauser's arrival and proposal, the discovery of the ESR of conduction electrons, the experiment confirming his idea in Li and Na metals, then in solutions of Na atoms in liquid ammonia, and the realization that a form of Overhauser effect might be seen when several isotopes were present in cases where atoms were mobile. Two other experiments at Illinois inspired by the Overhauser experiments are briefly described: measurement of the spin susceptibility of conduction electrons and measurement of spin–lattice relaxation times in superconductors.

Introduction

The editors have kindly invited me to write a short historical article about the discovery and first demonstration of dynamic polarization of nuclei: the Overhauser effect. My goal is to give some of the background to this themed issue of *PCCP* through some remarks about the delightful and close association my students and I had with Al Overhauser, the excitement his ideas unleashed, and especially the experiment my talented student Tom Carver and I did. To do so, I need to start with my time as a graduate student working for my PhD, where I first encountered magnetic relaxation, magnetic resonance, and concepts like spin temperature, since that time period had a profound effect on how we designed our test of Overhauser's amazing proposal. Because of Tom's untimely death in 1981, I write on behalf of us both. It is a joy to recall the very warm friendships and stimulating collaborations that were central to the story.

Graduate school

I entered graduate school at Harvard in the fall of 1946. My undergraduate

adviser had been J. H. Van Vleck, whom everyone referred to simply as "Van". He told me that Harvard had acquired an especially strong faculty in physics and urged me to stay at Harvard for graduate work. Of course, Van was famous for his pioneering work on the theory of electric and magnetic susceptibilities. One day in February of my first graduate school year, he stopped me in the hallway to say that he had a recommendation for a thesis adviser and a thesis topic. He suggested that I ask Professor Edward Purcell to be my adviser, and that I should propose using magnetic resonance of electron spins to study paramagnetic salts. He explained that Purcell was a great expert on these materials since, when a graduate student, he had coauthored a paper with Malcolm Hebb on their theory.[1]

I did not know who Purcell was, what magnetic resonance was, or much about paramagnetic salts, but I knew enough to take Van's advice! Of course, Purcell, Pound, and Torrey had discovered nuclear magnetic resonance just a little over a year previously. I went to Purcell, quite nervous, to ask him to be my thesis adviser to work on the topic Van had suggested. To my delight he immediately said yes. I now realize that almost surely Van had already discussed the whole idea of the project with Purcell and probably said that he had a suggestion

of someone to work on it before Van even mentioned it to me.

In the 1930s, Van had closely followed all the experiments on paramagnetic materials. He developed much of their theoretical understanding. In an important article in 1940 on the theory of the relaxation times of Cr and Ti alum,[23] he wrote: "A noteworthy series of experiments on magnetic absorption and dispersion in paramagnetic media at radio and lower frequencies have been performed within the last few years by Gorter and other Dutch physicists". Those experiments involved measuring the frequency-dependent real and imaginary parts of the magnetic susceptibility to deduce the spin–lattice and spin–spin relaxation properties of various materials. Van's paper was concerned with the theory of the spin–lattice relaxation time, τ, and the spin–spin relaxation time, τ_s, measured by these methods. Van goes on to say that "the most elegant way of determining τ is furnished by the thermodynamic treatment of Casimir and du Pré, the essence of which is as follows: if the impressed frequency ω is small compared with $1/\tau_s$, the spins are in thermodynamic equilibrium with each other so that it makes sense to talk of a "spin temperature" T_s. The latter, however, will not be the same as the lattice temperature if ω is comparable

Department of Physics, University of Illinois at Urbana-Champaign, Urbana, IL 61801-3080, USA.

Handbook of High Field Dynamic Nuclear Polarization, Edited by Vladimir K. Michaelis, Robert G. Griffin, Björn Corzilius and Shimon Vega, ISBN: 978-1-119-44164-9
Original Publication: Charles P. Slichter, *Physical Chemistry Chemical Physics* 2010, **12**, 5741-5751.
DOI: 10.1039/c003286g, Reproduced by permission of The Royal Society of Chemistry.

with, or larger than, $1/\tau$. Obviously the magnetic susceptibility will be of the isothermal variety in the region $\omega \ll 1/\tau$, where spin and lattice are in equilibrium, but becomes adiabatic when $\omega \sim 1/\tau$. The critical transition region $\omega \gg 1/\tau$ is characterized by dispersion, and by determining this region, τ can be evaluated experimentally".

I have found it impressive that so many electron spin–spin and spin–lattice relaxation times were being measured well before magnetic resonance was discovered. I studied the papers of Gorter and of Van Vleck as part of my thesis research, becoming familiar with such concepts as spin temperature, spin–spin and spin–lattice relaxation. In the summer of 1948, Gorter was a visiting professor at Harvard, and I was a student in the course he taught on these subjects. In his paper on Ti and Cr spin–lattice relaxation, Van gave a complete quantum mechanical discussion, presenting the Hamiltonians of these ions in solids with interactions with the crystalline electric fields. He treats the effect of the lattice vibrations in relaxing the electrons, considering processes in which a single phonon is absorbed or emitted and processes in which one phonon is absorbed and another emitted with an energy change (a so-called two-phonon or Raman process). Moreover, he shows how, if he describes the magnetic atoms as having a spin temperature, he is able to calculate the relaxation times without ever needing to solve their Hamiltonians to obtain the eigenfunctions and energies. This paper inspired my student Hebel and me in our work on superconductivity, where we calculated the NMR spin lattice relaxation of dipolar-coupled nuclei in zero magnetic field, for which we did not know the solution of the Hamiltonian.

For my thesis, I built two electron spin resonance (ESR) spectrometers. The first was a klystron-powered 3 cm ESR apparatus that I used to measure crystal splittings of Fe and Mn ions in things such as iron alum and manganese sulfate. I then built a high-power instrument using a pulsed magnetron to measure spin–lattice relaxation by studying the saturation of the resonance signal as a function of the power level (Fig. 1).

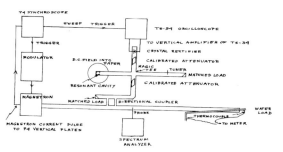

Fig. 1 Block diagram of microwave apparatus for measuring saturation of electron spin resonance signals as a function of incident power.

The magnetron produced 1 μs pulses with a peak power of 12 kW; 90 percent of the power was dissipated in a matched load consisting of a thin glass tube through which water flowed. I determined the power level of the magnetron by measuring the rise in temperature of the flowing water. About 10 percent of the power was routed *via* a directional coupler to the sample arm of the spectrometer. That enabled me to generate H_1 values of 50 G at the sample in the cavity. Fig. 2 (from my thesis) shows the saturation of the Mn^{++} ESR signal in $MnSO_4{\cdot}4H_2O$ at room temperature. The signal was cut in two at an H_1 of 52 G.

When I began work with Purcell, Bloembergen was just finishing his thesis research, and George Pake had a year yet to go before finishing his graduate studies. Thus I was on hand when he discovered the "Pake doublets", and when he and Gutowsky did their famous studies discovering the collapse of the dipolar splittings in molecules with CH_2 or CH_3 groups when they were warmed to temperatures at which the groups could reorient about a molecular symmetry axis. I was the only one of Purcell's students doing ESR.

One day Purcell came into the lab and asked me, "Charlie, why don't you detect an electron spin resonance from the electrons in the Cu cavity walls?" This was typical Purcell—he thought of something that, on reflection, seemed like an obvious question, but I had never thought of it! Of course, I had no answer. Then the next morning he came in with a big smile and told me he thought he knew why I saw no signs of the conduction electron spin resonance. He pointed out that only the electrons at the Fermi energy could give such a signal. They are moving at the Fermi velocity, v_F, quite a high speed. They can give rise to a resonance signal only when acted on by the microwave magnetic field. That field is confined to within the microwave skin depth δ of the surface, so they experience

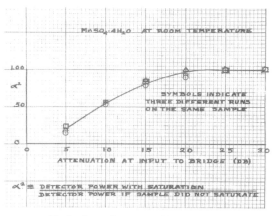

Fig. 2 Saturation curves of $MnSO_4{\cdot}4H_2O$.

these fields only when their trajectory brings them close to the surface. Given their high speed, they experience just a very short pulse of microwave field of duration approximately δ/v_F. Purcell pointed out that such a lifetime effect would give very broad and thus weak resonance lines.

I finished my thesis work and took the oral examination on my thesis in the spring of 1949. James Fisk, a visiting scientist at Harvard from Bell Labs that semester, was made a member of my examining committee. I remember vividly his question: "Charlie, tell me a bit more about this idea of spin temperature". Little did we know how important this issue would be for me within two years! Fisk later became president of Bell Labs and was my host in 1970, when I spent a sabbatical leave there.

Magnetic resonance at Illinois

In September 1949, I joined the physics department of the University of Illinois with the rank of instructor. Erwin Hahn was there. He had been a radar technician in the Navy during the war. He had received his PhD at Illinois also in June 1949. That summer, he discovered spin echoes. We had a wonderful collaboration for a year until he went to Stanford as a National Research Council postdoctoral fellow.

I was attracted to Illinois because Fred Seitz was launching a program in solid state physics there. I had originally planned to set up a microwave apparatus to continue work on ESR. My idea was to detect the ESR of the F center, thought to be a single electron trapped at the position of a missing halogen atom in materials such as KCl. But after arriving at Illinois, I heard that Beringer and Castle at Yale, who had a very sensitive apparatus with a bolometer detector, had been unable to detect the spin resonance of F centers. So I decided instead to focus on nuclear magnetic resonance. In addition to Erwin, there was another student, Dick Norberg, also a war veteran (Air Force), who was beginning work on NMR. Erwin had not really had a thesis adviser. I became Dick's informal adviser. Dick took the solid state physics course taught by Fred Seitz in the fall of 1949, where he learned about H in Pd and proposed studying

this system for a thesis.[2] The next year, two other students, Don Holcomb and Tom Carver, joined the group.

In the chemistry department, Herb Gutowsky had built a steady-state NMR apparatus. While studying Na metal, he found that the NMR line exhibited what looked to be motional narrowing at temperatures above 150 K, well below the melting point of Na (453 K). Herb proposed that he was observing motional narrowing and therefore the process of diffusion[3] in the solid phase. He could study the effect over a limited range of temperatures since the line narrowing ran into magnet inhomogeneity broadening above about 220 K, well below the melting temperature. Dick and I realized that with spin echoes we could overcome this handicap and study the motion all the way to the melting temperature, since echoes refocused the dephasing from inhomogeneous magnetic fields. This was my first experiment at Illinois[4] and perhaps the first use of spin echoes to study a problem in solid state physics. The most important thing was that this experiment launched us on our study of the alkali metals.

Overhauser at Illinois

In 1951, Al Overhauser (Fig. 3) finished his PhD at Berkeley, where his thesis adviser had been Charles Kittel. For his thesis, Al made a theoretical study of the spin–lattice relaxation of conduction electrons in metals. Kittel pointed out that Al needed a job. He asked Al to step outside his office, and a few minutes later called him back in again to say that he had just called Fred Seitz at Illinois, and in a short phone conversation got Al a postdoctoral position at Illinois to work on radiation defects in solids in the Seitz program.

In his thesis, Al had concluded that the spin–lattice relaxation time would be about a microsecond, leading to a very narrow ESR line about 0.1 G wide. This was a theoretical prediction, but there was no experimental value for comparison, since no one had as yet observed the conduction electron spin resonance in any metal. That such a narrow line had not shown up in my thesis experiments led credence to Purcell's calculation that there would be

enormous lifetime broadening of the resonance line at microwave frequencies.

My group soon got to know Al and learned about his thesis, and he learned about our studies of nuclear relaxation times in metals. In particular, he attended a talk that Dick Norberg gave to my group. Al wrote to me that "*what Norberg said in his seminar was: the free induction decay contains information. I believed that, during the decay, the system is out of equilibrium (of course). So what I did in the next two days was to find out what happens when the system is held (steadily) out of equilibrium. (The discovery followed quickly!)*".

In trying to reconstruct the exact date of Al's discovery, I have limited information. Dick's paper on the H–Pd system was received at *Physical Review* on January 21, 1952. My guess is that at this seminar Dick was talking about that study. The discovery was sometime before March 1952, when Overhauser gave a 10 minute talk about his PhD thesis. Al has told me that after the talk, Bloembergen came over to him and in their conversation, Overhauser told him about his idea for dynamic polarization, a topic not in his talk.

At that time, as mentioned above, no one had observed the conduction electron spin magnetic resonance, a prerequisite to producing Overhauser's effect! So we set out to find a conduction electron spin resonance.

Keeping in mind Purcell's thoughts about microwave ESR linewidths, and realizing that for our NMR studies of metals we employed metal powders in which the individual powder grains were smaller than the radio frequency (rf) skin depths, I realized that at NMR frequencies the Purcell-effect line broadening mechanism would be absent, since the electrons were under continuous radiation. Overhauser's theoretical prediction of a very narrow and intense ESR line should make for a very strong ESR signal at NMR frequencies, so I decided we should stay away from microwaves and instead work at frequencies typical of NMR at that time (10 to 40 MHz). I enlisted Don Holcomb for this project and used an NMR apparatus to search in Cu powders, Al powders, and even in an old sample of powdered Na. We were looking for a resonance line whose width was a few

This journal is © the Owner Societies 2010

Phys. Chem. Chem. Phys., 2010, **12**, 5741–5751 | 5743

Fig. 3 Al Overhauser. Picture from his 1951 application for a postdoctoral appointment at the University of Illinois.

tenths of gauss. For this, we built a solenoid with 6th-order end corrections to generate uniform magnetic fields. We tried and tried but had no success. Eventually I decided we should give up. Don joined Dick in spin echo NMR studies of the alkali metals. These studies were indeed rich. In the course of this work, they introduced the boxcar integrator that enabled them to signal average spin echoes, much as the lock-in amplifier enabled one to signal average steady-state NMR signals.

Then, late one morning in early December 1952, I was in the Physics Library reading the latest issue of *Physical Review* (volume 88), when I saw the article by Griswold and coworkers[5] reporting the observation of the conduction electron spin magnetic resonance of Na metal at 3 cm band microwaves. The line was quite broad but strong. I dashed up to the 4th floor where our labs were located under the low, sloping ceiling to find that my students had already left for lunch. Don and Dick had now made samples of Li

metal and were studying their NMR. I took the Li sample, stuck it in the coil of a little transitron NMR oscillator[6] that we used for calibration of the strength of our electromagnets, put that into the solenoid Don and I had made, attached the leads of a Variac to the solenoid to sweep the field back and forth by 20 G to 30 G, and put the transitron output on the scope screen. There was a gigantic signal from the conduction electron ESR! Fig. 4 shows such a signal.†

I kicked myself around for not having done just this simple experiment, sweeping the field over a large range, months earlier with our Na sample. I had been so focused on finding a very narrow line that I never thought just to try for a vastly broader one!

Don and Dick were now deeply immersed in alkali-metal NMR, so I asked Tom Carver (Fig. 5) if he would like to work with me to try to verify

---

† This particular figure, published in 1954, was taken from a paper by Schumacher and coworkers that I discuss below.[7]

Overhauser's remarkable proposal. Perhaps I should say "test" rather than "verify", but we understood his argument and believed it was correct!

The first issue was how to generate and detect nuclear polarization. Overhauser had proposed a method. It was based on his realization that the frequency of the electron spin resonance was affected by the polarization of the nuclei. So if the nuclear polarization grows as one saturates the electron resonance, it will shift the frequency of the electron spin resonance. Thus, the effect could be detected by use solely of the electron spin resonance signal. This method requires substantial nuclear polarization, thus a strong magnetic field such as used for microwave ESR and temperatures in the liquid He range. We had no apparatus for either the microwave ESR or the low-temperature work.

Tom and I decided that a less demanding experiment would be to observe the nuclear spin polarization directly *via* a nuclear resonance signal while simultaneously saturating the electron spin resonance. This method could be done at room temperature and at any magnetic field strength at which we could do both resonances. We did not need large nuclear polarization, simply a large change in the amount of nuclear polarization produced by saturating the electron spin resonance. Moreover, we either had or felt we could build the necessary apparatus.

Another consideration was that we needed to have the electron rf fields and the nuclear rf fields occupy the same physical space in the sample, so that all the nuclei we detected were subject to Overhauser enhancement. Note that this situation would not obtain if one used thick metal samples, since the penetration depths would be greater at the nuclear frequency than at the 1000-fold higher electron frequency. These length differences are macroscopic, too large to be bridged by spin diffusion. So both frequencies must give penetration depths that are greater than the diameter of the metal particles of the powder composing the sample. The size of the metal particles we were able to make limited the upper frequency at which the ESR could be set. Given the samples we could make, we felt that about 100 MHz was our upper frequency limit. That placed the

5744 | *Phys. Chem. Chem. Phys.*, 2010, **12**, 5741–5751

This journal is © the Owner Societies 2010

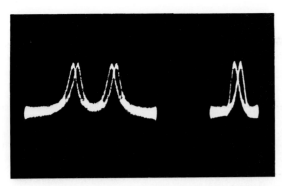

Fig. 4 Oscilloscope pictures of electron and nuclear resonance absorption in lithium *vs.* magnetic field. Resonance apparatus operating at 17.4 MHz. Left: conduction electron spin resonance. Zero magnetic field at the center of the sweep. The four peaks arise from a small phase shift between the horizontal and vertical axes. Sweep amplitude about 40 G from left to right, putting the resonances at an absolute value around 6 G. *Right*: ^{7}Li nuclear resonance at about 10 000 G. Sweep amplitude approximately 3 G.

Fig. 5 Tom Carver in 1951.

NMR frequency in the range of tens of kHz.

Of course, most nuclear magnetic resonance studies at that time were done above 10 MHz in order to get the biggest signals. However, we knew that Dick Brown, another of Purcell's students while I was at Harvard, had recently studied proton NMR in water at magnetic fields of 6 and 12 G, corresponding to Larmor frequencies of 25 kHz and 50 kHz respectively, using a sample of 30–40 in^3 volume.[8] It therefore seemed reasonable to try such frequencies.

We began by trying to find NMR signals at low frequencies. We chose proton NMR at 16 kHz. We built a bridge apparatus using a Wheatstone bridge, a Hewlett–Packard audio signal generator, and a harmonic wave analyzer amplifier for signal detection. We were unsuccessful. We then moved to 50 kHz, replacing the Wheatstone bridge (for which there cannot be a common ground for the input and output signals, making it impossible to shield the equipment) with a "twin-tee bridge"‡ that had a common ground and thus could be shielded. The size of the actual metal samples we used was small so that we could generate enough power to saturate the conduction electron resonance. Finally we made a simple tuned preamplifier, connected it to a mixer driven by a 550 kHz signal, and fed the 600 kHz output to a communication receiver. A twin-tee bridge has a null at a particular frequency. Therefore the signal generator for the nuclear resonance had to be highly stable in frequency and noise free. Tom built a battery-operated oscillator consisting of a 100 kHz crystal-controlled fixed frequency and a frequency divider circuit using integers from 2 to 10 as our signal generator. With this equipment, we found the proton NMR in glycerine at 50 kHz!

The ESR signals were enormous in the rf range. If the entire Li ESR line width of about 5 G arose from the Li spin–lattice relaxation rate, we knew we would need an alternating magnetic field H_1 of about 5 G to saturate the ESR. From my thesis work, I knew that in the 3 cm band region this would require a magnetron to generate enough power to produce such a field. We had no microwave equipment of any kind at our disposal. However, our choice of 50 kHz for the NMR frequency put the ESR frequency for Li at 84 MHz and for Na at 124 MHz and eventually for polarizing ^{1}H at 33 MHz. For these frequencies, there was well known technology from amateur radio. We were aided by use of the famous *American Radio Relay League Handbook*. A one-tube oscillator based on a 3E29 vacuum tube provided the ESR power. It could produce 50 W to 100 W of rf

---

‡ A T-network is a circuit with three circuit elements connected to look like a letter T. A "twin-tee" consists of two such T networks in which the two input terminals of one T circuit are connected in parallel with the two input terminals of the other T, and the two output terminals of each T are connected in parallel with those of the other T. Twin-tee's with properly chosen circuit elements are frequently used in electric circuits as tuned filters.

Fig. 6   Coils and samples used to demonstrate the Overhauser effect. Left to right starting with the back row: low-frequency coils for frequencies 33–50 kHz, 200 kHz, and 100 kHz; 30 cc samples of distilled water, glycerine, sodium, lithium; scale with 1 in marks; 19 cc samples of glycerine, lithium, sodium; a 124 MHz sample holder and coil; an 84 MHz sample holder and coil.

power. It had to be operated in a closed metal box to shield the sensitive NMR amplifier from its influence. To do a double resonance, we needed to be able to apply two alternating magnetic fields of widely differing frequencies simultaneously to the sample, with one of the fields strong enough to saturate the electron spin resonance. To do so, the solenoid, the NMR coil, and the ESR coil were oriented with their axes mutually perpendicular. Fig. 6 shows a number of the coils, and Fig. 7 shows the 3E29 inside its metal box, with the ESR coil across a transmission line, the NMR coil sitting properly oriented but not in place, and the solenoid sitting, also properly oriented, but also not in place.

To measure the frequency of the electron oscillator, we coupled the output loosely to a parallel wire transmission line and measured the wavelength

two parallel wires of the transmission line and the moveable short are visible on the far left side of the figure.

In summary, we had a largely homemade apparatus.

The output of the 3E29 vacuum tube could be turned off or on by means of its plate voltage, a term today's practitioners of semiconductor electronics do not encounter! We had a switch that could turn the plate voltage off or on. I remember vividly being across the room from the apparatus when Tom was at the controls on probably August 10, 1953, with everything assembled and tuned up when Tom let out a yell! I turned and there on the screen of the oscilloscope was a gigantic signal! Tom

Fig. 8   The apparatus assembled. Communication receiver on the top on the right. The parallel wire transmission line used to measure the 3E29 frequency foreground on the far left. The large metal box contains the elements shown in Fig. 7.

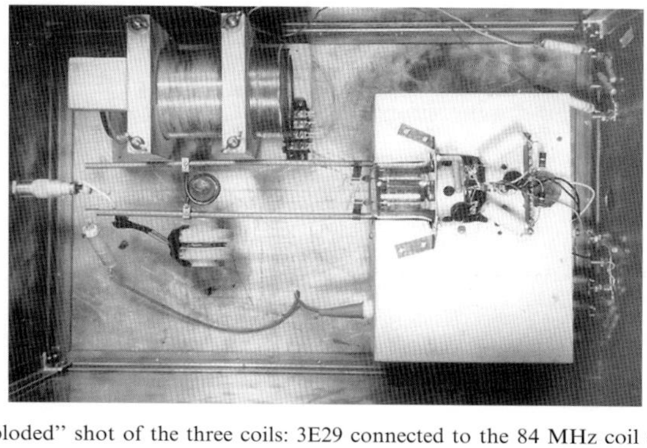

by adjusting the length of the transmission line until it lit a small light bulb connected across the wires as a moveable short. Fig. 8 shows our apparatus. The

explained that it was the Li NMR signal. With the plate voltage off, there was nothing but a flat baseline as the field was slowly swept. But the instant he threw the plate voltage switch, turning on the ESR oscillator, the Li NMR signal leapt into view. The Overhauser prediction was confirmed! Fig. 9 shows the images we made that afternoon. The top line is the NMR signal without saturating the electron spins; the middle line is the Li NMR signal with the ESR saturated. The bottom line shows the proton signal from a large reference sample.

In June, Overhauser had left Illinois to take up his duties as a young faculty member in the physics department at Cornell. We immediately sent him a telegram, followed by a short letter giving some details. Then we wrote a short paper announcing our result and sent it

Fig. 7   "Exploded" shot of the three coils: 3E29 connected to the 84 MHz coil containing Li sample. Solenoid above used to generate the static magnetic field; 35–50 kHz coil below. Walls of the metal box visible on all sides.

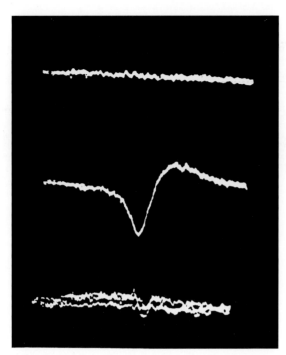

Fig. 9 The very first demonstration of the Overhauser effect. Oscilloscope pictures of 50 kHz nuclear resonance absorption *vs.* static magnetic field. Field excursion 0.2 G. Top line: ^{7}Li resonance (lost in noise). Middle line: ^{7}Li resonance enhanced by electron saturation. Bottom line: proton resonance at 50 kHz in a glycerine sample of 8 times the volume of the Li sample, giving an estimated Overhauser enhancement of a factor of 100.

saturation diminishes the population difference much as would a higher temperature. So, naively, I thought of saturation as raising the spin temperature. However, the spins are not the only degree of freedom of a conduction electron in a metal. With Al's help, we realized that one should make a diagram of the electrons in the quantum states of a metal. Because of the exclusion principle, the electrons' kinetic energy must be included, giving one the famous Fermi functions to describe the occupation of the electron states. Two figures from Tom's thesis tell the story. Fig. 11a shows the population of the up spins and down spins for a magnetic field pointing in the up direction when the system is in thermal equilibrium. There are more electrons in the down spin orientation, since the electron gyromagnetic ratio is negative. Saturating the electron spin resonance equalizes the number of spins in the two orientations, producing Fig. 11b. The nuclear T_1 process involving the electron–nuclear hyperfine coupling ($I \cdot S$) scatters electrons from one spin orientation to the other, simultaneously changing the electron kinetic energy, since the total energy gain (loss) of the electron must match the energy loss (gain) of the nucleus. Such

to *Physical Review*. It was received on August 17, 1953, Al Overhauser's 28th birthday.[9]

Several months earlier, Overhauser had presented a 10 minute paper at the meeting of the American Physical Society in Washington, DC. Fig. 10 shows the abstract, the first public announcement of Al's discovery. The abstract had attracted a good audience, including Purcell, Bloch, Rabi, Bloembergen, Ramsey, and Abragam. I was at the meeting but not at Al's talk. After the talk, Al had to leave because he had an appointment scheduled. Members of this group found me and quizzed me about the mechanism. They were quite skeptical, but soon fell to talking to each other when I tried to give an explanation.

I was not surprised at their skepticism, since when Al first told me his idea it did not seem reasonable to me either. I was used to thinking of the saturation of a bunch of individual spins. For a spin-$\frac{1}{2}$ system in a strong static magnetic field,

Q10. Polarization of Nuclei in Metals. ALBERT W. OVER-HAUSER, *University of Illinois.*—A new method for producing polarized nuclei is made possible by the spin paramagnetic resonance absorption of conduction electrons in metals. The important interaction between electrons in the conduction band and nuclear spins is the hyperfine structure coupling of an S state, proportional to $I \cdot s$, which gives rise to the predominant nuclear relaxation process in metals and contributes to the electron paramagnetic relaxation. If the electron resonance of a metal in a magnetic field is completely saturated, it follows from a *dynamical* study of the relaxation processes that the nuclei will be polarized to the same degree they would be if their gyromagnetic ratio were that of the electron spin. More generally, the degree of polarization can be described by an effective gyromagnetic ratio; $\gamma_{eff} = \gamma_n + s|\gamma_e|$. Here, γ_n and γ_e are the gyromagnetic ratios of the nuclei and electron spins, and s is the fractional saturation multiplied by T/T', where T is the nuclear relaxation time and T' is the nuclear relaxation time produced only by $I \cdot s$ interactions. At 2°K and fields of 10^4 oersteds, 50 percent polarization appears possible by present techniques, the major problem being the preparation of sufficiently finely powdered specimens. Some obvious applications are polarized neutron reactions (e.g., on Li), angular dependence of radioactive disintegration, and nuclear resonances resulting from rare isotopes.

Fig. 10 Abstract of Overhauser's talk that attracted a celebrated but skeptical audience.

thermal equilibrium with the lattice produces a population difference between the two energy levels described by the Boltzmann factors. The onset of

transitions are shown on the diagrams by the arrows between points A and B.

There is no temperature at which the electron system looks like Fig. 11b. So

This journal is © the Owner Societies 2010

Phys. Chem. Chem. Phys., 2010, **12**, 5741–5751 | 5747

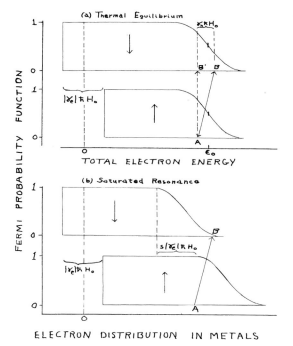

ELECTRON DISTRIBUTION IN METALS

Fig. 11 Electron distributions in metals (from Carver's thesis). Top: Fermi function for system in thermal equilibrium. Long vertical arrows specify electron spin orientation for a magnetic field pointing up. Arrows between A and B illustrate an electron scattering event resulting from the electron–nuclear hyperfine coupling. Note that the electron changes both Zeeman and kinetic energy. Bottom: electron distribution when the electron spin resonance is saturated, equalizing the number of electron spins pointing up or down. There is no temperature at which such a distribution arises naturally. Note that the fall-off of the tail corresponds to the lattice temperature.

one should not say that saturating the electron spin resonance produces a higher electron spin temperature. However, a simple argument enables one to see what happens.[10]

Contemplating Fig. 11b, we note that if we did not have the magnetic shifts of the electron distributions so that the bottoms of the distributions coincided, the tops of the distributions would also coincide and would look like a system in zero applied field in thermal equilibrium at the lattice temperature. We can remove this electron spin magnetic splitting by transforming the electron spins to a reference frame rotating at an angular frequency $\Omega = \gamma_e H$, where γ_e is the magnitude of the electron gyro-magnetic ratio (*i.e.*, the symbol γ_e is positive) and H is the applied static magnetic field. However, we also have to transform the nuclear spins to the same reference frame in order to keep

the hyperfine interaction $I \cdot S$ between the nuclear and electron spins independent of time. In this reference frame, the nucleus comes to the temperature of the lattice. Its Zeeman interaction in this frame is

$$\mathcal{H}_{\text{Zeeman}} = -\gamma_e \hbar (H + \Omega/\gamma_e).$$

It is acted on, therefore, by an effective magnetic field

$$H_{\text{effective}} = H(1 + (\gamma_e/\gamma_n)),$$

where γ_n is the nuclear gyromagnetic ratio, producing therefore a vastly larger nuclear polarization.

This relationship is just Overhauser's result. Tom and I demonstrated the Overhauser enhancement experimentally in both Li and Na metals, but it was many years before anyone succeeded in observing the electron spin resonance in any other metal with the exception of Be.

Moving beyond those metals was made possible by the invention of a new technique, spin transmission through a thin metal film, invented and demonstrated by Lewis and Carver[11] and also by Vander Ven and Schumacher.[12] Schultz and Latham used the new method to observe the conduction electron spin resonance in Cu metal.[13]

Feher and Kip made a thorough study of the line shape of conduction electron ESR,[14] testing the theory of Dyson.[15] They found that the line broadening predicted by Purcell did not in fact occur. The explanation can be understood easily in terms of the Carver/Schumacher spin transmission method. The electron motion in a metal is diffusive, with a mean free path that is short. The typical electron scattering event does not flip the electron spin. So a precessing electron spin maintains coherence despite many scattering collisions. Thus an electron can come to the surface multiple times precessing coherently, and thus the line width is not broadened.

Other systems

On achieving a successful verification of Overhauser's amazing prediction, we immediately wondered if such an effect is confined to metals. Clearly, one key feature was a mutual spin flip, but for it to occur, one needed to conserve energy. As one sees from Fig. 11b, in the metal this is made possible by a change in electron kinetic energy to compensate for the change in Zeeman energy from a mutual spin flip arising from the difference in the nuclear and electron gyromagnetic ratios. Another case where this energy compensation should be possible is in a liquid where the translational and rotational degrees of freedom play a role in conventional theories of nuclear relaxation. So Tom and I decided to try to achieve an Overhauser effect in a liquid. For that purpose we needed a liquid that had an electron spin resonance we could saturate. We settled on a system that had been recently and thoroughly studied by Hutchison and Pastor,[16] Na dissolved in liquid ammonia. The system had also been explored theoretically by Kaplan and Kittel.[17] When the Na atoms are dissolved, they give up their outer electron, giving rise to a ESR line only about 0.02 G wide, the

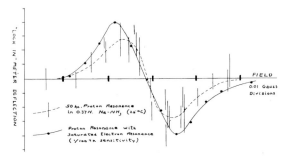

ENHANCEMENT IN NA-NH₃ SOLUTION

Fig. 12 Demonstration of the Overhauser effect in a non-metal: proton resonance signals from ammonia molecules in solutions of Na in liquid ammonia *vs.* applied magnetic field. The signals are obtained by modulating the static magnetic field and recording the signal with a lock-in amplifier. The points taken when the electron spin was saturated are recorded with a factor of 100 signal attenuation, showing that the Overhauser enhancement obtained is approximately a factor of 100.

narrowest ESR line of which we were aware. I will not here describe the whole experiment, since we have published the work.[18] The narrow electron resonance was easy to saturate. We replaced the 3E29 oscillator with a General Radio signal generator (visible on the left in Fig. 8), giving us easy control of the drive signal amplitude and frequency. Fig. 12 shows the lock-in signal of the ammonia protons with and without saturation of the electron resonance.

The plot of the signal under saturation is made with a factor of 100 scale change, showing that we achieved a factor of 100 increase in the proton absorption rate and thus enhancement. Since saturation increases proton absorption rather than turning it into emission, the coupling must involve terms such as $I^{+}S^{-}$ rather than terms such as $I^{+}S^{+}$ that might arise if conventional dipolar coupling dominated. This conclusion supports a picture in which the electron spin sits in an s-orbit on the H atoms, providing a contact interaction. We found that the degree of maximum enhancement was proportional to the Na concentration, suggesting that the rapid diffusion of the electrons meant that individual H atoms experienced the electron spin intermittently. All these results were in Tom's PhD thesis (June 1954) and in our major paper that we were very slow in publishing.[18]

While Tom and I were working on our experiments, Norberg (Fig. 13) and Holcomb (Fig. 14) were studying the NMR of the Li, Na, and Rb alkali

metals. All exhibited the motional narrowing at sufficiently high temperatures, indicating rapid diffusion at temperatures well below their melting points. These experiments firmly established NMR as a tool for studying diffusion in solids. It was well known from the work of Bloembergen, Purcell, and Pound that the fluctuation of the dipolar coupling between pairs of nuclear moments in liquids, resulting from rotational or diffusive motions of

the spins, produced relaxation effects. The fluctuations in dipolar coupling that accompanied the spatial motion of molecules in a liquid were responsible for the T_1 and T_2. In the Hamiltonian describing the process, there were terms in which only one spin is flipped and terms in which both spins flip. In the latter case, the two could both flip up (down), or one could flip up (down) the other flip down (up). The same should hold true if the spins were different species.

The Rb isotopes ^{85}Rb and ^{87}Rb have abundances of 73 percent and 27 percent respectively. Holcomb and Norberg wondered to what extent such phenomena played a role in the alkalis. In the case of Rb, with two quite abundant isotopes, they had a case of two families of coupled spins, *i.e.*, an I spin family and an S spin family. Some of these relaxation processes involved mutual spin flips. They also recognized that for a case such as Rb, where there were two fairly abundant isotopes, there would be mutual spin flip terms between the species and, therefore, that disturbing the population of one species from equilibrium would also disturb the population of the other species. In his thesis, Holcomb points out that this is

Fig. 13 Dick Norberg on entering graduate school at Illinois.

This journal is © the Owner Societies 2010

Phys. Chem. Chem. Phys., 2010, **12**, 5741–5751 | 5749

Fig. 14 Don Holcomb on entering graduate school at Illinois.

similar to the Overhauser effect. They concluded, however, that the effect in their particular case was too small to detect.[19]

Independently, Ionel Solomon, while visiting Harvard University the following year, examined the problem of coupled nuclear spins in a liquid belonging to two different nuclear species. He likewise found that one could have a nuclear equivalent of the Overhauser effect. Solomon published his treatment of the problem of two coupled spins and verified the theory by his beautiful experiments on the highly toxic molecule HF.[20]

The afterglow

In the summer of 1954, armed with their new PhDs, Carver went off to Princeton and Holcomb to Cornell to start their academic careers. Although we took no more data, Tom and I did not finish writing our second paper[18] until the fall of 1955, since Tom was busy setting up a new lab at Princeton and I was intensely involved in two new experiments (see below).

Don Holcomb was launched on a distinguished career of teaching and research at Cornell, where he held various important positions such as head of the physics department and director of their Laboratory of Atomic and Solid State Physics. He has served a term as president of the American Association of Physics Teachers.

Carver likewise had a distinguished career at Princeton. Tom continued highly innovative research closely related to magnetic resonance. I have already mentioned his invention of the spin transmission method through thin metal films that made possible detecting conduction electron spin resonance in Cu. He also utilized optical methods. One of particular interest in connection with dynamic polarization was his invention of a method of polarizing noble gases by spin-exchange optical pumping.[21]

Norberg joined the Physics Department at Washington University. He served many years as the head of the department. In 2004, Dick and his first PhD student, Irving Lowe, received the

ISMAR Prize for discovering the Fourier transform theorem for magnetic resonance and for the invention and demonstration of magic angle spinning.

I had moved on to new experiments that in many ways were the direct result of our work at low magnetic field and double resonance. The first of these with my student Bob Schumacher, aided initially by Tom Carver, was an experiment to measure the electron spin contribution to the magnetic susceptibility of metals, first calculated by Wolfgang Pauli. David Pines, John Bardeen's postdoc, pointed out to me that the spin susceptibility of metals had never been measured because there was also an orbital contribution to the magnetic susceptibility of metals, but measurements of magnetic susceptibility could not separate the two contributions. Pines had a great interest, since he had just calculated the many-body correction to the Pauli formula and was eager for an experimental test. As a result of the fact that we were looking at conduction electron spin resonances, I realized we could determine the spin paramagnetic susceptibility from the area under the ESR absorption line, using the Kramers–Krönig relationships. Moreover, I realized how we could use the metal's NMR signal to calibrate the ESR absorption scale. We built a bridgeless NMR apparatus for precise measurements of the imaginary part of magnetic susceptibilities. The ESR and NMR signals are shown in Fig. 4. On the left is the ESR of conduction electrons in Li metal, on the right the ^{6}Li NMR signal in the same sample and taken at the same resonance frequency. We first recorded the ESR signal in a low magnetic field, and then raised the magnetic field to produce a nuclear resonance in the sample at the same frequency. We submitted this paper to *Physical Review* in late June 1954, just before Tom left for Princeton.[7] I like to think of this experiment as a "time-delay double resonance" in which we observe two resonances sequentially. The whole experiment was a direct result of my work with Al, both because that led to our looking at conduction electron spin resonances, because we thought to do ESR at very low magnetic fields, and also because it jogged our minds out of any possible rut.

Also in 1954, I attended a talk given by my colleague John Bardeen about superconductivity. He had come to believe that when a metal was in the superconducting phase, its density of states was modified. In particular, he believed that there was a gap in the density of states right at the Fermi level. Because of the work of Holcomb and Norberg on relaxation times in metals, I realized that such a gap would have a major effect on the nuclear spin–lattice relaxation time of the metal nuclei. I realized it would be a terrific experiment to measure the relaxation time in the superconducting state. Then my spirits sank when I remembered that a superconductor is a perfect diamagnet and hence excludes magnetic fields from its interior. (This was before the discovery of type II superconductors.) How could one do magnetic resonance in a material that excludes the magnetic field! I believe that once again the experiment to test the Overhauser effect was key, augmented now by thinking of our "time-delayed double resonance" experiment with Schumacher and Carver, by my familiarity with spin temperature and its uses, by the work of Gorter and Van Vleck, and by the mind-freeing effect of associating with Al Overhauser. Application of a strong enough magnetic field suppresses superconductivity. So why not start with such a strong magnetic field to polarize the nuclear spins in the normal state, then lower the magnetic field adiabatically to zero, putting the sample in the superconducting state, but at a much lower spin temperature. Let the nuclear spins exchange energy with the lattice for a time t_{super}, then

raise the magnetic field adiabatically to its initial value, observing the NMR signal on the way up. Studying the signal as a function of t_{super} then gives one the spin–lattice relaxation time in the superconducting state. I enlisted the help of another brilliant student, Chuck Hebel, for this project. Again we built a brand-new NMR apparatus, operating at 400 kHz. To do the magnetic field cycle, we built a special magnet that we could turn from 500 G to 0 G in a millisecond. We got our first experimental results in the fall of 1956, a matter of a few months before Bardeen, Cooper, and Schrieffer (BCS) invented their famous theory that explained superconductivity. We published our report on this experiment and its analysis using the BCS theory in *Physical Review*.[22] Leon Cooper showed our NMR results as well as those of Redfield in his Nobel Prize lecture.

I believe the thought processes that led to this experiment are rather well described by the term "phase transition double resonance". It was directly stimulated by John Bardeen. But the ultimate inspiration for this experiment was our interaction with Al Overhauser.

Acknowledgements

Working with Al Overhauser was one of the most exciting and gratifying experiences of my life. I feel deeply the roots originating in working with Van Vleck, Purcell, and Gorter, and finally the inspiration and can-do spirit of my students Dick Norberg, Don Holcomb, Tom Carver, Bob Schumacher, and Chuck Hebel. Preparation of the figures in the article would not have been possible without the superb help of Celia Elliott.

References

1 M. H. Hebb and E. M. Purcell, *Phys. Rev.*, 1937, **5**, 338.
2 R. E. Norberg, *Phys. Rev.*, 1952, **86**, 745.
3 H. S. Gutowsky, *Phys. Rev.*, 1951, **83**, 1073.
4 R. E. Norberg and C. P. Slichter, *Phys. Rev.*, 1951, **83**, 1074.
5 T. W. Griswold, A. F. Kip and C. Kittel, *Phys. Rev.*, 1952, **88**, 951.
6 E. L. Hahn and H. W. Knoebel, *Rev. Sci. Instrum.*, 1951, **22**, 904.
7 R. T. Schumacher, T. R. Carver and C. P. Slichter, *Phys. Rev.*, 1954, **95**, 1089. See also: R. T. Schumacher and C. P. Slichter, *Phys. Rev.*, 1956, **101**, 58.
8 R. M. Brown, *Phys. Rev.*, 1950, **78**, 530.
9 T. R. Carver and C. P. Slichter, *Phys. Rev.*, 1953, **92**, 212.
10 C. P. Slichter, *Phys. Rev.*, 1955, **99**, 1822, erratum *Phys. Rev.* 1956, **103**, 1905.
11 R. B. Lewis and T. R. Carver, *Phys. Rev. Lett.*, 1964, **12**, 693.
12 N. S. Vander Ven and R. T. Schumacher, *Phys. Rev. Lett.*, 1964, **12**, 695.
13 S. Schultz and C. Latham, *Phys. Rev. Lett.*, 1965, **15**, 148.
14 G. Feher and A. F. Kip, *Phys. Rev.*, 1955, **98**, 337.
15 F. Dyson, *Phys. Rev.*, 1955, **98**, 349.
16 C. A. Hutchison, Jr. and R. C. Pastor, *Rev. Mod. Phys.*, 1953, **25**, 285.
17 J. Kaplan and C. Kittel, *J. Chem. Phys.*, 1953, **21**, 1429.
18 T. R. Carver and C. P. Slichter, *Phys. Rev.*, 1956, **102**, 975.
19 D. F. Holcomb and R. E. Norberg, *Phys. Rev.*, 1955, **98**, 1074 and D. F. Holcomb, *Nuclear Magnetic Resonance in the Alkali Metals*, PhD Thesis, University of Illinois, June, 1954.
20 I. Solomon, *Phys. Rev.*, 1955, **99**, 559.
21 M. A. Bouchiat, T. R. Carver and C. M. Varnum, *Phys. Rev. Lett.*, 1960, **5**, 373.
22 L. C. Hebel and C. P. Slichter, *Phys. Rev.*, 1957, **107**, 901; L. C. Hebel and C. P. Slichter, *Phys. Rev.*, 1959, **113**, 1504.
23 J. H. Van Vleck, *Phys. Rev.*, 1940, **57**, 426.

Chapter 2
DNP Mechanisms

Krishnendu Kundu[1], Frédéric Mentink-Vigier[2], Akiva Feintuch[1], and Shimon Vega[1]

[1]*Chemical and Biological Physics Department, Weizmann Institute of Science, Rehovot, Israel*
[2]*National High Magnetic Field Laboratory, Florida State University, Tallahassee, FL, USA*

2.1 INTRODUCTION

In this chapter, we have attempted to present a comprehensive description of the spin dynamics leading to the nuclear signal enhancements observed during dynamic nuclear polarization (DNP) experiments on glassy amorphous solid solutions of free radicals. Since Overhauser's[1] initial proposal of enhancing NMR signal intensities by microwave (MW)

Handbook of High Field Dynamic Nuclear Polarization.
Edited by Vladimir K. Michaelis, Robert G. Griffin, Björn Corzilius and Shimon Vega
© 2020 John Wiley & Sons, Ltd. ISBN: 978-1-119-44164-9
Also published in eMagRes (online edition)
DOI: 10.1002/9780470034590.emrstm1550

irradiation of samples containing free electrons, mechanisms responsible for DNP enhancements have been introduced and checked experimentally. As a result, we can find during the years a large collection of review articles and a recent book summarizing the different stages of development of these mechanisms.[2–8]

These reviews together with their cited original papers form the basis of our understanding of DNP. We must therefore always be aware of the pioneering contributions that created the base of all our modern DNP experiments and their applications. It is sometimes difficult to trace back the origin of ideas and concepts we are using today. Being aware of this, we apologize from the start for the lack of appropriate referencing, when discussing our present understanding of the DNP spin dynamics. In this chapter, we restrict ourselves to discussing the basic DNP mechanisms and do not deal with practical aspects, such as experimental conditions, DNP equipment, and choice of free radicals or even absolute enhancement issues. We rely on the other chapters to discuss these subjects and hope that our contribution may assist others in the interpretation and evaluation of their DNP data.

DNP was thus first predicted by Albert Overhauser[1] in 1953 and demonstrated by Carver and Slichter[9,10] on a sample of metallic lithium. The Overhauser effect was further applied on liquid solutions of free radicals and has become an important enhancement tool.[2,11,12] Shortly after the demonstration of the Overhauser effect, a new mechanism for DNP in the solid state was proposed[13,14] and experimentally demonstrated in glassy nonconducting solids by Jeffries.[15] In this case, the polarization enhancement of the nuclei appears due to effective MW irradiation on the zero-quantum (ZQ) or double-quantum (DQ) transitions of coupled electron-nuclear spin systems. The frequencies of these transitions are separated by twice the Larmor frequency of the nuclei. This 'solid-effect' (SE) polarization transfer mechanism results in positive or negative nuclear signal enhancements, depending on the DQ- or ZQ-MW irradiation, respectively. In amorphous samples, SE enhancements can be easily recognized, when the Larmor frequency of the nuclei coupled to the unpaired electrons is larger than the width of the EPR line.[16,17] In samples where this frequency is smaller than the width of the EPR line, the positive maximum and negative minimum of the enhancement frequency profile can be separated by less than twice the nuclear Zeeman frequency. These profiles were first demonstrated experimentally and explained by Hwang and Hill.[18,19] They proposed a theoretical model based on the work by Kessenikh *et al.,*[20,21] in which the EPR line was split into frequency bins, where the electrons are divided according to their bin frequencies. Then irradiation on an electron, which interacts with another electron with a bin frequency removed by the nuclear Zeeman frequency, can polarize efficiently their neighboring nuclei. This enhancement process was called the cross effect (CE).

An additional enhancement process predicted and extensively discussed by Abragam and Goldman[22,23] and by de Boer, Borghini, Wollan, and others,[24-26] is responsible for DNP enhancements, in samples containing relatively high concentrations of unpaired electrons. This process relies on a cross-relaxation[27-29] mechanism between the spins and was called thermal mixing (TM). Its enhancement mechanism was explained using the thermodynamic description of an interacting spin system, introduced by Redfield.[30] This thermal representation of these spin systems was extended profoundly and implemented into DNP by Provotorov.[31,32] It assigns different spin temperatures to different parts of the system, defined by the different interactions in the Hamiltonian describing the electron–nuclear spin system. These parts are treated as energy reservoirs, each having their own single temperature coefficient, that are in thermal contact with each other via spin interactions and with the lattice. A complete description of the heat flow between the reservoirs

and its consequences in terms of DNP phenomena are well summarized in the recently published book by Wenckebach[8] and further clarified in two of his recent publications.[33,34] The TM description of the spin system was first applied to electron systems with homogeneously broadened EPR spectra and later extended by de Boer[24] to inhomogeneously broadened systems. Very recently, the validity of the TM mechanism at very low temperatures was also proven.[35] In a variety of cases, the results of DNP experiments could be understood relying on this TM mechanism,[24,36,37] and detailed theoretical derivations based on spin thermodynamics can be found in the literature.[5,35]

Over the years, more sophisticated types of DNP experiments based on pulsed techniques, instead of continuous-wave irradiation, were introduced. It was demonstrated that coherence transfer of polarization from the electrons to the nuclei can be achieved via short pulses of strong MW irradiation, inducing a mechanism analogous to cross-polarization (CP) in solid-state NMR.[38] This pulse DNP experiment, known as nuclear orientation via electron spin locking (NOVEL),[39] and the integrated solid effect (ISE) method[40] were first introduced by Wenckebach[40,41] and later extended also by the group of Griffin.[39,42] NOVEL is based on the fact that the nutation of the nuclei in the laboratory frame can be matched to the Rabi frequency of the electron in the rotating frame. This Hartmann-Hahn type of matching condition allows nuclear polarizations to reach the values of electron polarizations. Other types of pulse DNP experiments are the nuclear rotating frame (NRF)[16] and the dressed state solid effect (DSSE)[43] techniques, which can operate at relatively low microwave power. Very recently, an additional pulse DNP technique was introduced based on recoupling concepts.[44]

Forty years after the discovery of Overhauser DNP, following the pioneering works by Wind, Yannoni, and Schaefer and coworkers,[45-47] Griffin and coworkers demonstrated that by using high-power, high-frequency MW sources, it is possible to obtain a significant sensitivity enhancement during MAS-NMR experiments.[48] Ten years later, Ardenkjaer-larsen and Golman *et al.*[49] introduced the idea of dissolution DNP that can enhance solution NMR signals by four orders of magnitude. These two achievements resulted in a broad spectrum of applications of DNP on static and rotating sample, including NMR of proteins, molecular adsorption on surfaces, materials research, and new contrast agents for MRI, as can be found in recent reviews.[7,50,51]

This revival of the interest in DNP also resulted in a renewed interest in theoretical aspects describing DNP mechanisms, while also taking advantage of modern computational abilities.[52–58] As part of this renewed interest in understanding DNP mechanisms, Vega and Goldfarb and coworkers introduced an experimental setup with combined NMR and EPR capabilities aimed at studying these mechanisms.[59] Temperature-dependent experiments performed on this setup demonstrated the complexity of the DNP process as a combination of different mechanisms.[60]

In this chapter, we focus on a theoretical approach that emphasizes the role of the electron spins in the DNP processes in static samples and in samples rotating at the magic angle. A better understanding of the role of the electrons in static samples can be achieved by relying on results of EPR experiments performed at DNP conditions. These types of experiments have been pioneered many years ago by Atsarkin,[61] who used cw-EPR to demonstrate the formation of an electron non-Zeeman spin temperature, following long MW irradiation. With the revival of interest in DNP, Kockenberger and coworkers[62] and Goldfarb and coworkers[63] pioneered the use of pulse EPR electron–electron double resonance (ELDOR) experiments at W-band, to probe the electron state after long MW irradiation. Following in their footsteps, ELDOR experiments have revealed that at high enough concentrations, even in the case of broad EPR spectra, the single-frequency MW irradiation can affect significantly the full spectrum by spectral diffusion[64] and, as a result, can increase the nuclear enhancement.[65,66] To evaluate and understand these effects, we use a model spin system recently introduced,[67] containing nine electrons and one nucleus, and quantum mechanical calculations to simulate the spin dynamics of this system under long MW irradiation. Then we introduce a mathematical approach for describing the spin evolution of macroscopic systems and use this model to analyze experimental ELDOR and DNP results.

In the rotating samples, the spin dynamics during DNP differs from the static case, because of the time dependence of the spin interactions. Therefore, the theoretical description of the static DNP processes must be modified, and in the last section, we summarize the MAS-DNP description, introduced by Thurber and Tycko[68] and Mentink-Vigier *et al.*[69] This description defines the different stages of the spin dynamics leading to DNP enhancements in small spin systems. Again, we will not discuss here technical details concerning MAS-DNP hardware.[6,70–73]

differentiation in the DNP efficiency of different radicals[74–79] and their actual enhancement values, and the excellent applications in biochemistry, surface chemistry, and materials research in general.

2.2 THE SPIN SYSTEMS

2.2.1 Glassy Amorphous Solid Solutions of Free Radicals

As mentioned in the introduction, we focus on frozen solutions of free radicals in glassy matrices. These samples are treated as spin systems composed of a network of interacting nuclei, with low concentrations of randomly distributed radicals with unpaired electrons imbedded between them. We will have in mind only three-dimensional systems and exclude two-dimensional electron systems representing surface or membrane samples. Crystalline samples with long-range structural order are also outside the scope of our discussions. The unpaired electrons of the free radicals have a spin $S = 1/2$ and are assumed to have anisotropic g-tensors. They are coupled to each other via electron–electron dipolar interactions and possibly via exchange interactions. We describe the electron spin system by a spin Hamiltonian, which for N_e coupled electrons will have 2^{N_e} eigenstates. Ignoring the electron–nuclear interactions, the frequencies of the SQ transitions between the eigenstates constitute the EPR spectrum of the system. These states are superpositions of the single-electron spin product states (i.e. the eigenstates of the z-components of the angular momentum operators), which are mixed due to the presence of the flip-flop terms in the dipolar and spin exchange Hamiltonian terms. This state mixing will result in a homogeneous broadening of the EPR spectrum, beyond the inhomogeneous width due to the g-tensor anisotropy. When the g-tensor frequency shifts are much larger than the dipolar frequencies, the EPR spectrum is mainly inhomogeneous; however, some state mixing still prevails. When reintroducing the hyperfine coupling, its electron–nuclear interactions in the spin system cause splitting of the electron transitions, which, in turn, can cause a significant broadening of the EPR spectrum. These interactions create also a mixing between the electron product states and the nuclear product states.

In a typical DNP experiment, thousands of nuclear spins, which all can experience a DNP enhancement, surround each electron spin. These polarization enhancements spread from nuclei, which are hyperfine coupled to the electrons, to all non-hyperfine-coupled nuclei in the sample. This polarization transfer process is governed by the spin diffusion mechanism, derived from the nuclear dipolar interactions. Thus, to understand the DNP process, we must first try to comprehend the enhancement process of the nuclei coupled to the electrons and then consider the spin diffusion process. Proceeding along these lines, we first discuss the direct DNP enhancement and make a distinction between nuclei that are strongly coupled to the electrons and those that are weakly coupled. The first are the 'core nuclei' that are polarized by the DNP process, but cannot transfer their polarization to the rest of the nuclei. This stagnation of the polarization transfer process to these 'bulk' nuclei is due to hyperfine truncation of the nuclear dipolar interactions (the spin diffusion barrier). The second are the 'local nuclei' that are polarized via the direct or dipolar assisted hyperfine interactions, but are able to transfer their polarization via spin diffusion to the bulk nuclei.[80]

In this section, we first introduce the spin Hamiltonian, defining the static interactions in the spin systems under consideration, and the spin relaxation processes necessary to support the DNP process. We then describe the spin dynamics of these systems, with the help of a set of rate equations for the populations of their eigenstates. These equations are derived from the master equation of the spin density matrix elements in Liouville space. Solving these population equations provides the basis for the description of the spin dynamics of the electrons and nuclei, leading to the DNP enhancements. Examples of solutions of these equations are first presented for small spin systems in Sections 2.3 and 2.4. Then we very briefly discuss the local-to-bulk nuclear polarization transfer, supported by the spin diffusion mechanism. Finally, in Section 2.5, we consider real spin systems and their DNP characteristics.

2.2.2   The Spin Hamiltonian

We begin by considering a spin system, which includes a finite number of electrons and nuclei, $n_e = 1, \ldots, N_e$ and $n_n = 1, \ldots, N_n$, respectively. The spin Hamiltonian that describes this system, when placed in an external magnetic field and exposed to MW irradiation, is composed of a set of terms that depend on the angular momentum operator components of the nuclei, $\widehat{I}_{n_n,p}$,

with $p = x, y, z$, allowing different types of nuclei, and of the electrons, $\widehat{S}_{n_e,p}$. Here we limit ourselves to cases with $I_{n_n} = 1/2$ and $S_{n_e} = 1/2$. As is common in magnetic resonance in general, we present the Hamiltonian of the system in the rotating frame with respect to a fixed MW frequency ω_{MW}. Doing so, we neglect any of its time-dependent terms oscillating at the frequency ω_{MW} or its multiples.

Thus, the total Hamiltonian in the rotating frame of the MW irradiation field only, excluding the MW contribution itself, is given by

$$\widehat{H}^{RoF} = \widehat{H}^{RoF}_{Z,e} + \widehat{H}^{RoF}_{Z,n} + \widehat{H}^{RoF}_{e-e} + \widehat{H}^{RoF}_{J} + \widehat{H}^{RoF}_{e-n} + \widehat{H}^{RoF}_{n-n} \tag{2.1}$$

where the different terms are given as follows:

2.2.2.1   *The Electron Zeeman Hamiltonian*

$$\widehat{H}^{RoF}_{Z,e} = \sum_{n_e} \Delta\omega_{n_e} \widehat{S}_{n_e,z} \tag{2.2}$$

where we define the individual off-resonance frequencies as $\Delta\omega_{n_e} = \omega_{n_e}(\alpha_{n_e}, \beta_{n_e}, \gamma_{n_e}) - \omega_{MW}$. The Zeeman frequency is $\omega_{n_e}(\alpha_{n_e}, \beta_{n_e}, \gamma_{n_e}) = g(\alpha_{n_e}, \beta_{n_e}, \gamma_{n_e})\beta_e B_0$, where β_e is the Bohr magneton and $(\alpha_{n_e}, \beta_{n_e}, \gamma_{n_e})$ are the Euler angles between the g-tensor principal axis system and the laboratory frame.

2.2.2.2   *The Nuclear Zeeman Hamiltonian*

$$\widehat{H}^{RoF}_{Z,n} = -\sum_{n_n} \omega_{n_n} \widehat{I}_{n_n,z} \tag{2.3}$$

with the nuclear Zeeman frequencies ω_{n_n}.

2.2.2.3   *The Electron–electron Dipolar Interaction Hamiltonian*

$$\widehat{H}^{RoF}_{e-e} = \sum_{n_e < n'_e} \omega_{n_e,n'_e} \left\{ 2\widehat{S}_{n_e,z}\widehat{S}_{n'_e,z} - \frac{1}{2}\left(\widehat{S}^+_{n_e}\widehat{S}^-_{n'_e} + \widehat{S}^-_{n_e}\widehat{S}^+_{n'_e} \right) \right\} \tag{2.4}$$

where we neglected all terms of the full dipolar Hamiltonian that do not commute with $\widehat{S}_z = \sum_{n_e} \widehat{S}_{n_e,z}$. The dipolar frequency of each interacting pair of electrons is

$$\omega_{n_e,n'_e} \equiv \omega_{n_e,n'_e}(\theta^D_{n_e,n'_e})$$
$$= g_{n_e} g_{n'_e} \beta_{e^2} \mu_0 / 4\pi\hbar r^3_{n_e,n'_e} \times 1/2(3\cos^2\theta^D_{n_e,n'_e} - 1) \tag{2.5}$$

with r_{n_e,n'_e} the distance between the electrons, n_e and n'_e, and $\theta^D_{n_e,n'_e}$ the angle between the distance vector r_{n_e,n'_e} and the external magnetic field.

2.2.2.4 The Exchange Interaction Hamiltonian

$$\hat{H}_J^{\mathrm{RoF}} = \sum_{n_e < n'_e} J_{n_e,n'_e} \left\{ \hat{S}_{n_e,z} \cdot \hat{S}_{n'_e,z} + \frac{1}{2} \left(\hat{S}^+_{n_e} \hat{S}^-_{n'_e} + \hat{S}^-_{n_e} \hat{S}^+_{n'_e} \right) \right\}$$

(2.6)

where J_{n_e,n'_e} are the Heisenberg spin exchange coupling constants between covalently interacting electrons.

2.2.2.5 The Nuclear–nuclear Dipolar Interaction Hamiltonian

$$\hat{H}_{n-n}^{\mathrm{RoF}} = \sum_{n_e < n'_e} \omega_{n_n,n'_n} \left\{ 2\hat{I}_{n_n,z} \hat{I}_{n'_n,z} - \frac{1}{2} \left(\hat{I}^+_{n_n} \hat{I}^-_{n'_n} + \hat{I}^-_{n_n} \hat{I}^+_{n'_n} \right) \right\}$$

(2.7)

where we truncated the total dipolar interaction Hamiltonian by leaving only those terms that commute with $\sum_{n_n} \hat{I}_{n_n,z}$. The dipolar frequency coefficients

$$\omega_{n'_n n'_n} \equiv \omega_{n_n,n'_n}(\theta^d_{n_n,n'_n})$$
$$= -\gamma_{n_n} \gamma_{n'_n} \mu_0 \hbar^2 / r^3_{n_n,n'_n} \times 1/2(3\cos^2\theta^d_{n_n,n'_n} - 1)$$

(2.8)

are dependent on the angle $\theta^d_{n_n,n'_n}$ between the distance vector between n_n and n'_n and the external magnetic field.

2.2.2.6 The Electron–nuclear Hyperfine Interaction Hamiltonian

$$\hat{H}_{e-n}^{\mathrm{RoF}} = \sum_{n_e,n_n} A_{n_e,n_n} \hat{S}_{n_e,z} \hat{I}_{n_n,z} + \frac{1}{2} B^+_{n_e,n_n} \hat{S}_{n_e,z} \hat{I}^+_{n_n}$$
$$+ \frac{1}{2} B^-_{n_e,n_n} \hat{S}_{n_e,z} \hat{I}^-_{n_n}$$

(2.9)

Here again we neglect terms that do not commute with $\hat{S}_{n_e,z}$ and we are left with the secular and pseudosecular terms. The coefficient of the secular term is $A_{n_e n_n} \equiv A^0_{n_e,n_n} + A_{n_e,n_n}(\theta^{\mathrm{HF}}_{n_e,n_n})$ where $A^0_{n_e,n_n}$ is the isotropic contribution to the electron–nuclear interaction, and $A_{n_e,n_n}(\theta^{\mathrm{HF}}_{n_e,n_n})$ is the dipolar contribution given by $A_{n_e,n_n}(\theta^{\mathrm{HF}}_{n_e,n_n}) = \frac{1}{2}\omega_{n_e,n_n}(3\cos^2\theta^{\mathrm{HF}}_{n_e,n_n} - 1)$ with $\omega_{n_e,n_n} = g_{n_e}\beta_e\gamma_{n_n}\mu_0\hbar/r^3_{n_e,n_n}$. $\theta^{\mathrm{HF}}_{n_e,n_n}$ is the angle

between the distance vector connecting electron n_e and nucleus n_n and the external magnetic field direction. The coefficient of the pseudosecular term is $B^\pm_{n_e,n_n}(\theta^{\mathrm{HF}}_{n_e,n_n}) = \frac{3}{2}\omega_{n_e,n_n} \sin\theta^{\mathrm{HF}}_{n_e,n_n} \cos\theta^{\mathrm{HF}}_{n_e,n_n}$. These terms present the static interactions that make the transfer of electron polarization to nuclear polarization possible.

2.2.3 The Spin Dynamics During MW Irradiation

The frequency of the MW irradiation applied on a spin system determines its rotating frame, and the MW Hamiltonian has the form

$$\hat{H}_{\mathrm{MW}}^{\mathrm{RoF}} = \omega_1 \sum_{n_e} \hat{S}_{n_e,x}$$

(2.10)

assuming, for simplicity, that during the upcoming discussions the MW field always points in the x direction. The irradiation field drives the system away from its thermal equilibrium, and the response of the system can result in DNP enhancement.

At this point, we have to introduce the theoretical approach we have chosen to describe the mechanisms responsible for the DNP enhancements. In this chapter, we base our description on spin dynamics calculations of small spin systems. The results of these calculations will then determine the way of describing DNP in real systems. The dynamics of the small spin systems will be presented here by solving rate equations for their eigenstate populations. In the eigenstate representation, the Hamiltonian $\hat{H}^{\mathrm{RoF}}$ becomes diagonal. Adding now the continuous MW Hamiltonian $\hat{H}_{\mathrm{MW}}^{\mathrm{RoF}}$ and assuming we can ignore all coherences generated by the MW irradiation, the diagonal elements of the spin density matrix in the same representation, i.e. the populations, are sufficient to describe the spin dynamics. In this representation, the temporal evolutions of the populations do not depend on the diagonal elements of the Hamiltonian in Hilbert space and thus also in Liouville space. The master equation therefore is a set of rate equations for the populations with rate constants, determined by irradiation parameters and the relaxation mechanisms. Thus with this assumption, in addition to the relaxation rate constants, we must introduce MW irradiation rates that are a function of the MW intensity, off-resonance values, and transverse relaxation times.

The general procedure for calculating the spin dynamics of a spin system during MW irradiation is composed of the following steps:

1. At first we generate $\hat{H}^{\text{RoF}}$ in its product state matrix representation and diagonalize it by a unitary transformation matrix $\hat{D}$. To simplify the forthcoming equations, we assume at this point that all nuclei have the same value $\omega_{n_n} = \omega_n$. The diagonalization

$$\hat{\Lambda}^{\text{RoF}} = \hat{D}^{-1}\hat{H}^{\text{RoF}}\hat{D} \qquad (2.11)$$

results in the eigenstates $|\psi_i\rangle$ with $i = 1, \ldots, 2^{N_e + N_n}$ with their rotating-frame energies $\varepsilon_i = \Lambda^{\text{RoF}}_{ii}$. These states have energies in the laboratory frame equal to $E_i = \langle\psi_i|\omega_{\text{MW}}\hat{S}_z|\psi_i\rangle + \varepsilon_i$, with $\hat{S}_z = \sum_{n_e} \hat{S}_{n_e,z}$.

2. Then we apply the diagonalization transformation to the matrix representations of the linear operators $\hat{S}_p = \sum_{n_e}\hat{S}_{n_e,p}$ and $\hat{I}_p = \sum_{n_n}\hat{I}_{n_n,p}, p = x, y, z$, resulting in the matrices $\{\hat{D}^{-1}\hat{S}_p\hat{D}\}$ and $\{\hat{D}^{-1}\hat{I}_p\hat{D}\}$.

3. At this stage, we define a population vector $\vec{p}(t)$, composed of all diagonal elements $p_i(t)$ of the spin density matrix in the $\{|\psi_i\rangle\}$ eigenstate representation, and determine their thermal equilibrium values, $p_i^{\text{eq}} = Z^{-1}e^{-E_i/k_B T_L}$, with Z the partition function, T_L the temperature of the lattice, i.e. the experimental temperature of the sample, and k_B the Boltzmann factor. We can now transfer the Liouville-von Neumann equation in Hilbert space, describing the dynamics of the spin system, to Liouville space, while considering only the diagonal $p_i(t)$ elements of the density matrix. The resulting rate equation for $\vec{p}(t)$ contains then only relaxation rate matrices and an MW rate matrix. We restrict ourselves to two types of relaxation processes, comprising the relaxation rate matrices. (i) The longitudinal relaxation rate matrix $\hat{R}_1$ that drives the elements of $\vec{p}(t)$ to their Boltzmann statistics values and (ii) the zero-quantum cross-relaxation rate matrix $\hat{R}_D$ that creates a Boltzmann distribution between the elements p_i of $\vec{p}(t)$, belonging to states with the same total z-component M_e of $\hat{S}_z$. Additional relaxation mechanisms, such as hyperfine-induced relaxation or double quantum dipolar supported relaxation, are only introduced when necessary.

4. To set the values of the elements of $\hat{R}_1$, we rely here on a semiempirical approach, in which we use measured values of T_1 and do not calculate their values from first principles, based on motional correlation times of the various relaxation mechanisms in our systems. During the upcoming discussions, we will for simplicity use only two longitudinal relaxation times, one for all electrons and one for all nuclei, respectively $T_{1,e}$ and $T_{1,n}$, also assuming that $T_{1,e}$ is isotropic. To simplify the calculations even further, we will not involve core nuclei with their short relaxation times. The $\hat{R}_1$ relaxation matrix has elements according to

$$\frac{\text{d}}{\text{d}t}\vec{p}(t) = -\hat{R}_1\vec{p}(t) \text{ with}$$

$$\frac{\text{d}}{\text{d}t}\begin{pmatrix} p_i \\ p_{i'} \end{pmatrix}(t) = \frac{R_{1,ii'}}{1+\eta_{ii'}}\begin{pmatrix} -1 & \eta_{ii'} \\ 1 & -\eta_{ii'} \end{pmatrix}\begin{pmatrix} p_i \\ p_{i'} \end{pmatrix}(t) \qquad (2.12)$$

where $\eta_{ii'} = e^{-(E_i - E_{i'})/k_B T}$ ensures a Boltzmann distribution at thermal equilibrium. For simplicity, we assume that the relaxation rates can be expressed as if they are caused by fluctuations of an effective magnetic field pointing in the x-direction perpendicular to the external magnetic field. Of course, the actual sources of these effective fluctuating fields are the dipolar, Raman, and direct relaxation processes.[81] As a result of this assumption, we use the following simple expression:

$$R_{1,ii'} = \frac{|\{2\hat{D}^{-1}\hat{S}_x\hat{D}\}_{ii'}|^2}{T_{1e}} + \frac{|\{2\hat{D}^{-1}\hat{I}_x\hat{D}\}_{ii'}|^2}{T_{1n}} \qquad (2.13)$$

It is important to note that by using this definition one obtains T_1 relaxation only between levels that are connected through linear angular momentum components. These connected levels also include the 'forbidden' ZQ and DQ (electron–nuclear) transitions after diagonalization of H^{RoF}. Because we are here not relying on explicit expressions for the relaxation rates, we ignored here again for simplicity fluctuations in the y- or z-direction. When necessary, variations in the relaxation times for different electrons and nuclei can easily be implemented in the calculations. All assumptions mentioned here are introduced to simplify the upcoming calculation.

5. In addition to the $\hat{R}_1$ relaxation mechanisms, we also take into account a zero-quantum cross-relaxation process between eigenstates with equal total z-component electron angular momentum values. This mechanism is a result of fluctuations of the electron dipolar interaction flip-flop terms $\omega_{n_e,n'_e}(\hat{S}^+_{n_e}\hat{S}^-_{n'_e} + \hat{S}^-_{n_e}\hat{S}^+_{n'_e})$ in equation (2.6). In our population rate equations, this is represented by the rate matrix $\hat{R}_D$, with elements that depend on the coefficients of the flip-flop

terms. During our calculations, we require that this relaxation process support the Boltzmann distribution of the populations at thermal equilibrium. We therefore have chosen the following form for the matrix elements of $\hat{R}_D$:

$$\frac{d}{dt}\overline{p}(t) = -\hat{R}_D\overline{p}(t) \quad \text{and}$$

$$\frac{d}{dt}\begin{pmatrix} p_i \\ p_{i'} \end{pmatrix}(t) = \frac{R_{D,ii'}}{1+\eta_{ii'}}\begin{pmatrix} -1 & \eta_{i,i'} \\ 1 & -\eta_{i,i'} \end{pmatrix}\begin{pmatrix} p_i \\ p_{i'} \end{pmatrix}(t) \tag{2.14}$$

with again $\eta_{ii'} = e^{-(E_i-E_{i'})/k_B T}$. Defining empirical expressions for the elements of $R_{D,ii'}$ that are consistent with experimental observations, we have chosen rate constants that decay according to $(\varepsilon_i - \varepsilon_{i'})^2$ for energy differences $|\varepsilon_i - \varepsilon_{i'}|$ significantly larger than $|\omega_{n_e,n'_e}|$. Assuming a common cross-relaxation time constant T_D for the whole electron system, we express the elements of $\hat{R}_D$ in the form

$$R_{D,ii'} = \sum_{n_e < n'_e} \frac{\omega^2_{n_e,n'_e}}{(|\omega_{n_e,n'_e}| + |\varepsilon_i - \varepsilon_{i'}|)^2}$$
$$\times \frac{|\{\hat{D}^{-1}(\hat{S}^+_{n_e}\hat{S}^-_{n'_e} + \hat{S}^-_{n_e}\hat{S}^+_{n'_e})\hat{D}\}_{ii'}|^2}{T_D} \tag{2.15}$$

We must realize that we can monitor the magnitude of T_D only indirectly, by measuring the spectral diffusion time constant T_{1D}.[82] Because $\hat{H}^{RoF}$ cannot cause any mixing between states with different total z-components of $\hat{S}_z$, the matrix elements in equation (2.15) do not connect states with different $\langle\psi_i|\hat{S}_z|\psi_i\rangle = M_e\hbar$ values. Therefore, this relaxation mechanism is indeed active only inside the manifold of states with the same M_e value.

6. In order to calculate the dynamics of the populations under MW irradiation, we introduce into the rate equation the effect of the MW irradiation in the form

$$\frac{d}{dt}\overline{p}(t) = \hat{W}_{MW}\overline{p}(t) \quad \text{and}$$

$$\frac{d}{dt}\begin{pmatrix} p_i \\ p_{i'} \end{pmatrix}(t) = W_{MW,ii'}\begin{pmatrix} -1 & 1 \\ 1 & -1 \end{pmatrix}\begin{pmatrix} p_i \\ p_{i'} \end{pmatrix}(t) \tag{2.16}$$

In this way, we ignore all coherences created by the MW irradiation matrix in the diagonalized frame. The elements of $\hat{W}_{MW}$ for each transition

$|\psi_i\rangle \leftrightarrow |\psi_{i'}\rangle$ are defined by[83]

$$W_{MW,ii'} = \frac{\omega^2_1|(D^{-1}S_x D)_{ii'}|^2 T_{2e}}{1 + (\varepsilon_i - \varepsilon_{i'})^2 T^2_{2e}} \tag{2.17}$$

where we introduced an effective electron transverse relaxation time T_{2e}, determining the off-resonance excitation efficiency.
We can now construct the overall rate equation for $\overrightarrow{p}(t)$ in the MW rotating frame

$$\frac{d}{dt}\overrightarrow{p}(t) = (\hat{W}_{MW} - \hat{R}_1 - \hat{R}_D)\overrightarrow{p}(t) \tag{2.18}$$

and solve it for different interaction coefficients and relaxation parameters.

2.2.4 EPR, ELDOR, and DNP Spectra

The solution for $\overrightarrow{p}(t)$, as a function of the MW irradiation period t, can now be used to calculate the time evolution of the intensities of the EPR and NMR spectra at different detection frequencies $v_{det} = \omega_{det}/2\pi$, for a fixed excitation frequency $v_{exc} = \omega_{MW}/2\pi$. The spectra are composed of signals originating from the different transitions of the spin system. These transitions $|\psi_i\rangle \leftrightarrow |\psi_{i'}\rangle$ are between levels with populations $p_i(t)$ and $p_{i'}(t)$, which contribute a value to the intensity of the EPR spectrum at frequency $v_{ii'} = (E_i - E_{i'})/2\pi$

$$S^e_{ii'}(t, v_{ii'}, [v_{exc}]) = (p_i(t) - p_{i'}(t))|\langle\psi_i|\hat{S}_x|\psi_{i'}\rangle|^2 \tag{2.19}$$

and to the intensity of the NMR spectrum

$$S^n_{ii'}(t, v_{ii'}, [v_{exc}]) = (p_i(t) - p_{i'}(t))|\langle\psi_i|\hat{I}_x|\psi_{i'}\rangle|^2 \tag{2.20}$$

The brackets $[v_{exc}]$ indicate that the v_{exc} frequency is fixed. The sum of the contributions of all transitions generates the normalized EPR spectrum

$$S^{EPR}(t, v_{det}, [v_{exc}]) = \frac{S_e(t, v_{det}, [v_{exc}])}{\{S_e(t, v_{det}, [v_{exc}])\}_{max}}$$
$$S_e(t, v_{det}, [v_{exc}]) = \sum_{i<i'} S^e_{i,i'} g(v_{ii'} - v_{det}) \tag{2.21}$$

where $g(v)$ is a narrow Gaussian function with $g(0) = 1$. The denominator is the maximum value of $S_e(t_{MW}, v_{det}, [v_{exc}])$ as a function of v_{det}. Because we are not dealing here with high-resolution NMR experiments, we are interested in the integrated and

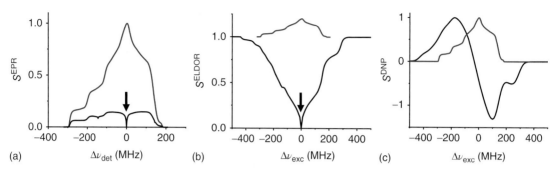

Figure 2.1. Examples of three types of simulated steady-state spectra, discussed in this chapter (in black): (a) A steady-state EPR spectrum during MW irradiation at the frequency assigned by the arrow. (b) A steady-state ELDOR spectrum, detected at the frequency of the arrow. (c) A normalized DNP spectrum, describing the normalized nuclear enhancement as a function of the irradiation frequency. The unperturbed equilibrium EPR spectrum in all three figures is shown in red. The parameters used for the simulation are the g-tensor elements of TEMPOL, an MW intensity of $v_1 = 0.1$ MHz, $T_{1e} = 10$ ms, $T_{2e} = 10$ μs, the electron spectral diffusion coefficient $\Lambda^{\text{eSD}} = 1000$ μs^{-3}, and the hyperfine parameter $A^{\text{SE}}/2\pi = 1$ MHz

normalized nuclear signals:

$$S^{\text{NMR}}(t, [v_{\text{exc}}]) = \frac{S_n(t, [v_{\text{exc}}])}{S_n^{\text{eq}}};$$

$$S_n(t, [v_{\text{exc}}]) = \sum_{i<i'} S_{i,i'}^n(t, v_{ii'}, [v_{\text{exc}}]) \qquad (2.22)$$

Here S_n^{eq} is obtained from equation (2.20), by inserting thermal equilibrium values for the populations $p_i(t) = p_i^{\text{eq}}$. We remind ourselves here that the EPR spectra and the normalized nuclear signal, i.e. the enhancement, are functions of the coefficients of the Hamiltonian of the relaxation rates and of the MW irradiation parameters. Keeping all these parameters unchanged, but changing the MW frequency v_{exc}, will result in different EPR spectra and must tell us about the MW frequency dependence of the DNP enhancement efficiencies.

In many circumstances, it is more practical to vary the excitation frequency while keeping the detector constant, as commonly used in EPR spectroscopy for measuring ELDOR-detected NMR spectra.[84] The alternative of keeping the excitation frequency constant and moving the detector, which gives us the EPR spectrum under constant irradiation as is typically done in DNP, is less practical. Normally, we normalize the ELDOR spectra as follows:

$$S^{\text{ELDOR}}(t, v_{\text{exc}}, [v_{\text{det}}]) = \frac{S_e(t, v_{\text{det}}, [v_{\text{exc}}])}{S_e(t, v_{\text{det}}, [v'_{\text{exc}}])} \qquad (2.23)$$

where for each frequency pair $\{v_{\text{det}}, v_{\text{exc}}\}$ the EPR signal is normalized by the signal obtained for the pair

$\{v_{\text{det}}, v'_{\text{exc}}\}$, with v'_{exc} far removed from the frequency range $[\Omega_{\text{min}}, \Omega_{\text{max}}]$ of the EPR spectrum.

Of course, both types of measurements contain the same information about the depolarization of the electrons in the sample during MW. To support our theoretical descriptions, in the next sections we will calculate the EPR spectra (see Figure 2.1a) to present the spin dynamics. In practice, however, we will rely mainly on experimental ELDOR data (see Figure 2.1b) to describe the electron spin dynamics.

The dependence of the enhancement as a function of the excitation frequency can also be plotted, resulting in, what we will call, a DNP spectrum:

$$S^{\text{DNP}}(t, v_{\text{exc}}) = S^{\text{NMR}}(t, [v_{\text{exc}}]) \qquad (2.24)$$

In the forthcoming text, we discuss the lineshapes of these DNP spectra and plot normalized spectra (see Figure 2.1c) or consider relative enhancements as a function of experimental parameters.

In the following section, we present a short discussion about electron polarizations in the case of coupled electron spin systems.

2.2.5 The Electron Polarizations

The mechanisms responsible for the enhancement of nuclear signals or the depletion of EPR signals are in many instances explained in terms of electron and nuclear polarizations. We define the polarization of an electron n_e as

$$P_{n_e}(t) = \text{Tr}(\hat{\rho}^{\text{RoF}}(t)\hat{S}_{z,n_e}) = \text{Tr}(\hat{\rho}^{\text{diag}}(t)\hat{D}^{-1}\hat{S}_{z,n_e}\hat{D}) \tag{2.25}$$

where $\hat{\rho}^{\text{RoF}}(t)$ and $\hat{S}_{z,n_e}$ are the spin density and angular momentum operators in the nondiagonalized frame. The diagonal elements of $\hat{\rho}^{\text{diag}}(t) = \hat{D}^{-1}\hat{\rho}^{\text{RoF}}(t)\hat{D}$ are in fact the elements of $\vec{p}(t)$.

When the dipolar flip-flop terms of $\hat{H}^{\text{RoF}}$ are ignored and we concentrate on the electrons only, its eigenstates are pure product states. Then the expressions for $P_{n_e}(t)$ can be written as a sum of population differences $(p_i(t) - p_{i'}(t))$ between states for which $\langle\psi_i|\hat{S}_{x,n_e}|\psi_{i'}\rangle \approx 1/2$, all with a transition frequency $\varepsilon_i - \varepsilon_{i'} = \nu_{i,i'} = \nu_{n_e}$. The intensity of the EPR spectrum at a frequency ν_{det} is thus proportional to the sum of all polarizations of the electrons with transition frequencies about equal to ν_{det}. MW irradiation at a frequency ν_{exc} can affect the polarization of the electrons resonating at a frequency ν_{det}, and as a result the EPR intensity at that frequency will be affected. By measuring the EPR intensity, we can extract the electron polarization by

$$P_e(t, \nu_{\text{det}}, [\nu_{\text{exc}}]) = \int_{\nu_{\text{det}}-\delta\nu}^{\nu_{\text{det}}+\delta\nu} \frac{S^{\text{EPR}}(t, \nu', [\nu_{\text{exc}}])}{S_{\text{eq}}^{\text{EPR}}(\nu')} P_e^{\text{eq}}(\nu')d\nu' \tag{2.26}$$

Of course, as soon as the electron dipolar interaction starts mixing the product states this correspondence between the EPR intensities and the single-electron polarizations is not valid anymore. Despite this fact, in the case of weak dipolar interactions we still consider different electron polarizations resonating at narrow frequency ranges that together compose the EPR lines. When the dipolar interaction increases significantly and we must consider the homogeneous broadening effects of inhomogeneous EPR spectra, we should refrain from correlating polarizations with well-defined electrons. However, we still consider 'polarizations of the electron spin system' with resonance frequencies in well-defined small frequency ranges. Thus, an intensity of the EPR line at a certain frequency can still be thought of as a sum of polarizations.

2.3 DNP SIMULATIONS OF A BASIC SPIN SYSTEM

2.3.1 The Solid Effect

The first DNP enhancement mechanism we discuss here is the 'solid effect'. It is the enhancement obtained during an MW excitation of electron–nucleus ZQ or DQ 'forbidden' transitions that become allowed due to the presence of the pseudosecular terms of the hyperfine interactions in equation (2.9), $B_{n_e,n_n}^{\pm}S_{n_e,z}I_{n_n}^{\pm}$. These terms cause a weak mixing of the electron–nucleus product states, when $B_{n_e,n_n}^{\pm} \ll \omega_n$. To demonstrate the SE enhancement mechanism, we consider a two-spin system $\{e-n\}$ containing one electron, with a resonance frequency ω_e, coupled to a nucleus with a Larmor frequency ω_n. The only interaction parameters of this system are the hyperfine interaction coefficients $A_{e,n}$ and $B_{e,n}^{\pm}$. The energy level diagram of this system is shown in Figure 2.2(b), and the thermal equilibrium EPR spectrum $S_{\text{eq}}^{\text{EPR}}(\Delta\nu_{\text{det}})$, with $\Delta\nu_{\text{det}} = \nu_{\text{det}} - \nu_e$, is shown in black in Figure 2.2(e). Adding MW irradiation and relaxation parameters to the system, we can solve the corresponding population rate equation in equation (2.18) and calculate a steady-state spectrum for the electron and nucleus. In Figure 2.2(e) the steady-state DNP spectrum $S^{\text{DNP}}(t, \Delta\nu_{\text{exc}})$ is shown in red, with $\Delta\nu_{\text{exc}} = \nu_{\text{exc}} - \nu_e$, for a set of interaction parameters given in the figure caption. As we see, NMR enhancement is obtained when $\Delta\nu_{\text{exc}} = \pm \nu_n$. The time evolutions of the electron and nuclear signals, which are proportional to their polarizations, are also shown as a function of MW irradiation time t in Figure 2.2(e) and (f). Some observations can be made: The buildup time of the nuclear polarization is somewhat longer than the electron T_{1e} and its steady-state value approaches the electron polarization. The electron polarization decays initially and returns almost to its original value also in a time determined by T_{1e}.

Here we should add some comments about the SE enhancement process. The strength of the irradiation experienced by the 'forbidden' ZQ and DQ transitions is represented in the population rate equation by the MW rate defined in equation (2.17). In this equation, the matrix elements of $\hat{S}_x$, in the eigenstate representation of H^{RoF}, are of the order of magnitude $(B_{e-n}^{\pm}/\omega_n) \cdot \omega_1$, assuming again $B_{e-n}^{\pm} \ll \omega_n$. Thus, an effectively weak irradiation $\omega_1^{\text{SE}} \approx \frac{1}{4}(B_{e-n}^{\pm}/\omega_n) \cdot \omega_1$ induces the enhancement. Whether or not this irradiation succeeds in equalizing the population of the two states of the ZQ and DQ transitions depends, of course, on the values of T_{2e} and on the spin–lattice relaxation times of these 'forbidden' transitions. Expressions for these times are in our model derived from equation (2.13) and are of the order of $(B_{e-n}^{\pm}/\omega_n)^{-2}T_{1e}$. Thus, a weak effective MW field and this relatively long effective spin–lattice relaxation time enable an efficient equilibration of the populations.

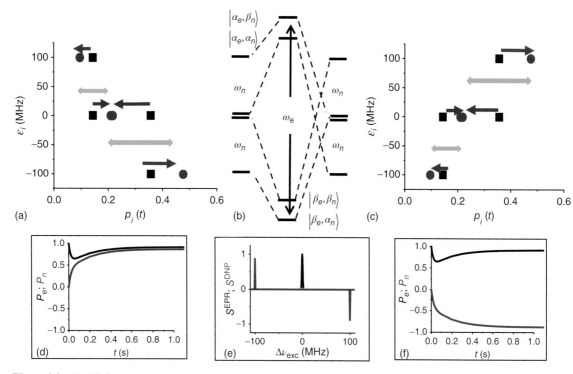

Figure 2.2. In this figure, a schematic representation of the SE-DNP mechanism is presented. In (b) the energy level diagram in the laboratory frame of an $\{e-n\}$ spin system is drawn. The states are added and the frequency scale is of course different for the nuclear Larmor frequency and the electron Larmor frequency. On the left-hand side of this diagram the rotating-frame energy diagram is shown, for an MW field excitation of the DQ transition and on the right-hand side, for an MW excitation of the ZQ transition. In (a) the energy population plot in the DQ rotating frame is shown and in (c) the same in the ZQ rotating frame. In both plots, the thermal equilibrium populations are shown as black squares. The result of the MW saturation is presented by blue arrows and the spin–lattice relaxation effects by red arrows. The steady-state populations p_i after a long MW irradiation are drawn as red dots. The gray arrow in (a) and (c) indicates the population differences determining the nuclear polarization. In (d) and (f) the normalized electron P_e (in black) and the nuclear polarization P_n (in red) during the polarization buildup are shown and in (e) the DNP spectrum is drawn in red and the EPR line in black. The parameters for these simulations are $v_n = 100\,\text{MHz}$; $B^{\pm}_{e,n}/2\pi = 3\,\text{MHz}$; $A_{e,n}/2\pi = 1\,\text{MHz}$; $T_{1n} = 1\,\text{s}$; $T_{1e} = 100\,\text{ms}$; $T_{2e} = 100\,\mu\text{s}$; $v_1 = 50\,\text{kHz}$

Another way of showing the SE enhancement process is by an energy versus population plot in the rotating frame, as in Figure 2.2(a) and (c). These energies are fixed, and in the rotating frame, the two energies of the irradiated transition are about equal. Initially, all populations, $p_{\chi\chi'} = p_{\chi_e\chi_n}$ with $\chi, \chi' = \alpha$, β, are determined by their thermal equilibrium values (black squares) and during irradiation the populations of the states of the irradiated transition become equal (blue arrows). This is followed by the action of the relaxation mechanism, modifying the populations of the nonirradiated transitions by trying to reach Boltzmann ratios between the populations (red arrows).

Altogether, this results in steady-state populations (red circles). The energy–population plots in Figure 2.2(a) and (c) are calculated for a set of interaction and relaxation parameters given in the caption. Realizing that the electron polarizations are about proportional to the sum of the population differences $p_{\beta a} - p_{\alpha a}$ and $p_{\beta\beta} - p_{\alpha\beta}$ and the nuclear polarization to the differences $p_{\beta a} - p_{\beta\beta}$ and $p_{\alpha a} - p_{\alpha\beta}$, respectively, it is easy to see that $P_n(t, \Delta v_{\text{exc}})$, with $\Delta v_{\text{exc}} = \pm v_n$, reaches about the value $\pm P_e^{\text{eq}}(\Delta v_{\text{det}})$ at $\Delta v_{\text{det}} = 0$.

In situations where the electron polarization $P_e^{\text{fix}}(0) < P_e^{\text{eq}}(0)$ is fixed, because of interactions of our $\{e-n\}$ system with its environment, the population differences between the electron transition states become fixed as well. Then the MW irradiation can still saturate the ZQ or DQ transitions and can reach an equilibration with $P_n(t, \Delta \nu_{\text{exc}}) \cong \pm P_e^{\text{fix}}(0)$.

2.3.2 The Cross Effect

The next spin system we consider is a three-spin system $\{e_2 - e_1 - n\}$. In this system, an electron e_1 is coupled to a nucleus n via the hyperfine interaction and to another electron e_2 via the electron dipolar interaction. This system will be used to demonstrate the action of the 'CE' process. Here we consider only systems where the dipolar and the pseudosecular hyperfine interactions are smaller than the nuclear Zeeman–Larmor frequency. The parameters of this system will be chosen such that we can discuss sequentially the spin dynamics of (i) the SE, (ii) the direct CE, (iii) the indirect CE, and (vi) the heteronuclear CE mechanisms.

We begin by defining the 'CE condition' for a nucleus with a Larmor frequency ω_n as

$$|\omega_{e_1} - \omega_{e_2}| \approx |\omega_n| \qquad (2.27)$$

Here ω_{e_1} and ω_{e_2} are the resonance frequencies of the two electrons e_1 and e_2, respectively, in the laboratory frame. A direct consequence of this condition is a degeneracy (see Figure 2.3) of the product state energies (before taking $\widehat{H}_{\text{D}}^{\text{RoF}}$ and $\widehat{H}_{\text{HF}}^{\text{RoF}}$ into account)

$$|\alpha_{e_1}, \beta_{e_2}, \beta_n\rangle \leftrightarrow |\beta_{e_1}, \alpha_{e_2}, \alpha_n\rangle \text{ or}$$
$$|\alpha_{e_1}, \beta_{e_2}, \alpha_n\rangle \leftrightarrow |\beta_{e_1}, \alpha_{e_2}, \beta_n\rangle \qquad (2.28)$$

and an overlap between a ZQ or DQ transition of one electron with an SQ transition of the other electron. For example, for the second case in equation (2.28), the SQ transition between $|\beta_{e_1} \beta_{e_2} \beta_n\rangle$ and $|\beta_{e_1} \alpha_{e_2} \beta_n\rangle$ of e_2 overlaps with the DQ transition $|\beta_{e_1} \beta_{e_2} \beta_n\rangle$ and $|\alpha_{e_1} \beta_{e_2} \alpha_n\rangle$ corresponding to $\{e_1 - n\}$. To follow the spin dynamics of the $\{e_1 - e_2 - n\}$ system under DNP conditions, we add to the Hamiltonian the electron–electron dipolar and hyperfine interaction parameters and transfer it to its diagonal form. At the CE conditions, the degenerate states can become fully mixed, despite the fact that the matrix representations of $\widehat{H}_{\text{D}}^{\text{RoF}}$ and $\widehat{H}_{\text{HF}}^{\text{RoF}}$ do not have any off-diagonal elements between these degenerate states. However, the combination of the pseudosecular hyperfine terms of $\widehat{H}_{\text{HF}}^{\text{RoF}}$, $(B_{e_1,n}^+ \widehat{S}_{z,e_1} \widehat{I}_n^+ + B_{e_1,n}^- \widehat{S}_{z,e_1} \widehat{I}_n^-)$, and the electron flip-flop dipolar terms of $\widehat{H}_{\text{D}}^{\text{RoF}}$, $\omega_{e_1,e_2}(\widehat{S}_{e_1}^+ \widehat{S}_{e_2}^- + \widehat{S}_{e_1}^- \widehat{S}_{e_2}^+)$, creates an effective

off-diagonal element between the degenerate states. Applying degenerate perturbation theory[85] and still assuming that $B_{e_1,n}^\pm, \omega_{e_1,e_2} \ll \omega_n$, these elements can be represented in $\widehat{H}^{\text{RoF}}$ as effective interaction terms of the form[86]

$$(B_{e_1,n}^+ \omega_{e_1,e_2}/8\omega_n)\widehat{S}_{e_1}^+ \widehat{S}_{e_2}^- \widehat{I}_n^+ + (B_{e_1,n}^- \omega_{e_1,e_2}/8\omega_n)$$
$$\times \widehat{S}_{e_1}^+ \widehat{S}_{e_2}^- \widehat{I}_n^- + c.c \qquad (2.29)$$

Although these terms are small, they can cause a state mixing of the form

$$2^{-1/2}(|\alpha_{e_1}, \beta_{e_2}, \beta_n\rangle \pm |\beta_{e_1}, \alpha_{e_2}, \alpha_n\rangle) \text{ or}$$
$$2^{-1/2}(|\alpha_{e_1}, \beta_{e_2}, \alpha_n\rangle \pm |\beta_{e_1}, \alpha_{e_2}, \beta_n\rangle) \qquad (2.30)$$

After the diagonalization, the MW rate constants $\widehat{W}_{\text{MW},ii'}$ in equation (2.17) become proportional to

$$\omega_1^{\text{CE}} \equiv |\langle 2^{-1/2}(\alpha_{e_1}, \beta_{e_2}, \beta_n \pm \beta_{e_1}, \alpha_{e_2}, \alpha_n)|$$
$$\times 2\widehat{S}_{x,e_2}|\alpha_{e_1}, \alpha_{e_2}, \beta_n\rangle|^2$$
$$\omega_1 = 2^{-1}\omega_1 \qquad (2.31)$$

This shows that the MW field can excite, very efficiently, transitions that include nuclear spin flips. Excitation of these transitions can thus become very efficient and the magnitude of ω_1^{CE} indicates that we expect broad $S^{\text{DNP}}(t, \nu_{\text{exc}}, [\nu_{\text{det}}])$ profiles because of efficient off-resonance saturation. We can now investigate the DNP enhancements that result from the CE conditions.

Introducing relaxation parameters to the system and choosing a set of interaction parameters, we can solve the population rate equation in equation (2.18) and calculate $S^{\text{EPR}}(t, \nu_{\text{det}}, [\nu_{\text{exc}}])$ and $S^{\text{DNP}}(t, \nu_{\text{exc}})$.

2.3.2.1 SE: $\omega_{e_1} - \omega_{e_2} \neq |\omega_n|$

We start by choosing the resonance frequencies of the two electrons such that they are not at a CE condition. In this case, we expect to observe only SE-DNP enhancements, when $\nu_{\text{exc}} = \nu_{e_1} \pm \nu_n$. Except for some dipolar splitting, we obtain the same results for $S^{\text{DNP}}(t, \nu_{\text{exc}})$ around ν_{e_1} as in the previous section.

2.3.2.2 Direct CE: $\omega_{e_1} - \omega_{e_2} = |\omega_n|$

Next we assume that the electrons e_1 and e_2 are at a CE condition. Then direct irradiation of one of these electrons will result in a nuclear enhancement. The effective irradiation strength of ω_1^{CE} is of the order of ω_1 itself, and the DNP spectrum becomes

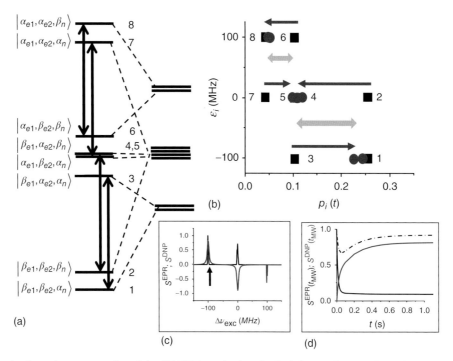

Figure 2.3. A schematic representation of the CE-DNP mechanism. In the left part of (a), a schematic energy level diagram of a three spin system $\{e_1 - e_2 - n\}$ at a CE condition in the laboratory frame is drawn, and the levels are assigned according to increasing energy as 1, ... 8. The CE condition sets the levels $\{4,5\}$ at the same energy. The MW irradiation on electron 2 is presented by four arrows. In the right-hand side of (a) the rotating-frame diagram is drawn, where the pairs of levels $\{1,3\}$, $\{2,4\}$, $\{5,7\}$, and $\{6,8\}$ become degenerate. In (b) the rotating-frame energy–population plot is presented with the black squares, representing the thermal equilibrium populations. The MW saturation process, together with the spin–lattice relaxation effect, is presented as red arrows. The steady-state populations are shown in red, and the nuclear polarization is determined by the population differences assigned by the gray arrows. In (c), the EPR (in black) and the DNP (in red) spectra are drawn. In (d), the buildup time of the normalized nuclear polarization is shown (in red), together with the polarizations of electron e_1 (the dotted black line) and of e_2 (the solid black line). The black arrow in (b) indicates the value of the MW frequency. The parameters used here are $v_{e_2} = -100\,\text{MHz}$; $v_{e_1} = 0\,\text{MHz}$; $B_{e,n}^{\pm}/2\pi = 3\,\text{MHz}$; $A_{e_1,n}/2\pi = 1\,\text{MHz}$ and $T_{1n} = 1\,\text{s}$; $T_{1e} = 100\,\text{ms}$; $T_{2e} = 100\,\mu\text{s}$; $\omega_{e_1,e_2}/2\pi = 1\,\text{MHz}$; $v_1 = 50\,\text{kHz}$

much broader than in the SE case, as is clearly seen in Figure 2.3(c). Here again we can investigate the DNP process by following the solution of equation (2.18) and drawing an energy–population plot. This is shown in Figure 2.3(b) for $v_{\text{exc}} = v_{e_2}$, assuming that the off-resonance saturation bandwidth covers the dipolar and hyperfine broadened single-electron spectra. Again the equilibrium populations (black squares) of the excited transitions become about equal and the MW irradiation and relaxation processes (red arrows) bring the system to its steady state (red circles). The major difference between the SE- and the CE-DNP processes is thus the effective power of the MW irradiation on the 'forbidden' transition, leading to

the enhancement of the nuclear spin polarization and as a direct consequence there will be a difference in the width of the off-resonance effect. For comparison, the buildup time of the nuclear polarization in the CE-DNP case is also shown in Figure 2.3(d).

For completeness, we show in Figure 2.4 $S^{\text{DNP}}(t, v_{\text{exc}})$ spectra of a three-spin system with $\omega_{e_1,e_2} \approx 1/2\omega_n$. In this case, the diagonalization of the Hamiltonian, before introducing the hyperfine interaction, results in eight eigenstates $|1^{\chi}\rangle = |\beta_{e_1}\beta_{e_2}\chi_n\rangle$, $|2^{\chi}\rangle = |\psi_e^{(2)}\chi_n\rangle$, $|3^{\chi}\rangle = |\psi_e^{(3)}\chi_n\rangle$, and $|4^{\chi}\rangle = |\alpha_{e_1}\alpha_{e_2}\chi_n\rangle$, with $\chi_n = \alpha_n, \beta_n$. The electron states $|\psi_e^{(2)}\rangle$ and $|\psi_e^{(3)}\rangle$ are linear combinations of $|\alpha_{e_1}\beta_{e_2}\rangle$ and $|\beta_{e_1}\alpha_{e_2}\rangle$ and their energies are separated

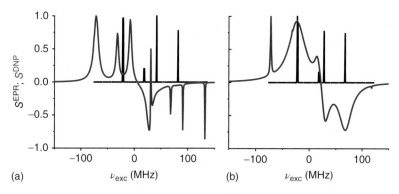

Figure 2.4. EPR stick spectra (black) and DNP spectra (red) of a system of two electrons and a nucleus $\{e_1 - e_2 - n\}$ with a strong electron–electron dipolar interaction $\omega_{e_1,e_2}/2\pi = 20$ MHz on the order of the nuclear Larmor: $\nu_n = 50$ MHz. The DNP calculations are performed for two values of the difference between the electron frequencies $\Delta\nu = \nu_{e_1} - \nu_{e_2}$; (a) $\Delta\nu = 60$ MHz, (b) $\Delta\nu = 45$ MHz. Clearly, in both cases, although we are far from the CE condition, $\Delta\nu = \nu_n$, we still see broad DNP spectra, as typical for the CE case. This is a signature of the strong state mixing in the spin system. Other parameters are $B_{e,n}^{\pm}/2\pi = 2$ MHz; $A_{e1,n}/2\pi = 1$ MHz; $T_{1e} = 100$ ms; $T_{2e} = 100$ μs and a Gaussian broadening of 0.5 MHz was applied

by a value $\{(\omega_{e_1} - \omega_{e_2})^2 + \omega_{e_1,e_2}^2\}^{1/2}$. Then adding the hyperfine interaction with $|B_{e-n}^{\pm}| \ll |\omega_n|, |\omega_{e_1,e_2}|$ to the diagonalized Hamiltonian can cause a mixing of the states $|\psi_e^{(2)}\alpha_n\rangle$ and $|\psi_e^{(3)}\beta_n\rangle$, when their energies become equal:

$$\{(\omega_{e_1} - \omega_{e_2})^2 + \omega_{e_1,e_2}^2\}^{1/2} \approx |\omega_n| \qquad (2.32)$$

Because of this mixing, the MW field can again excite transitions with mixed electron–nuclear states and results in nuclear enhancements. This can be clearly observed in Figure 2.4. We do not deal with systems with nuclear Larmor frequencies of the order of dipolar interactions, as it is not common in DNP.

2.3.2.3 Indirect CE: $|\omega_{e_1} - \omega_{e_2}| \approx |\omega_n|$

We saw in the *direct* CE case that MW saturation of one of the two coupled electrons, $P_{e_1}(t) = 0$, results in an increase in the nuclear polarization, reaching a value $P_n(t) = P_{e_2}^{eq}$. Two electrons at a CE condition that are part of a multielectron spin system can become depolarized and can reach different values of polarization. This can happen even when the spin system is irradiated at a frequency position in the EPR line removed from the transitions of the two electrons. These different polarizations are then a result of the spectral diffusion mechanism, which for long MW irradiation can result in an extremely broad excitation profile. We can mimic this situation in our small spin system's calculations by adding to our three-spin

system a third electron e_3, with dipolar coefficients $\omega_{e_1,e_3} \neq \omega_{e_2,e_3}$ in equation (2.4). Continuous irradiation at the resonance frequency of e_3 can then, because of the cross-relaxation mechanism defined in equation (2.15), result in steady-state polarizations that are fixed and nonequal, $P_{e_1}^{fix} \neq P_{e_2}^{fix}$. Putting this constraint in the population rate equation for the three-spin system $\{e_2 - e_1 - n\}$, we can calculate $S^{DNP}(t, \nu_{exc})$. The steady-state solution for the nuclear polarization for this case, after a long irradiation period t, has been derived already more than forty years ago and is given by[87]

$$P_n(t) = \frac{P_{e_1}^{fix} - P_{e_2}^{fix}}{1 - P_{e_1}^{fix} P_{e_2}^{fix}} \qquad (2.33)$$

where $\nu_{e_1} < \nu_{e_2}$. This expression is consistent with the populations at thermal equilibrium, as can easily be shown by inserting the expressions for the equilibrium electron polarizations and by using trigonometric relations:

$$P_n(t) = \frac{\tanh(\hbar\omega_{e_1}/2k_B T_L) - \tanh(\hbar\omega_{e_1}/2k_B T_L)}{1 - \tanh(\hbar\omega_{e_1}/2k_B T_L) \times \tanh(\hbar\omega_{e_1}/2k_B T_L)}$$
$$= \tanh(\hbar(\omega_{e_1} - \omega_{e_2})/2k_B T_L) \qquad (2.34)$$

The right-hand side of this equality is, at the CE condition, just equal to the polarization of the nucleus.

To demonstrate the solution of the population rate equation for fixed electron polarizations, we can again plot its solution in the form of an energy–population plot as in Figure 2.5. In this representation, at the high temperature approximation, fixed polarizations mean

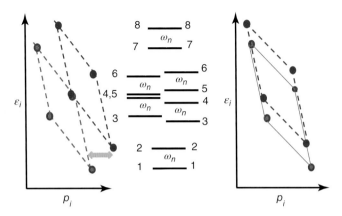

Figure 2.5. A schematic presentation of the iCE enhancement mechanism. In the center, the energy level diagram is shown for a $\{e_1 - e_2 - n\}$ spin system in an arbitrary rotating frame. The state assignments of the levels are the same as in Figure 2.3. Levels 3 to 6 on the right represent a system without degeneracy of levels 4 and 5, and on the left with degeneracy of levels 4 and 5, corresponding to the CE condition. On the right and left, the corresponding energy–population plots of these systems are drawn, where the blue dots present population of states with α_n and the red with β_n. The position of the dots corresponds to the level positions in the center diagram. In both cases, we assume that the electron polarizations are fixed, i.e. for e_2 the differences $p_1 - p_3 = p_2 - p_4$ and $p_5 - p_7 = p_6 - p_8$ are fixed and are not equal to the fixed differences $p_1 - p_5 = p_2 - p_6$ and $p_3 - p_7 = p_4 - p_8$ for e_1. In the nondegenerate case drawn on the right, the values of p_3 and p_4, as well as p_5 and p_6, are the same, because of the T_{1n} relaxation. In the degenerate case, due to the state mixing between levels 4 and 5, p_4 and p_5 become equal, overriding T_{1n}, while keeping the population differences fixed as in the nondegenerate case. This results in a shift of all red dots away from the blue dots, thus creating an enhancement of the nuclear polarization, indicated by the gray arrow

fixed population differences indicated by the dotted lines. The energy degeneracy results in a population equilibrium without changing these differences, and as a result for long t we see that $P_n(t) = P_{e_1}^{\text{fix}} - P_{e_2}^{\text{fix}}$, as explained in the figure caption.

Introducing more nuclei with the same Larmor frequency to the system can result in additional CE conditions of higher order[24]:

$$|\omega_{e_1} - \omega_{e_2}| = k|\omega_n| \qquad (2.35)$$

with k an integer. These conditions are, of course, narrow because the effective off-diagonal elements derived from degenerate perturbation theory are very small. An extension of these higher-order conditions is the heteronuclear CE conditions.

2.3.2.4   Heteronuclear CE:
$$\omega_{n_1} \neq \omega_{n_2}; \, |\,\omega_{e_1} - \omega_{e_2}\,| \approx |\,\omega_{n_1} \pm \omega_{n_2}\,|$$

In most samples, more than one type of nuclear spins are present and can be polarized by DNP. Considering only two types of nuclei n_1 and n_2 with Larmor frequencies ω_{n_1} and ω_{n_2}, we recognize, in addition to the individual CE conditions $|\omega_{e_1} - \omega_{e_2}| \approx |\omega_{n_{1/2}}|$, the

existence of the conditions:

$$|\,\omega_{e_2} - \omega_{e_1}\,| \approx |\,\omega_{n_1} + \omega_{n_2}\,| \text{ or}$$
$$|\,\omega_{e_2} - \omega_{e_1}\,| \approx |\,\omega_{n_2} - \omega_{n_1}\,| \qquad (2.36)$$

These equalities are called the heteronuclear CE conditions (hn-CE).[88] In a spin system $\{e_1 - e_2 - n_1 - n_2\}$, they cause degeneracies between the product states

$$|\alpha_{e_1}\beta_{e_2}\beta_{n_1}\beta_{n_2}\rangle \leftrightarrow |\beta_{e_1}\alpha_{e_2}\alpha_{n_1}\alpha_{n_2}\rangle \text{ or}$$
$$|\alpha_{e_1}\beta_{e_2}\beta_{n_1}\alpha_{n_2}\rangle \leftrightarrow |\beta_{e_1}\alpha_{e_2}\alpha_{n_1}\beta_{n_2}\rangle \qquad (2.37)$$

When we again assume that the electron polarizations are fixed and not equal, $P_{e_1}^{\text{fix}} < P_{e_2}^{\text{fix}}$, then the steady-state values of the nuclear polarizations satisfy the following equalities:

$$P_{n_1}^{\text{ss}} + P_{n_2}^{\text{ss}} = \frac{P_{e_1}^{\text{fix}} - P_{e_2}^{\text{fix}}}{1 - P_{e_1}^{\text{fix}} P_{e_2}^{\text{fix}}} \text{ or } P_{n_1}^{\text{ss}} - P_{n_2}^{\text{ss}} = \frac{P_{e_1}^{\text{fix}} - P_{e_2}^{\text{fix}}}{1 - P_{e_1}^{\text{fix}} P_{e_2}^{\text{fix}}}$$
$$(2.38)$$

where we assumed that $\omega_{e_1} < \omega_{e_2}$; $\omega_{n_1} < \omega_{n_2}$. The individual values of the two nuclear polarizations $P_{n_1}^{\text{ss}}$ and $P_{n_2}^{\text{ss}}$ are dependent on the relative magnitudes of the two nuclear relaxation times T_{1n_1} and T_{1n_2}.[88] Figure 2.6 shows an example of the calculation, following a time evolution of a $\{e_1 - e_2 - n_1 - n_2\}$ system fulfilling

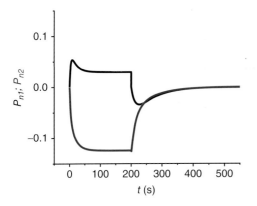

Figure 2.6. An example of the evolution of the intensity of the normalized nuclear polarizations P_{n_i} with $i = 1, 2$ of a $\{e_1 - e_2 - n_1 - n_2\}$ spin system at the hn-CE condition, as in equation (2.36), during MW irradiation at the frequency of electron e_1. For this specific system, the MW irradiation results in opposite polarizations for the two nuclei. After a time $t = 200\,$s the MW is shut off and the polarization of n_1, represented by the black curve, is set to zero. The calculation demonstrates the approach of the two nuclear polarizations toward each other, before reaching thermal equilibrium. The parameters of the simulation are $v_{e1} = -109.93\,$MHz; $v_{e2} = 0\,$MHz; $v_{n1} = 150\,$MHz; $v_{n2} = 40\,$MHz; $B_{e,n_1}^{\pm}/2\pi = 2\,$MHz; $A_{e,n_1}/2\pi = 1\,$MHz; $B_{e,n_2}^{\pm}/2\pi = 3\,$MHz; $A_{e,n_2}/2\pi = 1\,$MHz; $T_{1n_1} = 10\,$s; $T_{1n_2} = 100\,$s; $T_{1e} = 100\,$ms; $T_{2e} = 100\,\mu$s; $\omega_{e_1,e_2}/2\pi = 2\,$MHz; $v_1 = 200\,$kHz

the hn-CE condition $(\omega_{e_2} - \omega_{e_1}) \approx (\omega_{n_1} - \omega_{n_2})$. The parameters of the system are given in the figure caption.

The experiment we have in mind starts with a long MW irradiation on electron e_1, creating two different nuclear steady-state polarizations. In this system, the nuclear polarizations have opposite signs and different in amplitudes. This result is dictated by the difference between the T_{1n} values of the nuclei, which also causes the polarizations to reach their steady-state values in the time scale of the T_{1n}s. After reaching the steady state, the MW field is turned off and the polarization of one of the nuclei n_1 is saturated. Continuing the calculation, without turning the MW field on again, shows that the saturated nucleus equalizes its polarization to the polarization of the second nucleus, before they both reach their thermal equilibrium values. This can be understood by realizing that after about T_{1e} the difference between the electron polarizations becomes about $P_{e_1}^{eq} - P_{e_2}^{eq} \approx 0$, and insertion in equation (2.38) results in $P_{n_1}^{ss}(t) \approx P_{n_2}^{ss}(t)$.

2.3.3 Overhauser DNP

The earliest DNP mechanism, introduced by Overhauser[1] and demonstrated experimentally by Carver and Slichter,[10] is now known as Overhauser DNP. This mechanism is relevant mainly to samples with free conducting electrons and liquid solutions of free radicals.[2] However, recently it has been shown that the OE DNP enhancement process can also play an important role in amorphous insulating solids containing radicals with narrow EPR spectra.[89,90]

For completeness, we show here a calculation demonstrating the Overhauser enhancement DNP (OE-DNP) mechanism in a small spin system. This mechanism originates from thermal fluctuations of the isotropic and dipolar hyperfine interactions. OE-DNP in liquid solutions containing free radicals is discussed extensively in a separate chapter and in a large set of articles,[11,91] in which the source of the fluctuations are introduced and their resulting relaxation processes derived. Following the approach we present in this chapter, we do not dwell on these issues but rather assume the presence of these relaxation processes in our simple model.

The simplest way of showing the OE-DNP mechanism is again to consider a two-spin system $\{e - n\}$. This system is again defined by its Hamiltonian parameters in

$$\hat{H}^{\text{RoF}} = \Delta\omega_e \hat{S}_z - \omega_n \hat{I}_z + A\hat{S}_z \hat{I}_z + B^+ \hat{S}_z \hat{I}^+ + B^- \hat{S}_z \hat{I}^- \tag{2.39}$$

and experiences $T_{1,n}$ and $T_{1,e}$ relaxation. To describe the OE enhancement mechanism by solving equation (2.18), it is necessary to introduce two additional relaxation processes, with time constants $T_{1,ZQ}$ and $T_{1,DQ}$ for the zero- and double-quantum transitions, respectively. These relaxation processes originate from the fluctuations of the coefficients in the nontruncated hyperfine Hamiltonian, where the isotropic interaction terms dominate the ZQ relaxation and the dipolar interaction terms the DQ relaxation.[2] The eigenstates $\{|\psi_i\rangle\}_{i=1,4}$ of this Hamiltonian with $|B^{\pm}| \ll |\omega_n|$ are the weakly mixed (indicated by the approximate equal sign $\cong$) product spin states

$$|\psi_1\rangle \cong |\beta_e, \alpha_n\rangle; \ |\psi_2\rangle \cong |\beta_e, \beta_n\rangle;$$
$$|\psi_3\rangle \cong |\alpha_e, \alpha_n\rangle; \ |\psi_4\rangle \cong |\alpha_e, \beta_n\rangle \tag{2.40}$$

with their populations p_i, $i = 1, 4$. The rate equations for these four populations, in analogy to Solomon's equations,[92] are similar to equation (2.18), containing the spin lattice relaxation rate matrix with elements of

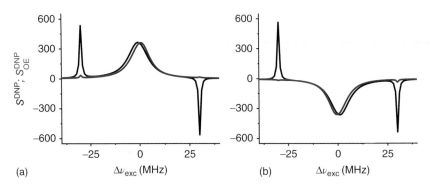

Figure 2.7. Calculated DNP spectra (black) as a result of the Overhauser DNP mechanism in a $\{e-n\}$ spin system, by solving equation (2.41). The values of the DQ and ZQ relaxation times are for the spectrum in (a) $T_{1,\text{DQ}} = 500\,\text{ms}$; $T_{1,\text{ZQ}} = 100\,\text{ms}$; and in (b) $T_{1,\text{DQ}} = 100\,\text{ms}$; $T_{1,\text{ZQ}} = 500\,\text{ms}$. In red we show the same spectra obtained using equation (2.44). The small shift between the two spectra is a result of the simulations. Other parameters used in the calculation are $B^{\pm}_{e,n}/2\pi = 1\,\text{MHz}$; $A_{e,n}/2\pi = 3\,\text{MHz}$; $T_{1n} = 100\,\text{s}$; $T_{1e} = 10\,\text{ms}$; $T_{2e} = 100\,\mu\text{s}$; $\omega_{e_1,e_2}/2\pi = 1\,\text{MHz}$; $v_1 = 500\,\text{kHz}$; $v_n = 30\,\text{MHz}$.

$\hat{R}_1$ as in equation (2.15) and the MW matrix elements of $\hat{W}_{\text{MW}}$ as in equation (2.17). Adding the matrix $R_{1,\text{HF}}$ and presenting the ZQ and DQ relaxation, we get

$$\frac{\text{d}}{\text{d}t}\vec{p}(t) = (\hat{W}_{\text{MW}} - \hat{R}_1 - \hat{R}_{1,\text{HF}})\vec{p}(t) \qquad (2.41)$$

The matrix elements of $\hat{R}_{\text{HF}}$ are in our simple spin system defined according to

$$\frac{\text{d}}{\text{d}t}\begin{pmatrix} p_i \\ p_{i'} \end{pmatrix}(t) = \frac{R_{1,ii'} + R_{1,ii'}^{\text{HF}}}{1+\eta_{ii'}}\begin{pmatrix} -1 & \eta_{ii'} \\ 1 & -\eta_{ii'} \end{pmatrix}\begin{pmatrix} p_i \\ p_{i'} \end{pmatrix}(t) \qquad (2.42)$$

with

$$R_{1,ii'}^{\text{HF}} = \frac{|\langle\psi_i|(\hat{S}_{n_e}^+\hat{I}_{n_n}^+ + \hat{S}_{n_e}^-\hat{I}_{n_n}^-)|\psi_{i'}\rangle|^2}{T_{1,\text{DQ}}}$$
$$+ \frac{|\langle\psi_i|(\hat{S}_{n_e}^+\hat{I}_{n_n}^- + \hat{S}_{n_e}^-\hat{I}_{n_n}^+)|\psi_{i'}\rangle|^2}{T_{1,\text{ZQ}}} \qquad (2.43)$$

Inserting these elements into the rate equation of $\overline{p}(t)$ and choosing values for the interaction and relaxation parameters, we can obtain a solution for the enhancement $S^{\text{DNP}}(t)$. Radiating close to the SQ transitions $|\psi_1\rangle \leftrightarrow |\psi_3\rangle$ and $|\psi_2\rangle \leftrightarrow |\psi_4\rangle$ and ignoring state mixing, the expression for the steady-state enhancement during irradiation becomes

$$S^{\text{DNP}}(t) = 1 - \frac{P_e^{\text{eq}}}{P_n^{\text{eq}}} \times s_e \times \chi \qquad (2.44)$$

with the parameters

$$s_e(t) = \frac{P_e^{\text{eq}} - P_e}{P_e^{\text{eq}}} \quad \text{and} \quad \chi = \frac{T_{1,\text{DQ}}^{-1} - T_{1,\text{ZQ}}^{-1}}{T_{1,\text{DQ}}^{-1} + T_{1,\text{ZQ}}^{-1} + 2T_{1,n}^{-1}} \qquad (2.45)$$

Here, $s_e(t)$ is the saturation factor and χ the enhancement factor. In the OE-DNP literature, the enhancement factor is expressed in terms of a coupling factor ξ and a leakage factor f via $\chi = f \cdot \xi$.[2] In Figure 2.7, we show typical DNP spectra of the $\{e-n\}$ spin system for two relative values for the DQ and ZQ relaxation times, $T_{1,\text{DQ}} < T_{1,\text{ZQ}}$ and $T_{1,\text{DQ}} > T_{1,\text{ZQ}}$.

2.3.4 Pulsed DNP

The population rate equation in equation (2.18) was introduced, assuming that the coherences in the diagonalized rotating-frame Hamiltonian H_T^{RoF} can be ignored. As mentioned earlier, this assumption fails when we apply an MW irradiation that locks the electron polarization in the direction of the MW field or when we have to deal with adiabatic level crossings. An important DNP experiment based on electron spin locking, introduced by Wenckebach, is the NOVEL pulse sequence.[93] This DNP experiment has the important advantage that the electron-to-nuclear polarization transfer time between the electrons and their core and local nuclei can be made very short. Another important method also initiated by Wenckebach is the integrated SE (ISE) pulse DNP technique[94] that depends on population exchange processes of energy-level anticrossings in the rotating frame. Without discussing any experimental aspects of these experiments, and in the framework of our efforts to present only basic DNP concepts, we again turn to the simple two-spin system $\{e - n\}$ for explaining the NOVEL and ISE spin dynamics. Thus, here we present short summaries only and refer to detailed descriptions elsewhere.[93,95–97]

The rotating-frame Hamiltonian of the two-spin system contains the off-resonance term of the electron, $\Delta\omega S_z$, the nuclear Zeeman term, $\omega_n I_z$, the hyperfine interaction terms $(AS_z I_z + BS_z I_x)$, where we assumed for simplicity that $B \equiv B^+ = B^-$ is real, and the MW term, $\omega_1 S_x$, assuming the MW irradiation to point in the x-direction in the rotating frame. To simplify the upcoming discussion, in an effort to provide some insights into the essence of the spin evolution during NOVEL and ISE, it is sufficient to continue the discussion without taking the secular hyperfine term of H^{RoF} into account:

$$H^{\text{RoF}} = \Delta\omega S_z - \omega_n I_z + \omega_1 S_x + B S_z I_x \quad (2.46)$$

In this representation, the reduced spin density operator[98] has the form

$$\sigma^{\text{eq}} = s_e^{\text{eq}} S_z \quad (2.47)$$

where we neglected its small initial nuclear contribution. At this point, it is convenient to represent H^{RoF} and σ^{eq} in terms of the fictitious spin-half operators I_p^{i-j}[99] with $p = x, y, z$ and $i < j = 1, \ldots, 4$, in the basis set:

$$|1\rangle = |\beta_e, \alpha_n\rangle; \quad |2\rangle = |\beta_e, \beta_n\rangle;$$
$$|3\rangle = |\alpha_e, \alpha_n\rangle; \quad |4\rangle = |\alpha_e, \beta_n\rangle \quad (2.48)$$

with increasing energies when $\Delta\omega > \omega_n > 0$:

$$H^{\text{RoF}} = -\Delta\omega(I_z^{1-3} + I_z^{2-4}) - \omega_n(I_z^{1-2} + I_z^{3-4})$$
$$+ \omega_1(I_x^{1-3} + I_x^{2-4}) - \frac{1}{2}B(I_x^{1-2} - I_x^{3-4}) \quad (2.49)$$

$$\sigma^{\text{eq}} = -s_e^{\text{eq}}(I_z^{1-3} + I_z^{2-4}) \quad (2.50)$$

Here, we have expressed the operator determining the electron polarization as $S_z = -I_z^{1-3} - I_z^{2-4}$ and the nuclear polarization operator as $I_z = I_z^{1-2} + I_z^{3-4}$.[99] Using this representation, we now derive the spin evolution during the two DNP experiments. Both experiments are composed of a repetition of a short MW irradiation period followed by a delay. As we deal here only with the irradiation period, we follow the time evolution of $\sigma(t)$ in Hilbert space.

2.3.4.1 NOVEL

The NOVEL type experiments start with a short on- or off-resonance MW prepulse that rotates the electron spin in the rotating frame over some angle for aligning its polarization in the direction of the effective MW field. According to H^{RoF} in equation (2.49), this field makes an angle θ with the magnetic field direction, determined by $\tan\theta = \omega_1/\Delta\omega$. As in CP experiments,[100] the prepulse maximizes the component of the electron magnetization in the direction of the effective MW field. To simplify the description of the spin dynamics after the prepulse, we diagonalize the terms $-\Delta\omega(I_z^{1-3} + I_z^{2-4}) + \omega_1(I_x^{1-3} + I_x^{2-4})$ of H^{RoF}, by applying a rotation operator $\exp\{i\theta(I_y^{1-3} + I_y^{2-4})\}$. This results in

$$H_T^{\text{RoF}} = -\omega_{\text{eff}}(I_z^{1-3} + I_z^{2-4}) - \omega_n(I_z^{1-4} - I_z^{2-3})$$
$$- \frac{1}{2}B\cos\theta(I_x^{1-2} - I_x^{3-4})$$
$$+ \frac{1}{2}B\sin\theta(I_x^{1-4} + I_x^{2-3}) \quad (2.51)$$

where the first two terms are diagonal and the pseudo-hyperfine terms contain all four I_x^{i-j} operators. These terms are a result of the rotating transformation, where we used the commutation relation $[I_y^{1-3} + I_y^{2-4}, I_x^{1-2} - I_x^{3-4}] = i(I_x^{1-4} + I_x^{2-3})$. Here the effective MW field equals $\omega_{\text{eff}} = \sqrt{\Delta\omega^2 + \omega_1^2}$. The spin density matrix following the prepulse can now be presented as

$$\sigma_T(0^+) = -s_{0^+}(I_z^{1-3} + I_z^{2-4}) = -s_{0^+}(I_z^{1-4} + I_z^{2-3}) \quad (2.52)$$

where the magnitude of s_{0^+} is determined by the yield of the prepulse and $t = 0^+$ is the time immediately after the prepulse. In equations (2.51, 2.52) we used $(I_z^{i-j} + I_z^{k-l}) = (I_z^{i-l} + I_z^{k-j})$. When the MW is applied at resonance with $\Delta\omega = 0$ and $I^{i-j} = I^{j-i}$ the effective field equals ω_1 and $\sin\theta = 1$. We must notice that the nuclear I_z operator after the transformation is still of the form $(I_z^{1-2} + I_z^{3-4}) = (I_z^{1-4} - I_z^{2-3})$. Comparing this with equation (2.52), we realize already that the purpose of the NOVEL experiments is to transfer $\sigma_T(0^+)$ to $\sigma_T(\tau) = \pm s_{0^+}(I_z^{1-4} - I_z^{2-3})$, by inverting I_z^{2-3} or I_z^{1-4}, after a time τ.

To show this, we simplify the above expression for H_T^{RoF} in equation (2.51) even further, by ignoring its third term, because with $B \ll \omega_n$ this off-resonance term hardly influences the spin evolution during the MW irradiation. It is easy to show that the remaining terms can then be rewritten as

$$H_T^{RoF} = -\frac{1}{2}(\omega_{eff} + \omega_n)I_z^{1-4} - \frac{1}{2}(\omega_{eff} - \omega_n)I_z^{2-3}$$
$$+ \frac{1}{2}B\sin\theta(I_x^{1-4} - I_x^{2-3}) \qquad (2.53)$$

This Hamiltonian must reveal the NOVEL spin dynamics. The off-resonance term in H_T^{RoF} is the only term that can cause the aforementioned I_z^{2-3} or I_z^{1-4} inversion. However, this is only possible when the coefficient of I_z^{2-3} or I_z^{1-4} in H^{RoF} becomes about zero. This condition defines the NOVEL condition for the DNP polarization exchange

$$\omega_{eff} = \sqrt{\Delta\omega^2 + \omega_1^2} = |\omega_n| \qquad (2.54)$$

which can be looked upon as the Hartmann–Hahn condition of NOVEL. At this condition, and looking for example at the case $\omega_n > 0$, the $1/2B\sin\theta I_x^{2-3}$ term of H_T^{RoF} causes the oscillation

$$\sigma_T(t) = -s_{0^+}\left(\left(I_z^{1-4} + I_z^{2-3}\right)\cos\left\{\frac{1}{2}B\sin\theta t\right\}\right.$$
$$\left. - I_y^{2-3}\sin\left\{\frac{1}{2}B\sin\theta t\right\}\right) \qquad (2.55)$$

leading to the oscillating polarization exchange between the electron to the nucleus:

$$\langle I_z\rangle(t) = \langle\hat{I}_z\rangle_T(t) = \langle\hat{I}_z^{1-4} - \hat{I}_z^{2-3}\rangle(t)$$
$$= -\frac{1}{2}s_{0^+}\left(1 - \cos\left\{\frac{1}{2}B\sin\theta t\right\}\right) \qquad (2.56)$$

Thus, the nuclear polarization will oscillate with a frequency, determined by the pseudohyperfine interaction coefficient B and the off-resonance value, via $\sin\theta t$. For our two-spin system, the length of the MW irradiation should be chosen to be $\tau = 2\pi/(B\sin\theta)$. Wenckebach's original NOVEL experiments were performed at $\Delta\omega = 0$ and $\sin\theta = 1$, where the oscillation frequency is maximum and it is possible to set $s_{0^+} = s^{eq}$. Jain et al.[101] repeated this experiment by irradiating off-resonance and reduced the required magnitude of ω_1 in equation (2.54). In this case, the oscillation frequency decreases and in general $s_{0^+} < s^{eq}$, but the yield of nuclear polarization stays the same as for $\Delta\omega = 0$.

In real solid samples, the electron spins polarize all their neighboring core and local nuclei simultaneously. The polarization transfer during the MW irradiation can then become inefficient because the oscillations of the polarizations of the individual nuclei cause a depletion of the electron polarization. After a single NOVEL pulse sequence, the local nuclear polarizations must be transferred to the bulk nuclei via spin diffusion.

Thus, to benefit from the NOVEL type polarization transfer experiments, the basic pulse sequence is repeated many times, with time delays that allow spin diffusion to polarize the bulk nuclei and spin–lattice relaxation to recover the electron polarization. The repetition rate of the sequence must therefore be chosen in accordance with the spin–lattice relaxation times of the electron and the nuclei. The actual experimental details of NOVEL and its correlated DNP experiments can be found in the literature.[42,93,96,101]

2.3.4.2 The Integrated Solid Effect

We now turn our attention to the ISE experiment, again restricting ourselves to the small two-spin system $\{e-n\}$. This DNP technique consists of a set of short MW pulses separated by longer delay times. During these pulses, the magnetic field[94] or the frequency[97] is incremented or decremented. The range of these scans is chosen such that the instantaneous MW covers a spectral width somewhat bigger than $2\omega_n$. Here, we again restrict ourselves to discuss the spin evolution of one ISE pulse only – all other aspects of real ISE experiments can be found in the literature.[41,95,97]

To describe the spin evolution in our two-spin system during the ISE pulse, we should go back to the Hamiltonian in equation (2.49). In this case, the rotating frame Hamiltonian becomes time dependent, and choosing the field-swept ISE version, the off-resonance value $\Delta\omega$ varies from $\pm(\omega_n + \delta\omega_{ISE})$ to $\mp(\omega_n + \delta\omega_{ISE})$, depending on the sweep direction. The following derivation can be easily transferred to the frequency-swept version of ISE.

To derive the spin evolution, it is preferable to first diagonalize the terms $-\omega_n(I_z^{1-2} + I_z^{3-4}) - \frac{1}{2}B(I_x^{1-2} - I_x^{3-4})$ in H^{RoF} by applying a transformation operator of the form $\exp\{-i\phi(I_y^{1-2} - I_y^{3-4})\}$. This results in

$$H_T^{RoF}(t) = -\Delta\omega(t)(I_z^{1-3} + I_z^{2-4}) - \omega_{T,n}(I_z^{1-2} + I_z^{3-4})$$
$$+ \omega_1\cos\phi(I_x^{1-3} + I_x^{2-4})$$
$$+ \omega_1\sin\phi(I_x^{1-4} + I_x^{2-3}) \qquad (2.57)$$

with $\omega_{T,n} = \sqrt{\omega_n^2 + 1/4B^2}$ and $\tan\phi = B/2\omega_1$. When $B << \omega_n$, the $\sin\phi$ is small and the nuclear signal can still be considered as proportional to $\langle I_x\rangle_T \approx \langle I_z^{12} + I_z^{3-4}\rangle$. The first two terms of $H_T^{RoF}(t)$ describe the time dependence of the energy levels with constant eigenstates. These levels experience four level crossings during the sweep of the off-resonance value. Because of the relatively small value of ω_1, we can expect the off-resonance terms in $H_T^{\mathrm{RoF}}(t)$ to have an influence on $\sigma_T(t)$ only close to these crossings. When the off-resonance value decreases, starting with $\Delta\omega(0) > \omega_{T,n}$, the order of the level crossings is *I*: $|2\rangle \overset{x}{\longleftrightarrow} |3\rangle$ (when $\Delta\omega \approx \omega_{T,n}$), *II*: $|1\rangle \overset{X}{\longleftrightarrow} |3\rangle$ and $|2\rangle \overset{X}{\longleftrightarrow} |4\rangle$ (when $\Delta\omega \approx 0$), *III*: $|1\rangle \overset{x}{\longleftrightarrow} |4\rangle$ (when $\Delta\omega \approx -\omega_{T,n}$). Here we added the factors x and X, representing the efficiency of the exchange of the populations at the *I*-DQ and *III*-ZQ crossings, x, and at the *II*-SQ crossings, X, assuming only population exchange. The values of these anti-crossing parameters x and X can vary from 1, when no population change occurs, to -1, when the MW fully exchanges the populations. The values of these parameters are a function of the MW intensity and the sweep rate $\Delta\omega(t)$, and can be calculated using the Landau–Zener formula.[102] The effective MW power at the first and last crossings is much smaller than the two SQ crossings.

To follow the effects of the crossings, we can represent the spin density operators by their four diagonal elements only in the eigenstate representation of the diagonal elements of H_T^{RoT}. Before the crossing *I*, the initial spin density operator is given by

$$\sigma_T(I^{\mathrm{before}}) = \sigma^{\mathrm{eq}} = -\frac{s^{\mathrm{eq}}}{2}\begin{pmatrix} 1 & & & \\ & 1 & & \\ & & -1 & \\ & & & -1 \end{pmatrix} \quad (2.58)$$

After this crossing, $\sigma_T(I^{\mathrm{after}})$ stays constant until the next crossing and $\sigma_T(II^{\mathrm{before}}) = \sigma_T(I^{\mathrm{after}})$, assuming no relaxation effects between the crossings. Then we can also write $\sigma_T(III^{\mathrm{before}}) = \sigma_T(II^{\mathrm{after}})$, and thus we have only to calculate four time points in the evolution of the diagonal elements of the density matrix: $\{\sigma_T(I^{\mathrm{before}}) \to \sigma_T(I^{\mathrm{after}}) \to \sigma_T(II^{\mathrm{after}}) \to \sigma_T(III^{\mathrm{after}})\}$. Following the order of the level crossings, we then get for the diagonal elements of σ_T:

$$-\frac{s^{\mathrm{eq}}}{2} \times \left\{ \begin{pmatrix} 1 \\ 1 \\ -1 \\ -1 \end{pmatrix} \overset{I}{\to} \begin{pmatrix} 1 \\ x \\ -x \\ -1 \end{pmatrix} \overset{II}{\to} \begin{pmatrix} (1-x)+(1+x)X \\ -(1-x)+(1+x)X \\ (1-x)-(1+x)X \\ -(1-x)-(1+x)X \end{pmatrix} \right.$$
$$\left. \overset{III}{\to} \frac{1}{2} \begin{pmatrix} (1-x)x+(1+x)Xx \\ -(1-x)+(1+x)X \\ (1-x)-(1+x)X \\ -(1-x)x-(1+x)Xx \end{pmatrix} \right\} \quad (2.59)$$

We can now write $\sigma_T(III^{\mathrm{after}})$ using the fictitious spin operators as

$$\sigma(III^{\mathrm{after}}) = -\frac{s^{\mathrm{eq}}}{4}(1-x^2)(1-X)(I_z^{1-2}+I_z^{3-4})$$
$$- \{(1-x)^2 - (1+x)^2 X\}(I_z^{1-3}+I_z^{2-4}) \quad (2.60)$$

The nuclear signal at the end of the pulse will then be given by

$$<I_z^{1-2}+I_z^{3-4}> = -\frac{s^{\mathrm{eq}}}{4}\{(1-x^2)(1-X)\} \quad (2.61)$$

When the sweep is performed in the other direction (frequency sweep vs field sweep), $\Delta\omega(t)$ changes during the MW irradiation in the reverse order, $III \to II \to I$, the diagonal elements as in equation (2.59) become now

$$-\frac{s^{\mathrm{eq}}}{2} \times \left\{ \begin{pmatrix} 1 \\ 1 \\ -1 \\ -1 \end{pmatrix} \overset{III}{\to} \begin{pmatrix} x \\ 1 \\ -1 \\ -x \end{pmatrix} \overset{II}{\to} \frac{1}{2}\begin{pmatrix} -(1-x)+(1+x)X \\ (1-x)+(1+x)X \\ -(1-x)-(1+x)X \\ (1-x)-(1+x)X \end{pmatrix} \right.$$
$$\left. \overset{I}{\to} \frac{1}{2} \begin{pmatrix} -(1-x)+(1+x)X \\ (1-x)x+(1+x)Xx \\ -(1-x)x-(1+x)Xx \\ (1-x)-(1+x)X \end{pmatrix} \right\} \quad (2.62)$$

and the nuclear signal is given by

$$<I_z^{1-2}+I_z^{3-4}> = -\frac{s^{\mathrm{eq}}}{2}\{-(1-x^2)(1-X)\} \quad (2.63)$$

These results show that for adiabatic DQ and ZQ anticrossings, $x = -1$, the nuclear polarization $<I_z^{1-2}+I_z^{3-4}>$ becomes zero, while for full saturation at the DQ and ZQ anticrossings, $x = 0$, this polarization depends on the value of X. An ideal combination could then be SQ adiabatic passages, $X = -1$, and DQ and ZQ saturations, $x = 0$. Then the nuclear signals become $\pm s^{\mathrm{eq}}/2$ for increasing and decreasing $\Delta\omega(t)$. Thus, the MW intensity, the strength of B and the time derivative of $\Delta\omega(t)$ all influence the end nuclear polarization.

Again, in real experiments, multiple nuclei are polarized by each electron, and spin diffusion must be taken into account for polarizing the bulk nuclei. Repetitions of the ISE pulse followed by delays make this technique rather efficient. For additional discussions on ISE, refer to the literature Refs. 42, 93.

2.4 SMALL SPIN SYSTEMS AS MODEL SYSTEMS FOR MACROSCOPIC SAMPLES

At this stage, we must extend the description of the SE- and CE-DNP mechanisms toward real coupled electron–nuclear spin systems. We first present the mechanisms bringing about DNP effects in small spin systems. On grounds of the results from these investigations, we then suggest possible computational models that make it possible to analyze the spin dynamics in large systems. The DNP mechanisms in real samples are governed by free electrons that are interacting with each other and that are each surrounded by their own interacting nuclei. To present the DNP process in these spin systems, we realize that the MW irradiation has a direct effect on the electrons composing the EPR spectrum, in terms of an on- and off-resonance saturation of their polarizations. Indirectly, it can cause a severe depolarization of the rest of the electrons due to the cross-relaxation process. In addition, SE-DNP enhancements occur around electrons that are removed by ω_n from the MW frequency ω_{MW} and iCE-DNP enhancements around all interacting electron pairs satisfying a CE condition. We can therefore imagine our samples as being composed of a dipolar coupled electron network of DNP centers, imbedded in a dipolar coupled nuclear spin system. These centers can be the free electrons of mono-, bi-, or tri-radicals. To simplify the description of the mechanisms involved in the DNP enhancement of the nuclear polarization, we discuss their actions in three stages. First, we consider the spin dynamics of the electron network during MW excitation, then we discuss the polarization enhancement of the nuclei coupled to the individual electrons, and at the end, we concentrate briefly on the spin diffusion mechanism polarizing the bulk nuclei. For this reason, we divide this section into three subsections and discuss first the electron evolution during MW irradiation in a system of nine coupled electrons. Then we add nuclei to this system and show their DNP enhancement during this irradiation, and finally we briefly discuss the spin diffusion process in real systems.

2.4.1 A Coupled Electron Spin System

Here we consider first a spin system of nine electrons, without nuclei, that are dipolar coupled to each other and that should resemble, in some way, the electrons in a real system. This is, of course, rather difficult, mainly because of the amorphous nature of our samples. The electrons can have large anisotropic g-tensors that are randomly oriented; thus the interacting electrons can have very different resonance frequencies. For these reasons, we decided to construct an electron spin system that cannot exist in reality but that has some typical features of a real system: multiple spin pair interactions with varying dipolar strengths and resonance frequency differences. Thus, for computational reasons, we consider a nine-electron system with each electron having a different resonance frequency $\omega_{n_e} = \omega_e + (n_e - 5)\delta\omega_e$, with $n_e = 1, \ldots, 9$. In our calculations, we set the frequency differences equal to $\delta\omega_e/2\pi = 8$ MHz. The electrons are dipolar coupled with interaction coefficients ω_{n_e, n'_e}, as in equation (2.4), equal to $x_{n_e-n'_e}D$. The value of D represents the overall dipolar interaction strength in frequency units and could be correlated to the radical concentration in real systems. The factors $x_{n_e-n'_e}$ are chosen randomly with the restrictions $-1/2 < x_{n_e-n'_e} < 1$ and $\sum_{n_e < n'_e} x_{n_e-n'_e} = 0$.

First we follow the response of this system to an MW irradiation at $\nu_{exc} = \nu_{MW}$ after a period t, in the absence of cross-relaxation, i.e $T_D = \infty$. Using the population equation in the diagonalized frame in equation (2.18) for a given set of values for $\{D; \delta\omega_e; T_{1e}; T_2; \omega_1\}$, we can evaluate EPR spectra $S^{EPR}(t, \nu_{det}, [\nu_{exc}])$ of the system from its eigenstate populations $p_i(t)$. Here we should notice again that during the simulations the whole Hamiltonian $\hat{H}^{RoF}$ is taken into account, including the flip-flop terms of the dipolar interactions. Furthermore, the relaxation processes are restricted to a uniform T_{1e} value for all electrons. Possible DQ cross-relaxation mechanisms are not taken into account. The assumption we make, which should be repeated here, is the fact that we ignore any generation of coherences in the diagonalized frame by the MW excitation, for example, as a result of spin locking.

Before presenting the results of these simulations, we must remind ourselves that the Hamiltonian H^{RoF} of the system commutes with the total z-component of the electron spin angular momentum, $\hat{S}_z = \sum_{n_e} \hat{S}_{n_e, z}$. Its eigenstates can therefore be subdivided into subsets of states $\{|\psi_{k_M}^{M_e}\rangle\}$ with constant M_e that satisfy $S_z|\psi_{k_M}^{M_e}\rangle = M_e|\psi_{k_M}^{M_e}\rangle$ and that have energies $\varepsilon_{k_M}^{M_e}$ in the rotating frame. The subscript is $k_M = 1, \ldots, K_{M_e}$, where K_{M_e} is the total number of states with the same M_e value. The energies of the states in the laboratory frame are, of course, equal to $E_{k_M}^{M_e} = M_e\omega_{MW} + \varepsilon_{k_M}^{M_e}$. For this system, we can plot an EPR spectrum by assigning to each transition between levels in adjacent M_e manifolds $|\psi_{k_M}^{M_e}\rangle \leftrightarrow |\psi_{k_{M+1}}^{M_e+1}\rangle$ a Gaussian, centered at the transition frequency $\nu_{k_M,k_{M+1}}^{\mathrm{RoF}} = (\varepsilon_{k_M}^{M_e} - \varepsilon_{k_{M+1}}^{M_e+1})/2\pi$, with some width $\delta_G = 1\,\mathrm{MHz}$, and an intensity given by $|\langle\psi_{k_{M+1}}^{M+1}|S^+|\psi_{k_M}^M\rangle|^2(p_{k_M}^M(t) - p_{k_{M+1}}^{M+1}(t))$. Each EPR spectrum will then be a sum of these Gaussians. In the following, all EPR spectra are plotted as a function of the detection frequency, in terms of $\Delta\nu_{\mathrm{det}}$ with $\Delta\nu_{\mathrm{det}} = \nu_{\mathrm{det}} - \nu_e$, for a fixed ν_{exc}.

In Figure 2.8, we show the results of the calculations for two D values. Figure 2.8(a) and (d) show the energy–population plots in the rotating frame for $D/2\pi = 6\,\mathrm{MHz}$ and $D/2\pi = 2\,\mathrm{MHz}$, respectively. As this system has 2^9 states, in Figure 2.8(b) and (e) we show only populations of states with energies $\varepsilon_{k_M}^{M_e}$ with $-3/2 < M_e < 3/2$. Points corresponding to the same M_e value are given the same color. The square symbols

are the populations at thermal equilibrium and appear almost as a vertical distribution of points. Starting with the case $D/2\pi = 6\,\mathrm{MHz}$, the majority of the populations in Figure 2.8(a) and (d), after a period t of MW irradiation longer than T_{1e}, follow approximately a single exponential function of the form

$$p_{k_M}^{M_e}(t) = Z^{-1}e^{-\beta^{\mathrm{RoF}}\varepsilon_{k_M}^{M_e}} \qquad (2.64)$$

This is an indication that the spin system can be characterized by a common spin temperature in the rotating frame, defined by a temperature coefficient $\beta^{\mathrm{RoF}} = 1/k_B T^{\mathrm{RoF}}$ with k_B the Boltzmann factor. A direct consequence of this is the peculiar shape of the EPR spectrum, shown in Figure 2.8(c). It shows a crossing through zero at the MW frequency, where the electron transitions are saturated. This characteristic shape is well known and indeed corresponds to electron spin systems that are described relying on the

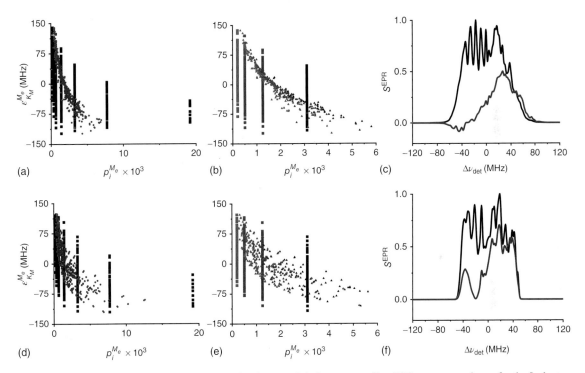

Figure 2.8. Energy–population plots in the rotating frame and their corresponding EPR spectra are shown for the 9-electron spin system. The top row is for $D/2\pi = 6\,\mathrm{MHz}$ and the bottom row for $D/2\pi = 2\,\mathrm{MHz}$. The thermal equilibrium populations are drawn as black squares and the values $p_i^{M_e}$ after a long MW irradiation as colored triangles. In (a) and (d) we show all the 2^9 populations, while in (b) and (e) we show only the populations with $-3/2 < M_e < 3/2$. The different colors correspond to populations with different M_e values. In (c) and (f) we show the EPR spectra of the systems at thermal equilibrium (black) and following MW irradiation at $\Delta\nu_{\mathrm{exc}} = -20\,\mathrm{MHz}$ (red). Other parameters of the system are $\delta\omega_e/2\pi = 8\,\mathrm{MHz}$; $T_{1e} = 100\,\mathrm{ms}$; $T_{2e} = 100\,\mathrm{\mu s}$; $T_D = \infty$; $\nu_1 = 500\,\mathrm{kHz}$; $\nu_n = 30\,\mathrm{MHz}$

TM model.[35,103,104] This model describes the status of the electron spin system in the laboratory frame by the Zeeman spin temperature T_{Ze}^L, of a Zeeman interaction thermal spin bath, and a non-Zeeman spin temperature T_{nZe}^L, of an inhomogeneously broadened dipolar interaction spin bath. An experimental example of this has been shown early on by Atsarkin[105] In the rotating frame, these two temperatures, $T_{Ze}^{RoF} = (\omega_{MW}/\omega_e)T_{Ze}^L$ and $T_{nZe}^{RoF} = T_{nZe}^L$, are about equal and correspond to the value of T^{RoF} for $D/2\pi = 6$ MHz. Here we should comment that, despite the fact that the electron frequency differences are larger than the dipolar interactions, a single spin temperature in the rotating frame can be reached.

For $D/2\pi = 2$ MHz, the populations can again be calculated using equation (2.18) and are shown for our system in Figure 2.8(d) and (e). Comparing this energy–population plot with Figure 2.8(a) clearly shows that we cannot talk about one common spin temperature in this case. This is also apparent when we compare the EPR spectra for the two D values, in Figure 2.8(c) and (f), obtained

after a time of MW irradiation $t > T_{1e}$. In the last case, the EPR profile shows a broad hole around $\Delta v_{exc} = v_{exc} - v_e = -20$ MHz. The differences in the EPR spectra are, of course, a direct result of the distribution of the population differences for $\Delta M = \pm 1$. We should emphasize that the unusual shapes of the EPR lines at thermal equilibrium in Figure 2.6 are a consequence of the small size of the spin system.

The differences between the energy–population plots for the two dipolar interaction values are dictated by the relative magnitudes of these values with respect to the differences between the electron resonance frequencies. This indicates that in practice we can only expect a spin temperature behavior of the electron spin system, when its EPR linewidth is reduced and the radial concentration increased, like comparing TEMPOL with trityl samples at high radical concentrations.

We now add cross-relaxation to the system, by setting $T_D = 5$ ms, and show again the energy–population plots and the EPR spectra in Figure 2.9. Comparing the results with those in Figure 2.8 shows very similar

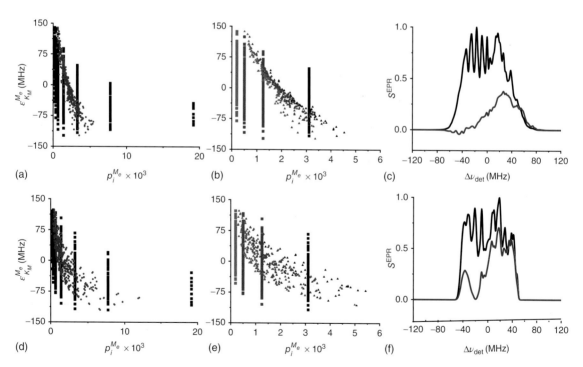

Figure 2.9. Energy–population plots and EPR spectra as in Figure 2.8, but this time including also a cross-relaxation process. In the top row $D/2\pi = 6$ MHz and the cross-relaxation time equals $T_D = 5$ ms. In the bottom row, these parameters are $D/2\pi = 2$ MHz and $T_D = 5$ ms

behaviors. We recognize again a TM spin temperature picture for $D/2\pi = 6$ MHz, with its typical EPR lineshape, and for 2 MHz, a non-TM picture with a broad hole in its EPR spectrum at $\Delta v_{exc} = -20$ MHz. The addition of T_D to the calculation equalizes neighboring populations and should therefore result in a flattening of the EPR lineshape. In Figure 2.9(c), this is manifested in a lowering of the slope of the EPR profile, while in Figure 2.9(f) we hardly see any effect. In Ref. 106, similar calculations were presented for $T_D = 100$ μs. There, in the strong dipolar case the spectrum approaches zero at all frequencies, and in the weak dipolar case, we obtain a significant depletion of the whole spectrum.

Here, we should mention a major difference between systems without and with cross-relaxation, following a long MW irradiation. In the first case, all polarizations (i.e. EPR intensities) return to their equilibrium value in a time scale determined by T_{1e}, while in the second case the cross-relaxation causes an equilibration of the polarizations in a time scale determined by T_D and then a collective return of all polarizations (i.e. the whole EPR lineshape) to its equilibrium shape in a time T_{1e}. This is discussed in Ref. 106 and is demonstrated in Figure 2.10.

The results of this section reflect the electronic response to an MW irradiation of real systems. Their EPR and ELDOR spectra should also show differences that correspond to the spectral features appearing in Figures 2.8 and 2.9. Thus, these features can be indicators whether the system corresponds to a weak or a strong dipolar coupled electron spin system. As already mentioned, the weak and strong regimes depend on the concentration of the electrons in the sample and on their inhomogeneously broadened EPR linewidth. Therefore, in practice, it is *a priori* not clear in which regime the electrons find themselves. A sample can also be in an intermediate regime, which due to lack of space will not be discussed here.

To be able to utilize the above-presented simulations for analyzing real experiments, it is necessary to simplify the calculations. To accomplish this, we introduce computational models that will be discussed in Section 2.5. However, before doing so, we must first introduce nuclei to our small spin system in order to investigate their influence on the electron polarization and to calculate DNP spectra.

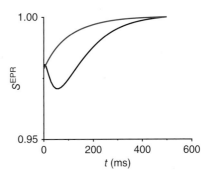

Figure 2.10. The response of the normalized EPR line intensity at $\Delta v_{det} = 42$ MHz of the 9-electron spin system, after a short MW excitation of 1 ms at $\Delta v_{exc} = 20$ MHz. The red curve is calculated without the presence of cross-relaxation, when all electron polarizations return to thermal equilibrium in a timescale of T_{1e}. The black curve is calculated in the presence of cross-relaxation with a time constant $T_D = 1$ ms. In this case, the EPR intensity initially decays, due to equalization of polarizations in the system, and then returns, together with the other intensities, to equilibrium. The initial values of the two curves are dictated by the presence of the dipolar interaction during the off-resonance MW excitation. The parameters used for these calculations are $T_{1e} = 100$ ms; $T_{2e} = 100$ μs; $D/2\pi = 2$ MHz; $T_D = 2$ ms; $v_1 = 0.5$ MHz

2.4.2 Electron–nuclear Spin Systems

After discussing the response of the pure electron spin system to MW irradiation, we now extend the system by adding nuclei. To make the following calculations also relevant for understanding real systems, we concentrate now on the second stage of the DNP process. During this stage, the electrons polarize their neighboring nuclei during MW irradiation, still ignoring the spin diffusion stage. Thus, we use a simplified view in which each nucleus is coupled only to its neighbor electron in the coupled electron spin system and that it gets polarized independent of the presence of other nuclei and other electrons. Because of this notion, we can calculate the nuclear enhancement by considering a set of coupled electron spin systems, where in each system only one electron is hyperfine coupled to nuclei. Applying our nine-electron spin system, we must thus consider nine spin systems where in each system a different electron experiences hyperfine couplings. To minimize the computational effort, we simplify the calculations by adding to each system only one nucleus, representing the interacting local nuclei. Furthermore, we do not consider here the presence of core

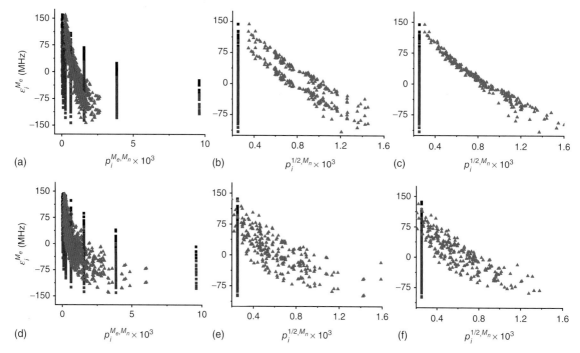

Figure 2.11. Energy–population plots in the rotating frame are shown for a 10-spin system with 9 electrons and one nucleus. The results in the top row are calculated for $D/2\pi = 6$ MHz and the bottom row for $D/2\pi = 2$ MHz. The thermal equilibrium values are drawn as black and red squares for $M_n = -1/2$ and $M_n = 1/2$, respectively. The steady-state populations after long MW irradiation at $\Delta v_{MW} = -20$ MHz are drawn in green and magenta for $M_n = -1/2$ and $M_n = 1/2$, respectively. The populations of the whole system are shown in (a) and (d), while in (b) and (e) only the populations belonging to the states $M_e = 1/2$ are plotted. In (c) and (f), a hyperfine interaction of the nucleus with electron e_2 is added to the system with coefficients: $A_{2,n}/2\pi = 3$ MHz; $B_{2,n}^{\pm}/2\pi = 1$ MHz

nuclei, which in real systems can have a pronounced effect on local polarizations. With these restrictions, we can simulate in a reasonable CPU time, together with the EPR and ENDOR spectra, also the nuclear signals of each system in the form of DNP spectra.

In Figure 2.11 we show energy–population plots of a ten-spin system, with $D/2\pi = 6$ MHz and $D/2\pi = 2$ MHz, the 'strong' and 'weak' dipolar interaction cases, respectively. The nuclear Larmor frequency v_n was chosen to be smaller than the EPR linewidth, as given in the figure caption. The energy–population plot of the system with the nucleus bound to electron $n_e = 2$ shows again a clear difference between the weak and strong dipolar interactions. When $D/2\pi = 6$ MHz, we recognize the single spin temperature for the whole system, indicating that $T_{Ze}^{RoT} = T_{nZe}^{RoT} = T_{Zn}^{RoT}$, with T_{Zn}^{RoT} the spin temperature defining the nuclear population differences. For $D/2\pi = 2$ MHz, these temperatures cannot be defined

and we cannot consider nuclear and electron thermal spin baths with equal spin temperatures. Comparing the ELDOR and DNP spectra in Figure 2.12, for the weak and strong cases of the 10-spin system, reveals the local polarization effects in the weak case and the global polarization effects in the strong case. Already realizing that in real systems the overall bulk nuclear polarization after the spin diffusion process is proportional to the average of the local polarizations, we show here also the average nuclear polarization of the individual polarizations in the form of DNP spectra. These spectra are the average nuclear polarization as a function of the MW irradiation frequency.

These DNP spectra are in fact the result of a combination of SE and CE processes, which are discussed in Section 2.3. Irradiation at a frequency v_{exc} results, due to the SE mechanisms, in localized nuclear enhancements in the DNP spectra and polarization depletions in the ELDOR spectra at $v_{exc} \pm v_n$. The nuclear

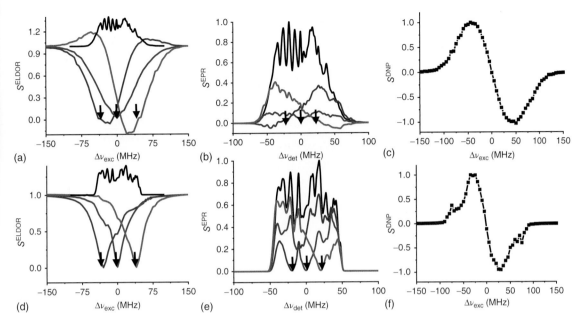

Figure 2.12. In (a) and (b) ELDOR and EPR spectra of the 10-spin system, with nine electrons and one nucleus, are shown for $D/2\pi = 6$ MHz and in (d) and (e) the same for $D/2\pi = 2$ MHz. The equilibrium EPR spectrum is plotted in black. The steady-state ELDOR spectra in (a) and (d) are plotted for $\Delta\nu_{det} = -20, 0, 20$ MHz in red, blue, and green, respectively. The EPR spectra in (b) and (e) are plotted for $\Delta\nu_{exc} = -20, 0, 20$ MHz in red, blue, and green, respectively. In (c) and (f) the normalized DNP spectrum of the 10-spin system is shown. This spectrum is derived by calculating the DNP spectra of nine systems, with the nucleus interacting each time with one of the same nine coupled electrons. The spectra in the figure are the average of these nine spectra. In all figures, the arrows indicate the excitation or detection frequencies used in the calculations. The following parameters are used here : $T_{1e} = 100$ ms; $T_{2e} = 100$ µs; $T_D = 5$ ms; $A_{2,n}/2\pi = 1$ MHz; $B_{2,n}^{\pm}/2\pi = 3$ MHz; $D/2\pi = 2$ MHz; $\nu_1 = 500$ kHz

enhancements are, however, mainly attributed to the iCE mechanism (*vide infra*), derived from the electron polarization gradients created by the electron spectral diffusion. While the SE leads to some SE-induced electron depolarization effects, the iCE mechanism does not cause visible changes in the depolarization profile. Another difference between the two mechanisms originates from the fact that irradiation outside the frequency range of the EPR line can directly result in SE enhancement and some SE depolarization. This depolarization together with the cross-relaxation can again create polarization gradients leading to iCE enhancement.

The source of the DNP enhancement in real systems, composed of a solution of free electrons with arbitrarily oriented g-tensors, also originates from the SE and iCE mechanisms. The former depends on the number of electrons at $\nu_{exc} \pm \nu_n$ and their polarizations and the latter depends on the number of electron pairs at the CE condition and their polarization differences.

For sufficiently large free electron concentrations, the number of iCE inducing electron pairs and the number of SE inducing electrons can be of the same order of magnitude. The iCE contribution to the overall DNP enhancements can still be dominant. At low concentrations or when ν_n is larger than the EPR linewidth, the SE contributions will dominate.

2.4.3 Spin Diffusion

Before presenting the spin dynamics in realistic spin systems, we briefly discuss here the polarization transfer process from the local nuclei to the bulk nuclei. These local nuclei are polarized by their neighboring electrons either directly via their hyperfine interaction with the electron, or indirectly by the combined action of the hyperfine and the dipolar interaction with an intermediate nucleus.[107] These local nuclei differ from the core nuclei that cannot directly transfer polarization to the bulk because of the truncation of the nuclear dipolar interaction due to their large hyperfine. On the basis of the assumption that these local nuclei are polarized via the SE or iCE mechanism, their polarization is transferred to the bulk nuclei by the spin diffusion process. As theoretical and experimental aspects of this process are extensively discussed in the literature,[4,80,108] we decided not to dwell on these aspects. However, we should mention some factors relevant to the DNP enhancement in real systems.

The nuclei interacting with the unpaired electrons are localized around these electrons and are kept polarized by DNP processes during the MW irradiation. We could say that the nuclei surrounding the electrons form polarized 'cold spots' that are separated from each other and that can have different levels of polarization. They form the origins of the nuclear polarization that is spread over the bulk nuclei. The value of the bulk polarization depends on a set of parameters, some of which include the average number of bulk nuclei per cold spot, the relaxation times of the electrons, the

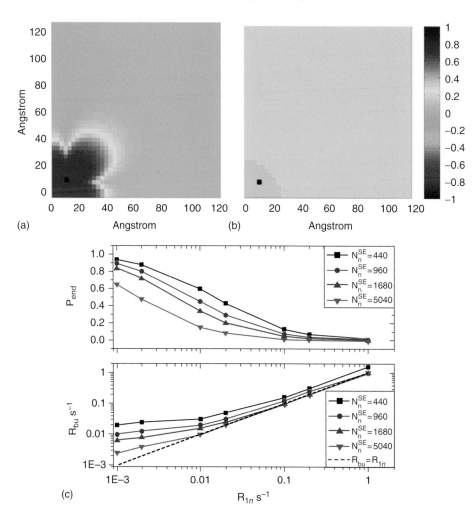

Figure 2.13. In (a) and (b), the steady-state normalized nuclear polarization as a function of position is drawn for a 2D network of interacting nuclei that are polarized by one electron (the black dot) via SE-DNP. In (a) the spin diffusion rate constant is set at $R_{sd} = 0$, while in (b) $R_{sd} \gg R_{1n}$. In (c) the enhancement and the buildup rate of the average nuclear polarization are calculated as a function of N_n^{SE} and R_{1n}. (Daphna Shimon, Akiva Feintuch, Daniella Goldfarb and Shimon Vega Phys. Chem. Chem. Phys., 2014, 16, 6687. Reproduced with permission from Royal Society of Chemistry)

core and the bulk nuclei, and the spin diffusion rate constant. Without going into more details, we should emphasize two features that are important for calculating average bulk polarization. At first, we assume that the value of the overall bulk nuclear polarization at steady state is proportional to the average polarizations of the nuclei at each cold spot. This allows us to estimate the value of the bulk polarization by calculating the nuclear polarizations around all polarizing electrons separately. Typical values of the concentration in glassy samples, used during DNP experiments, are tens of millimolar for the electrons and several up to tens of molar for the nuclei. Secondly, it is known that the buildup time of the bulk polarization has as its upper limit the spin–lattice relaxation time of the bulk nuclei.[109] This time, as well as the characteristic time of the spin diffusion process itself, is in general much longer than T_{1e}. Among others, three important factors influencing the buildup time are (i) the MW irradiation power, (ii) the relaxation time of the core nuclei, and (iii) the ratio between the average number of bulk nuclei and local nuclei per electron. Of course, the average number of bulk nuclei per electron pair at a CE condition or per SE-DNP polarizing electron is much larger than this number. The effective MW irradiation strength is larger for the CE electrons than the SE electrons. These factors, together with the relaxation parameters, make it possible that the total bulk enhancement originates from both DNP processes simultaneously.[110]

To investigate the influence of the aforementioned parameters on the bulk enhancements, a simple two-dimensional SE-DNP model, containing one electron and many nuclei, was constructed.[109] Taking the DNP enhancement of the core and local nuclei into account, and considering a two-dimensional network of interacting nuclei, we can simulate some of the main characteristics of the bulk polarization buildup time, as mentioned earlier. In Figure 2.13(a) and (b), an electron (in the bottom-left corner) polarizes a network of $N_n^{SE} = 960$ coupled nuclei, with nearest neighbor distances of 3.1 A, via the SE-DNP mechanism and spin diffusion. In Figure 2.13(a), in the absence of spin diffusion, only the core and local nuclei are polarized, while in Figure 2.13(b), in the presence of spin diffusion, all nuclei are polarized. The interaction parameters relaxation and spin diffusion rate constants of this spin system can be found in Ref. 109. To demonstrate some of the aforementioned characteristics of the bulk enhancement process, a dependence of the bulk polarization buildup rate R_{bu}

and the end polarization P_{end} on the number of bulk nuclei N_n^{SE} and their nuclear relaxation rate R_{1n} for the range $N_n^{SE} = 400 - 5040$ is shown in Figure 2.13(c).

2.5 LARGE SPIN SYSTEMS

We would now like to extend the simulated results, concerning small spin systems, to real spin systems, which typically can have 10^{3-4} nuclear spins per unpaired electron. We must therefore use a different approach for analyzing the EPR, ELDOR, and DNP spectra of these samples. For this, we will use a computational model that, on the one hand, can mimic the calculations in Section 2.4 and, on the other hand, can be applied to analyze real data. In the following text, we briefly introduce this model and show how its results resemble that of the exact calculations. Some examples of experimental results and their analysis are also presented.

Because the dimension of the matrix representations of the Hamiltonian of real systems is too large to handle numerically, we replace its population rate equations by a small dimensional set of polarization rate equations by using the frequency bin model and introducing a set of coupled equations for the polarization of the electrons in each bin.[26,58,65] Following the results of Section 2.4, we first evaluate the electron polarization and then derive from those the nuclear polarization. The EPR spectra of these electrons are in general inhomogeneously broadened by the *g*-tensor anisotropy and have a width larger than the electron dipolar interactions. These spectra are composed of the intensities of the allowed transition between mixed product states. As mentioned earlier for the weak dipolar case, we can assign single-electron polarizations to the transitions in narrow frequency bands. For the strong dipolar case, this is in fact not possible because of state mixing. However, we will still consider electron polarization in each bin.

2.5.1 The Polarization Rate Equation – The eSD Model

To reduce the dimensionality of the population calculations, we introduce a set of coupled rate equations for the electron bin polarizations. These equations still must include MW irradiation, spin relaxation, and cross-relaxation processes. To do so, we subdivide the

EPR spectrum of the system under investigation into N_b frequency bins, each with an average frequency v_j with $j = 1, \ldots, N_b$. All bins have the same width δv_b. We assign to the electrons in each bin j a polarization $P_{e,j}(t)$, which at thermal equilibrium is determined by the Boltzmann distribution. To take into account the different intensities of the bins, composing the equilibrium EPR spectrum, we introduce parameters f_j that represent the relative number of electron spins contributing to each bin j. These parameters are derived from the intensities S_j^{EPR} at v_j of the equilibrium EPR spectrum, by generating an equivalent spectrum with intensities $f_j P_{e,j}^{\mathrm{eq}}$ at v_j and with $\sum_j f_j = 1$, such that for all j and a constant c_f factor $f_j P_{e,j}^{\mathrm{eq}} = c_f S_j^{\mathrm{EPR}}$. So far, we do not determine the value of δv_b, although it should not be very much narrower than the dipolar homogeneous linewidth. With this subdivision of the EPR line, we now introduce a mathematical model for the time evolution of the $P_{e,j}(t)$'s. The values of $P_{e,j}(t)$ can then be used to simulate EPR line intensities $f_j P_{e,j}(t)$, during MW irradiation. This model has been described in more detail in previous publications.[65,88]

The coupled rate equations for the $P_{e,j}(t)$'s can be written in the form

$$\frac{\mathrm{d}}{\mathrm{d}t}\vec{P}_e(t) = (\hat{W}_{\mathrm{MW}}^b - \hat{R}_1^b - \hat{R}_D^b - \hat{R}_{\mathrm{TM}}^b + \hat{W}_{\mathrm{SE}}^b)\vec{P}_e(t)$$
(2.65)

where $\vec{P}_e(t)$ is the vector composed of all $P_{e,j}(t)$'s and the rate matrices W_{MW}^b and $\hat{W}_{\mathrm{SE}}^b$, and the three, R^b, relaxation matrices are defined below. In the following text, we still make a distinction between weak and strong dipolar interaction cases. The samples in the weak interaction regime have EPR spectra that exhibit hole burning during MW irradiation. Those in the strong dipolar interaction regime have EPR spectra that show the typical TM features. For simplicity, we do not discuss the intermediate case.

As introduced in Section 2.2, when MW irradiation is sufficiently strong, it can be represented by a rate matrix $\hat{W}_{\mathrm{MW}}^b$ that in the polarization equations is diagonal with elements:

$$W_{\mathrm{MW},jj}^b = -\left(\frac{\omega_1^2 T_2}{1 + \left(\omega_j - \omega_{\mathrm{MW}}\right)^2 T_2^2}\right)$$
(2.66)

where T_2 is an effective electron–electron relaxation time determining the off-resonance efficiency of MW irradiation. For the spin–lattice relaxation rate matrices, we need to extend $\vec{P}_{e,j}$ by an additional element

equal to 1.[111] The matrix $\hat{R}_{1e}$ is then defined such that

$$\frac{\mathrm{d}}{\mathrm{d}t}\begin{pmatrix} 1 \\ P_{e,j}(t) \end{pmatrix} = \begin{pmatrix} 0 & 0 \\ \frac{P_{e,j}^{\mathrm{eq}}}{T_1} & -\frac{1}{T_1} \end{pmatrix}\begin{pmatrix} 1 \\ P_{e,j}(t) \end{pmatrix}$$
(2.67)

where $P_{e,j}^{\mathrm{eq}} = \tanh(\omega_j/2k_{\mathrm{B}}T)$.

The next rate matrix $\hat{R}_D^b$ is responsible for electron spectral diffusion and replaces the cross-relaxation rate matrix in the population calculations. We realize that in the case of a spectrum with two discrete lines of two dipolar-coupled electrons, the zero-quantum cross-relaxation exchange process equalizes the intensity of the lines, while keeping the spectral integral unchanged. In our systems, we therefore assume that the cross-relaxation between two bins keeps their intensities constant $\frac{\mathrm{d}}{\mathrm{d}t}(N_b f_i P_{e,j}(t) + N_b f_{j'} P_{e,j'}(t)) = 0$ and tries to make $f_j P_{e,j}(t)/f_{j'} P_{e,j'}(t)$ equal to $f_j P_{e,j}^{\mathrm{eq}}(t)/f_{j'} P_{e,j'}^{\mathrm{eq}}(t)$. This assumption makes sure that at thermal equilibrium, and in the presence of the exchange process, the polarizations follow the Boltzmann statistics. Based on these considerations, we can write the rate equation for this process, from which we derive the elements of R_D^b:

$$\frac{\mathrm{d}}{\mathrm{d}t}\begin{pmatrix} P_{e,j}(t) \\ P_{e,j'}(t) \end{pmatrix} = \frac{1}{2}R_{j,j'}^D\begin{pmatrix} -\eta_{j,j'}f_j & f_{j'} \\ \eta_{j,j'}f_j & -f_j \end{pmatrix}\begin{pmatrix} P_{e,j}(t) \\ P_{e,j'}(t) \end{pmatrix}$$
(2.68)

with

$$\eta_{j,j} = \frac{P_{e,j}^{\mathrm{eq}}}{P_{e,j'}^{\mathrm{eq}}} = \exp(-(\omega_j - \omega_{j'})\hbar/k_{\mathrm{B}}T)$$
(2.69)

We define here a single-electron spectral diffusion (eSD) coefficient, Λ^{eSD}, that expresses the empirically determined dependence of the rate constants $R_{j,j'}^D$ on the square of the bin frequency differences $(\omega_j - \omega_{j'})$:

$$R_{j,j'}^D = \frac{\Lambda^{\mathrm{eSD}}}{(\omega_j - \omega_{j'})^2}.$$
(2.70)

Solving equation (2.65), after inserting all necessary parameters from equations (2.66–2.70), gives $\vec{P}_e(t)$ values, enabling the simulation of EPR and ELDOR spectra that correspond to the weak interaction case. Indeed, with the single eSD coefficient, Λ^{eSD}, together with the inverse square dependence on the bin frequency differences in equation (2.70), we succeeded to simulate EPR and ELDOR spectra that fit the experimental ones in the weak interaction case. For the moment, we ignored the rate matrices $\hat{R}_{\mathrm{TM}}^b$ and $\hat{W}_{\mathrm{SE}}^b$.

The validity of the eSD model, defined by the polarization rate equation in equation (2.65), can be verified by comparing the shapes of the calculated

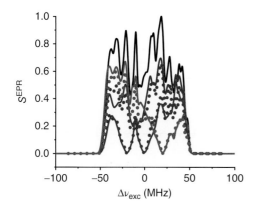

Figure 2.14. A comparison between EPR spectra of the nine-electron spin system, with $D/2\pi = 2$ MHz, obtained by an exact calculation (solid lines) and by using the eSD model (dotted lines). The parameters for the exact calculation are the same as in Figure 2.11, with for the eSD calculation $\Lambda^{eSD} = 1\,\mu s^{-3}$

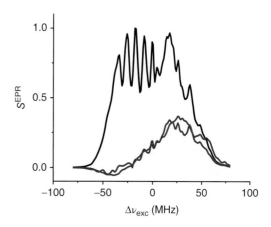

Figure 2.15. The EPR spectrum (blue) of a nine-electron spin system calculated, using $D/2\pi = 6$ MHz, is compared with the spectrum obtained by using the extended eSD model, including the TM relaxation term in equation (2.71). This eSD calculation relies on the equilibrium EPR spectrum in black. The parameters used for the simulations are $T_{1e} = 100$ ms; $T_{2e} = 100\,\mu s$; $\nu_1 = 500$ KHz; $\nu_{exc} = 20$ MHz; $R_{TM} = 0.01\,\mu s^{-1}$; $T_{TM}^{ROF} = 5$ mK; $\Lambda^{eSD} = 100\,\mu s^{-3}$

EPR spectra of our 9-electron system, as obtained from full quantum calculations, with those derived from the eSD model. Choosing a set of $x_{i,i'}$ values and a small D, and in addition values for $\{\omega_1; T_{1e}; T_{2e}; T_D\}$, we can calculate the equilibrium and steady-state EPR spectra. Using the equilibrium spectrum and choosing different values of Λ^{eSD}, we can then calculate a set of steady-state spectra using the eSD model. Comparing these last spectra with the above-calculated spectrum, we can determine a best-fit value for Λ^{eSD}. Figure 2.14 shows an example of the result of such a fitting procedure, for a 9-electron spin system, with parameters summarized in the figure caption.

In the strong dipolar interaction regime, the eSD model in equation (2.65) is not able to generate EPR spectra of electron spin systems that exhibit spin temperatures T_{Ze}^{RoF} and T_{nZe} in the rotating frame. As shown in the population calculations, the appearance of a single-electron spin temperature $T_{TM}^{RoT} = T_{Ze}^{RoF} = T_{nZe}$ is a result of the joint action of the dipolar interaction, the MW field, and the cross-relaxation mechanism. We realize that the eSD model relies on the presence of polarizations of electrons resonating in narrow frequency regimes. In the strong dipolar case, because of state mixing, this is formally not anymore the case. Still, we can define collective electron polarizations $P_{e,j}(t)$ that are determined by the EPR spectrum and represent a fraction of the strongly coupled electron system. Thus, to simulate EPR spectra of real samples in this case, we can use the basic structure of

equation (2.65). However, we must add to it an additional rate matrix $\hat{R}_{TM}$ responsible for the TM features in the EPR spectra. The elements of this rate matrix are defined by two parameters – the spin temperature in the rotating-frame T_{TM}^{RoF} and a rate constant R_{TM} that determines whether or not this TM process dominates the depolarization process:

$$\frac{d}{dt}\begin{pmatrix} 1 \\ P_{e,j}(t) \end{pmatrix} = R_{TM}\begin{pmatrix} 0 & 0 \\ P_{e,j}^{TM} & -1 \end{pmatrix}\begin{pmatrix} 1 \\ P_{e,j}(t) \end{pmatrix} \quad (2.71)$$

Here the spin temperature polarizations are defined as $P_{e,j}^{TM} = \tanh(-(\omega_j - \omega_{MW})\hbar/2k_B T_{TM}^{RoF})$. To verify the validity of this addition to equation (2.65), we show in Figure 2.15 a comparison between an EPR spectrum, simulated by solving equation (2.65) for an 9-electron spin system with $D/2\pi = 6$ MHz, and a spectrum, obtained by solving equation (2.65). The values of R_{TM} and T_{TM}^{RoF} are given in the figure caption. Early on Atsarkin showed experimentally an EPR spectrum exhibiting negative intensities,[105] corresponding to the strong dipolar case. Recent examples of experimental ELDOR and EPR spectra showing the TM features are under investigation.

At this point, we present spectra that are results from actual experiments of systems in the weak interaction regime. These experimental spectra are plotted as a function of $\Delta v_{det} = v_{det} - v_e$ or $\Delta v_{exc} = v_{exc} - v_e$, where v_e is now the frequency of the maximum intensity of their corresponding EPR spectra. In many instances, we measure a set of ELDOR spectra $E^{ELDOR}(t, \Delta v_{exc}, [\Delta v_{det}])$ for different Δv_{det} values, in order to determine the corresponding EPR spectra $E^{EPR}(t, \Delta v_{det}, [\Delta v_{exc}])$ with different Δv_{exc} values. This is done by fitting the ELDOR spectra, using equation (2.65), for determining a best-fit Λ^{eSD} value, followed by simulating EPR spectra using the fitted value. Figure 2.16 shows a set of ELDOR spectra obtained from 20 to 40 mM TEMPOL samples at different temperatures and their best-fit ELDOR spectra.

To account for the last rate matrix in equation (2.65), we must remind ourselves that the interaction of the electrons, with their neighboring nuclei, can also cause frequency-localized depolarization in the steady-state EPR and ENDOR spectra. These depolarizations are caused by the SE-DNP processes. To take these effects into account, we add to equation (2.65) a rate matrix $\widehat{W}_{SE}$ that results in appropriate depolarizations of the electrons.[112] In our model, this is done by defining its elements as

$$\frac{d}{dt} P_{e,j}(t) = -W_{jj}^{SE} P_{e,j}(t) \text{ with}$$

$$W_{jj}^{SE} = \frac{(A^{SE}\omega_1)^2 T_2}{1 + (\omega_j - \omega_n)^2 T_2^2} + \frac{(A^{SE}\omega_1)^2 T_2}{1 + (\omega_j + \omega_n)^2 T_2^2} \quad (2.72)$$

where the SE parameter A^{SE} can be determined via data fitting and can vary with t. An example showing the need for a time-dependent SE parameter is shown in Figure 2.17. Similar experiments were performed by the Han group at 200 GHz also demonstrating these effects.[113]

2.5.2 The DNP Spectra – The iCE Model

The equivalence between the results of the extended eSD approach and the population calculations supports our basic description of the electron spin dynamics laid out in Section 2.4. The eSD model enables us to derive, from a set of experimental ELDOR spectra, the lineshapes of the EPR spectra for different MW excitation frequencies v_{exc}. As we will now discuss, we will need these lineshapes in order to derive DNP spectra.

The nuclear enhancements are mainly a consequence of the two DNP mechanisms, the SE and CE processes. In the weak interaction case, these two mechanisms must be sufficient to account for the enhancement process, while in the strong dipolar case the TM process must also be taken into account. At this stage, we ignore OE-DNP effects. As mentioned earlier, we assume that the bulk nuclei reach a common polarization that is proportional to the average value of the local polarizations. The actual value of the bulk polarization depends on the spin diffusion process that is, among others, dependent on the structural and relaxation properties of the bulk. For this reason, we do here not dwell on the actual values of the enhancements.

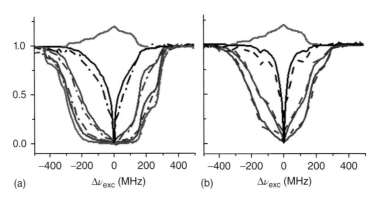

(a)   (b)

Δv_{exc} (MHz)

Figure 2.16. Experimental steady-state ELDOR spectra (dotted lines) obtained at different temperatures, 3K in magenta, 5K in blue, 10K in red, and 20K in black, from a solid solution with 40 mM TEMPOL in (a) and with 20 mM TEMPOL in (b). The solid lines are the spectra simulated using the eSD model. (Y. Hovav, D. Shimon, I. Kaminker, A. Feintuch, D. Goldfarb and S. Vega Phys. Chem. Chem. Phys., 2015, 17, 6053. Reproduced with permission from Royal Society of Chemistry)

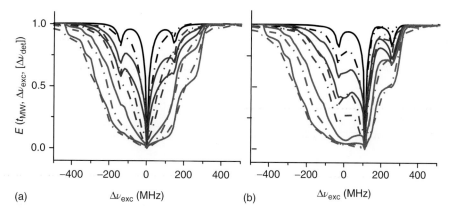

Figure 2.17. Experimental (dotted lines) and simulated (solid lines) ELDOR spectra of a 40mM sample of TEMPOL at 10K. The spectra are measured and simulated for five different values of the MW irradiation times 0.15, 0.6, 1.4, 20 ms in black, red, blue, green, and magenta, respectively. The detection frequencies were set at $\Delta v_{det} = 0$ MHz in (a) and at $\Delta v_{detr} = 100$ MHz in (b). The simulated solid lines are obtained using the eSD model by considering a time-dependent value for A^{SE}. (Krishnendu Kundu, Marie Ramirez Cohen, Akiva Feintuch, Daniella Goldfarb and Shimon Vega Physical Chemistry Chemical Physics. 21(1):478–489. Reproduced with permission from Royal Society of Chemistry)

From the start we realize that the SE and CE mechanisms are dependent on the nuclear Larmor frequency v_n. The first relies on the excitation of ZQ and DQ transitions and the second on the CE condition. When the overall EPR linewidth Ω_{EPR} of the radicals is smaller than v_n, SE-DNP enhancement is only possible when the MW irradiation is applied outside the EPR frequency range. The DNP spectrum shows then two, positive and negative, enhancement patterns that are removed from the EPR line by $\pm v_n$ and have a width about equal to the EPR linewidth Ω_{EPR}. The corresponding ELDOR spectrum shows two polarization depletions at $\Delta v_{exc} = \Delta v_{det} \pm v_n$, as long as Δv_{det} is inside the EPR line. Here again $\Delta v_{exc} = v_{exc} - v_e$ and $\Delta v_{det} = v_{det} - v_e$, where v_e is still the frequency at the maximum of the EPR spectrum. An example is shown in Figure 2.18. When v_n is smaller than the EPR linewidth, the SE process happens at all excitation frequencies $|\Delta v_{exc}| < |1/2\Omega_{EPR} + v_n|$. Roughly speaking, the SE-DNP spectra as a function of Δv_{exc} can almost be decomposed into two spectra of opposite polarity, removed from each other by $2v_n$, with EPR lineshapes. The SE process results, in addition to the nuclear enhancement, also in some depletion of the electron polarization.

In cases where v_n is smaller than the EPR linewidth, the main source of the nuclear polarization enhancement is the iCE-DNP process. For real systems, the probability for two interacting free electrons to fulfill

the CE condition can be rather low. Therefore, only a fraction of the local nuclei get polarized by the iCE process. In a steady-state situation, the MW irradiation at some frequency Δv_{exc}, together with the spin–lattice relaxation and the spectral diffusion mechanism, determine the values of the electron $P_{e,j}$s of bin j with a frequency $\Delta v_j = v_j - v_e$. The polarizations of the nuclei interacting with the electrons in each bin j can then be evaluated using the iCE expression:

$$P_{n,j}(\Delta v_{exc})$$
$$= \left[f_j f_{j+\Delta j_n} \frac{\left(P_{e,j+\Delta j_n} - P_{e,j}\right)}{1 - P_{e,j} P_{e,j+\Delta j_n}} + f_j f_{j'-\Delta j_n} \frac{(P_{e,j} - P_{e,j-\Delta j_n})}{1 - P_{e,j} P_{e,j'-\Delta j_n}} \right]$$

(2.73)

where $\Delta j_n \times \delta v_b = v_n$. $f_j f_{j'}$ stands for the probability of the electrons in bin j and bin j' to interact significantly enough for the iCE process to take place. The choice of the expression for this probability is, of course, not rigorously derived, and we should therefore expect small deviations between experimental and simulated iCE-DNP spectra. Considering all N_b bins with a width δv_b, assuming again that the bulk nuclear polarization after the spin diffusion stage equals to the average polarizations of the iCE enhanced local nuclei, we can

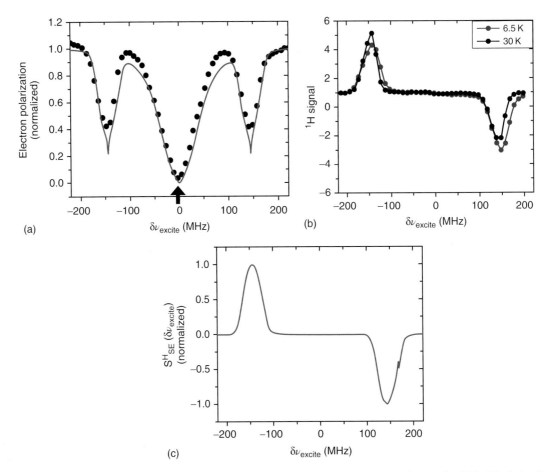

Figure 2.18. (a) Experimental (dotted lines) and simulated (solid lines) ELDOR spectra of a sample of 15 mM trityl radicals (OX63) in a solution of 50/50 (% v/v) of $H_2O/1,3$-$^{13}C_2$-glycerol. The ZQ and DQ transitions of the proton result in the two dips on the two sides of the main central line depolarization. (b) The DNP spectrum for two temperatures of the same sample, demonstrating the result of the SE mechanism. (c) A simulated DNP spectrum for the same sample. (Daphna Shimon, Yonatan Hovav, Ilia KaminkerAkiva Feintuch, Daniella Goldfarb and Shimon Vega: Phys. Chem. Chem. Phys., 2015, 17, 11868. Reproduced with permission from Royal Society of Chemistry)

calculate a bulk nuclear polarization:

$$P_n(\Delta v_{\text{exc}}) = \frac{1}{N_b - \Delta j_n} \sum_{j=1,N_b-\Delta j_n} f_j f_{j+\Delta j_n} \frac{P_{e,j} - P_{e,j+\Delta j_n}}{1 - P_{e,j}P_{e,j+\Delta j_n}}$$

(2.74)

Deriving the values of $P_n(\Delta v_{\text{exc}})$ from the EPR spectra for different MW frequencies Δv_{exc} enables us to construct normalized iCE-DNP enhancement spectra. These spectra must then coincide with experimentally obtained DNP spectra. The validity of equation (2.73) was recently evaluated numerically, by comparing DNP spectra of 10-spin (9 electrons and 1 nucleus)

systems, obtained by full Liouville space calculations, with DNP results obtained from applying the iCE model.[67] This study has shown that the iCE approach is valid in the weak dipolar regime as well as in the strong dipolar regime, where the EPR spectra exhibit the above-discussed TM characteristics.

Two examples of iCE-DNP spectra are shown in Figures 2.19 and 2.20 and compared with experimental DNP spectra. These iCE-DNP spectra were constructed from the EPR spectra, derived from experimental ENDOR spectra, using the eSD model. The spectra were obtained from samples containing

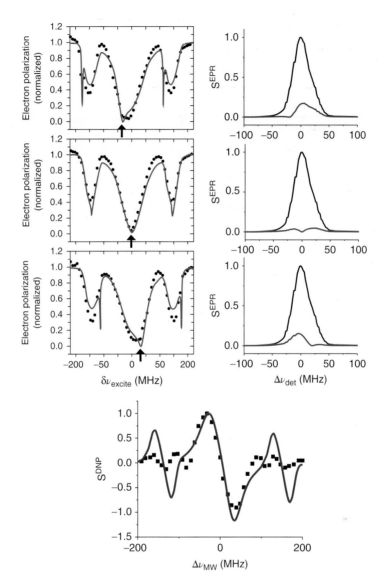

Figure 2.19. On the left are shown experimental and simulated ELDOR spectra of a 15mM trityl sample at 6.5K (note that the *x*-axis is equivalent to what we termed in this review as $\Delta\nu_{exc}$). The detection frequency was positioned at -32 MHZ (top), 0 MHz (middle), 32 MHz (bottom) relative to the center of the EPR line, as indicated by the arrows. The simulated spectra are obtained by fitting eSD-derived spectra to the experimental ones. The central column shows EPR spectra (in red) simulated, using the eSD parameters obtained from the ELDOR fits. The equilibrium EPR spectrum is drawn in black. The spectrum on the right shows the experimental DNP spectrum (black dotted) and a calculation spectrum, using the iCE model relying on the calculated EPR spectra. (Daphna Shimon, Yonatan Hovav, Ilia KaminkerAkiva Feintuch, Daniella Goldfarb and Shimon Vega: Phys. Chem. Chem. Phys., 2015, 17, 11868. Reproduced with permission from Royal Society of Chemistry)

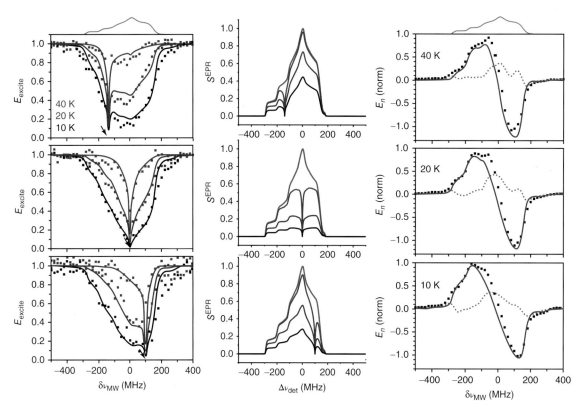

Figure 2.20. On the left are shown experimental and simulated ELDOR spectra of 40 mM TEMPOL solution in a 50 : 50 (% v/v) mixture of dimethyl sulfoxide (DMSO) and H_2O, at three different temperatures, 40 K in black, 20 K in blue, and 10 K in red (again the x-axis is equivalent to what we termed in this review as $\Delta\nu_{exc}$). The detection frequency was positioned at 100 MHZ (top), 0 MHz (middle), 100 MHz (bottom) relative to the maximum intensity of the TEMPOL EPR line. The simulated spectra are obtained by fitting eSD-derived spectra to the experimental ones. The central column shows EPR spectra simulated, using the eSD parameters obtained from the ELDOR fits, with the MW excitation frequencies of 100 MHZ (top), 0 MHz (middle), 100 MHz (bottom). The equilibrium EPR spectrum is each time drawn in green. The spectra on the right show the experimental DNP spectra (black dots) and calculated spectra, using the iCE model relying on the calculated equilibrium EPR lineshape. The dotted curve shows the difference between the experimental and simulated spectra. Details of the sample and ELDOR simulations can be found in Yonatan Hovav, Daphna Shimon, Ilia Kaminker, Akiva Feintuch, Daniella Goldfarb and Shimon Vega Phys. Chem. Chem. Phys., 2015, 17, 6053

trityl[114] and TEMPOL radicals.[115] The experimental and numerical parameters of these experiments and simulations are summarized in the figure captions. Some deviations between the experimental and simulated spectra are apparent, but the overall lineshapes, in particular their widths and the ratio between the maxima and minima, overlap reasonably well.

Another example concerning the simultaneous enhancement of protons and deuterium in a sample of a 40 mM TEMPOL radical dissolved in $^1H_2O/DMSO\text{-}d_6$ is shown in Figure 2.21.[116] The

^{1}H-DNP and ^{2}H-DNP spectra at long irradiation times, shown in Figure 2.21(a), overlap not only in shape but also in the enhancement. These spectra have a shape that is close to the ^{1}H-iCE-DNP lineshape derived from the experimental EPR lineshapes. For short irradiation times, the ^{2}H-DNP spectrum is clearly different from the ^{1}H-DNP spectrum, as shown in Figure 2.21(b), and its shape is close to that of the ^{2}H-iCE-DNP spectrum, derived from the same EPR spectrum. The result at long irradiation would be an indication of the action of the TM process, despite

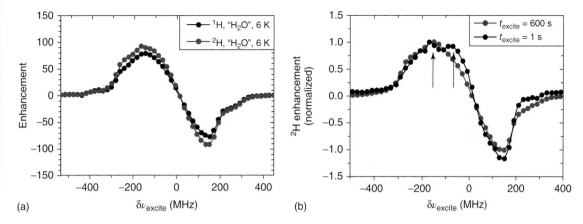

Figure 2.21. (a) Normalized steady-state ^{1}H- and ^{2}H-DNP spectra measured on a sample with 40 mM TEMPOL in ^{1}H$_2$O/DMSO-d$_6$, showing their complete overlap. (b) Normalized ^{2}H-DNP spectra obtained after MW irradiation of 1 and 600 s, showing the change in the shape of ^{2}H-DNP spectrum over time. (I. Kaminker, D. Shimon, Y. Hovav, A. Feintuch and S. Vega Phys. Chem. Chem. Phys., 2016, 18, 11017. Reproduced with permission from Royal Society of Chemistry)

the fact that the EPR spectra do not show the typical TM features. A possible explanation could be related to the hn-CE processes in the sample under study. A consequence of the presence of these processes was demonstrated by following the spontaneous appearance of ^{2}H enhancement, after its saturation, in the presence of enhanced ^{1}H polarization.[116] Further analysis of this phenomena is still under investigation.

What summarizes our understanding and these examples is the fact that the shape of the EPR line determines the enhancement. Of course, this is right as long as the iCE-DNP mechanism dominates the polarizations of the local nuclei in the amorphous samples, and the spin diffusion distributes these polarizations over the bulk equally. This spin diffusion process can become insufficient, for example, when the concentration of the bulk nuclei is low or when their spin–lattice relaxation rates become too fast. Another factor, not discussed here, disturbing the bulk enhancement is a very short relaxation time of the core nuclei. For ^{1}H-DNP, this time can be influenced by partial deuteration.

In the case of iCE-DNP, in practice we must try to generate EPR lineshapes that are favorable for DNP.

Factors influencing these shapes are intrinsic and experimental parameters of the sample, such as the radical concentration–dependent Λ^{eSD} parameter, the MW intensity ω_{1_e} and the T_{1e} relaxation time. Increasing values for Λ^{eSD} and decreasing values for T_{1e} can broaden the MW burned hole in the spectra and thus modify the electron polarization gradients.

Influencing these effects requires changes in the radical concentrations or T_{1e} values, for example, by adding Gd-ions to the sample.[117]

An additional example, where modifying the EPR lineshape results in an increase in the iCE-DNP enhancement, is application of frequency modulation of the MW irradiation. Significant gains in DNP enhancements due to such modulations have been reported.[118,119] In particular, Kaminker *et al.*[120] showed that changes in the ELDOR spectra, as a result of broadband excitation, go together with increasing enhancements up to a factor of 5. As an example, we choose to show here the results of this frequency modulation approach by Hovav *et al.*,[118] applied to a set of samples containing TEMPOL radicals at different concentrations and temperatures. Figure 2.22 summarizes the results of this study for a constant MW frequency, $\Delta\nu_{exc}$, and a modulated MW frequency, centered at $\Delta\nu_{exc}$ with an amplitude of modulation $\pm\delta\nu_{exc}$. In this figure, the ratio between the constant frequency and modulated frequency enhancements, E_{fm}/E_{cf}, is plotted as a function of the TEMPOL concentration and temperature of the sample. The frequency $\Delta\nu_e$ was set at the frequency of the maximum of the DNP profile. E_{fm}/E_{cf} shows a maximum at 10 mM and an increase for increasing temperature. It should be noted that the modulation directly broadens the MW saturated range of the EPR line. If the hole in the spectrum is already wide for constant MW, this broadening does not have a significant effect, as we observe at 40 mM. On the other hand, when T_{1e}

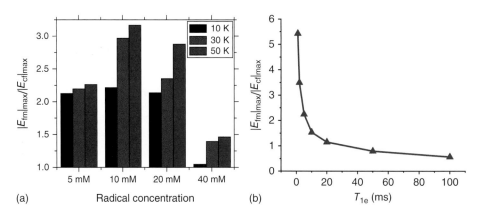

Figure 2.22. (a) Experimental results demonstrating the ratio between the proton NMR signal (defined here as P^{mf}/P^{cf}), obtained without and with modulation of the MW field, of samples with 5, 10, 20, and 40 mM TEMPOL in 56/44 wt% of DMSO/H$_2$O, at three temperatures. (b) The simulated results as a function of T_{1e} of the same ratio from the spin system discussed in the main text. (Yonatan Hovav; Akiva Feintuch; Shimon Vega; Daniella Goldfarb, Journal of Magnetic Resonance. 2014, 238:94–105. Reproduced with permission from Royal Society of Chemistry)

is short, and the hole before modulation is narrow, the modulation broadens the hole, increasing the number of CE electron pairs with large polarization differences, as is probable the reason for the high ratio at 10 mM. As we did not measure the EPR spectra corresponding to the data in Figure 2.22, we show here only a simulated result. Assuming that the modulated MW saturates the spectra over the whole range $\Delta \nu_{exc} \pm \delta \nu_{exc}$, we plot in Figure 2.22(b) the value of E^{cf}_{mf} for an EPR line belonging to TEMPOL as a function of T_{1e} in the range 1-100 ms and a constant value for $\Lambda^{eSD} = 200\,\mu s^{-3}$ all for a modulation frequency range $\delta \nu_{exc} = 67$ MHz. For these parameters, the E^{cf}_{mf} ratio varies between 1 and 5.

2.6 THE SPIN DYNAMICS LEADING TO MAS-DNP

After presenting a description of the spin dynamics leading to the DNP enhancement in static glassy samples containing free radicals, we now concentrate on samples that are spinning at the magic angle (MAS) in the magnetic field. The terms of the Hamiltonian defining the interactions of the spins in these rotating samples, as well as the spin–lattice relaxation mechanisms, are identical to the terms of the static case, except that the interaction coefficients and the relaxation times become time dependent. Thus, the same interactions are responsible for the enhancements in static and rotating samples, but their DNP spin

dynamics differ because of the time dependence. For example, the steady state of a rotating spin system, even after a long MW irradiation, remains periodically time dependent. Hence, we call this state a quasiperiodic (qp) steady state. Consequently, the previously derived static eSD, TM, and iCE models for analyzing EPR and DNP spectra in large spin systems cannot be directly applied.

To reveal the unique MAS-DNP enhancement processes, we follow the evolution of the spins by numerical simulations. Computational requirements limit the calculations of the spin dynamics to small spin systems if no approximations are considered. Only at the last stage of the text, we consider larger spin systems and address DNP-related phenomena that are correlated to spectral and spin diffusion processes. As in the previous sections, we will not dwell on the actual enhancement values achieved in practice or on comparisons between the DNP efficiency of different types of free radicals. All these aspects are discussed in other chapters.

To simplify the discussions in the following sections, the reader must keep in mind that in all cases we assume that the typical spin–lattice relaxation times of the electron spins are much shorter than those of the nuclei. For example, the electron spin–lattice relaxation time of free radicals in glassy solids at 100 K is of the same order of magnitude as the MAS rotor period, while the nuclear relaxation times are, in general, of longer orders of magnitude. As we will see in the following sections, this has an important consequence on the MAS-DNP processes.

2.6.1 MAS-DNP on a Two-electron and One-nucleus Spin System

In 2012 two research initiatives, of Kent Thurber and Robert Tycko[68] and of Frederic Mentink-Vigier *et al.*,[69] approached the MAS-DNP problem by considering a three-spin system, composed of two interacting electrons and a nucleus $\{e_a - e_b - n\}$. In most of the MAS-DNP experiments, the free electrons in the solid samples come in the form of biradicals.[74] Without the presence of the multispin spectral and spin diffusion effects, the three-spin system $\{e_a - e_b - n\}$ is an excellent model for investigating the spin physics of the MAS-DNP mechanism. The anisotropic interactions of this system, when placed in an external magnetic field, are defined by the time-dependent coefficients $g_i(t)$, $A_{i,n}(t)$, $B_{i,n}^{\pm}(t)$ and $\omega_{a,b}(t)$, with $i = a$, b, in the Zeeman, hyperfine, and dipolar terms of the Hamiltonian in equation (2.1):

$$\hat{H}^{\text{RoF}}(t) = \hat{H}_Z(t) + \hat{H}_{\text{HF}}(t) + \hat{H}_D(t) + \hat{H}_J + \hat{H}_{\text{MW}}$$
$$= \hat{H}_0(t) + \hat{H}_{\text{MW}} \qquad (2.75)$$

Here the *MW* Hamiltonian is included in $\hat{H}^{\text{RoF}}(t)$. In the following, we again assume that the hyperfine coefficients are smaller than the nuclear Larmor frequency, $|A_{i,n}|$, $|B_{i,n}^{\pm}| \ll |\omega_n|$.

Introducing the spin density matrix of this system in the MW rotating frame, $\hat{\rho}(t)$, we can derive the spin dynamics by solving the full master equation in Liouville space. After presenting $\hat{H}^{\text{RoF}}(t)$ in this space and introducing a relaxation superoperator, we can obtain the solution of this equation by a small time-step integration approach. As the Hamiltonian is periodic with a repetition time equal to $t_r = 2\pi/\omega_r$, where ω_r is the spinning frequency of the sample in sec^{-1} units, we divide each rotor period into k_r equal-length time intervals Δt with $t_r = k_r\Delta t$. Then, for each time interval $Nt_r + (k-1)\Delta t \rightarrow Nt_r + k\Delta t$, where N is an integer and $k = 1, \ldots, k_r$, we aim at computing its evolution operator $\hat{U}_k \equiv \hat{U}((k-1)\Delta t, k\Delta t)$, for all k, in Liouville space.

For a set of randomly oriented equivalent spin systems, imbedded in an amorphous solid forming a spin powder, each system has a propagator that can be calculated by a four-step procedure:

1. Define for each time interval, with $k = 1, \ldots, k_r$ a rotating-frame Hamiltonian with constant interaction coefficients that are equal to the coefficients of the Hamiltonian $\hat{H}_0(t_k)$ at $t_k = (k - 1/2)\Delta t$.

2. Diagonalize these Hamiltonians $\hat{\Lambda}_k = \hat{D}_k^{-1}\hat{H}_0(t_k)\hat{D}_k$ for all k time intervals and derive the k_r Liouvillians $\hat{L}_k$ of $\hat{\Lambda}_k + \hat{D}_k^{-1}\hat{H}_{\text{MW}}\hat{D}_k$.

3. Introduce $T_{1,e}$, $T_{2,e}$, and $T_{1,n}$ relaxation times, evaluate the spin–lattice and spin–spin relaxation rates in the diagonalized representation of $\hat{\Lambda}_k$, following equation (2.13) using the operators $\hat{S}_x$, $\hat{S}_z$, and $\hat{I}_x$, respectively, and derive the relaxation superoperator $\hat{R}_k$.

4. Calculate for each time period of length Δt an evolution superoperator in Liouville space $e^{\hat{L}_k + \hat{R}_k}$ in the diagonalized MW-rotating frame, and transfer it to the nondiagonalized MW-rotating frame: $\hat{U}_k$.

The evolution operator of a single rotor period can then be calculated, $\hat{U}_r = \Pi\hat{U}_k$, and the spin density vector $\vec{\rho}(t)$ in Liouville space at $t = Nt_r + k\Delta t$ becomes

$$\vec{\rho}(t) = \hat{U}(t)\vec{\rho}(0) \; ; \hat{U}(t) = (\hat{U}_r)^N\Pi\hat{U}_k \qquad (2.76)$$

It is then possible to extract relevant spin parameters, such as the time-dependent electron and nuclear polarizations. This procedure is repeated for each of the powder orientations and their NMR signals can be calculated. Their mean values can then be compared to experimentally obtained signals. A similar approach based on simulations in Liouville space can be found in the work of Mance *et al.*[121]

2.6.2 The Rotor Events of the Three-spin System

To gain insight into the spin dynamics during MAS-DNP and understand the interplay between the spin interactions and the experimental results, we focus on the time course of the three-spin system. In particular, we calculate the spin evolution during a single rotor period, after the system has reached a qp steady state. In Figure 2.23 we show a typical energy–time profile during one rotor period of a three-spin system at high field, with interaction parameters equal to those of a nitroxide biradical, widely used under MAS-DNP conditions.[77,79] The large anisotropic *g*-tensors of the two electrons (their interaction parameters are summarized in the figure caption) create frequency differences between the instantaneous eigenstate energies that are most of the time significantly larger than any of the other interactions in the system. In

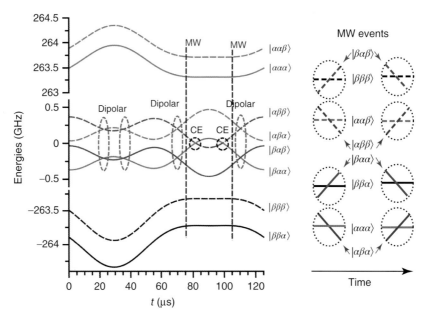

Figure 2.23. The time dependence of the energy levels in the laboratory frame of a single three-spin system composed of two electrons and a nucleus $\{e_a - e_b - n\}$ during a single rotor period. The g-tensor principle elements $[g_x, g_y, g_z]$ of the two electrons are $[2.0092, 2.006, 2.0021]$ and the Euler angles between the two g-tensors are $(90°, 90°, 90°)$. The orientation of the g-tensor of e_a with respect to the magnetic field at $t = 0$ is defined by the angles $(2.5°, 51.8°, 80°)$. The electron dipolar interaction was set equal to 23 MHz and the $e_a - e_b$ bond direction makes an angle of 70° with the magnetic field at $t = 0$. The strength of the hyperfine interaction between e_a and n was set to 3 MHz. The magnetic field was set to 9.394 T, the MW frequency to 263.45 GHz, and the corresponding nutation frequency was 0.35 MHz. The energies of this system are assigned according to their product states $|\chi, \chi', \chi''\rangle = |\chi_{e_a}, \chi'_{e_b}, \chi''_n\rangle$. The DJ and CE rotor events are indicated by their appropriate energy level crossings. The vertical dotted lines indicate the moments of MW events in this system. These MW moments are also dictated by energy level crossing, but this time in the rotating frame. To emphasize this fact, we added at the right side of the figure the MW event energy level crossings in the rotating frame

other words, most of the time the eigenstates of $\hat{H}^{\text{RoF}}(t)$ are determined by the $\hat{S}_{z,a}$, $\hat{S}_{z,b}$, and $\hat{I}_{z,n}$ operators of the electron and nuclear Zeeman terms and the hyperfine interaction. These eigenstates, however, vary when two energy levels approach each other in the rotating frame. Their off-diagonal elements in the matrix representation of $\hat{H}^{\text{RoF}}(t)$ create state mixings and cause energy level anticrossings. More precisely, in the matrix representation defined by the product states $|\chi, \chi', \chi''\rangle \equiv |\chi_a, \chi_b', \chi_n''\rangle$ with energies $\varepsilon_{\chi\chi'\chi''}(t)$ and $\chi, \chi', \chi'' = \alpha, \beta$, the rotating-frame Hamiltonian has in addition to its diagonal terms

$$\hat{H}^{\text{diag}}(t) = \sum_{i=a,b} \Delta\omega_e(t)\hat{S}_{z,i} + A_{z,i,n}(t)\hat{S}_{z,i}\hat{I}_{z,n}$$
$$- \omega_n \hat{I}_{z,n} - (\omega_{a,b}(t) - 2J_{a,b})\hat{S}_{z,a}\hat{S}_{z,b} \quad (2.77)$$

with $\Delta\omega_e = (g_i(t)\beta_e B_0 - \omega_{\text{MW}})$, $\omega_{a,b}(t)$ the anisotropic electron dipole-dipole coefficient and $J_{a,b}$ the isotropic Heisenberg exchange coefficient, three types of off-diagonal terms:

1. The electron dipolar and exchange flip-flop terms, connecting the states $|\alpha, \beta, \chi\rangle \leftrightarrow |\beta, \alpha, \chi\rangle$

$$\hat{H}_{\text{DJ}}^{\text{off}}(t) = -1/2(\omega_{a,b}(t) - 2J_{a,b})(\hat{S}_a^+\hat{S}_b^- + \hat{S}_a^-\hat{S}_b^+)$$
$$(2.78)$$

2. The MW irradiation term, connecting the states $|\alpha, \chi, \chi'\rangle \leftrightarrow |\beta, \chi, \chi'\rangle$ and $|\chi, \alpha, \chi'\rangle \leftrightarrow |\chi, \beta, \chi'\rangle$

$$\hat{H}_{\text{MW}}^{\text{off}} = \omega_1(\hat{S}_{x,a} + \hat{S}_{x,b}) \quad (2.79)$$

For simplicity, we assume here that the MW irradiation with amplitude ω_1 is applied in the x-direction of the laboratory frame.

3. The pseudohyperfine terms, assuming, for simplicity, $B_{i,n} = B_{i,n}^{\pm}$ to be real, connecting pairs of states of the form $|\chi, \chi', \alpha\rangle \leftrightarrow |\chi, \chi', \beta\rangle$

$$\hat{H}_{e,n}^{\text{off}}(t) = 1/2 \sum_{i=a,b}(B_{i,n}(t)\hat{S}_{z,i}\hat{I}_n^+ + B_{i,n}(t)\hat{S}_{z,i}\hat{I}_n^-)$$

(2.80)

As discussed in the earlier sections, the direct influence of the pseudosecular hyperfine terms $\hat{H}_{e,n}^{\text{off}}(t)$ on the eigenstates is minor when the energy difference ω_n between the states ($|\chi, \chi', \alpha\rangle \leftrightarrow |\chi, \chi', \beta\rangle$) is much larger than $B_{i,n}$. Still, their influence becomes very significant when we can derive, by using degenerate perturbation theory[85] involving these pseudohyperfine terms, effective Hamiltonian terms that directly connect pairs of degenerate states in the rotating frame.

A. When the ZQ pairs of states $\{|\alpha, \chi, \beta\rangle, |\beta, \chi, \alpha\rangle\}$ or $\{|\chi, \alpha, \beta\rangle, |\chi, \beta, \alpha\rangle\}$ or the DQ pairs of states $\{|\alpha, \chi, \alpha\rangle, |\beta, \chi, \beta\rangle\}$ or $\{|\chi, \alpha, \alpha\rangle, |\chi, \beta, \beta\rangle\}$ become degenerate in the rotating frame at a time t_{SE}, the simultaneous presence of the $\hat{H}_{e,n}^{\text{off}}(t_{\text{SE}})$ and $\hat{H}_{\text{MW}}^{\text{off}}$ leads to effective off-diagonal terms in the Hamiltonian of the form

$$\hat{H}_{\text{SE}}^{\text{off}}(t_{\text{SE}})$$
$$= \sum_{i=a,b}\{1/2(B_{i-n}(t_{\text{SE}})/\omega_n)\omega_1(\hat{S}_i^+\hat{I}^+ + \hat{S}_i^-\hat{I}^-)$$
$$+ 1/2(B_{i-n}(t_{\text{SE}})/\omega_n)\omega_1(\hat{S}_i^-\hat{I}^+ + \hat{S}_i^-\hat{I}^+)\}$$

(2.81)

These terms are the source of the SE enhancement mechanism during MAS-DNP.

B. When the pairs of states $\{|\alpha, \beta, \beta\rangle, |\beta, \alpha, \alpha\rangle\}$ or $\{|\alpha, \beta, \alpha\rangle, |\beta, \alpha, \beta\rangle\}$ become degenerate at a time t_{CE}, the simultaneous presence of $\hat{H}_{e,n}^{\text{off}}(t_{\text{SE}})$ and $\hat{H}_{\text{DJ}}^{\text{off}}(t_{\text{CE}})$ leads to effective off-diagonal terms of the form

$$\hat{H}_{\text{CE}}^{\text{off}}(t_{\text{CE}}) = 1/4(\omega_{a,b}^{\text{DJ}}(t_{\text{CE}}) \times \Delta B(t_{\text{CE}})/\omega_n)$$
$$\times (\hat{S}_a^+\hat{S}_b^-\hat{I}^+ + \hat{S}_a^-\hat{S}_b^+\hat{I}^-)$$
$$+ 1/4(\omega_{a,b}^{\text{DJ}}(t_{\text{CE}}) \times \Delta B(t_{\text{CE}})/\omega_n)$$
$$\times (\hat{S}_a^+\hat{S}_b^-\hat{I}^- + \hat{S}_a^-\hat{S}_b^+\hat{I}^+) \quad (2.82)$$

with $\omega_{a,b}^{\text{DJ}}(t) \equiv \omega_{a,b}(t) - 2J_{a,b}$ and $\Delta B(t) = B_{a,n}(t) - B_{b,n}(t)$:

These terms form the source of the CE enhancement mechanisms during MAS-DNP.

Since the duration of these events is very short, most of the time the spin system is exposed only to relaxation effects. However, when the energies of these states in the rotating frame become equal, and they are connected via one of the four effective off-diagonal terms in equations (2.78, 2.79) and (2.81, 2.82), the populations and coherences of the spin system will experience profound changes. From now on, we call these energy anticrossings 'rotor events'. The four types of rotor events are thus

1. the DJ rotor events, when $\varepsilon_{\beta\alpha\chi} \approx \varepsilon_{\alpha\beta\chi}$, induced by the off-diagonal terms in equation (2.78);
2. the MW rotor events, when $\varepsilon_{\beta\chi\chi'} \approx \varepsilon_{\alpha\chi\chi'}$ and $\varepsilon_{\chi\beta\chi'} \approx \varepsilon_{\chi\alpha\chi'}$, induced by the off-diagonal terms in equation (2.79);
3. the DQ- and ZQ-SE rotor events, when $\varepsilon_{\beta\chi\beta} \approx \varepsilon_{\alpha\chi\alpha}$ and $\varepsilon_{\beta\chi\alpha} \approx \varepsilon_{\alpha\chi\beta}$, induced by the off-diagonal terms of equation (2.81); and
4. the CE rotor events, when $\varepsilon_{\beta\alpha\beta} \approx \varepsilon_{\alpha\beta\alpha}$ and $\varepsilon_{\beta\alpha\alpha} \approx \varepsilon_{\alpha\beta\beta}$, induced by the off-diagonal terms in equation (2.82).

Each of these four types of rotor events can occur 0, 2, or 4 times per rotor period and together are responsible for the DNP enhancement. The actual effects of these events on the elements of $\vec{\rho}(t)$ must be calculated using the small-time-step integration approach for an accurate evaluation. However, to get a better insight into the spin dynamics during MAS-DNP, we concentrate here on the time evolution of the product state populations only.

2.6.3 The Time Evolution of the Populations and the Landau–Zener Coefficient

To analyze the time dependence of the system's populations, we should make a distinction between their time evolution at the rotor events and in-between. During the events, we use the Landau–Zener (LZ) formula[102] as suggested first by Thurber and Tycko,[68] while during the rest of the rotor cycle the evolution of the populations depends on spin–lattice relaxation mechanisms. The LZ formula enables us to calculate changes in the populations caused by the rotor event. The validity of this approach depends on the magnitudes of the off-diagonal elements, causing the anticrossings, and on the spin–lattice relaxation rates.[122–124] When these are relatively small, the formula can be safely used and when not, the stepwise integration or the Bloch-type approach, involving coherences, must be applied[69] to calculate the shifts in the population from before to after the events.

In short, the LZ formula correlates the values of the populations $p_1(t)$ and $p_2(t)$ of a two-level system after an anticrossing event with those before the event. To present the formula, we consider a two-level system with a Hamiltonian $\hat{H}_{1,2}(t)$ that, in the representation of the pair of states $|\psi_1\rangle$ and $|\psi_2\rangle$, has diagonal elements, $\langle\psi_1|\hat{H}_{1,2}(t)|\psi_1\rangle = \varepsilon_1(t)$ and $\langle\psi_2|\hat{H}_{1,2}(t)|\psi_2\rangle = \varepsilon_2(t)$, and off-diagonal elements, $\langle\psi_1|\hat{H}_{1,2}(t)|\psi_2\rangle = \varepsilon_{12}(t)$ and its complex conjugate. When at a time t_x the diagonal elements become equal, $\varepsilon_1(t_x) = \varepsilon_2(t_x)$, while at $t_{\text{before}} < t_x$ $\varepsilon_1(t_{\text{before}}) < \varepsilon_2(t_{\text{before}})$, and at $t_{\text{after}} > t_x$ $\varepsilon_1(t_{\text{after}}) > \varepsilon_2(t_{\text{after}})$, the LZ formula correlates the populations after t_x, $p_1^{\text{after}} = p_1(t_{\text{after}})$ and $p_2^{\text{after}} = p_2(t_{\text{after}})$ with those before t_x, $p_1^{\text{before}} = p(t_{\text{before}})$ and $p_2^{\text{before}} = p(t_{\text{before}})$. Assuming that at $t = t_{\text{before}}$ and $t = t_{\text{after}}$ $|\varepsilon_1(t) - \varepsilon_2(t)| \gg |\varepsilon_{12}(t)|$:

$$(p_1^{\text{after}} + p_2^{\text{after}}) = (p_1^{\text{before}} + p_2^{\text{before}})$$
$$(p_1^{\text{after}} - p_2^{\text{after}}) = (2e^{-\pi\xi_{\text{mix}}^2/2\xi_t} - 1)(p_1^{\text{before}} - p_2^{\text{before}})$$
$$= X_{\text{LZ}}(p_1^{\text{before}} - p_2^{\text{before}}) \quad (2.83)$$

Here $\xi_{\text{mix}}^2 = \varepsilon_{12}^2(t_x)$ and $\xi_t = \{d(\varepsilon_1(t) - \varepsilon_2(t))/dt\}_{t_x}$ are the square of the strength of the state mixing matrix elements and the crossing rate, respectively. In this equation, we also defined the LZ factor, $X_{\text{LZ}} = (2e^{-2\pi\xi_{\text{mix}}^2/\xi_t} - 1)$ with $1 \leq X_{\text{LZ}} \leq -1$, correlating the differences of the populations before and after the crossing. The actual change in each of the populations (the step size) can also be expressed in terms of X_{LZ}

$$(p_1^{\text{after}} - p_1^{\text{before}}) = -(p_2^{\text{after}} - p_2^{\text{before}})$$
$$= \frac{1}{2}(1 - X_{\text{LZ}})(p_1^{\text{before}} - p_2^{\text{before}}) \quad (2.84)$$

and we can define the LZ step factor by $s_{\text{LZ}} = 1/2(1 - X_{\text{LZ}})$. A step factor equal to 0 corresponds to $X_{\text{LZ}} = 1$ and a step factor equal to 1 corresponds to an inversion of the populations with $X_{\text{LZ}} = -1$. A discussion about the validity of these equations as a function of the relaxation parameters can be found in the literature.[125,126]

Important conclusions derived from equation (2.83) are that for an increase of ξ_{mix} the value X_{LZ} decays and s_{LZ} increases. On the other hand, when the crossing rate increases X_{LZ} approaches 1 and s_{LZ} declines.

Furthermore, for $\xi_{\text{mix}} \gg \xi_t$ the X_{LZ} approaches -1 and s_{LZ} reaches its maximum value, and for $\xi_t \gg \xi_{\text{mix}}$ the X_{LZ} is close to 1 and s_{LZ} is about 0.

Using this approach later on, we assume that the time duration of the event is very short with respect to other time scales in the system. When we cannot use the LZ approach, the small-step integration approach still correlates the populations before and after the crossing. We can thus define a factor X similarly to that in equation (2.83), as long as there is no significant loss of coherence. A step factor can then also be defined in this case, using the definition of s_{LZ}, replacing X_{LZ} by X. The duration of the population exchange process of these events can take a significant portion of the rotor period.

2.6.4 The Quasiperiodic Steady-state Polarizations

In our three-spin systems, we can now concentrate on the time evolution of the populations $p_{\chi\chi'\chi''} \equiv p_{\chi\chi'\chi''}(t)$ of the eight product states $|\chi, \chi', \chi''\rangle$, again with $\chi, \chi', \chi'' = \alpha, \beta$. Their time dependence is determined by T_{1n} and T_{1e} between the rotor events and their exchange during the events. The values of $p_{\chi\chi'\chi''}$ determine the polarization coefficients, which are defined from now on as the polarizations $P_i \equiv P_i(t)$ of the two electrons $(i = a, b)$ and $P_n \equiv P_n(t)$ of the nucleus:

$$P_a = \sum_{\chi,\chi'} p_{\beta\chi\chi'} - p_{\alpha\chi\chi'}; P_b = \sum_{\chi,\chi'} p_{\chi\beta\chi'} - p_{\chi\alpha\chi'};$$
$$P_n = \sum_{\chi,\chi'} p_{\chi\chi'\alpha} - p_{\chi\chi'\beta} \quad (2.85)$$

Each partial exchange of two populations at a rotor event can be translated to a change in the values of these polarizations. After a certain MW irradiation time, the system reaches a 'quasiperiodic steady state', which will be the focus of this section.

A typical time evolution of the polarization of a three-spin system during a rotor period at the qp steady state is shown in Figure 2.24.

By focusing on the effects of the rotor events on the populations, we can infer the impact of the DJ, the CE and MW rotor events, on the polarizations. Because, in most practical cases, the CE events dominate the SE events, these last events are, for simplicity, ignored in the forthcoming discussion. The polarizations can be expressed as a sum of population differences as

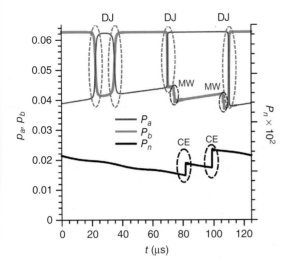

Figure 2.24. Simulated time dependences of three polarizations, P_a in red, P_b in green, and P_n in black, of a single three-spin system $\{e_a - e_b - n\}$, with parameters as in Figure 2.23 and $v_r = 8\,\text{kHz}$, at its quasiperiodic steady state. The relaxation parameters during the calculation were $T_{1,e} = 0.3\,\text{ms}$, $T_{2,e} = 2\,\mu\text{s}$ and $T_{1,n} = 4\,\text{s}$, $T_{2,n} = 20\,\text{ms}$. The average value of $\overline{P}_n$ is equal to $\overline{|P_a - P_b|}$, the average value of the electron polarization difference

follows:

$$P_a = (p_{\beta\beta\alpha} - p_{\alpha\alpha\alpha}) + (p_{\beta\beta\beta} - p_{\alpha\alpha\beta})$$
$$+ (p_{\beta\alpha\alpha} - p_{\alpha\beta\alpha}) + (p_{\beta\alpha\beta} - p_{\alpha\beta\beta})$$
$$P_b = (p_{\beta\beta\alpha} - p_{\alpha\alpha\alpha}) + (p_{\beta\beta\beta} - p_{\alpha\alpha\beta})$$
$$- (p_{\beta\alpha\alpha} - p_{\alpha\beta\alpha}) - (p_{\beta\alpha\beta} - p_{\alpha\beta\beta})$$
$$P_n = (p_{\beta\beta\alpha} - p_{\beta\beta\beta}) + (p_{\alpha\alpha\alpha} - p_{\alpha\alpha\beta})$$
$$+ (p_{\beta\alpha\alpha} - p_{\beta\alpha\beta}) + (p_{\alpha\beta\alpha} - p_{\alpha\beta\beta}) \quad (2.86)$$

In these expressions for P_a and P_b, the first two terms are equal and they change only during MW events, while the third and fourth terms have opposite signs and vary at all three events. The four terms of P_n determine the four contributions leading to the enhancement and their values get modified only at the CE events. We rewrite these four population differences as

$$\Delta_{\chi\chi'} = p_{\chi\chi'\alpha} - p_{\chi\chi'\beta} \quad (2.87)$$

and the nuclear polarization gets the form

$$P_n = \Delta_{\beta\beta} + \Delta_{\alpha\alpha} + \Delta_{\beta\alpha} + \Delta_{\alpha\beta} \quad (2.88)$$

At high temperatures, the four terms of P_n at thermal equilibrium are equal. At a static equilibrium, the polarizations P_a^{eq} and P_b^{eq} are determined by the

Boltzmann distribution of the populations $p_{\chi\chi'\chi''}^{\text{eq}}$ at the sample temperature. In the laboratory frame, the energy differences between the states $|\alpha\alpha\chi\rangle$ and $|\beta\beta\chi\rangle$ are significantly larger than those of $|\beta\alpha\chi\rangle$ and $|\alpha\beta\chi\rangle$, and therefore the first two terms of P_a^{eq} and P_b^{eq} largely contribute to their values. Because of the equality of the first two terms in equation (2.86), the last terms determine the value of the polarization difference:

$$(P_a - P_b) = 2(p_{\beta\alpha\alpha} - p_{\alpha\beta\alpha}) + 2(p_{\beta\alpha\beta} - p_{\alpha\beta\beta}) \quad (2.89)$$

In a previous publication,[127] we demonstrated that this electron polarization difference is an essential parameter in determining P_n. There we showed that, when the nuclear relaxation times are relatively long (> 1 s for a typical biradical), the maximum value of $|P_a - P_b|$, during a qp steady-state rotor period, determines the magnitude of P_n:

$$|P_n| = |P_a - P_b|_{\text{max}} \quad (2.90)$$

To assist clarifying the interplay between the three main mechanisms leading to MAS-DNP enhancement, we describe now each type of rotor events sequentially, starting with DJ events, followed by the combination of CE and DJ events and finally introducing the MW events.

2.6.4.1 The DJ Events

First, we describe the DJ events and their influence on the polarization. At these events, the third and fourth terms of P_a and P_b in equation (2.68) change their values and the first two terms stay unchanged. The sum of the two electronic polarizations also stays constant and therefore their changes are opposite:

$$(P_a^{\text{after}} - P_a^{\text{before}}) = -(P_b^{\text{after}} - P_b^{\text{before}}) \quad (2.91)$$

At the zero-quantum DJ events, the changes in $(p_{\beta\alpha\chi} - p_{\alpha\beta\chi})$, determining the value of $(P_a^{\text{after}} - P_a^{\text{before}})$, can become very substantial. Typically, the dipolar interaction strength requires an explicit calculation of the population exchange of the DJ events. These calculations show that in most cases, when the anti-crossings are not very fast and the $\omega_{a,b}$ not very small, $(p_{\beta\alpha\chi}^{\text{after}} - p_{\alpha\beta\chi}^{\text{after}}) \approx X_{\text{LZ}}^{\text{DJ}}(p_{\beta\alpha\chi}^{\text{before}} - p_{\alpha\beta\chi}^{\text{before}})$ according to equation (2.83) with $X_{\text{LZ}}^{\text{DJ}} \approx -1$ and thus $P_a^{\text{after}} \approx P_b^{\text{before}}$ and $P_b^{\text{after}} \approx P_a^{\text{before}}$, indicating a full exchange between P_a and P_b.

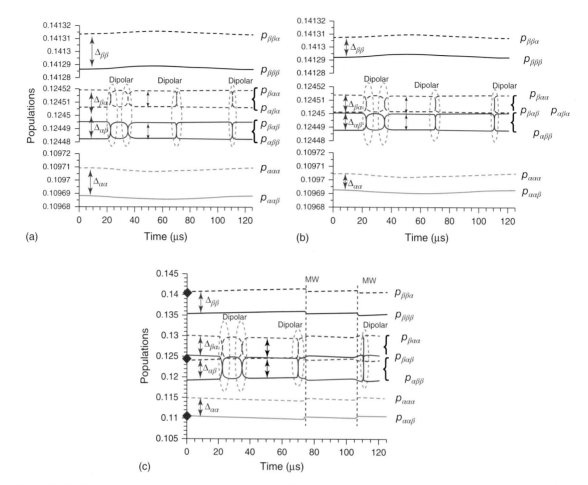

Figure 2.25. The simulated time dependence of the populations for a single three-spin system $\{e_a - e_b - n\}$ at its quasiperiodic steady state. The parameters of this spin system are the same as in Figures 2.23 and 2.24. In (a) the hyperfine coupling is set to 0 MHz (no cross-effect) and the MW irradiation is off. In (b) no MW is applied, but the pseudohyperfine coupling is set to be 3 MHz. In (c) the sample is irradiated and the hyperfine coupling is present. The four red arrows in each population diagram are equal to the $\Delta_{\chi\chi'}$ contributions to the nuclear polarization. The black arrows indicate the values of $|p_{\beta\alpha\alpha} - p_{\alpha\beta\alpha}|$ and $|p_{\beta\alpha\beta} - p_{\alpha\beta\beta}|$, whose sum equals to the difference between the polarization of the electrons $1/2\,|\,P_a - P_b|$. Figures (a) and (b) have equal vertical scales, while (c) has a different scale. To emphasize this difference, we added in (c) three blue diamonds at the values of the three ranges of population in (a) and (b). The height of the diamonds corresponds to the ranges of values of the populations drawn in (a) and (b)

In case the electrons are not interacting with the nucleus and there is no MW irradiation present, only the DJ events and the T_{1e} relaxation influence the values of the populations during the rotor periods. Hence, at the qp steady-state evolution, the overall changes in the values of $(p_{\beta\alpha\alpha} - p_{\alpha\beta\alpha})$ and $(p_{\beta\alpha\beta} - p_{\alpha\beta\beta})$ at these events must be compensated by the changes caused by the electron relaxation time T_{1e}.

Thus at the quasiperiodic steady state the difference between the populations, exchanging during the DJ events, depends on a delicate balance between the values of the X_{LZ}^{DJ} factors and the electron relaxation time T_{1e}.

This is demonstrated in Figure 2.25(a) for a $\{e_a - e_b - n\}$ system at its qp steady state. The absolute differences between the exchanging populations

$p_{\beta\alpha\chi}$ and $p_{\alpha\beta\chi}$ reach a value that is smaller than their averaged Boltzmann equilibrium differences. The polarization difference $|P_a - P_b|$ away from the DJ events, according to equation (2.88), equals twice the sum of the two black arrows in Figure 2.25(a). This difference varies for different crystallite orientations.

Their values are particularly sensitive to the values of the dipolar interactions, ω_r and $T_{1,e}$. When the dipolar coupling is weak (e.g. <10 MHz at 8 kHz), the electron relaxation time T_{1e} is long (e.g. >1 ms at 8 kHz) or when the sample rotation frequency becomes very high, we observe a significant reduction in the population difference and thus of $|P_a - P_b|$.

In addition, Figure 2.25(a) confirms that $p_{\beta\beta\chi}$ and $p_{\alpha\alpha\chi}$ are unaffected by the DJ events and hence $\Delta_{\beta\beta}$ and $\Delta_{\alpha\alpha}$ (the upper and lower red arrows) stay equal to their equilibrium values. The two population pairs $\{p_{\beta\alpha\chi}, p_{\alpha\beta\chi}\}$ experience their DJ events independently, and the difference between these two pairs determines the values of $\Delta_{\beta\alpha}$ and $\Delta_{\alpha\beta}$ (the middle red arrows). Since the equivalent DJ events of $\{p_{\beta\alpha\alpha}, p_{\alpha\beta\alpha}\}$ and $\{p_{\beta\alpha\beta}, p_{\alpha\beta\beta}\}$ happen at the same time, the two $\Delta_{\beta\alpha}$ and $\Delta_{\alpha\beta}$ also maintain their thermal equilibrium values, and P_n is not affected by these events.

2.6.4.2 The DJ and CE Events Leading to Depolarization

When the hyperfine interaction is added to the spin system, the CE events can influence the populations significantly. The shifts of the populations at these events are in general rather small [see equation (2.83)], because with $\xi_{mix}^{CE} \ll \xi_t^{CE}$ their LZ factors, X_{LZ}^{CE}, are slightly below 1. Thus, the CE events cause each time a small depletion of the polarization differences $(p_{\beta\alpha\alpha} - p_{\alpha\beta\alpha})$ and $(p_{\beta\alpha\beta} - p_{\alpha\beta\beta})$. However, these depletions accumulate over time to a significant decrease between the two population pairs $\{p_{\beta\alpha\alpha}, p_{\alpha\beta\alpha}\}$ and $\{p_{\beta\alpha\beta}, p_{\alpha\beta\beta}\}$. This reorganization of the populations, during the qp steady-state rotor periods, is this time compensated by the nuclear spin–lattice relaxation T_{1n}. Thus, at quasiperiodic steady state the proximity of the populations, exchanging during the CE event, depends on a delicate balance between the X_{LZ}^{CE} factors and the nuclear relaxation time $T_{1,n}$. In the system considered here, with parameters compatible with a bis-nitroxide, this results in a close approach of the $\{p_{\beta\alpha\alpha}, p_{\alpha\beta\alpha}\}$ and $\{p_{\beta\alpha\beta}, p_{\alpha\beta\beta}\}$ pairs, as shown in Figure 2.25(b). The population differences of these pairs (the black arrows), which are determined by the

DJ events, are equal to the values of $\Delta_{\beta\alpha} = p_{\beta\alpha\alpha} - p_{\beta\alpha\beta}$ and $\Delta_{\alpha\beta} = p_{\alpha\beta\alpha} - p_{\alpha\beta\beta}$ (the middle red arrows). At qp steady state, the overall changes in the population, as a result of the DJ and CE events, are thus compensated by both relaxation mechanisms, T_{1e} and T_{1n}. A direct consequence of this T_{1e} dependence is the depletion of the difference between $p_{\beta\beta\chi}$ and $p_{\alpha\alpha\chi}$, as a result of the change in $p_{\beta\alpha\chi}$ and $p_{\alpha\beta\chi}$. This, in turn, causes a depletion of the $\Delta_{\beta\beta}$ and $\Delta_{\alpha\alpha}$ values, with respect to their equilibrium values. Thus, overall, all four $\Delta_{\chi\chi'}$ values are reduced, and as a consequence, the polarization of the nucleus gets depolarized. This is clearly visible in the population plot in Figure 2.25(b), when comparing it with Figure 2.25(a). Inspection of Figure 2.25(b) also reveals that the sum of the lengths of the four red arrows, equal to $|P_n|$, is equal to twice the sum of the lengths of the black arrows, equal to $|P_a - P_b|$, in accordance with equation (2.90). This equality can be examined further by considering a powder of randomly oriented three-spin systems and by calculating the qp steady-state polarizations. This has been done for a system, similar to the one presented in the previous figures, by putting it in the magnetic field at different orientations. Using the 144-ZCW orientations,[128,129] the calculated values of P_n can then be plotted as a function of the maximum absolute difference between P_a and P_b, as in Figure 2.26(a).[127,130] First, we see in this figure the expected distribution of $|P_a - P_b|_{max}$ values for different crystallite orientations. Furthermore, we see that the expected dependence of P_n on $|P_a - P_b|_{max}$ is not far from being correct for most orientations.

In powders each individual three-spin system will experience a different depolarization factor $\varepsilon_{Depo} = P_n/P_n^{eq}$. When comparing the depolarization in real samples with the ε_{Depo} results of the three-spin powder, we should evaluate their average value $\bar{\varepsilon}_{Depo}$. In practice, spin diffusion takes care of this averaging process, determining the bulk polarization.

The nuclear depolarization was first discussed and shown by the Tycko group at very low temperatures[131] and later extensively studied by the group of De Paepe, who correlated it to the nature of the biradicals.[130] Thurber and Tycko.[131] and Mentink-Vigier et al.[75,125,130] provided a detailed descriptions of the depolarization effect, and the reader is referred to the above references for much more information about this effect.

As is clear from the foregoing discussion, the depolarization of P_n is a complicated function of all parameters of the spin system. Therefore, we should

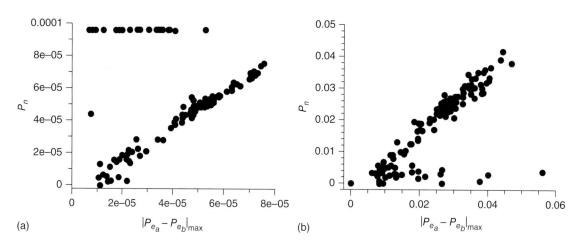

Figure 2.26. Nuclear polarizations P_n of a set of 144 $\{e_a - e_b - n\}$ spin systems, with parameters defined in the figure caption of Figures 2.23 and 2.24, oriented differently with respect to the magnetic field direction at $t = 0$. The orientations are chosen according to the 144 ZCW angle data set. The nuclear polarizations are plotted as a function of the maximum difference between the electron polarizations for each orientation $|P_a - P_b|_{max}$. In (a) the quasiperiodic steady-state polarizations are calculated in the absence of MW irradiation and in (b) with MW irradiation. Notice the difference of the scales in (a) and (b), indicating a maximum nuclear enhancement of up to $0.04/0.0001 = 400$

mention here that in practice most biradicals exhibit nuclear depolarizations as discussed earlier, but for some biradicals the depolarization effect is absent.[75] In certain circumstances, even some enhancement of the nuclear polarization is possible in the absence of MW irradiation.[131]

2.6.4.3 The DJ, MW, and CE Events Leading to Enhancement

At this point, we add the MW events to the calculations. These events induce changes in the population differences $(p_{\beta\chi\chi'} - p_{\alpha\chi\chi'})$ or $(p_{\chi\beta\chi'} - p_{\chi\alpha\chi'})$. These changes do, of course, not affect P_n directly, except when they correspond to the SE events or in the very rare situation when the MW and CE events coincide. In the following, we will use the LZ formula to characterize the MW events, despite the fact that in many calculations the small-step integration method has been used. The MW events, of course, influence the values of P_a or P_b directly. Similar to the DJ events, the overall changes during the MW events at qp steady state are compensated by T_{1e} relaxation. *Thus, at the quasiperiodic steady state, differences between the populations, exchanging during the*

MW events, depend on a delicate balance between the values of the X_{LZ}^{MW} factors and the electron relaxation time T_{1e}.

The addition of the MW events to the DJ events induces depletions of the differences $(p_{\beta\chi\chi'} - p_{\alpha\chi\chi'})$ or $(p_{\chi\beta\chi'} - p_{\chi\alpha\chi'})$. This in turn causes a distancing of the DJ-exchanging population pairs $\{p_{\beta\alpha\alpha}, p_{\alpha\beta\alpha}\}$ and $\{p_{\beta\alpha\beta}, p_{\alpha\beta\beta}\}$, because the CE events maintain their population proximities between $p_{\beta\alpha\alpha}$ and $p_{\alpha\beta\beta}$, and between $p_{\beta\alpha\beta}$ and $p_{\alpha\beta\alpha}$. In Figure 2.25(c) the MW irradiation is added to the simulation, at a frequency away from the middle of the static EPR line, and in comparison with Figure 2.25(a), we observe that the two black arrows, determining the absolute electron polarization difference, and the four red arrows, determining the nuclear polarization, became significantly larger. This immediately indicates a simultaneous increase of $|P_a - P_b|$ and P_n. An easy inspection of the population plot in Figure 2.25(c) shows that P_n is positive. To analyze the nature of this plot, we present here two schematic population schemes: on the left corresponding to Figure 2.25(c) where $P_n > 0$ and on the right a different schematic where $P_n < 0$:

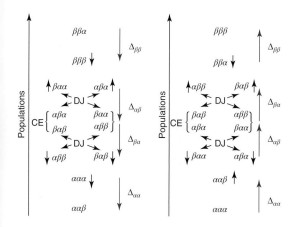

In these schemes, the crossing arrows represent the strong population exchange of the DJ events. The populations between the parentheses indicate the population proximities as a result of the CE events, and the up and down arrows show the increase or depletion of the populations $\chi\chi'\chi'' \equiv p_{\chi\chi'\chi''}$. On the left, all $p_{\chi\chi'\alpha} > p_{\chi\chi'\beta}$ and thus $P_n > 0$. On the right, all $p_{\chi\chi'\alpha} < p_{\chi\chi'\beta}$ and $P_n < 0$. The two time-domain patterns are thus a combined result of the three types of events and the relaxation compensations. A more elaborate analysis of the interplay between all of those is, of course, possible, but is outside the scope of this presentation. More elaborate calculations have shown that for many spin systems

$$P_n > 0 : P_n \cong (P_a - P_b)_{\max};$$
$$P_n < 0 : P_n \cong (P_a - P_b)_{\min} \qquad (2.92)$$

or, for very long T_{1n} values, at least

$$| P_n | \leq |P_a - P_b|_{\max} \qquad (2.93)$$

These equations are particularly useful for gaining insight into the MAS-DNP mechanism, as long as the electron–electron state mixing remains moderate up to a few MHz; otherwise, the dependences on P_a and P_b are not of practical use.

In Figure 2.26(b) we show the dependence of the P_n values of a three-spin system at 144 different orientations as a function of their $|P_a - P_b|_{\max}$, including MW irradiation at a frequency set on the low frequency side of the static powder EPR spectrum of the system. For this irradiation frequency all P_n's are positive. First, we see that the different orientations result in a large spread of $|P_a - P_b|_{\max}$ values. Furthermore, equation (2.92) is satisfied for most of the orientations.

The present analysis of the time dependence of the populations prevents us from approaching the MAS-DNP problem as 'static'. One of the more troubling aspects of understanding DNP under MAS is the consequence of the CE events in the absence of MW irradiation. As mentioned in the case of bis-nitroxides, these events can lead to a nuclear polarization at quasiperiodic steady state that is significantly lower than its Boltzmann equilibrium value. Thus, the actual MAS-DNP enhancement, ε_B, must therefore be defined as the ratio between the quasiperiodic steady-state value, in the presence of MW irradiation, P_n^{on}, and the nonspinning thermal Boltzmann equilibration value P_n^{eq}. This enhancement is in general not equal to the ratio, $\varepsilon_{\mathrm{on/off}}$, between P_n^{on} and the depolarized value, without MW irradiation, P_n^{off}. To stress this difference, we use the following definitions:

$$\varepsilon_B = P_n^{\mathrm{on}}/P_n^{\mathrm{eq}}; \; \varepsilon_{\mathrm{Depo}} = P_n^{\mathrm{off}}/P_n^{\mathrm{eq}}; \; \varepsilon_{\mathrm{on/off}} = P_n^{\mathrm{on}}/P_n^{\mathrm{off}} \qquad (2.94)$$

and thus

$$\varepsilon_B = \varepsilon_{\mathrm{on/off}} \times \varepsilon_{\mathrm{Depo}} \qquad (2.95)$$

As a consequence, $\varepsilon_{\mathrm{on/off}}$ does not quantify the polarization gain achieved under MAS-DNP conditions, and it depends on the nature of the biradicals.[75,130] It should therefore not be used to compare the DNP efficiency of different biradicals. These issues are covered in greater detail in another chapter.

In Figure 2.27 we show a powder $\bar{\varepsilon}_B$ spectrum (here as a function of the main magnetic field B_0 rather than the MW frequency), calculated for a powder of $\{e_a - e_b - n\}$ systems. This shows the expected negative–positive nuclear polarization.

2.6.5 Overhauser DNP in a Spinning Sample

As in nonspinning insulating sample, Overhauser DNP enhancement was also observed in spinning samples.[89,132] In particular, in insulating samples with narrow line monoradicals, such as BDPA and SA-BDPA, and even at very high magnetic fields, it was shown that significant enhancements can be reached, when the MW frequency was set on-resonance at the EPR frequency of the radicals. As already discussed, the OE-DNP mechanism depends on the imbalance between electron–nuclear ZQ and DQ relaxation processes, necessarily governed by fluctuations of the ZQ and DQ terms in the full hyperfine interaction Hamiltonian.[13,133] The

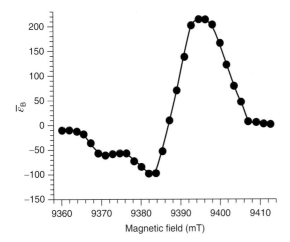

Figure 2.27. Calculated MAS-DNP enhancement spectrum, i.e. the average enhancement of 144 ZCW orientations of a three-spin system with parameters summarized in the caption of Figures 2.23 and 2.24. The spectrum is plotted as a function of magnetic field B_0 with MW irradiation at 263.45 GHz

requirement for a significant positive OE-DNP enhancement is again the presence of strong isotropic hyperfine coefficients that result in $T_{1,ZQ} < T_{1,DQ}$ as was verified experimentally. In the framework of this publication, we again do not discuss the mechanisms responsible for the interaction fluctuations, and refer to the literature for further follow-up of OE-DNP during MAS.[134]

Simulations of the OE-MAS-DNP enhancement during the qp steady state of a two-spin system $\{e-n\}$ can be performed by following the small-step integration procedure presented in Sections 2.6.1 and 2.6.3, after the addition of ZQ and DQ relaxation matrices as in equation (2.43). Figure 2.28 shows experimental results with the radicals BDPA and SA-BDPA in Figure 2.28(a) in comparison to the simulations shown in Figure 2.28(b).[135]

2.6.6 Spinning Frequency and Magnetic Field Dependence

In this section, we discuss the dependence of the enhancement ε_B and the depolarization factor ε_{Depo} of a bis-nitroxide like $\{e_a - e_b - n\}$ spin system on the spinning frequency ω_r, the relaxation time T_{1e}, and the external magnetic field B_0. In equation (2.83)

we introduced the LZ step factor s_{LZ} expressing the change in the populations during the events. For the CE and the MW rotor events, we can thus define s_{LZ}^{RE} with RE = CE, MW and for the DJ rotor event s^{DJ} [see the discussion following equation (2.84)].

An increase of ω_r raises the crossing rates of all events and thus lowers their step factors. In addition, it shortens the rotor period and thus the time for the relaxation to compensate for the changes caused by the rotor events. An increase in the magnetic field generates larger off-resonance shifts $g(\Omega; t)\beta_e B_0$, and thus also increases the crossing rates and thus a depletion of the step factors. For the CE events, the field increase lowers the effective mixing coefficients, because of their inverse proportionality to ω_n [see equation (2.82)]. Finally, a change in the relaxation times modifies the compensation ability of the relaxation process to reach the qp steady-state conditions $p_{\chi\chi'\chi''}(t) = p_{\chi\chi'\chi''}(t-t_r)$. Thus, to rationalize the changes in the $\Delta_{\chi\chi'}$ values and thus in ε_B, as a function of ω_r, B_0, or $T_{1,e}$, is not trivial. Numerical simulations of these changes must therefore assist us in this process.[75,76,127,136]

To demonstrate some of these effects, we show in Figure 2.28(a)–(c) the dependence of the powder averaged $\bar{\varepsilon}^{Depo}$ and $\bar{\varepsilon}_B$ values of the $\{e_a - e_b - n\}$ spin system on ω_r, $T_{1,e}$, and B_0. Although we are dealing with a powder averaging, it is useful to look at Figure 2.25(c) for explaining some of the trends shown in Figure 2.28. We mainly discuss effects that are a result of changes in the MW and CE events. Most of the DJ events have a s^D factor close to 1 and therefore cause a full exchange in the $\{p_{\beta\alpha\chi}, p_{\alpha\beta\chi}\}$ population pairs. Increasing ω_r results at first in an increase and then a decay of $\bar{\varepsilon}_B$. The first increase arises from the shortening of t_r. The CE events maintain the population proximities, because $T_{1,n}$ is much longer than the rotor period. However, the MW events must lower the values of $s_{LZ}^{MW}(p_{\alpha\chi\chi'} - p_{\beta\chi\chi'})$ or $s_{LZ}^{MW}(p_{\chi\alpha\chi'} - p_{\chi\beta\chi'})$ [see equation (2.84)], by lowering their population differences (i.e. an increased saturation), to allow $T_{1,e}$ compensation in a time t_r. As a result P_n increases, as can be best understood by inspecting Figure 2.25(c), where by the making the differences between $p_{\alpha\chi\chi'}$ and $p_{\beta\chi\chi'}$ or $p_{\chi\alpha\chi'}$ or $p_{\chi\beta\chi'}$ smaller the black and red arrows all increase. $\bar{\varepsilon}_B$ reaches its maximum value when the rotor period is of the same order of magnitude as the relaxation time $T_{1,e}$. Beyond that point, $\bar{\varepsilon}_B$ decays

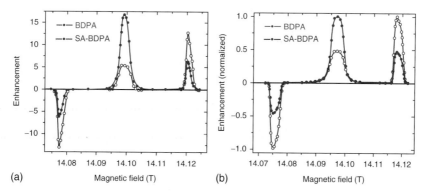

Figure 2.28. (a) ^{1}H DNP enhancement field profile of BDPA (full circles) and SA-BDPA (open circles). (b) Simulations of the DNP profiles shown in (a). (T. V. Can, M. A. Caporini, F. Mentink-Vigier, B. Corzilius, J. J. Walish, M. Rosay, W. E. Maas, M. Baldus, S. Vega, T. M. Swager, and R. G. Griffin, J. Chem. Phys. 141, 1 (2014). Reproduced with permission from Royal Society of Chemistry)

for increasing ω_r, which is a direct consequence of the decay of the step factors s_{LZ}^{MW}, and eventually of s_{LZ}^{CE} and s^D. A decrease in s_{LZ}^{MW} requires an increase of its population difference in order to maintain the value of $s_{LZ}^{MW}(p_{\alpha\chi\chi'} - p_{\beta\chi\chi'})$. The last argument is also responsible for the decay of $\bar{\varepsilon}_{Depo}$ for increasing ω_r. The dependence of $\bar{\varepsilon}_B$ against T_{1e} in Figure 2.29(a) shows a higher $\bar{\varepsilon}_B$ for longer T_{1e}. This can be understood by realizing that for increasing $T_{1,e}$ the compensation efficiency goes down, which again causes a decay in the differences between the populations involved in the MW events and a gain in $|P_a - P_b|_{max}$ and P_n.[127] This trend has been experimentally verified by Lund et al.[137]

Regarding the decay of $\bar{\varepsilon}_B$ for increasing B_0, this dependence is a combined effect of all rotor events, having reduced step factors and thus a lowering of P_n, as explained earlier. Thus, at higher magnetic field, it becomes harder to create an electron polarization difference $|P_a - P_b|_{max}$ and an enhancement.

There exists no simple field dependence of $\bar{\varepsilon}_B$ vs B_0, like a simple power law as initially suggested.[6] The general trends are of course observed: very high magnetic fields usually lead to poorer performance of nitroxide biradicals,[121,138] while other biradicals, such as Trityl-Tempo and BDPA-Tempo, remain good performers.[75,139,140] One of the reasons for this important observation is the large difference between the g-tensor anisotropies of the two monoradicals composing the bis-radical.

Because of the combined dependence of $\bar{\varepsilon}_B$ and $\bar{\varepsilon}_{off}$, the field dependence of $\bar{\varepsilon}_{on/off}$ seems particularly drastic as illustrated in Figure 2.28(c). Nonetheless, we should keep in mind that $\bar{\varepsilon}_B$ is the enhancement quantifying the MAS-DNP efficiency.

2.6.7   The Fifth Type of Rotor Event

It is thus evident that the MAS-DNP enhancement is a direct consequence of the four rotor events, where two states anticross in the rotating frame. A similar type of event can occur when two crossing energy levels have states that are coupled by the flip-flop term of the nuclear dipolar interaction. This event can of course only occur in a system with at least two nuclei. To appreciate the consequences of such a nuclear dipolar (dn) rotor event, let us consider a four-spin system $\{e_a - e_b - n_1 - n_2\}$ with product state energies $\varepsilon_{\chi_{e_1}\chi_{e_2}\chi_{n_1}\chi_{n_2}}$ and populations $p_{\chi_{e_1}\chi_{e_2},\chi_{n_1}\chi_{e_2}}$, with χ, χ', χ'', $\chi''' = \alpha, \beta$. Furthermore, let us assume that the two electrons, e_a and e_b, are dipolar coupled and the two magnetically equivalent nuclei, n_1 and n_2, are also dipolar coupled. In addition one nucleus, n_1, is hyperfine coupled to e_a and the other nucleus, n_2, is not coupled to the electrons. Without the $n_1 - n_2$ interaction, the four standard rotor events determine the values of all populations $p_{\chi\chi',\chi''\alpha}$ and $p_{\chi\chi',\chi''\beta}$, in particular, the nuclear polarizations:

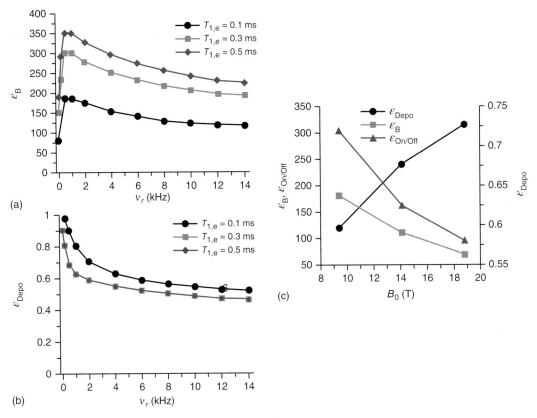

(a)

(b)

(c)

Figure 2.29. Rotation frequency dependence of the polarization gain ε_B in (a) and the depolarization ε_{Depo} in (b) of a three-spin system $\{e_a - e_b - n\}$, with parameters defined in the figure caption of Figures 2.23 and 2.24, for different $T_{1,e}$ values. In (c) the values of ε_B, ε_{Depo}, and $\varepsilon_{On/Off}$ for $T_{1,e} = 0.3$ ms are plotted as a function of the external magnetic field B_0

$$
\begin{aligned}
P_{n_1} &= \sum_{\chi,\chi'=\beta,\alpha} (p_{\chi\chi',\alpha\alpha} - p_{\chi\chi',\beta\alpha}) \\
&\quad + \sum_{\chi,\chi'=\beta,\alpha} (p_{\chi\chi',\alpha\beta} - p_{\chi\chi',\beta\beta}) \\
&= \sum_{\chi,\chi'=\beta,\alpha} (p_{\chi\chi',\alpha\alpha} - p_{\chi\chi',\beta\beta}) \\
&\quad + \sum_{\chi,\chi'=\beta,\alpha} (p_{\chi\chi',\alpha\beta} - p_{\chi\chi',\beta\alpha}) \equiv \Sigma_{\alpha\alpha}^{\beta\beta} + \Sigma_{\alpha\beta}^{\beta\alpha} \\
P_{n_2} &= \sum_{\chi,\chi'=\beta,\alpha} (p_{\chi\chi',\alpha\alpha} - p_{\chi\chi',\alpha\beta}) \\
&\quad + \sum_{\chi,\chi'=\beta,\alpha} (p_{\chi\chi',\beta\alpha} - p_{\chi\chi',\beta\beta}) \\
&= \sum_{\chi,\chi'=\beta,\alpha} (p_{\chi\chi',\alpha\alpha} - p_{\chi\chi',\beta\beta}) \\
&\quad + \sum_{\chi,\chi'=\beta,\alpha} (p_{\chi\chi',\alpha\beta} - p_{\chi\chi',\beta\alpha}) \equiv \Sigma_{\alpha\alpha}^{\beta\beta} - \Sigma_{\alpha\beta}^{\beta\alpha}
\end{aligned}
\tag{2.96}
$$

Ignoring the thermal equilibrium nuclear polarization $P_{n_2}^{eq}$, and without the $n_1 - n_2$ interaction the values of the populations of all $p_{\chi\chi'\chi''\alpha}$ and $p_{\chi\chi'\chi''\beta}$ are equal so that $P_{n_2} = 0$. In that case, the values of $\Sigma_{\alpha\alpha}^{\beta\beta}$ and $\Sigma_{\alpha\beta}^{\beta\alpha}$, as defined in equation (2.96), are equal. The energies $\varepsilon_{\chi\chi'\chi''\alpha}$ and $\varepsilon_{\chi\chi'\chi''\beta}$ are all separated by ω_n, but the energies $\varepsilon_{\chi\chi',\alpha\beta}$ and $\varepsilon_{\chi\chi',\beta\alpha}$ are separated by the hyperfine interaction coefficient $\pm\frac{1}{2}A_{z,e_1,n_1}(t)$, which can become zero. Introducing the $n_1 - n_2$ interaction, its flip-flop matrix element connects the states $|\chi\chi'\alpha\beta\rangle$ and $|\chi\chi'\beta\alpha\rangle$ and can result in dn rotor events, when at a time t_{nd} $A_{z,e_1,n_1}(t_{nd}) = 0$ and $\omega_{n1,n2}(t_{nd}) \neq 0$. At these events, an exchange between the $p_{\chi\chi'\alpha\beta}$ and $p_{\chi\chi'\beta\alpha}$ populations can take place, making their overall difference $\Sigma_{\alpha\beta}^{\beta\alpha}$ smaller. Although the populations $p_{\chi\chi'\alpha\alpha}$ and $p_{\chi\chi'\beta\beta}$ do not experience any dn events, they can be slightly influenced via $T_{1,n}$ relaxation. Still, we must expect that the new value of $\Sigma_{\alpha\beta}^{\beta\alpha}$ becomes smaller

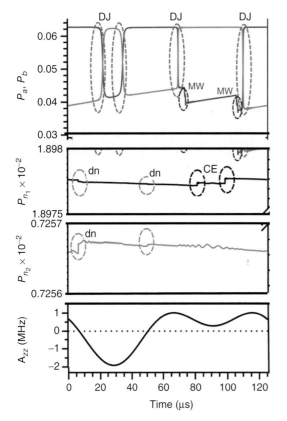

Figure 2.30. Evolution of the polarizations at quasiperiodic steady state of a four-spin system $\{e_a - e_b - n_1 - n_2\}$. Except for n_2, the spin system has parameters as summarized in the captions of Figures 2.23 and 2.24. The second nucleus n_2 is dipolar coupled with $\omega_{n_1,n_2}/2\pi = 1\,\text{kHz}$ to n_1 only. The $n_1 - n_2$ bond makes an angle of 45^0 with the g_z axis of the electron e_a. The plots of the polarizations have each their own scale. At the bottom, the secular hyperfine coefficient $A_{zz} = A_{e_a,n_1}$ is plotted as a function of time

than $\Sigma_{\beta\beta}^{\alpha\alpha}$. Then the value $(\Sigma_{\beta\beta}^{\alpha\alpha} - \Sigma_{\alpha\beta}^{\beta\alpha})$ of P_{n_2} becomes different from zero and the value $(\Sigma_{\beta\beta}^{\alpha\alpha} + \Sigma_{\alpha\beta}^{\beta\alpha})$ of P_{n_1} changes as well. To show these effects, we draw in Figure 2.30 the calculated qp steady-state polarizations of the $\{e_a - e_b - n_1 - n_2\}$ spin system, with parameters given in the figure caption.

In the nonspinning case, when the difference between the hyperfine interactions of the two nuclei is much larger than that of the dipolar interaction between them, this nuclear interaction is quenched. This effect is the source of the well-known spin diffusion barrier. As a result, the strongly hyperfine-coupled

'core' nuclei as defined in Section 2.4.3 do get hyperpolarized, but they cannot (or very slowly) transfer their populations to the 'bulk' nuclei. However, in the magic angle spinning case, almost all strongly coupled nuclei can pass their polarization to weakly coupled nuclei, when the difference between their secular hyperfine coefficients becomes zero at a time that the flip-flop term of the nuclear dipolar interaction differs from zero. In that case, the strongly coupled nuclei can transfer their populations to their neighboring, weakly coupled, nuclei, and we do not have to consider spin diffusion barriers. This polarization process can be compared with the ''indirect'' polarization of the local nuclei in the static case as described in Section 2.4.3.

Inspecting Figures 2.23 and 2.29, we see that for the system under study and without the $n_2 - n_2$ interaction $\Sigma_{\beta\beta}^{\alpha\alpha} = \Sigma_{\alpha\beta}^{\beta\alpha} \approx 0.0098$ and with this interaction $\Sigma_{\beta\beta}^{\alpha\alpha} \approx 0.013$ and $\Sigma_{\alpha\beta}^{\beta\alpha} \approx 0.006$. It is worth noting that the final polarization of nucleus n_1 (the black line), $0.013 + 0.006 = 0.019$, is almost equal to its value without the $(n_1 - n_2)$ interaction: $2 \times 0.0098 = 0.0196$. Presumably, the CE events, being much stronger than the dn events, overall compensate for the loss of P_{n_1} polarization due to the polarization transfer of P_{n_1} to P_{n_2}. This is an important observation, as it shows how the polarization of the biradical is transferred to the strongly coupled nuclei and from them to the weakly coupled nuclei, without losing their own polarization. In the next section, we further discuss some of these features.

2.6.8 Multispin Systems

Finally, we will discuss some aspects dealing with the analogs of the static spectral and spin diffusion mechanisms in large spin systems rotating at the magic angle. To perform simulations of real spin systems, one would need to perform full Liouville space calculations for systems containing together a few hundred electron spins and a few thousand nuclei. Unfortunately, even the best computer/algorithm today cannot do so, let alone store the resulting density matrix. Thus, a simplified model has been developed that can handle systems with hundreds of spins, because it scales linearly with the number of spins. For the exact details of this model, we refer to a paper by Mentink-Vigier *et al.*[125] It assumes, for instance, that high-order effects due to multiple coupled spins are negligible and that state mixing away from the rotor events is weak. Furthermore, no cross-relaxation effects are taken into account, and all polarization transfers are derived from rotor events, except spin diffusion. The simulation code again employs the Landau–Zener formula to calculate the effects of the rotor events on the Liouville operators, while Thurber and Tycko[131] used populations. We should mention here that additional groups have been working on multispin problems. For instance, to name two examples, Pinon *et al.*[141] evaluated how to polarize bulky samples and Perras and Pruski[142] investigated the impact of the MAS frequency on the enhancement and the level of deuteration level on the nuclear spin diffusion process using the full Liouville equation.

To demonstrate the capability of this model in calculating the flow of polarization in such large systems, we consider two specific systems: (i) the system containing a single pair of electrons coupled to a large set of dipolar coupled nuclei and (ii) the system composed of many coupled three-spin $\{e_a - e_b - n\}$ systems. The latter is similar to the 3000-spin model reported by Tycko's lab.[131]

In addition to the DJ, CE, and MW events that occur inside the three-spin systems themselves, the model must also take into account the following events:

- The dn rotor events that enable the transfer of the polarization from the hyperfine-coupled nuclei to the bulk.
- DJ rotor events between electrons belonging to different three-spin systems. We refer to these events as intermolecular DJ events. We should already notice that for a three-spin concentration of the order of 10 mM, the average dipolar interaction between intermolecular electrons is about 3 MHz, which is not sufficiently strong to lead to significant intermolecular DJ events.
- CE rotor events can, of course, also occur between the electron pair of one three-spin system with the nucleus of another three-spin system. Although these events can be taken into account, their effect on the nuclear polarizations is rather weak.

Without repeating the details of the simulations, as introduced in Ref. 125, we summarize here their main results. We start with a multinuclear system, containing one pair of electrons with one of its electrons hyperfine coupled to 60 nuclei, as pictured in Figure 2.31(a). These 'local' nuclei are dipolar coupled to an increasing number of bulk nuclei, the last experiencing spin diffusion.

The main conclusions, derived from the average result of the simulations on 144-ZCW oriented multinuclear systems, can be summarized as follows:

- The strongly coupled local nuclei have a significant influence on the overall enhancement $\bar{\varepsilon}_B$. They tend to determine the maximum polarization gain that can be reached. The relaxation times of both the strongly coupled nuclei and the bulk nuclei determine the buildup time T_B and the final value of $\bar{\varepsilon}_B$.
- There exists no spin diffusion barrier that prevents polarization transfer between strongly and weakly hyperfine-coupled nuclei. A slow-down of the polarization transfer may nonetheless exist.
- The buildup time T_B of the enhancement is shorter than the nuclear relaxation time. It is determined in part by the efficiency of the CE mechanism. For instance, in samples with AMUPol[79] radicals and more recently with TEMTriPol,[139] and ASYMPolPOK[76] radicals, this buildup time is much faster than in samples with TOTAPOL[74] radicals.
- Increasing the average number of nuclei, polarized by a single radical, results in a decrease of $\bar{\varepsilon}_B$ and a lengthening of T_B in agreement with the experiments.

Figure 2.31 illustrates the dependence of $\bar{\varepsilon}_B$ and T_B on the relaxation times of the most strongly coupled nucleus and the bulk nuclei and further on the number of bulk nuclei, polarized via spin diffusion. Overall, these results are intuitive, but they remain to be proven experimentally. For instance, it has been observed that lowering the biradical concentration in amorphous solids leads to a lengthening of the polarization buildup times in homogenous samples.[143–146]

Concentrating now on multi-biradical spin systems, we present here some results also reported by Fred Mentink-Vigier *et al.*[125] They calculated the biradical concentration dependence of the polarization gain $\bar{\varepsilon}_B$ and the depolarization factor $\bar{\varepsilon}_{Depo}$ in a system of 40 randomly oriented three-spin systems $\{e_a - e_b - n\}$ (referred to as the box model). All structural and interaction details can be found in the reference. In addition to the CE and DJ events in each three-spin system, intermolecular DJ events are also taken into account. The collective effect of these DJ events leads to what we can call a 'MAS-induced spectral diffusion' process. This process seems to lower the values of the individual polarization differences $|P_a - P_b|_{max}$

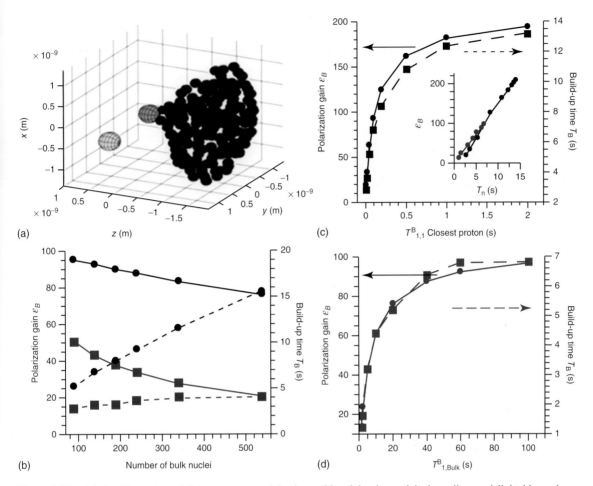

Figure 2.31. (a) An illustration of the geometry used in the multinuclei spin model; the yellow and light blue spheres correspond to the biradical, the blue ones to the surrounding protons. (b) The effect of the number of bulk nuclei per biradical on the polarization gain ε_B and polarization time T_B for two cases: $T_{1,n,\text{Bulk}} = 5$ s (red squares) and $T_{1,n,\text{Bulk}} = 60$ s (black circles squares). In (c) and (d) the effect of the relaxation time of the closest and bulk nuclei, respectively, on the polarization gain ε_B and polarization time T_B. 400 nuclei were considered, including 60 treated as locals. (From Mentink-Vigier, F., Vega, S., & De Paëpe, G. (2017). Fast and accurate MAS–DNP simulations of large spin ensembles. *Physical Chemistry Chemical Physics*, 19(5), 3506–3522. Reproduced with permission from Royal Society of Chemistry)

and, therefore, of the P_ns of the three-spin systems, both in the presence and in the absence of MW irradiation (see Section 2.6.4). Figure 2.32 shows the dependence of $\bar{\varepsilon}_B$ and $\bar{\varepsilon}_{\text{Depo}}$ of the box model on the biradical concentration, and we can conclude that as long as the $T_{1,n}$ of the bulk nuclei is large, an increase in the radical concentration in a sample beyond about 10 mM can reduce the gain of the nuclear polarization in homogeneously distributed biradicals.

In practice, this observation is particularly true for bis-nitroxides, and as a consequence low biradical

concentrations should be used to obtain maximum polarizations for homogenous samples. It is noteworthy that the trends for inhomogeneous samples are not simple.[147,148]

The multispin system simulations are thus a powerful tool for supporting our understanding of the MAS-DNP phenomena. They help us strengthen our intuition about how to prepare samples with high nuclear polarizations or fast buildup times. For example, these simulations have led to the development of a new bis-nitroxide family called ASYMPol,

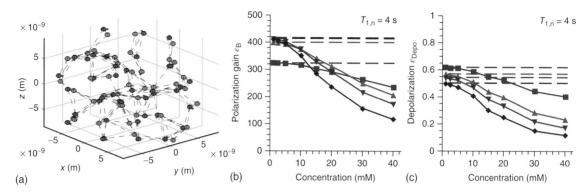

Figure 2.32. (a) An illustration of a multi-spin model in which 40 biradicals are randomly dispersed in a box, such that their concentration matches 15 mM. Green and yellow spheres corresponds to the electron spins, while orange spheres corresponds to the nuclei. The dotted lines represent the dipolar vectors connecting the spins. In (b) and (c) the polarization gain ε_B and depolarization parameter ε_{Depo} calculated and plotted as a function of the concentration of the bi-radicals for four different electron relaxation times $T_{1,e} = 0.1$ ms (blue squares), $T_{1,e} = 0.3$ ms (green up-pointing triangles), $T_{1,e} = 0.5$ ms (red down-pointing triangles) and $T_{1,e} = 1$ ms (purple diamonds). (Mentink-Vigier, F., Vega, S., & De Paëpe, G. (2017). Fast and accurate MAS–DNP simulations of large spin ensembles. Physical Chemistry Chemical Physics, 19(5), 3506–3522. Reproduced with permission from Royal Society of Chemistry)

with fast polarization characteristics and moderate depolarization factors.[76] They have recently been used to successfully predict the properties of AMUPol and TEKPol.[149]

ACKNOWLEDGMENTS

We are extremely grateful to Daniella Goldfarb for the important role she played in our joint DNP journey and for her help in writing this chapter. This research was made possible in part by the historic generosity of the Harold Perlman Family. The National High Magnetic Field Laboratory is supported by the National Science Foundation through NSF/DMR-1644779 and the State of Florida.

REFERENCES

1. A. W. Overhauser, *Phys. Rev.*, 1953, **92**, 411.

2. K. H. Hausser and D. Stehlik, *Adv. Magn. Opt. Reson.*, 1968, **3**, 79.

3. W. T. Wenckebach, T. J. B. Swanenburg, and N. J. Poulis, *Phys. Rep.*, 1974, **14**, 181.

4. V. A. Atsarkin, *Sov. Phys. Uspekhi*, 1978, **21**, 725.

5. A. Abragam and M. Goldman, *Rep. Prog. Phys*, 1978, **41**, 395.

6. T. Maly, G. T. Debelouchina, V. S. Bajaj, K.-N. Hu, C.-G. Joo, M. L. Mak–Jurkauskas, J. R. Sirigiri, P. C. A. van der Wel, J. Herzfeld, R. J. Temkin, and R. G. Griffin, *J. Chem. Phys.*, 2008, **128**, 052211.

7. A. S. Lilly Thankamony, J. J. Wittmann, M. Kaushik, and B. Corzilius, *Prog. Nucl. Magn. Reson. Spectrosc.*, 2017, **102–103**, 120.

8. T. Wenckebach, Essentials of Dynamic Nuclear Polarisation, Spindrift Publications: The Netherlands, 2016.

9. T. R. Carver and C. P. Slichter, *Phys. Rev.*, 1956, **102**, 975.

10. T. R. Carver and C. P. Slichter, *Phys. Rev.*, 1953, **92**, 212.

11. C. Griesinger, M. Bennati, H. M. Vieth, C. Luchinat, G. Parigi, P. Hofer, F. Engelke, S. J. Glaser, V. Denysenkov, and T. F. Prisner, *Prog. Nucl. Magn. Reson. Spectrosc.*, 2012, **64**, 4.

12. J. M. Franck, A. Pavlova, J. A. Scott, and S. Han, *Prog. Nucl. Magn. Reson. Spectrosc.*, 2013, **74**, 33.

13. A. Abragam, *Phys. Rev.*, 1955, **98**, 1729.

14. C. D. Jeffries, *Phys. Rev.*, 1960, **117**, 1056.

15. T. J. Schmugge and C. D. Jeffries, *Phys. Rev.*, 1965, **138**, A1785.

16. R. A. Wind, L. Li, H. Lock, and G. Maciel, *J. Magn. Reson.*, 1988, **79**, 577.

17. D. Shimon, Y. Hovav, I. Kaminker, A. Feintuch, D. Goldfarb, and S. Vega, *Phys. Chem. Chem. Phys.*, 2015, **17**, 11868.

18. C. Hwang and D. Hill, *Phys. Rev. Lett.*, 1967, **18**, 110.

19. C. F. Hwang and D. A. Hill, *Phys. Rev. Lett.*, 1967, **19**, 1011.

20. A. V. Kessenikh, A. A. Manenkov, and G. I. Pyatnitskii, *Sov. Phys. Solid State*, 1964, **6**, 641.

21. A. V. Kessenikh, V. I. Lushchikov, A. A. Manenkov, and Y. V. Taran, *Sov. Phys. Solid State*, 1963, **5**, 321.

22. M. Goldman, Spin Temperature and Magnetic Resonance in Solids, Oxford University Press: London, 1970.

23. A. Abragam, Principles of Nuclear Magnetism, Clarendon Press: Oxford, 1961.

24. W. de Boer, *J. Low Temp. Phys.*, 1976, **22**, 185.

25. M. Borghini, *Phys. Rev. Lett.*, 1968, **20**, 419.

26. D. Wollan, *Phys. Rev. B*, 1976, **13**, 3686.

27. N. Bloembergen, S. Shapiro, P. S. Pershan, and J. O. Artman, *Phys. Rev.*, 1959, **114**, 445.

28. V. A. Atsarkin and F. S. Dzheparov, *Zeitschrift Fur Phys. Chemie*, 2017, **231**, 545.

29. V. A. Atsarkin and A. V. Kessenikh, *Appl. Magn. Reson.*, 2012, **43**, 7.

30. A. G. Redfield, *Phys. Rev.*, 1955, **98**, 1787.

31. B. N. Provotorov, *Sov. Phys. JETP*, 1962, **14**, 1126.

32. B. N. Provotorov, *Sov. Phys. JETP*, 1962, **15**, 611.

33. W. T. Wenckebach, *J. Magn. Reson.*, 2019, **299**, 124.

34. W. T. Wenckebach, *J. Magn. Reson.*, 2019, **299**, 151.

35. W. T. Wenckebach, *J. Magn. Reson.*, 2017, **277**, 68.

36. S. Jannin, A. Comment, F. Kurdzesau, J. A. Konter, P. Hautle, B. Van Den Brandt, and J. J. Van Der Klink, *J. Chem. Phys.*, 2008, **128**, 241102.

37. S. Jannin, A. Comment, and J. J. van der Klink, *Appl. Magn. Reson.*, 2012, **43**, 59.

38. A. Pines, M. G. Gibby, and J. S. Waugh, *J. Chem. Phys.*, 1972, **56**, 1776.

39. G. Mathies, S. Jain, M. Reese, and R. G. Griffin, *J. Phys. Chem. Lett.*, 2016, **7**, 111.

40. D. J. van den Heuvel, A. Henstra, T.-S. Lin, J. Schmidt, and W. T. Wenckebach, *Chem. Phys. Lett.*, 1992, **188**, 194.

41. A. Henstra and W. T. Wenckebach, *Mol. Phys.*, 2014, **112**, 1761.

42. T. V. Can, J. J. Walish, T. M. Swager, and R. G. Griffin, *J. Chem. Phys.*, 2015, **143**, 1.

43. V. Weis, M. Bennati, M. Rosay, and R. G. Griffin, *J. Chem. Phys.*, 2000, **113**, 6795.

44. K. O. Tan, C. Yang, R. T. Weber, G. Mathies, and R. G. Griffin, *Sci. Adv.*, 2019, **5**, 1.

45. R. A. Wind, M. J. Duijvestijn, C. van der Lugt, A. Manenschijn, and J. Vriend, *Prog. Nuc. Mag. Res. Sp.*, 1985, **17**, 33.

46. M. Afeworki and J. Schaefer, *Macromolecules*, 1992, **25**, 4092.

47. H. C. Dorn, J. Gu, D. S. Bethune, R. D. Johnson, and C. S. Yannoni, *Chem. Phys. Lett.*, 1993, **203**, 549.

48. L. R. Becerra, G. J. Gerfen, R. J. Temkin, D. J. Singel, and R. G. Griffin, *Phys. Rev. Lett.*, 1993, **71**, 3561.

49. J. H. Ardenkjær-larsen, B. Fridlund, A. Gram, G. Hansson, L. Hansson, M. H. Lerche, R. Servin, M. Thaning, and K. Golman, *PNAS*, 2003, **100**, 10158.

50. G. Zhang and C. Hilty, *Magn. Reson. Chem.*, 2018, **56**, 566.

51. E. M. Serrao and K. M. Brindle, *Porto Biomed. J.*, 2017, **2**, 71.

52. K. N. Hu, G. T. Debelouchina, A. A. Smith, and R. G. Griffin, *J. Chem. Phys.*, 2011, **134**.

53. A. Karabanov, G. Kwiatkowski, C. U. Perotto, D. Wisniewski, J. McMaster, I. Lesanovsky, and W. Köckenberger, *Phys. Chem. Chem. Phys.*, 2016, **18**, 30093.

54. A. Karabanov, A. Van Der Drift, L. J. Edwards, I. Kuprov, and W. Köckenberger, *Phys. Chem. Chem. Phys.*, 2012, **14**, 2658.

55. D. Wišniewski, A. Karabanov, I. Lesanovsky, and W. Köckenberger, *J. Magn. Reson.*, 2016, **264**, 30.

56. A. De Luca and A. Rosso, *Phys. Rev. Lett.*, 2015, **115**, 1.

57. A. De Luca, I. Rodríguez-Arias, M. Müller, and A. Rosso, *Phys. Rev. B*, 2016, **94**, 014203.

58. S. C. Serra, A. Rosso, and F. Tedoldi, *Phys. Chem. Chem. Phys.*, 2013, **15**, 8416.

59. A. Feintuch, D. Shimon, Y. Hovav, D. Banerjee, I. Kaminker, Y. Lipkin, K. Zibzener, B. Epel, S. Vega, and D. Goldfarb, *J. Magn. Reson.*, 2011, **209**, 136.

60. D. Shimon, Y. Hovav, A. Feintuch, D. Goldfarb, and S. Vega, *Phys. Chem. Chem. Phys.*, 2012, **14**, 5729.

61. V. A. Atsarkin, M. I. Rodak, *Sov. Phys. Uspekhi*, 1972, **15**, 251.

62. J. Granwehr and W. Kockenberger, *Appl. Magn. Reson.*, 2008, **34**, 355.

63. V. Nagarajan, Y. Hovav, A. Feintuch, S. Vega, and D. Goldfarb, *J. Chem. Phys.*, 2010, **132**, 214504.

64. S. A. Dzuba and A. Kawamori, *Concepts Magn. Reson.*, 1996, **8**, 49.

65. Y. Hovav, I. Kaminker, D. Shimon, A. Feintuch, D. Goldfarb, and S. Vega, *Phys. Chem. Chem. Phys.*, 2015, **17**, 226.

66. A. Leavesley, D. Shimon, T. A. Siaw, A. Feintuch, D. Goldfarb, S. Vega, I. Kaminker, and S. Han, *Phys. Chem. Chem. Phys.*, 2017, **19**, 3596.

67. K. Kundu, A. Feintuch, and S. Vega, *J. Phys. Chem. Lett.*, 2019, **108**, 1769.

68. K. R. Thurber and R. Tycko, *J. Chem. Phys.*, 2012, **137**, 084508.

69. F. Mentink-Vigier, U. Akbey, Y. Hovav, S. Vega, H. Oschkinat, and A. Feintuch, *J. Magn. Reson.*, 2012, **224**, 13.

70. R. G. Griffin, *J. Magn. Reson.*, 2011, **213**, 410.

71. A. B. Barnes, E. Markhasin, E. Daviso, V. K. Michaelis, E. A. Nanni, S. K. Jawla, E. L. Mena, R. Derocher, A. Thakkar, P. P. Woskov, J. Herzfeld, R. J. Temkin, and R. G. Griffin, *J. Magn. Reson.*, 2012, **224**, 1.

72. F. J. Scott, E. P. Saliba, B. J. Albert, N. Alaniva, E. L. Sesti, C. Gao, N. C. Golota, E. J. Choi, A. P. Jagtap, J. J. Wittmann, M. Eckardt, W. Harneit, B. Corzilius, S. T. Sigurdsson, and A. B. Barnes, *J. Magn. Reson.*, 2018, **289**, 45.

73. S. R. Chaudhari, P. Berruyer, D. Gajan, C. Reiter, F. Engelke, D. L. Silverio, C. Copéret, M. Lelli, A. Lesage, and L. Emsley, *Phys. Chem. Chem. Phys.*, 2016, **18**, 10616.

74. C. Song, K. N. Hu, C. G. Joo, T. M. Swager, and R. G. Griffin, *J. Am. Chem. Soc.*, 2006, **128**, 11385.

75. F. Mentink-Vigier, G. Mathies, Y. Liu, A. L. Barra, M. A. Caporini, D. Lee, S. Hediger, R. Griffin, and G. De Paëpe, *Chem. Sci.*, 2017, **8**, 8150.

76. F. Mentink-Vigier, I. Marin-Montesinos, A. P. Jagtap, T. Halbritter, J. Van Tol, S. Hediger, D. Lee, S. T. Sigurdsson, and G. De Paëpe, *J. Am. Chem. Soc.*, 2018, **140**, 11013.

77. D. J. Kubicki, G. Casano, M. Schwarzwälder, S. Abel, C. Sauvée, K. Ganesan, M. Yulikov, A. J. Rossini,

G. Jeschke, C. Copéret, A. Lesage, P. Tordo, O. Ouari, and L. Emsley, *Chem. Sci.*, 2016, **7**, 550.

78. M. Kaushik, M. Qi, A. Godt, and B. Corzilius, *Angew. Chemie – Int. Ed.*, 2017, **56**, 4295.

79. C. Sauvée, M. Rosay, G. Casano, F. Aussenac, R. T. Weber, O. Ouari, and P. Tordo, *Angew. Chemie – Int. Ed.*, 2013, **52**, 10858.

80. C. Ramanathan, *Appl. Magn. Reson.*, 2008, **34**, 409.

81. S. S. Eaton and G. R. Eaton, *EMagRes*, 2016, **5**, 1543.

82. M. Romanelli and L. Kevan, *Concepts Magn. Reson.*, 1997, **9**, 403.

83. Y. Hovav, A. Feintuch, and S. Vega, *J. Chem. Phys.*, 2011, **134**, 074509.

84. D. Goldfarb, *EMagRes*, 2017, **6**, 101.

85. C. P. Slichter, Principles of Magnetic Resonance, 3rd edn, Springer: Berlin, Heidelberg, 1990.

86. Y. Hovav, A. Feintuch, and S. Vega, *J. Magn. Reson.*, 2012, **214**, 29.

87. D. Wollan, *Phys. Rev. B*, 1976, **13**, 3671.

88. D. Shimon, Y. Hovav, I. Kaminker, A. Feintuch, D. Goldfarb, and S. Vega, *Phys. Chem. Chem. Phys.*, 2015, **17**, 11868.

89. T. V. Can, M. A. Caporini, F. Mentink-Vigier, B. Corzilius, J. J. Walish, M. Rosay, W. E. Maas, M. Baldus, S. Vega, T. M. Swager, and R. G. Griffin, *J. Chem. Phys.*, 2014, **141**.

90. X. Ji, T. V. Can, F. Mentink-Vigier, A. Bornet, J. Milani, B. Vuichoud, M. A. Caporini, R. G. Griffin, S. Jannin, M. Goldman, and G. Bodenhausen, *J. Magn. Reson.*, 2018, **286**, 138.

91. T. Prisner and M. J. Prandolini, in *Multifrequency Electron Paramagn. Reson. Theory Appl.* (2011).

92. I. Solomon, *Phys. Rev.*, 1955, **99**, 559.

93. A. Henstra, P. Dirksen, J. Schmidt, and W. T. Wenckebach, *J. Magn. Reson.*, 1988, **77**, 389.

94. A. Henstra, P. Dirksen, and W. T. Wenckebach, *Phys. Lett. A*, 1988, **134**, 134.

95. A. Henstra and W. T. Wenckebach, *Mol. Phys.*, 2014, **112**, 1761.

96. T. V. Can, J. J. Walish, T. M. Swager, and R. G. Griffin, *J. Chem. Phys.*, 2015, **143**, 054201.

97. T. V. Can, R. T. Weber, J. J. Walish, T. M. Swager, and R. G. Griffin, *Angew. Chemie – Int. Ed.*, 2017, **56**, 6744.

98. A. Feintuch and S. Vega, *EMagRes*, 2017, **6**, 427.

99. S. Vega, *J. Chem. Phys.*, 1978, **68**, 5518.

100. A. Pines, M. G. Gibby, and J. S. Waugh, *J. Chem. Phys.*, 1973, **59**, 569.

101. S. K. Jain, G. Mathies, and R. G. Griffin, *J. Chem. Phys.*, 2017, **147**, 164201.

102. C. Zener, *Proc. R. Soc.*, 1932, **137**, 696.

103. W. T. Wenckebach, *J. Magn. Reson.*, 2017, **284**, 104.

104. W. T. Wenckebach, Essentials of Dynamic Nuclear Polarisation, Spindrift publications: The Netherlands, 2016.

105. V. A. Atsarkin, *Sov. Phys. JETP*, 1971, **32**, 421.

106. K. Kundu, A. Feintuch, and S. Vega, *J. Phys.Chem. Lett.*, 2018, **9**, 1793.

107. Y. Hovav, A. Feintuch, and S. Vega, *Phys. Chem. Chem. Phys.*, 2013, **15**, 188.

108. D. S. Wollan, *Phys. Rev. B*, 1976, **13**, 3671.

109. D. Shimon, A. Feintuch, D. Goldfarb, and S. Vega, *Phys. Chem. Chem. Phys.*, 2014, **16**, 6687.

110. D. Shimon, Y. Hovav, A. Feintuch, D. Goldfarb, and S. Vega, *Phys. Chem. Chem. Phys.*, 2012, **14**, 5729.

111. M. H. Levitt and L. Di Bari, *Phys. Rev. Lett.*, 1992, **69**, 3124.

112. Y. Hovav, D. Shimon, I. Kaminker, A. Feintuch, D. Goldfarb, and S. Vega, *Phys. Chem. Chem. Phys.*, 2015, **17**, 6053.

113. A. Leavesley, D. Shimon, T. A. Siaw, A. Feintuch, D. Goldfarb, S. Vega, I. Kaminker, and S. Han, *Phys. Chem. Chem. Phys*, 2017, **19**, 3596.

114. D. Shimon, Y. Hovav, I. Kaminker, A. Feintuch, D. Goldfarb, and S. Vega, *Phys. Chem. Chem. Phys.*, 2015, **17**, 11868.

115. Y. Hovav, I. Kaminker, D. Shimon, A. Feintuch, D. Goldfarb, and S. Vega, *Phys. Chem. Chem. Phys.*, 2015, **17**, 6053.

116. I. Kaminker, D. Shimon, Y. Hovav, A. Feintuch, and S. Vega, *Phys. Chem. Chem. Phys.*, 2016, **18**, 11017.

117. E. Ravera, D. Shimon, A. Feintuch, D. Goldfarb, S. Vega, A. Flori, C. Luchinat, L. Menichetti, and G. Parigi, *Phys. Chem. Chem. Phys.*, 2015, **17**, 26969.

118. Y. Hovav, A. Feintuch, S. Vega, and D. Goldfarb, *J. Magn. Reson.*, 2014, **238**, 94.

119. A. Bornet, J. Milani, B. Vuichoud, A. J. Perez Linde, G. Bodenhausen, and S. Jannin, *Chem. Phys. Lett.*, 2014, **602**, 63.

120. I. Kaminker, R. Barnes, and S. Han, *J. Magn. Reson.*, 2017, **279**, 81.

121. D. Mance, P. Gast, M. Huber, M. Baldus, and K. L. Ivanov, *J. Chem. Phys.*, 2015, **142**, 234201.

122. F. Lucas, and K. Hornberger, *Phys. Rev. Lett.*, 2013, **110**, 240401.

123. N. V. Vitanov, and B. M. Garraway, *Phys. Rev. A*, 1996, **54**, 5458.

124. N. V. Vitanov, *Phys. Rev. A*, 1999, **59**, 988.

125. F. Mentink-Vigier, S. Vega, and G. De Paëpe, *Phys. Chem. Chem. Phys.*, 2017, **19**, 3506.

126. N. V. Vitanov and B. M. Garraway, *Phys. Rev. A - At. Mol. Opt. Phys.*, 1996, **53**, 4288.

127. F. Mentink-Vigier, Ü. Akbey, H. Oschkinat, S. Vega, and A. Feintuch, *J. Magn. Reson.*, 2015, **258**, 102.

128. M. Edén, *Concepts Magn. Reson. Part A Bridg. Educ. Res.*, 2003, **18**, 24.

129. V. B. Cheng, H. H. Suzukawa, and M. Wolfsberg, *J. Chem. Phys.*, 1973, **59**, 3992.

130. F. Mentink-Vigier, S. Paul, D. Lee, A. Feintuch, S. Hediger, S. Vega, and G. De Paëpe, *Phys. Chem. Chem. Phys.*, 2015, **17**, 21824.

131. K. R. Thurber and R. Tycko, *J. Chem. Phys.*, 2014, **140**, 184201.

132. M. Lelli, S. R. Chaudhari, D. Gajan, G. Casano, A. J. Rossini, O. Ouari, P. Tordo, A. Lesage, and L. Emsley, *J. Am. Chem. Soc.*, 2015, **137**, 14558.

133. K. H. Hausser and D. Stehlik, Dynamic Nuclear Polarization in Liquids, Academic Press Inc., 1968.

134. S. Pylaeva, K.L. Ivanov, M. Baldus, D. Sebastiani, and H. Elgabarty, *J. Phys. Chem. Lett.*, 2017, **8**, 2137–2142.

135. T. V. Can, M. A. Caporini, F. Mentink-Vigier, B. Corzilius, J. J. Walish, M. Rosay, W. E. Maas, M. Baldus, S. Vega, T. M. Swager, and R. G. Griffin, *J. Chem. Phys.*, 2014, **141**, 1.

136. F. Mentink-Vigier, S. Vega, and G. De Paëpe, *Phys. Chem. Chem. Phys.*, 2017, **19**, 3506.

137. A. Lund, A. Equbal, and S. Han, *Phys. Chem. Chem. Phys.*, 2018, **20**, 23976.

138. E. J. Koers, E. A. W. Van Der Cruijsen, M. Rosay, M. Weingarth, A. Prokofyev, C. Sauvée, O. Ouari, J. Van Der Zwan, O. Pongs, P. Tordo, W. E. Maas, and M. Baldus, *J. Biomol. NMR*, 2014, **60**, 157.

139. G. Mathies, M. A. Caporini, V. K. Michaelis, Y. Liu, K. N. Hu, D. Mance, J. L. Zweier, M. Rosay, M. Baldus, and R. G. Griffin, *Angew. Chemie – Int. Ed.*, 2015, **54**, 11770.

140. D. Wisser, G. Karthikeyan, A. Lund, G. Casano, H. Karoui, M. Yulikov, G. Menzildjian, A. C. Pinon, A. Purea, F. Engelke, S. R. Chaudhari, D. Kubicki, A. J. Rossini, I. B. Moroz, D. Gajan, C. Copéret, G. Jeschke, M. Lelli, L. Emsley, A. Lesage, and O. Ouari, *J. Am. Chem. Soc.*, 2018, **140**, 13340.

141. A. C. Pinon, U. Skantze, J. Viger-Gravel, S. Schantz, and L. Emsley, *J. Phys. Chem. A*, 2018, **122**, 8802.

142. F. A. Perras and M. Pruski, *J. Chem. Phys.*, 2018, **149**, 154202.

143. S. Lange, A. H. Linden, Ü. Akbey, W. Trent Franks, N. M. Loening, B. J. Van Rossum, and H. Oschkinat, *J. Magn. Reson.*, 2012, **216**, 209.

144. H. Takahashi, C. Fernández-De-Alba, D. Lee, V. Maurel, S. Gambarelli, M. Bardet, S. Hediger, A. L. Barra, and G. De Paëpe, *J. Magn. Reson.*, 2014, **239**, 91.

145. A. J. Rossini, A. Zagdoun, M. Lelli, D. Gajan, F. Rascón, M. Rosay, W. E. Maas, C. Copéret, A. Lesage, and L. Emsley, *Chem. Sci.*, 2012, **3**, 108.

146. C. Sauvée, G. Casano, S. Abel, A. Rockenbauer, D. Akhmetzyanov, H. Karoui, D. Siri, F. Aussenac, W. Maas, R. T. Weber, T. Prisner, M. Rosay, P. Tordo, and O. Ouari, *Chem. – A Eur. J.*, 2016, **22**, 5598.

147. F. A. Perras, L. L. Wang, J. S. Manzano, U. Chaudhary, N. N. Opembe, D. D. Johnson, I. I. Slowing, and M. Pruski, *Curr. Opin. Colloid Interface Sci.*, 2018, **33**, 9.

148. P. Thureau, M. Juramy, F. Ziarelli, S. Viel, and G. Mollica, *Solid State Nucl. Magn. Reson.*, 2019, **99**, 15.

149. F. Mentink-Vigier, A. L. Barra, J. Van Tol, S. Hediger, D. Lee, and G. De Paëpe, *Phys. Chem. Chem. Phys.*, 2019, **21**, 2166.

Chapter 3

Pulsed Dynamic Nuclear Polarization

Kong Ooi Tan[1], Sudheer Jawla[2], Richard J. Temkin[2], and Robert G. Griffin[1]

[1]*Department of Chemistry, Francis Bitter Magnet Laboratory, Massachusetts Institute of Technology, Cambridge, MA, USA*
[2]*Plasma Science and Fusion Center, Massachusetts Institute of Technology, Cambridge, MA, USA*

3.1 INTRODUCTION

Dynamic nuclear polarization[1,2] (DNP) is one of the hyperpolarization techniques used to enhance the sensitivity of NMR experiments by transferring polarizations from unpaired electrons to nearby nuclei via microwave irradiation. For transfers involving electrons and ^1H nuclei, the theoretical maximum enhancement factor is $\varepsilon \sim 658$. However, in practical situations, the enhancement factor ε varies with

temperature (see Chapter 9), magnetic fields, the type(s) of radical(s) employed (see Chapter 5), and other experimental conditions. There are two main categories of DNP mechanisms that differ in the type of microwave irradiation employed, namely continuous-wave (CW) DNP[2–5] and pulsed DNP.[6–10] The main difference in the underlying physics between the two DNP mechanisms is that the CW-DNP techniques rely on saturating the transitions between the energy levels, while the pulsed-DNP polarization transfer is facilitated by generating coherences (transverse magnetization). Thus, in order to achieve efficient CW-DNP transfer, the saturation rate, which depends on the Rabi frequency ω_{1s}, must be larger than the rate at which the system returns to thermal equilibrium (spin-lattice/T_1 relaxation), as described by equation (3.1). Similarly, the pulsed-DNP techniques require that the excitation rate of the coherence (transverse magnetization) exceeds the rate at which it decays due to the spin-spin/T_2 relaxation:

$$\omega_{1s} \gg 2\pi/T_1 \ \text{(CW-DNP)}$$
$$\omega_{1s} \gg 2\pi/T_2 \text{(Pulsed DNP)} \qquad (3.1)$$

Figure 3.1 shows the results of numerical simulations performed on a single-spin system that illustrates these constraints on the Rabi frequency. For an electron with relaxation times of $T_1 = 1\,ms$ and $T_2 = 1\,\mu s$, the plot shows that one would need $\sim 100\,kHz$ of Rabi frequency applied continuously

Handbook of High Field Dynamic Nuclear Polarization.
Edited by Vladimir K. Michaelis, Robert G. Griffin, Björn Corzilius and Shimon Vega
© 2020 John Wiley & Sons, Ltd. ISBN: 978-1-119-44164-9
Also published in eMagRes (online edition)
DOI: 10.1002/9780470034590.emrstm1551

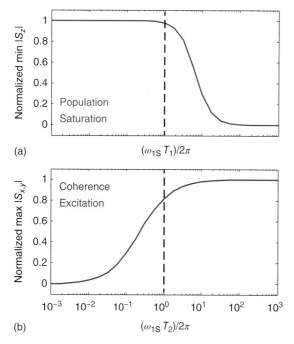

(a)

(b)

Figure 3.1. Numerical simulations of the population and coherence of a single electron spin system with $T_1 = 1$ ms and $T_2 = 1$ μs. The plot shows the minimum absolute population difference $|\hat{S}_z|$ (a) and maximum coherence $|\hat{S}_{x,y}| = \sqrt[2]{\hat{S}_x^2 + \hat{S}_y^2}$ excited (b), as a function of the Rabi frequency ω_{1s}

to completely saturate the electron spin, which in turn facilitates the CW-DNP mechanism. On the other hand, a minimum Rabi frequency of ~5 MHz is required to excite the coherence by over 95% following an application of a simple $\pi/2$ pulse. It is easy to notice that the pulsed-DNP techniques demand a more stringent microwave power requirement than the CW-DNP technique since $T_1 > T_2$ in theory. Note that this is only the minimum requirement needed to implement the pulsed DNP sequences. Further requirements such as higher microwave power, the ability to gate the microwave radiation and manipulation of the microwave phase, amplitude and frequency are often desired in some pulsed DNP experiments. The stringent technical and power requirements demanded by the pulsed DNP technique are two reasons why most DNP spectrometers at higher fields (≥5 T/140 GHz/211 MHz) built to date are designed only to perform CW-DNP experiments. However, it was reported that the enhancement factors of most CW-DNP techniques decrease at higher magnetic

fields, i.e., the solid effect (SE) and cross effect (CE) mechanisms in the CW-DNP techniques have transfer efficiencies that scale with the magnetic field as B_0^{-2} and B_0^{-1}, respectively.[11] Nevertheless, an exception to this trend was found recently for Overhauser effects (OE) in insulating solids,[12] where the enhancement factor increases with the strength of the Zeeman magnetic fields. The underlying mechanism[13] that governs the polarization transfer in the OE remains to be established.

In contrast, pulsed DNP techniques mediate polarization transfer via a coherent pathway, which in theory can have an enhancement factor that is field-independent. In addition, it has been demonstrated experimentally that some pulsed DNP experiments yield significant enhancement factors at room temperature[14–16] at a magnetic field of 0.35 T. Both the field-independent enhancement and viability at higher temperatures are features that suggest that pulsed DNP experiments can outperform CW techniques when applied to biological or materials samples. In Section 3.2, we will discuss two main categories of the pulsed DNP techniques, namely those with and without simultaneous radio frequency (rf) irradiation on the nuclei involved. We shall see that the implementation, terms of hyperfine interactions used, and the effects of them are rather different in these two categories.

3.2 THEORY

3.2.1 Pulsed DNP with rf Irradiation of the Nuclear Spins

Two pulse sequences will be discussed here, namely DNP in the nuclear rotating frame (NRF-DNP)[17,18] and the dressed-state solid effect (DSSE).[19] There are two features that distinguish these two experiments from other DNP techniques. First, as shown in Figure 3.2 rf irradiation is applied to the nuclear spins during the polarization period. Second, both NRF-DNP and DSSE exploit the secular hyperfine coupling A_{zz} rather than the non-secular contributions B_{zx} to mediate polarization transfer. The main consequence of using the secular contribution is that the technique can possibly be applied to liquid-state samples. In addition, the theory does not indicate any unfavorable scaling of the enhancement with the static magnetic field, assuming ideal conditions. Although

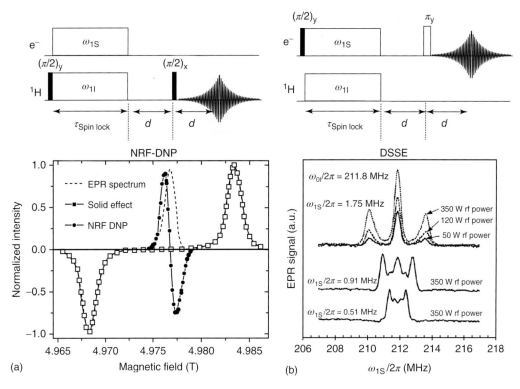

Figure 3.2. (a) Schematic diagram of the pulse sequences (a) NRF-DNP with NMR detected using solid echo and (b) DSSE with EPR signal detected using spin echo, at ~5 T. An enhancement factor of less than 1 is obtained for NRF-DNP while the nuclear enhancement was not directly observable in the DSSE case. (Reprinted with permission from Refs 17, 18, 20)

there is limited experimental evidence supporting this theoretical result, its validity is worthy of further investigation.

We will begin with a general Hamiltonian for a two-spin system before considering individual sequences:

$$\widehat{\mathcal{H}} = \Omega\widehat{S}_z + \omega_{1s}\widehat{S}_x - \omega_{0I}\widehat{I}_z + 2\omega_{1I}\cos(\omega_{rf}t)\widehat{I}_x \\ + A_{zz}\widehat{S}_z\widehat{I}_z + B_{zx}\widehat{S}_z\widehat{I}_x \tag{3.2}$$

where A_{zz} and B_{zx} are the secular and non-secular components of the hyperfine interaction, ω_{0I} is the Larmor frequency of the nuclear spin, Ω is the microwave off-set frequency, ω_{rf} is the rf carrier frequency, ω_{1S} and ω_{1I} are the nutation frequencies for the electron and nuclear spins, respectively. The Hamiltonian is rotated by $\widehat{U}_t = \exp(-i\theta\widehat{S}_y)$ with an angle $\theta = \tan^{-1}(\omega_{1s}/\Omega_s)$,

so that the z-axis is along the effective field $\omega_{eff}^{(S)} = \sqrt{\omega_{1s}^2 + \Omega_s^2}$:

$$\widehat{U}_t^{-1}\widehat{\mathcal{H}}\widehat{U}_t = \omega_{eff}^{(S)}\widehat{S}_z - \omega_{0I}\widehat{I}_z + 2\omega_{1I}\cos(\omega_{rf}t)\widehat{I}_x \\ + (A_{zz}\widehat{I}_z + B_{zx}\widehat{I}_x)(\widehat{S}_z\cos\theta - \widehat{S}_x\sin\theta) \tag{3.3}$$

We then transform to an rf interaction-frame with $\widehat{U}_{rf} = \exp(-i(-\omega_{rf}\widehat{I}_z)t)$:

$$\widehat{U}_{rf}^{-1}\widehat{U}_t^{-1}\widehat{\mathcal{H}}\widehat{U}_t\widehat{U}_{rf} = \omega_{eff}^{(S)}\widehat{S}_z + \Omega_I\widehat{I}_z + \omega_{1I}\widehat{I}_x \\ + A_{zz}\cos\theta\widehat{S}_z\widehat{I}_z - A_{zz}\sin\theta\widehat{S}_x\widehat{I}_z \tag{3.4}$$

where $\Omega_I = \omega_{0I} - \omega_{rf}$ is the rf offset frequency. Note that the non-secular term B_{zx} vanishes due to the high-field approximation. Similarly, if we now tilt the Hamiltonian such that the z-axis of the nuclear frame is directed along the effective field $\omega_{\text{eff}}^{(I)}$, we find:

$$\hat{\hat{\mathcal{H}}} = A_{zz}(\cos\theta\cos\varphi\hat{S}_z\hat{I}_z - \cos\theta\sin\varphi\hat{S}_z\hat{I}_x$$
$$- \sin\theta\cos\varphi\hat{S}_x\hat{I}_z + \sin\theta\sin\varphi\hat{S}_x\hat{I}_x)$$
$$+ \omega_{\text{eff}}^{(S)}\hat{S}_z + \omega_{\text{eff}}^{(I)}\hat{I}_z \qquad (3.5)$$

where $\varphi = \tan^{-1}(\omega_{1\mathrm{I}}/\Omega_{\mathrm{I}})$. We have now obtained a simplified Hamiltonian in the double-rotating frame, which can be used to discuss the both NRF-DNP and DSSE experiments. The NRF-DNP experiments utilize off-resonance microwave irradiation together with on-resonance spin-locking rf pulses. In contrast, DSSE employs the opposite scenario, i.e., on-resonance spin-locking microwave pulses with off-resonance rf irradiation. The consequences of the two different implementations are discussed in the following sections.

3.2.1.1   Nuclear Rotating Frame-Dynamic Nuclear Polarization

The NRF-DNP experiment uses on-resonance rf irradiation ($\varphi = \pi/2$) (Figure 3.2a) and far off-resonance microwave irradiation (small θ for $|\omega_{1\mathrm{s}}/\Omega_{\mathrm{s}}| \ll 1$). Hence, equation (3.5) simplifies to the form

$$\hat{\hat{\mathcal{H}}} \approx \omega_{\text{eff}}^{(S)}\hat{S}_z + \omega_{1\mathrm{I}}\hat{I}_z - A_{zz}\hat{S}_z\hat{I}_x. \qquad (3.6)$$

By inspecting the resulting Hamiltonian in equation (3.6), one can see that it has a similar form as the lab-frame Hamiltonian required to describe the SE ($\hat{\mathcal{H}} = \omega_{0S}\hat{S}_z - \omega_{0I}\hat{I}_z + B_{zx}\hat{S}_z\hat{I}_x$) except that the off-diagonal term here is contributed by the secular A_{zz} term instead of the pseudo-secular term B_{zx}. Hence, a similar state-mixing mechanism to the one that occurs in the SE is also expected to drive the double-quantum (DQ) and zero-quantum (ZQ) transitions in this tilted frame, with the degree of mixing $\eta = \tan^{-1}(A_{zz}/2\omega_{1\mathrm{I}})$.[21] Following that, it is straightforward to diagonalize the Hamiltonian [equation (3.6)] with $\hat{U} = \exp(-i(-\eta 2\hat{S}_z\hat{I}_y)t)$ and obtain the eigenvalues needed to determine the matching condition:

$$\omega_{\text{eff}}^{(S)} = \pm\sqrt{(A_{zz}/2)^2 + \omega_{1\mathrm{I}}^2}$$
$$\Omega_s \sim \pm\sqrt{(A_{zz}/2)^2 + \omega_{1\mathrm{I}}^2} \qquad (3.7)$$

where in the second line we have considered the case for $|\omega_{1\mathrm{s}}/\Omega_{\mathrm{s}}| \ll 1$. The matching conditions imply that the separation between the DQ and ZQ conditions is much smaller than the case expected for the SE,

i.e., $2\omega_{0\mathrm{I}}$ (Figure 3.2a). An enhancement factor of ~ 0.89 was measured.[18] Although this enhancement factor may be considered to be small, the experimental advantage comes indirectly from the short T_1 of electrons relative to nuclei. The low enhancement factor could be rationalized by the fact that the polarization is built up along the spin-lock axis, which has a relaxation time $T_{1\rho}$ (~ 27 ms) several orders of magnitude smaller than T_1 (~ 1056 s) of the proton at 11 K. In the case that $T_{1\rho}$ becomes comparable or smaller than the spin diffusion timescale, the bulk nuclei can no longer be polarized efficiently by the spin diffusion mechanism.

3.2.1.2   Dressed-State Solid Effect

We mentioned that the DSSE sequence has an opposite pulsing regime compared to NRF-DNP, i.e., $\theta = \pi/2$ (Figure 3.2b) and small φ for the case $|\omega_{1\mathrm{I}}/\Omega_{\mathrm{I}}| \ll 1$. Hence, we simplify equation (3.5) to yield

$$\hat{\hat{\mathcal{H}}} \approx \omega_{\text{eff}}^{(I)}\hat{I}_z + \omega_{1\mathrm{S}}\hat{S}_z - A_{zz}\hat{S}_x\hat{I}_z. \qquad (3.8)$$

This is similar to the Hamiltonian for NRF-DNP [equation (3.6)], except that the spin part of the off-diagonal term here is $\hat{S}_x\hat{I}_z$ instead of $\hat{S}_z\hat{I}_x$. Likewise, the Hamiltonian [equation (3.8)] can be diagonalized with $\hat{U} = \exp(-i(-\eta 2\hat{S}_y\hat{I}_z)t)$ with $\eta = \tan^{-1}(A_{zz}/2\omega_{1\mathrm{S}})$, and the matching conditions are given as

$$\omega_{\text{eff}}^{(I)} = \pm\sqrt{(A_{zz}/2)^2 + \omega_{1\mathrm{S}}^2}$$
$$\Omega_{\mathrm{I}} \sim \pm\omega_{1\mathrm{S}} \qquad (3.9)$$

where we have considered the case for $|\omega_{1\mathrm{I}}/\Omega_{\mathrm{I}}| \ll 1$ and $A_{zz} \ll \omega_{1\mathrm{S}}$ in the second line. Although the enhanced nuclear signal in the DSSE has not yet been observed directly, there are indications of polarization transfer via monitoring the loss of electron polarization (Figure 3.2b) and also via numerical simulations.[19] Nonetheless, the fact that the microwave irradiation is applied directly on-resonance on the EPR line allows the possibility of employing electron decoupling during NMR detection.[22]

3.2.2   Pulsed DNP without rf Irradiation of the Nuclear Spins

There are in general three different types of pulsed DNP sequences that employ only microwave irradiation (no excitation of NMR transitions), namely nuclear orientation via electron spin locking (NOVEL), the frequency-swept integrated solid effect (FS-ISE), and time-optimized pulsed dynamic nuclear polarization (TOP-DNP). These three 'longitudinal' pulsed DNP sequences are in general more efficient at enhancing nuclear polarization than the 'transverse' sequences discussed above and have other desirable features. First, any NMR polarization built up in the transverse plane will be subjected to nuclei transverse relaxation which has a shorter lifetime than when it is along the z-axis. Second, the precession frequencies of the nuclei might be shifted away from the Larmor frequency due to distance-dependent hyperfine interactions. This causes an additional complication when choosing the frequency for the rf pulses. Therefore, it is not surprising that the pulsed DNP sequences without rf irradiation have been experimentally demonstrated to be more robust than the transverse versions.

3.2.2.1 Nuclear Orientation via Electron Spin-Locking (NOVEL)

As shown in Figure 3.3(a), the NOVEL experiment uses a strong microwave field to spinlock the electrons at a Rabi frequency that matches the nuclear Larmor frequency ($\omega_{1S} = \omega_{0I}$). If the condition is fulfilled, the energy levels become degenerate in the rotating frame and allow for a resonant transfer of polarization from the electrons to the nearby nuclei via dipolar couplings. This idea is similar to the Hartmann–Hahn cross-polarization (CP) technique,[24] and thus NOVEL is a lab frame-rotating frame version of the ubiquitous CP experiment. The NOVEL technique was first demonstrated experimentally by Henstra *et al.*[6] and Henstra and Wenckebach[25] on a single crystal of silicon doped with boron, which under uniaxial stress forms an effective electron spin-1/2 system. The derivation of the Hamiltonian governing NOVEL starts with a Hamiltonian similar to equation (3.2) except that the microwave irradiation is applied on resonance

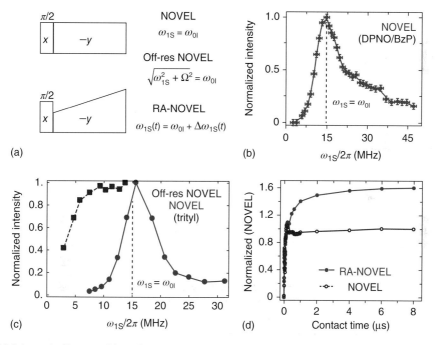

(a)

(b)

(c)

(d)

Figure 3.3. (a) Schematic diagram of the pulse sequences NOVEL, off-resonance NOVEL, and RA-NOVEL. (b) Experimental results of NOVEL on a single crystal of diphenyl nitroxide (DPNO) doped benzophenone (BzP), with optimum transfer at the NOVEL condition ($\omega_{1S} = \omega_{0I}$). (c) Field profiles of the off-resonance (blue) and on-resonance (red) NOVEL sequence. (d) A linear ramp of 18 MHz applied in RA-NOVEL outperforms the constant-amplitude NOVEL by a factor of ~1.6. All experiments were performed at 0.35 T. (Reprinted with permission from Refs 7, 10, 23)

$\Omega = 0$ and there is no rf irradiation $\omega_{1I} = 0$ during the DNP:

$$\widehat{\mathcal{H}} = -\omega_{0I}\widehat{I}_z + A_{zz}\widehat{S}_z\widehat{I}_z + B_{zx}\widehat{S}_z\widehat{I}_x + \omega_{1S}\widehat{S}_x \quad (3.10)$$

Next, we perform an interaction-frame transformation with the propagator $\widehat{U}_1 = \exp(-i(\omega_{1S}\widehat{S}_x - \omega_{0I}\widehat{I}_z)t)$ that yields

$$\widehat{\widetilde{H}}(t) = \widehat{U}_1^{-1}(t)\widehat{H}\widehat{U}_1(t) - (\omega_{1S}\widehat{S}_x - \omega_{0I}\widehat{I}_z)$$
$$= (A_{zz}\widehat{I}_z + B_{zx}(\widehat{I}_x \cos\omega_{0I}t + \widehat{I}_y \sin\omega_{0I}t))$$
$$\times (\widehat{S}_z \cos\omega_{1S}t + \widehat{S}_y \sin\omega_{1S}t) \quad (3.11)$$

This results in a time-dependent Hamiltonian that upon integration becomes time-independent at the conditions $\omega_{1S} = \pm\omega_{0I}$:

$$\widehat{\widehat{\mathcal{H}}} = \frac{\omega_{0I}}{2\pi}\int_0^{(2\pi)/\omega_{0I}} \widehat{\widetilde{\mathcal{H}}}(t)\mathrm{d}t = \frac{B_{zx}}{2}(\widehat{S}_z\widehat{I}_x \pm \widehat{S}_y\widehat{I}_y)$$
$$\text{if } \omega_{1S} = \pm\omega_{0I} \quad (3.12)$$

where the Hamiltonian [equation (3.12)] dictates a DQ condition if the electron polarization is spin locked along the microwave field ($\omega_{1S} = \omega_{0I}$), or ZQ condition if they are antiparallel ($\omega_{1S} = -\omega_{0I}$). The sequence has been shown to be robust and yields enhancement factors >300 in experiments at 0.35 T.[23,26] A microwave field profile (enhancement vs. $\omega_{1S}/2\pi$) for diphenyl-nitroxide (DPNO) doped into benzophenone (BzP) is illustrated in Figure 3.3b. The enhancement is maximized at $\omega_{1S}/2\pi = 15\,\text{MHz}$ and shows additional enhanced intensity at 30 and 45 MHz. This is likely due to the strong couplings of the electron on DPNO to the ^{1}H's on BzP.

In addition, it was demonstrated recently that the enhancement factor can be increased further by a factor of 1.6 by making the sequence adiabatic (Figure 3.3a and d), i.e., sweeping the microwave amplitude $\omega_{1S}(t)$ across the matching condition using an arbitrary waveform generator (AWG).[10] There are two main reasons that the ramped-amplitude nuclear orientation via electron spin locking (RA-NOVEL) sequence improves the efficiency significantly. First, the transfer efficiency of an adiabatic sequence is theoretically higher (close to 100%) as it compensates for different optimum mixing times in each crystallite, whose magnitudes of dipolar couplings components along the static magnetic field are orientation-dependent. This feature is implied in the relatively long build-up curve of RA-NOVEL (Figure 3.3d) since the crystallites with fast transfer have to wait for those with slow transfer in an adiabatic sequence for maximum

efficiency. Second, the adiabatic sequence has higher tolerance towards the mismatch of the matching condition ($\omega_{1S} = \omega_{0I}$) and power inhomogeneity across the sample. Following that, it was reported that the tangent ramp is more efficient than the linear ramp at shorter mixing times.[10]

One of the main challenges of implementing NOVEL at higher fields is the requirement of high microwave Rabi frequencies. It was demonstrated that this issue can be mitigated by exploiting the off-resonance NOVEL condition ($\sqrt{\omega_{1S}^2 + \Omega^2} = \omega_{0I}$),[7] inspired by the SPECIFIC-CP sequence in solid-state NMR.[27] Figure 3.3(c) shows that the off-resonance NOVEL sequence has an enhancement factor ~70% of on-resonance NOVEL but requires only ~11% of the microwave power.

3.2.2.2 *Frequency-swept Integrated Solid Effect (FS-ISE)*

Matching the NOVEL condition at high fields is technically demanding, and this fact provided the impetus to explore the FS-ISE as an alternative approach that can be implemented at lower Rabi frequencies. The integrated solid effect (ISE) sequence was first demonstrated by sweeping the magnetic field through both SE matching conditions (Figure 3.4a), e.g., from $\Omega \ll -\omega_{0I}$ to $\Omega \gg \omega_{0I}$ within a short time period ($\sim T_{1e}$).[29-31] This approach might appear to be counter-intuitive because the ZQ and DQ matching conditions yield nuclear polarizations of opposite signs. However, the electron polarization is also inverted during the adiabatic process and hence the polarization transfer adds constructively. In order to perform the ISE at higher fields – for example, in a 9.4 T/400 MHz magnet – the magnetic field must be swept over ~28 mT within 1 ms. This is technically challenging and the field drift and inhomogeneity induced by the sweep makes the sequence inappropriate for NMR applications at higher fields. Alternatively, the offset frequencies Ω can be manipulated by changing the microwave frequency while keeping the magnetic field constant. Similar frequency modulation schemes were demonstrated to improve the enhancement by a factor of ~2–3 at both 3.34 T (95 GHz) and 6.7 T (188 GHz) on TEMPOL at temperatures (10–50 K) favoring thermal mixing (TM).[32,33] The frequency modulation was performed by changing the microwave carrier frequencies directly via a voltage-controlled oscillator (VCO). Alternatively, the frequency modulation can

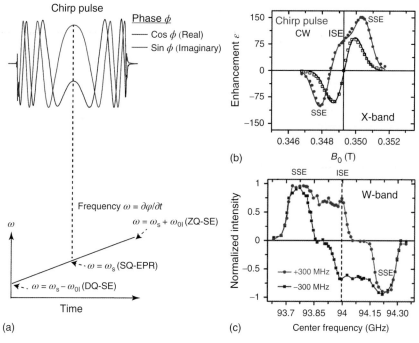

Figure 3.4. (a) Schematic diagram of the frequency-swept integrated solid effect (FS-ISE) sequence, which performs a sweep in microwave frequency via phase-modulated pulses using an AWG. Experimental field profiles with the positions of the ISE and SSE marked at (b) X-band/9 GHz/0.35 T and (c) W-band/94 GHz/3.35 T. It was observed that the optimum SSE enhancement shows a displacement from the normal solid effect condition. (Reprinted with permission from Refs 9, 28)

also be implemented by generating a microwave pulse train of varying phases (Figure 3.4a) using an AWG, while having the same carrier frequency in all pulses. It was demonstrated recently that such FS-ISE experiments performed using $\omega_{1S}/2\pi = 1.5$ MHz ($\sim 0.1\omega_{0I}$) can yield similar enhancements compared to the ISE at the NOVEL condition ($\omega_{1S}/2\pi = 15$ MHz) 0.35 T (Figure 3.4b).[9]

While investigating the FS-ISE using an AWG, we observed a more efficient polarization-transfer condition, but at microwave offset frequencies that are larger than the normal SE conditions (Figure 3.4b and c), i.e., a stretched solid effect (SSE). Such an effect has been exploited experimentally at 0.35 and 3.35 T, and the enhancement factor in SSE is ~ 1.4 times larger than that of an ISE.[9,28] The underlying mechanism that causes the displacement observed in the SSE field profile is currently being investigated in our laboratory. Nevertheless, with the availability of gyro-amplifiers (*vide infra*) or other high-power controllable microwave sources, the low-power frequency-swept ISE and SSE sequences could be

practically implemented at higher fields. In addition, the FS-ISE/SSE experiments have been performed with static samples only and its effect under the MAS condition remains to be investigated in the future.

3.2.2.3 Time-optimized Pulsed DNP (TOP-DNP)

We have discussed pulsed DNP sequences that achieve DNP matching conditions by exploiting the Rabi (NOVEL) and/or offset (SE/ISE/SSE) frequencies. A new class of pulsed DNP sequence was recently introduced in our laboratory, which mediates the polarization transfer via a train of microwave pulses, τ_p, and delays, d, of optimal timings (Figure 3.5a), i.e., TOP-DNP.[8] The design of the sequence was inspired by MAS dipolar recoupling experiments,[34] where averaged anisotropic interactions can be reintroduced via rotor-synchronized pulses. A prominent example is radio frequency-driven recoupling (RFDR),[35] which consists of a train of π-pulses applied during the mixing period to transfer polarization.

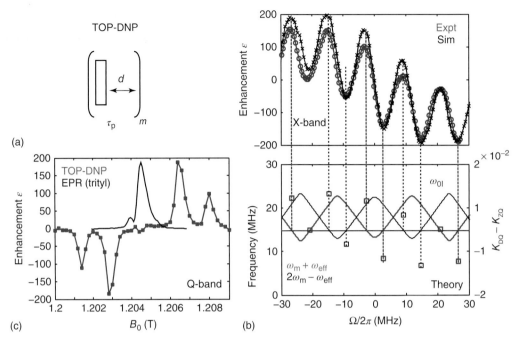

Figure 3.5. (a) Schematic diagram of the TOP-DNP sequence, which is comprised of a train of microwave pulses interleaved with delays. (b) Experimental and simulated field profile (top) of TOP-DNP at 0.35 T. The bottom plot shows the calculated characteristic frequencies and the matching conditions of TOP-DNP at the intersections. (c) TOP-DNP field profile measured experimentally at 1.2 T (Q-band). (Reprinted with permission from Ref. 8)

Similarly, the averaging of the pseudo-secular term B_{zx} by the Larmor interaction $\omega_{0I}\hat{I}_z$ is inhibited and restored to transfer polarization in TOP-DNP, i.e., hyperfine recoupling. A brief theoretical analysis of the pulse sequence using Floquet theory[36,37] is presented here.

There are three characteristic frequencies[38–40] involved in the TOP-DNP sequence, namely the nuclear Larmor frequency, ω_{0I}, the modulation frequency $\omega_m = 2\pi/(\tau_p + d)$, and the effective field, ω_{eff}. ω_{eff} is defined as the net rotation angle, β_{eff}, over the cycle time, τ_m, of a periodic sequence. An effective Hamiltonian is obtained if there exists a linear combination of the characteristic frequencies that satisfies the resonance conditions (Figure 3.5b):

$$\omega_{0I} + k\omega_m \pm \omega_{eff} = 0 \tag{3.13}$$

where k is an integer number. There are in principle multiple resonance conditions exploiting different values of k. It can be shown that the effective Hamiltonian for one of the matching conditions, $\omega_m + \omega_{eff} = \omega_{0I}$, is

given by

$$\hat{\mathscr{H}}^{(1,-1,-1)} + \hat{\mathscr{H}}^{(-1,1,1)}$$
$$= \frac{B}{4}(a_-^{(-1,-1)}\hat{S}^-\hat{I}^- + a_+^{(-1,-1)}\hat{S}^+\hat{I}^-$$
$$+ 2a_z^{(-1,-1)}\hat{S}_z\hat{I}^-) + c.c. \tag{3.14}$$

where $c.c.$ stands for complex conjugate, $\hat{S}^\pm = \hat{S}_x \pm i\hat{S}_y$ are the ladder operators, and $a_\pm^{(-1,-1)}$ are the Fourier coefficients. The first two terms in the effective Hamiltonian [equation (3.14)] are responsible for driving the DQ and ZQ transfers, respectively. It was shown that the amplitudes of the Fourier coefficients $a_{+,-}^{(-1,-1)}$ calculated theoretically are directly correlated with the performance at the matching condition in the field profile (Figure 3.5b). This sequence yields a broadband DNP field profile spanning a bandwidth of $4\omega_{0I}$ with a maximum enhancement of $\sim$200 at 0.35 T (Figure 3.5b). The result is promising as it yields $\sim$15% higher signal intensity than NOVEL with only 7% of the microwave power. Furthermore, the effective Hamiltonian [equation (3.14)] suggests that the enhancement factor is field-independent as long as the Rabi frequency is scaled proportionally. This statement was confirmed to be valid at Q-band/1.2 T/33 GHz (Figure 3.5c), which is the highest frequency available for such experiments at the time of writing.

3.3 NUMERICAL SIMULATIONS

Despite the fact that DNP has been studied extensively since its first demonstration in the 1950s, the underlying factors and optimum conditions required to yield efficient DNP experiments are not yet fully understood. For instance, the choice of solvent, type of radical, degree of deuteration, temperature, and even the protocol to freeze the sample all affect the enhancement factor.[41] In situations like this, numerical simulations and theoretical modeling can provide useful insight and enable DNP mechanisms to be studied at

reduced complexity to achieve enhancements closer to the theoretical maximum of ~658 for e-^{1}H transfers.

In principle, DNP can be a multistep mechanism where in the first step, the polarization is transferred from an unpaired electron to the nearby nucleus coupled directly via the hyperfine interaction.[42] Then, the nearby nucleus spreads its polarization among the bulk nuclei via dipolar couplings. Alternatively, the electron can polarize the bulk nuclei directly without a relayed transfer via the nearby nuclei. The first stage is the main process where the actual DNP between an electron and a few nuclei take place, while the second stage is governed by the spin-diffusion mechanism, i.e., an NMR process. In the context of numerical simulations for DNP, the first stage of the DNP process, in the case of a small and isolated spin system, can usually be simulated sufficiently well using a full quantum-mechanical description in the numerical simulation. In addition to the simulated TOP-DNP field profile shown in Figure 3.5(b), Figure 3.6 shows

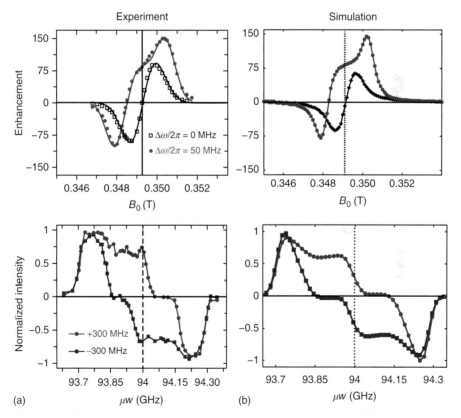

Figure 3.6. Field profiles of frequency-swept stretched solid effect (FS-SSE) obtained via (a) experiments and by using (b) numerical simulations at 0.35 T (top) and 3.35 T (bottom). (Reprinted with permission from Refs 9, 28)

that the main features observed in the experimental FS-ISE field profiles can also be reproduced reliably in the numerical simulations (unpublished results). These simulated field profiles were calculated using a custom simulation package developed in our laboratory.[8] There are other custom-designed programs written in other laboratories.[43,44] In particular, Mentink-Vigier *et al.* have demonstrated the effect of depolarization, as well as simulating CE field profiles via numerical simulations.[45,46] Simulation packages that are available for public users include *Spinach*[47] and SPINEVOLUTION.[48] Note that the former program is free and an open-source package in MATLAB while the latter is a excellent standalone commercial software optimized for rapid calculations.

The general features of the simulated field profiles are usually comparable to the experimental results, and qualitative arguments can be made on the performance of the DNP process. However, the simulated enhancement factors do not necessarily resemble the observed value because factors like spin diffusion, the number of protons surrounding each radical, and effect of temperature are beyond the capability of the full quantum-mechanical calculations.

3.4   INSTRUMENTATION

In order to implement the pulsed DNP sequences, particularly at high fields, where the spin dynamics that govern the polarization transfer are fundamentally different from the CW-DNP techniques, it is necessary to develop new instrumentation. In general, the spin coherences can be generated or manipulated efficiently in pulsed DNP if the excitation rate is faster than the coherence decay rate, i.e., $\omega_{1s} \gg 2\pi/T_{2e}$ [equation (3.1)]. This inequality can be satisfied by increasing the Rabi frequency ω_{1s} and/or lengthening the electron transverse relaxation time T_{2e}. The former parameter is governed by the available microwave power P and/or quality factor Q at the sample, i.e., $\omega_{1s} \propto \sqrt{PQ}$. Although the T_{2e} can be lengthened by performing the experiment at lower temperatures (see Chapter 9), the benefit can be offset by a longer T_1 of both the electrons and the nuclei of interest, leading to longer experimental acquisition times. We review the instrumentation needed to generate high ω_{1s} microwave pulses.

3.4.1   High-power Microwave Sources

There are in general two strategies for generating high-power microwave irradiation: (i) direct output from a high-power oscillator or (ii) use a lower-power source to generate microwave pulses, which are then amplified via a microwave amplifier. Although the ability to generate intense microwave pulses is one of the main requirements to perform the pulsed DNP experiments, other considerations including the manipulation of pulse widths, phases, and amplitudes are often advantageous in the development of pulsed DNP sequences. Thus, the strategy of generating low power/frequency pulses in a first step followed by amplification and upconversion to the final intended frequency is a preferred approach, primarily due to the cost and availability of microwave components, e.g., switches, attenuators, phase shifters, etc. at lower frequencies. It is worth noting that modern heterodyne NMR spectrometers employ a similar architecture. The generation of the pulses is best performed using a fast AWG (see Chapter 10), which is instrumental in implementing the FS-ISE experiments.[9,28] In addition, other practical specifications like frequency bandwidth and the duty cycle of a microwave source could have significant impacts on the overall efficiency of DNP experiments.

For the pulsed DNP experiments performed at 9 GHz/0.35 T (X-band) and 33 GHz/1.2 T (Q-band), the microwave pulses are first synthesized at low power, and then amplified via a traveling wave tube (TWT) amplifier, which has a maximum output power in the range of 0.15–1 kW.[8,23,26] Since the maximum output power of TWT's decreases at higher frequencies,[28,49] it is necessary to use an extended interaction klystron (EIK) operating as an amplifier at 94 GHz/3.35 T (W-band). The klystron-type amplifiers/oscillators are commercially available in the frequency range of 90–300 GHz and have been used in DNP experiments. For instance, a 300 W extended interaction oscillator (EIO) was used for pulsed DNP and EPR at 140 GHz,[50,51] a 9 W EIK was used for CW experiments at 187 GHz,[52] and a 1.5 W EIO at 263 GHz.[53] However, these are slow-wave devices and their structures for high frequencies become difficult to fabricate because of their small sizes. In addition, because of the reduced dimensions, the power density escalates and this limits the maximum output power at frequencies above ~250 GHz. In contrast, gyrotrons are fast-wave devices and do not suffer from these issues.[54] Hence, gyrotrons were developed for

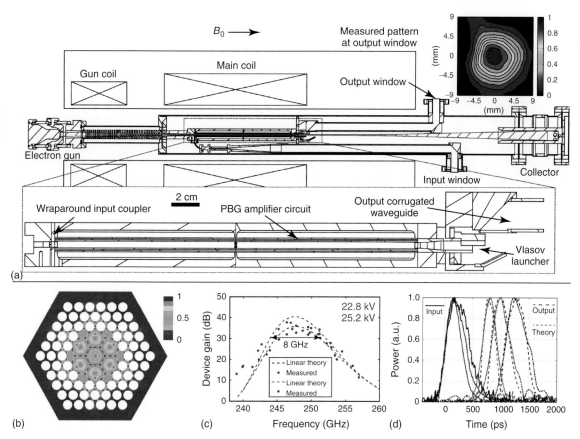

Figure 3.7. (a) Schematic diagram of a 250 GHz gyro-amplifier with the measured microwave beam at the output (upper right) with (b) the electric-field map inside a PBG waveguide operating in a TE_{03}-like mode. (c) The device gain-bandwidth from the gyro-amplifier for two different operating conditions 675 mA/22.8 kV (red) and 511 mA/25.2 kV (blue). A maximum circuit gain of 55 dB was achieved while the device observed 17 dB losses in the input and output components leaving a net device gain of 38 dB. (d) The amplified short pulses of variable lengths 250–800 ps. Input pulses (solid lines), measured amplified output pulses (- -), and, theoretically modeled pulses (--). The shortest amplified pulse (red) is 260 ps long and maintaining the temporal shape. (Reprinted with permission from Refs 57, 58)

DNP experiments at frequencies ranging from 140 to 593 GHz (see Chapter 8), and currently remain the preferred state-of-the-art microwave sources for CW-DNP.

One of the disadvantages of gyrotrons is that they are relatively fixed-frequency oscillators and therefore it is difficult to control the timing, amplitude, and phase of pulses during an experiment. Scott *et al.* have partially addressed this issue by modulating the accelerating voltage of the gyrotron, to modulate the microwave frequency directly.[55] Another potential approach would be gating the microwave output from a high-power gyrotron using a laser-driven

semiconductor switch.[56] Nevertheless, complete control over the pulse amplitude, phase, and width remains a challenging issue for oscillators operating at high power/frequencies. Accordingly, gyro-amplifiers were developed at 140 and 250 GHz (Figure 3.7a), so that the pulsed DNP experiments demonstrated at lower frequencies (X, and Q-band) can be extended to higher fields.[57,59] Recently, a 250 GHz short pulse (2 μs) wideband gyro-amplifier was built at MIT using a novel metallic photonic bandgap (PBG) based resonator cavity operating in a TE_{03}-like mode (Figure 3.7b). The amplifier achieved a record interaction circuit gain of >55 dB with a 8 GHz instantaneous

3 dB-bandwidth. The measured gain–bandwidth is shown in Figure 3.7(c) for two optimized operating voltages, 22.8 kV (red) and 25.2 kV (blue), with beam currents of 675 and 511 mA, respectively. The amplifier operated in zero drive stable condition for these operating parameters. These amplified microwave pulses are as short as 260 ps without any temporal or spatial distortion to the shape of the output pulse.[58] The measured and theoretically simulated input and output picosecond pulses at 249.7 GHz are shown in Figure 3.7(d). This amplifier will be used for generating high-power microwave pulses needed for various pulsed DNP sequences such as NOVEL and TOP-DNP. We foresee that this will be a more robust and promising approach to performing pulsed DNP at high fields.

3.4.2 High-*Q* Resonant Cavities

A resonant cavity is a structure in which the microwave irradiation redistributes its magnetic and electric field components in the closed region, and forms a standing wave with maximum amplitude where the microwave power is concentrated. There are different strategies in designing a cavity to confine the microwave power, and the efficiencies are characterized by the quality (*Q*) factor, which determines the Rabi frequency for a given microwave power. It is clear that having a high-*Q* resonant cavity is desireable as it relaxes the need of high-power microwave sources, and secondly, with careful design, it can minimize the electric-field components at the sample position so that the dielectric heating is minimized. However, a high-*Q* cavity inherently limits the frequency bandwidth of the system, which is a disadvantage for broadband frequency-swept sequences like FS-ISE. In addition, a high-*Q* cavity requires a longer time to dissipate the microwave power (longer dead time) and thus limits time resolution in transient EPR experiments.

The pulsed DNP experiments reported at X- and Q-band were performed using the commercial ENDOR probes equipped with dielectric resonators, which support a TE_{011} mode. The cavities are then wound around by rf coils for NMR spectroscopy.[8,23,26] A similar design but with the rf coil placed inside an oversized microwave cavity was developed by Wind *et al.*[60] at 1.4 T. The same group also fabricated a DNP-CPMAS probe equipped with a horn antenna with a reflector approach. Despite being able to preserve a high NMR filling factor, the system does not have a high-quality factor.

At higher frequencies, it becomes increasingly challenging to fabricate the microwave cavity due to smaller wavelengths, e.g., $\lambda/2 \sim 0.57$ mm at 263 GHz/400 MHz. An additional complexity is the incorporation of a microwave cavity with an rf coil without perturbing the resonant structure. A high-*Q* helix-cylindrical TE_{011} coiled resonator was designed and constructed at 140 GHz, and later adapted for 263 GHz (Figure 3.8a).[61,63,64] Despite the high *Q* factor achieved in such a design, it does limit the sample volume available for NMR spectroscopy. One way to circumvent this issue is to use a Fabry–Pérot cavity together with an rf coil or a stripline structure

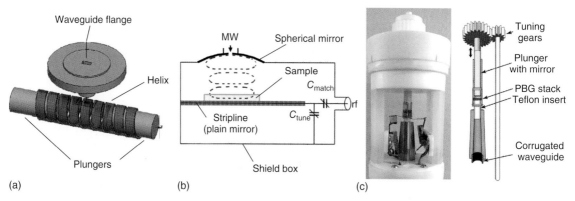

Figure 3.8. Various types of double-resonance structures: (a) helix-cylindrical TE_{011} coiled resonator, (b) stripline-Fabry–Pérot resonance structure, (c) and one-dimensional PBG in a saddle rf coil. (Reprinted with permission from Refs 61, 62)

(Figure 3.8b).[61,65] In a Fabry–Pérot resonator, two reflecting mirrors are placed in parallel to achieve a partial confinement of the microwave irradiation. Such a configuration restricts only one dimension and it allows two-dimensional samples like films or surfaces to be investigated at larger volumes while retaining a moderate value for Q factor. An alternate scheme to implement a double-resonance structure was reported recently by Nevzorov *et al.*,[62] where a one-dimensional PBG microwave resonator is incorporated within a saddle coil (Figure 3.8c). Such a regime yields a Q-factor of ~252 on microliter volume samples.

It is evident that the fabrication of a double-resonance structure is technically challenging at higher frequencies, and it becomes more demanding for it to be compatible for MAS. The state-of-the-art MAS-DNP probes developed for the CW-DNP techniques do not have a Q-factor as large as the resonant cavities discussed earlier, and it is an ongoing effort to improve the microwave coupling in the MAS stator. We expect that the assembly of a double-resonance structure integrated with a MAS stator will play a significant role in driving pulsed DNP applications at higher fields.

3.5 RADICALS FOR PULSED DNP

As is the case for the CW-DNP techniques, some factors including the stability of the radical, tendency of the radical to bind to the target molecule,[66] and the method of dispersion across the sample, e.g., the solubility of the radical in the solvent (if any), have to be considered when choosing the most suitable radical for DNP applications. Besides that, we have emphasized in the introduction that some of the commonly used radicals for the CW-DNP techniques (see Chapter 5) have not yet been tested for efficiency in pulsed experiments. This is because the spectral width of the radical, both homogeneous (Δ) and inhomogeneous (Δ) due to the g-anisotropy contributions, was not considered thoroughly in the development of the pulsed DNP sequences. In addition, the theory assumes that all electron spin packets can be excited simultaneously by the intense microwave pulse, an idea that led to the development of FT-NMR.[67] Moreover, the interaction-frame transformation with respect to the microwave irradiation, a convenient mathematical tool frequently used to describe the spin dynamics, remains a valid theoretical description for all electron spins in

such a regime. In short, at this stage of the development of pulsed DNP, we assume that the linewidth of a radical should fulfill the criteria $\Delta < \omega_{1S}^{(max)}, \omega_{0I}$.

Another important consideration is the relaxation times of both the electrons and the nuclei. As we have discussed earlier, a radical polarizes a small number of nearby nuclei each cycle of microwave irradiation, and the NMR polarization will spread across the bulk nuclei via spin diffusion. Thus, the amount of electron polarization that recovers, which is governed by T_{1e}, before the next microwave cycle begins can affect the DNP efficiency. Hence, contrary to some of the CW-DNP techniques where longer T_{1e} is desirable for easier saturation, a shorter T_{1e} is preferred for pulsed DNP experiments as long as the correlated T_{2e} remains sufficiently long. Besides that, the bulk nuclei relaxation time T_{1n}, which is affected by the type of radicals employed, can limit the rate of spin diffusion across the bulk nuclei. Secondly, the nuclear T_{1n} also determines the maximum number of cycles that a nuclear spin can be polarized by each radical before reaching the equilibrium value. Thus, we introduce the polarization cycle defined as $C_{pol} = T_{1e}/T_{1n}$, to be an empirical but yet universal parameter that can be used to estimate the efficacy of a radical for pulsed DNP applications. Some factors that one can manipulate to control C_{pol} include the concentration of the radical employed, temperature, presence of another paramagnetic species,[68] etc.

It was shown that trityl and bis-diphenylene-phenyl-allyl (BDPA) (Figure 3.9a), along with their derivatives[69,70] are efficient for pulsed DNP experiments at magnetic fields ranging from 0.35 to 3.35 T.[8,9,26] Although trityl has a relatively narrow linewidth at low magnetic fields, the EPR spectrum broadens as the magnetic field increases due to the g-anisotropy (inhomogeneous broadening). BDPA, on the other hand, is mostly broadened by the hyperfine interactions (homogenous broadening) instead of the g-anisotropy.[71] Hence, we foresee that BDPA and its derivatives will be potential candidates for pulsed DNP at higher magnetic fields, i.e., the linewidths of trityl OX063 and BDPA are 42 and 20 MHz, respectively, at 5 T.[72] This effect was also implied in the numerical simulations (Figure 3.9b), where a lower ω_{1S} is needed for BDPA to achieve the same enhancement compared to trityl.[8]

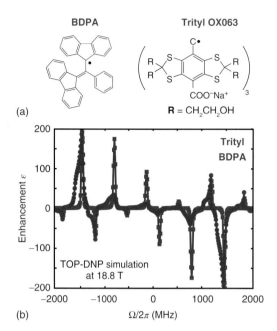

(a)

(b)

Figure 3.9. (a) Structures of BDPA and trityl OX063 and (b) their simulated field profiles at 18.8 T using TOP-DNP sequence. (Reprinted with permission from Ref. 8)

3.6 SUMMARY AND OUTLOOK

In this chapter, we have outlined the properties and the key differences between the pulsed- and CW-DNP techniques. One of the key features of pulsed DNP is the field-independent enhancement factor compared to CW-DNP, whose efficiency decreases at higher magnetic fields. We have summarized the theory and performances of pulsed DNP sequences that currently exist – namely NRF-DNP, DSSE, NOVEL, FS-ISE, and TOP-DNP. Following that, we discussed the instrumentation required to implement pulsed DNP, particularly the strategy to generate high-intensity microwave pulses using a combination of high-Q resonant cavities and microwave amplifiers, at magnetic fields ranging from 0.35 to 8.9 T. Then, we briefly discussed the desired features and examples of radicals that will be suited for pulsed DNP.

At present, there are limited applications of pulsed DNP techniques at lower fields because the resolution in the NMR spectrum is suboptimal. Nevertheless, the instrumentation available at these frequencies permits us to design and test ideas and a number of new intellectual avenues have emerged – RA-NOVEL, FS-ISE, the SSE, and TOP-DNP.[8–10,28] Thus, we anticipate

that these low-frequency experiments will remain a fertile intellectual proving ground for the near future. However, we also anticipate that the integration of the gyro-amplifiers into DNP instruments will permit implementation of pulsed experiments at higher frequencies. An example of this direction is the integration of the 250 GHz microwave gyro-amplifier with the 380 MHz CW DNP spectrometer in our laboratory. This instrumentation, together with the development of high-frequency pulsed DNP MAS probes, will open a new avenue for spin physics and applications that are currently inaccessible by the CW methods.

ACKNOWLEDGMENTS

We would like to thank Michael Mardini for proof-reading the manuscript. K.O.T. is supported by an Early Postdoc Mobility grant from the Swiss National Science Foundation (Grant no. 165285). This work was supported by the National Institute of General Medical Sciences (GM132997 and GM132079), the National Institute for Biomedical Imaging and Bioengineering (NIBIB) (EB001965 and EB004866), and formerly by NIBIB under grants EB002804 and EB002026.

RELATED ARTICLE IN EMAGRES

Overhauser, Albert W.: Dynamic Nuclear Polarization

REFERENCES

1. A. Abragam and M. Goldman, *Rep. Prog. Phys.*, 1978, **41**, 395.

2. T. R. Carver and C. P. Slichter, *Phys. Rev.*, 1953, **92**, 212.

3. C. F. Hwang and D. A. Hill, *Phys. Rev. Lett.*, 1967, **18**, 110.

4. A. Abragam and W. G. Proctor, *C. R. Acad. Sci.*, 1958, **246**, 2253.

5. M. Borghini, *Phys. Rev. Lett.*, 1968, **20**, 419.

6. A. Henstra, P. Dirksen, J. Schmidt, and W. T. Wenckebach, *J. Magn. Reson.*, 1988, **77**, 389.

7. S. K. Jain, G. Mathies, and R. G. Griffin, *J. Chem. Phys.*, 2017, **147**, 164201.

8. K. O. Tan, C. Yang, R. T. Weber, G. Mathies, and R. G. Griffin, *Sci. Adv.*, 2019, **5**, eaav6909.

9. T. V. Can, R. T. Weber, J. J. Walish, T. M. Swager, and R. G. Griffin, *Angew. Chem. Int. Ed.*, 2017, **56**, 6744.

10. T. V. Can, R. T. Weber, J. J. Walish, T. M. Swager, and R. G. Griffin, *J. Chem. Phys.*, 2017, **146**, 154204.

11. K.-N. Hu, G. T. Debelouchina, A. A. Smith, and R. G. Griffin, *J. Chem. Phys.*, 2011, **134**, 125105.

12. T. V. Can, M. A. Caporini, F. Mentink-Vigier, B. Corzilius, J. J. Walish, M. Rosay, W. E. Maas, S. Vega, T. M. Swager, and R. G. Griffin, *J. Chem. Phys.*, 2014, **141**, 064202.

13. S. Pylaeva, K. L. Ivanov, M. Baldus, D. Sebastiani, and H. Elgabarty, *J. Phys. Chem. Lett.*, 2017, **8**, 2137.

14. A. Henstra, T.-S. Lin, J. Schmidt, and W. T. Wenckebach, *Chem. Phys. Lett.*, 1990, **165**, 6.

15. K. Tateishi, M. Negoro, S. Nishida, A. Kagawa, Y. Morita, and M. Kitagawa, *Proc. Natl. Acad. Sci. U. S. A.*, 2014, **111**, 7527.

16. S. Fujiwara, M. Hosoyamada, K. Tateishi, T. Uesaka, K. Ideta, N. Kimizuka, and N. Yanai, *J. Am. Chem. Soc.*, 2018, DOI: 10.1021/jacs.8b10121.

17. R. A. Wind, L. Li, H. Lock, and G. E. Maciel, *J. Magn. Reson.*, 1988, **79**, 577.

18. C. T. Farrar, D. A. Hall, G. J. Gerfen, M. Rosay, J.-H. Ardenkjær-Larsen, and R. G. Griffin, *J. Magn. Reson.*, 2000, **144**, 134.

19. V. Weis, M. Bennati, M. Rosay, and R. G. Griffin, *J. Chem. Phys.*, 2000, **113**, 6795.

20. T. V. Can, Q. Z. Ni, and R. G. Griffin, *J. Magn. Reson.*, 2015, **253**, 23.

21. B. Corzilius, A. A. Smith, and R. G. Griffin, *J. Chem. Phys.*, 2012, **137**, 054201.

22. E. P. Saliba, E. L. Sesti, F. J. Scott, B. J. Albert, E. J. Choi, N. Alaniva, C. Gao, and A. B. Barnes, *J. Am. Chem. Soc.*, 2017, **139**, 6310.

23. T. V. Can, J. J. Walish, T. M. Swager, and R. G. Griffin, *J. Chem. Phys.*, 2015, **143**, 054201.

24. S. R. Hartmann and E. L. Hahn, *Phys. Rev.*, 1962, **128**, 2042.

25. A. Henstra and W. T. Wenckebach, *Mol. Phys.*, 2008, **106**, 859.

26. G. Mathies, S. Jain, M. Reese, and R. G. Griffin, *J. Phys. Chem. Lett.*, 2016, **7**, 111.

27. M. Baldus, A. T. Petkova, J. Herzfeld, and R. G. Griffin, *Mol. Phys.*, 1998, **95**, 1197.

28. T. V. Can, J. E. McKay, R. T. Weber, C. Yang, T. Dubroca, J. van Tol, S. Hill, and R. G. Griffin, *J. Phys. Chem. Lett.*, 2018, **9**, 3187.

29. A. Henstra, P. Dirksen, and W. T. Wenckebach, *Phys. Lett. A*, 1988, **134**, 134.

30. A. Henstra and W. T. Wenckebach, *Mol. Phys.*, 2013, **112**, 1.

31. T. R. Eichhorn, B. van den Brandt, P. Hautle, A. Henstra, and W. T. Wenckebach, *Mol. Phys.*, 2013, **112**, 1773.

32. A. Bornet, J. Milani, B. Vuichoud, A. J. Perez Linde, G. Bodenhausen, and S. Jannin, *Chem. Phys. Lett.*, 2014, **602**, 63.

33. Y. Hovav, A. Feintuch, S. Vega, and D. Goldfarb, *J. Magn. Reson.*, 2014, **238**, 94.

34. E. R. Andrew, A. Bradbury, and R. G. Eades, *Nature*, 1958, **182**, 1659.

35. A. E. Bennett, R. G. Griffin, J. H. Ok, and S. Vega, *J. Chem. Phys.*, 1992, **96**, 8624.

36. I. Scholz, J. D. van Beek, and M. Ernst, *Solid State Nucl. Magn. Reson.*, 2010, **37**, 39.

37. M. Leskes, P. K. Madhu, and S. Vega, *Prog. Nucl. Magn. Reson. Spectrosc.*, 2010, **57**, 345.

38. K. O. Tan, M. Rajeswari, P. K. Madhu, and M. Ernst, *J. Chem. Phys.*, 2015, **142**, 065101.

39. K. O. Tan, V. Agarwal, B. H. Meier, and M. Ernst, *J. Chem. Phys.*, 2016, **145**, 094201.

40. K. O. Tan, A. B. Nielsen, B. H. Meier, and M. Ernst, *J. Phys. Chem. Lett.*, 2014, **5**, 3366.

41. A. Leavesley, C. B. Wilson, M. Sherwin, and S. Han, *Phys. Chem. Chem. Phys.*, 2018, **20**, 9897.

42. A. A. Smith, B. Corzilius, A. B. Barnes, T. Maly, and R. G. Griffin, *J. Chem. Phys.*, 2012, **136**, 015101.

43. F. Mentink-Vigier, S. Vega, and G. De Paëpe, *Phys. Chem. Chem. Phys.*, 2017, **19**, 3506.

44. K. R. Thurber and R. Tycko, *J. Chem. Phys.*, 2012, **137**, 084508.

45. F. Mentink-Vigier, G. Mathies, Y. Liu, A.-L. Barra, M. A. Caporini, D. Lee, S. Hediger, R. G. Griffin, and G. De Paëpe, *Chem. Sci.*, 2017, **8**, 8150.

46. F. Mentink-Vigier, A. L. Barra, J. van Tol, S. Hediger, D. Lee, and G. De Paëpe, *Phys. Chem. Chem. Phys.*, 2019, **21**, 2166.

47. H. J. Hogben, M. Krzystyniak, G. T. P. Charnock, P. J. Hore, and I. Kuprov, *J. Magn. Reson.*, 2011, **208**, 179.

48. M. Veshtort and R. G. Griffin, *J. Magn. Reson.*, 2006, **178**, 248.

49. P. A. S. Cruickshank, D. R. Bolton, D. A. Robertson, R. I. Hunter, R. J. Wylde, and G. M. Smith, *Rev. Sci. Instrum.*, 2009, 80. DOI: 10.1063/1.3239402.

50. T. F. Prisner, S. Un, and R. G. Griffin, *Isr. J. Chem.*, 1992, **32**, 357.

51. S. Un, T. Prisner, R. T. Weber, M. J. Seaman, K. W. Fishbein, A. E. McDermott, D. J. Singel, and R. G. Griffin, *Chem. Phys. Lett.*, 1992, **189**, 54.

52. T. F. Kemp, H. R. W. Dannatt, N. S. Barrow, A. Watts, S. P. Brown, M. E. Newton, and R. Dupree, *J. Magn. Reson.*, 2016, **265**, 77.

53. K. Thurber and R. Tycko, *J. Magn. Reson.*, 2016, **264**, 99.

54. K. L. Felch, B. G. Danly, H. R. Jory, K. E. Kreischer, W. Lawson, B. Levush, and R. J. Temkin, *Proc. IEEE*, 1999, **87**, 752.

55. F. J. Scott, E. P. Saliba, B. J. Albert, N. Alaniva, E. L. Sesti, C. Gao, N. C. Golota, E. J. Choi, A. P. Jagtap, J. J. Wittmann, M. Eckardt, W. Harneit, B. Corzilius, S. T. Sigurdsson, and A. B. Barnes, *J. Magn. Reson.*, 2018, **289**, 45.

56. J. F. Picard, S. C. Schaub, G. Rosenzweig, J. C. Stephens, M. A. Shapiro, and R. J. Temkin, *Appl. Phys. Lett.*, 2019, **114**, 5.

57. E. A. Nanni, S. M. Lewis, M. A. Shapiro, R. G. Griffin, and R. J. Temkin, *Phys. Rev. Lett.*, 2013, **111**, 1.

58. E. A. Nanni, S. Jawla, S. M. Lewis, M. A. Shapiro, and R. J. Temkin, *Appl. Phys. Lett.*, 2017, **111**, 233504.

59. A. V. Soane, M. A. Shapiro, S. Jawla, and R. J. Temkin, *IEEE Trans. Plasma Sci.*, 2017, **45**, 2835.

60. R. A. Wind, R. A. Hall, A. Jurkiewicz, H. Lock, and G. E. Maciel, *J. Magn. Reson. Ser. A*, 1994, **110**, 33.

61. T. Prisner, V. Denysenkov, and D. Sezer, *J. Magn. Reson.*, 2016, **264**, 68.

62. A. A. Nevzorov, S. Milikisiyants, A. N. Marek, and A. I. Smirnov, *J. Magn. Reson.*, 2018, **297**, 113.

63. V. Weis, M. Bennati, M. Rosay, J. A. Bryant, and R. G. Griffin, *J. Magn. Reson.*, 1999, **140**, 293.

64. A. A. Smith, B. Corzilius, J. A. Bryant, R. DeRocher, P. P. Woskov, R. J. Temkin, and R. G. Griffin, *J. Magn. Reson.*, 2012, **223**, 170.

65. D. J. Singel, H. Seidel, R. D. Kendrick, and C. S. Yannoni, *J. Magn. Reson.*, 1989, **81**, 145.

66. M. Nagaraj, T. W. Franks, S. Saeidpour, T. Schubeis, H. Oschkinat, C. Ritter, and B. J. van Rossum, *ChemBioChem*, 2016, 1308.

67. R. R. Ernst, *Angew. Chem. Int. Ed. Engl.*, 1992, **31**, 805.

68. L. Lumata, M. Merritt, C. Khemtong, S. J. Ratnakar, J. van Tol, L. Yu, L. Song, and Z. Kovacs, *RSC Adv.*, 2012, **2**, 12812.

69. O. Haze, B. Corzilius, A. A. Smith, R. G. Griffin, and T. M. Swager, *J. Am. Chem. Soc.*, 2012, **134**, 14287.

70. O. Y. Rogozhnikova, V. G. Vasiliev, T. I. Troitskaya, D. V. Trukhin, T. V. Mikhalina, H. J. Halpern, and V. M. Tormyshev, *Eur. J. Org. Chem.*, 2013, **2013**, 3347.

71. N. S. Dalal, D. E. Kennedy, and C. A. McDowell, *J. Chem. Phys.*, 1974, **61**, 1689.

72. T. Maly, G. T. Debelouchina, V. S. Bajaj, K.-N. Hu, C.-G. Joo, M. L. Mak-Jurkauskas, J. R. Sirigiri, P. C. A. van der Wel, J. Herzfeld, R. J. Temkin, and R. G. Griffin, *J. Chem. Phys.*, 2008, **128**, 052211.

Chapter 4

MAS-DNP Enhancements: Hyperpolarization, Depolarization, and Absolute Sensitivity

Sabine Hediger[1], Daniel Lee[1], Frédéric Mentink-Vigier[2], and Gaël De Paëpe[1]

[1]*Univ. Grenoble Alpes, CEA, CNRS, INAC-MEM, Grenoble, France*
[2]*National High Magnetic Field Laboratory, Florida State University, Tallahassee, FL, USA*

4.1 INTRODUCTION

Since the earliest days of solid-state dynamic nuclear polarization (DNP) and the first experiment by Carver and Slichter on lithium,[1] the ability of this technique to enhance the polarization of nuclear spins has been quantified by comparing the NMR signal obtained under suitable microwave (μwave) irradiation of the electron spins with the same signal in absence of this irradiation. This is a very direct and easy experiment, which immediately reflects the gain in sensitivity obtained on the NMR experiment by making use of the larger

Handbook of High Field Dynamic Nuclear Polarization.
Edited by Vladimir K. Michaelis, Robert G. Griffin, Björn Corzilius and Shimon Vega
© 2020 John Wiley & Sons, Ltd. ISBN: 978-1-119-44164-9
Also published in eMagRes (online edition)
DOI: 10.1002/9780470034590.emrstm1559

electron spin polarization. This DNP enhancement factor, normally denoted ε with various indices (depending on the polarization transfer mechanism, on the targeted isotope, on the permanently on-going developments, etc.) has accompanied all the instrumental and technical developments that have been required to make DNP compatible with experimental conditions of modern high-resolution high-field solid-state NMR.[2-4] Whereas the step-wise evolution to higher magnetic fields impacted each time negatively on the enhancement, as expected from the inverse field dependence of the ubiquitous DNP mechanisms, the enhancement factor, called in this contribution $\varepsilon_{\mathrm{on/off}}$, was able to recover somewhat through improvements to the source of electron spins, known as the polarizing agent (PA). A first move from BDPA and the solid-effect (SE) mechanism[5] to nitroxide radicals in frozen aqueous solutions using the cross effect (CE)[6] led to the introduction of bis-nitroxide PAs[7] and a water soluble version TOTAPOL.[8] With the arrival of the first commercial DNP spectrometer at 9.4 T in 2010,[9] a growing number of research groups have become involved in these modern DNP developments, and values of $\varepsilon_{\mathrm{on/off}}$ have increased regularly with the introduction of new bis-nitroxide radicals or modifications of existing ones,[10-17] designed to improve the DNP efficiency and/or to broaden the applicability of the technique by modifying solubility conditions. At the time of this writing, the 'gold standards' in magic-angle spinning (MAS) DNP under the most common conditions of 9.4 T and 100 K are AMUPol[15] for aqueous frozen solutions and TEKPol[14] for organic solvents, both leading to proton enhancement factors of more than 200 in optimal samples.[16,17] Considering that the theoretical limit for the enhancement of protons via DNP under continuous μwave irradiation is 658, given by the ratio between electron and proton gyromagnetic ratios, these recent experimental results suggest that about 1/3 of the electron polarization could effectively be transferred to the surrounding proton spins.

Going further in the optimization of the DNP process itself, it was shown that the incorporation of dielectric solid particles and removal of dissolved paramagnetic oxygen through degassing of the frozen solution amplifies the DNP enhancement by improving μwave propagation and increasing spin relaxation times, respectively, and $\varepsilon_{on/off}$ of 515 has been observed.[18] Lowering the sample temperature is also known to slow down the relaxation of both the electron and nuclear spins thus improving the transfer of polarization. In this context, a DNP enhancement factor of 677, astonishingly above the 'theoretical limit' of 658, has been reported for MAS-DNP using AMUPol for the CE at a sample temperature of 55 K.[19] This required using a closed-loop recirculating helium system for spinning and cooling.[19,20]

These latter enhancement factors, of course, look very promising at first sight, but they require a critical review. Considering the CE mechanism, which is now the most established for current standard MAS-DNP conditions and used in the aforementioned studies, the theoretical maximum value of 658 could only then be achieved, if one electron spin of the biradical can be completely saturated by the continuous μwave irradiation, without impacting the second electron at all, and the difference of polarization between the two electron spins can entirely be transferred to proton spins. Under MAS, both steps happen successively in so-called rotor events (see the following section)[21–23] corresponding to crossing or anti-crossing of some of the energy levels of the 3-spin system, which are modulated through the MAS. The μwave field strength for electron irradiation has been reported to be <1 MHz for an input power of 5 W at high magnetic field,[24,25] a realistic value for the actual instrumentation. It is of the same order of magnitude as the inverse of the phase memory times T_m (or T_{2e}) reported for bis-nitroxide radicals used in MAS-DNP conditions (around 0.6–4 μs).[14,16,17] (Note: In EPR, the phase memory

time T_m is the characteristic time describing the decay of the electron-spin Hahn-echo intensity with increasing echo time. It corresponds to the refocused transverse relaxation time in solid-state NMR, denoted T_2'. For homogeneously broadened EPR spectra, T_m corresponds to the transverse relaxation time T_{2e}.) However, even if the μwave irradiation is performed in a continuous-wave modus, attempted saturation of single electron spins happens via the short rotor events that are periodically repeated according to the MAS frequency. As such, this saturation process is in competition with the longitudinal electron relaxation T_{1e}, which is on the order of the rotor period for bis-nitroxides at the temperature and spinning frequencies standardly used for MAS-DNP (about 100 K and 5–40 kHz).[14,16] Under such conditions, and even without considering the losses through other nonideal types of rotor events, it is then questionable that full saturation of one electron spin can occur and a hyperpolarization close to 80–100%, as suggested by the above reported record $\varepsilon_{on/off}$ enhancement factors, can experimentally be obtained with hyperpolarization agents known so far. (Note: In dissolution DNP (see Chapter 22), 91% of polarization on ^1H can be achieved at the more favorable conditions of 1.2 K and 6.7 T. The buildup time for the hyperpolarization is however 150 s.[26]) It is, therefore, crucial to understand the origin of the potential discrepancy between $\varepsilon_{on/off}$ and the real polarization obtained in a MAS-DNP experiment. This will be the focus of the following section.

Another perturbing aspect of the enhancement ratio $\varepsilon_{on/off}$ consists in its lack of information about the actual NMR sensitivity from a DNP experiment. This is nicely illustrated in the simple example on glycine by Takahashi *et al.*, which is reproduced in Figure 4.1.[27] Whereas an encouraging $\varepsilon_{on/off}$ of 20 was obtained using TOTAPOL for a frozen solution of 0.1 M [2-^{13}C]-glycine, one of the standard ways to prepare a DNP sample, the sensitivity of the MAS-DNP enhanced signal, defined as the signal-to-noise ratio per unit square root of the experimental time, $(S/N)_{\sqrt{t}}$, was almost 50× less than the sensitivity obtained with standard solid-state MAS-NMR on a powdered sample. A high $\varepsilon_{on/off}$ value may, therefore, be informative on a well-working hyperpolarization mechanism but cannot be used alone to estimate the gain in sensitivity and experimental time one can obtain by using MAS-DNP. Several groups have addressed this problem of quantifying correctly the sensitivity

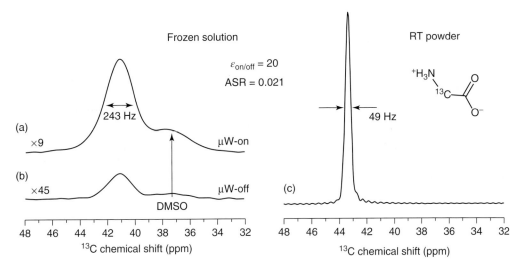

Figure 4.1. ^{13}C CPMAS spectra of [2-^{13}C]-glycine recorded at 9.4 T using a MAS frequency of 8 kHz. (a–b) Frozen solution of 0.1 M [2-^{13}C]-glycine in d_6-DMSO/D$_2$O/H$_2$O (6 : 3 : 1 v/v/v) with 20 mM TOTAPOL, recorded at 105 K with (a) and without (b) µwave irradiation for CE DNP. (c) Powder sample of [2-^{13}C]-glycine recorded at RT using conventional solid-state NMR. ASR (absolute sensitivity ratio) is the ratio of $(S/N)_{\sqrt{t}}$ between the spectra obtained under DNP and standard ssNMR conditions. (Reproduced with permission from Ref. 27. © John Wiley and Sons, 2012)

enhancement of MAS-DNP experiments, and a survey of the different solutions proposed will be presented.

In the jungle of different ways that have been proposed to measure the efficiency of MAS-DNP to enhance NMR spectra, the DNP investigator has good reasons to be lost, not knowing which factor would be the most relevant to assert one's findings in terms of DNP sensitivity. We will, therefore, in the conclusion section go through different types of DNP studies and developments to highlight the corresponding relevant measure of the DNP-enhanced NMR sensitivity.

4.2 THE FORCE OF $\varepsilon_{\text{on/off}}$ AND ITS DARK SIDE

To better understand the dark side of $\varepsilon_{\text{on/off}}$ without going into the deepest details of the CE theory under MAS, we will first consider a simplified thermodynamical picture of the CE mechanism under static conditions, which will then be extended to MAS. Thorough theoretical descriptions of the CE at high magnetic field in static[3,28] and under MAS[21,22] conditions, inclusive of depolarization,[29,30] can be found elsewhere.[31]

4.2.1 CE DNP in Static Samples

Hyperpolarization by CE happens in a three-spin system composed of two coupled electron spins, with respective EPR transition frequencies ω_{e1} and ω_{e2}, and one coupled nuclear spin with Larmor frequency ω_n. The CE condition requires the difference between the two electron-spin frequencies to match the nuclear Larmor frequency, $\omega_{e1} - \omega_{e2} = \pm \omega_n$ (see Figure 4.2a). This condition can be experimentally observed for PAs whose EPR transition is inhomogeneoulsy broadened by the g-anisotropy such that the frequency range of the EPR line is larger than the Larmor frequency of the considered nuclear spin (see Figure 4.2a). Coupled electron spins whose respective g-tensor orientations are such that the CE condition is met can then produce hyperpolarization of the nuclear spin when µwave irradiation is applied to the EPR transition of one of the two electron spins.

Let us consider an arbitrary irradiation at the frequency of ω_{e1} (the same reasoning can be done at ω_{e2}). The populations of the four energy levels of the two-electron spin system (see Figure 4.2b), originally at their Boltzmann equilibrium (denoted ρ_{ij}^{eq} with $i, j = \beta$ or α, the two spin–1/2 states), will be perturbed such to approach population equilibration of the energy levels corresponding to the spin transition of the

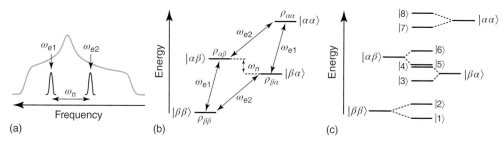

Figure 4.2. (a) Schematic representation of the EPR line of a nitroxide radical, inhomogeneoulsy broadened by the g-anisotropy, with the selection of two crystallite orientations whose respective EPR frequencies ω_{e1} and ω_{e2} are separated by the Larmor frequency of the nuclear spin ω_n. (b) Energy level diagram for the two coupled electron spins of (a). The change in energy due to the dipolar coupling is neglected. The energy levels are labeled according to the spin state α or β of each electron. (c) Energy level diagram for a three-spin system composed of the two previous electron spins coupled to one nuclear spin. The energy levels are numbered from $|1\rangle$ to $|8\rangle$

first electron:

$$\rho_{\alpha\beta} = \frac{1}{2}\left[(2-s)\,\rho_{\alpha\beta}^{eq} + s\rho_{\beta\beta}^{eq}\right]$$

$$\rho_{\alpha\alpha} = \frac{1}{2}\left[(2-s)\,\rho_{\alpha\alpha}^{eq} + s\rho_{\beta\alpha}^{eq}\right]$$

$$\rho_{\beta\beta} = \frac{1}{2}\left[(2-s)\,\rho_{\beta\beta}^{eq} + s\rho_{\alpha\beta}^{eq}\right]$$

$$\rho_{\beta\alpha} = \frac{1}{2}\left[(2-s)\,\rho_{\beta\alpha}^{eq} + s\rho_{\alpha\alpha}^{eq}\right] \quad (4.1)$$

with s, the saturation factor, which takes values between 0 in absence of irradiation and 1 for irradiation able to completely saturate the ω_{e1} transition. The polarization P_{e1} and P_{e2} for the two electron spins, obtained from the energy-levels' population difference is then:

$$P_{e1} = (1-s)P_{e1}^{eq}$$

$$P_{e2} = P_{e2}^{eq} \quad (4.2)$$

with P_{e1}^{eq}, P_{e2}^{eq} the thermal equilibrium Boltzmann polarizations of the two electron spins. Owing to the (partial) saturation of the first electron spin under μwave irradiation, its polarization is reduced toward 0 depending on the saturation factor, whereas the polarization of the second electron is maintained at its thermal equilibrium. If we now consider that the two-electron spins system is coupled to a nuclear spin, each of the four electron energy levels is split into two levels separated by an energy corresponding to ω_n (Figure 4.2c). Their respective populations ρ_{ij} are spread on the two levels with a difference corresponding to the thermal equilibrium nuclear polarization P_n^{eq}. When the CE condition is met, levels $|4\rangle$ and $|5\rangle$, with respective populations of $\rho_4 = \rho_{\beta\alpha} - \frac{P_n^{eq}}{2}$ and $\rho_5 = \rho_{\alpha\beta} + \frac{P_n^{eq}}{2}$ to first approximation, are degenerate, leading to strong state mixing,

with the resulting mixed eigenstates $|4\rangle_m = \frac{1}{\sqrt{2}}(|4\rangle + |5\rangle)$ and $|5\rangle_m = \frac{1}{\sqrt{2}}(|4\rangle - |5\rangle)$. The coupling element between these two levels (composed of the dipolar and hyperfine couplings) will lead to an equilibration of the populations, and the mixed states will end up with the populations

$$\rho_4^{mixed} = \rho_5^{mixed} = \frac{1}{2}(\rho_{\alpha\beta} + \rho_{\beta\alpha}) \quad (4.3)$$

The nuclear spin polarization obtained from the population difference between levels $|3\rangle$ and $|4\rangle_m$ as well as $|5\rangle_m$ and $|6\rangle$ is then (provided that two-spin or three-spin order are not created)

$$P_n^{on} = \frac{1}{2}(P_{e1} - P_{e2} + P_n^{eq}) = \frac{1}{2}(\Delta P_e + P_n^{eq}) \quad (4.4)$$

with P_{e1} and P_{e2} given in equation (4.2). At the CE condition, we can consider that the nuclear polarization is put into contact with the electron polarization difference created by the μwave irradiation. Of course, we have to keep in mind that the electron transition ω_{e1} (or ω_{e2}) is continuously irradiated during the whole process. This maintains the electron polarization difference ΔP_e at the same level despite continuous transfer of polarization to the nuclear spin. P_n^{on} increases, therefore, above the value given in equation (4.4) until it reaches, in the limit of infinite nuclear longitudinal relaxation T_{1n}, the same value as the electron polarization difference (quasi-equilibrium state),

$$P_n^{on} \rightarrow \Delta P_e \quad (4.5)$$

The nuclear spin hyperpolarization is transferred to distant coupled nuclear spins via spin-diffusion.[32,33] This process is, however, not described further here. In absence of irradiation ($s = 0$), the reasoning leading to equation (4.5) still applies as long as the CE condition is met. In this case, the difference in polarization of the two electron spins ΔP_e corresponds to the Boltzmann polarization difference, which is proportional to the difference in frequencies of the two electrons, $\Delta P_e(s = 0) \propto (\omega_{e1} - \omega_{e2})$, and, therefore, equal to the nuclear Boltzmann polarization (as expected for the CE condition). The nuclear spin polarization is thus not modified and

$$P_n^{\text{off}} = \Delta P_e^{\text{eq}} = P_n^{\text{eq}} \qquad (4.6)$$

This apparent trivial result will highlight its importance once MAS is introduced.

In the case of a substantial irradiation, $s > 0$, the difference in electron polarization ΔP_e is negative for irradiation at ω_{e1} and positive for irradiation at ω_{e2}, corresponding to negative and positive enhancement of the nuclear polarization, respectively. In the optimum case of complete saturation, $s = 1$, the nuclear spin polarization will be enhanced up to the electron polarization, $P_n^{\text{on}}(s = 1) = P_e^{\text{eq}}$, leading to the already mentioned upper limit for the enhancement factor:

$$\max\left(\varepsilon_{\text{on/off}}^{\text{static}}\right) = \frac{P_n^{\text{on}}(s = 1)}{P_n^{\text{off}}} = \frac{P_e^{\text{eq}}}{P_n^{\text{eq}}} = \frac{\gamma_e}{\gamma_n} \qquad (4.7)$$

For protons ($n = {}^1$H), this ratio is 658 (Note: In the situation where the polarization of one electron spin could be inverted, instead of saturated, the theoretical maximum enhancement would be 1316. This would require coherent control on the electron spins by i.e. pulsed DNP.[34]). In practice, several factors prevent obtaining the maximal value: obviously an incomplete saturation factor (depends on the μwave power and the homogeneous linewidth, and therefore on T_m) but also the longitudinal electron and nuclear relaxation times T_{1e} and T_{1n}, which continuously fight against the out-of-equilibrium situation encountered under μwave irradiation.

4.2.2 CE DNP in Magic-angle Spinning Samples

Under MAS, the frequencies ω_{e1} and ω_{e2} of the two coupled electron spins (see Figure 4.2a) are not fixed any more, but move across the EPR line, due to the relative reorientation of the respective g-tensors of the electron spins. As a result, the eight energy levels of the two-electrons-one-nucleus spin system are

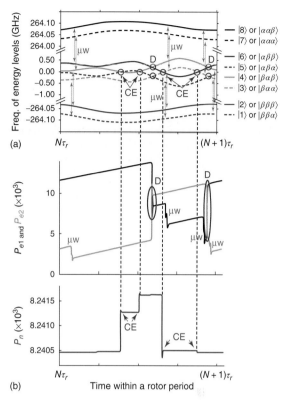

Figure 4.3. (a) Time evolution during a steady-state rotor period of the frequencies of the eight energy levels of Figure 4.2(c) under MAS for one crystallite. μwave rotor events (μw) are indicated by orange vertical arrows, CE and dipolar (D) rotor events by red and blue circles, respectively. Color and line code for the different levels is given next to the figure. (b) Corresponding evolution of the polarization for the first electron spin in black, for the second electron spin in green, and for the nuclear spin in red

modulated with the MAS frequency ω_r, as can be seen in Figure 4.3(a).

During the course of the rotor period, the different energy levels can occasionally and sequentially fulfill some different conditions, called rotor events.[21,22] It is the periodic succession of these discrete rotor events that brings the spin system to a quasi-periodic out-of-equilibrium state. There are four types of rotor events: (i) μwave rotor events, when the resonant frequency of one electron spin matches the μwave frequency, $\omega_{e,i} = \omega_{\mu w}$ ($i = 1$, 2); (ii) CE rotor events, when the electron frequencies match the CE condition $|\omega_{e1} - \omega_{e2}| = \omega_n$; (iii) dipolar events, when the two electron spin frequencies are identical $\omega_{e1} = \omega_{e2}$; (iv) SE events when the SE condition is met $|\omega_{e,i} - \omega_{\mu w}| \approx \omega_n$ (see Figure 4.3).

Rotor events (i) and (ii) are essential to obtain CE hyperpolarization under MAS, and they act exactly the same as in the static thermodynamical picture given above, except that their effect is (normally) sequential. Each time the modulated energy levels experience a µwave rotor event, the polarization of the corresponding electron spin will be decreased by a partial saturation. This modifies the difference in electron polarization, ΔP_e, which later will partially exchange with the nuclear polarization P_n whenever a CE rotor event is encountered. The dipolar coupling (besides the J exchange interaction) between the two electrons leads to an exchange of polarization between the two electrons during the so-called dipolar rotor events (iii). The SE rotor events (iv) will be neglected here as their effect is marginal under conditions chosen to optimize CE for protons using nitroxide-based PAs. The transfer efficiency of all rotor events depends on the effective strength of the active interaction, as well as on the relaxation times and spinning frequency, which all together define the degree of adiabaticity for the fast passage crossings or anticrossings. Thus, dipolar events will entirely exchange polarization, and therefore keep the already built difference in electron polarization, $|\Delta P_e|$, only if the effective dipolar (or J exchange) coupling element is strong enough. If the coupling element is too small, the polarization exchange will be incomplete resulting in a reduction of $|\Delta P_e|$. This is, for example one reason hyperpolarization gets less efficient when the biradical concentration is too high and the intermolecular dipolar coupling to a third electron spin can no longer be neglected.[30,35]

The buildup of nuclear spin polarization under MAS is, therefore, complex and requires proper computational tools to obtain the result of the successive different rotor events for all orientations encountered in a solid sample.[23,35] We now focus on two aspects that have been shown to be relevant for sensitivity considerations: the electron and nuclear spin polarizations in absence of µwave irradiation and the buildup time of nuclear hyperpolarization.

4.2.2.1 Electron and Nuclear Spin Polarization in Absence of µwave Irradiation

In absence of µwave irradiation, which is the situation encountered when the 'off' signal is measured, the energy levels of the spin system are modulated exactly the same way as under irradiation, and all rotor events are active except the µwave rotor events. For instance, at each CE rotor event, the nuclear polarization will still equilibrate with the difference of electron spin polarization. However, under MAS, the thermal electron spin polarizations are (partially) averaged by the MAS rotation toward values proportional to the isotropic frequency of their respective g-tensors. For nitroxides and bis-nitroxides, the two coupled electron spins can, therefore, have very similar polarization under MAS. In absence of irradiation, the difference of polarization $|\Delta P_e|$ can, therefore, become very small and even tend toward 0 when the T_{1e} gets much longer than the rotor period. In such a case, the Boltzmann nuclear polarization can be larger than the difference of electron polarization, and each CE rotor event will lead to a decrease of nuclear polarization to compensate for the smaller difference of electron polarization, until a new quasi-equilibrium is reached, below the nuclear Boltzmann thermal equilibrium. Therefore, under MAS, the 'off' signal reflects a depolarized nuclear state compared to the thermal value measured for static conditions, $P_n^{\text{off,MAS}} = \varepsilon_{\text{depo}} \cdot P_n^{\text{eq}}$, with $\varepsilon_{\text{depo}}$ (also called χ_{depo} in Ref. 36) taking values between 0 and 1 for protons with the biradicals investigated so far. As soon as irradiation is turned 'on', a large difference of electron polarization will be created, allowing for hyperpolarization of the nuclear spins. The traditional $\varepsilon_{\text{on/off}}$ measured as the ratio between the NMR signal intensity with and without µwave irradiation needs, therefore, to be corrected under MAS by the depolarization factor $\varepsilon_{\text{depo}}$ to reflect the real polarization gain with respect to Boltzmann equilibrium, $\varepsilon_B = \varepsilon_{\text{on/off}} \cdot \varepsilon_{\text{depo}}$.

As shown in Ref. 30, nuclear spin depolarization increases for longer T_{1e}, as the averaging of the g-tensor by MAS is then more efficient, and in addition, the recovery of polarization difference after an inefficient dipolar rotor event will be minor (see Figure 4.4b). It explains, at least in part, why AMUPol is found to depolarize more than TOTAPOL.[30] It is important to realize that it is the same CE mechanism that leads to nuclear depolarization in absence of irradiation as for nuclear hyperpolarization when irradiation is turned on. It is, therefore, not astonishing that there is a correlation between the capacity of bis-nitroxide orientations to depolarize and hyperpolarize, as shown in the simulations of Figure 4.4(a). The enhancement ratio $\varepsilon_{\text{on/off}}$ under MAS reflects, therefore, not only the ability of a biradical to hyperpolarize, but the sum of its depolarization and hyperpolarization capacities, with respect to the Boltzmann equilibrium, as sketched

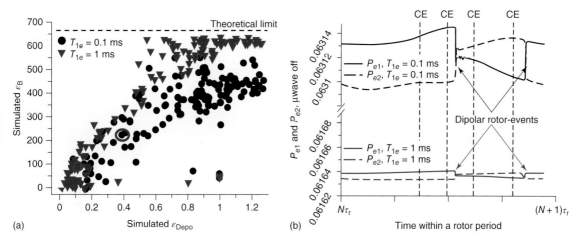

Figure 4.4.   (a) Correlation of the simulated ^{1}H hyperpolarization obtained in presence of μwave irradiation and normalized by Boltzmann thermal equilibrium (simulated ε_{B}), with the simulated ^{1}H depolarization observed in absence of irradiation (simulated ε_{depo}) for 144 different crystallite orientations of a bis-nitroxide, considering two different electron spin relaxation time constants T_{1e} of 0.1 ms (in black) and 1 ms (in red) and a MAS frequency set at 12.5 kHz. (b) Simulation of the electron polarization P_{e1} and P_{e2} (full and dashed lines) over one rotor period τ_{r} after $N\tau_{r} = 8.33$ s of evolution without μwave irradiation (quasi-equilibrium reached) for the crystal orientation circled in blue in (a), using the same color code as in (a). (Reproduced from Ref. 30 with permission from the PCCP Owner Societies)

in Figure 4.5. This renders the quantitative comparison of biradicals for CE DNP on the basis of their $\varepsilon_{on/off}$ alone seriously flawed. The introduction in 2013 of the rigid biradical AMUPol[15] provided a huge jump in $\varepsilon_{on/off}$ of a factor of more than 4 (up to 6 at 10 kHz MAS frequency) compared to TOTAPOL.[30] However, in reality, a longer T_{1e} resulting from a more rigid biradical can enhance depolarization even more than hyperpolarization, and, for AMUPol, the real gain with respect to Boltzmann equilibrium is only up to 3 at 10 kHz MAS frequency and 110 K.[30] Until recently, all studies to improve the efficiency of biradicals were based on $\varepsilon_{on/off}$, and we can, therefore, expect that some PAs may have been wrongly rejected.

Two recent studies have highlighted that efficient biradicals for CE under MAS, which present reduced and even no depolarization, are possible. TEMtriPol-1[38] is a TEMPO-trityl-based mixed biradical. The different nature of the two coupled electron spins is such that their isotropic frequencies differ by about the proton nuclear Larmor frequency. Thanks to this particular spectral signature, the MAS-averaged difference of electron spin polarization in absence of irradiation matches approximatively the ^{1}H Boltzmann equilibrium nuclear spin polarization, thus avoiding depolarization for low radical concentration. Note that the same behavior would be expected for

BDPA-TEMPO biradicals.[39] Despite a much smaller $\varepsilon_{on/off}$, the enhancement ratio with respect to Boltzmann thermal equilibrium, ε_{B}, is only slightly less than for AMUPol at 9.4 T. The main advantage of TEMTriPol-1 is at higher fields where it becomes much more efficient than AMUPol (four times more at 18.8 T[36]). AsymPolPOK[37] is an asymmetric biradical as well, which links together five-membered and six-membered ring nitroxides. It is the first biradical that has been specifically designed to optimize ε_{B}, and not $\varepsilon_{on/off}$. This was achieved by combining these rings by a short, electron rich linker, resulting in an increased dipolar electron interaction and a large J exchange interaction. These strong electron–electron interactions promote fast hyperpolarization while keeping a reduced depolarization.

4.2.2.2   Buildup Time of Hyperpolarized Nuclear Spin Polarization

Except for the special case of single-scan spectroscopy, modern Fourier-transform NMR relies on signal averaging of several transients. The sensitivity of all NMR experiments, therefore, strongly depends on the ability to recover rapidly the initial polarization state after its use in the previous transient. Whereas in

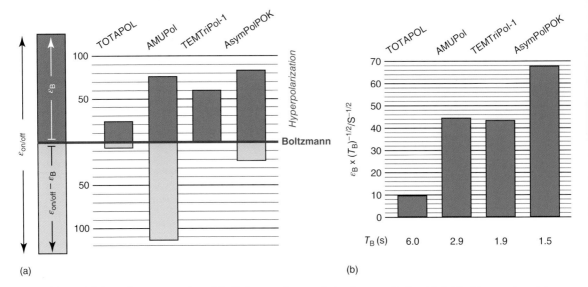

Figure 4.5. (a) Schematic representation of the Boltzmann hyperpolarization contribution (dark blue) to experimental $\varepsilon_{\text{on/off}}$ at 9.4 T, 10 kHz MAS frequency, and 110 K for different water-soluble biradicals: TOTAPOL, AMUPol, TEMTriPol-1, and AsymPolPOK. Note that the size of the light blue bars only represents the discrepancy between $\varepsilon_{\text{on/off}}$ and ε_{B}. It is not proportional to the amount of depolarization, which cannot be larger than the nuclear Boltzmann polarization, $P_n^{\text{eq}} - P_n^{\text{off,MAS}} = (1 - \varepsilon_{\text{depo}}) P_n^{\text{eq}}$. Experimental values for $\varepsilon_{\text{on/off}}$ and ε_{B} (the enhancement ratio with respect to Boltzmann equilibrium) are taken from Ref. 30 for TOTAPOL and AMUPol, Ref. 36 for TEMTriPol-1 and Ref. 37 for AsymPolPOK. The biradical concentration was 12 mM for TOTAPOL and AMUPol, and 10 mM for TEMTriPol-1 and AsymPolPOK, in a standard DNP glassy matrix. Other experimental details can be found in the corresponding references. (b) Schematic representation of the sensiti*vity, express*ed as the enhancement factor ε_{B} divided by the square root of the hyperpolarization buildup time constant T_{B}, for the same biradicals as in (a). Experimental values for T_{B} are given below each bar and are taken from the same references as data for (a)

'standard' liquid- and solid-state NMR experiments the return to equilibrium of nuclear polarization is induced by the incoherent fluctuation of surrounding magnetic interactions and is dictated by the longitudinal spin–lattice relaxation time constant T_{1n}, in experiments using hyperpolarization it will be dominated by the time required to create the hyperpolarized state. Indeed, the best biradical we can imagine, with the highest ε_{B}, will be quite inefficient in terms of sensitivity if the time required to build the hyperpolarized nuclear spin state with this PA is very long. As seen previously, under conditions of MAS-CE, the buildup time constant for nuclear hyperpolarization, T_{B}, depends particularly on the efficiency of the individual rotor events to polarize the coupled nuclear spins and of spin diffusion to transfer this polarization to all further homonuclei in the sample. It is, therefore, based among others on coherent interactions and in this sense cannot be considered as relaxation induced by incoherent averaging. Note that paramagnetic-induced

relaxation can affect T_{1n} of surrounding nuclei and as such only indirectly impact the level of hyperpolarization ε_{B} and the buildup time T_{B}.[35] Owing to the complexity of the CE process in a spin-system large enough to account correctly for homonuclear spin–diffusion as well, the exact T_{B} is very difficult to predict with simulations, although satisfactory trends can still be obtained (see Figure 4.6).[35] In contrast, it can easily be measured with a saturation-recovery experiment. It is interesting and important to note that as the mechanism leading to the depolarized nuclear spin state in absence of μwaves is exactly the same as for hyperpolarization, namely CE and homonuclear spin-diffusion, the same buildup time (or apparent T_1) is observed in the 'off' as in the 'on' experiment (for homogeneously distributed PAs).

The hyperpolarization performance of a radical is thus entirely (and therefore best) characterized by both the real enhancement ratio with respect to Boltzmann equilibrium, ε_{B}, which should be as high as possible,

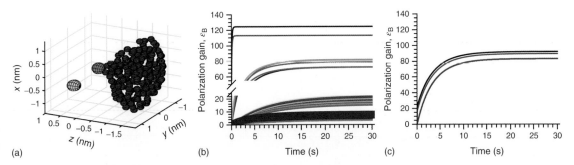

Figure 4.6. (a) Example of the spin system used in the bulk model at the base of simulations in (b) and (c). The orange and yellow spheres correspond to electron 1 and 2, and the blue spheres to 182 nuclear spins. (b,c) Bulk model simulations of the polarization buildup ε_B for the nuclear spins without (b), and with (c) nuclear-dipolar rotor events (i.e. nuclear spin-diffusion). The black curve corresponds to the first proton (the one closest to the electrons), the blue curve to the second proton, and all other colored curves to further protons. In (b), the thick blue curve represents the mean nuclear polarization buildup. In (c), the pink curve represents the common polarization buildup of the further protons. Simulations were performed using the TOTAPOL geometry with $\omega_{1e}/2\pi = 0.85$ MHz, $T_{1e} = 0.3$ ms, $T_{2e} = 1$ ms, hyperfine coupling $= 3$ MHz, $\omega_{\mu w}/2\pi = 263.45$ GHz, $B_0 = 9.394$ T, and $v_r = 8$ kHz. The bulk relaxation time of the nuclear spins was $T_{1n,\text{Bulk}} = 10$ s, the closest proton relaxation time was $T_{1,n1} = 0.15$ s. (Adapted from Ref. 35 with permission from the PCCP Owner Societies)

and the buildup time constant T_B, which needs to be as short as possible. For best comparison, they can be combined in the expression $\varepsilon_B/\sqrt{T_B}$, which respects their relative weight toward experimental sensitivity. As an example, Figure 4.5(b) compares the performances for four different water-soluble biradicals. Whereas the ε_B values are quite similar for AMUPol, TEMTriPol-1 and AsymPolPOK, the efficiency of AsymPolPOK expressed as $\varepsilon_B/\sqrt{T_B}$ is much higher thanks to its very short relative buildup time constant.

It should be noted that the performance of PAs has also been expressed using the 'practical sensitivity gain', $E = \varepsilon_{\text{abs}} \cdot \sqrt{T_{1n,\text{undoped}}/T_B}$,[36,40] with ε_{abs} being the absolute enhancement ratio, which takes into account depolarization and quenching effects due to the presence of paramagnetic species inside the sample, and $T_{1n,\text{undoped}}$, the nuclear longitudinal relaxation time constant measured under identical conditions but without radicals inside the sample. In addition, first to be more complicated to determine and second to exaggerate practical sensitivity gains due to usually large $T_{1n,\text{undoped}}$ at the cryogenic temperatures still required for DNP, it introduces additional factors that do not depend on the nature of the radical itself, but more on its presence (quenching) and on the quality of the glassy state of the sample ($T_{1n,\text{undoped}}$). Moreover, misconceptions will arise when comparing biradicals dissolved in different solvents with dissimilar $T_{1n,\text{undoped}}$. We recommend, therefore, to use

$\varepsilon_B/\sqrt{T_B}$ for the characterization of PAs. Other factors impacting the overall DNP sensitivity will be treated in the following section.

4.3 WHAT ELSE SHOULD WE CONSIDER?

Whereas the nature of the radical will have a direct impact on the sensitivity of the DNP experiment through its efficiency in hyperpolarizing, expressed by the parameters ε_B and T_B, many other factors play a role in the absolute sensitivity of a DNP-enhanced NMR spectrum, defined as the signal-to-noise ratio per the square-root of the experimental time, $(S/N)_{\sqrt{t}}$. All these factors are linked to the particular physical and chemical experimental conditions used to perform DNP experiments. It is important to keep these factors in mind, as the best DNP enhancement can be counterbalanced by unfavorable conditions, such that a study using standard solid-state NMR without DNP may be finally more appropriate.[41] The main experimental factors which strongly impact the sensitivity of DNP experiments are reviewed here.

4.3.1 The Temperature

DNP experiments are standardly performed at temperatures around 100 K. Lowering the temperature produces a gain in sensitivity which is threefold:

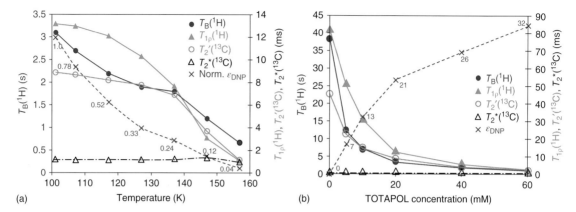

Figure 4.7. Temperature (a) and TOTAPOL concentration (b) dependence of DNP enhancement $\varepsilon_{on/off}$ (red crosses) and different lifetimes (T_B in blue full circles, $T_{1\rho}$ in green full triangles, T_2' in orange open circles, T_2^* in black open triangles) on 2 M ^{13}C-urea in d_6-DMSO/D$_2$O/H$_2$O (6 : 3 : 1 by volume) at 9.4 T. For (a), the TOTAPOL concentration was 20 mM. For (b), the temperature was 105 K. (Reproduced with permission from Ref. 42. © Elsevier, 2014)

an increase of magnetization (Boltzmann distribution), a decrease of the thermal noise due to the cooling of some of the detection devices, and an improved efficiency of the CE DNP mechanism thanks to the slowdown of the spins' relaxation (see Figure 4.7a). However, temperature can have an indirect effect on the spectral linewidths as well, in particular for systems that can present fast and large amplitude motions. Indeed, whereas dynamics at higher temperatures may help to obtain narrow resonances by averaging out structural heterogeneity, decrease of the temperature below the glass transition will freeze the system in a distribution of conformations characterized by inhomogeneously broadened resonances of lower intensity. The opposite effect may occur with an increase in sensitivity at lower temperature for highly flexible sites that are difficult to observe at room temperature. Sample temperature has also an effect on all other spin lifetimes, e.g. T_B, $T_{1\rho}$, or T_2', as can be seen in Figure 4.7(a). This will be commented further below.

4.3.2 The Effective (or Detectable) Sample Amount

In solid-state MAS NMR, the sample is usually in the form of a powder which is directly inserted inside the MAS rotor. The limitation in detectable sample originates solely from the rotor volume, chosen as a function of the available probes, the desired MAS frequency, and the sample availability. For

DNP, several aspects have to be considered. First, the presence of paramagnetic PAs inside the sample broadens and potentially shifts the resonances of the closest neighboring nuclear spins beyond detection. Part of the sample becomes invisible (bleaching/quenching effect), reducing thus the effective amount of sample which gives rise to signal. The amount of bleaching depends on the radical concentration,[40,43–45] the type of PA and of the directly hyperpolarized nuclear spin. Second, the PA is best uniformly distributed inside the sample. This can require specific sample preparation techniques, which may result in a dilution of the sample in the fixed volume of the rotor, compared to standard solid-state NMR. The original sample preparation consisting in the use of a glassy frozen solution[6] is quite inefficient in terms of effective detectable number of spins and sensitivity. It is still used for the evaluation of PA performances or DNP mechanism investigations, but for applications other methods that minimize the dilution of the sample, such as impregnation,[46] film casting,[47] or matrix-free protocols[27,48,49] have been developed and are preferred. It is to note that removal of the solvent in the matrix-free approach may be deleterious for the DNP enhancement through aggregation or phase separation of the hyperpolarizing agent. This can be avoided by the use of a direct or indirect affinity of the PA with the analyte.[27,40,48] In particular, chemical grafting of DNP radicals on the system of interest itself or its specific ligand has been proposed in biomolecular applications to target the signal enhancement (Figure 4.8).[50–58]

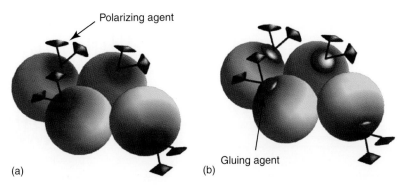

Figure 4.8. Schematic illustration of the two matrix-free preparation strategies using either a direct (a) or an indirect (b) affinity of PAs to the system of interest. Targeted DNP obtained by grafting the PA directly onto the system or to a ligand of the system can be seen as special cases of (a) or (b). (Reproduced from Ref. 48 with permission from The Royal Society of Chemistry)

4.3.3 The Coherence Lifetimes

The sensitivity of an experiment depends on the type of experiment, and more specifically on the lifetime of the coherences present during the course of the pulse sequence, e.g. during mixing times, echoes, *J*-evolution periods, acquisition (see Figure 4.9b), etc. The temperature, as well as the amount of paramagnetic PA used for DNP, has a strong effect on all coherence lifetimes, as can be seen in Figure 4.7.[42] This effect has to be properly investigated in order to choose the optimal PA concentration, which may be different for one type of experiment to the other. For instance, 20 mM TO-TAPOL reduces $^{13}C T_2'$ from 45 to 10 ms for ^{13}C-urea in DMSO/water (see Figure 4.7b). This would reduce the efficiency of a *J*-based homonuclear correlation experiment such as refocused INADEQUATE[60] or SARCOSY[61] by approximately a factor of 2.

4.3.4 Magnetic Field, MAS Frequency, Rotor Size, etc.

Even if the available magnetic fields and MAS probes for DNP are diversifying (see Figure 4.9), the range of possibilities are still much more reduced for DNP compared to conventional solid-state NMR. Notably, rotors of the same size, and therefore of the same theoretical capacity in terms of sample amount, can achieve only reduced MAS frequencies at 100 K due to the change in density of nitrogen gas (Note: A solution to that problem is to use helium gas instead, which however requires recovery of the turbine driving

and cooling gas through a closed-loop system to be sustainable).[19,20] As known from standard NMR, all these instrumental aspects do impact on the sensitivity and resolution of NMR experiments.

The fact that the sensitivity of the DNP experiment is affected by many other experimental parameters in addition to the DNP enhancement has been addressed by various groups, and different ways of correcting $\varepsilon_{on/off}$ (or ε_B) have been proposed to take into account some of these parameters. Thus, a quenching factor and the square-root of the ratio of longitudinal buildup times have been introduced by Rossini et al.,[43] for example, to obtain a global enhancement factor. Similarly, the 'practical sensitivity gain', $E = \varepsilon_{abs} \cdot \sqrt{T_{1n,undoped}/T_B}$,[36,40] takes also into account the change in apparent recovery time and the bleaching effect. The impact of the presence of solvent, the change in temperature and in some coherence lifetimes has been additionally considered by Kobayashi et al. in their global sensitivity enhancement.[62]

Correct identification and estimation of these various contributions for a proper correction of $\varepsilon_{on/off}$ is important to further develop DNP but is very time consuming and impossible to envisage for each application. This is the reason we introduced in 2012 the absolute sensitivity ratio, ASR, which expresses the measured experimental sensitivity gain brought by DNP.[27,41] Instead of correcting $\varepsilon_{on/off}$ for the separate contributions affecting sensitivity, the empirical ASR is obtained by comparing the *S/N* per unit square root of time from a spectrum obtained under optimal DNP conditions with one obtained under optimal standard NMR conditions. The advantage of the ASR

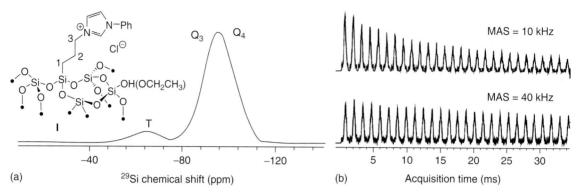

(a) ^{29}Si chemical shift (ppm)   (b)   Acquisition time (ms)

Figure 4.9. (a) DNP-enhanced $\{^1\text{H-}\}^{29}\text{Si}$ CPMAS NMR spectrum of **I** impregnated with 10 mM AMUPol in 90 : 10 $\text{D}_2\text{O}/\text{H}_2\text{O}$, recorded at 18.8 T and ~115 K, using a MAS rate of 40 kHz and CPMG acquisition (summation of 60 whole echoes). (b) Free induction decays of experiment from (a) recorded at 10 and 40 kHz MAS rate. (Source: Chaudhari,[59] http://pubs.rsc.org/en/content/articlehtml/2016/cp/c6cp00839a.

is that it intrinsically takes into account all possible experimental contributions that may impact on the sensitivity of the DNP experiment compared to a reference solid-state NMR experiment, including potential differences in available equipment (magnetic field, type of probe, etc.), with the acquisition of only two 1D spectra. Depending on the study, the standard NMR reference spectrum may be taken under conditions which are in part similar to the DNP spectrum, such as at low temperature, leading to a so-called *reduced ASR*. The disadvantage of the ASR is that sometimes DNP is so effective that a conventional reference spectrum is impossible to acquire.[63,64] The same limitation is obviously present for $\varepsilon_{\text{on/off}}$, ε_{B}, E, etc.

4.4 CONCLUSIONS

Realizing that the goal of MAS-DNP is generally to increase the sensitivity of NMR experiments to address systems and questions that were beyond reach, e.g. detection and correlation of low-gamma and/or low isotopic abundance spins, a proper characterization of the sensitivity increase is, therefore, essential. In the jungle of factors and ratios, DNP enhancement, depolarization, quenching, bleaching, global enhancement, practical, global, overall, or absolute sensitivity, it may be difficult to choose the correct way of doing it. This task is even more difficult considering that all these factors actually contain some information, but none of them is the perfect candidate to choose in any particular case. In addition, some of them are much easier to obtain than others, and we may be tempted

to favor those even if the information content is not the most appropriate. The chosen factor should reflect the correct information content and therefore match the type of investigation.

The DNP enhancement factor $\varepsilon_{\text{on/off}}$, the easiest one to measure, does not give by itself any information about the sensitivity of the experiment. However, it is a good indication to evaluate whether the DNP process is working in a particular sample. Indeed, the DNP efficiency relies very much on the quality of the DNP sample preparation in terms e.g. of glass properties, concentration, and distribution of the hyperpolarizing agent. Progress in sample preparation can easily be followed by $\varepsilon_{\text{on/off}}$.

Investigating the efficiency of a PA, or of different DNP mechanisms, is at the heart of many studies aiming at improving the DNP technique itself in order to broaden its applicability to any kind of systems and conditions (higher magnetic fields, higher or lower temperatures, faster spinning, etc.). In such cases, it is important to take into account the efficiency in both the enhancement and the buildup time. A particular attention has to be paid to potential depolarization effects whose importance is dependent on the radical, the DNP mechanism and the experimental conditions. For such studies, it should be absolutely mandatory to report both the Boltzmann enhancement ε_{B} and the DNP buildup time T_{B}.

For applications, the ASR is very convenient to highlight the pertinence of using DNP-enhanced NMR for experiments that would be very time consuming under standard NMR conditions. It nicely underlines the broadened range of experimental possibilities that are offered by MAS-DNP for the structural investigation of various systems. For example, in the case of functionalized silica nanoparticles, an ASR value of 25 has been reported for the DNP-enhanced ^{29}Si CPMAS experiment, corresponding to an experimental time-saving factor of 625 (ASR^2).[65] This huge gain in sensitivity allowed the acquisition of ^{29}Si–^{29}Si correlation experiment within a few hours, despite the low natural abundance (NA) of ^{29}Si (4.7%). This experiment, which would have been impossible without DNP, was key to understanding the type of organosiloxane polymerization at the surface of the nanoparticles. ^{13}C–^{13}C and ^{15}N–^{13}C correlations at NA are also extremely challenging considering the isotopic abundance of 1.1% for ^{13}C and 0.4 % for ^{15}N. A ^{13}C–^{13}C correlation experiment on cellulose at NA was shown to be possible with DNP in only 20 min instead of several days thanks to an ASR of 47 (time-saving of more than 2000),[27] and the first ^{15}N–^{13}C correlation at NA was demonstrated on a guanosine derivative within 25 h (ASR of more than 10).[66]

MAS-DNP has slowly reached a regime where new methodology can be developed such that it can be used to answer important structural questions. In such cases, the only relevant question is whether the sensitivity of a particular sample is high enough to envisage the use of a certain methodology, without worrying about where the sensitivity is coming from (DNP, temperature, short buildup times, etc.). This requires that authors of new methodologies developed at the limit of the NMR sensitivity take the habit to give an indication of the signal-to-noise ratio per square root of the experimental time $(S/N)/\sqrt{t_{exp}}$ for the system used to demonstrate their methodology. This has not been done so far, but would definitely help scientists to decide on the best strategy to unravel structural information from their system using NMR, including DNP-enhanced NMR. Thus, having access to the original data from the few examples given above, we obtained values for $(S/N)_{\sqrt{t}}$ of 20 s$^{-1/2}$ for the $\{^1$H-$\}^{29}$Si CPMAS spectrum measured on the nanoparticle sample used for the ^{29}Si–^{29}Si correlation experiments of Ref. 65, 130 s$^{-1/2}$ for the $\{^1$H-$\}^{13}$C CPMAS spectrum of the cellulose sample which allowed the acquisition of a ^{13}C–^{13}C correlation experiment in 20 min,[27] 15 s$^{-1/2}$ for the $\{^1$H-$\}^{13}$C CPMAS spectrum of the guanosine sample on which the ^{15}N–^{13}C correlation experiment at NA was demonstrated.[66] As a further example, the measurement of ^{13}C–^{13}C distances at NA that allowed to predict the crystal structure of cyclo-FF

nanotubes[67,68] relied on a $(S/N)_{\sqrt{t}}$ value of 80 s$^{-1/2}$ (measured on the $\{^1$H-$\}^{13}$C CPMAS spectrum).

Even if MAS-DNP has now developed so far that it can be considered as a complementary powerful tool for the structural investigation of a multitude of interesting and complex systems, there is still much room for further improvements of the technique itself. Instrumentation, methodology, and application will all continue to evolve in parallel and the description of their various impacts on the NMR sensitivity will still need to be best described using the appropriate factor. We hope that this overview on DNP enhancements for solid-state NMR has given a good indication of how best to proceed with this in mind.

ACKNOWLEDGMENTS

This work was supported by the French National Research Agency (ANR-16-C81L-0030-03, ANR-12-BS08-0016-01, ANR-11-LABX-0003-01, and RTB) and the European Research Council (ERC-CoG-2015, No. 682895). The National High Magnetic Field Laboratory is supported by National Science Foundation through NSF/DMR-1644779 and the State of Florida.

RELATED ARTICLES IN EMAGRES

Overhauser, Albert W.: Dynamic Nuclear Polarization

Dynamic Nuclear Polarization and High-Resolution NMR of Solids

High-Frequency Dynamic Nuclear Polarization

Pulse EPR

Spin Diffusion in Solids

REFERENCES

1. T. R. Carver and C. P. Slichter, *Phys. Rev.*, 1953, **92**, 212.

2. V. S. Bajaj, C. T. Farrar, M. K. Hornstein, I. Mastovsky, J. Vieregg, J. Bryant, B. Eléna, K. E. Kreischer, R. J. Temkin, and R. G. Griffin, *J. Magn. Reson.*, 2003, **160**, 85.

3. T. Maly, G. T. Debelouchina, V. S. Bajaj, K.-N. Hu, C.-G. Joo, M. L. Mak-Jurkauskas, J. R. Sirigiri, P. C. A. van der Wel, J. Herzfeld, R. J. Temkin, and R. G. Griffin, *J. Chem. Phys.*, 2008, **128**, 52211.

4. R. G. Griffin and T. F. Prisner, *Phys. Chem. Chem. Phys.*, 2010, **12**, 5737.

5. L. R. Becerra, G. J. Gerfen, R. J. Temkin, D. J. Singel, and R. G. Griffin, *Phys. Rev. Lett.*, 1993, **71**, 3561.

6. G. J. Gerfen, L. R. Becerra, D. A. Hall, R. G. Griffin, R. J. Temkin, and D. J. Singel, *J. Chem. Phys.*, 1995, **102**, 9494.

7. K.-N. Hu, H. Yu, T. M. Swager, and R. G. Griffin, *J. Am. Chem. Soc.*, 2004, **126**, 10844.

8. C. Song, K. Hu, C. Joo, T. M. Swager, and R. G. Griffin, *J. Am. Chem. Soc.*, 2006, **128**, 11385.

9. M. Rosay, L. Tometich, S. Pawsey, R. Bader, R. Schauwecker, M. Blank, P. M. Borchard, S. R. Cauffman, K. L. Felch, R. T. Weber, R. J. Temkin, R. G. Griffin, and W. E. Maas, *Phys. Chem. Chem. Phys.*, 2010, **12**, 5850.

10. Y. Matsuki, T. Maly, O. Ouari, H. Karoui, F. Le Moigne, E. Rizzato, S. Lyubenova, J. Herzfeld, T. Prisner, P. Tordo, and R. G. Griffin, *Angew. Chem. Int. Ed.*, 2009, **48**, 4996.

11. E. L. Dane, B. Corzilius, E. Rizzato, P. Stocker, T. Maly, A. A. Smith, R. G. Griffin, O. Ouari, P. Tordo, and T. M. Swager, *J. Org. Chem.*, 2012, **77**, 1789.

12. M. K. Kiesewetter, B. Corzilius, A. A. Smith, R. G. Griffin, and T. M. Swager, *J. Am. Chem. Soc.*, 2012, **134**, 4537.

13. A. Zagdoun, G. Casano, O. Ouari, G. Lapadula, A. J. Rossini, M. Lelli, M. Baffert, D. Gajan, L. Veyre, W. E. Maas, M. Rosay, R. T. Weber, C. Thieuleux, C. Coperet, A. Lesage, P. Tordo, and L. Emsley, *J. Am. Chem. Soc.*, 2012, **134**, 2284.

14. A. Zagdoun, G. Casano, O. Ouari, M. Schwarzwälder, A. J. Rossini, F. Aussenac, M. Yulikov, G. Jeschke, C. Copéret, A. Lesage, P. Tordo, and L. Emsley, *J. Am. Chem. Soc.*, 2013, **135**, 12790.

15. C. Sauvée, M. Rosay, G. Casano, F. Aussenac, R. T. Weber, O. Ouari, and P. Tordo, *Angew. Chem. Int. Ed.*, 2013, **52**, 10858.

16. D. J. Kubicki, G. Casano, M. Schwarzwälder, S. Abel, C. Sauvée, K. Ganesan, M. Yulikov, A. J. Rossini, G. Jeschke, C. Copéret, A. Lesage, P. Tordo, O. Ouari, and L. Emsley, *Chem. Sci.*, 2016, **7**, 550.

17. C. Sauvée, G. Casano, S. Abel, A. Rockenbauer, D. Akhmetzyanov, H. Karoui, D. Siri, F. Aussenac, W. Maas, R. T. Weber, T. Prisner, M. Rosay, P. Tordo, and O. Ouari, *Chem. A Eur. J.*, 2016, **22**, 5598.

18. D. J. Kubicki, A. J. Rossini, A. Purea, A. Zagdoun, O. Ouari, P. Tordo, F. Engelke, A. Lesage, and L. Emsley, *J. Am. Chem. Soc.*, 2014, **136**, 15711.

19. E. Bouleau, P. Saint-Bonnet, F. Mentink-Vigier, H. Takahashi, J.-F. Jacquot, M. Bardet, F. Aussenac, A. Purea, F. Engelke, S. Hediger, D. Lee, and G. De Paëpe, *Chem. Sci.*, 2015, **6**, 6806.

20. D. Lee, E. Bouleau, P. Saint-Bonnet, S. Hediger, and G. De Paëpe, *J. Magn. Reson.*, 2016, **264**, 116.

21. K. R. Thurber and R. Tycko, *J. Chem. Phys.*, 2012, **137**, 84508.

22. F. Mentink-Vigier, U. Akbey, Y. Hovav, S. Vega, H. Oschkinat, and A. Feintuch, *J. Magn. Reson.*, 2012, **224**, 13.

23. F. Mentink-Vigier, Ü. Akbey, H. Oschkinat, S. Vega, and A. Feintuch, *J. Magn. Reson.*, 2015, **258**, 102.

24. E. A. Nanni, A. B. Barnes, Y. Matsuki, P. P. Woskov, B. Corzilius, R. G. Griffin, and R. J. Temkin, *J. Magn. Reson.*, 2011, **210**, 16.

25. E. P. Saliba, E. L. Sesti, F. J. Scott, B. J. Albert, E. J. Choi, N. Alaniva, C. Gao, and A. B. Barnes, *J. Am. Chem. Soc.*, 2017, **139**, 6310.

26. A. Bornet and S. Jannin, *J. Magn. Reson.*, 2016, **264**, 13.

27. H. Takahashi, D. Lee, L. Dubois, M. Bardet, S. Hediger, and G. De Paëpe, *Angew. Chem. Int. Ed.*, 2012, **51**, 11766.

28. Y. Hovav, O. Levinkron, A. Feintuch, and S. Vega, *Appl. Magn. Reson.*, 2012, **43**, 21.

29. K. R. Thurber and R. Tycko, *J. Chem. Phys.*, 2014, **140**, 184201.

30. F. Mentink-Vigier, S. Paul, D. Lee, A. Feintuch, S. Hediger, S. Vega, and G. De Paëpe, *Phys. Chem. Chem. Phys.*, 2015, **17**, 21824.

31. A. S. Lilly Thankamony, J. J. Wittmann, M. Kaushik, and B. Corzilius, *Prog. Nucl. Magn. Reson. Spectrosc.*, 2017, **102–103**, 120.

32. Y. Hovav, A. Feintuch, and S. Vega, *J. Magn. Reson.*, 2012, **214**, 29.

33. D. Wiśniewski, A. Karabanov, I. Lesanovsky, and W. Köckenberger, *J. Magn. Reson.*, 2016, **264**, 30.

34. S. Un, T. Prisner, R. T. Weber, M. J. Scaman, K. W. Fishbein, A. E. McDermott, D. J. Singel, and R. G. Griffin, *Chem. Phys. Lett.*, 1992, **189**, 54.

35. F. Mentink-Vigier, S. Vega, and G. De Paëpe, *Phys. Chem. Chem. Phys.*, 2017, **19**, 3506.

36. F. Mentink-Vigier, G. Mathies, Y. Liu, A.-L. Barra, M. A. Caporini, D. Lee, S. Hediger, R. G. Griffin, and G. De Paëpe, *Chem. Sci.*, 2017, **8**, 8150.

37. F. Mentink-Vigier, I. Marin-Montesinos, A. P. Jagtap, J. Van Tol, S. Hediger, D. Lee, S. T. Sigurdsson, and G. De Paëpe, *J. Am. Chem. Soc.*, 2018, **140**, 11013.

38. G. Mathies, M. A. Caporini, V. K. Michaelis, Y. Liu, K.-N. Hu, D. Mance, J. L. Zweier, M. Rosay, M. Baldus, and R. G. Griffin, *Angew. Chem.*, 2015, **127**, 11936.

39. E. L. Dane, T. Maly, G. T. Debelouchina, R. G. Griffin, and T. M. Swager, *Org. Lett.*, 2009, **11**, 1871.

40. B. Corzilius, L. B. Andreas, A. A. Smith, Q. Z. Ni, and R. G. Griffin, *J. Magn. Reson.*, 2014, **240**, 113.

41. D. Lee, S. Hediger, and G. De Paëpe, *Solid State Nucl. Magn. Reson.*, 2015, **66–67**, 6.

42. H. Takahashi, C. Fernández-de-Alba, D. Lee, V. Maurel, S. Gambarelli, M. Bardet, S. Hediger, A.-L. Barra, and G. De Paëpe, *J. Magn. Reson.*, 2014, **239**, 91.

43. A. J. Rossini, A. Zagdoun, M. Lelli, D. Gajan, F. Rascón, M. Rosay, W. E. Maas, C. Copéret, A. Lesage, and L. Emsley, *Chem. Sci.*, 2012, **3**, 108.

44. S. Lange, A. H. Linden, Ü. Akbey, W. Trent Franks, N. M. Loening, B.-J. van Rossum, and H. Oschkinat, *J. Magn. Reson.*, 2012, **216**, 209.

45. R. Rogawski, I. V. Sergeyev, Y. Zhang, T. H. Tran, Y. Li, L. Tong, and A. E. McDermott, *J. Phys. Chem. B*, 2017, **121**, 10770.

46. A. Lesage, M. Lelli, D. Gajan, M. A. Caporini, V. Vitzthum, P. Miéville, J. Alauzun, A. Roussey, C. Thieuleux, A. Mehdi, G. Bodenhausen, C. Coperet, and L. Emsley, *J. Am. Chem. Soc.*, 2010, **132**, 15459.

47. D. Le, G. Casano, T. N. T. Phan, F. Ziarelli, O. Ouari, F. Aussenac, P. Thureau, G. Mollica, D. Gigmes, P. Tordo, and S. Viel, *Macromolecules*, 2014, **47**, 3909.

48. H. Takahashi, S. Hediger, and G. De Paëpe, *Chem. Commun.*, 2013, **49**, 9479.

49. C. Fernández-de-Alba, H. Takahashi, A. Richard, Y. Chenavier, L. Dubois, V. Maurel, D. Lee, S. Hediger, and G. De Paëpe, *Chem. A Eur. J.*, 2015, **21**, 4512.

50. V. Vitzthum, F. Borcard, S. Jannin, M. Morin, P. Miéville, M. A. Caporini, A. Sienkiewicz, S. Gerber-Lemaire, and G. Bodenhausen, *ChemPhysChem*, 2011, **12**, 2929.

51. T. Maly, D. Cui, R. G. Griffin, and A. F. Miller, *J. Phys. Chem. B*, 2012, **116**, 7055.

52. A. N. Smith, M. A. Caporini, G. E. Fanucci, and J. R. Long, *Angew. Chem. Int. Ed.*, 2015, **54**, 1542.

53. M. Kaushik, T. Bahrenberg, T. V. Can, M. A. Caporini, R. Silvers, J. Heiliger, A. A. Smith, H. Schwalbe, R. G. Griffin, and B. Corzilius, *Phys. Chem. Chem. Phys.*, 2016, **18**, 27205.

54. M. A. Voinov, D. B. Good, M. E. Ward, S. Milikisiyants, A. Marek, M. A. Caporini, M. Rosay, R. A. Munro, M. Ljumovic, L. S. Brown, V. Ladizhansky, and A. I. Smirnov, *J. Phys. Chem. B*, 2015, **119**, 10180.

55. E. A. W. van der Cruijsen, E. J. Koers, C. Sauvée, R. E. Hulse, M. Weingarth, O. Ouari, E. Perozo, P. Tordo, and M. Baldus, *Chem. A Eur. J.*, 2015, **21**, 12971.

56. B. J. Wylie, B. G. Dzikovski, S. Pawsey, M. Caporini, M. Rosay, J. H. Freed, and A. E. McDermott, *J. Biomol. NMR*, 2015, **61**, 361.

57. T. Viennet, A. Viegas, A. Kuepper, S. Arens, V. Gelev, O. Petrov, T. N. Grossmann, H. Heise, and M. Etzkorn, *Angew. Chem. Int. Ed.*, 2016, **55**, 10746.

58. R. Rogawski and A. E. McDermott, *Arch. Biochem. Biophys.*, 2017, **628**, 102.

59. S. R. Chaudhari, P. Berruyer, D. Gajan, C. Reiter, F. Engelke, D. L. Silverio, C. Copéret, M. Lelli, A. Lesage, and L. Emsley, *Phys. Chem. Chem. Phys.*, 2016, **18**, 10616.

60. A. Lesage, M. Bardet, and L. Emsley, *J. Am. Chem. Soc.*, 1999, **121**, 10987.

61. D. Lee, J. Struppe, D. W. Elliott, L. J. Mueller, and J. J. Titman, *Phys. Chem. Chem. Phys.*, 2009, **11**, 3547.

62. T. Kobayashi, O. Lafon, A. S. Lilly Thankamony, I. I. Slowing, K. Kandel, D. Carnevale, V. Vitzthum, H. Vezin, J.-P. Amoureux, G. Bodenhausen, and M. Pruski, *Phys. Chem. Chem. Phys.*, 2013, **15**, 5553.

63. F. Blanc, L. Sperrin, D. Lee, R. Dervişoğlu, Y. Yamazaki, S. M. Haile, G. De Paëpe, and C. P. Grey, *J. Phys. Chem. Lett.*, 2014, **5**, 2431.

64. D. Lee, C. Leroy, C. Crevant, L. Bonhomme-Coury, F. Babonneau, D. Laurencin, C. Bonhomme, and G. De Paëpe, *Nat. Commun.*, 2017, **8**, 14104.

65. D. Lee, G. Monin, N. T. Duong, I. Z. Lopez, M. Bardet, V. Mareau, L. Gonon, and G. De Paëpe, *J. Am. Chem. Soc.*, 2014, **136**, 13781.

66. K. Märker, M. Pingret, J. M. Mouesca, D. Gasparutto, S. Hediger, and G. De Paëpe, *J. Am. Chem. Soc.*, 2015, **137**, 13796.

67. K. Märker, S. Paul, C. Fernández-de-Alba, D. Lee, J.-M. Mouesca, S. Hediger, and G. De Paëpe, *Chem. Sci.*, 2017, **8**, 974.

68. K. Märker, Atomic-level structure determination of organic assemblies by dynamic nuclear polarization enhanced solid-state NMR, PhD thesis from Univ. Grenoble Alpes, 2017.

Chapter 5

Polarizing Agents: Evolution and Outlook in Free Radical Development for DNP

Gilles Casano, Hakim Karoui, and Olivier Ouari

CNRS, ICR, Aix Marseille University, Marseille, France

5.1 INTRODUCTION

Dynamic nuclear polarization (DNP) is a technique that relies on the transfer of Boltzmann polarization between unpaired electron and nuclear spins induced by microwave irradiation. Due to electron spins being 660 times more polarized than protons, an nuclear magnetic resonance (NMR) signal enhancement of several orders of magnitudes can be expected, solving (at least partially) the long-lasting limitation of NMR sensitivity.

DNP technique has now a long and rich history since its inception in the early 1950s, with a landmark renaissance for the broad scientific community in the 1990s with the use of high-frequency/high-power microwave sources by Griffin and collaborators.[1,2] This seminal work demonstrated the successful use of DNP/solid-state nuclear magnetic resonance (ssNMR) technique at high magnetic fields in studies where sensitivity was a critical issue, highlighting the potential of the technique and opening new areas of applications and new perspectives. In 2003, Ardenkjaer-Larsen and collaborators introduced dissolution DNP for MRI applications.[3] Quickly, this renaissance received a good deal of interest and stimulated new developments and other approaches (shuttle DNP, high-frequency liquid DNP) in the magnetic resonance community but also in materials science and in metabolic/clinical imaging.[4] Rapidly, the polarizing agents (PAs) have shown to be an essential ingredient for successful DNP experiments, each DNP approach requiring a type of paramagnetic PA fulfilling criteria ruled by the related polarizing mechanism. Here, a special focus is given to the design and development of PAs for magic angle spinning (MAS) ssNMR/DNP. For a discussion on PAs of other DNP techniques, excellent chapters and reviews are available in this handbook but also in the literature for dissolution DNP,[5–11] high-field liquid DNP,[12,13] and Shuttle DNP.[14,15]

Handbook of High Field Dynamic Nuclear Polarization.
Edited by Vladimir K. Michaelis, Robert G. Griffin, Björn Corzilius and Shimon Vega
© 2020 John Wiley & Sons, Ltd. ISBN: 978-1-119-44164-9
Also published in eMagRes (online edition)
DOI: 10.1002/9780470034590.emrstm1547

5.2 PRINCIPLE

In 1953, Overhauser[16] predicted that a polarization transfer can occur between the conduction electrons

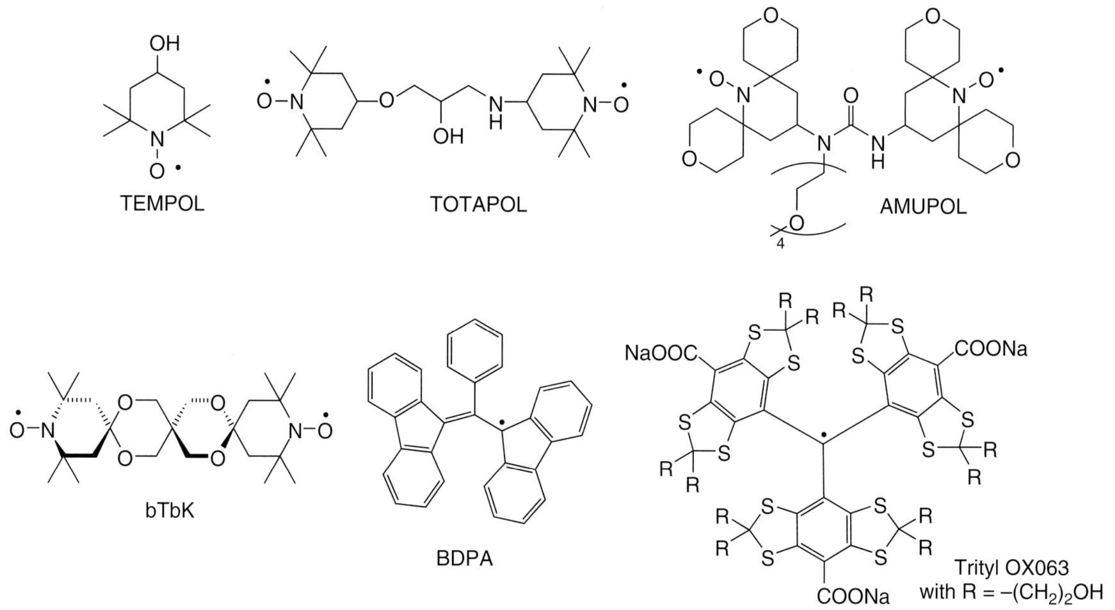

Figure 5.1. Structures of polarizing agents

and the nuclei in a metal and shortly after Slichter and Carver[17] demonstrated the phenomenon on a powdered Li sample, with an enhancement of the ^{7}Li NMR signal of nearly 100. However, such an experiment was possible only in systems containing conduction electrons having hyperfine interaction with nuclear spins, such as metals. In the following years, DNP technique was used in nuclear physics to produce highly polarized solid targets at low temperatures for neutron scattering, revealing the multiple possible polarizing transfer mechanisms.

In the decades that followed, samples doped with a wide range of paramagnetic species (porphyrexide, TEMPOL, BDPA, trityls, Frémy's salt, chromium (III), gadolinium (III) and manganese (II) complexes, optically excited pentacene and fullerene derivatives, UV photo-generated radicals, homo- and heterobiradicals, diamonds, etc.) as PAs have demonstrated to produce proton or carbon NMR signal enhancement, highlighting the fabulous potential of DNP for NMR and MRI applications.

Collaborative efforts in experimental and theoretical studies moved DNP far beyond the proof of concept and enabled to push the DNP factors of NMR signal enhancement every time closer to the theoretical maximum value, defined as the ratio between the gyromagnetic ratios of the electron spins and the

nuclear spin (γ_e/γ_n), i.e., nearly 660 for ^{1}H and 2620 for ^{13}C.

Two types of samples can be investigated by DNP. The polarizing source (unpaired electron) is endogenous to the sample, such as in flavodoxin protein,[18] in RNA complex with Mn(II),[19] in diamonds,[20] and in thermally carbonized porous silicon.[21] The second case relies on doping the sample with stable paramagnetic species (Figure 5.1); this alternative has quickly shown to be very efficient, enabling the study of numerous diamagnetic systems and expanding considerably the range of DNP applications and NMR possibilities. In the latter case, the PAs are usually divided into two classes depending of their EPR properties, such as the homogeneous linewidth (δ), the inhomogeneous breadth of their EPR spectrum (Δ), and g-anisotropy.

Several possible polarizing mechanisms (solid effect (SE), cross effect (CE), thermal mixing (TM), and Overhauser effect) can be active and their efficiency will depend on parameters dictated by the experimental approach, such as temperature, sample rotation, magnetic field, microwave power, continuous wave or pulsed microwave, and type of nucleus to polarize.[22-24] The design of PAs will have to fulfill specific criteria for each mechanism in order to take the full benefit of the DNP process.

The SE is a two-spin process, involving one electron spin and one nuclear spin. The polarization transfer relies on the microwave excitation of forbidden electron–nuclear transitions (zero- and double-quantum coherences, ZQ and DQ, respectively) that are partially allowed by the mixing of spin states induced by anisotropic hyperfine interactions. The matching conditions are satisfied when the frequency of the microwave irradiation corresponds to the sum or difference of the electron and nuclear Larmor frequencies ($\omega_{\mu w} = \omega_{0S} \pm \omega_{0I}$). Because ZQ and DQ transitions lead to opposite nuclear enhancement and cancellation of the SE when excited equally, the PA has to satisfy specific EPR requirements. Thus, PAs exhibiting a narrow EPR line and a weak *g*-anisotropy relative to the nuclear Larmor frequency are the most efficient agents for SE, meaning a small homogeneous linewidth (δ) and inhomogeneous breadth of their EPR spectrum (Δ) in order to satisfy $\delta, \Delta < 2 \, \omega_{0I}$. Carbon-centered radicals such as trityl-based and BDPA have been mostly considered for SE DNP experiments.

The CE can be modeled as a three-spin process, involving two coupled electron spins and a nuclear spin. This effect is based on allowed transitions. A peculiar set of conditions has to be satisfied for an efficient CE, with the difference of the Larmor frequency of the two electron spins matching the nuclear Larmor frequency ($\omega_{0I} = \omega_{0S2} - \omega_{0S1}$) while the microwave irradiation frequency saturating one of the two electron spins ($\omega_{\mu w} = \omega_{0S1}$ or ω_{0S2}). A pair of paramagnetic species fulfilling the CE matching conditions and with isotropic Zeeman interactions and small breadth of the EPR spectrum relative to the nuclear Larmor frequency would be an ideal system ($\delta, \Delta > \omega_{0I}$). Unfortunately, no such system is available at the moment. However, nitroxides have shown to be good candidates despite their broad EPR line due to their large *g*-anisotropy, enabling to satisfy the CE matching conditions.[25,26] A further important point for the CE process arises when the sample is not static but under MAS. As nicely described by Thurber *et al.* and then by Mentink-Vigier *et al.*, the spin system can periodically undergo level anti-crossing (LAC) events, where polarization can be exchanged between the spins, leading to a MAS-induced depolarization process.[27–29]

The TM mechanism is mainly active at very low temperatures (<5 K) and involved a strongly coupled electron and nuclear spin system and slow relaxation.[30,31] The unpaired electrons are usually added at high concentrations to the sample as monoradical, mostly derived from tetrathiatriarymethyl (TAM) or TEMPO radicals. The EPR spectrum of the PAs has to be homogeneously broadened, satisfying $\delta, \Delta > \omega_{0I}$.

The Overhauser effect is a polarizing process well described for liquids and conductive solids where fluctuations of hyperfine interactions induce electron–nuclear cross-relaxation, but its existence in insulating solids has been surprisingly observed too.[32–34] Even if moderate DNP enhancement factors of around 80 have been obtained so far via OE ssNMR/DNP at 9, its unique field dependence displaying larger signal enhancement at higher MAS rate and higher magnetic fields opens promising perspectives for DNP at 18.8 T and above.[35] Fluctuations of hyperfine interactions at a frequency close to the Larmor frequency in BDPA have been rationalized to be responsible for the OE in insulating solids.[34,35] The nature of the fluctuations inducing OE in a solid sample is still unclear.

5.2.1 Absolute Sensitivity Ratio, Bleaching/Quenching Effect, and Depolarization Process

The quantification of the efficiency of MAS ssNMR/DNP is usually described through the DNP enhancement factor (ε, defined as the ratio of signal intensities of spectra acquired with and without microwave irradiation). However, it is clear now that ε is not sufficient to describe the performances of a PA. Many experimental parameters have shown to have a significant effect on the DNP factors, such as sample temperature, magnetic field, microwave power, glass-forming solvent, MAS rate, and PA concentration. Several authors proposed different methods to quantify the overall/absolute sensitivity ratio, taking into account the bleaching effect, the acceleration factor, the Boltzmann factor, the cross-polarization step, and line broadening.[36–38]

However, the sample is modified by the introduction of a radical-containing solution. The introduction of exogenous radicals to polarize the sample will therefore affect the sensitivity with respect to a conventional NMR experiment, especially for CE DNP under MAS. In fact, CE does not require microwave irradiation to transfer polarization between electron and nuclear spins. Under MAS, the nuclear polarization and the difference in electron polarization between the two electron spins fulfilling the CE matching conditions are strongly linked. This phenomenon causes CE-induced depolarization under MAS in the absence of microwave irradiation,[27,29] and can produce overestimated DNP factors because the microwave-on spectrum is compared to a microwave-off spectrum that is not at thermal equilibrium because of the presence of the electron spin pairs satisfying the CE matching conditions. At the opposite, in the case of paramagnetic bleaching/quenching, the NMR signal intensity is partially reduced even though with large DNP factors can be measured, because of the reduced number of observable nuclear spins. Interestingly, the depolarization process has shown to be present for dinitroxide PAs[29,39] but absent for heterobiradicals CE PAs.[40]

5.3 POLARIZING AGENTS (PAs) FOR MAS ssNMR/DNP

The design of PAs for MAS ssNMR/DNP is a growing area of research and tremendous progress has been made in the last few years. Taking all other parameters to be the same (9.4 T, 100 K, 10 KHz MAS, 10 mM in DNP juice), the DNP performances were improved by more than 10 from the first-used PA (TEMPOL, $\varepsilon = 20$) to the most efficient at the moment (AMUPOL, $\varepsilon = 240$, Figure 5.1). As the time of signal acquisition scales inversely with the square of the enhancement factor (ε), tremendous time savings and previously unfeasible NMR studies have been achieved. Additionally, lowering the sample temperature to 55 K allowed reaching a signal enhancement of 677 with AMUPOL at 9.4 T and 10 kHz, which matches the theoretical maximum value.[41] Interestingly, DNP factors up to 500 have been obtained with TEKPOL (Figure 5.3) at 100 K in mixtures containing dielectric particles such as sapphire microparticles.[42]

PAs are usually added to the sample in the form of an exogenous mono- or biradical (Figure 5.1) dissolved in a glass-forming solvent mixture, mostly glycerol-d_8/D$_2$O/H$_2$O mixture 60/30/10 (DNP juice) and 1,1,2,2-tetrachloroethylene (TCE), and the experiment is performed with MAS at temperatures of about 100 K. The solvent matrix is highly deuterated with reduced proton abundance (ca 10%) to optimize signal enhancement factors and transfer of polarization to the analyte through a spin-diffusion process. In this experimental approach, the DNP experiment is usually driven by the CE and the SE. The PAs discussed below will be classified according to these two polarizing

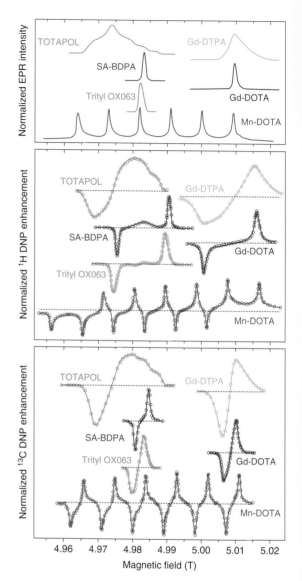

Figure 5.2. Comparison of the EPR spectra and the corresponding ^{1}H and ^{13}C DNP field profiles for different PAs discussed in the text. (Source: Kaushik,[43] http://pubs.rsc.org/en/content/articlehtml/2016/cp/c6cp04623a. Licensed under CC BY 3.0)

transfer mechanisms (CE or SE), and subclassified depending on their EPR properties, including their homogeneous linewidth, the inhomogeneous breadth of their EPR spectrum, their *g*-anisotropy, and their Larmor frequency (Figure 5.2).

5.3.1 Nitroxides, Dinitroxides, and Heterobiradicals for CE MAS ssNMR/DNP

Currently, the most efficient DNP mechanism in MAS ssNMR/DNP at 100 K is the CE. As discussed above, an ideal PA for CE DNP would be two EPR narrow-line radicals whose Larmor frequencies are separated by the nuclear Larmor frequency. However, such a system is not available. Another option is to have a radical with an inhomogeneous linewidth larger than the nuclear Larmor frequency. Nitroxide radicals such as TEMPO present g-anisotropy on the order of the ^{1}H Larmor frequency, making them efficient PA to polarize protons. The principal values of the g tensor are $g_{xx} = 2.0090$, $g_{yy} = 2.0062$, $g_{zz} = 2.0025$, and the difference between g_{xx} and g_{yy} or between g_{yy} and g_{zz} satisfies ideally the CE matching conditions whatever the magnetic field is applied (Figure 5.3). As the anisotropy of nitroxides scales with the field, at higher fields the adiabaticity of microwave excitation events under MAS is reduced due to larger spread in frequency during the rotor cycle.[27,44–46]

Polarizing proton nuclei offers several benefits in DNP experiments. Protons are very common nuclei, they are good nuclei for cross-polarization experiments enabling to detect low gamma nuclei, they exhibit relatively short longitudinal relaxation times relative to other nuclei, and proton spin diffusion is efficient to generate homogeneous polarization through the sample.

5.3.1.1 Nitroxides, Dinitroxides, and Polynitroxides

There has been much interest in improving the existing PAs to make the CE transfer more efficient. A series of key steps have been reached to this end. After the seminal experiment using TEMPOL in ssNMR/DNP,[1] the following step was to fulfill the required electron–electron coupling by switching from a high concentration of monomeric nitroxides to tethered dinitroxides. The study of the bTnE series (Figure 5.4), in which the length of the tether was changed, demonstrated that the size of the electron–electron dipolar coupling drives the CE efficiency and that fixing the distance between the two unpaired electrons by a tether is crucial (Figure 5.4).[47] Solubility in an aqueous glass-forming matrix is one of the multiple criteria required for an efficient PA,

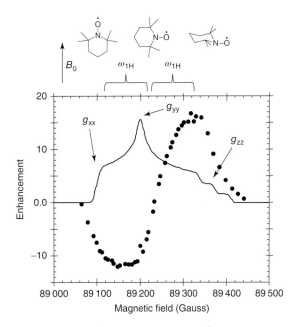

Figure 5.3. Dependence of the DNP ^{1}H enhancement on principal electron resonance values superimposed with TEMPO EPR spectrum in frozen solution. Bottom: Orientation of TEMPO radical at principal electron resonance values (g_{xx}, g_{yy}, g_{zz}) relative to applied magnetic field

and the limited solubility of BT2E dinitroxide in a glycerol/water mixture led to the design of TOTAPOL that has been used in numerous DNP studies.[48]

TOTAPOL-based tri (DOTOPA-TEMPO, DOTOPA-4OH) and tetraradicals[49–51] (Figure 5.5) have been prepared to evaluate the possible CE improvement by adding interacting unpaired electrons spatially close to the one saturated by the microwave irradiation. Their evaluation has been mainly performed in low microwave power, low temperatures (25–30 K), and static and MAS ssNMR/DNP experiments at 9.4 T. Best results were obtained with triradical derivatives, possibly due to too fast spectral diffusion process limiting electron saturation in tetraradicals. DNP performances were improved relative to TOTAPOL at 25–30 K, but the trend was opposite at higher temperatures (100 K) with larger microwave power.

At a point, it has been assumed that the CE matching conditions were obtained for the largest number of orientations of the PA when the two nitroxide moieties are locked in an orthogonal mutual orientation. As a result, the dinitroxide bTbK (Figure 5.1) was

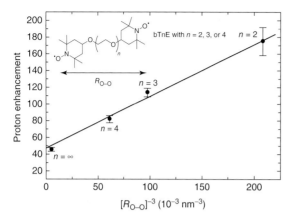

Figure 5.4. ^{1}H DNP factors in the bTnE series as a function of dipole coupling estimated via electron–electron distance $(R_{O-O})^{-3}$. (Reprinted with permission from K. N. Hu, H. H. Yu, T. M. Swager, R. G. Griffin, J. Am. Chem. Soc., 2004, 126, 10844. Copyright (2004) American Chemical Society)

designed where the two TEMPO moieties are held orthogonal by a rigid bisketal tether, leading to the situation where the y- and z-axes of the g-tensor of electrons 1 and 2, respectively, are nearly parallel.[52] DNP enhancement (ε) was improved by a factor of 1.4 relative to TOTAPOL under identical conditions in a DMSO/water mixture. Electron–electron dipolar interactions (around 25–30 MHz) in bTbK and TOTAPOL were measured to be very similar, but the rigid tether in bTbK possibly contributes also to its higher efficiency. Remarkably, when the orientation of the axes of the g-tensors of the two nitroxide moieties does not fulfill the CE matching conditions, the DNP factors are low. This is illustrated with bTOxa and dCdO dinitroxides (Figure 5.5) where the axes are collinear and the g_{xx} component of one TEMPO is parallel to the g_{zz} component of the other, respectively.[53–55]

The limited solubility of bTbK in aqueous glass-forming solvent mixture prompted to the search of water-soluble bTbK derivative, either by introducing polar groups on the skeleton,[56,57] or by making a host guest complex with a β-cyclodextrin derivative.[58] Interestingly, aqueous glassy matrices were formed in the presence of micelles, and good DNP enhancements were obtained using water-insoluble PAs in the presence of deuterated sodium octylsulfate[59] and non-deuterated Tween-80 surfactants.[60]

Beyond the geometry, the electron relaxation properties are another key factor affecting the DNP process. The CE process depends on the saturation of

the electron spins in resonance with the microwave irradiation, with longer the electron relaxation time, higher the DNP efficiency. Capitalizing on the structure of bTbK, a twofold longer T_{1e} relaxing derivative (bCTbK, Figure 5.5) has been prepared by removing the methyl groups in β-position of the nitroxide and replacing them by larger spirocyclohexyl groups.[61] The DNP performance of bCTbK in 1,1,2,2-tetrachloroethane (TCE) was nearly two times higher than bTbK under identical conditions.

The impact of electron relaxation times on DNP factors has been examined in a series of bCTbK derivatives (Figure 5.6), and demonstrated that longer inversion recovery and phase memory relaxation times provide larger DNP factors, as illustrated with TEKPOL ($\varepsilon = 200$, 16 mM in TCE at 9.4 T, 100 K, 10 kHz, Figure 5.5) that outperformed the previous PAs used so far in DNP experiments on materials.[62] DNP factors in TCE of 33 at 180 K and 12 at 200 K were obtained with TEKPOL, suggesting that with further improvements, larger signal enhancement may be reached at higher temperatures. For instance, using *ortho*-terphenyl (OTP) as glass-forming medium, DNP enhancements of 15–20 have been observed at room temperature using TEKPOL at 9.4 T. OTP forms a homogeneous glass phase up to its glass transition temperature (243 K), higher relative to TCE, and DNP enhancement factors of about 60–80 are obtained at 200 K.[35]

In 2013, capitalizing on the gained knowledge on long electron relaxation times, sufficiently larger electron–electron interaction and rigid tether, the water-soluble AMPUPOL and PyPol dinitroxides (Figures 5.1 and 5.5) have been introduced, yielding unprecedented ^{1}H enhancements (ε) of 240 and 200, respectively, at 9.4 T and 100 K in a water/glycerol solvent mixture.[63] The DNP factors for AMUPOL are 3.5–4 times larger than TOTAPOL at 9.4 and 14.1 T at 100 K, under identical conditions. At 18.8 T, the DNP factor of AMUPOL is ca 35. The electron–electron dipole interaction and spin exchange interaction in AMUPOL have been evaluated to be ca 35 and 45 MHz, respectively, supporting the view that larger magnetic interactions (but not too large in regard to proton Larmor frequency) can be beneficial for improved CE DNP at high fields.[64] Recently, a highly water-soluble (>50 mM) PA (bCtol) has been reported with similar DNP performances to AMUPOL at 20 mM and 9.4 T.[65]

Detailed studies on the rational design of series of more than 40 dinitroxides have been reported for

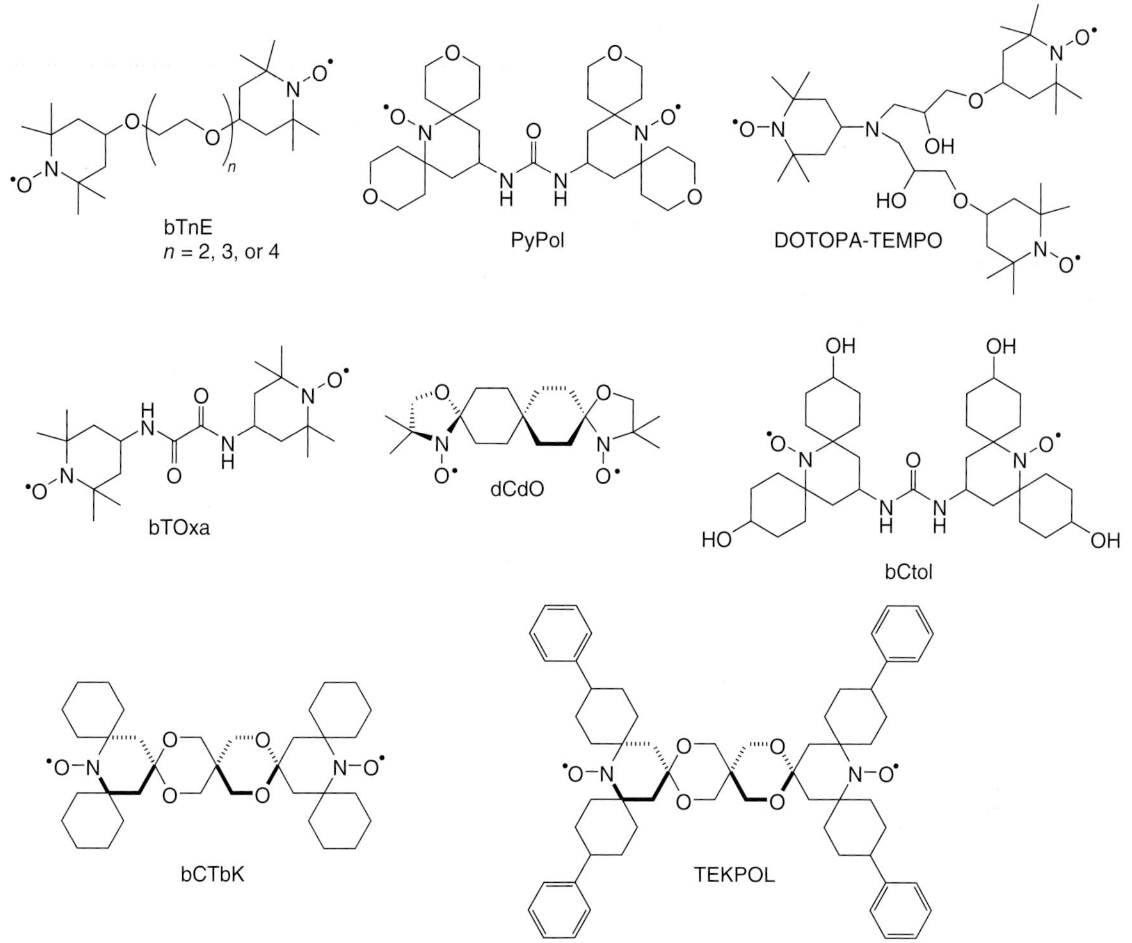

Figure 5.5. Structures of dinitroxide PAs

TCE and glycerol/water mixture, with, for instance, discussion on the impact of deuteration of the PA skeleton, overall sensitivity enhancements, and crowding around the unpaired electrons. It is hypothesized that the optimum relaxation parameters in the TEKPOL and PyPol series have been reached at 9.4 T and 100 K.[39,66] TEKPOL2 and PyPolPEG2OH provide the larger DNP enhancements so far at 9.4 T and 100 K, with $\varepsilon = 220$ in TCE and 300 in the glycerol/water mixture, respectively.[39,66]

When MAS ssNMR/DNP experiments required higher fields, such as 18.8 T or above, for structural analysis at higher resolution,[67–70] the performances of the current dinitroxide PAs drop tremendously. The observed decrease when measuring ε at 400, 600, and 800 MHz is steeper than what theory predicts when all other parameters are kept the same.[46] For instance, for AMUPOL (in DNP juice) and TEKPOL (in TCE), the values decrease from $\varepsilon = 240$ at 9.4 T to 35 at 18.8 T and $\varepsilon = 200$ at 9.4 T to 15 at 18.8 T, respectively.

The decrease in DNP efficiency is related to a combination of various factors, mainly driven by the change of the width of the EPR signal at higher fields, leading to the excitation of smaller spin packets, but also due to microwave irradiation at lower power, increase of the sample temperature, and change in relaxation and diffusion properties. The design of new dinitroxide PAs exhibiting larger electron–electron interactions is expected to limit the drop of DNP efficiency induced by the use of higher magnetic fields.[64] However, at the moment, increasing the magnetic interactions between

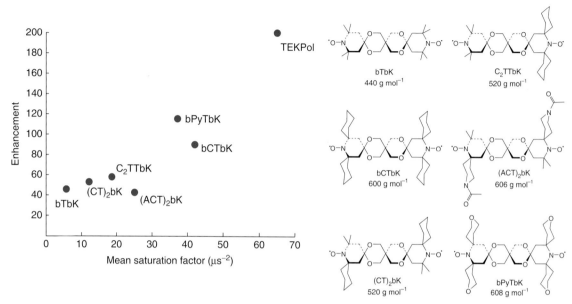

Figure 5.6. DNP enhancement as a function of the mean saturation factor ($T_{1e} \times T_{2e}$) for 16 mM bulk solutions of a series of bTbK-based dinitroxides in TCE. (Reprinted with permission from A. Zagdoun, G. Casano, O. Ouari, M. Schwarzwälder, A. J. Rossini, F. Aussenac, M. Yulikov, G. Jeschke, C. Copéret, A. Lesage, P. Tordo, L. Emsley, J. Am. Chem. Soc., 2013, 135, 12790. Copyright (2013) American Chemical Society)

the nitroxide moieties have not brought solutions for very high field (800 MHz) DNP experiments, as illustrated with data obtained for PyPoldiPEG,[66] TEMPOXAZO, and TalCyO (Figure 5.7).[54,55]

These results pointed out the need to design new dinitroxides exhibiting larger magnetic interactions than in AMUPOL, where the dipolar and spin exchange interactions are ca 35 and 45 MHz, respectively[66,71] but not too large relative to the nuclear Larmor frequency. It has been established that the mixing of the spin states for the three-spin CE process is optimized when the following equation[64] is satisfied:

$$\omega_{0S1} - \omega_{0S2} \approx \sqrt{(\omega_{0I}^2 - D_0^2)} \approx \omega_{0I} - \frac{D_0^2}{2\,\omega_{0I}} \approx \omega_{0I}$$
$$(5.1)$$

with $D_0 = -(d + J)$, where d and J are the magnetic dipole and exchange couplings, respectively.

Perspectives to tackle the limitations of high-field ssNMR/DNP can come from heterobiradicals, as recently reported in a series of biradicals (TEMTriPols, see below) where a trityl moiety was tethered to a nitroxide using a linker of different length.[72] A second approach to surpass the high-field limitation of

DNP efficiency at 100 K is attributed to the Overhauser effect (see below).

5.3.1.2 Heterobiradicals

As early as 2007, the strategy of combining two free radical types having different EPR properties and satisfying the CE conditions for ^{1}H polarization has been reported.[73] Using a mixture of deuterated Finland trityl radical (pseudo-isotropic narrow EPR line) and TEMPO (inhomogeneous broad EPR line), a DNP factor of more than threefold larger was obtained as compared to TEMPO in a sample of similar unpaired electron concentration (40 mM).

The improvement was observed only if the microwave irradiation was fixed at the frequency of the narrow-line trityl radical that exhibits longer electron relaxing properties. Later, a heterobiradical BDPA–TEMPO (Figure 5.8) where a nitroxide was covalently linked to the BDPA phenyl ring was successfully prepared.[74] The BDPA radical exhibits a narrow homogenous EPR linewidth (similar to trityl radicals), a very small g-anisotropy, and slow relaxation. However, the BDPA–TEMPO biradical is shown to have limited DNP efficiency (DNP factor

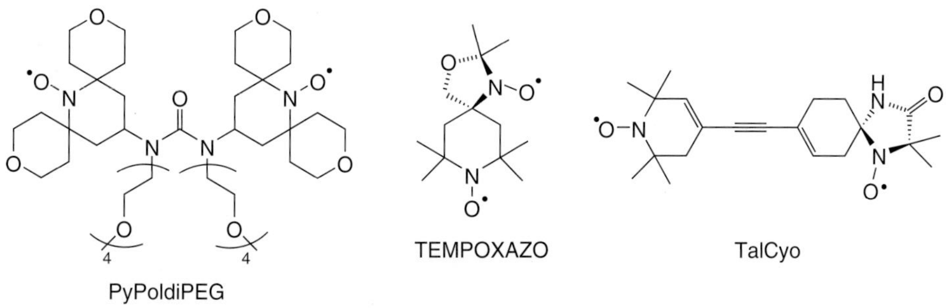

Figure 5.7. Structures of dinitroxides with strong spin exchange interactions

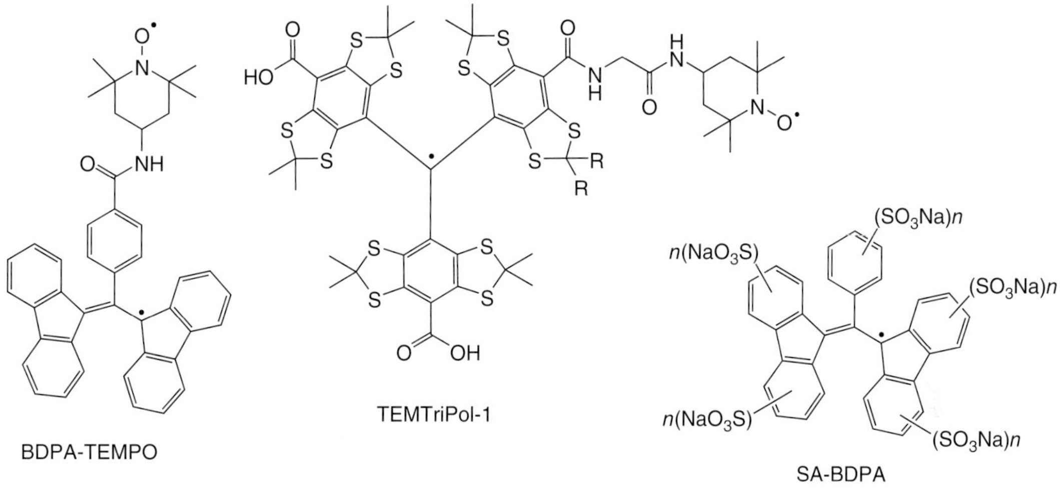

Figure 5.8. Structures of heterobiradicals

not reported), possibly because of the too large spin exchange interactions. More recently, a series of Finland trityl radical covalently tethered to TEMPO (Figure 5.8) with linkers of various sizes has been synthesized and their DNP properties were evaluated in the glycerol/water mixture at various fields (4.7, 14.1, and 18.8 T).[72] Interestingly, the DNP factors showed a field-dependence not observed for dinitroxides, and the best compound TEMTriPol-1 gave DNP factors (ε of 50, 85, and 65 at 5, 14.1, and 18.8 T, respectively. The DNP enhancement observed with TEMTriPol-1 (Figure 5.8) at 18.8 T is the largest enhancement achieved by the CE mechanism at this magnetic field, and this result is consolidated by the absence of depolarization phenomenon that is observed usually for dinitroxides.[40] The remarkable field dependence of the CE efficiency observed in the series of TEMTriPol

biradicals has been related to the strength of the spin exchange interaction between the trityl and nitroxide moieties. Other derivatives of the TEMTriPol series exhibiting smaller or larger exchange interactions led to poor DNP efficiency. As discussed before for the magnetic dipole interactions, the exchange interaction contributes to the state mixing required for CE DNP, and the fine-tuning of these interactions is crucial for the design of improved PAs.

A series of BDPA-nitroxide biradicals soluble in organic solvents has been investigated. By tuning the distance between the two electrons and the substituents at the nitroxide moiety, correlations between the electron-electron interactions and the electron spin relaxation, and the DNP enhancement factors are established. In a 1.3 mm prototype DNP probe, HyTEK2 yields enhancements of up to 185 at 18.8 T

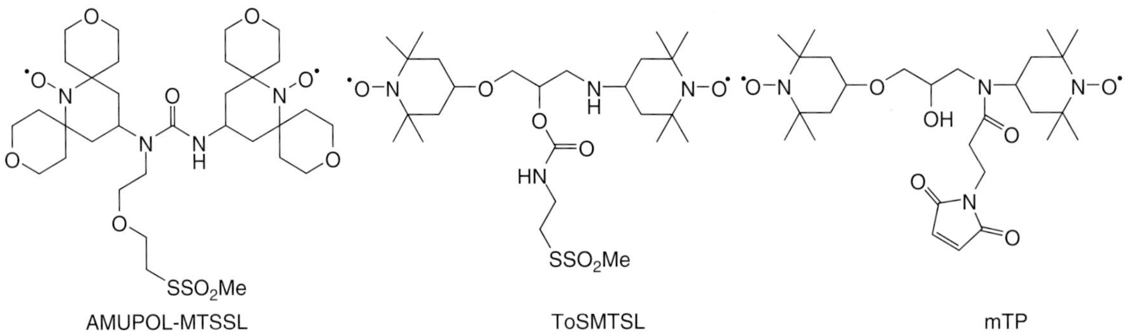

Figure 5.9. Site-specific targeted PAs

(800 MHz ^{1}H resonance frequency) and 40 kHz MAS.[75] Comparably, the direct polarization of ^{13}C at 5 T and 82 K (without the cross-polarization step from ^{1}H) has been achieved successfully using a mixture of two narrow-linewidth radicals (OX063 trityl and SA-BDPA, Figures 5.1 and 5.8) whose Larmor frequencies are separated by approximately the ^{13}C nuclear Larmor frequency.[76] An enhancement factor (ε) of 620, e.g., nearly 25% of the theoretical maximum enhancement ($\gamma_e/\gamma^{13}C \approx 2620$), has been obtained using a total unpaired electron concentration of 40 mM. The hypothesis of a beneficial difference of electron relaxation has been discussed to support the effective CE enhancement, with T_{1eBDPA} and $T_{1etrityl}$ of 3.6 and 1.4 ms, respectively, in the mixture (20 mM each) but with T_{1eBDPA} and $T_{1etrityl}$ of 28.9 and 1.4 ms, respectively, in 40 mM solution of the corresponding monoradical. A combination of CE and SE enhancements has been observed, highlighting the potential of this approach for the study of sample containing no or few active ^{1}H for CP and for dissolution DNP experiments. A few years before, a detailed study was performed to compare the efficiency of the direct and indirect ^{13}C polarization using TOTAPOL dinitroxide as PA at 5 T and 90 K.[77]

5.3.1.3 CE PAs for Specific Application

DNP experiments usually involve the addition of soluble PAs to a sample dissolved or dispersed in a glass-forming solution to achieve a homogeneous distribution of the unpaired electrons through the sample. However, in some cases, it can be of benefit to target and localize the PAs at specific spots.

For instance, in the study of heterogeneous systems, such as membrane-embedded peptides, proteins, and large protein assemblies, site-specific PAs have shown to usually bring additional structural information and improved signal enhancement. Two different approaches have been mainly employed, based on the previous covalent labeling of the proteins using PAs (AMUPOL-MTSSL, ToSMTSL, mTP, Figure 5.9) bearing a cysteine site-directed reactive group (methanethiosulfonate and maleimide groups)[78–81] (or also using a PA-tethered ligand for non-covalent labeling)[82] or based on amphipathic mono- and dinitroxide PAs (Figure 5.10) that homogeneously disperse in the membrane bilayers.[83–87] In the former case, the well-defined spatial localization of the PA enabled to gain unique information by DNP but also to use the paramagnetic properties of the PA to study the signal quenching and paramagnetic relaxation enhancement (PRE) effects.[81] In the latter approach, the DNP factors and resonance linewidth have shown to significantly depend on the structure of the lipid and free radical parts, but also on the membrane lipid composition and biradical-to-lipid ratio.[83]

Polarization silica matrices made of mesoporous hybrid materials containing homogeneously dispersed mono- or dinitroxide PAs linked on and in the walls have been developed and investigated to polarize solutions containing various analytes.[88–90] Some of these mesostructured polarization sources (HYPSOs) have been successfully employed for dissolution DNP experiments at 1.2 K, enabling straightforward separation of the polarization source from the hyperpolarized solution by filtration.[91,92]

Dynamic nuclear polarization surface-enhanced NMR spectroscopy (DNP SENS)[93] has shown to be an effective method to characterize solid surfaces, such as metal oxides, mesoporous silica, hybrid

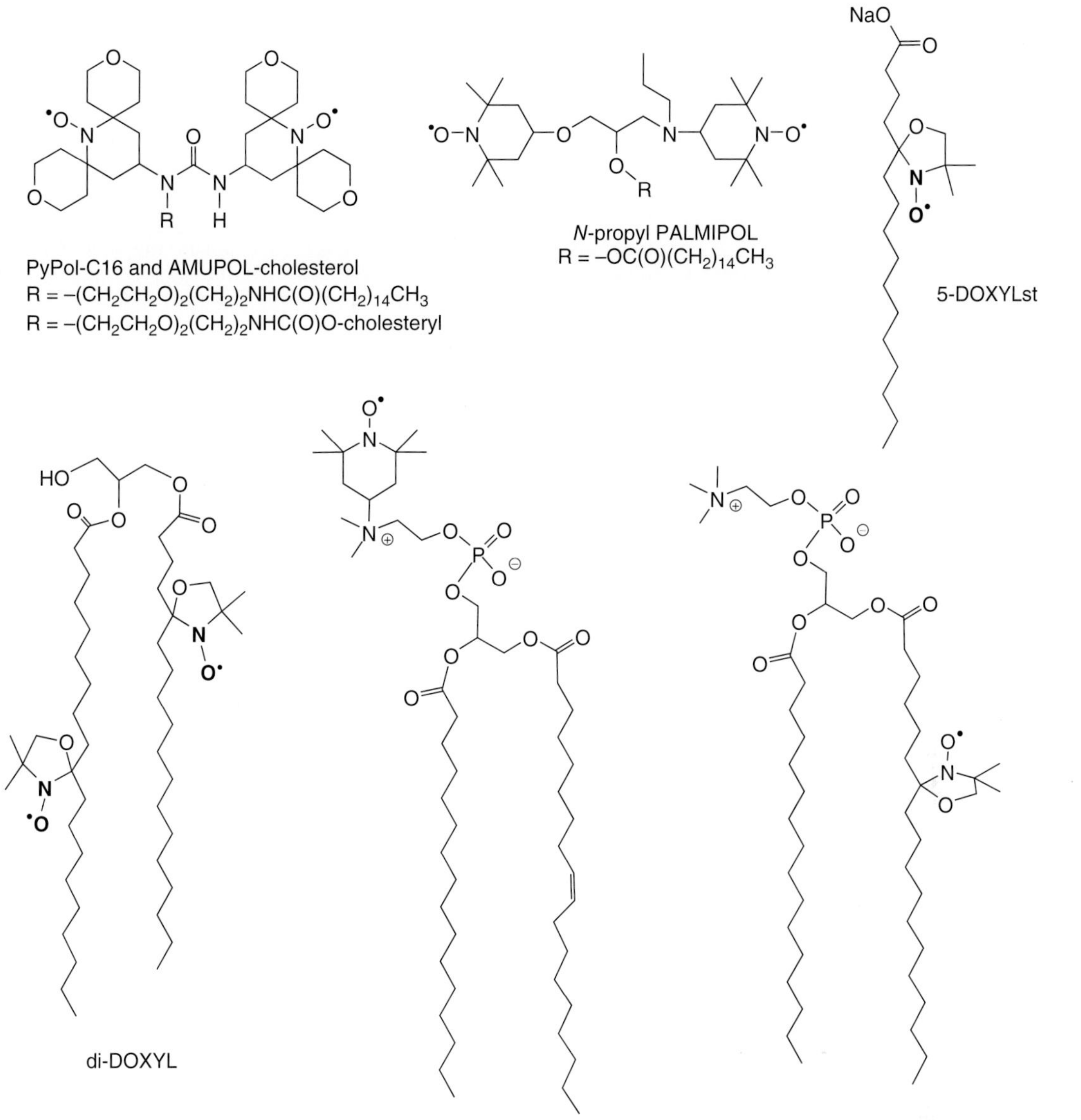

Figure 5.10. Amphipathic PA for membrane protein CE DNP studies

materials, and nanoparticles. However, the nitroxide moiety of the PAs can react with sensitive substrate, such as supported heterogeneous metathesis catalysts. A carbosilane-based dendritic PA where the nitroxide moieties are partially shielded has been prepared and successfully used to characterize a highly reactive single-site cationic alkene metathesis W@SiO$_2$ heterogeneous catalyst, while no DNP enhancement was observed with TEKPOL.[94]

5.3.2 Narrow EPR Linewidth Free Radicals for SE MAS ssNMR/DNP

Free radicals with a narrow homogeneous EPR linewidth (δ) and small inhomogeneous breadth (Δ) relative to the nuclear Larmor frequency (δ, $\Delta < 2$ ω_{0I}) can be efficient PAs for SE polarization at high magnetic fields. SE relies on the excitation of the 'forbidden' electron-nuclear ZQ and DQ transitions by the microwave irradiation, leading to an experimentally less efficient (but theoretically equivalent)[25] polarizing mechanism relative to CE at 100 K and with a theoretically less favorable field-dependence. Trityl-based and BDPA radicals (Figure 5.1) satisfy these requirements and have shown to generate SE-induced ^{1}H,[95] ^{2}H,[96] ^{13}C,[76] and ^{17}O[97] polarization in MAS ssNMR/DNP. Additionally, with the development of pulsed DNP, trityl OX063 and BDPA (Figure 5.1) has shown to be efficient PAs for NOVEL experiments, due to their narrow EPR line allowing efficient excitation of all the unpaired electrons with very short pulses, and their appropriate electron relaxation time enabling efficient spin-locking but also fast repetition of the locking sequence.[98,99]

The chemical stability and its narrow EPR line at high magnetic fields made α,-bisdiphenylene-β-phenylallyl radical (BDPA) an attracting PA in the early developments of MAS ssNMR/DNP.[100–104] BDPA linewidth ($\Delta = 26$ and 28 MHz at 9 GHz and 140 GHz, respectively) is dominated by unresolved hyperfine couplings with protons and the line broadens insignificantly at higher fields as a result of a very small g-anisotropy. Moreover, longitudinal electron relaxation of BDPA (T_{1e}) is longer than for trityl-based radicals, leading to an efficient saturation at lower microwave power but also to longer build-up times. More recently, a water-soluble randomly sulfonated-BDPA derivative (SA-BDPA, Figure 5.8), retaining the EPR properties of the parent BDPA, has been synthesized.[105] The SE DNP enhancement factor of SA-BDPA has shown to be superior to OX063 trityl radical in ^{1}H-enhanced MAS ssNMR/DNP, but at the expense of longer build-up times.

Trityl (also triarylmethyl, TAM) radicals such as CT-03 and OX063 (Figure 5.1) exhibit narrow linewidth (50 MHz at 140 GHz) that increases linearly with the external field because g-anisotropy is the dominant broadening mechanism. Thus, the smaller g-anisotropy makes BDPA an interesting PA for SE DNP at higher fields relative to trityl-based radicals.[105] CT-03 and OX063 exhibit high chemical stability, water solubility, and longer relaxation times relative to nitroxides. OX063 is usually the best PA for the direct polarization of low gyromagnetic ratio nuclei, such as ^{2}H, ^{13}C, and ^{17}O, mainly because its EPR linewidth is of similar value to the nuclear Larmor frequencies, enabling to have both SE and CE operative. But trityl OX063 is most famous for its role as PA in dissolution DNP for the direct polarization of ^{13}C at temperature below 2 K.[3]

5.3.3   Free Radical PAs for Overhauser Effect

The origin of the OE effect in ssNMR/DNP experiment on frozen samples is still unclear, due to the need of intermolecular dynamic processes causing fluctuations of the hyperfine interactions at relevant Larmor frequency that generate subsequently electron–nuclear cross-relaxation.[22,34] No such phenomenon was expected to be observed in insulating dielectrics until its observation with BDPA-doped samples, while d$_{21}$-BDPA and trityl OX063 have shown to be poorly efficient.[32,33] Interestingly, OE DNP generates larger DNP factors at higher fields, and BDPA provides among the largest signal enhancement at 800 MHz in OE DNP experiments.[32] New PAs have to be designed in order to gain better understanding of the parameters driving the OE in frozen dielectric systems.[32,106]

5.3.4   Paramagnetic Metal Ions for MAS ssNMR/DNP

Since the early developments of DNP, several paramagnetic metal ions (Ce^{3+}, Cr^{3+}, Cr^{5+}, Gd^{3+}, and Mn^{2+}) have shown to be effective PAs, but with moderate efficiency in regard to dinitroxide PAs.[43,107–109] However, the rich DNP properties of metal ions have not been fully evaluated and makes them interesting PA candidates for high-field DNP and specific applications, notably for the study of metalloproteins or for *in cell* DNP. The DNP evaluation of paramagnetic metal ions as PAs has been hampered by limitations related to the complex spin systems and multiple spin transitions with different properties and the use of DNP spectrometers with fixed-frequency microwave source (gyrotron) and limited field-sweep capability of the NMR magnet. Indeed, paramagnetic species with g-factors significantly deviating from the value of the free electron ($g_e = 2.0023$), and most of the organic free radicals, such as nitroxide, BDPA, and trityl-based, are not accessible. Many metal ions with low spin quantum numbers ($S = 1/2$), such as Cu^{2+}, Co^{2+}, and Ce^{3+}, are characterized by strong spin orbit coupling (SOC), generating important g-anisotropy. Among metal ions with high spin numbers ($S > 1/2$), Gd^{3+} and Mn^{2+} complexes have shown to have peculiar electron spin properties, with narrow and isotropic central transition as well as favorable field-dependence of its linewidth that allow ^{1}H polarization by SE and direct low-γ nuclei polarization by CE (^{13}C and ^{15}N).[19,43,108,110]

5.4 OUTLOOK

DNP techniques have attracted a tremendous and large interest in the scientific community in the last decade due to their ability to routinely enable studies that were not previously feasible or considered. The continuous progress of the DNP techniques relies on the synergetic and interdisciplinary efforts, such as theoretical modeling of the polarizing mechanisms, hardware engineering, synthesis of tailored PAs, and methodological developments. This effort has boosted the renaissance of DNP at high fields, stimulated renewed interests, and brought closer the EPR and NMR magnetic resonance communities. Creativity and inspiration from other techniques but also from other DNP approaches sparked interest and make DNP a continuous evolving technique. The design and development of new and improved PAs have contributed to this success. Still, further improvements are required and high-field and high-temperature DNP pose rewarding challenges. Further detailed studies and improvement of PA families (dinitroxides, heterobiradicals, narrow-line radicals, and PAs for OE) but also completely new PAs are required to tackle the actual limitations and open new possibilities in magnetic resonance.

ACKNOWLEDGMENTS

The authors are most grateful to Anne Lesage, Lyndon Emsley, Christophe Copéret (and their respective collaborators), Aaron Rossini, Moreno Lelli, Guido Pintacuda, Maxim Yulikov, Gunnar Jeschke, Burkhard Bechinger, Evgeniy Salnikov, Marc Baldus, Melanie Rosay, Fabien Aussenac, Ralph Weber, Werner Maas, and R. G. Griffin for fruitful collaborations and discussions. Financial support from ANR-15-CE29-0022-01 and ANR-12-BSV5-01 is gratefully acknowledged.

RELATED ARTICLES IN EMAGRES

Overhauser, Albert W.: Dynamic Nuclear Polarization

REFERENCES

1. D. A. Hall, D. C. Maus, G. J. Gerfen, S. J. Inati, L. R. Becerra, F. W. Dahlquist, and R. G. Griffin, *Science*, 1997, **276**, 930.

2. T. Maly, G. T. Debelouchina, V. S. Bajaj, K.-N. Hu, C.-G. Joo, M. L. Mak-Jurkauskas, J. R. Sirigiri, P. C. A. van der Wel, J. Herzfeld, R. J. Temkin, and R. G. Griffin, *J. Chem. Phys.*, 2008, **128**, 052211.

3. J. H. Ardenkjaer-Larsen, B. Fridlund, A. Gram, G. Hansson, L. Hansson, M. H. Lerche, R. Servin, M. Thaning, and K. Golman, *Proc. Natl. Acad. Sci. U. S. A.*, 2003, **100**, 10158.

4. R. G. Griffin and T. F. Prisner, *Phys. Chem. Chem. Phys.*, 2010, **12**, 5737.

5. L. L. Lumata, M. E. Merritt, C. R. Malloy, A. D. Sherry, J. van Tol, L. Song, and Z. Kovacs, *J. Magn. Reson.*, 2013, **227**, 14.

6. L. Lumata, M. Merritt, C. Khemtong, S. J. Ratnakar, J. van Tol, L. Yu, L. Song, and Z. Kovacs, *RSC Adv.*, 2012, **2**, 12812.

7. L. Lumata, M. E. Merritt, C. R. Malloy, A. D. Sherry, and Z. Kovacs, *J. Phys. Chem. A*, 2012, **116**, 5129.

8. A. Capozzi, T. Cheng, G. Boero, C. Roussel, and A. Comment, *Nat. Commun.*, 2017, **8**, 15757.

9. T. R. Eichhorn, Y. Takado, N. Salameh, A. Capozzi, T. Cheng, J. N. Hyacinthe, M. Mishkovsky, C. Roussel, and A. Comment, *Proc. Natl. Acad. Sci. U. S. A.*, 2013, **110**, 18064.

10. A. Bornet and S. Jannin, *J. Magn. Reson.*, 2016, **264**, 13.

11. X. Ji, A. Bornet, B. Vuichoud, J. Milani, D. Gajan, A. J. Rossini, L. Emsley, G. Bodenhausen, and S. Jannin, *Nat. Commun.*, 2017, **8**, 13975.

12. C. Griesinger, M. Bennati, H. M. Vieth, C. Luchinat, G. Parigi, P. Höfer, F. Engelke, S. J. Glaser, V. Denysenkov, and T. F. Prisner, *Prog. Nucl. Magn. Reson. Spectrosc.*, 2012, **64**, 4.

13. G. Liu, M. Levien, N. Karschin, G. Parigi, C. Luchinat, and M. Bennati, *Nat. Chem.*, 2017, **9**, 676.

14. J. Leggett, R. Hunter, J. Granwehr, R. Panek, A. J. Perez-Linde, A. J. Horsewill, J. McMaster, G. Smith, and W. Köckenberger, *Phys. Chem. Chem. Phys.*, 2010, **12**, 5883.

15. M. Sharma, G. Janssen, J. Leggett, A. P. M. Kentgens, and P. J. M. van Bentum, *J. Magn. Reson.*, 2015, **258**, 40.

16. A. W. Overhauser, *Phys. Rev.*, 1953, **92**, 411.

17. T. R. Carver and C. P. Slichter, *Phys. Rev.*, 1953, **92**, 212.

18. T. Maly, D. Cui, R. G. Griffin, and A.-F. Miller, *J. Phys. Chem. B*, 2012, **116**, 7055.

19. P. Wenk, M. Kaushik, D. Richter, M. Vogel, B. Suess, and B. Corzilius, *J. Biomol. NMR*, 2015, **63**, 97.

20. C. O. Bretschneider, Ü. Akbey, F. Aussenac, G. L. Olsen, A. Feintuch, H. Oschkinat, and L. Frydman, *ChemPhysChem*, 2016, **17**, 2691.

21. J. Riikonen, S. Rigolet, C. Marichal, F. Aussenac, J. Lalevée, F. Morlet-Savary, P. Fioux, C. Dietlin, M. Bonne, B. Lebeau, and V.-P. Lehto, *J. Phys. Chem. C*, 2015, **119**, 19272.

22. A. S. Lilly Thankamony, J. J. Wittmann, M. Kaushik, and B. Corzilius, *Prog. Nucl. Magn. Reson. Spectrosc.*, 2017, **102−103**, 120.

23. F. Mentink-Vigier, S. Vega, and G. De Paëpe, *Phys. Chem. Chem. Phys.*, 2017, **19**, 3506.

24. Q. Z. Ni, E. Daviso, T. V. Can, E. Markhasin, S. K. Jawla, T. M. Swager, R. J. Temkin, J. Herzfeld, and R. G. Griffin, *Acc. Chem. Res.*, 2013, **46**, 1933.

25. K. N. Hu, G. T. Debelouchina, A. A. Smith, and R. G. Griffin, *J. Chem. Phys.*, 2011, **134**, 125105.

26. Y. Hovav, A. Feintuch, and S. Vega, *J. Magn. Reson.*, 2012, **214**, 29.

27. K. R. Thurber and R. Tycko, *J. Chem. Phys.*, 2014, **140**, 184201.

28. S. R. Chaudhari, P. Berruyer, D. Gajan, C. Reiter, F. Engelke, D. L. Silverio, C. Copéret, M. Lelli, A. Lesage, and L. Emsley, *Phys. Chem. Chem. Phys.*, 2016, **18**, 10616.

29. F. Mentink-Vigier, S. Paul, D. Lee, A. Feintuch, S. Hediger, S. Vega, and G. De Paëpe, *Phys. Chem. Chem. Phys.*, 2015, **17**, 21824.

30. A. V. Kessenikh, V. I. Lushchikov, A. A. Manenkov, and Y. V. Taran, *Sov. Phys. Solid State*, 1963, **5**, 321.

31. M. Borghini, *Phys. Rev. Lett.*, 1968, **20**, 419.

32. T. V. Can, M. A. Caporini, F. Mentink-Vigier, B. Corzilius, J. J. Walish, M. Rosay, W. E. Maas, M. Baldus, S. Vega, T. M. Swager, and R. G. Griffin, *J. Chem. Phys.*, 2014, **141**, 064202.

33. S. R. Chaudhari, D. Wisser, A. C. Pinon, P. Berruyer, D. Gajan, P. Tordo, O. Ouari, C. Reiter, F. Engelke, C. Copéret, M. Lelli, A. Lesage, and L. Emsley, *J. Am. Chem. Soc.*, 2017, **139**, 10609.

34. S. Pylaeva, K. L. Ivanov, M. Baldus, D. Sebastiani, and H. Elgabarty, *J. Phys. Chem. Lett.*, 2017, **8**, 2137.

35. M. Lelli, S. R. Chaudhari, D. Gajan, G. Casano, A. J. Rossini, O. Ouari, P. Tordo, A. Lesage, and L. Emsley, *J. Am. Chem. Soc.*, 2015, **137**, 14558.

36. A. J. Rossini, A. Zagdoun, M. Lelli, D. Gajan, F. Rascon, M. Rosay, W. E. Maas, C. Coperet, A. Lesage, and L. Emsley, *Chem. Sci.*, 2012, **3**, 108.

37. T. Kobayashi, O. Lafon, A. S. Lilly Thankamony, I. I. Slowing, K. Kandel, D. Carnevale, V. Vitzthum, H. Vezin, J.-P. Amoureux, G. Bodenhausen, and M. Pruski, *Phys. Chem. Chem. Phys.*, 2013, **15**, 5553.

38. H. Takahashi, C. Fernández-de-Alba, D. Lee, V. Maurel, S. Gambarelli, M. Bardet, S. Hediger, A.-L. Barra, and G. De Paëpe, *J. Magn. Reson.*, 2014, **239**, 91.

39. D. J. Kubicki, G. Casano, M. Schwarzwälder, S. Abel, C. Sauvée, K. Ganesan, M. Yulikov, A. J. Rossini, G. Jeschke, C. Copéret, A. Lesage, P. Tordo, O. Ouari, and L. Emsley, *Chem. Sci.*, 2016, **7**, 550.

40. F. Mentink-Vigier, G. Mathies, Y. Liu, A.-L. Barra, M. A. Caporini, D. Lee, S. Hediger, R. G. Griffin, and G. De Paëpe, *Chem. Sci.*, 2017, **8**, 8150.

41. E. Bouleau, P. Saint-Bonnet, F. Mentink-Vigier, H. Takahashi, J.-F. Jacquot, M. Bardet, F. Aussenac, A. Purea, F. Engelke, S. Hediger, D. Lee, and G. De Paëpe, *Chem. Sci.*, 2015, **6**, 6806.

42. D. J. Kubicki, A. J. Rossini, A. Purea, A. Zagdoun, O. Ouari, P. Tordo, F. Engelke, A. Lesage, and L. Emsley, *J. Am. Chem. Soc.*, 2014, **136**, 15711.

43. M. Kaushik, T. Bahrenberg, T. V. Can, M. A. Caporini, R. Silvers, J. Heiliger, A. A. Smith, H. Schwalbe, R. G. Griffin, and B. Corzilius, *Phys. Chem. Chem. Phys.*, 2016, **18**, 27205.

44. F. Mentink-Vigier, Ü. Akbey, H. Oschkinat, S. Vega, and A. Feintuch, *J. Magn. Reson.*, 2015, **258**, 102.

45. K. R. Thurber and R. Tycko, *J. Chem. Phys.*, 2012, **137**, 084508.

46. D. Mance, P. Gast, M. Huber, M. Baldus, and K. L. Ivanov, *J. Chem. Phys.*, 2015, **142**, 234201.

47. K. N. Hu, H. H. Yu, T. M. Swager, and R. G. Griffin, *J. Am. Chem. Soc.*, 2004, **126**, 10844.

48. C. Song, K. N. Hu, C. G. Joo, T. M. Swager, and R. G. Griffin, *J. Am. Chem. Soc.*, 2006, **128**, 11385.

49. K. R. Thurber and R. Tycko, *J. Magn. Reson.*, 2010, **204**, 303.

50. A. Potapov, K. R. Thurber, W.-M. Yau, and R. Tycko, *J. Magn. Reson.*, 2012, **221**, 32.

51. W.-M. Yau, K. R. Thurber, and R. Tycko, *J. Magn. Reson.*, 2014, **244**, 98.

52. Y. Matsuki, T. Maly, O. Ouari, H. Karoui, F. LeMoigne, E. Rizzato, S. Lyubenova, J. Herzfeld, T. Prisner, P. Tordo, and R. G. Griffin, *Angew. Chem. Int. Ed. Engl.*, 2009, **48**, 4996.

53. K. N. Hu, C. Song, H. H. Yu, T. M. Swager, and R. G. Griffin, *J. Chem. Phys.*, 2008, **18**, 052302.

54. C. Ysacco, E. Rizzato, M. A. Virolleaud, H. Karoui, A. Rockenbauer, F. LeMoigne, D. Siri, O. Ouari, R. G. Griffin, and P. Tordo, *Phys. Chem. Chem. Phys.*, 2010, **12**, 5841.

55. C. Ysacco, H. Karoui, G. Casano, F. LeMoigne, S. Combes, A. Rockenbauer, M. Rosay, W. Maas, O. Ouari, and P. Tordo, *Appl. Magn. Reson.*, 2012, **43**, 251.

56. E. L. Dane, B. Corzilius, E. Rizzato, P. Stocker, T. Maly, A. A. Smith, R. G. Griffin, O. Ouari, P. Tordo, and T. M. Swager, *J. Org. Chem.*, 2012, **77**, 1789.

57. M. K. Kiesewetter, B. Corzilius, A. A. Smith, R. G. Griffin, and T. M. Swager, *J. Am. Chem. Soc.*, 2012, **134**, 4537.

58. J. Mao, D. Akhmetzyanov, O. Ouari, V. Denysenkov, B. Corzilius, J. Plackmeyer, P. Tordo, T. F. Prisner, and C. Glaubitz, *J. Am. Chem. Soc.*, 2013, **135**, 19275.

59. M. K. Kiesewetter, V. K. Michaelis, J. J. Walish, R. G. Griffin, and T. M. Swager, *J. Phys. Chem. B*, 2014, **118**, 1825.

60. M. Lelli, A. J. Rossini, G. Casano, O. Ouari, P. Tordo, A. Lesage, and L. Emsley, *Chem. Commun.*, 2014, **50**, 10198.

61. A. Zagdoun, G. Casano, O. Ouari, G. Lapadula, A. J. Rossini, M. Lelli, M. Baffert, D. Gajan, L. Veyre, W. E. Maas, M. Rosay, R. T. Weber, C. Thieuleux, C. Copéret, A. Lesage, P. Tordo, and L. Emsley, *J. Am. Chem. Soc.*, 2012, **134**, 2284.

62. A. Zagdoun, G. Casano, O. Ouari, M. Schwarzwälder, A. J. Rossini, F. Aussenac, M. Yulikov, G. Jeschke, C. Copéret, A. Lesage, P. Tordo, and L. Emsley, *J. Am. Chem. Soc.*, 2013, **135**, 12790.

63. C. Sauvée, M. Rosay, G. Casano, F. Aussenac, R. T. Weber, O. Ouari, and P. Tordo, *Angew. Chem. Int. Ed.*, 2013, **52**, 10858.

64. T. V. Can, Q. Z. Ni, and R. G. Griffin, *J. Magn. Reson.*, 2015, **253**, 23.

65. A. P. Jagtap, M.-A. Geiger, D. Stoppler, M. Orwick-Rydmark, H. Oschkinat, and S. T. Sigurdsson, *Chem. Commun.*, 2016, **52**, 7020.

66. C. Sauvée, G. Casano, S. Abel, A. Rockenbauer, D. Akhmetzyanov, H. Karoui, D. Siri, F. Aussenac, W. Maas, R. T. Weber, T. Prisner, M. Rosay, P. Tordo, and O. Ouari, *Chem. Eur. J.*, 2016, **22**, 5598.

67. R. Gupta, M. Lu, G. Hou, M. A. Caporini, M. Rosay, W. Maas, J. Struppe, C. Suiter, J. Ahn, I. J. Byeon, W. T. Franks, M. Orwick-Rydmark, A. Bertarello, H. Oschkinat, A. Lesage, G. Pintacuda, A. M. Gronenborn, and T. Polenova, *J. Phys. Chem. B*, 2016, **120**, 329.

68. P. Fricke, D. Mance, V. Chevelkov, K. Giller, S. Becker, M. Baldus, and A. Lange, *J. Biomol. NMR*, 2016, **65**, 121.

69. E. J. Koers, E. A. van der Cruijsen, M. Rosay, M. Weingarth, A. Prokofyev, C. Sauvée, O. Ouari, J. van der Zwan, O. Pongs, P. Tordo, W. E. Maas, and M. Baldus, *J. Biomol. NMR*, 2014, **60**, 157.

70. M. Kaplan, A. Cukkemane, G. C. P. van Zundert, S. Narasimhan, M. Daniëls, D. Mance, G. Waksman, A. M. J. J. Bonvin, R. Fronzes, G. E. Folkers, and M. Baldus, *Nat. Methods*, 2015, **12**, 649.

71. P. Gast, D. Mance, E. Zurlo, K. L. Ivanov, M. Baldus, and M. Huber, *Phys. Chem. Chem. Phys.*, 2017, **19**, 3777.

72. G. Mathies, M. A. Caporini, V. K. Michaelis, Y. Liu, K. N. Hu, D. Mance, J. L. Zweier, M. Rosa, M. Baldus, and R. G. Griffin, *Angew. Chem. Int. Ed.*, 2015, **54**, 11770.

73. K. N. Hu, V. S. Bajaj, M. Rosay, and R. G. Griffin, *J. Chem. Phys.*, 2007, **126**, 044512.

74. E. L. Dane, T. Maly, G. T. Debelouchina, R. G. Griffin, and T. M. Swager, *Org. Lett.*, 2009, **11**, 1871.

75. D. Wisser, G. Karthikeyan, A. Lund, G. Casano, H. Karoui, M. Yulikov, G. Menzildjian, A. C. Pinon, A. Purea, F. Engelke, S. R. Chaudhari, D. Kubicki, A. J. Rossini, I. B. Moroz, D. Gajan, C. Coperet, G. Jeschke, M. Lelli, L. Emsley, O. Ouari, J. Am. Chem. Soc., 2018, **140**, 13340.

76. V. K. Michaelis, A. A. Smith, B. Corzilius, O. Haze, T. M. Swager, and R. G. Griffin, *J. Am. Chem. Soc.*, 2013, **135**, 2935.

77. T. Maly, A.-F. Miller, and R. G. Griffin, *ChemPhysChem*, 2010, **11**, 999.

78. E. A. W. van der Cruijsen, E. J. Koers, C. Sauvée, R. E. Hulse, M. Weingarth, O. Ouari, E. Perozo, P. Tordo, and M. Baldus, *Chem. Eur. J.*, 2015, **21**, 12971.

79. M. A. Voinov, D. B. Good, M. E. Ward, S. Milikisiyants, A. Marek, M. A. Caporini, M. Rosay, R. A. Munro, M. Ljumovic, L. S. Brown, V. Ladizhansky, and A. I. Smirnov, *J. Phys. Chem. B*, 2015, **119**, 10180.

80. T. Viennet, A. Viegas, A. Kuepper, S. Arens, V. Gelev, O. Petrov, T. N. Grossmann, H. Heise, and M. Etzkorn, *Angew. Chem. Int. Ed.*, 2016, **55**, 10746.

81. B. J. Wylie, B. G. Dzikovski, S. Pawsey, M. Caporini, M. Rosay, J. H. Freed, and A. E. McDermott, *J. Biomol. NMR*, 2015, **61**, 367.

82. R. Rogawski, I. V. Sergeyev, Y. Zhang, T. H. Tran, Y. Li, L. Tong, and A. E. McDermott, *J. Phys. Chem. B*, 2017, **121**, 10770.

83. E. S. Salnikov, S. Abel, G. Karthikeyan, H. Karoui, F. Aussenac, P. Tordo, B. Bechinger, and O. Ouari, *ChemPhysChem*, 2017, **18**, 2103.

84. E. S. Salnikov, C. Aisenbrey, F. Aussenac, O. Ouari, H. Sarrouj, C. Reiter, P. Tordo, F. Engelke, and B. Bechinger, *Sci. Rep.*, 2016, **6**, 20895.

85. A. N. Smith, M. A. Caporini, G. E. Fanucci, and J. R. Long, *Angew. Chem. Int. Ed.*, 2015, **54**, 1542.

86. A. N. Smith, U. T. Twahir, T. Dubroca, G. E. Fanucci, and J. R. Long, *J. Phys. Chem. B*, 2016, **120**, 7880.

87. C. Fernández-de-Alba, H. Takahashi, A. Richard, Y. Chenavier, L. Dubois, V. Maurel, D. Lee, S. Hediger, and G. De Paëpe, *Chem. Eur. J.*, 2015, **21**, 4512.

88. D. Gajan, M. Schwarzwälder, M. P. Conley, W. R. Grüning, A. J. Rossini, A. Zagdoun, M. Lelli, M. Yulikov, G. Jeschke, C. Sauvée, O. Ouari, P. Tordo, L. Veyre, A. Lesage, C. Thieuleu, L. Emsley, and C. Copéret, *J. Am. Chem. Soc.*, 2013, **135**, 15459.

89. E. Besson, F. Ziarelli, E. Bloch, G. Gerbaud, S. Queyroy, S. Viel, and S. Gastaldi, *Chem. Commun.*, 2016, **52**, 5531.

90. D. L. Silverio, H. A. van Kalkeren, T.-C. Ong, M. Baudin, M. Yulikov, L. Veyre, P. Berruyer, S. Chaudhari, D. Gajan, D. Baudouin, M. Cavaillés, V. Basile, A. Bornet, G. Jeschke, G. Bodenhausen, A. Lesage, L. Emsley, S. Jannin, C. Thieuleux, and C. Copéret, *Helv. Chim. Acta*, 2017, **100**, e1700101.

91. D. Gajan, A. Bornet, B. Vuichoud, J. Milani, R. Melzi, H. A. van Kalkeren, L. Veyre, C. Thieuleux, M. P. Conley, W. R. Gruning, M. Schwarzwalder, A. Lesage, C. Coperet, G. Bodenhausen, L. Emsley, and S. Jannin, *Proc. Natl. Acad. Sci.*, 2014, **111**, 14693.

92. D. Baudouin, H. A. van Kalkeren, A. Bornet, B. Vuichoud, L. Veyre, M. Cavaillès, M. Schwarzwälder, W.-C. Liao, D. Gajan, G. Bodenhausen, L. Emsley, A. Lesage, S. Jannin, C. Copéret, and C. Thieuleux, *Chem. Sci.*, 2016, **7**, 6846.

93. A. J. Rossini, A. Zagdoun, M. Lelli, A. Lesage, C. Copéret, and L. Emsley, *Acc. Chem. Res.*, 2013, **46**, 1942.

94. W. C. Liao, T. C. Ong, D. Gajan, F. Bernada, C. Sauvée, M. Yulikov, M. Pucino, R. Schowner, M. Schwarzwälder, M. R. Buchmeiser, G. Jeschke, P. Tordo, O. Ouari, A. Lesage, L. Emsley, and C. Copéret, *Chem. Sci.*, 2017, **8**, 416.

95. B. Corzilius, A. A. Smith, and R. G. Griffin, *J. Chem. Phys.*, 2012, **13**, 054201.

96. T. Maly, L. B. Andreas, A. A. Smith, and R. G. Griffin, *Phys. Chem. Chem. Phys.*, 2010, **12**, 5872.

97. V. K. Michaelis, B. Corzilius, A. A. Smith, and R. G. Griffin, *J. Phys. Chem. B*, 2013, **117**, 14894.

98. G. Mathies, S. Jain, M. Reese, and R. G. Griffin, *J. Phys. Chem. Lett.*, 2016, **7**, 111.

99. T. V. Can, R. T. Weber, J. J. Walish, T. M. Swager, and R. G. Griffin, *Angew. Chem. Int. Ed.*, 2017, **56**, 6744.

100. R. A. Wind, F. E. Anthonio, M. J. Duijvestijn, J. Smidt, J. Trommel, and G. M. C. de Vette, *J. Magn. Reson.*, 1983, **52**, 424.

101. R. A. Wind, M. J. Duijvestijn, C. van der Lugt, A. Manenschijn, and J. Vriend, *Prog. Nucl. Magn. Reson. Spectrosc.*, 1985, **17**, 33.

102. M. Afeworki, R. A. McKay, and J. Schaefer, *Macromolecules*, 1992, **25**, 4084.

103. M. Afeworki, S. Vega, and J. Schaefer, *Macromolecules*, 1992, **25**, 4100.

104. L. R. Becerra, G. J. Gerfen, B. F. Bellew, J. A. Bryant, D. A. Hall, S. J. Inati, R. T. Weber, S. Un, T. F. Prisner, A. E. McDermott, K. W. Fishbein, K. E. Kreischer, R. J. Temkin, D. J. Singel, and R. G. Griffin, *J. Magn. Reson. Ser. A*, 1995, **117**, 28.

105. O. Haze, B. Corzilius, A. A. Smith, R. G. Griffin, and T. M. Swager, *J. Am. Chem. Soc.*, 2012, **134**, 14287.

106. B. A. DeHaven, J. T. Tokarski, A. A. Korous, F. Mentink-Vigier, T. M. Makris, A. M. Brugh, M. D. E. Forbes, J. van Tol, C. R. Bowers, and L. S. Shimizu, *Chem. Eur. J.*, 2017, **23**, 8315.

107. M. Abraham, M. A. H. McCausland, and F. N. H. Robinson, *Phys. Rev. Lett.*, 1959, **2**, 449.

108. B. Corzilius, A. A. Smith, A. B. Barnes, C. Luchinat, I. Bertini, and R. G. Griffin, *J. Am. Chem. Soc.*, 2011, **133**, 5648.

109. B. Corzilius, *Phys. Chem. Chem. Phys.*, 2016, **18**, 27190.

110. M. Kaushik, M. Qi, A. Godt, and B. Corzilius, *Angew. Chem. Int. Ed.*, 2017, **56**, 4295.

Chapter 6

Paramagnetic Metal Ions for Dynamic Nuclear Polarization

Björn Corzilius

University of Rostock, Rostock, Germany

6.1 INTRODUCTION

Since the inception of dynamic nuclear polarization (DNP) immediately following the advent of magnetic resonance, it has become clear that a huge variety of associated technological aspects have to be considered for its further development.[1] Even though DNP has become somewhat fractionated during the

Handbook of High Field Dynamic Nuclear Polarization.
Edited by Vladimir K. Michaelis, Robert G. Griffin, Björn Corzilius and Shimon Vega
© 2020 John Wiley & Sons, Ltd. ISBN: 978-1-119-44164-9
Also published in eMagRes (online edition)
DOI: 10.1002/9780470034590.emrstm1593

renaissance of high-field methods – giving rise to several more streamlined but more or less independent application directions[2] – each resulting flavor (i.e., Overhauser/solution DNP, dissolution DNP, and solid-state/MAS DNP) is in constant developmental flux. However, the individual degree of this flux might differ for each method. For example, dissolution DNP is well established with approved protocols in place for clinical applications while methodological advances are pursued and step-wise implemented into the standardized instrumentation and methods. In contrast, high-field solution-based methods are still in their infancy where exciting shifts in paradigm are frequently encountered, but no 'standard' application or instrumentation is yet available. MAS DNP, however, is situated somewhere in between these 'extremes': while an impressive set of rather well-established applications in biological and materials NMR has been developed over the last few years, there still exists an overwhelming rate of development, from an instrumental, methodological, and theoretical viewpoint. This is apparent, for example, when following the evolution of commercially built hardware on the one hand or the paradigm shifts frequently encountered in theoretical understanding of the underlying, complex DNP mechanisms on the other hand.

One crucial aspect related to DNP which has shown tremendous developments over the last few years, is the polarizing agent (PA). Its importance is emphasized by a central role in DNP research; the PA provides the large electron spin polarization, which is subsequently transferred to the nuclear spins exploiting one of the known DNP mechanisms [i.e., solid effect (SE), cross effect (CE), Overhauser effect, or thermal mixing]. In order to optimize this crucial step in DNP, not only MAS NMR is naturally involved, but also EPR and synthetic chemistry are indispensable tools. This has led to the recent closing of the gap between the NMR and EPR communities – which has before been widening over several decades – and also to the highly interdisciplinary nature of research related to DNP.

The most commonly encountered PAs are persistent radicals based on nitroxides (e.g., TEMPO) or carbon-centered radicals (e.g., trityl, BDPA). These have gone through several developmental stages, giving rise to highly efficient biradicals such as AMUPol[3] or TEKPol[4] aiming toward 'general application', or PAs tailored toward a specific problem. One of the most recent advances is based on heterodimeric biradicals, for example, constituent of a TEMPO and a trityl moiety, which bring together the advantages of each PA and circumvent problems associated with bisnitroxides, such as nuclear depolarization and a detrimental dependence of the DNP enhancement on the external magnetic field.[5]

Besides persistent radicals, open-shell metal ions feature one or more unpaired electron spins available for potential polarization transfer to nuclei (see Section 6.2). In fact, among the earliest DNP experiments, metal ions such as Ce^{3+} and Cr^{5+} played an important role in early fundamental studies, and for scattering experiments on polarized targets; also ferric (Fe^{3+}) impurities have been used to demonstrate direct DNP of ^{59}Co within $K_3Co(CN)_6$.[1] All of these metal ions have spin doublet ($S = 1/2$) ground states; however, strong spin-orbit coupling (SOC) leads in most cases to very large effective **g** anisotropy (e.g., $\Delta g/g_{iso} \approx 0.8$ for $K_3Co(CN)_6$:Fe^{3+}).[6] Whereas this still allows for DNP on single crystals or at low magnetic field, the absolute resonance frequency dispersion in disordered systems at high magnetic field impedes their use in contemporary MAS DNP (see Section 6.4). Cr^{5+} in organic complexes might be an exemption since it has shown a near isotropic g factor[7]; however, a small but significant shift of $g_{iso} = 1.981$ with respect to the free electron pushes the expected DNP matching conditions outside the reachable sweep width of most available DNP instrumentation (see Section 6.5).

High-spin metal ions have found their way into the field of DNP rather late – except for one early demonstration using an impurity of Cr^{3+} for hyperpolarization of ^{27}Al within synthetic sapphire.[8] The first use of Gd^{3+} and Mn^{2+} complexes as PAs

for general MAS DNP application was published in 2011[9] and several studies followed recently, demonstrating various potential fields of applications for these intriguing PA species (see Section 6.6). Particularly, targeted DNP by an endogenously bound Mn^{2+} within an RNA complex,[10] as well as by site-specific labeling of a protein with a Gd^{3+} chelate complex was demonstrated.[11] While Mn^{2+} is biologically relevant and plays a crucial role in many biomolecular systems, such as in enzymes and metalloproteins,[12] Gd^{3+} fulfills no known essential function but has a rather high toxicity.[13] Nevertheless, due to high redox-stability of the ion, as well as the extraordinarily large binding affinity to chelators such as 1,4,7,10-tetraazacyclododecane-1,4,7,10-tetraacetic acid (DOTA),[14] the utilization as PA for in-cell DNP – where nitroxide-based PAs may be quickly reduced inside the cytoplasm[15,16] – is an intriguing possibility, as has been recently demonstrated in the field of dipolar EPR spectroscopy.[17,18]

6.2 EPR PROPERTIES OF PARAMAGNETIC METAL IONS

Since the general theory of DNP in light of the EPR properties of the (doublet) PAs have already been explained in other chapters in this handbook, we focus on the peculiarities encountered for paramagnetic metal ions in this chapter. The most striking differences between metal ions and typical radical PAs manifest in the electron Zeeman (EZ) interaction due to SOC and – for the case of high-spin systems featuring more than one unpaired electron – in zero-field splitting (ZFS). In the following, we provide an abridged overview about the concept and conclusions, while for an in-depth derivation and discussion, the reader is referred to the excellent classical works by Abragam and Bleaney as well as more modern aspects compiled by Schweiger and Jeschke.[6,19]

6.2.1 Electron Zeeman Interaction and Spin-orbit Coupling

6.2.1.1 Atomic and Atom-like States: Isotropic Zeeman Interaction

For a free electron, the ideal electron Zeeman interaction can be described by

$$\hat{H}_{EZ} = \frac{\mu_B g_e}{\hbar} \mathbf{B}^T \hat{\mathbf{S}} \tag{6.1}$$

where **B** is the external magnetic field vector (comprised of a static component $\mathbf{B}_0$ and an oscillatory component $\mathbf{B}_1$), $g_e = 2.002319$ is the free electron g factor, and $\hat{\mathbf{S}} = (\hat{S}_x, \hat{S}_y, \hat{S}_z)^T$ is the (pure) electron spin vector operator of spin angular momentum **S**; superscript 'T' denotes the vectors' transpose. μ_B and $\hbar$ are used according to their usual definition for the Bohr magneton and the reduced Planck constant, respectively. Equation (6.1) is strictly valid only for an electron in vacuum; as soon as the electron is bound within an atomic state, an orbital angular momentum **L** with vector operator $\hat{\mathbf{L}} = (\hat{L}_x, \hat{L}_y, \hat{L}_z)^T$ has to be considered. For the sake of simplicity, we first discuss a system in the strong-field limit. In this case, the Paschen-Back effect allows us to treat spin and orbital angular momentum separately because both being conserved quantities each, while the SOC is a perturbation and results in mixing between spin and orbital angular states. This coupling is described by the last term in the Zeeman Hamiltonian:

$$\hat{H}_Z = \frac{\mu_B}{\hbar} \mathbf{B}^T (\hat{\mathbf{L}} + g_e \hat{\mathbf{S}}) + \lambda \hat{\mathbf{L}}^T \hat{\mathbf{S}} \qquad (6.2)$$

As a relativistic effect, the SOC is depending on the kinetic energy of the electron. For hydrogen or hydrogen-like ions, the SOC constant λ is proportional to the atomic number Z of the nucleus forcing the electron on its orbit and inversely proportional to the expectation value of the cubic electron–nuclear distance r:

$$\lambda = \frac{\mu_0 \mu_B^2}{2\pi\hbar} \frac{Z}{\langle r^3 \rangle} \qquad (6.3)$$

Although equation (6.3) is strictly valid only for systems consisting of one electron orbiting an otherwise bare nucleus, the qualitative behavior can be transferred to more complex electronic systems. Here, it becomes clear that SOC is typically small for radicals based on light elements such as carbon, nitrogen, and oxygen, whereas it can become substantial for transition metal ions. For heavy elements with severely contracted open shells – a situation typically encountered with lanthanide ions due to their compact f-orbitals – this interaction is particularly strong. In this case, the above-described Paschen-Back effect is not valid, and an intermediate or weak-field situation is encountered, i.e., SOC competes with or dominates over the interactions of spin and orbital momenta with the magnetic field. **S** and **L** are not conserved anymore individually, but so is the total angular momentum:

$$\mathbf{J} = \mathbf{L} + \mathbf{S} \qquad (6.4)$$

Within the Russell-Saunders approximation (*LS* coupling), the possible values of the total angular momentum quantum number J can be constructed from the respective Clebsch–Gordan (CG) series and can take values between $J = L + S$ and $J = |L - S|$. For atomic states, an isotropic Zeeman interaction is observed:

$$\hat{H}_Z = \frac{\mu_B g_J}{\hbar} \mathbf{B}^T \hat{\mathbf{J}} \qquad (6.5)$$

where $\hat{\mathbf{J}} = (\hat{J}_x, \hat{J}_y, \hat{J}_z)^T$ is now the total angular momentum vector operator and the Landé g factor can be derived from the contributions of L and S to the total J:

$$g_J = g_L \frac{J(J+1) + L(L+1) - S(S+1)}{2J(J+1)} \\ + g_S \frac{J(J+1) - L(L+1) + S(S+1)}{2J(J+1)} \qquad (6.6)$$

g_L and g_S are the orbital and spin g factors, which usually take the values $g_L = 1$ and $g_S = g_e = 2.002319$, respectively. However, relativistic effects can lead to a slight deviation from these values if heavy elements are involved. For the lanthanide series of tripositive ions, the various possibilities to combine L and S give rise to g_J values between 0 (e.g., $^7F_0$ state of Eu^{3+}) and ~ 2 (e.g., $^8S_{7/2}$ state of Gd^{3+}).[6]

If SOC is such strong that the spin and orbital momenta of individual electrons – before being combined to the respective atomic quanta – couple and form an individual total angular momentum vector $\mathbf{j}_i$ for each electron, the so-called *jj* coupling case has to be considered. This can lead to significant deviations from the simpler *LS* coupling scheme. While Russel-Saunders approximation is typically valid for the first row of transition metal ions, contributions from strong (*jj*) coupling have to be accounted for particularly in the case of lanthanides and heavier elements.

6.3 NONDEGENERATE GROUND STATES: ANISOTROPIC ZEEMAN INTERACTION

In many practical cases encountered in EPR (or DNP), the electronic system exists in a nondegenerate ground state. This can be caused by occupation of every d- or f-shell atomic orbital with a single unpaired electron each, or by a strong, nonspherical electrostatic potential lifting the degeneracy of atomic orbitals, which is typically encountered in radical centers due to molecular bonding environment. As a consequence, the orbital momentum can be mostly quenched, resulting in an effective $L = 0$; nevertheless, partial admixture of excited electronic state(s) with $L > 0$ can lead to significant SOC. In these cases, it has become customary to describe the behavior of such a system by the *effective* electron Zeeman Hamiltonian:

$$\hat{H}_{EZ} = \frac{\mu_B}{\hbar} \mathbf{B}^T \mathbf{g} \hat{\mathbf{S}}' \qquad (6.7)$$

Here, the *effective* spin vector operator $\hat{\mathbf{S}}' = (\hat{S}_x', \hat{S}_y', \hat{S}_z')^T$ acts on the *effective* electron spin basis (including orbital admixture due to SOC) the same way the *pure* spin and orbital operators act on the *pure* basis described by the spin and orbital quantum numbers S and L as described by equation (6.2). As a result, the **g** tensor, **g**, now describes the deviations of this effective Zeeman effect from the pure spin behavior and therefore also includes the anisotropic contribution of the SOC tensor $\mathbf{\Lambda}$:

$$\mathbf{g} = g_e \mathbf{1} + 2\lambda \mathbf{\Lambda} \qquad (6.8)$$

If the perturbation due to SOC is small (i.e., excited electronic states are high in energy with respect to the nondegenerate ground state), the matrix elements Λ_{ij} of the SOC tensor $\mathbf{\Lambda}$ can be adequately obtained by perturbation theory:

$$\Lambda_{ij} = \sum_{n \neq 0} \frac{\langle \psi_0 | \hat{L}_i | \psi_n \rangle \langle \psi_n | \hat{L}_j | \psi_0 \rangle}{\varepsilon_n - \varepsilon_0} \qquad (6.9)$$

Here, ψ_0 and ψ_n are the electronic wave functions of the ground state and the n-th excited state, whereas ε_0 and ε_n are their respective eigenenergies; the indices i and j count over all Cartesian axes x, y, and z. This admixture of orbital angular momentum to the spin has two (tightly connected) consequences: a significant anisotropy of the electron Zeeman interaction is introduced and the isotropic g factor can be significantly offset with respect to that of the free electron. The latter aspect can be quantified by the trace of **g** following equations (6.8) and (6.9):

$$g = g_e + \frac{2\lambda}{3} \sum_{i = \{x,y,z\}} \sum_{n \neq 0} \frac{\langle \psi_0 | \hat{L}_i | \psi_n \rangle \langle \psi_n | \hat{L}_i | \psi_0 \rangle}{\varepsilon_0 - \varepsilon_n} \qquad (6.10)$$

For DNP, a certain Zeeman or **g** anisotropy might be desirable, because this is crucial for allowing the highly efficient CE under MAS due to level-anticrossings (LACs) induced upon sample rotation inside the magnetic field.[20-22] However, if the anisotropy is exceeding the magnitude of the nuclear Larmor frequency, CE efficiency is reduced due to vanishing adiabaticity of the associated LACs. An even simpler argument emphasizes the detrimental impact of a significant offset of the isotropic g factor: EPR/DNP excitation may be prohibited per se if a fixed-frequency microwave (μw) source is used in combination with a nonsweepable NMR magnet, or if the deviation exceeds any sweep range of the magnetic field or frequency source.

In general, organic radicals feature rather high-lying excited states and as a consequence only a minor shift in the isotropic g factor. Nevertheless, the **g** anisotropy is often significant and, for the typically used PAs, can be on a spectrum between vanishingly small even at highest magnetic fields (e.g., BDPA),[11,23,24] or dominating the EPR spectrum already at moderate fields. The most prominent PA-type in MAS DNP, a nitroxide-based biradical, falls in the latter category with anisotropy in the range of ~0.3%. Many open-shell metal ions, on the other hand, feature huge anisotropies of ~20% (e.g., Cu^{2+} with $g_\parallel \approx 2.5$ and $g_\perp \approx 2.1$ in $La_2Mg_3(NO_3)_{12} \cdot 24D_2O$) and much larger (e.g., low-spin Fe^{3+} with $g_{xx} \approx 2.4$, $g_{yy} \approx 2.1$, and $g_{zz} \approx 0.9$ in $K_3Co(CN)_6$, or low-spin Mn^{2+} with $g_{xx} \approx 2.6$, $g_{yy} \approx 2.2$, and $g_{zz} \approx 0.6$ in $K_4Fe(CN)_6 \cdot 3H_2O$).[6] For large SOC, rather extreme cases such as Ce^{3+} can be found with $g_\parallel \approx 1.8$ and $|g_\perp| \leq 0.1$ in $La_2Mg_3(NO_3)_{12} \cdot 24H_2O$.[25] Any of these cases results in a diminishing fraction of the EPR spectrum being accessible in a DNP setup with monochromatic irradiation and/or narrow sweep range. In some cases, the resulting shift in isotropic g factor is even such strong that the full inhomogeneously broadened EPR spectrum is shifted and no spin packet resonates at the Larmor frequency of the free electron as is often encountered for Cu^{2+}.

6.3.1   High-spin Systems and Fine Structure

6.3.1.1   The Electronic Structure of Many-electron Metal Centers

For metal ions containing more than one unpaired electron in the open shell, the situation can become much more complex. If a (nonquenched) orbital momentum **L** strongly interacting with the spin is present – as found in most paramagnetic lanthanide ions except the f^7 configuration – the analytical description of the magnetic states including fine structure (i.e., the interaction within the system of strongly coupled unpaired electrons) is tedious. In general, the electronic configuration starkly depends on the magnitude and geometry of the crystal or ligand field and significant deviations from Russell-Saunders (*LS*) coupling may occur for heavy elements.

Nevertheless, in lanthanide ions the deeply buried 4f orbitals are strongly shielded by the surrounding orbitals, resulting in an atom-like configuration. This makes the description on the basis of the total angular momentum quantum number comprised of coupled orbital and spin angular momenta rather intuitive. In short, for half-integer J (i.e., in Kramers systems), magnetic resonance transitions can usually only be observed within Kramers doublets. The separation of these doublets is typically on the order of $10-100\,\text{cm}^{-1}$ ($0.3-3\,\text{THz}$) and therefore transitions between states of different Kramers doublets are rarely accessible for currently available (tunable) μw sources. Because the spin angular momentum **S** is not conserved due to strong SOC, 'forbidden' EPR transitions violating the $\Delta m_S = \pm 1$ selection rule are possible; however, often only the Kramers doublet lowest in energy is thermally populated and can give rise to an EPR signal. This transition can be conveniently described by a *fictitious spin* with a large anisotropy of the electron Zeeman interaction and principal axis components of the effective **g** tensor span over a wide range of values. For non-Kramers ions, the situation becomes more complicated. Within a ligand field, the total angular momentum states form a series of doublets and singlets between which magnetic transitions can only be excited by an oscillating magnetic field component $\mathbf{B}_1 \parallel \mathbf{B}_0$, i.e., parallel to the static external field (whereas typically EPR transitions are excited by a transverse μw field $\mathbf{B}_1 \perp \mathbf{B}_0$). A fictitious spin treatment may be desirable as well, and again effective g components $\gg 2$ are often observed for thermally accessible transitions. As a conclusion, we only note that these properties are incompatible with the requirements associated with efficient PAs for MAS DNP. Nevertheless, we refer the interested reader to the excellent treatment by Abragam and Bleaney for a more detailed description.

In open-shell transition metal complexes, the d electrons are strongly influenced by the ligand field and actively participate in the formation of molecular orbitals. Thus, the electronic states cannot be described by atomic term symbols, and symmetry-adapted molecular term symbols are more adequate. As a consequence of this, spin and orbital momenta are not conserved individually (if the latter is present), and the total angular momentum generally has to be considered. The variety of properties due to different electronic ground-state terms, low-lying excited states, as well as ligand fields of varying magnitude and symmetry lead to a rather overwhelming set of possibilities, and each case requires individual consideration. This is even more emphasized by the fact that in many cases, small changes in ligand field can have significant impact on the EPR properties, which impedes intuitive prediction.

The situation considerably simplifies when a half-filled electron subshell is encountered. This can be generally found in high-spin d^5 and f^7 systems such as Mn(II) and Gd(III), respectively. In this case, the orbital momentum of the many-electron system is naturally quenched because every electron with an individual orbital momentum is opposed by another electron with opposite orbital magnetic quantum number m_l and the ground state is strictly nondegenerate. Furthermore, due to the stability of this peculiar ground state with maximum spin multiplicity, the energy separation to the nearest electronic excited state is large. Then, admixture of orbital momentum can be neglected in many cases, particularly if the ligand field is relatively weak and of high symmetry. In the case of the deeply buried f^7 system of Gd(III) (as well as the chemically less stable Eu^{2+} and Tb^{4+}), this leads to an ^8S electronic ground state; the ^6S atomic ground state of the d^5 system in Mn(II) complexes is typically expressed as an $^6A_1$ molecular term due to the larger participation of d electrons within molecular bonding to the ligands. A similar situation is found for Cr(III) in a highly symmetric, near-octahedral ligand field: even though a ^4F atomic ground state with $L = 3$ is found in the free ion, the d^3 system forms a $^4A_2$ molecular ground state upon introduction of a ligand field. This can be intuitively understood because the ligand–field interaction lifts the degeneracy of the t_{2g} and e_g orbitals. Consequently, the triply occupied t_{2g} orbitals form a nondegenerate ground state with quartet spin multiplicity and quenched orbital momentum. However, under a distortion from octahedral symmetry, the rather easily accessible $^4T_2$ excited state can lead to significant admixture of orbital momentum, resulting in a significant deviation from the free electron g factor due to SOC.

6.3.1.2 *Zero-field Splitting*

Even in cases where SOC is only a perturbation, ZFS may still become an important or even the dominant interaction besides electron Zeeman interaction. In the most general sense, ZFS is the interaction of the higher-order spin moment of a high-spin system ($S \geq 1$) with a nonspherical electronic distribution. For $1 \leq S \leq 2$, only quadrupolar moment may be present in environments of noncubic symmetry, whereas for $S > 2$ higher-order moments (i.e., hexadecapolar) can additionally occur even in environments of cubic symmetry groups. However, the latter are usually small and not to be concerned with respect to DNP applications. Therefore, we can limit our description – in the absence of any SOC – to the quadrupolar interaction in the high-spin system expressed by the second-rank ZFS tensor $\mathbf{D}$:

$$\hat{H}_{ZFS} = \hat{\mathbf{S}}^{T} \mathbf{D} \hat{\mathbf{S}} \qquad (6.11)$$

which can be expressed in its principal axis system as:

$$\hat{H}_{ZFS} = D\left[\hat{S}_z^2 - \frac{S(S+1)}{3}\hat{1} \right] + E(\hat{S}_+^2 + \hat{S}_-^2) \qquad (6.12)$$

Here, D and E are the ZFS anisotropy and asymmetry parameters, respectively, in their common form derived from the principal axis ZFS tensor elements: $D = 3(D_{33}/2)$; $E = (D_{22} - D_{11})/2$; $0 \leq E \leq D/3$. $\hat{S}_+$ and $\hat{S}_-$ are the electron spin raising and lowering operators, respectively, following $\hat{S}_\pm = \hat{S}_x \pm i\hat{S}_y$. A more general description of ZFS (also allowing for treatment of interactions of higher than quadrupolar order) is based on the rotational symmetry-adapted Stevens operators but beyond the scope of our treatment.

ZFS arises from the dipolar interaction between the electron spins within a nonspherical electron distribution due to a low-symmetric electrostatic potential (i.e., nonspherical electric field gradient). Therefore, it can be increased by raising the strength and/or reducing the symmetry of the ligand field that acts on the high-spin metal center. The field's symmetry group is therefore directly influencing the symmetry properties of $\mathbf{D}$: for a cubic symmetry group, the (quadrupolar) ZFS vanishes ($D = 0$, $E = 0$), for tetragonal symmetry the tensor acquires axial structure ($D \neq 0$, $E = 0$), whereas for lower than tetragonal symmetry an (ortho)rhombic ZFS tensor is observed ($D \neq 0$, $0 < |E| \leq |D/3|$).

As long as ZFS is relatively small with respect to the electron Zeeman interaction, the eigenenergies of such a high-spin system can be derived by perturbation theory.[19] For this, it is most convenient to define the orientation between the tensor frame and the laboratory frame (defined by the external magnetic field) by the longitudinal and azimuthal angles, θ and φ, respectively. All satellite transitions (ST) – where $|m_S|$ is changing by one unit – can be sufficiently described by only considering first-order ZFS terms.

This is especially true for line shape analysis of frozen solutions at high magnetic field where typically the distribution of ZFS parameters is much larger than second-order effects (see below). In this case, the eigenenergies are given by

$$\frac{\Delta E_{ZFS}^{(1)}(m_S)}{\hbar} = \frac{D(3\cos^2\theta - 1) + 3E\sin^2\theta(2\sin^2\varphi - 1)}{6}$$
$$\times [3m_S^2 - S(S+1)] \qquad (6.13)$$

and the EPR frequency shifts for a general transition $|m_S\rangle \leftarrow |m_S - 1\rangle$ are

$$\omega_{ZFS}^{(1)}(m_S, \theta, \varphi) = [D(3\cos^2\theta - 1)$$
$$+ 3E\sin^2\theta(2\sin^2\varphi - 1)]\left(m_S - \frac{1}{2} \right) \qquad (6.14)$$

The resulting eigenstate and transition frequency shifts are depicted in Figure 6.1 for an $S = 7/2$ system. For the EPR central transition (CT) – where $|m_S| = 1/2$ is conserved – the first-order ZFS vanishes and the second-order term dominates the line shape.

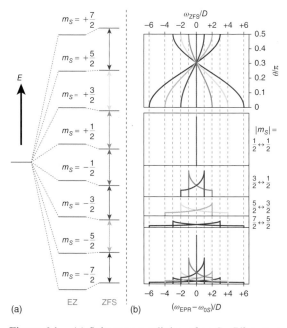

Figure 6.1. (a) Spin energy splitting of an $S = 7/2$ system (e.g., Gd^{3+}) with axially symmetric ZFS ($E = 0$). Note that EZ and ZFS interaction are not to scale. (b) Angular dependence of EPR transitions shifted by ZFS (first order) and resulting shape of the EPR line as deduced from individual EPR transitions. (Source: Corzilius,[22] https://pubs.rsc.org/en/content/articlehtml/2016/cp/c6cp04621e. Licensed under CC BY 3.0)

Therefore, the transition frequency can be expressed as

$$\omega_{ZFS}^{CT}\left(m_S = \frac{1}{2}, \theta, \varphi\right) = \frac{4S(S+1)-3}{16\omega_{0S}}$$
$$\times [p_1(\theta)D^2 - 2p_2(\theta, \varphi)DE + p_3(\theta, \varphi)E^2] \quad (6.15)$$

with

$$p_1(\theta) = \sin^2\theta(1 - 9\cos^2\theta)$$
$$p_2(\theta, \varphi) = \sin^2\theta(9\cos^2\theta + 1)(2\cos^2\varphi - 1)$$
$$p_3(\theta, \varphi) = 4 - 12\sin^2\theta + 9\sin^4\theta(2\cos^2\varphi - 1)^2$$
$$(6.16)$$

This second-order ZFS scales inversely with the Zeeman frequency – and thus the external magnetic field – and is therefore most important at lower magnetic fields usually employed in low-field EPR. Nevertheless, due to the oftentimes large ZFS constants in combination with the absence of other broadening mechanisms, the CT line shape is in many cases dominated by ZFS even at DNP-relevant magnetic fields of 9.4 T and higher.

Besides shifts in the eigenstates, high-spin properties affect the transition moments of connected (allowed) EPR transitions. For a transition $|m_S\rangle \leftarrow |m_S - 1\rangle$, this can be deduced from the matrix elements $\langle m_S|\omega_{1S}\hat{S}_x|m_S - 1\rangle$ and can be calculated as[26]

$$\omega_N^{EPR}(m_S) = [S(S+1) - m_S(m_S - 1)]^{1/2}\omega_{1S} \quad (6.17)$$

with the Rabi frequency $\omega_{1S} = \mu_B g B_1/\hbar$. For $S = 7/2$, this leads to ~2.65–4 times larger transition moments or 7–16 times larger transition probabilities compared to $S = 1/2$.

As a result, several typical ZFS line shape functions have to be considered based on the abovementioned ZFS tensor symmetry. In Figure 6.2, these are shown for the ideal axial ($E = 0$), intermediate orthorhombic ($0 < E < D/3$), and ideal orthorhombic case ($E = D/3$). Under the typical situation observed for the metal ions with quenched SOC, D is much smaller than the electron Zeeman frequency, and as such, a Pake-like multiplet structure is observed. The CT is virtually unaffected by ZFS and by far the most intense feature; nevertheless, a significant second-order effect dominates the CT line shape if other broadening mechanisms are negligible.

In this context, it is important to consider the populations of states in thermal equilibrium as well. Because neither of the states connected by the CT is the energetic ground state of the electron spin system under dominant electron Zeeman interaction, these states

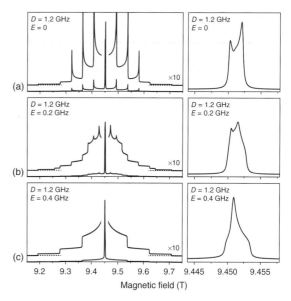

Figure 6.2. Simulations of 263 GHz EPR line shape under high-temperature approximation of Gd^{3+} with three different ZFS parameter sets representing axial (a), intermediate orthorhombic (b), and the ideal orthorhombic (c) case. The slightly thinner sets of lines are the same spectra multiplied by a factor of 10; their baseline is indicated by horizontal, dashed black lines at the edges of the spectrum. The right column shows horizontal magnifications of the respective CTs in a narrow field range. (Source: Corzilius,[22] https://pubs.rsc.org/en/content/articlehtml/2016/cp/c6cp04621e. Licensed under CC BY 3.0)

will be depopulated at sufficiently low temperatures (Figure 6.3). At high fields relevant to MAS DNP (i.e., neglecting the absolute population changes due to ZFS shifts), the maximum population difference for the CT can be estimated to occur at $T_{max}[K] \approx 0.69 \times B_0[T] \times \sqrt{S(S+1)}$. Thus, for Gd^{3+} and Mn^{2+} at 9.4 T, maximum CT intensity is observed at ~26 and ~19 K, respectively.

In solutions of Gd^{3+} complexes, the ligands maintain a rather large flexibility within the first coordination sphere of the lanthanide ion. As a consequence, the D and E parameters are subject to large distributions upon vitrification, with standard deviations being of the same order of magnitude as the expectation value itself. The situation has been modeled by Raitsimring *et al.* using a bimodal standard distribution of D symmetric to 0, with a significant distribution width of each lobe being half of its mean $|D|$ value; the E/D ratio takes up a quadratic distribution with a maximum

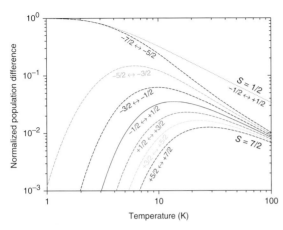

Figure 6.3. Relative population difference of neighboring m_S states of a $S = 1/2$ system (dotted line) and a $S = 7/2$ system (STs dashed, CT solid line), exemplarily shown for a field of 5 T

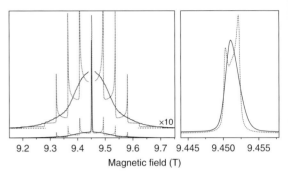

Figure 6.4. Simulations of 263 GHz EPR line shape of Gd^{3+} with axial ZFS of $D = 1.2$ GHz (dashed red line) in comparison with a bimodal ZFS distribution with maximum occurrences at $|D| = 1.2$ GHz and $E/D = 0.25$ (solid blue line), following the model of Raitsimring *et al.*[27] The slightly thinner sets of lines are the same spectra multiplied by a factor of 10; their baseline is indicated by horizontal dashed black lines. The right plot shows a horizontal magnification of the CT in a narrow field range. (Source: Corzilius,[22] https://pubs.rsc.org/en/content/articlehtml/2016/cp/c6cp04621e. Licensed under CC BY 3.0)

occurrence near 0.25 and vanishing probability toward $E/D = 0.$[27] A slightly different model has been introduced by Benmelouka *et al.*[28]; however, it has been recently shown that both models indeed lead to very similar effective distributions of D and E parameters and comparable fitting quality for EPR spectra of Gd^{3+} complexes.[29] As a result, no resolved features can be detected in a field/frequency-swept EPR spectrum; the CT shows a typical asymmetric broadening while the STs are only visible as a rather broad feature of relatively weak intensity (Figure 6.4).

6.4 DNP THEORY OF HIGH-SPIN SYSTEMS

6.4.1 Solid Effect

The SE description can be adopted to metal ion PAs by combination of SE theory for $S = 1/2$ and the above-described peculiarities of high-spin systems. A detailed derivation of the theoretical background has been published earlier and is the basis for the explanations here.[22] Shifts in eigenenergies due to ZFS cause inhomogeneous broadening, with effects similar to significant **g** anisotropy. The periodic modulation of the eigenstate energy during MAS will cause a comparable situation including transient SE LACs as has been discussed for nitroxide PAs;[21] however,

the consequences with respect to a static sample situation is not completely clear and requires thorough experimental investigation of SE dependence on MAS frequency.

Basis for further discussion is provided by the general pseudo-high-field Hamiltonian for a two-spin (electron–nuclear) system:

$$\widehat{H}_{SE} = \frac{\mu_B}{\hbar} \mathbf{B}^T \mathbf{g} \widehat{\mathbf{S}} + \widehat{\mathbf{S}}^T \mathbf{D} \widehat{\mathbf{S}} - \frac{\mu_n g_n}{\hbar} \mathbf{B}^T \widehat{\mathbf{I}} + \widehat{\mathbf{S}}^T \mathbf{A} \widehat{\mathbf{I}} \quad (6.18)$$

Here, the terms are the electron Zeeman interaction, ZFS, nuclear Zeeman (NZ) interaction, and hyperfine interaction (HFI), in that order; **A** is the HFI tensor. Under µw irradiation and in the pseudo-high-field approximation, we can simplify this to

$$\widehat{H}_{SE} = (\omega_{0S} - \omega_{\mu w}) \widehat{S}_z + D_{zz} \left(\widehat{S}_z^2 - \frac{1}{3} \widehat{\mathbf{S}}^2 \right)$$
$$- \omega_{0I} \widehat{I}_z + A \widehat{S}_z \widehat{I}_z + B \widehat{S}_z \widehat{I}_x + \omega_{1S} \widehat{S}_x \quad (6.19)$$

where ω_{0S} and ω_{0I} are the electron and nuclear Zeeman frequencies, ω_{1S} is the electron Rabi frequency, $\omega_{\mu w}$ is the frequency of the incident µw irradiation, and $A = A_{zz}$ as well as $B = \sqrt{A_{zx}^2 + A_{zy}^2}$ are the secular and pseudo-secular HFI coupling constants, respectively.

Next, we want to analyze shifts in eigenenergy by first-order perturbation theory:

$$\frac{E^{(1)}(m_S, m_I)}{\hbar} = m_S\omega_{0S} + \left(m_S^2 - \frac{S(S+1)}{3}\right)D_{zz}$$
$$- m_I\omega_{0I} + m_Sm_IA \qquad (6.20)$$

We can obtain the general SE matching conditions by deriving the energy differences for the ZQ and DQ transition, respectively, i.e. $|m_S, m_I\rangle \leftarrow |m_S - 1, m_I \pm 1\rangle$, where m_S and m_I denote the *final* state:

$$\omega_{\mu w}^{ZQ/DQ} = \omega_{0S} + (2m_S - 1)D_{zz} \pm \omega_{0I} \pm (1 - m_S \pm m_I)A \qquad (6.21)$$

A unique situation that is encountered only when dealing with high-spin systems is that the SE matching condition not only depends on the ZFS [which can also be described by the respective constant, ω_{ZFS}, as defined in equations (6.14) or (6.15)], but additionally on the secular HFI to the nuclear spin to be polarized within the electron–nucleus pair. For the $S = 1/2$ case, in contrast, the electron and nuclear Zeeman interactions are the only relevant terms as the HFI completely cancels within the ZQ and DQ transitions. Here, however, this HFI shift occurs exclusively when STs are considered and vanishes for the CT.

In a general sense, the ZFS acts on the ZQ and DQ SE matching conditions the same way as on the (single quantum) EPR resonance condition, i.e., the inhomogeneous ZFS broadening of the EPR spectrum can be projected simply onto the SE field profile. By restricting the treatment to first-order ZFS, the same matching conditions would be obtained for the CT as for $S = 1/2$ due to the only possible final state with $m_S = 1/2$. If a significant second-order effect is encountered, the effective ZFS shift can be used as given in equation (6.15) and a simplified CT SE matching condition is obtained for the typical $I = 1/2$ case:

$$\omega_{\mu w}^{ZQ/DQ} = \omega_{0S} + \omega_{ZFS}^{CT} \pm \omega_{0I} \qquad (6.22)$$

By introducing the effective EPR frequency $\omega_{EPR} = \omega_{0S} + \omega_{ZFS}^{CT}$, which combines the electron Zeeman and ZFS terms, a form analogous to the well-known and simple SE matching condition is obtained[30]:

$$\omega_{\mu w}^{ZQ/DQ} = \omega_{EPR} \pm \omega_{0I} \qquad (6.23)$$

In fact, any interaction except the HFI to the nucleus to be polarized may be contained in ω_{EPR} such that, for example, also the large sextet splitting from ^{55}Mn HFI can be effectively treated within equation (6.23). As a result, the SE field/frequency profile can be interpreted the same way as for a typical narrow-line radical such as trityl,[31] except that the inhomogeneous broadening is caused by (second-order) ZFS instead of **g** anisotropy. Since the first-order ZFS acting on the STs is typically of the same magnitude or larger than the nuclear Larmor frequency ($D \geq \omega_{0I}$) even at the highest magnetic fields, it is expected that any contribution from the broad ST 'background' results in differential SE and thus leads to very small DNP enhancements.[8,32]

Another theoretical peculiarity can be found in the SE transition moment, ω_N^{SE}, by which electron polarization is transformed into electron–nuclear DQ or ZQ coherence under µw irradiation. This is not only enhanced due to the generally larger CG coefficients obtained for high-spin systems [see equation (6.17)], but also depends on m_S due to the branching of nuclear spin quantization under pseudo-secular HFI with the high-spin electronic system:

$$\omega_N^{SE}(m_S) = [S(S+1) - m_S(m_S - 1)]^{1/2}\omega_{1S} \sin\eta_{m_S}^- \qquad (6.24)$$

Here, $\eta_{m_S}^-$ is the differential branching angle between the initial state $m_S - 1$ and the final state m_S:

$$\eta_{m_S}^- = \frac{\eta_{m_S} - \eta_{m_{S-1}}}{2} \qquad (6.25)$$

The individual branching angles can be obtained from geometrical considerations as depicted in Figure 6.5:

$$\eta_{m_S} = \arctan\left(\frac{m_SB}{\omega_{0I} + m_SA}\right) \quad \text{with} \quad -\frac{\pi}{2} \leq \eta_{m_S} < \frac{\pi}{2} \qquad (6.26)$$

In the high-field case, this can often be simplified to

$$\eta_{m_S} \approx \arctan\left(\frac{m_SB}{\omega_{0I}}\right) \approx \frac{m_SB}{\omega_{0I}} \quad \text{for} \quad \omega_{0I} \gg A, B \qquad (6.27)$$

Here, we find that for neighboring m_S states

$$\eta_{m_S}^- \approx \frac{m_SB - (m_S - 1)B}{2\omega_{0I}} = \frac{B}{2\omega_{0I}} \qquad (6.28)$$

and consequently

$$\omega_N^{SF}(m_S) \approx [S(S+1) - m_S(m_S - 1)]^{1/2}\frac{\omega_{1S}B}{2\omega_{0I}} \qquad (6.29)$$

Except for the larger CG coefficient, this result is again equal to what would be obtained for the $S = 1/2$ case. This guides us to the conclusion that no qualitative differences have to be considered under high magnetic fields besides the obviously increased high-spin transition moments, leading to a 16-fold increase in

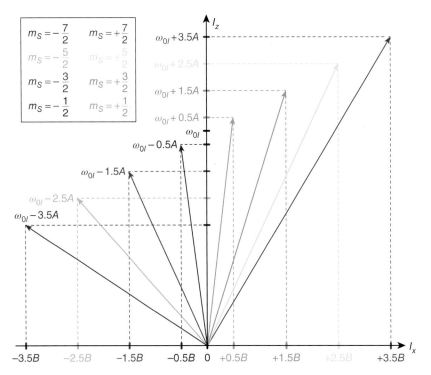

Figure 6.5.   Branching of nuclear spin quantization for the different m_S substates of an $S = 7/2$, $I = 1/2$ spin system. (Source: Corzilius,[22] https://pubs.rsc.org/en/content/articlehtml/2016/cp/c6cp04621e. Licensed under CC BY 3.0)

SE efficiency for the CT. Particularly if the STs are significantly broadened and do not effectively contribute to the SE field/frequency profile, the same semiquantitative models can be applied as already described for 'conventional' radical PAs, for example, the prediction of field profiles on the basis of the CT EPR spectrum as introduced by Vega and coworkers.[33]

6.4.2   Cross Effect

In the simplest general case, the analytical description of the CE is based on a Hamiltonian describing two electron spins (S_1, S_2) and one nuclear spin (I):

$$
\hat{H}_{CE} = \frac{\mu_B}{\hbar}\mathbf{B}^T(\mathbf{g}_1\hat{\mathbf{S}}_1 + \mathbf{g}_2\hat{\mathbf{S}}_2) - \frac{\mu_n g_n}{\hbar}\mathbf{B}^T\hat{\mathbf{I}} + \hat{\mathbf{S}}_1^T\mathbf{D}_1\hat{\mathbf{S}}_1
$$
$$
+ \hat{\mathbf{S}}_2^T\mathbf{D}_2\hat{\mathbf{S}}_2 + \hat{\mathbf{S}}_1^T\mathbf{D}^{ee}\hat{\mathbf{S}}_2 + (\hat{\mathbf{S}}_1^T\mathbf{A}_1 + \hat{\mathbf{S}}_2^T\mathbf{A}_2)\hat{\mathbf{I}}
$$
$$(6.30)$$

Here, nomenclature is the same as in equation (6.18); the additional numerical indices refer to the first or second electron spin, respectively. Furthermore,

interaction between the two electron spins is described by the electron–electron (e–e) coupling tensor $\mathbf{D}^{ee}$. Note that in contrast to equation (6.19), a rotating-frame treatment is not required because µw-induced coherences do not directly play a role in CE transfer.

Nevertheless, this confronts us with a situation that is very tedious, if not impossible to treat in an elegant and general analytic fashion for a general electron spin. For the $S_1 = S_2 = 1/2$ case, the analytical treatment is based on dividing the 8×8 Hamiltonian in Hilbert space into its DQ and ZQ subspace of dimensionality 4 (for each), where every population state is only connected to one other state via a DQ or a ZQ coherence, respectively, in combination with a change in nuclear magnetic spin state (i.e., $\langle\alpha\alpha\alpha|\beta\beta\beta\rangle$ or $\langle\alpha\alpha\beta|\beta\beta\alpha\rangle$ in the DQ, as well as $\langle\alpha\beta\alpha|\beta\alpha\beta\rangle$ or $\langle\alpha\beta\beta|\beta\alpha\alpha\rangle$ in the ZQ subspace).[34] For the general high-spin case, this elegant transformation is impossible. Even for the simplest half-integer high-spin system ($S_1 = S_2 = 3/2$), the dimensionality of the full Hilbert space is already 32, and a clear separation of the DQ and ZQ subspaces is not possible, because several eigenstates are connected to more than one relevant state via DQ or ZQ transitions at the same time. Therefore, a perturbation treatment on the basis of eigenstate analysis for several distinct matching conditions seems adequate to understand the implications of the unique impact of the high spin on the CE. For this, we can describe equation (6.30) in the principal axis system of the e–e interaction while using the standard notation for the respective diagonal interaction tensor with the elements $D_{xx}^{ee} = D_{yy}^{ee} = -d - 2J$ and $D_{zz}^{ee} = 2d - 2J$ (with d and J being the dipolar and exchange coupling constants, respectively). Furthermore, we initially reduce the problem in the high-field approximation, i.e., we discard all off-diagonal (nonsecular and pseudo-secular) elements and assume that the Zeeman interactions of both electrons are isotropic and equal, thus they can be described by a common ω_{0S}. The latter assumption is valid for systems consisting of two ions of the same metal and ligand geometry and where SOC is sufficiently quenched. This is well fulfilled for the systems under consideration such as Gd(III) and Mn(II). Then we are left with the following truncated three-spin Hamiltonian containing only on diagonal elements:

$$\hat{H}_{CE}^{on} = \omega_{0S}(\hat{S}_{1z} + \hat{S}_{2z}) - \omega_{0I}\hat{I}_z + D_{1zz}\hat{S}_{1z}^2 + D_{2zz}\hat{S}_{2z}^2$$
$$+ D_d^{ee}\hat{S}_{1z}\hat{S}_{2z} + (A_{1zz}\hat{S}_{1z} + A_{2zz}\hat{S}_{2z})\hat{I}_z \quad (6.31)$$

$D_d^{ee} = 2(d - J)$ is the effective e–e coupling constant. Note that in order to obey conservation of energy, the ZFS terms need to be expressed by the square spin operators $\hat{S}_{iz}^2 - \frac{1}{3}\hat{S}_i^2$ [see equation (6.19)]; however, we have dropped the last term for simplicity since it leads to a general energy shift of all magnetic spin states. In first-order approximation, we find

$$\frac{E^{(1)}(m_{S_1}, m_{S_2}, m_I)}{\hbar} = M_S\omega_{0S} + m_{S_1}^2 D_{1zz} + m_{S_2}^2 D_{2zz}$$
$$- m_I\omega_{0I} + m_{S_1}m_{S_2}D_d^{ee}$$
$$+ m_I(m_{S_1}A_{1zz} + m_{S_2}A_{2zz}) \quad (6.32)$$

with $M_S = m_{S_1} + m_{S_2}$ and again neglecting the overall center of weight shift due to $-\frac{D_{1zz}}{3}S_1(S_1 + 1) - \frac{D_{2zz}}{3}S_2(S_2 + 1)$. When considering the multitude of possible CE matching conditions between different m_S states of each electron spin center, three distinct cases can be identified: (i) matching between the CTs of each spin center, (ii) matching between the CT of one spin center and an ST of the other, and (iii) matching between two STs of each center. Up to the date of preparation of this handbook, only case (i) has been observed unambiguously in experiment so far[11,35]; nevertheless, contributions from cases (ii)

and (iii) cannot be excluded and may be the cause for several unexplained observations.[35,36]

Generally, a CE matching condition is achieved when two states connected by an e_1-e_2-n flip–flop–flip (or flip–flop–flop for inverse sign of nuclear hyperpolarization) transition are degenerate in energy. In high-spin systems, many or even most states can be reached by either raising or lowering the respective magnetic spin quantum numbers; nevertheless, it is sufficient to consider the transitions $|m_{S_1}, m_{S_2}, m_I\rangle \leftarrow |m_{S_1} - 1, m_{S_2} + 1, m_I \pm 1\rangle$ from which all possible matching conditions can be derived by variation of m_{S_1} and m_{S_2}. The matching conditions are effectively described by

$$\pm\omega_{0I} = (2m_{S_1} - 1)D_{1zz} - (2m_{S_2} + 1)D_{2zz}$$
$$+ (1 - m_{S_1} + m_{S_2})D_d^{ee} \quad (6.33)$$

where the variation in relative sign indicates opposing directions of resulting nuclear hyperpolarization (i.e., positive or negative enhancement). In equation (6.33), we have neglected any potential shift due to differential HFI between the nucleus and the two electron spins because this is typically very small in comparison with the difference in effective EPR transition frequencies dominated by differential ZFS between the two electron spins as well as their coupling D_d.

If the CE is confined to occur within the CTs of both metal centers, case (i), we conveniently use the effective second-order shift due to the ZFS according to equation (6.15) and the simplified CT CE matching condition

$$\pm\omega_{0I} = \omega_{ZFS,1}^{CT} - \omega_{ZFS,2}^{CT} \quad (6.34)$$

is achieved, which does not depend on the e–e coupling in first-order approximation. In a system lacking local order (e.g., vitrified solution of individual metal complexes or bis-complexes with sufficient bending or rotational degrees of freedom within the tether unit), CE matching is thus possible between the CTs of two metal complex sites. An interesting peculiarity arises if the PA sites are embedded in a crystalline matrix, or if rigidly interconnected bis-complexes are used. The restricted orientational space of the two centers' molecular frames with respect to their connecting vector causes concerted shifts of the energy levels and may partially or completely prevent the CE matching condition from occurring. Such a situation has been discussed in the context of both doped crystalline solids and rigid bis-complexes based on indicative data[35,36]; however, additional experiments are required in order to fully understand the implications of these effects.

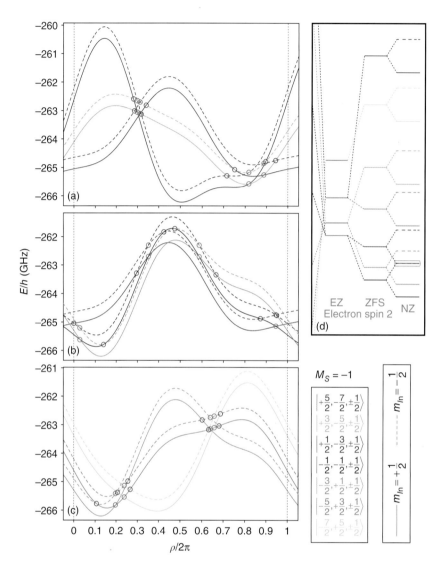

Figure 6.6. (a–c) Evolution of eigenstates under one period of MAS due to ZFS (up to second order) within the $M_S = -1$ subspace. For better visibility, the levels have been separated into three graphs in a way that all relevant LACs can be identified. Different m_I states are shown with solid and dashed lines, respectively. CE LACs (red circles) occur between lines of different type, purely e–e dipolar LACs (green circles) between lines of same type. (d) Level splitting scheme including EZ, ZFS, and NZ for a system of two Gd^{3+} and one proton showing only the $M_S = -1$ subspace including NZ splitting. CE-enabling degeneracy is fulfilled within the red dotted box. (Source: Corzilius,[22] https://pubs.rsc.org/en/content/articlehtml/2016/cp/c6cp04621e. Licensed under CC BY 3.0)

Note that in equation (6.34), any effects leading to e–e state mixing introduced by the nonsecular $D_0^{ee} = -(d + 2J)$ have been neglected. These effects are expected to lead to an effective matching condition with $\omega_{0I} = \pm\sqrt{\Delta\omega_{0S}^2 + (D_0^{ee})^2}$ in the spin-1/2 case.[5] The moderate e–e couplings typically encountered with a bisnitroxide PA are on the order of $D_d^{ee}, D_0^{ee} \leq 2\pi \times 100\,\text{MHz}$ and thus, changes in effective CE matching condition due to D_0^{ee} are rather small. In contrast, for a high-spin system, the direct occurrence of D_d^{ee} in matching condition (6.33) leads to a situation where the e–e coupling can directly influence CE matching. For any CE transition $|m_{S_1}, m_{S_2}, m_I\rangle \leftarrow |m_{S_1} - 1, m_{S_2} + 1, m_I \pm 1\rangle$ except where $m_{S_1} = m_{S_2} + 1$ or $m_{S_2} = m_{S_1} - 1$, the secular e–e interaction is active in first order, leading to potentially large contributions to the CE matching condition. This discussion about secular e–e shifts might seem to be somewhat academic in nature. For D_d^{ee} to persist in the matching condition (6.33) a ST has to be strictly involved, which typically are subject to large ZFS with $D \gg D_d^{ee}$. Nevertheless, this topic has barely been discussed in the context of current research. In particular, the conditional contribution of e–e dipolar broadening in high-spin electron spin pairs to their CE field profile has been overlooked in relevant analysis.[35]

Coming back to the general problem of CE matching in high-spin systems, equation (6.33) now allows us to derive a similar picture of (partially) avoided level-crossings under MAS (Figure 6.6) as has been described by Thurber and Tycko as well as Mentink-Vigier *et al.* for bisnitroxides.[20,21] A detailed description of the $S_1 = S_2 = 7/2$ case has been presented under the assumption of vanishing level shifts due to e–e interaction.[22] Analysis of the MAS-frequency-dependent adiabaticity of the respective LACs induced by ZFS indicate that all relevant events can be of similar magnitude as expected with bisnitroxides at intermediate magnetic fields (i.e., 9.4 T). While the larger transition moments within the high-spin system may compensate for larger absolute crossing rates due to significant first-order ZFS, this interaction is independent of external magnetic field, suggesting a relative advantage over bisnitroxides particularly at highest fields.

6.5 PRACTICAL ASPECTS OF DNP WITH HIGH-SPIN SYSTEMS

6.5.1 Spin Diffusion and Matrix Proton Concentration

The faster electron spin relaxation and stronger paramagnetic effects on surrounding nuclei lead to a distinctly different interplay between DNP transfer, relaxation, and spin diffusion for high-spin metals compared to radical PAs such as trityl. While the fast recovery of electron spin polarization prevents

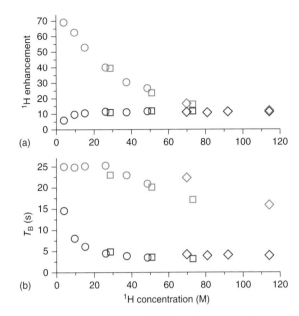

Figure 6.7. Comparison of DNP enhancement factor (a) and build-up time constant (b) between 40 mM trityl Ox063 (green) and 10 mM Gd-DOTA (blue) for various matrix ^{1}H concentrations. For symbol code and experimental details, see original publication. (Reproduced with permission from Ref. 31 © AIP Publishing, 2012)

significant saturation under DNP,[23] the larger effective magnetic moment of the high-spin system causes a stronger quenching of detectable resonances as well as suppression of spin-diffusion in the vicinity of the PA.[37] Consequently, the effective transfer of magnetization from the electron to the nuclei is limited by the spreading of enhanced polarization from the inner (core) nuclei to the bulk.[31] For $S = 1/2$ systems, in contrast, the initial DNP transfer between the electron and the core nuclei is rate limiting, as has been indicated by matrix-proton concentration dependence (Figure 6.7).

For practical considerations, experiments with Gd-DOTA with fully protonated solvent matrix (^{13}C$_3$-glyceral/H$_2$O, 60/40% v/v) have thus yielded equal and slightly better ^{1}H and ^{13}C DNP performance, respectively, compared to the typically used 'DNP juice' (d$_8$-^{13}C$_3$-glycerol/D$_2$O/H$_2$O, 60/30/10% v/v).[11] However, when targeted DNP is sought after, perdeuteration of the matrix may support localization of the enhanced polarization on the analyte molecules that bind the metal ion PA.[10,11]

6.5.2 Limitations of Metal Ion DNP with Current Instrumentation

The efficiency of metal ion PAs falls significantly short when directly compared with well-optimized bisnitroxides in model systems. Therefore, metal ions will most likely not be used as general purpose PAs but will fulfill important roles where doping with nitroxides or other radical PA is not preferential or impossible. For example, long-term stable complexes such as Gd-DOTA may be used for in-cell DNP, endogenously metal-binding (bio)molecules may be targeted by Mn^{2+} or Gd^{3+},[10,11] or metal ion PAs may be isostructurally doped into materials.[36,38]

In any case, the DNP enhancement has thus far been limited by the available DNP transition moment. By increasing this factor – either by providing higher μw power or by utilizing a resonant μw structure – enhancement factors would be improved several-fold and may even approach those obtained by bisnitroxides operating on the CE. While the virtually linear power dependence during MAS DNP allows for promising extrapolation,[11] low-power μw fields amplified within a cavity resonator directly provide evidence for large obtainable ^{1}H DNP efficiency: at 80 K, a DNP enhancement factor of 109 has been measured; at 20 K, this factor increased to 157 (Figure 6.8).[39] Including the advantage from a significant reduction of the DNP built-up time constant, T_B, with respect to the nuclear spin lattice relaxation time constant, T_{1l}, the largest enhancement in signal-to-noise has been recorded as 234-fold. These improvements are not limited to ^{1}H DNP: under similar conditions, a direct ^{13}C enhancement factor in excess of 400 has been observed.[11] These results suggest that given considerable advancements in DNP instrumentation, similar improvements may be achieved under MAS. In this context, it is noteworthy to mention the advent of high-field pulsed DNP, which would allow to manipulate the high-spin system prior to or during the DNP transfer in order to maximize the utilizable electron polarization, as has been already demonstrated for EPR spectroscopy.[40–42]

6.6 DEMONSTRATIONS OF MAS DNP WITH PARAMAGNETIC METAL IONS

6.6.1 Gd^{3+} and Mn^{2+} Bound within Dissolved Chelate Complexes

6.6.1.1 *Monomeric Complexes*

Paramagnetic metal ions have been first introduced as PAs for MAS DNP in 2011. DOTA and diethylene-triaminepentaacetate (DTPA) complexes of Gd^{3+} and Mn^{2+} have been initially chosen due to their excellent binding stability as well as highly symmetric ligand field in aqueous solutions[28,43,44]; furthermore, they are ubiquitously used as contrast agents for MRI (e.g.,

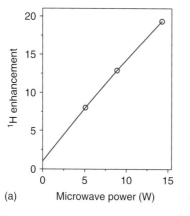
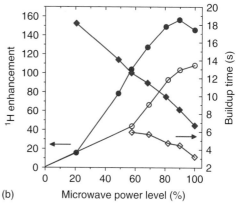

(a)   Microwave power (W)   (b)   Microwave power level (%)

Figure 6.8. Microwave field strength and temperature dependence of ^{1}H DNP enhancements by Gd-DOTA at 140 GHz (5 T). (a) Data obtained under MAS at ~86 K and using a gyrotron. (b) Data obtained under static conditions inside a μw resonator driven by a 120 mW low-power source at 20 K (filled symbols) and 80 K (open symbols). Left ordinate (red circles): SE enhancement factor; right ordinate (blue diamonds): DNP buildup time constants (T_B). (Source: Data taken from (MAS)[11] and (static)[39])

gadoteric acid, gadopentetic acid).[45,46] The acyclic ligand in the DTPA complex of Gd^{3+} induces a larger ZFS and causes much larger inhomogeneous broadening of the CT.[28] This set of samples has allowed not only to demonstrate that high-spin metal complexes are able to generally hyperpolarize ^1H by SE DNP but also to deduce the effects of the strong hyperfine (Fermi contact) interaction to ^{55}Mn as well as the ZFS broadening on the DNP enhancement quantitatively (Figure 6.9).[9] The study was subsequently extended

to include the aqua complex of Gd^{3+},[11] which had already been investigated by electron–electron double resonance (ELDOR) in the laboratory of D. Goldfarb (Weizmann Institute of Science, Rehovot, Israel) in light of potential DNP.[47] For the 'free' Gd^{3+} ion, the existence of more than one structurally different aqua complex seems to result in an unproportionally large reduction of the DNP efficiency; however, the exact cause for this effect has not yet been identified. The direct DNP of ^{13}C and ^{15}N has been demonstrated

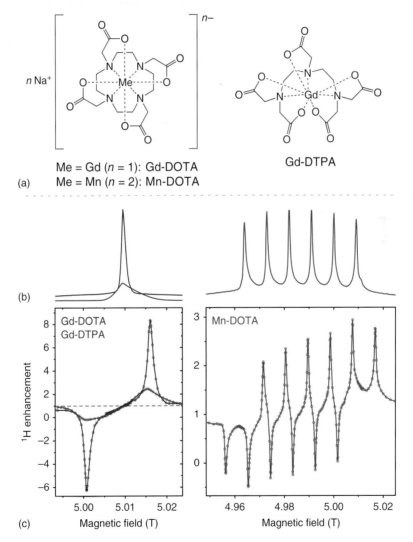

Figure 6.9. (a) Chemical structures, (b) EPR spectra, and (c) ^1H DNP field profiles of Gd-DOTA (blue), Gd-DTPA (red), and Mn-DOTA (green). EPR and DNP profiles were recorded with 140 GHz µw. For further details, see original places of publication[9,11]

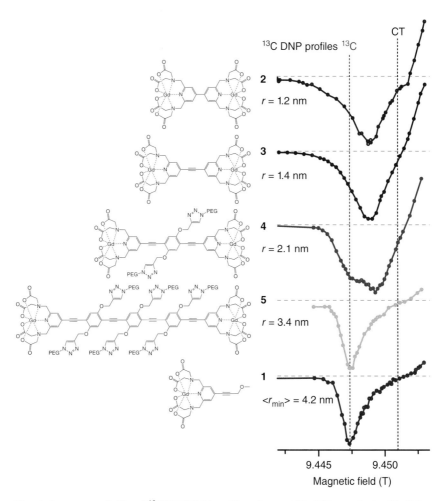

Figure 6.10. Chemical structures (left) and ^{13}C DNP field profiles of several bis-(Gd-complex)es (**2–5**) in comparison with a mono-Gd-complex (**1**). At Gd–Gd distances of $r \leq 2.1$ nm the CE contribution can be seen at a field position in between the ^{13}C SE matching maximum (marked by red vertical line) and the EPR CT maximum (marked by black vertical line). For further details, see original publication[35]

as well and enhancement factors on the order of 100 have been obtained. The PA concentration dependence of DNP field profiles has furthermore indicated the occurrence of CE within the CTs of two neighboring Gd^{3+} complexes.[11]

6.6.1.2 *Bis-complexes of Gd³⁺*

The abovementioned contribution of CE to the DNP enhancement in solutions of Gd-DOTA with moderate-to-large (i.e., 10–20 mM) concentrations led to the design of tailored bis-Gd-complexes in which

two chelate moieties are linked by a rigid tether. The defined but tunable intramolecular Gd–Gd distance allows the disentanglement of dipolar electron–electron coupling from complex concentration. As a result, it was possible to prove the transition from SE to CE DNP by stepwise reduction of the linker length and thus increase of dipolar e–e interaction (Figure 6.10). Interestingly, this transition occurred at shorter Gd–Gd distances with larger nuclear gyromagnetic ratios (i.e., ^{13}C vs ^{15}N); in either case, the onset of CE was accompanied by a significant increase in DNP enhancement. For ^{1}H DNP, the analysis was inconclusive; however, several features which may indicate the occurrence of unconventional polarization transfer mechanisms have been observed in the DNP field profile.[35]

6.6.1.3 DNP by Endogenously Bound Metal Ions

We have also demonstrated the direct DNP of a (full-length) hammerhead ribozyme (HHRz) using endogenously bound Mn^{2+} as PA.[10] The HHRz family consists of catalytically active ribonucleic acids with a highly conserved central structural element, which can (self-)cleave a nucleic acid strand at a very specific position if a divalent metal ion (e.g., Mg^{2+} or Mn^{2+}) is present.[48] Upon occupation of the specific metal ion binding site with paramagnetic Mn^{2+}, it was possible to hyperpolarize the ^{13}C nuclei within the uniformly isotope-labeled Mn^{2+}-HHRz complex (Figure 6.11). The biomolecule was dissolved in a vitrified solution, and intermolecular transfer could be experimentally excluded from the direct intracomplex DNP transfer.[10]

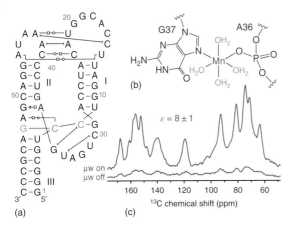

Figure 6.11. DNP of a full-length hammerhead ribozyme by endogenously bound Mn^{2+}: (a) Tertiary structure following Martick and Scott,[62] (b) proposed geometry of the Mn^{2+} binding site following Morrissey *et al.*[63] as well as Wang *et al.*,[64] and (c) ^{13}C DNP-enhanced direct polarization MAS NMR spectrum of the uniformly ^{13}C,^{15}N-labeled Mn^{2+}–HHRz complex at 9.4 T under Mn^{2+}–^{13}C SE matching. For further details, see original publication.[10] (Reproduced with permission from Ref. 10. © Springer Science+ Business Media Dordrecht, 2015)

6.6.1.4 Site-directed Spin Labeling of a Protein with Chelate Tags

Diamagnetic proteins that do not feature a metal binding site suitable for complexation of a DNP-active paramagnetic metal ion may be targeted by site-directed spin labeling with a chelate tag. Several tags have been introduced toward paramagnetic NMR or EPR applications based on ethylenediaminetetraacetic acid (EDTA), DOTA, dipicolinic acid (DPA), or PyMTA ligand moieties[18,49–55]; in these cases, coupling to cysteine residues is performed by disulfur- or maleimido-linkage. We have demonstrated this approach with single-site cysteine mutants of ubiquitin, where we have observed the impact of varying linker lengths and thus varying Gd–protein distances on ^{13}C DNP enhancement and paramagnetic resonance quenching/broadening (Figure 6.12). Current experiments are focused on the suppression of intrinsic ^{13}C relaxation and hence polarization drainage by partial or full deuteration of the protein; furthermore, the direct DNP of ^{15}N has shown promising results with DNP enhancements >100 and polarization gradients between different side-chain resonances. Results will be published elsewhere shortly.

6.6.1.5 DNP by isostructural doping of (poly-)crystalline materials

The utilization of paramagnetic metal ion dopants within (poly-)crystalline materials has been demonstrated already in the early days of DNP.[8] Regarding MAS DNP at high fields, the first demonstration was performed on a $[Co^{III}(en)_3Cl_3]_2 \cdot NaCl \cdot 6H_2O$ (en = ethylenediamine) polycrystalline powder isostructurally doped with Cr^{3+}.[36] Due to the high symmetry of the ethylenediamine complex, the ZFS of the $S = 3/2$ system is small, and DNP of ^1H, ^{13}C, and ^{59}Co was shown. The increase of ^{13}C and ^{59}Co DNP enhancement factors with larger doping ratios and deviations in their DNP field profile from the simple SE model indicated that CE between two neighboring Cr^{3+} might significantly contribute to the observed net enhancement (Figure 6.13).

Such a general approach may be extremely useful for nonporous materials lacking a dense proton network capable of transferring hyperpolarization into its bulk. For example, the use of metal ions (Mn^{2+}) doped into functional electrode materials is currently investigated in the laboratory of M. Leskes (Weizmann

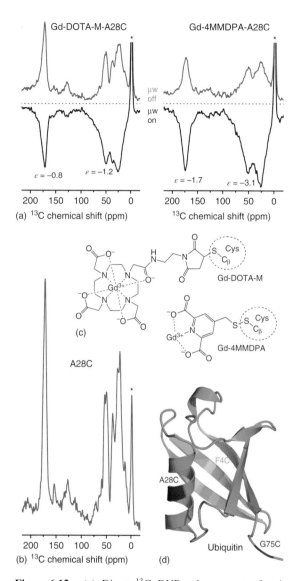

Figure 6.12. (a) Direct ^{13}C DNP enhancement of uniformly [^{13}C,^{15}N]-labeled ubiquitin mutant A28C using site-directed spin labeling with Gd-DOTA-M and Gd-4MMDPA tags at 9.4 T. The field was optimized for Gd^{3+} ^{13}C SE resulting in negative signal enhancement. (b) ^{13}C MAS spectrum of A28C ubiquitin (without attached spin label). Asterisks mark signals from silicone plugs. (c) Chemical structures of Gd-4MMDPA and Gd-DOTA-M spin labels connected to cysteine residues. (d) Ribbon structure of ubiquitin (PDB ID 1UBQ),[65] with F4C, A28C, and G75C single-site mutations marked in yellow, red, and blue, respectively. For further details, see original publication[11]

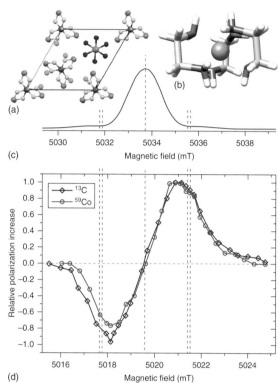

Figure 6.13. DNP of Cr^{3+}-doped [Co(en)$_3$Cl$_3$]$_2$·NaCl·6H$_2$O (en: ethylenediamine, C$_2$H$_8$N$_2$): (a) Crystallographic unit cell; projection of the crystal structure with the c axis perpendicular to the plane of the drawing. (b) Molecular unit with hydrogen atoms. (c) Field-swept, echo-detected 140 GHz EPR spectrum of 3% Cr^{3+} doping at a temperature of 80 K. (d) ^{13}C and ^{59}Co DNP field profiles recorded under 8 W of 140 GHz μw irradiation. (Reprinted with permission from B. Corzilius, V. K. Michaelis, S. A. Penzel, E. Ravera, A. A. Smith, C. Luchinat, and R. G. Griffin, J. Am. Chem. Soc., 2014, 136, 11716. Copyright (2014) American Chemical Society. Further permissions related to the material excerpted should be directed to the ACS)

Institute of Science) for DNP-enhanced NMR studies of battery materials.[38]

6.7 SUMMARY

As laid out in the above sections, DNP with paramagnetic metal ions offers exciting prospects for specialized applications such as in-cell DNP, targeted DNP of (metallo-)biomolecules, as well as DNP on isostructurally doped materials. However, the general applicability is limited compared to the more efficient bisnitroxides. This situation may change in the intermediate future, when methodological advances such as (ultra-)low-temperature MAS DNP, μw-field amplification, or time-domain (pulsed) DNP are introduced for mainstream applications.

Nevertheless, the available space of paramagnetic metal ions offers a large variety of different species – each featuring unique properties – and is mostly uncharted with respect to DNP. Due to the limited knowledge and understanding of potential DNP mechanisms within the complicated spin systems, new methods might still be developed based on radically new approaches. For example, sample rotation of lanthanides with large SOC such as Yb^{3+}, Dy^{3+}, or Ce^{3+} inside a magnetic field can cause a 'nuclear spin refrigerator' effect due to electron–nuclear cross relaxation caused by the large anisotropies of the electron Zeeman effect and spin relaxation.[56-60] This concept has recently been revived in the context of high-field MAS DNP by Bodenhausen and coworkers.[61] Given the significant advancements in instrumentation expected in the next years combined with the outstanding creativity of the research teams working within this interdisciplinary field, similarly groundbreaking concepts might arise in the near to intermediate future. This might bring many metal ions that are currently not amenable to 'conventional' MAS DNP mechanisms but are highly relevant within molecular biology or materials research into the focus of high-field DNP.

ACKNOWLEDGMENTS

Financial support for own work featured in this chapter has been provided by the Deutsche Forschungsgemeinschaft through Emmy Noether grant CO802/2 as well as collaborative research center SFB902. Further support has been provided by the Center for Biomolecular Magnetic Resonance (BMRZ). I am highly indebted to Robert G. Griffin (MIT, Cambridge, MA) for the fantastic support provided during and after my research stay at the Francis Bitter Magnet Laboratory. Special credits also belong to the late Ivano Bertini as well as to Claudio Luchinat (CERM, Florence) for triggering the initial work on MAS DNP with metal ions. I thank Albert A. Smith (ETH Zürich) for the close support during the early work on that topic, as well as Enrico Ravera (CERM) and Vladimir Michaelis (University of Alberta) for the inspiring collaboration on the Cr^{3+} project. I am grateful to Marc Vogel and Beatrix Suess (TU Darmstadt) as well as Erhan Cetiner and Harald Schwalbe (GU Frankfurt) for providing high-quality samples of HHRz and ubiquitin, respectively. Finally, the largest contribution to the work featured in Section 6.6 was made by my outstanding students, in particular Thorsten Bahrenberg, Diane Daube, Jörg Heiliger, Monu Kaushik, and Patricia Wenk, without whom none of these demonstrations would have been possible.

RELATED ARTICLES IN EMAGRES

EPR Interactions – *g*-Anisotropy

EPR Interactions – Zero-field Splittings

EPR Interactions – Coupled Spins

EPR Interactions – Hyperfine Couplings

Very-high-frequency EPR

Overhauser, Albert W.: Dynamic Nuclear Polarization

Dynamic Nuclear Polarization and High-Resolution NMR of Solids

High-Frequency Dynamic Nuclear Polarization

REFERENCES

1. A. S. Lilly Thankamony, J. J. Wittmann, M. Kaushik, and B. Corzilius, *Prog. Nucl. Magn. Reson. Spectrosc.*, 2017, **102–103**, 120.

2. R. G. Griffin and T. F. Prisner, *Phys. Chem. Chem. Phys.*, 2010, **12**, 5737.

3. C. Sauvée, M. Rosay, G. Casano, F. Aussenac, R. T. Weber, O. Ouari, and P. Tordo, *Angew. Chem. Int. Ed.*, 2013, **52**, 10858.

4. A. Zagdoun, G. Casano, O. Ouari, M. Schwarzwalder, A. J. Rossini, F. Aussenac, M. Yulikov, G. Jeschke, C. Coperet, A. Lesage, P. Tordo, and L. Emsley, *J. Am. Chem. Soc.*, 2013, **135**, 12790.

5. G. Mathies, M. A. Caporini, V. K. Michaelis, Y. Liu, K.-N. Hu, D. Mance, J. L. Zweier, M. Rosay, M. Baldus, and R. G. Griffin, *Angew. Chem. Int. Ed.*, 2015, **54**, 11770.

6. A. Abragam, and B. Bleaney, Electron Paramagnetic Resonance of Transition Metal Ions, 1st edn, Oxford University Press, Oxford, 1970.

7. H. Glättli, M. Odehnal, J. Ezratty, A. Malinovski, and A. Abragam, *Phys. Lett. A*, 1969, **29**, 250.

8. M. Abraham, M. A. H. McCausland, and F. N. H. Robinson, *Phys. Rev. Lett.*, 1959, **2**, 449.

9. B. Corzilius, A. A. Smith, A. B. Barnes, C. Luchinat, I. Bertini, and R. G. Griffin, *J. Am. Chem. Soc.*, 2011, **133**, 5648.

10. P. Wenk, M. Kaushik, D. Richter, M. Vogel, B. Suess, and B. Corzilius, *J. Biomol. NMR*, 2015, **63**, 97.

11. M. Kaushik, T. Bahrenberg, T. V. Can, M. A. Caporini, R. Silvers, J. Heiliger, A. A. Smith, H. Schwalbe, R. G. Griffin, and B. Corzilius, *Phys. Chem. Chem. Phys.*, 2016, **18**, 27205.

12. D. C. Weatherburn, in 'Handbook on Metalloproteins', 1st edn, eds I. Bertini, A. Sigel, and H. Sigel, Marcel Dekker, Inc.: New York, 2001, Chap. 8, p. 193.

13. M. Rogosnitzky and S. Branch, *BioMetals*, 2016, **29**, 365.

14. R. H. Knop, J. A. Frank, A. J. Dwyer, M. E. Girton, M. Naegele, M. Schrader, J. Cobb, O. Gansow, M. Maegerstadt, M. Brechbiel, L. Baltzer, and J. L. Doppman, *J. Comput. Assist. Tomogr.*, 1987, **11**, 35.

15. M. Azarkh, O. Okle, P. Eyring, D. R. Dietrich, and M. Drescher, *J. Magn. Reson.*, 2011, **212**, 450.

16. A. P. Jagtap, I. Krstic, N. C. Kunjir, R. Hänsel, T. F. Prisner, and S. T. Sigurdsson, *Free Radic. Res.*, 2015, **49**, 78.

17. M. Qi, A. Groß, G. Jeschke, A. Godt, and M. Drescher, *J. Am. Chem. Soc.*, 2014, **136**, 15366.

18. A. Martorana, G. Bellapadrona, A. Feintuch, E. Di Gregorio, S. Aime, and D. Goldfarb, *J. Am. Chem. Soc.*, 2014, **136**, 13458.

19. A. Schweiger, and G. Jeschke, Principles of Pulse Electron Paramagnetic Resonance, 1st edn, Oxford University Press, Oxford, 2001.

20. K. R. Thurber and R. Tycko, *J. Chem. Phys.*, 2012, **137**, 084508.

21. F. Mentink-Vigier, Ü. Akbey, Y. Hovav, S. Vega, H. Oschkinat, and A. Feintuch, *J. Magn. Reson.*, 2012, **224**, 13.

22. B. Corzilius, *Phys. Chem. Chem. Phys.*, 2016, **18**, 27190.

23. O. Haze, B. Corzilius, A. A. Smith, R. G. Griffin, and T. M. Swager, *J. Am. Chem. Soc.*, 2012, **134**, 14287.

24. T. V. Can, M. A. Caporini, F. Mentink-Vigier, B. Corzilius, J. J. Walish, M. Rosay, W. E. Maas, M. Baldus, S. Vega, T. M. Swager, and R. G. Griffin, *J. Chem. Phys.*, 2014, **141**, 064202.

25. K. D. Bowers and J. Owen, *Rep. Prog. Phys.*, 1955, **18**, 304.

26. A. V. Astashkin and A. Schweiger, *Chem. Phys. Lett.*, 1990, **174**, 595.

27. A. M. Raitsimring, A. V. Astashkin, O. G. Poluektov, and P. Caravan, *Appl. Magn. Reson.*, 2005, **28**, 281.

28. M. Benmelouka, J. Van Tol, A. Borel, M. Port, L. Helm, L. C. Brunel, and A. E. Merbach, *J. Am. Chem. Soc.*, 2006, **128**, 7807.

29. J. A. Clayton, K. Keller, M. Qi, J. Wegner, V. Koch, H. Hintz, A. Godt, S. Han, G. Jeschke, M. S. Sherwin, and M. Yulikov, *Phys. Chem. Chem. Phys.*, 2018, **20**, 10470.

30. A. Abragam and M. Goldman, *Rep. Prog. Phys.*, 1978, **41**, 395.

31. B. Corzilius, A. A. Smith, and R. G. Griffin, *J. Chem. Phys.*, 2012, **137**, 054201.

32. W. T. Wenckebach, *Appl. Magn. Reson.*, 2008, **34**, 227.

33. D. Shimon, Y. Hovav, A. Feintuch, D. Goldfarb, and S. Vega, *Phys. Chem. Chem. Phys.*, 2012, **14**, 5729.

34. K. N. Hu, G. T. Debelouchina, A. A. Smith, and R. G. Griffin, *J. Chem. Phys.*, 2011, **134**, 19.

35. M. Kaushik, M. Qi, A. Godt, and B. Corzilius, *Angew. Chem. Int. Ed.*, 2017, **56**, 4295.

36. B. Corzilius, V. K. Michaelis, S. A. Penzel, E. Ravera, A. A. Smith, C. Luchinat, and R. G. Griffin, *J. Am. Chem. Soc.*, 2014, **136**, 11716.

37. B. Corzilius, L. B. Andreas, A. A. Smith, Q. Z. Ni, and R. G. Griffin, *J. Magn. Reson.*, 2014, **240**, 113.

38. T. Chakrabarty, N. Goldin, A. Feintuch, L. Houben, and M. Leskes, *ChemPhysChem*, 2018, **19**, 2139.

39. A. A. Smith, B. Corzilius, J. A. Bryant, R. DeRocher, P. P. Woskov, R. J. Temkin, and R. G. Griffin, *J. Magn. Reson.*, 2012, **223**, 170.

40. I. Kaminker, A. Potapov, A. Feintuch, S. Vega, and D. Goldfarb, *Phys. Chem. Chem. Phys.*, 2009, **11**, 6799.

41. A. Doll, M. Qi, N. Wili, S. Pribitzer, A. Godt, and G. Jeschke, *J. Magn. Reson.*, 2015, **259**, 153.

42. P. E. Spindler, P. Schöps, A. M. Bowen, B. Endeward, and T. F. Prisner, in 'eMagRes', eds R. K. Harris and R. L. Wasylishen, Wiley: New York, 2016, DOI: 10.1002/9780470034590.emrstm1520.

43. J. C. Bousquet, S. Saini, D. D. Stark, P. F. Hahn, M. Nigam, J. Wittenberg, and J. T.Ferrucci Jr., *Radiology*, 1988, **166**, 693.

44. H. Stetter and W. Frank, *Angew. Chem. Int. Ed.*, 1976, **15**, 686.

45. P. Hermann, J. Kotek, V. Kubicek, and I. Lukes, *Dalton Trans.*, 2008, 3027.

46. D. Pan, A. H. Schmieder, S. A. Wickline, and G. M. Lanza, *Tetrahedron*, 2011, **67**, 8431.

47. V. Nagarajan, Y. Hovav, A. Feintuch, S. Vega, and D. Goldfarb, *J. Chem. Phys.*, 2010, **132**, 13.

48. W. G. Scott, L. H. Horan, and M. Martick, *Prog. Mol. Biol. Transl. Sci.*, 2013, **120**, 1.

49. D. Banerjee, H. Yagi, T. Huber, G. Otting, and D. Goldfarb, *J. Phys. Chem. Lett.*, 2012, **3**, 157.

50. X.-C. Su and G. Otting, *J. Biomol. NMR*, 2010, **46**, 101.

51. H. Yagi, D. Banerjee, B. Graham, T. Huber, D. Goldfarb, and G. Otting, *J. Am. Chem. Soc.*, 2011, **133**, 10418.

52. A. Potapov, H. Yagi, T. Huber, S. Jergic, N. E. Dixon, G. Otting, and D. Goldfarb, *J. Am. Chem. Soc.*, 2010, **132**, 9040.

53. A. M. Raitsimring, C. Gunanathan, A. Potapov, I. Efremenko, J. M. L. Martin, D. Milstein, and D. Goldfarb, *J. Am. Chem. Soc.*, 2007, **129**, 14138.

54. A. Collauto, A. Feintuch, M. Qi, A. Godt, T. Meade, and D. Goldfarb, *J. Magn. Reson.*, 2016, **263**, 156.

55. D. Goldfarb, in 'Structural Information from Spin-Labels and Intrinsic Paramagnetic Centres in the Biosciences', eds C. R. Timmel and J. R. Harmer, Springer-Verlag: Berlin, 2014, Vol. **152**, p. 163.

56. C. D. Jeffries, *Cryogenics*, 1963, **3**, 41.

57. A. Abragam, *Cryogenics*, 1963, **3**, 42.

58. K. H. Langley and C. D. Jeffries, *Phys. Rev.*, 1966, **152**, 358.

59. J. Lubbers and W. J. Huiskamp, *Physica*, 1967, **34**, 193.

60. H. B. Brom and W. J. Huiskamp, *Physica*, 1972, **60**, 163.

61. A. J. Perez Linde, S. Chinthalapalli, D. Carnevale, and G. Bodenhausen, *Phys. Chem. Chem. Phys.*, 2015, **17**, 6415.

62. M. Martick and W. G. Scott, *Cell*, 2006, **126**, 309.

63. S. R. Morrissey, T. E. Horton, and V. J. DeRose, *J. Am. Chem. Soc.*, 2000, **122**, 3473.

64. S. Wang, K. Karbstein, A. Peracchi, L. Beigelman, and D. Herschlag, *Biochemistry*, 1999, **38**, 14363.

65. S. Vijay-Kumar, C. E. Bugg, and W. J. Cook, *J. Mol. Biol.*, 1987, **194**, 531.

Chapter 7

Instrumentation for High-field Dynamic Nuclear Polarization NMR Spectroscopy

Guy M. Bernard and Vladimir K. Michaelis

Gunning-Lemieux Chemistry Centre, University of Alberta, Edmonton, Alberta, Canada

7.1 INTRODUCTION

The dynamic nuclear polarization nuclear magnetic resonance (DNP NMR) phenomenon was predicted by Overhauser in 1953[1] and confirmed experimentally by Carver and Slichter.[2,3] Initially, the technique was restricted to low-field measurements and thus considered a niche application due to its limited practical applications. However, the advent of high-frequency microwave sources in the past 30 years has allowed the use of the technique for high-field NMR magnets, leading to explosive developments for the technique.[4] In this chapter, we present an overview of the important developments in the instrumentation for this rapidly developing spectroscopic field with applications

in numerous areas, such as analytical chemistry,[5,6] biological science,[7-11] and materials science.[5,8,12,13]

Figure 7.1 illustrates a modern custom-built high-field DNP NMR installation. The key components discussed in detail in the following sections are (1) a microwave source with its associated controller and magnet, (2) a transmission line to deliver high-power microwaves from the microwave source to the NMR probe, (3) the DNP NMR probe, (4) the magic-angle spinning (MAS) controller along with the special rotors required for DNP and the cryogenics gas source, and finally (5) the NMR magnet and console. With the exception of the latter, the hardware illustrated here is either not found in a typical NMR lab or is significantly different. Hence, in the ensuing sections, these components are discussed in detail. Finally, we conclude with a discussion on advances that one may anticipate in the near future.

Since the focus of this chapter is on instrumentation, a detailed discussion of the background theory is beyond the scope of this chapter; readers are encouraged to consult associated chapters within this book as well as several detailed reviews on DNP theory[4,14-20] and related concepts, such as enhancement factors[21-23] and polarizing agents.[4,15,17,23,24] Readers are assumed to understand the fundamentals of solid-state NMR[25-30] and electron paramagnetic resonance (EPR)[31-36] spectroscopy and should consult the many texts on the topics for further details. Note that this chapter discusses the instrumentation required for DNP NMR of solid samples. DNP NMR

Handbook of High Field Dynamic Nuclear Polarization.
Edited by Vladimir K. Michaelis, Robert G. Griffin, Björn Corzilius and Shimon Vega.
© 2020 John Wiley & Sons, Ltd. ISBN: 978-1-119-44164-9
Also published in eMagRes (online edition)
DOI: 10.1002/9780470034590.emrstm1560

Figure 7.1. Modern custom-built DNP NMR installation (700 MHz/460 GHz, FBML-MIT). (1) Gyrotron and its magnet; (2) waveguide; (3) N_2 gas separation system; (4) NMR magnet; (5) cryogen dewar; (6) refrigeration (chiller) unit; (7) VT controller and (8) vacuum-jacketed cryogenic gas transfer lines (drive, bearing, VT)

for solutions is possible, although the method poses additional challenges.[14,17,37] In dissolution NMR, one hyperpolarizes a frozen sample as for solid samples, then dissolves it; the NMR spectrum is then acquired.[38–41] The hardware requirements for this[39,42] and other solution DNP NMR[43] techniques have been discussed.

7.2 COMPONENTS OF A DNP NMR SYSTEM

7.2.1 The Microwave Source

Combining the improved resolution of high magnetic fields with the greatly improved sensitivity afforded by DNP has been a long-term goal for NMR spectroscopists.[44] The fundamental DNP experiment entails the saturation of EPR resonances via microwave irradiation. Since the ratio of the Larmor frequency for the electron to that for ^1H is approximately 660, the microwave frequencies must be in the gigahertz range for virtually all NMR applications (an interesting exception, not discussed further here, is DNP NMR spectroscopy undertaken at Earth's field, where the electron-^{14}N hyperfine coupling of 131.5 MHz, in a polarizing agent at 10 mT, was used to polarize ^1H).[45] For many years, the development of microwave sources capable of generating the required frequencies lagged behind advances in high-field NMR magnet technology, precluding the routine application of DNP at high fields,[46,47] although some workaround solutions were proposed. For example, Dorn *et al.* enhanced the NMR signal *via* DNP at 0.34 T (i.e., requiring a 9.4 GHz microwave source), then, to improve resolution, transferred the sample to a 4.7 T NMR instrument.[48] Although the method provided the desired enhancement, technical difficulties hindered a general application of these and related techniques. Other researchers obtained high-frequency microwave sources, albeit at low power and at very low temperatures. For example, in 2002, Brill proposed an ENDOR-DNP (electron nuclear double resonance-dynamic nuclear polarization) approach, achieving 50% polarization at 1 K and $B_0 = 5$ T.[49] Using a similar approach, Morley *et al.* achieved an enhancement of 10^3 for ^{15}N NMR spectroscopy of $^{15}N@C_{60}$ at $B_0 = 8.6$ T.[50] Tycko and coworkers[51] obtained an enhancement factor of 80 for a glycerol/H_2O sample using a 30 mW microwave system effectively operating at 264 GHz in conjunction with a 9.39 T magnet (400 MHz ^1H).

In the 1990s, extended interaction oscillators (EIOs) or extended interaction klystrons (EIKs) were applied for the desired high-frequency microwaves. These are linear-beam devices, but unfortunately, the power required for DNP NMR spectroscopy at high magnetic fields reduces the lifetimes of these devices,[15] and extending their frequency range beyond those required for 5 T magnets remains a challenge. Despite some drawbacks, they remain the preferred option for certain systems. For example, Dupree and coworkers recently described a tunable 187 GHz EIK amplifier as a microwave source for a DNP NMR system.[52] Likewise, Thurber and Tycko obtained enhancement factors in the 100–200 range for ^{13}C NMR signals using an EIO with an output power of 1.5 W as the microwave source.[53] In addition, a 263 GHz EIK is offered by Bruker as an option for their 263 GHz/400 MHz DNP NMR system.

A major breakthrough in high-field DNP NMR spectroscopy came from the Francis Bitter Magnet Laboratory at MIT,[20,44,54,55] where a cyclotron-resonance maser (i.e., a gyrotron) was adapted to operate at the times and power levels required for high-frequency, high-power microwave radiation that enabled DNP NMR at a ^{1}H frequency of 211 MHz (i.e., $B_0 = 5$ T). The gyrotron is used to stimulate cyclotron radiation, generated within a superconducting magnet,[8] as the microwave source. Although the gyrotron was invented over 50 years ago,[56] use of this device, which typically operated at higher powers to generate microsecond pulses, required modifications to permit continuous wave (CW) operation.[54] DNP NMR devices operating at up to 460 GHz have since been developed by Griffin and coworkers and by Idehara *et al.*[57] allowing spectroscopy at a 700 MHz ^{1}H frequency. In addition, Temkin and coworkers developed a 20 W, 527 GHz gyrotron suitable for DNP NMR at a ^{1}H frequency of 800 MHz.[58] The success of these and other research groups prompted Bruker Biospin, in collaboration with Communications & Power Industries (CPI), to market DNP NMR systems, with gyrotrons operating at frequencies up to 593 GHz (i.e., 900 MHz ^{1}H). Other commercial sources for gyrotrons have been used (see, for example, Refs. 37 and 59), but these are not marketed as complete DNP NMR instruments. Further details about high-frequency microwave devices can be found elsewhere in this book.

The DNP NMR experiment requires matching of the EPR and NMR frequencies, a significant complication since EPR signals at a given field strength cover a large frequency range, depending on the polarizing agent. This issue has generally been addressed through the use of sweep coils (vide infra), but there has been recent progress in implementing frequency-tunable gyrotrons, which would alleviate the necessity for sweep coils and the complications associated with these. Relatively small frequency ranges (approximately 100 MHz) have been achieved by adjusting the temperature in the cavity circuit.[60] Larger tuning ranges (1.2 GHz) have been achieved,[61,62] but challenges, such as varying output power, remain.[4] Tuning ranges greater than 2 GHz have recently been reported by the Griffin laboratory,[63,64] who suggested that combining a tunable NMR magnet with a tunable gyrotron would enhance efficiency, and by Matsuki *et al.*[65] and Idehara *et al.*[66]; the former also reported that the output power was not constant over the gyrotron frequency range.[65]

The hyperfine interaction arising from the polarizing agent may have undesirable effects on the NMR spectra of samples acquired under MAS conditions, leading to paramagnetic quenching (i.e., loss of intensity).[22] Similar to the situation for routine cross-polarization NMR spectroscopy, where ^{1}H decoupling leads to increased resolution in the NMR spectra,[67] decoupling of the hyperfine interaction offers the potential of increased resolution and of longer nuclear relaxation times in DNP NMR spectra acquired with MAS and may improve the polarization enhancement.[68] Thus, the Barnes research group has been working extensively in the area of hyperfine decoupling.[69–72] Development is ongoing but promising results have been reported. For example, ^{13}C NMR linewidths were reduced by 11% and the signal intensity increased by 14% using microwave frequency sweeps through the EPR pattern (Figure 7.2).[69]

Optimum signal enhancement is achieved when all electrons in the polarizing agent are irradiated.[73] Some agents with large *g*-anisotropy have large inhomogeneous EPR linewidths, which increase at higher magnetic fields.[74] Hence, if DNP NMR measurements are conducted with CW irradiation of the polarizing agent at a constant field, only a fraction of the electrons are irradiated, particularly in the absence of MAS. Frequency sweeps have been used but the method frequently is inefficient.[73] Hovav *et al.* used microwave frequency modulation to irradiate a broader bandwidth,[75] obtaining a three-fold improvement on the enhancement, compared to that obtained in the absence of frequency modulation. Recently, Pines and coworkers demonstrated that improved polarization of agents with broad bandwidths may be achieved through the use of microwave frequency combs.[73] The method involves simultaneously sweeping the entire electron linewidth with the 'teeth' of the comb (Figure 7.3).

7.2.2 The Waveguide

Another significant challenge in DNP NMR spectroscopy is the transmission of the microwaves, generated as discussed above, to the sample. Since current technology requires that the gyrotron magnet be situated at a distance from the NMR magnet, the microwaves must be transmitted over several meters while minimizing losses.[20,44,47,59,62,76–78] For example, the waveguide used at the Francis Bitter Magnet Laboratory is a 4.65 m corrugated waveguide

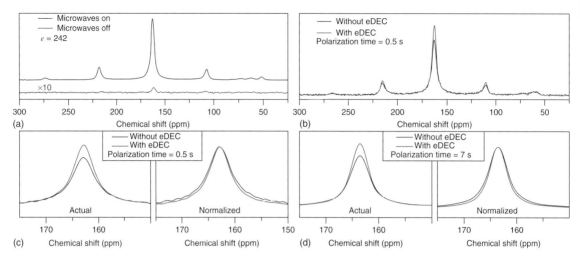

Figure 7.2. Comparison of NMR spectra of urea acquired with and without DNP (a) and with DNP but with or without electron decoupling (b–d). (Reprinted with permission from E. P. Saliba, E. L. Sesti, F. J. Scott, B. J. Albert, E. J. Choi, N. Alaniva, C. Gao and A. B. Barnes, *J. Am. Chem. Soc.*, 2017, **139**, 6310. Copyright (2017) American Chemical Society)

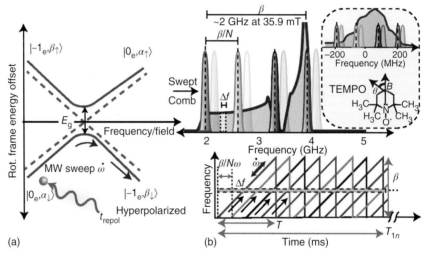

Figure 7.3. Frequency comb used for DNP enhancement. In (a), the hyperpolarization process using the frequency/field swept techniques is illustrated. (b): The red trace illustrates the inhomogeneously broadened electron spectrum (with line width of B) arising from nitrogen-vacancy center defects in diamond. The frequency comb, allowing repeated polarization transfers, is illustrated at lower right. The inset illustrates the TEMPO ((2,2,6,6-tetramethylpiperidine-1-yl)oxyl) polarization agent as well as its electron spectrum, another good candidate for the technique. (Reproduced with permission from A. Ajoy, R. Nazaryan, K. Liu, X. Lv, B. Safvati, G. Wang, E. Druga, J. A. Reimer, D. Suter, C. Ramanathan, C. A. Meriles and A. Pines, *PNAS*, 2018, **115**, 10576. © National Academy of Sciences, 2018)

with an inner diameter of 19.05 mm.[47,62] Currently, the favored technology is the use of corrugated over-moded waveguides,[62,79] which are more efficient than fundamental-mode rectangular waveguides.[47] Part of the output from the waveguide must be diverted to determine the output frequency.[59,62] Hill and coworkers have described a waveguide equipped with a splitter, permitting simultaneous DNP NMR measurements on two 14.1 T NMR magnets.[77] It, of course, is not possible to transmit the microwaves in a direct line from the gyrotron to the sample within the NMR magnet, so a series of high precision miter bends are required within the waveguide[8]; reducing the number of bends is highly desirable to minimize complications and potential losses in power and beam structure.[59]

Quasioptical transmission systems have been proposed as an alternative to the corrugated waveguide described above.[8,80,81] In this setup, the microwaves are transmitted via mirrors. The method allows more flexibility, such as power attenuation and beam splitting.[8]

7.2.3 The NMR Probe

In addition to the functionality typically associated with conventional MAS NMR probes (e.g., stable high-frequency spinning, accurate setting at the magic angle), the corresponding DNP NMR probes have

additional requirements.[4,59] They must allow much lower-temperature operation and hence gas lines within the probe must be insulated, typically with a vacuum jacket. In fact, some DNP probes are vacuum jacketed to maintain a cool internal temperature while protecting the NMR magnet bore from these low temperatures. The stator housing must accommodate the microwave transmission line (vide supra). In addition, the ability to eject and insert samples with the probe at low temperature is important for high-throughput studies, since several hours are required to cool the probe to operating temperatures of between 80 and 110 K and then to heat it back up to room temperature.[79,82] Figure 7.4 illustrates a modern custom-built DNP NMR probe and stator.

In a typical DNP NMR setup, the microwaves are delivered to the sample from the bottom of the probe, and then directed through a series of miter bends such that they strike the sample orthogonal to the MAS rotor axis (Figure 7.4). In this configuration, the microwaves must traverse the NMR radio frequency coil, adversely

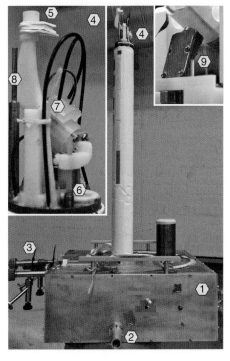
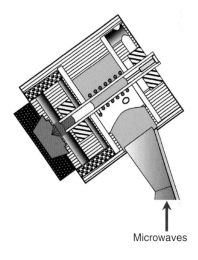

Microwaves

Figure 7.4. Photographs of an NMR probe and expanded view of some of its components (left). (1) NMR probe; (2) waveguide; (3) insulated bearing and drive gas lines; (4) probe head, (5) sample insert/eject; (6) magic angle adjust; (7) stator; (8) variable temperature (VT) gas line; and (9) waveguide to the sample. At right, a schematic representation of a rotor and stator. (Adapted from Barnes 2009 and Nanni 2011)

impacting their enhancement effects.[83] Nanni *et al.* have shown that these effects can be minimized by adjusting the coil spacing.[83] Alternate designs have been proposed.[59] For example, the systems developed by Horii and coworkers[84] and by Pike *et al.*[81] deliver the microwaves from the top of the probe and direct them to the top of the rotor. This configuration allows the probe to be quickly removed or installed and saves space within the probe body since the waveguide is external to it.[81] With the latter schemes, the microwave irradiation is parallel to the rotor axis and is sensitive to constraints imposed by the MAS system.[59] In addition, sample penetration is imperfect when the microwaves are introduced from the top of the rotor. Other designs are currently being investigated (see Ref. 59).

7.2.4 Magic Angle Spinning: Gas Sources, Rotors, and Controllers

As for other solid-state NMR experiments, MAS has the potential to greatly improve the resolution of the spectra that are obtained with DNP. Unfortunately, in the latter studies, MAS may also induce depolarization, negating some of the advantages of DNP.[85–87] Careful selection of polarizing agents have mitigated the problem.[85,88] A further concern is that the problem may worsen as the spinning frequency increases,[89,90] thus further radical development is on-going. In their investigation of this effect, Emsley and coworkers obtained indirect ^{29}Si DNP NMR spectra in the 10–40 kHz range and found that, although the depolarization does indeed increase with spinning frequency, the benefits of DNP and rapid MAS outweigh the disadvantages: a significant fraction of the sample still contributes and yields an enhancement and the higher spinning frequency increased the coherence lifetimes of the material under investigation.[89] With the assistance of computer simulation programs, De Paëpe and coworkers recently prepared a series of polarizing agents that show little or no depolarization at high MAS frequencies.[90]

Another important requirement of a DNP NMR lab is the source of MAS and variable temperature (VT) gases. DNP experiments are typically performed at low temperatures to improve electron and nuclear relaxation behavior and hence to improve the transfer of polarization to the nuclei.[15,91] Sustained low-temperature MAS imposes numerous challenges; in particular, gas flow rates must be varied depending on the desired spinning frequency and the mass of

the sample, rendering temperature regulation more difficult.[59,92,93] Continuous operation at temperatures near 90 K has been achieved with the use of nitrogen gas as the bearing and drive gas sources while simultaneously providing the sample temperature regulation. The gas may be sourced from large (150 or 220 L) dewars; two such dewars ensure that data acquisition is not disrupted when one dewar's supply is exhausted,[92] but a disadvantage of this system is that the high pressures required for NMR at high MAS frequencies means that large quantities of liquid nitrogen are consumed, driving costs; a typical DNP NMR setup may use several hundred liters per day. Matsuki *et al.*[94] as well as Griffin and coworkers[46] have used an N_2 gas separator in conjunction with an electric chiller to generate low-temperature gas which is then chilled by liquid N_2. The authors report that this strategy reduces nitrogen consumption by a factor of four compared to the consumption experienced under the same conditions using N_2 boil-off as a gas source.[94]

Another complication with N_2 as a VT gas is that its viscosity increases as the liquefaction point is approached, precluding its use for MAS NMR spectroscopy at lower temperatures.[4] Hence, lower temperature MAS has been achieved with helium as the VT gas source and as the liquid heat-exchange medium,[93,95] although this comes with additional challenges.[15] To minimize the prohibitive costs of helium, Thurber and Tycko developed an MAS system whereby the bearing and drive gases were obtained from N_2 while helium was used as the VT gas; longer rotors were used to isolate the VT helium gas stream from that for nitrogen.[53,96] Closed-loop helium recirculation systems have been designed.[97–99]

In room-temperature MAS NMR spectroscopy, stable spinning is achieved through regulation of the bearing and drive gas pressures, typically using MAS controllers one routinely finds in solid-state NMR labs. This hardware is suitable for DNP NMR applications since the drive and bearing gases are cooled down the line from the MAS controller.[4] After exiting the MAS unit, the gases, along with the VT gas, are cooled to the target temperature inside a pressurized heat exchanger.[4] To maintain temperature stability, the gas transfer lines from the heat exchanger to the probe are vacuum jacketed. Control of the cooling capacity is achieved by a heat exchanger that permits the replenishment of liquid nitrogen during operation.[82] A more recent design uses separate heat exchangers for the drive, bearing and VT gas streams, allowing greater flexibility in experimental conditions.[4] As discussed above, liquid helium must be used as the heat exchange medium for temperatures below approximately 90 K.[4]

The NMR rotors must permit effective transmission of microwaves into the sample while being robust enough to handle low temperatures as well as the excessive forces associated with high spinning frequencies (>5 kHz). Typically, these are made of sapphire[60,82,100] or of zirconia (i.e., the material typically used for solid-state NMR rotors). While the latter are more robust,[15] sapphire rotors are transparent to microwave radiation and have high thermal conductivity, an important consideration for samples that must be cooled to 100 K or lower.[82] In addition, greater signal enhancements have been reported when sapphire rotors were used.[60,100] Until recently, DNP NMR measurements were undertaken primarily with 3.2 or 4.0 mm outer diameter rotors, but, as discussed above, faster spinning DNP NMR probes are being developed.[89] A commercial (Bruker) 1.9 mm DNP NMR probe permits sample spinning frequencies of up to 25 kHz at 100 K and allows sample changes while the sample is cold. This manufacturer also offers a 1.3 mm probe, permitting MAS experiments at 40 kHz, at approximately 115 K.[101] These rotors, of course, have smaller sample volumes, but the loss of signal from this factor is mitigated by the greater microwave penetration and hence the improved DNP enhancement. For non-spinning applications, quartz EPR tubes have been used as well.[50,55]

Recently, Barnes and coworkers have demonstrated the benefits of using spinning spheres rather than cylindrical rotors for DNP NMR samples.[102] The authors demonstrate that such spheres provide stable sample spinning and improved signal to noise ratios while conserving space within the probe head. Another benefit of spinning spheres is that a single gas stream is required for the bearing, drive and VT gases.

7.2.5 The NMR Spectrometer

The basic NMR spectrometer, with the exception of the probe discussed above, is that used for solid-state NMR spectroscopy, but a wide-bore (i.e., 89 mm) magnet must currently be used to accommodate the hardware, such as the waveguide and electronics, insulation, variable temperature lines, and vacuum jacketed dewar associated with DNP applications at cryogenic temperatures.[15] Likewise, the NMR console is that used for solid-state NMR applications.

As discussed above, there has been significant progress in frequency-tunable gyrotrons. Nevertheless, because of continuing challenges with these methods, sweepable NMR magnets are often used to achieve the desired electron/nucleus frequency ratio.[4] This is commonly accomplished by coupling the NMR magnet with a superconducting sweep coil[103–105]; however, challenges remain for this technique as well. For example, the sweep rates are limited and may lead to temporary field instability and excess cryogen consumption through He boil-off. In addition, the sweepable range remains limited; that reported by Kaushik *et al.* has a sweep range of ±75 mT[105] and hence there are still cases where it is not possible to achieve the matching conditions for some polarizing agents, such as some metal ions.[4]

7.3 FUTURE DEVELOPMENTS

Despite the extensive recent developments, many avenues of research in DNP NMR remain open. For example, the behavior underlying certain mechanisms causing the DNP phenomenon is not fully understood.[106] While the complex interplay of factors governing the DNP efficiency means that the theoretical maximum enhancement factors are unachievable,[23] there surely is much room for improvement, as attested by the quantity of chapters published in recent months addressing these issues.[23,90,107–110]

Developments in other areas are also promising. As discussed above, the requirement for cryogens is a major driver of costs for DNP NMR. In addition to recent developments discussed above, Scott *et al.*[111] have developed a cryostat-transfer line system that increases the efficiency of DNP NMR measurements below 6 K. Alternatives are being investigated. Eichhorn *et al.* have presented preliminary results whereby the electron spins needed for the DNP process are obtained through a photoexcitation process.[112] Results are promising, albeit at a relatively low field (0.3 T).

The requirement for a gyrotron source located at a distance from the NMR magnet is a significant complication for a DNP NMR lab, particularly since another superconducting magnet is required, for the gyrotron. In the hope of avoiding the necessity for a second superconducting magnet, Bratman *et al.* described a DNP NMR system that incorporates the gyrotron within the NMR magnet.[113] Since the presence of a standard power gyrotron (10–20 kV) within an NMR magnet would compromise the field homogeneity, the authors developed a low-voltage gyrotron.[114] Although development work remains, the approach is promising. Likewise, Ryan *et al.* also recently discussed advances in the goal of integrating a gyrotron into an NMR magnet.[115] The gyrotron is located immediately above the NMR probe; to avoid the requirement for slightly different field strengths for the gyrotron and NMR probe, a ferroshim assembly is proposed for the latter. Although still under development, the method offers the hope that the infrastructure required for a DNP lab may in the future be significantly reduced.

A major feature of modern NMR spectroscopy is the advent of magnets with increasing magnetic field strengths. Commercial magnets up to 23.5 T (1.0 GHz, ^{1}H) are currently available, with 1.1 and 1.2 GHz instruments on the horizon. This begs the question, can DNP keep up? Combining the high-resolution benefits of the high-field NMR magnets with the potential sensitivity of DNP is very appealing, but challenges must be overcome. In particular, the cross effect, which is the major mechanism used in current DNP NMR applications, scales inversely with field.[87] Another challenge is that it is desirable to spin samples at higher frequencies when obtaining NMR spectra at higher magnetic field strengths, but, as discussed above, this introduces further complications in the polarization transfer mechanism.[89,90] A further challenge is the requirement for higher frequency gyrotrons,[4,24,59,116,117] and of course, as with NMR spectroscopy in general, higher magnetic fields come at a higher cost. Despite such challenges, DNP NMR results at 21.1 T have been reported.[101,118]

7.4 CONCLUDING REMARKS

The promise of DNP NMR is highlighted by the high interest shown in the technique and by the fact that several chapters, either using the technique or proposing to improve the technique, appear each week. Despite the numerous DNP NMR labs now found, particularly in Europe and North America, it is surely too early to describe DNP NMR as a mature research field, since the pace of rapid development in instrumentation and methodologies observed over the past two decades shows no sign of abating, as demonstrated by the number of recently published chapters cited in this chapter. The method will one day reach a point where development is incremental at which point it may be considered a mature field, taking its

place in the experimentalist's toolkit for unraveling the mysteries of nature. However, that point has not been reached yet: the near future promises many more exciting developments!

RELATED ARTICLES IN EMAGRES

Griffin, Robert G.: Perspectives of Magnetic Resonance

Dynamic Nuclear Polarization and High-Resolution NMR of Solids

Electron–Nuclear Hyperfine Interactions

Electron–Nuclear Interactions

Magic Angle Spinning

Sensitivity of the NMR Experiment

Dynamic Nuclear Polarization: Applications to Liquid-State NMR Spectroscopy

High-Frequency Dynamic Nuclear Polarization

Chemically Induced Dynamic Nuclear Polarization

Hyperfine Spectroscopy – ENDOR

Pulse EPR

EPR Interactions – Hyperfine Couplings

EPR Spectroscopy of Nitroxide Spin Probes

REFERENCES

1. A. W. Overhauser, *Phys. Rev.*, 1953, **92**, 411.

2. T. R. Carver and C. P. Slichter, *Phys. Rev.*, 1953, **92**, 212.

3. T. R. Carver and C. P. Slichter, *Phys. Rev.*, 1956, **102**, 975.

4. A. S. L. Thankamony, J. J. Wittmann, M. Kaushik, and B. Corzilius, *Prog. Nucl. Magn. Reson. Spectrosc.*, 2017, **102–103**, 120.

5. A. C. Pinon, U. Skantze, J. Viger-Gravel, S. Schantz, and L. Emsley, *J. Phys. Chem. A*, 2018, **122**, 8802.

6. A. Masion, A. Alexandre, F. Ziarelli, S. Viel, and G. M. Santos, *Sci. Rep.*, 2017, **7**, 3430.

7. J.-H. Ardenkjaer-Larsen, G. S. Boebinger, A. Comment, S. Duckett, A. S. Edison, F. Engelke,

C. Griesinger, R. G. Griffin, C. Hilty, H. Maeda, G. Parigi, T. Prisner, E. Ravera, J. van Bentum, S. Vega, A. Webb, C. Luchinat, H. Schwalbe, and L. Frydman, *Angew. Chem. Int. Ed.*, 2015, **54**, 9162.

8. A. N. Smith and J. R. Long, *Anal. Chem.*, 2016, **88**, 122.

9. K. Jaudzems, T. Polenova, G. Pintacuda, H. Oschkinat, and A. Lesage, *J. Struct. Biol.*, 2019, **206**, 90, DOI: 10.1016/j.jsb.2018.09.011.

10. Ü. Akbey and H. Oschkinat, *J. Magn. Reson.*, 2016, **269**, 213.

11. M. R. Elkins, I. V. Sergeyev, and M. Hong, *J. Am. Chem. Soc.*, 2018, **140**, 15437.

12. F. Pourpoint, A. S. L. Thankamony, C. Volkringer, T. Loiseau, J. Trébosc, F. Aussenac, D. Carnevale, G. Bodenhausen, H. Vezin, O. Lafon, and J.-P. Amoureux, *Chem. Commun.*, 2014, **50**, 933.

13. F. A. Perras, T. Kobayashi, and M. Pruski, *J. Am. Chem. Soc.*, 2015, **137**, 8336.

14. P. Niedbalski, A. Kiswandhi, C. Parish, Q. Wang, F. Khashami, and L. Lumata, *J. Phys. Chem. Lett.*, 2018, **9**, 5481.

15. M. Ha and V. K. Michaelis, in 'Modern Magnetic Resonance', ed G. Webb, Springer: Cham, Switzerland, 2018, p 1.

16. B. Corzilius, *Phys. Chem. Chem. Phys.*, 2016, **18**, 27190.

17. Ü. Akbey, W. T. Franks, A. Linden, M. Orwick-Rydmark, S. Lange, and H. Oschkinat, *Top. Curr. Chem.*, 2013, **338**, 181.

18. T. F. Prisner, in 'NMR of Biomolecules', eds I. Bertini, K. S. McGreevy, and G. Parigi, Wiley-VCH: Weinheim, 2012, Chapter 25, p 421.

19. K. H. Sze, Q. Wu, H. S. Tse, and G. Zhu, *Top. Curr. Chem.*, 2012, **326**, 215.

20. T. Maly, G. T. Debelouchina, V. S. Bajaj, K.-N. Hu, C.-G. Joo, M. L. Mak-Jurkauskas, J. R. Sirigiri, P. C. A. van der Wel, J. Herzfeld, R. J. Temkin, and R. G. Griffin, *J. Chem. Phys.*, 2008, **128**, 052211.

21. M. Ha and V. K. Michaelis, in 'Modern Magnetic Resonance', ed G. Webb, Springer: Cham, Switzerland, 2018, p 1207.

22. B. Corzilius, L. B. Andreas, A. A. Smith, Q. Z. Ni, and R. G. Griffin, *J. Magn. Reson.*, 2014, **240**, 113.

23. A. Leavesley, S. Jain, I. Kamniker, H. Zhang, S. Rajca, A. Rajca, and S. Han, *Phys. Chem. Chem. Phys.*, 2018, **20**, 27646.

24. Q. Z. Ni, E. Daviso, T. V. Can, E. Markhasin, S. K. Jawla, T. M. Swager, R. J. Temkin, J. Herzfeld, and R. G. Griffin, *Acc. Chem. Res.*, 2013, **46**, 1933.

25. M. J. Duer (ed), Solid State NMR Spectroscopy. Principles and Applications, Blackwell Science: Oxford, 2002.

26. D. C. Apperley, R. K. Harris, and P. Hodgkinson, Solid State NMR: Basic Principles and Practice, New York: Momentum Press, 2012.

27. D. E. Axelson, Solid State Nuclear Magnetic Resonance, Create Space Independent Publishing: A Practical Introduction, Kingston, 2012.

28. R. E. Wasylishen, S. E. Ashbrook, and S. Wimperis (eds), NMR of Quadrupolar Nuclei in Solid Materials, Wiley: Chichester, 2012.

29. J. C. C. Chan (ed), Solid State NMR. Topics in Current Chemistry, Springer-Verlag: Berlin, 2012, Vol. 306.

30. K. Müller and M. Geppi, Solid State NMR: Principles, Methods and Applications, Wiley-VCH: Chichester, 2018.

31. F. Gerson and W. Huber, Electron Spin Resonance Spectroscopy of Organic Radicals, Wiley-VCH: Weinheim, 2003.

32. J. A. Weil and J. R. Bolton, Electron Paramagnetic Resonance. Elementary Theory and Practical Applications, 2nd edn, Wiley: Hoboken, USA, 2007.

33. G. R. Eaton, S. S. Eaton, D. P. Barr, and R. T. Weber, Qauntitative EPR, Springer-Verlag: Wein, 2010.

34. S. K. Misra (ed), Multifrequency Electron Paramagnetic Resonance. Theory and Applications, Wiley-VCH: Weinheim, 2011.

35. S. K. Misra (ed), Multifrequency Electron Paramagnetic Resonance. Data and Techniques, Wiley-VCH: Weinheim, 2014.

36. D. Goldfarb and S. Stoll (eds), EPR Spectroscopy. Fundamentals and Methods, Wiley: Chichester, 2018.

37. V. Denysenkov and T. Prisner, *J. Magn. Reson.*, 2012, **217**, 1.

38. G. Zhang and C. Hilty, *Magn. Reson. Chem.*, 2018, **56**, 566.

39. J.-N. Dumez, *Magn. Reson. Chem.*, 2017, **55**, 38.

40. J. H. Ardenkjaer-Larsen, *J. Magn. Reson.*, 2016, **264**, 3.

41. A. Comment, *J. Magn. Reson.*, 2016, **264**, 39.

42. C. Griesinger, M. Bennati, H. M. Vieth, C. Luchinat, G. Parigi, P. Höfer, F. Engelke, S. J. Glaser, V. Denysenkov, and T. F. Prisner, *Prog. Nucl. Magn. Reson. Spectrosc.*, 2012, **64**, 4.

43. T. Prisner, V. Denysenkov, and D. Sezer, *J. Magn. Reson.*, 2016, **264**, 68.

44. A. B. Barnes, G. De Paëpe, P. C. A. van der Wel, K.-N. Hu, C.-G. Joo, V. S. Bajaj, M. L. Mak-Jurkauskas, J. R. Sirigiri, J. Herzfeld, R. J. Temkin, and R. G. Griffin, *Appl. Magn. Reson.*, 2008, **34**, 237.

45. M. E. Halse and P. T. Callaghan, *J. Magn. Reson.*, 2008, **195**, 162.

46. A. B. Barnes, E. Markhasin, E. Daviso, V. K. Michaelis, E. A. Nanni, S. K. Jawla, E. L. Mena, R. DeRocher, A. Thakkar, P. P. Woskov, J. Herzfeld, R. J. Temkin, and R. G. Griffin, *J. Magn. Reson.*, 2012, **224**, 1.

47. V. K. Michaelis, T.-C. Ong, M. K. Kiesewetter, D. K. Frantz, J. J. Walish, E. Ravera, C. Luchinat, T. M. Swager, and R. G. Griffin, *Isr. J. Chem.*, 2014, **54**, 207.

48. H. C. Dorn, R. Gitti, K. H. Tsai, and T. E. Glass, *Chem. Phys. Lett.*, 1989, **155**, 227.

49. A. S. Brill, *Phys. Rev. A*, 2002, **66**, 043405.

50. G. W. Morley, J. van Tol, A. Ardavan, K. Porfyrakis, J. Zhang, and G. A. D. Briggs, *Phys. Rev. Lett.*, 2007, **98**, 220501.

51. K. R. Thurber, W.-M. Yau, and R. Tycko, *J. Magn. Reson.*, 2010, **204**, 303.

52. T. F. Kemp, H. R. W. Dannatt, N. S. Barrow, A. Watts, S. P. Brown, M. E. Newton, and R. Dupree, *J. Magn. Reson.*, 2016, **265**, 77.

53. K. R. Thurber and K. Tycko, *J. Magn. Reson.*, 2016, **264**, 99.

54. L. R. Becerra, G. J. Gerfen, R. J. Temkin, D. J. Singel, and R. G. Griffin, *Phys. Rev. Lett.*, 1993, **71**, 3561.

55. L. R. Becerra, G. J. Gerfen, B. F. Bellew, J. A. Bryant, D. A. Hall, S. J. Inati, R. T. Weber, S. Un, T. F. Prisner, A. E. McDermott, K. W. Fishbein, K. E. Kreischer, R. J. Temkin, D. J. Singel, and R. G. Griffin, *J. Magn. Reson. Ser. A*, 1995, **117**, 28.

56. G. S. Nusinovich, M. K. A. Thumm, and M. I. Petelin, *J. Infrared Milli. Terhz. Waves*, 2014, **35**, 325.

57. T. Idehara, E. M. Khutoryan, Y. Tatematsu, Y. Yamaguchi, A. N. Kuleshov, O. Dumbrajs, Y. Matsuki, and T. Fujiwara, *J. Infrared Milli. Terhz. Waves*, 2015, **36**, 819.

58. S. Jawla, M. Shapiro, W. Guss, and R. Temkin, International Vaccum Electronics and Vaccum Electron Sources Conference, Monterey, CA, USA, 2012. https://ieeexplore.ieee.org/abstract/document/6264178.

59. M. Rosay, M. Blank, and F. Engelke, *J. Magn. Reson.*, 2016, **264**, 88.

60. M. Rosay, L. Tometich, S. Pawsey, R. Bader, R. Schauwecker, M. Blank, P. M. Borchard, S. R. Cauffman, K. L. Felch, R. T. Weber, R. J. Temkin, R. G. Griffin, and W. E. Maas, *Phys. Chem. Chem. Phys.*, 2010, **12**, 5850.

61. A. C. Torrezan, M. A. Shapiro, J. R. Sirigiri, R. J. Temkin, and R. G. Griffin, *IEEE Trans. Electron. Dev.*, 2011, **58**, 2777.

62. E. A. Nanni, S. K. Jawla, M. A. Shapiro, P. P. Woskov, and R. J. Temkin, *J. Infrared Milli. Terahz. Waves*, 2012, **33**, 695.

63. A. B. Barnes, E. A. Nanni, J. Herzfeld, R. G. Griffin, and R. J. Temkin, *J. Magn. Reson.*, 2012, **221**, 147.

64. S. Jawla, Q. Z. Ni, A. Barnes, W. Guss, E. Daviso, J. Herzfeld, R. Griffin, and R. Temkin, *J. Infrared Milli. Terahz. Waves*, 2013, **34**, 42.

65. Y. Matsuki, K. Ueda, T. Idehara, R. Ikeda, K. Kosuga, I. Ogawa, S. Nakamura, M. Toda, T. Anai, and T. Fujiwara, *J. Infrared Milli. Terahz. Waves*, 2012, **33**, 745.

66. T. Idehara, Y. Tatematsu, Y. Yamaguchi, E. M. Khutoryan, A. N. Kuleshov, K. Ueda, Y. Matsuki, and T. Fujiwara, *J. Infrared Milli. Terahz. Waves*, 2015, **36**, 613.

67. F. Engelke and S. Steuernagel, in 'Encyclopedia of NMR', eds R. K. Harris and R. E. Wasylishen, John Wiley & Sons: Chichester, 2012, Vol. 2, p 818.

68. D. E. M. Hoff, B. J. Albert, E. P. Saliba, F. J. Scott, E. J. Choi, M. Mardini, and A. B. Barnes, *Solid State Nucl. Magn. Reson.*, 2015, **72**, 79.

69. E. P. Saliba, E. L. Sesti, F. J. Scott, B. J. Albert, E. J. Choi, N. Alaniva, C. Gao, and A. B. Barnes, *J. Am. Chem. Soc.*, 2017, **139**, 6310.

70. E. L. Sesti, E. P. Saliba, N. Alaniva, and A. B. Barnes, *J. Magn. Reson.*, 2018, **295**, 1.

71. E. P. Saliba, E. L. Sesti, N. Alaniva, and A. B. Barnes, *J. Phys. Chem. Lett.*, 2018, **9**, 5539.

72. F. J. Scott, E. P. Saliba, B. J. Albert, N. Alaniva, E. L. Sesti, C. Gao, N. C. Golota, E. J. Choi, A. P. Jagtap, J. J. Wittmann, M. Eckardt, W. Harneit, B. Corzilius, S. T. Sigurdsson, and A. B. Barnes, *J. Magn. Reson.*, 2018, **289**, 45.

73. A. Ajoy, R. Nazaryan, K. Liu, X. Lv, B. Safvati, G. Wang, E. Druga, J. A. Reimer, D. Suter, C. Ramanathan, C. A. Meriles, and A. Pines, *PNAS*, 2018, **115**, 10576.

74. K.-N. Hu, V. S. Bajaj, M. Rosay, and R. G. Griffin, *J. Chem. Phys.*, 2007, **126**, 044512.

75. Y. Hovav, A. Feintuch, S. Vega, and D. Goldfarb, *J. Magn. Reson.*, 2014, **238**, 94.

76. O. Rybalko, S. Bowen, V. Zhurbenko, and J. H. Ardenkjær-Larsen, *Rev. Sci. Instrum.*, 2016, **87**, 054705.

77. T. Dubroca, A. N. Smith, K. J. Pike, S. Froud, R. Wylde, B. Trociewitz, J. McKay, F. Mentink-Vigier, J. van Tol, S. Wi, W. Brey, J. R. Long, L. Frydman, and S. Hill, *J. Magn. Reson.*, 2018, **289**, 35.

78. A. C. Torrezan, S.-T. Han, I. Mastovsky, M. A. Shapiro, J. R. Sirigiri, R. J. Temkin, A. B. Barnes, and R. G. Griffin, *IEEE Trans. Plasma Sci.*, 2010, **38**, 1150.

79. M. L. Mak-Jurkauskas and R. G. Griffin, in 'Solid-State NMR Studies of Biopolymers', eds A. E. McDermott and T. Polenova, John Wiley & Sons: Chichester, 2010, Chapter 9, p 159.

80. A. A. Bogdashov, V. I. Belousov, A. Y. Chirkov, G. G. Denisov, V. V. Korchagin, S. Yu. Kornishin, and E. M. Tai, *J. Infrared Milli. Terahz. Waves*, 2011, **32**, 823.

81. K. J. Pike, T. F. Kemp, H. Takahashi, R. Day, A. P. Howes, E. V. Kryukov, J. F. MacDonald, A. E. C. Collis, D. R. Bolton, R. J. Wylde, M. Orwick, K. Kosuga, A. J. Clark, T. Idehara, A. Watts, G. M. Smith, M. E. Newton, R. Dupree, and M. E. Smith, *J. Magn. Reson.*, 2012, **215**, 1.

82. A. B. Barnes, M. L. Mak-Jurkauskas, Y. Matsuki, V. S. Bajaj, P. C. A. van der Wel, R. DeRocher, J. Bryant, J. R. Sirigiri, R. J. Temkin, J. Lugtenburg, J. Herzfeld, and R. G. Griffin, *J. Magn. Reson.*, 2009, **198**, 261.

83. E. A. Nanni, A. B. Barnes, Y. Matsuki, P. P. Woskov, B. Corzilius, R. G. Griffin, and R. J. Temkin, *J. Magn. Reson.*, 2011, **210**, 16.

84. F. Horii, T. Idehara, Y. Fujii, I. Ogawa, A. Horii, G. Entzminger, and F. D. Doty, *J. Infrared Milli. Terahz. Waves*, 2012, **33**, 756.

85. F. Mentink-Vigier, G. Mathies, Y. Liu, A.-L. Barra, M. A. Caporini, D. Lee, S. Hediger, R. G. Griffin, and G. De Paëpe, *Chem. Sci.*, 2017, **8**, 8150.

86. K. R. Thurber and R. Tycko, *J. Chem. Phys.*, 2014, **140**, 184201.

87. K. R. Thurber and R. Tycko, *J. Chem. Phys.*, 2012, **137**, 084508.

88. G. Mathies, M. A. Caporini, V. K. Michaelis, Y. Liu, K.-N. Hu, D. Mance, J. L. Zweier, M. Rosay, M. Baldus, and R. G. Griffin, *Angew. Chem. Int. Ed.*, 2015, **54**, 11770.

89. S. R. Chaudhari, P. Berruyer, D. Gajan, C. Reiter, F. Engelke, D. L. Silverio, C. Copéret, M. Lelli, A. Lesage, and L. Emsley, *Phys. Chem. Chem. Phys.*, 2016, **18**, 10616.

90. F. Mentink-Vigier, I. Marin-Montesinos, A. P. Jagtap, T. Halbritter, J. van Tol, S. Hediger, D. Lee, S. Th. Sigurdsson, and G. De Paëpe, *J. Am. Chem. Soc.*, 2018, **140**, 11013.

91. R. Tycko, *Acc. Chem. Res.*, 2013, **46**, 1923.

92. P. J. Allen, F. Creuzet, H. J. M. de Groot, and R. G. Griffin, *J. Magn. Reson.*, 1991, **92**, 614.

93. M. Concistrè, O. G. Johannessen, E. Carignani, M. Geppi, and M. H. Levitt, *Acc. Chem. Res.*, 2013, **46**, 1914.

94. Y. Matsuki, H. Takahashi, K. Ueda, T. Idehara, I. Ogawa, M. Toda, H. Akutsu, and T. Fujiwara, *Phys. Chem. Chem. Phys.*, 2010, **12**, 5799.

95. A. Samoson, T. Tuherm, J. Past, A. Reinhold, T. Anupõld, and I. Heinmaa, *Top. Curr. Chem.*, 2004, **246**, 15.

96. K. R. Thurber and R. Tycko, *J. Magn. Reson.*, 2008, **195**, 179.

97. E. Bouleau, P. Saint-Bonnet, F. Mentink-Vigier, H. Takahashi, J.-F. Jacquot, M. Bardet, F. Aussenac, A. Purea, F. Engelke, S. Hediger, D. Lee, and G. De Paëpe, *Chem. Sci.*, 2015, **6**, 6806.

98. Y. Matsuki, T. Idehara, J. Fukazawa, and T. Fujiwara, *J. Magn. Reson.*, 2016, **264**, 107.

99. D. Lee, E. Bouleau, P. Saint-Bonnet, S. Hediger, and G. De Paëpe, *J. Magn. Reson.*, 2016, **264**, 116.

100. Ü. Akbey, W. T. Franks, A. Linden, S. Lange, R. G. Griffin, B.-J. van Rossum, and H. Oschkinat, *Angew. Chem. Int. Ed.*, 2010, **49**, 7803.

101. D. Wisser, G. Karthikeyan, A. Lund, G. Casano, H. Karoui, M. Yulikov, G. Menzildjian, A. C. Pinon, A. Purea, F. Engelke, S. R. Chaudhari, D. Kubicki, A. J. Rossini, I. B. Moroz, D. Gajan, C. Copéret, G. Jeschke, M. Lelli, L. Emsley, A. Lesage, and O. Ouari, *J. Am. Chem. Soc.*, 2018, **140**, 13340.

102. P. Chen, B. J. Albert, C. Gao, N. Alaniva, L. E. Price, F. J. Scott, E. P. Saliba, E. L. Sesti, P. T. Judge,

E. W. Fisher, and A. B. Barnes, *Sci. Adv.*, 2018, **4**, eaau1540.

103. S. Un, J. Bryant, and R. G. Griffin, *J. Magn. Reson. Ser. A*, 1993, **101**, 92.

104. T. Maly, J. Bryant, D. Ruben, and R. G. Griffin, *J. Magn. Reson.*, 2006, **183**, 303.

105. M. Kaushik, T. Bahrenberg, T. V. Can, M. A. Caporini, R. Silvers, J. Heiliger, A. A. Smith, H. Schwalbe, R. G. Griffin, and B. Corzilius, *Phys. Chem. Chem. Phys.*, 2016, **18**, 27205.

106. T. A. Siaw, M. Fehr, A. Lund, A. Latimer, S. A. Walker, D. T. Edwards, and S.-I. Han, *Phys. Chem. Chem. Phys.*, 2014, **16**, 18694.

107. J. Soetbeer, P. Gast, J. J. Walish, Y. Zhao, C. George, C. Yang, T. M. Swager, R. G. Griffin, and G. Mathies, *Phys. Chem. Chem. Phys.*, 2018, **20**, 25506.

108. A. Lund, A. Equbal, and S. Han, *Phys. Chem. Chem. Phys.*, 2018, **20**, 23976.

109. M.-A. Geiger, A. P. Jagtap, M. Kaushik, H. Sun, D. Stöppler, S. T. Sigurdsson, B. Corzilius, and H. Oschkinat, *Chem. Eur. J.*, 2018, **24**, 13485.

110. C. Parish, P. Niedbalski, A. Kiswandhi, and L. Lumata, *J. Chem. Phys.*, 2018, **149**, 054302.

111. F. J. Scott, N. Alaniva, N. C. Golota, E. L. Sesti, E. P. Saliba, L. E. Price, B. J. Albert, P. Chen, R. D. O'Connor, and A. B. Barnes, *J. Magn. Reson.*, 2018, **297**, 23.

112. T. R. Eichhorn, M. Haag, B. van den Brandt, P. Hautle, W. Th. Wenckebach, S. Jannin, J. J. van der Klink, and A. Comment, *J. Magn. Reson.*, 2013, **234**, 58.

113. V. L. Bratman, A. E. Fedotov, Yu. K. Kalynov, P. B. Makhalov, and A. Samoson, *J. Infrared Milli. Terahz. Waves*, 2013, **34**, 837.

114. V. L. Bratman, A. E. Fedotov, A. P. Fokin, M. Y. Glyavin, V. N. Manuilov, and I. V. Osharin, *Phys. Plasmas*, 2017, **24**, 113105.

115. H. Ryan, J. van Bentum, and T. Maly, *J. Magn. Reson.*, 2017, **277**, 1.

116. E. A. Nanni, A. B. Barnes, R. G. Griffin, and R. J. Temkin, *IEEE Trans. Terahz. Sci. Tech.*, 2011, **1**, 145.

117. T. Idehara and S. P. Sabchevski, *J. Infrared Milli Terahertz Waves*, 2017, 2017, **38**, 62.

118. S. Björgvinsdóttir, B. J. Walder, A. C. Pinon, J. R. Yarava, and L. Emsley, *J. Magn. Reson.*, 2018, **288**, 69.

Chapter 8

Millimeter-wave Sources for DNP-NMR

Monica Blank and Kevin L. Felch

Communications and Power Industries (CPI), Palo Alto, CA, USA

8.1 INTRODUCTION

The past several decades have seen a renewed and growing interest in solid-state dynamic nuclear polarization (DNP) enhanced nuclear magnetic resonance (NMR) spectroscopy for a variety of applications in structural biology and materials science.[1] Although solid-state NMR is a widely used and powerful spectroscopic method, its inherent low sensitivity can limit its effectiveness for some applications. DNP has emerged as a promising technique for improving NMR's low sensitivity and, thus, reducing the time required to obtain meaningful spectra. With DNP, the relatively large spin-polarization of electrons is transferred to the surrounding nuclear spins through the irradiation of the electron nuclear transitions at, or near, the electron paramagnetic resonance (EPR)

frequency.[2-4] This polarization transfer results in an increase in NMR signal by several orders of magnitude, which, in turn, results in greatly improved data acquisition rates. Depending on the parameters of the DNP experiment and the sample preparation, power ranging from 10s of mW to 100 W at frequencies above 200 GHz is needed for modern DNP-NMR systems.[5]

DNP-NMR was first conceived and demonstrated in the 1950s at low magnetic fields.[2-4] However, for some time, the efficacy of the technique was limited by the lack of high-frequency sources with sufficient power. In the early 1990s, the ground-breaking work by the Griffin Group at MIT in the development of DNP in solids at low temperatures made use of a 140 GHz gyrotron to achieve enhancement factors greater than 10.[6] This seminal work renewed interest in high-field DNP and ushered in the modern era of DNP research, which led, ultimately, to the present-day commercially available gyrotron-based DNP systems for NMR spectrometers at 400 MHz and above.[1,7]

The gyrotron, or cyclotron-resonance maser, is a device-type that was first discussed in the literature in the 1950s, with reports of early experimental demonstration efforts following shortly thereafter.[8-13] The results from the initial experimental demonstrations were not particularly promising and fell far short of the capabilities of conventional vacuum-electron devices (VEDs), and by the mid-1960s, many groups abandoned gyrotron work. However, in Russia, research and development of gyrotrons continued at a steady pace through the 1960s and 1970s. In the late 1970s and 1980s, with impressive gyrotron results reported in Russia,[14-16] interest in the gyrotron as a

Handbook of High Field Dynamic Nuclear Polarization.
Edited by Vladimir K. Michaelis, Robert G. Griffin, Björn Corzilius and Shimon Vega
© 2020 John Wiley & Sons, Ltd. ISBN: 978-1-119-44164-9
Also published in eMagRes (online edition)
DOI: 10.1002/9780470034590.emrstm1582

high-power source of high-frequency electromagnetic radiation grew and research began, or began again, in the United States, Europe, Japan, and around the world. For decades, gyrotron research and development was aimed at producing high-power (100s of kW to 1 MW), continuous-wave (cw) sources in the 28–200 GHz frequency range for electron cyclotron-resonance heating (ECRH) and electron cyclotron-resonance current drive (ECCD) in fusion plasmas.[17] However, as the development progressed, gyrotrons began to be considered superior VED sources for a variety of applications that require high power at high frequencies, including industrial processing, such as ceramic sintering and metal joining, high-frequency radar, accelerators, nonlethal weapons, and DNP-NMR.[17–19]

Although gyrotrons are widely viewed as the most promising source for modern high-field DNP-NMR, other classes of sources have been used in the past and are still of great interest. Despite being somewhat limited in power at frequencies above 200 GHz, solid-state devices, in particular Impatt and Gunn diodes, as well Schottky-diode-based solid-state multiplier chains, are viable DNP sources. In addition, 'slow-wave' VEDs, such as extended interaction klystrons (EIKs and EIOs) or clinotrons, offer many advantages as sources for high-field DNP.

In the following sections, the present state-of-the-art DNP sources will be discussed. In Section 8.2, the capabilities of solid-state DNP sources and examples of their use in DNP experiments will be detailed. In Section 8.3, the advantages and disadvantages of slow-wave VEDs, such as EIOs, EIKs and clinotrons, will be described. In Section 8.4, the features and design characteristics of gyrotron DNP sources will be discussed. A survey on worldwide gyrotron development for DNP applications will be made and details of the Bruker-CPI gyrotrons for commercial DNP-NMR systems will be provided. In addition, state-of-the-art research on gyrotron amplifiers for DNP will be highlighted. In the final section a summary will be provided.

8.2 SOLID-STATE SOURCES

Solids-state microwave and terahertz devices are viable sources for high-field DNP. In particular, Gunn and Impatt diodes, often used in amplifier frequency-multiplying chains, are commercially available at frequencies above 200 GHz.[20–22] Such

solid-state sources have several advantages, including ease of use, compactness, relatively low cost, and the lack of requirements for special laboratory services, such as cooling water or pressurized gases. However, the main disadvantage is that at high-field DNP relevant frequencies, in excess of 200 GHz, output power from solid-state sources and multiplier chains is currently limited to 200 or 300 mW.[23] The most promising solid-state sources for DNP consist of multiplier amplifying chains in which the signal from a low-frequency synthesizer, normally operating at 9–20 GHz, is multiplied and amplified to reach the target frequency and power. An example of one of the most advanced solid-state sources for DNP is a multiplying chain, based on Schottky diode multipliers. Such amplifier chains have produced 180 mW at 224 GHz and 120 mW at 250 GHz.[22,23]

Many DNP measurements have been made using solid-state microwave sources. For example, a solid-state diode source that produced 17 mW at 140 GHz was successfully used for DNP at 5 T.[24] More recently, a frequency-multiplying chain, which produced 30 mW at 264 GHz, was used in DNP experiments at 9.4 T.[25] However, in both cases, lower temperature DNP-NMR measurements were made to maximize the DNP enhancement at the reduced source power levels.

Extensive research and development of high-frequency solid-state sources continues, and improvements in power capabilities of devices applicable to DNP are being made. For example, Virginia Diodes has reported 3 μW at 2.7 THz.[22] Certainly, solid-state sources will remain of interest for a variety of lower temperature and lower field DNP-NMR work.[26,27] But, for higher power (10–100 W) at frequencies in the 200 GHz to 1 THz range, vacuum electronic sources, discussed in the following sections, are required.

8.3 SLOW-WAVE SOURCES

VEDs can be divided into two categories, fast-wave and slow-wave. In all VEDs, energy is transferred from an electron beam to an electromagnetic wave. For efficient energy transfer, some level of synchronism between the electron beam and electromagnetic wave must be achieved. For slow-wave devices, which include traditional microwave sources such as klystrons, EIOs and EIKs, traveling wave tubes (TWTs), and clinotrons, which are a special type of backward wave oscillators (BWOs), the synchronism is achieved by an interaction circuit designed to reduce the phase velocity of the electromagnetic wave to a value below the speed of light.[28] This reduction in the phase velocity of a slow-wave device is achieved by special structures and features of the interaction circuit, and circuit sizes are typically on the order of one to several wavelengths. For higher frequencies, wavelengths and, therefore, circuit sizes of slow-wave devices decrease, and it becomes difficult to produce high power due to thermal management issues related to cooling the ohmic power lost in the circuit.

One slow-wave device that has demonstrated impressive output power at frequencies above 200 GHz and has been successfully used as a source for high-field DNP is the EIO, or the amplifier version, the EIK.[29] A photograph of a commercially available 263 GHz EIO capable of producing up to 5 W with a mechanical tuning range of up to 2 GHz is shown in Figure 8.1. A 264 GHz EIO that produced up to 1.5 W was used for DNP at 25 K where enhancement factors of over 100 were achieved.[30] In addition, an EIK producing 3 W was used in DNP-NMR experiments at 187 GHz/284 MHz with enhancements of over 100 demonstrated.[31]

Another slow-wave device that could potentially generate enough power at frequencies beyond 200 GHz to be of interest for DNP-NMR is the clinotron, a type of backward wave oscillator. Clinotrons have delivered up to 1 W at 231 GHz, and designs for devices capable of generating up to 100 mW at 545 GHz have been made.[32]

Slow-wave vacuum electronic sources of several varieties offer many advantages for DNP applications. Although often larger than Impatt or Gunn diode systems and solid-state harmonic multiplying chains, slow-wave device systems are still relatively compact. In addition, slow-wave devices for DNP applications are generally very easy to use, with relatively few laboratory requirements, such as cooling water or compressed gasses. Certain slow-wave devices such as the EIO offer mechanical frequency tuning that can be advantageous in DNP applications. And, most importantly, slow-wave devices offer order-of-magnitude increases in power capabilities at DNP frequencies over solid-state sources.

However, slow-wave VEDs often suffer from some potential disadvantages. Despite the ability to produce higher power than have been demonstrated with solid-state sources in the 200 GHz–1 THz range, slow-wave sources are still limited in power at higher

Figure 8.1. Photograph of a commercially available EIO, which produces 5 W at 264 GHz. (Reprinted with permission from www.cpii.com)

frequencies. Because of the physics of the interaction, at higher frequencies, the beam duct and circuit structures become increasingly small and have reduced power-handling capabilities. In addition, the cost for high-performance slow-wave source systems for DNP such as that used in reference[30] can be quite high. Finally, the advertised and predicted lifetimes of state-of-the-art EIOs, such as the tunable 264 GHz device, is on the order of 20 000 h.[33] Despite a few distinct disadvantages, the potential of slow-wave VED sources for DNP-NMR, especially at 264 GHz/400 MHz and below, is strong, and it is likely that these slow-wave devices will be of interest in DNP for the foreseeable future.

8.4 GYROTRONS

8.4.1 Advantages and Disadvantages of Gyrotrons for DNP Systems

In contrast to conventional VEDs in the slow-wave category, gyrotrons, or cyclotron-resonance masers, are in the fast-wave category of sources in which the phase velocity of the electromagnetic wave is equal to or greater than the speed of light.[17,34] Gyrotrons take advantage of the cyclotron-resonance maser instability to transfer energy from an electron beam to an electromagnetic wave. In the interaction, an annular electron beam, comprised of mildly relativistic electrons traveling in helical paths, interacts with the electromagnetic field of a circuit or cavity in the presence of an externally applied axial magnetic field. The synchronism between the electron beam and the electromagnetic field of the circuit is achieved by choosing the applied magnetic field such that the cyclotron frequency of the electrons is nearly equal to the desired frequency of the electromagnetic wave. Gyrotrons can also operate at harmonics of the cyclotron frequency, allowing the axial magnetic field to be reduced by a factor equal to the harmonic number.

The benefits of gyrodevices, in terms of the ability to produce high power at high frequencies, are derived from the combination of the cyclotron-resonance interaction and the fast-wave interaction circuit. In gyrotron circuits, or cavities, the strength of the electromagnetic field, which can be quite high, is not dependent on the proximity of the metallic waveguide structure. This enables the electron beam to be situated in regions of high field, for optimum coupling, without necessarily placing the beam too close to the circuit walls. The interaction mode is selected by the magnitude of the applied magnetic field, which determines the electron cyclotron frequency, and the placement of the electron beam. If a high-order waveguide or cavity mode is selected, the transverse dimensions of the interaction structure can be several times, or even tens of times larger than the free-space wavelength. The larger interaction structure enhances the ability of the device to generate high output power at increasingly greater frequencies.

As discussed above, the natural ability, due to the interaction physics, of a gyrotron to produce very high power at high frequencies is one of the main advantages of the device for DNP applications. A related advantage is the relatively long lifetime of the gyrotron. Slow-wave device circuits at higher frequencies are made of delicate structures that are in proximity to high-energy electron beams. The requirements on the electron beam and damage to the circuits that can occur due to beam impact or heating from the beam or electromagnetic waves contribute to the reduced lifetime of slow-wave devices. In general, the gyrotron is predicted to have a longer life (>50 000 h)

due to its more robust circuit structures and larger beam clearances.

However, the gyrotron does have a few disadvantages that render slow-wave devices and solid-state DNP sources, especially below 300 GHz, of interest despite the gyrotron's superior power capabilities. For example, the external magnetic field required to generate power at higher frequencies necessitates the use of a superconducting magnet, which renders the gyrotron system large. Although many modern high-frequency gyrotrons use superconducting magnets with closed-cycle refrigerators that do not require liquid cryogens, some superconducting magnets still use liquid helium and nitrogen. The higher output power produced by gyrotrons means higher input voltages, often in the 15–20 kV range. Higher output power also leads to more cooling requirements, normally both water and air, for gyrotrons. And finally, gyrotron systems, which include superconducting magnets, are normally more expensive than slow-wave or solid-state competitors.

Despite some disadvantages (lack of compactness, special laboratory requirements, high voltage, cooling or cryogens, and relatively high cost), the gyrotron is still considered the best source for DNP at high frequencies and perhaps the only real source for DNP at frequencies above 300 GHz. Not all of the potential disadvantages are as severe as they might initially seem. For instance, a high-field NMR system already requires a laboratory environment with some amount of space and the availability of liquid cryogens. And all VEDs, fast-wave or slow-wave, will require relatively high-voltage supplies to make relevant microwave or THz power. Finally, although gyrotron systems are nearly always more expensive than solid-state or slow-wave systems, the cost per watt of a gyrotron system is generally the lowest of the three. A gyrotron system that can generate 100 W at 395 GHz can be priced in the range of $1000–$4000 per watt. This cost per watt is certainly an order of magnitude lower than the cost of a solid-state source system and less than half the cost per watt of an EIK source. Finally, the gyrotron is the only source with the demonstrated ability to produce tens of watts or hundreds of watts in the 300 GHz–1 THz range.

8.4.2 Features of High-frequency Gyrotrons for DNP

Key components of a modern DNP gyrotron include an electron gun, an interaction cavity, a mode converter,

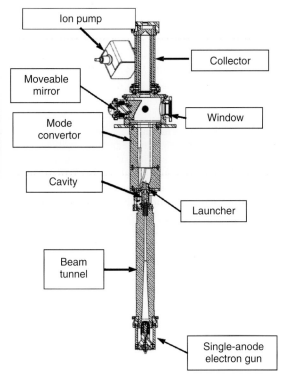

Figure 8.2. Schematic diagram of a typical gyrotron for DNP, the Bruker-CPI 50 W 395 GHz device

a vacuum window, a collector, and a superconducting magnet. A schematic diagram of the Bruker-CPI 50 W 395 GHz gyrotron, which is similar to most modern DNP gyrotrons, is shown in Figure 8.2.

In the figure, the superconducting magnet is not shown. An annular electron beam is produced by a single-anode magnetron injection gun operating at cathode voltages normally in the 10–20 kV range, with cathode currents up to several hundred milliamperes. The electron beam is accelerated toward the anode, which is at ground potential, and travels in a tapered beam tunnel region as its diameter is compressed by the increasing magnetic field. A superconducting magnet, either cryogen-free or liquid-helium-cooled, is used to produce the required magnetic field.

The interaction cavity is located at the peak of the axial magnetic field where the electron beam diameter is at its minimum. Because the larger part of the energy in the electron beam is in the perpendicular, not axial, direction, gyrotrons interact with transverse electric (TE) modes rather than transverse magnetic (TM) modes

that are usually used for slow-wave VEDs. The mode is carefully selected based on several criteria including the required electron beam diameter in the cavity for optimal beam-wave coupling, the power loading on the cavity walls, and competition with other cavity modes. For higher power and higher frequency gyrotrons, high-order cavity modes are normally used. Low-order cylindrical cavity modes, such as the TE_{11} or TE_{01} modes, are often not practical because of the small-diameter electron beam that would be optimal for the interaction, as well as the small diameter cavity, which would lead to large ohmic power densities on the cavity walls and would be difficult to effectively cool. Higher order modes, a unique feature of gyrotrons, offer the possibility of larger cavities with lower power densities on the cavity walls excited by larger diameter electron beams.

Gyrotrons are able to operate in modes with frequencies at, or near, the electron cyclotron frequency and at harmonics of the cyclotron frequency. Although operation at higher harmonics reduces the magnetic field requirement, and, therefore, the cost and complexity of a gyrotron system, the interaction strength at higher harmonics, and the resulting power and efficiency capabilities of a harmonic gyrotron are also reduced.[17]

Because high-order interaction modes, whether at fundamental or higher harmonic frequencies, are not suitable for efficient transmission in smooth or corrugated waveguides, modern gyrotrons typically make use of a converter, often in the vacuum envelope as shown in Figure 8.2, to transform the interaction mode to a Gaussian beam.[35] The mode converter normally consists of a helically cut launcher with specially designed wall perturbations to increase the Gaussian mode content of the launched beam, as well as several mirrors that serve to both shape and steer the Gaussian beam through a vacuum window. The internal converter also serves to separate the generated electromagnetic wave from the electron beam, which is guided by the magnetic field and continues along the gyrotron axis until it is incident on the walls of the collector. A gyrotron's vacuum is typically maintained by a vacuum ion pump, as shown in Figure 8.2.

8.4.3 Survey of Gyrotrons Used for DNP

Since the original demonstration of gyrotron-based DNP at MIT,[6] worldwide development of gyrotrons for DNP applications has taken place. Although not an exhaustive list, a survey of representative development programs and results from some of the key institutions where DNP gyrotron research and fabrication has been carried out are detailed below.

Work has continued at MIT, where several gyrotron oscillators have been developed and used in DNP experiments. In addition to the 140 GHz gyrotron oscillator described in the original article,[6] which was a fundamental frequency oscillator producing about 14 W at 140 GHz for a 210 MHz NMR,[36] MIT has developed gyrotrons at 250, 330, and 460 GHz. The 250 GHz gyrotron, which also operates at the fundamental cyclotron frequency, was designed for DNP experiments with a 380 MHz NMR system. The 250 GHz gyrotron produced 10–15 W, which resulted in 3–4 W of power at the sample, continuously for 72–96 h.[37] An updated version of this 250 GHz gyrotron produced 15–20 W cw output power and, with improved transmission efficiency of about 50%, resulted in approximately 8 W at the sample.[38,39] The updated gyrotron operated continuously for months without fault using proportional–integral derivative (PID) feedback control.[38,39] Gyrotrons that oscillate at double the cyclotron frequency, commonly called second-harmonic gyrotrons, were also developed at MIT. A second-harmonic, 330 GHz gyrotron capable of producing 18 W at 0.9% efficiency was developed for DNP and used with a 500 MHz NMR.[40] Through a combination of variation in voltage and magnetic field, a 1.2 GHz continuous tuning range was achieved for the 330 GHz gyrotron. Frequency tuning is a feature that can be useful in optimizing DNP enhancements. In addition, a second-harmonic 460 GHz gyrotron with a smooth tuning range of 1 GHz was developed for DNP with a 700 MHz NMR system.[41] Finally, a second-harmonic 527 GHz gyrotron for an 800 MHz DNP-NRM system is under development at MIT.

A long-standing and highly productive program in the development of gyrotrons for DNP and other applications at Fukui University, often in conjunction with team members from the Institute of Applied Physics of the Russian Academy of Sciences (IAP-RAS), has resulted in a series of DNP-relevant gyrotrons with impressive power and frequency capabilities.[42] The FU CW II series of gyrotrons, operating in second-harmonic modes at 395.6 GHz, produce up to 200 W of power and use PID feedback control of the voltage to achieve extremely low power fluctuations.[43] Several gyrotrons, including the FU CW II and FU CW VI gyrotrons, which also operate at 395 GHz, are currently in use at Osaka University in DNP

experiments on a 600 MHz NMR system.[44] The FU CW III gyrotron, which is designed to operate in a pulsed 20 T superconducting magnet, achieved the milestone frequency of 1.08 THz at the second harmonic of the cyclotron frequency.[45] The FU CW GO I and II gyrotrons, designed for 460 GHz/700 MHz DNP-NMR at Osaka University, have high-quality Gaussian output beams and produce power up to 100 W in the second harmonic.[46] At Warwick University, the FU CW VII gyrotron, operating at 395 GHz in a second-harmonic mode for DNP-NMR at 600 MHz, produces up to 50 W cw output power.[47]

Independent of collaborative work with Fukui University, research and development on gyrotrons for DNP has been carried out at IAP-RAS in Russia. For example, a second-harmonic gyrotron operating at 260 GHz with 100 W of output power has demonstrated very stable output power and frequency for long time periods, making it especially useful for the NMR spectrometer at the Institute of Biophysical Chemistry of Goethe University, where it is installed.[48,49]

DNP gyrotron work is also underway in Switzerland,[50] China,[51] and the United States.[7,52] At the Ecole Polytechnique Federal de Lausanne (EPFL) in Switzerland, a frequency tunable 260 GHz gyrotron was developed. Over a frequency tuning range of 1.2 GHz, the output power was measured to be at least 1.5 W, and a maximum power of 150 W was measured at 260.5 GHz.[50] At the School of Electronics Engineering and Computer Science, Peking University in Beijing, a 330 GHz gyrotron capable of producing more than 1 W over a 2 GHz tuning range has been designed. Bridge-12, a commercial company in the United States, designed and built a 395 GHz gyrotron intended for 600 MHz NMR.[52] And finally, in the United States, a successful large-scale program to develop gyrotrons for commercial DNP-NMR systems has be carried out and is described in the next section.

8.4.4   Commercially Available Gyrotrons for DNP

In the mid-2000s, Bruker, a leading supplier of NMR instruments, and CPI, a manufacturer of VEDs in general, and gyrotrons in particular, began an effort to develop commercially available gyrotron-based DNP-NMR systems.[7] A series of cw gyrotron oscillators operating at 263, 395, 527, and 593 GHz were developed by CPI for DNP enhancement of Bruker's

400, 600, 800, and 900 MHz NMR spectrometers, respectively. The gyrotrons at all four frequencies were designed to produce at least 50 W of cw output power at efficiencies greater than 1% using an electron beam in the 15–20 kV range at beam currents up to 200 mA. The electron beam is transported from the gun to the interaction cavity through a beam tunnel, which includes alternate rings of lossy ceramic and copper to inhibit any possible RF generation. As the electron beam travels along the beam tunnel toward the interaction cavity, it is compressed by the increasing magnetic field of the superconducting magnet. The size of the emitter and the beam compression ratio, which are not the same for the four gyrotrons, are carefully selected to ensure that the beam size in the interaction cavity is optimal for the chosen mode.

Early versions of the Bruker-CPI 263 GHz gyrotron operated at the fundamental cyclotron frequency in a helium-cooled 10 T magnet. However, the more modern 263 GHz gyrotron operates in the second-harmonic $TE_{11,2}$ mode at 4.9 T. The 395 GHz gyrotron operates in the second-harmonic $TE_{10,3}$ at 7.3 T; the 527 and 593 GHz gyrotrons both operate in the $TE_{14,3}$ mode at 9.7 and 10.9 T, respectively. In all three cases, the operating modes are converted to high-quality Gaussian beams with a launcher and a series of mirrors. The shape of the launcher, which has shallow wall deformations designed to create a mode mix that transforms the operating mode to a beam with a Gaussian field profile, is numerically optimized to maximize the desired Gaussian mode content of the launched beam. The mirrors are then used to shape the launched beam into a beam with waist size and waist position optimized for transport through the window and injection into a corrugated waveguide. The inner diameter of the corrugated waveguide is 19 mm for the 263 GHz gyrotron and 16 mm for the 395, 527, and 593 GHz gyrotrons. The final mirror in the system is adjustable so that the output beam can be positioned in the center of the Al_2O_3 output window and aligned for optimal coupling to the corrugated waveguide during operation.

The spent electron beam is then deposited on the walls of the collector, which is identical for the three lowest frequency gyrotrons and slightly longer for the 527 GHz device, to accommodate the slower expansion of the spent beam due to the higher magnetic field. A 2 ls^{-1} ion pump is used during gyrotron operation. The orientation and position of the ion pump are designed to use the fringing field of the

Figure 8.3. Photograph of the Bruker-CPI 593 GHz gyrotron

superconducting magnet, not a permanent magnet, for proper pump operation.

A photograph of the 593 GHz gyrotron, including one section of a 16-mm-diameter corrugated waveguide, is shown in Figure 8.3. The gyrotrons at the three lower frequencies are very similar and look nearly identical to the 593 GHz device. Figure 8.4 shows a second-harmonic 527 GHz gyrotron positioned in the 9.7 T superconducting magnet in test at CPI. A calorimetric load, located at the end of one section of the corrugated waveguide, is used to measure power during the tests.

Typical measured parameter variation curves for the gyrotrons at 263, 395, 527, and 593 GHz are plotted in Figures 8.5 and 8.6. As shown in Figure 8.5, for each of the four gyrotrons, the output power can be smoothly varied from over 40 W or more to less than 5 W by varying voltage. For the measurements, the beam current is held fixed at 100, 140, 130, and 180 mA for the 263, 395, 527, and 593 GHz gyrotrons, respectively. At 17.4 kV and 100 mA, the 263 GHz gyrotron

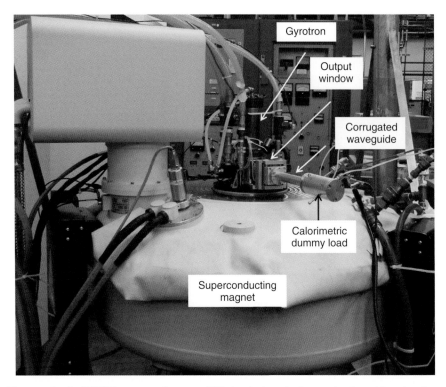

Figure 8.4. Photograph of a 527 GHz gyrotron in test at CPI. A calorimetric dummy load, positioned at the end of a section of a 16-mm-diameter corrugated waveguide, is used for power measurements

generates 109 W, corresponding to 10.9% efficiency. The efficiencies of the 395 and 527 GHz gyrotrons at the highest power points in Figure 8.5 are 2.5% and 2.2%, respectively. The efficiency of the 593 GHz maximum power point on Figure 8.5 is 1.4%.

In Figure 8.6, typical values of measured output power versus beam current with beam voltage held fixed are plotted for each of the four gyrotrons. As shown in the figure, for the 263 GHz gyrotron output power varies from 89 to 7 W as the beam current is reduced from 178 to 71 mA with the beam voltage held constant at 16.5 kV. Similarly, the 395 GHz gyrotron output power varies from 80 to 11 W at 17.4 kV as the beam current is reduced from 178 to 93 mA, and the 527 GHz gyrotron power at 17.5 kV beam voltage varies from 60 to 8 W as the beam current varies from 178 to 106 mA. In the most recent experimental results, power for the 593 GHz gyrotron operating at 18.7 kV was varied from 54 to 30 W as cathode current was reduced from 214 to 157 mA.

In addition to parameter variation curves, infrared images of the output beam are made for each gyrotron to determine the quality of the Gaussian beam. Figure 8.7 shows the setup for the infrared image measurement. The beam is incident on a paper target, which is laser-aligned with the section of the corrugated waveguide. The target is moved in a direction perpendicular to the waveguide, and images of the beam at several points along the propagation path are made. The infrared data is then analyzed, with the assumption that the measured temperature is proportional to the square of the electric field, and comparisons to the desired Gaussian beam shape are made. Figure 8.8 shows typical images of the beam along the propagation path for a 395 GHz gyrotron. In the figure, the distance from the output guide is indicated at the top of each image. The crosshairs represent the position of the waveguide center and the circle is 5.08 cm in diameter. As is evident from the figure, the gyrotron produces a high-quality Gaussian beam, which is centered with and travelling

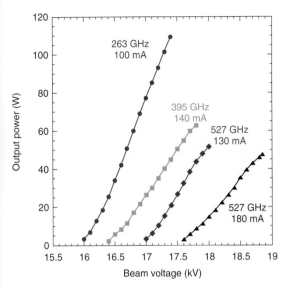

Figure 8.5. Measured output power versus beam voltage for the 263 GHz gyrotron operating at 100 mA beam current (filled circles), the 395 GHz gyrotron operating at 140 mA (filled squares), the 527 GHz gyrotron operating at 130 mA (filled diamonds), and the 593 GHz operating at 180 mA (filled triangles)

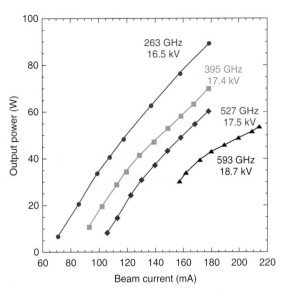

Figure 8.6. Measured output power versus beam current for the 263 GHz gyrotron operating at 16.5 kV beam voltage (filled circles), the 395 GHz gyrotron operating at 17.4 kV (filled squares), the 527 GHz gyrotron operating at 17.5 kV (filled diamonds), and the 593 GHz gyrotron operating at 18.7 kV (filled triangles)

perpendicular to the corrugated waveguide. Figure 8.9 shows the measured horizontal (filled squares) and vertical (filled circles) beam radii at which the electric field is 1/e of the peak value as functions of distance from the guide end compared to the theoretical waists for the four gyrotrons. The measured beam radii agree quite well with the theoretical values. Mathematical modal decompositions of the measured beam profiles show, typically, greater than 95% overlap between the measured beams and an ideal Gaussian beam from a 19-mm-diameter (263 GHz) or 16-mm-diameter (395, 527, and 593 GHz) corrugated waveguide.

In addition to parameter variation curves and in-frared images of the output beam, several other tests intended to demonstrate each gyrotron's stability and reliability are typically performed. For example, long-term stable operation of the gyrotron is demon-strated, wherein the gyrotron is operated without fault continuously for greater than 72 h with less than 1 MHz frequency variation and less than 0.3 W power variation over the run period. Also, operation at the nominal output power with 100% reflection is demonstrated by placing a piece of copper at the end

of the waveguide. The gyrotron is then operated at the nominal point with 100% reflection for at least 1 h without fault or damage.

To date, 35 Bruker-CPI gyrotrons are installed in DNP-enhanced NMR spectrometers around the world at Bruker customer sites. Bruker has delivered seventeen 263 GHz gyrotrons for 400 MHz NMR, twelve 395 GHz gyrotrons for 600 MHz NMR, five 527 GHz gyrotrons for 800 MHz NMR, and one 593 GHz gyrotron for a 900 MHz NMR spectrom-eter. These systems represent a major advance in the availability of commercial high-performance DNP-NMR systems, and bode well for the future of gyrotron-based DNP.

8.4.5 Gyrotron Amplifiers for DNP

A number of studies have shown the advantages of pulsed, or time-domain DNP-NMR.[53,54] The work suggests that the degradations in enhancements at higher frequencies observed for cw DNP-NMR

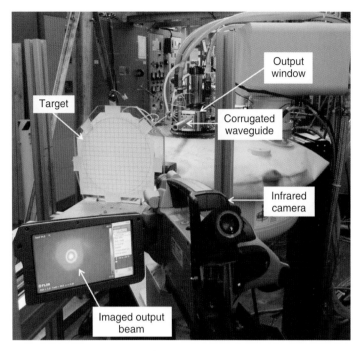

Figure 8.7. Photograph showing the test setup for infrared measurements of the output beam. The beam is incident on a paper target that is laser-aligned with the waveguide

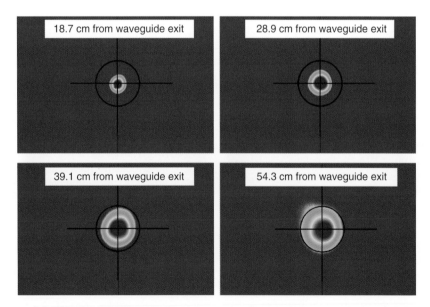

Figure 8.8. Infrared images of a 395 GHz gyrotron output beam. The crosshairs represent the position of the waveguide center and the circle is 5.08 cm in diameter. The target distance from the corrugated waveguide exit is indicated on each image

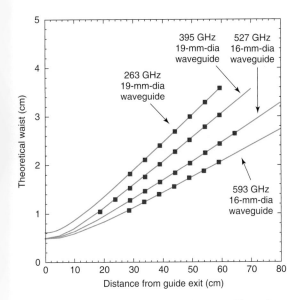

Figure 8.9. Theoretical and measured beam radii as a function of distance for the output beams from 263, 395, 527, and 593 GHz gyrotrons. For each of the four, the solid line shows the theoretical beam radii, the filled squares show the measured horizontal radii, and the filled circles show the measured vertical radii

can be at least partially countered by pulsing the high-frequency DNP source.[53] However, unless and intermediate anode is used, achieving short pulses of electromagnetic waves with a gyrotron oscillator would require pulsing the high-voltage source, which is normally in the 10–20 kV voltage range, which would, in turn, require a complex and costly power supply/modulator system. On the contrary, a gyrotron amplifier source can be easily pulsed by allowing the high-voltage power supply, and electron beam, to operate continuously while the amplified drive signal is pulsed. In addition, like the tunable gyrotron oscillators described above, a gyrotron amplifier can operate over a band of frequencies, which is often useful in achieving the highest enhancement for DNP-NMR.

At MIT, several gyrotron amplifiers for DNP have been developed and demonstrated. A 140 GHz gyrotron amplifier, using a unique confocal interaction circuit, has achieved 550 W peak output power with a bandwidth of 1.2 GHz.[55] The amplifier was demonstrated to be zero-drive-stable, meaning that no electromagnetic waves are generated when the drive source is off even when the gyrotron electron beam is on. Most impressively, the gyrotron amplifier was operated at pulses as short as 4 ns with no discernible pulse broadening.

In addition, a 250 GHz gyrotron amplifier with a photonic bandgap circuit has been developed and tested at MIT.[56] The gyrotron produced 45 W with a total device gain of 38 dB, a full-width-half-maximum instantaneous bandwidth of 8 GHz and an operational bandwidth of 16 GHz. The amplifier can operate at very short pulses, as short as 260 ps, without any pulse distortion or broadening.[56]

These two successful demonstrations of high-performance gyrotron amplifiers for DNP applications represent important steps in the continued advancement of high-field DNP performance.

8.5 SUMMARY

The recent availability of sources in the 200 GHz–1 THz range capable of producing 10–100 W has been a key factor in the renewed interest in solid-state DNP-NMR witnessed over the past several decades. Solid-state microwave sources, which are inexpensive, compact, easy to operate, and readily available at power levels in the tens of milliwatts at frequencies above 200 GHz, have been used in many DNP experiments. However, the decreased power capabilities at frequencies in the hundreds of GHz range limit their applicability to low-temperature DNP at lower fields. Conventional or slow-wave vacuum electronic sources, such as the EIO or EIK, offer higher power than solid-state sources at frequencies up to 260 GHz. Although they require high-voltage power supplies, EIOs and EIKs are still quite compact and easy to use. However, their limited lifetimes, relatively high cost, and decreasing power capabilities are hindrances to their use in DNP experiments for NMR at fields above 10 T. Following the initial demonstration of gyrotron-based DNP-NMR at MIT, ongoing development of DNP gyrotrons around the world has occurred. Gyrotrons, originally developed for fusion heating and current drive applications, have been shown to generate tens, hundreds, or even thousands of watts at frequencies between 200 GHz and 1 THz, which paved the way for Bruker to develop and introduce the first widely available, commercial DNP-NMR systems at 263 GHz/400 MHz, 395 GHz/600 MHz, 527 GHz/800 MHz, and recently 593 GHz/900 MHz. Finally, research and development of gyrotron amplifiers for use in time-domain or pulsed DNP is ongoing. Through the use of higher order

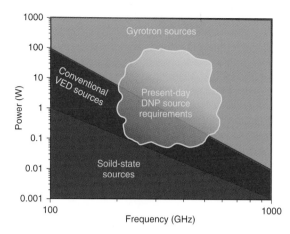

Figure 8.10. Schematic diagram of output power capabilities versus frequency for solid-state, conventional VED, and gyrotron DNP sources

modes and higher harmonics, gyrotron oscillators and, potentially gyrotron amplifiers, have the ability to meet the ever-increasing frequency needs of DNP-NMR.

In Figure 8.10, the rough output power capabilities of solid-state, conventional VED, and gyrotron sources are shown as a function of frequency from 200 GHz to 1 THz. All are capable of meeting some present-day DNP needs, but only gyrotrons show promise for producing the required power, and more, at the higher frequencies of interest.

RELATED ARTICLES IN EMAGRES

Overhauser, Albert W.: Dynamic Nuclear Polarization

Dynamic Nuclear Polarization and High-Resolution NMR of Solids

Dynamic Nuclear Polarization: Applications to Liquid-State NMR Spectroscopy

High-Frequency Dynamic Nuclear Polarization

REFERENCES

1. A. S. Lilly Thankamony, J. J. Wittmann, M. Kaushik, and B. Corzilius, *Prog. Nucl. Magn. Reson. Spectrosc.*, 2017, **102–103**, 120.

2. A. W. Overshauser, *Phys. Rev.*, 1953, **92**, 411.

3. T. R. Carver and C. P. Slichter, *Phys. Rev.*, 1953, **92**, 212.

4. T. R. Carver and C. P. Slichter, *Phys. Rev.*, 1956, **102**, 975.

5. R. G. Griffin and T. F. Prisner, *Phys. Chem. Chem. Phys.*, 2010, **22**, 5737.

6. L. R. Becerra, G. J. Gerfen, R. J. Temkin, D. J. Singel, and R. G. Griffin, *Phys. Rev. Lett.*, 1993, **71**, 3561.

7. M. Rosay, M. Blank, and F. Engelke, *J. Magn. Reson.*, 2017, **264**, 88.

8. R. Q. Twiss, *Aust. J. Phys.*, 1958, **11**, 564.

9. J. Schneider, *Phys. Rev. Lett.*, 1959, **2**, 504.

10. A. V. Gaponov, *Izv. VUZ. Radiofizika*, 1959, **2**, 450.

11. R. H. Pantell, *Proc. IRE*, 1959, **47**, 1146.

12. I. B. Bott, *Proc. IEEE*, 1964, **5**, 330.

13. J. L. Hirshfield and J. M. Wachtel, *Phys. Rev. Lett.*, 1964, **12**, 533.

14. N. I. Zaytsev, T. B. Pankratova, M. I. Petelin, and V. A. Flyagin, *Radiotekhnika I Electronica*, 1974, **19**, 1056.

15. D. V. Kisel, G. S. Korablev, V. G. Navel'yev, M. I. Petelin, and S. Y. Tsimring, *Radio Eng. Electron. Phys.*, 1974, **19**, 95.

16. A. A. Andronov, V. A. Flyagin, A. V. Gaponov, A. L. Gol'denberg, M. I. Petelin, V. G. Usov, and V. K. Yulpatov, *Infrared Phys.*, 1978, **18**, 385.

17. K. L. Felch, B. G. Danly, H. R. Jory, K. E. Kreischer, W. Lawson, B. Levush, and R. J. Temkin, *Proc. IEEE*, 1999, **87**, 752.

18. U.S. Department of Defense Non-Lethal Weapons Program, Active Denial System FAQs, 2018. http://jnlwp.defense.gov/About/Frequently-Asked-Questions/Active-Denial-System-FAQs/

19. M. Blank, P. Borchard, S. Cauffman, and K. Felch, *Terahertz Sci. Technol.*, 2016, **4**, 177.

20. Terasense, 2018. terasense.com

21. Sage Millimeter, 2018. www.sagemillimeter.com

22. Virigina Diodes, Inc., 2018. www.vadiodes.com

23. J. Hesler, 2nd Workshop on Terahertz: Opportunities for Industry, February 2014, Lausanne, Switzerland.

24. V. Weis, M. Bennati, M. Rosay, J. A. Bryant, and R. G. Griffin, *J. Magn. Reson.*, 1999, **140**, 293.

25. K. R. Thurber, W.-M. Yau, and R. Tycko, *J. Magn. Reson.*, 2010, **204**, 303.

26. T. A. Siaw, A. Leavesley, A. Lund, I. Kaminker, and S. Han, *J. Magn. Reson.*, 2016, **264**, 131.

27. A. Leavesley, D. Shimon, T. A. Siwa, A. Feintuch, D. Goldfarb, S. Vega, I. Kaminker, and S. Han, *Phys. Chem. Chem. Phys.*, 2017, **5**, 3596.

28. A.S. Gilmour, Jr., Microwave Tubes, Norwood, MA: Artech House, 1986.

29. D. Berry, H. Deng, R. Dobbs, P. Horoyski, M. Hyttinen, A. Kingsmill, R. MacHattie, A. Roitman, E. Sokol, and B. Steer, *IEEE Trans. Electron Devices*, 2014, **61**, 1830.

30. W.-M. Yau, K. R. Thurber, and R. Tycko, *J. Magn. Reson.*, 2014, **244**, 98.

31. T. F. Kemp, H. R. Dannatt, N. S. Barrow, A. Watts, S. P. Brown, M. E. Newton, and R. Dupree, *J. Magn. Reson.*, 2016, **265**, 77.

32. T. Idehara, A. Kuleshov, E. Khutoryan, S. Ponomarenko, S. Kishko, Yu. Kovshov, and A. Likhachev, 6th International Workshop on Far-Infrared Technologies (IW-FIRT 2017) and The 2nd International Symposium on Development of High Power Terahertz Science and Technology, March 2017, Fukui, Japan.

33. R. McHattie, 3rd Annual Meeting of the COST MPNS Action TD1103 "European Network for Hyperpolariztion Physics and Methodology in NMR and MRI," June 2014, Zurich, Switzerland.

34. G. S. Nusinovich, Introduction to the Physics of Gyrotrons, Baltimore, MD: Johns Hopkins University Press, 2004.

35. A. A. Bogdashov, A. V. Chirkov, G. G. Denisov, A. N. Kuftin, Y. V. Rodin, E. A. Soluyanova, and V. E. Zapevalov, *Int. J. Infrared Millim. Waves*, 2005, **26**, 771.

36. R. J. Temkin, *Terahertz Sci. Technol.*, 2014, **7**, 1.

37. V. S. Bajaj, C. T. Farrar, M. K. Hornstein, I. Mastovsky, J. Vieregg, J. Bryant, B. Eléna, K. E. Kreischer, R. J. Temkin, and R. G. Griffin, *J. Magn. Reson.*, 2003, **160**, 85.

38. A. B. Barnes, E. A. Nanni, J. Herzfeld, R. G. Griffin, and R. J. Temkin, *J. Magn. Reson.*, 2012, **221**, 147.

39. S. Jawla, Q. Z. Ni, A. Barnes, W. Guss, E. Daviso, J. Herzfeld, R. Griffin, and R. J. Temkin, *J. Infrared Millim. Terahertz Waves*, 2013, **34**, 42.

40. A. C. Torrezan, M. A. Shapiro, J. R. Sirigiri, R. J. Temkin, and R. G. Grifin, *IEEE Trans. Electron Dev.*, 2011, **58**, 2777.

41. A. C. Torrezan, S.-T. Han, I. Mastovsky, M. A. Shapiro, J. R. Sirigiri, R. J. Temkin, A. B. Barnes, and R. G. Griffin, *IEEE Trans. Plasma Sci.*, 2010, **38**, 1150.

42. M. Glyavin, T. Idehara, and S. Sabchevski, *IEEE Trans. Terahertz Sci. Technol.*, 2015, **5**, 788.

43. T. Idehara, A. Kuleshov, K. Ueda, and E. Khutoryan, *J. Infrared Millim. Terahertz Waves*, 2014, **35**, 159.

44. Y. Matsuki, K. Ueda, T. Idehara, R. Ikeda, K. Kosuga, I. Ogawa, S. Nakaura, M. Toda, T. Anai, and T. Fujiwara, *J. Infrared Millim. Terahertz Waves*, 2012, **33**, 745.

45. T. Idehara, I. Ogawa, H. Mori, S. Kobayashi, S. Mitsudo, and T. Saito, *J. Plasma Fusion Res. Ser.*, 2009, **8**, 1508.

46. T. Idehara, Y. Tatematsu, Y. Yamaguchi, E. M. Khutoryan, A. N. Kuleshov, K. Ueda, Y. Matsuki, and T. Fujiwara, *J. Infrared Millim. Terahertz Waves*, 2015, **36**, 613.

47. K. J. Pike, T. F. Kemp, H. Takahashi, R. Day, *et al.*, *J. Magn. Reson.*, 2012, **215**, 1.

48. A. A. Bogdashov, V. I. Belousov, A. V. Chirkov, G. G. Denisov, V. V. Korchagin, S. Y. Kornishin, and E. M. Ta, *J. Infrared Millim. Terahertz Waves*, 2011, **32**, 823.

49. V. Denysenkov, M. J. Prandolini, M. Gafurov, D. Sezer, B. Endeward, and T. F. Prisner, *Phys. Chem. Chem. Phys.*, 2010, **12**, 5786.

50. J. P. Hogge, F. Braunmueller, S. Alberti, and J. Genoud, 38th International Conference on Infrared, Millimeter, and Terahertz Waves (IRMMW-THz), Mainz, Germany, 2013.

51. Z. D. Li, C. H. Du, X. B. Qi, L. Luo, and P. K. Liu, *Chin. Phys. B*, 2016, **25**.

52. J. R. Sirigiri, T. Maly, and L. Tarricone, 36th International Conference on Infrared, Millimeter, and Terahertz Waves, Houston, TX, 2011.

53. S. Un, T. Prisner, R. T. Weber, M. J. Seaman, K. W. Fishbein, A. E. McDermott, D. J. Singel, and R. G. Griffin, *Chem. Phys. Lett.*, 1992, **189**, 54.

54. A. A. Smith, B. Corzilius, J. A. Bryant, R. DeRocher, P. P. Woskov, R. J. Temkin, and R. G. Griffin, *J. Magn. Reson.*, 2012, **223**, 170.

55. A. V. Soane, M. A. Shapiro, S. Jawla, and R. J. Temkin, *IEEE Trans. Plasma Sci.*, 2017, **45**, 2835.

56. E. A. Nanni, S. Jawla, S. M. Lewis, M. A. Shapiro, and R. J. Temkin, *Appl. Phys. Lett.*, 2017, **111**, 233504.

Chapter 9

Cryogenic Platforms and Optimized DNP Sensitivity

Yoh Matsuki and Toshimichi Fujiwara

Institute for Protein Research, Osaka University, Suita, Osaka, Japan

9.1 INTRODUCTION

Magic-angle spinning nuclear magnetic resonance (MAS NMR) spectroscopy is one of the most powerful tools for studying structure and dynamics of not only crystalline but also amorphous molecular systems such as polymers,[1,2] inorganic materials,[3] and biomacromolecules[4] at atomic resolution, but suffers from its inherently low sensitivity. Dynamic nuclear polarization (DNP) enables orders of magnitude sensitivity improvement of MAS NMR by transferring the large electron spin polarization to the nuclei of interest with an irradiation of strong microwaves at the frequency near the electron spin

Handbook of High Field Dynamic Nuclear Polarization.
Edited by Vladimir K. Michaelis, Robert G. Griffin, Björn Corzilius and Shimon Vega
© 2020 John Wiley & Sons, Ltd. ISBN: 978-1-119-44164-9
Also published in eMagRes (online edition)
DOI: 10.1002/9780470034590.emrstm1553

resonance (ESR). Although the phenomenon of DNP itself has been known since the 1950s and extensively studied,[5–8] prospects for its application to analysis of complex chemical systems have turned bright only in 1993[9] when a gyrotron, a vacuum electron device, was introduced as a new radiation source for high-frequency microwaves. This has opened up a way for the DNP technique, for the first time, to the 'high-field condition' in the modern NMR sense ($B_0 = 5$–21.1 T). The following two decades have witnessed rapid advances both in the instruments and methodologies of high-field DNP that include the development of cross-effect (CE)-based DNP methods,[10] biradical polarizing agents for efficiently driving CE-DNP,[11–13] advanced gyrotrons,[14–21] and cryogenic DNP MAS NMR probe systems.[22–24] As a result, high-field DNP MAS NMR spectrometers are now operative at $B_0 = 5$ T (212 MHz ^{1}H frequency), 6.7 T (284 MHz), 7 T (300 MHz), 11.7 T (500 MHz), 14.1 T (600 MHz), 16.4 T (700 MHz), 18.7 T (800 MHz), and 21 T (900 MHz) with many commercial spectrometers up and running worldwide.[25]

Modern high-field DNP NMR spectrometers are typically based on a cryogenic MAS capability that realizes a sample temperature T of ~100 K. The benefits of employing low temperatures are as follows: first of all, the thermal-equilibrium nuclear polarization, also known as the *static* nuclear polarization, increases inversely proportional to the sample temperature (Curie's law). On top of this, the DNP efficiency itself is improved through the better electron saturation as well as better accumulation of the *dynamically* produced nuclear polarization, thanks to the slower electron and nuclear spin relaxation, respectively. Moreover, freezing of the cryoprotectant solvents into a glassy solid (often occurring at $100-140\,K$) enables an efficient distribution of the locally produced nuclear hyperpolarization to the entire sample (including the solute and the matrix) through spin diffusion. The latter property ensures the generality of the method being able to polarize a wide range of sample setups involving the dissolved/suspended bio/organic/inorganic molecular systems such as supramolecular protein assemblies and surface-wetted polymers and materials.[26–35]

A sample temperature $T \sim 100\,K$ is readily obtained using a liquid nitrogen (LN_2)-based heat-exchanging system for a relatively low operational cost, while achieving substantial microwave-on versus -off signal enhancement ε_{DNP} in the order of 200 at moderate field conditions such as $B_0 = 9.4\,T$ or lower. Today, however, the need for performing MAS DNP at much lower sample temperatures, $T \ll 100\,K$, is receiving a growing attention. The reason is related to the observed inefficiency of the CE-based DNP process at very high-field conditions $B_0 > 10\,T$: in fact, ε_{DNP} for a standard sample using the TOTAPOL biradical at $T \sim 100\,K$ was scaled as, $\varepsilon_{DNP} \sim 180$ at 5 T, ~100 at 9.4 T, ~60 at 14.1 T, ~40 at 16.4 T, and ~10 at 18.8 T, being proportional to $\sim 1/B_0^p$, where $p \sim 2$.[36] This evokes a serious concern as the high-field condition is a prerequisite for the spectral resolution, and thus for targeting with DNP interesting chemical/biological systems with ever-increasing size and complexity. Meanwhile, it was shown that the ultralow temperature MAS DNP settings can dramatically improve the CE DNP-based sensitivity gain at $T \ll 100\,K$ by benefiting from a combined effect of increased static as well as dynamic nuclear polarization and from the suppressed radio frequency (RF) circuit thermal noise.[24,37] Clearly, the ultralow temperature MAS NMR capability is a powerful remedy to the loss of efficiency and is the crucial component for a success of 'very high-field DNP' in the coming years.

Besides the improved sensitivity gain, the benefits of establishing low/ultralow temperature MAS NMR are many fold: firstly, it enables the study of phenomena accessible only at ultralow temperatures, such as the Hall effect in two-dimensionally confined electrons,[38] super conduction of metallic boron compounds,[39] and ultralow energy-barrier molecular/molecular segmental motion,[40,41] to name a few. In biological

applications, a quantitative mapping of methyl group rotational barriers over a protein assembly, readily determined from the NMR linewidth variation at low temperature,[42] may provide a new and useful way to characterize the protein folding or interprotein interactions. Furthermore, together with visible light delivery to the sample, it has been shown that the low-temperature MAS platform enables sequential excitation and freeze trapping of the photoexcited intermediates of retinal proteins for structure analysis without relying on amino-acid mutations,[29,43,44] which is still a far-reaching goal for X-ray crystallography.

Important considerations in the implementation of low-temperature MAS NMR setups include how to produce, cool, and transfer sufficient spinner gas (for bearing and driving the sample rotor) and variable temperature (VT) gas streams to the NMR probe with the highest long-term stability and the minimum operational cost. Ease of system operation represents another important factor. In the following sections, we describe three contemporary cryogenic MAS NMR platforms: one of them is based on the use of cryogenic nitrogen gas (GN_2) for a sample temperature $T \sim 100\,K$, and the other two on liquid or gas helium (LHe/GHe) for $T \ll 100\,K$. The most detailed accounts are given on the systems developed and operated by the authors' laboratory in Osaka University, dedicated to the high-field (14.1 and 16.4 T) DNP NMR spectrometers. We first describe the prototype N_2-MAS system (see Section 9.2), then boiled-off LHe MAS systems (see Section 9.3.1), and finally a more sophisticated closed-cycle He MAS system relying only on GHe for sample cooling and -spinning (see Section 9.3.2). A section discussing the experimental DNP enhancements and effective sensitivity gain obtained with the described systems follows (see Section 9.4), and a brief summary (see Section 9.5) concludes this chapter.

9.2 N_2-BASED MAS NMR PLATFORM

Cryogenic MAS at $T \sim 100\,K$ can be achieved by cooling the spinner gas streams by heat exchanging them against relatively inexpensive LN_2 as previously implemented.[10,45,46] However, even based on the 'cheap' LN_2, the accumulative operational cost for DNP MAS NMR measurements can occupy a significant portion of the running cost of an NMR lab, especially for studies on biological systems that often rely on lengthy multidimensional data acquisition.

In many cases, people pay duly attention to the introduction and initial installation costs for a DNP spectrometer, but less attention to the day-to-day running cost for the system. However, the operational cost is equally important for constantly generating data out of the installed system; thus, inventing and implementing appropriate measures to reduce the unit time measurement cost deserves serious consideration.

9.2.1 Design and Operation of N_2-MAS System

Our N_2-cooling and -spinning MAS DNP NMR probe system incorporates four novel features for improving the long-term stability, the cost efficiency, and the readiness of the system's operation. Firstly, the VT and the spinner gas streams are produced by separating

pure and dry GN_2 from the atmospheric air, rather than by boiling LN_2 as conventionally done. Secondly, thus-produced GN_2 is precooled with an electric gas chiller before the heat-exchanging step against liquid cryogen to reduce the cryogen evaporation. Thirdly, for heat exchange, a liquid argon (LAr) bath was used instead of the conventional LN_2 bath to avoid liquefaction of the pressurized spinner GN_2 in the heat exchanger, increasing the MAS reliability/stability and the ease of operation. And lastly, the small amount of liquid oxygen (LO_2) remaining in the VT gas stream is utilized to achieve efficient cooling.

Figure 9.1 shows a schematic diagram of the N_2-MAS system dedicated to our custom-built 14.1 T MAS DNP NMR spectrometer.[30,37,46] It consists of the GN_2-production and -cooling systems and the dedicated NMR probes interconnected via vacuum-insulated flexible gas transfer lines. Two

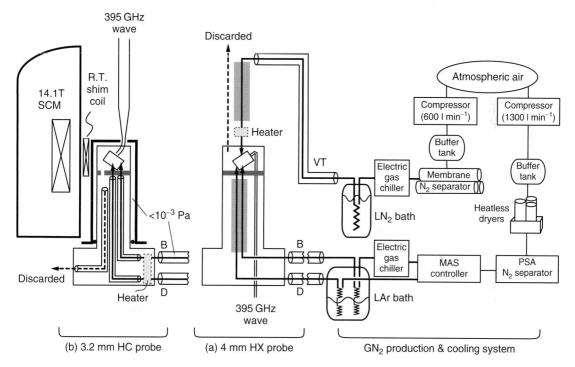

Figure 9.1. A schematic diagram of the open-circuit N_2-cooling and -spinning MAS NMR probe system dedicated to a 14.1 T DNP NMR spectrometer. The system consists of the gas production and cooling systems and the dedicated NMR probes (a,b). For the 4 mm HX probe (a), the flexible transfer lines delivering the VT and the spinner gas from the heat exchanger toward the probe are vacuum insulated as indicated by a cylinder enclosing the arrow of the gas flow, whereas the rigid gas pipes in the magnet bore or in the probe are only wrapped with a foam insulator and Mylar film depicted as gray shades. For the 3.2 mm HC probe (b), all the flexible as well as the rigid lines are vacuum insulated throughout. The spent GN_2 used for the sample cooling and spinning is simply discarded without further use

DNP-NMR probes are operative on this spectrometer: one is a modified Varian T3 probe with a MAS module for a 4 mm diameter rotor (4 mm HX probe), being doubly tunable to ^{1}H and X nuclei, where X covers the nuclear frequencies of ^{31}P down to ^{15}N. The other with a 3.2 mm MAS module was designed and built from scratch and tunes to ^{1}H and ^{13}C (3.2 mm HC probe). Both probes fit inside the room-temperature shim stack, and achieve regular high-resolution MAS NMR.

The 4 mm HX probe provides a path to deliver a VT gas stream to the sample in addition to the two paths for spinner gases (bearing and driving gas streams). The VT gas line comes down the magnet bore and is input to the probe from the top, while the spinner gases are input at the bottom of the probe, go through the probe's cylindrical body, and are joined to the MAS module. None of the three gas transfer lines running in the magnet bore or in the probe body is vacuum insulated, but are wrapped with a foam insulator (aerogel blanket, $\kappa \sim 0.02\,W\,m^{-1}K^{-1}$, Takumi Sangyo, Inc.) and Mylar^{TM} film (Dupont Teijin Films, Ltd). In addition, the probe outer surface is not treated for heat insulation, and instead a continuous flow of room-temperature GN_2 is maintained through the probe body to mitigate the frost developed on the probe surface. A 395 GHz microwave or, more properly defined, submillimeter wave (SMMW) is also input from the bottom of the probe box, transmitted up toward the MAS module, allowing it to irradiate the rotor axially after a tight turn made using a set of flat copper mirrors. An overmoded, smooth-wall waveguide was used for the wave transmission throughout.

The 3.2 mm HC probe incorporates three vacuum-jacketed paths: two for delivering the spinner gas streams, and the third for returning the spent GN_2 down the probe. On the way to the MAS module, the transfer lines have a thermal contact with a 50 W copper block heater proportional–integral–derivative (PID) regulated (Lakeshore Cryotronics Inc., Model 335) from the sample chamber temperature. The sample temperature was widely controllable at a fixed MAS rate by adjusting the heater and the driving gas flow rate; thus, the VT gas line was unnecessary for this probe. After the temperature control and a 90° turn, the stainless-steel rigid transfer line is joined to the vacuum-jacketed line leading to the MAS module using a He-tight indium sealing technique. The cylindrical part of the probe body is covered with a vacuum-tight hollow cylindrical probe jacket made of G10 glass fiber, and the space between the probe and

the jacket is evacuated ($<10^{-3}$ Pa) for heat insulation. The waveguide comes down the magnet bore, and is input to the probe from the top to illuminate the sample rotor transversely. This design of the 3.2 mm HC probe is mostly shared with the one employed for the closed-cycle He MAS system described below (see Section 9.3.2).

The spinner N_2 gas is first separated from the atmospheric air as follows. An oil-free air compressor (Hitachi Industries Equipment Systems, Ltd., SRL-11DMA) produces a $\sim$1300 l min^{-1} air stream at a pressure of 0.65–0.8 MPa. The produced compressed air is roughly dried to a dew point (d.p.) of $\sim$10 °C, buffered with first a 400 l, followed by another 55 l surge tank, filtered (CKD Corp., F8000, 5 μm mesh) to remove dust and oil mist, pressure regulated to 0.3 MPa, and then thoroughly dried using two series-connected pressure-swing adsorption (PSA)-type heatless dryers (CKD Corp., HD-9). The clean and dry air thus produced is finally input to the PSA-type N_2 separator (Kuraray Chemical Co. Ltd., MA2-5.5-7K) to produce pure (99.9%) and dry (d.p. ≈ -100 °C) GN_2 at a rate of $\sim$180 l min^{-1}. The oxygen content was typically less than 10 ppm. The GN_2 stream is then transferred to the NMR lab using stainless-steel tubing to avoid contamination from atmospheric moisture and impurity gases.

For the production of the VT N_2 gas stream for the 4 mm HX probe, another air compressor (Hitachi Industrial Equipment Systems, SRL-5.5DB6, 600 l min^{-1}) was operated. Similar to the spinner gas production, the air is compressed (0.65–0.78 MPa), roughly dried, buffered with a 100 l surge tank, filtered (CKD Corp., F4000), and then input into a now simpler N_2 separator based on a polyimide hollow fiber membrane filter (UBE Industries Ltd., NM-C05A). Depending on the *required* oxygen (O_2) level in the VT GN_2 stream, two or three units of the membrane filters are connected in series: the typical O_2 content is rated as $\sim$5% (with one filter unit) down to 0.01% (with five connected in series).

Both air-separated VT and the bearing gas streams are independently cooled to $\sim$190 K using an electric gas chiller (Polycold Systems International, PGC-152) before the full heat-exchanging step against the liquid cryogen.[46] To cool the VT gas, a 10 turn heat-exchanger coil wound from a 6 mm diameter copper tube was directly dipped into a LN_2 bath. Although the pressure on the VT gas line is not much higher than 1 bar, this easily liquefies the 'impurity' O_2 in the VT gas stream due to its boiling point

(~90 K) being much higher than that for LN_2 (~77 K). This intentional O_2 liquefaction was necessary for obtaining $T \sim 100$ K at the sample by overcoming the insufficient heat insulation for the 4 mm HX probe (results are described in Section 9.2.2). The spinner gas, on the other hand, is cooled using a heat exchanger immersed into a LAr bath, rather than LN_2 bath. The boiling point of LAr (~87 K) being ~10 K higher than that of LN_2 completely avoids liquefaction of the pressurized ($\lesssim 2$ bar) spinner GN_2 at the heat exchanger. The latter heat exchanger has a two-staged copper-tube coil structure. The first-stage five-turn coil is located above the LAr level and cools the gas stream to ~100 K primarily relying on the evaporated cold GAr flow on its way out the dewar, helping to reduce the LAr consumption. The second-stage six-turn heat-exchanger coil, connected via a 280 mm long low-thermal conduction stainless-steel pipe, is located below the liquid level for full heat exchange. The gas-transfer tube after the main exchanger is vacuum jacketed so that the gas temperature at the outlet (~88 K) is minimally affected by the LAr level in the dewar. The LAr dewar is equipped with an automatic refilling system in which a solenoid operated valve is triggered from a liquid level sensor, to regulate the nominal liquid level between 40% and 60% so that the liquid level locates between the heat-exchanger coils.

To avoid liquefaction of the spinner gas, conventionally, a pressurizable canister structure surrounding the heat-exchanging copper coil has been in use.[47] The internal pressure of the canister, i.e., the LN_2 level in the canister, is maintained using a N_2 gas cylinder or 'house N_2 gas' source with a regulator. As the MAS rate and sample temperature depend on the liquid level in the canister, the canister pressure has to be carefully supervised throughout the measurements that may last for days or a week. Instead, the use of LAr simplifies the heat-exchanging structure, eliminates the additional N_2 pressure supply, as well as the continuous supervision required during the measurement, and thus reduces the trouble and increases the ease of the system operation.

9.2.2 Performance of N_2-MAS System

For both 3.2 mm HC and 4 mm HX probes, the sample temperature of $T = 100$ K ± 1 K and the MAS rate v_R of 7 kHz (± 5 Hz) could be maintained for >10 h. No drop in the VT and spinner gas flow rate due to the internal

ice formation was observed, and even longer runs seem to be possible. The maximum MAS rate is mainly limited by the available amount of the driving gas flow and the turbine efficiency, both being the subject of further improvement.

The N_2 gas production from the air separation eliminated the conventional LN_2 boiling procedure typically required for generating the N_2 spinner gas, dramatically reducing the operational cost of the system. For example, 10 kHz MAS for a 3.2 mm rotor typically consumes ~150 l min^{-1} of spinner gases, which amounts to ~300 l day^{-1} of LN_2 boiled in a conventional procedure. On the other hand, the principal cost for the air separation method is the electricity expense required for the air compressor: ~260 kWh day^{-1}. On the basis of the local dispense prices for LN_2 and for the electricity in the authors' lab, the air-separation approach suppresses the gas production cost by a factor of ~2. In addition, the precooling procedure reduced the cryogen cost: it almost halved the liquid cryogen consumption at the heat exchanger (e.g., 400 to 200 l day^{-1}). This precooling strategy[46] has been adopted by the groups at the Massachusetts Institute of Technology (MIT)[45] and Washington University,[48] with similar cost reductions being reported.

The intentional use of LO_2 from impurities in the N_2 VT gas is intriguing. Empirically, three units of the membrane-filter N_2 separator connected in series (for the O_2 content of ~0.2%) yielded an optimum result. The VT gas flow of ~60 l min^{-1} measured after the N_2 separators enabled to achieve a minimum sample temperature of $T \sim 95$ K at $v_R = 7$ kHz for a fully loaded 4 mm zirconia rotor. Presumably, microdrops of LO_2 travel a significant fraction of the gas-transfer line and vaporize somewhere in between the heat exchanger and the input to the MAS module. The specific heat of LO_2 (53.8 J mol^{-1} K^{-1}) is almost twice that of GN_2 (29.4 J mol^{-1} K^{-1}), and when it evaporates it removes the heat of vaporization (~300 J g^{-1}) from GN_2. As a result, the VT gas temperature when it has reached the MAS module is significantly lower than that obtained with a completely pure GN_2 stream. In fact, with five units of the N_2 separators connected in series for ~0.01% of O_2 remaining in the VT flow, a minimum sample temperature of only ~140 K was obtained at the same VT flow (~60 l min^{-1}) after the N_2 separator. On the other hand, with a smaller number of filters (1 or 2), a part of the LO_2 reaches the MAS module and made the sample temperature highly unstable. It was also important to use a sample rotor made of zirconia for its low heat conductivity κ of ~ 3 W m^{-1}K^{-1}, rather than the more popular sapphire ($\kappa \sim 40$ W m^{-1}K^{-1}) or Si_3N_4 ($\kappa \sim 30$ W m^{-1}K^{-1}) for DNP to prevent the higher temperature spinner gas (~ 140 K) to affect the temperature of the center-packed sample although a standard-length rotor ($l = 52$ mm) was good enough for obtaining $T < 100$ K unlike the He-cooled, N_2-spinning probe previously reported.[24,49] Although zirconia, which has high dielectric constant ($\varepsilon' \sim 30$) and loss tangent (tan $\delta = 0.1$ at 50 GHz), is less transparent to the SMMW than sapphire or Si_3N_4 ($\varepsilon' \sim 10$, tan $\delta \sim 0.0003-0.002$), it did not obstruct the SMMW irradiated along the rotor axis as implemented in our 4 mm HX probe. The effect of the rotor material on the SMMW sample heating and on the DNP enhancement was previously investigated.[37] Although intriguing and useful, this trick of utilizing LO_2 cannot be applied to the spinner gas as even a partial liquefaction within the gas flow easily disturbs the stability of the rotor spinning.

The use of LAr for heat exchange completely abolished the spinner gas liquefaction without using the canister structure. The commercial price of LAr is only a few times that of LN_2. In addition, the typical evaporation rate of LAr for cooling two streams of the spinner gas (for bearing and driving) was only ~ 1.3 l h^{-1} as compared with ~ 2.5 l h^{-1} reported recently for a heat-exchanger system using a LN_2 bath,[48] partly due to the $\sim 40\%$ higher heat of vaporization for LAr than for LN_2. The small additional cost, if any, can be tolerated considering the ease of the system operation and the stability of the temperature/MAS rate obtained in exchange.

9.3 He-BASED MAS NMR PLATFORMS

For a sample temperature of $T \sim 100$ K or higher, a heat exchange with LN_2 or LAr sufficed for cooling the VT and spinner gas streams. To achieve lower temperatures, $T \ll 100$ K, GHe produced from[50–54] or cooled by LHe (b.p. ~ 4.2 K)[55] has been previously used. Although liquid hydrogen also has a viable heat of vaporization (30 kJ l^{-1}), the boiling point (~ 20 K) and attractive commercial price ($\sim \$1$ l^{-1}) for a purpose of cooling GHe, it is not suitable to a routine use due to its explosivity. As LHe is much more expensive than LN_2, a lot more careful treatment will be required in the design and operation of the system. In particular, an application of the vacuum heat-insulation technique is highly desirable everywhere including the

gas-transfer lines outside as well as inside the NMR probe. It is also essential to take measures to reduce He consumption by recovering/recycling the GHe exiting from the probe, looking toward routine application of He temperature MAS to lengthy biological NMR spectroscopy.

9.3.1 Liquid He Boil-off MAS NMR Probe System

9.3.1.1 Design and Operation of LHe Boil-off MAS System

Low-temperature He spinner gas streams are simply produced by boiling LHe. This method has historically been adopted by a number of groups,[50,52–54] as we have implemented this also for our first He MAS probe system illustrated in Figure 9.2.[56] We employed a 100 l pressurizable (<5 bar) dewar, in which the stored LHe is evaporated using a boiler heater (50 Ω, ~ 1 A max). The produced compressed (typically ≤ 3 bar) GHe stream (~ 7 K at the output of the dewar) is transferred with a C-shaped vacuum-jacketed rigid transfer line to a neighboring flow-regulator box that encases two cryogenic needle valves in vacuum

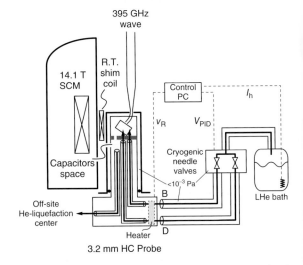

Figure 9.2. A schematic diagram of an open-circuit He-cooling and -spinning DNP NMR probe system based on LHe boil-off. The same 3.2 mm HC probe used for the open-circuit N_2-MAS experiment is used. The spent GHe is collected and recycled off-site

($<10^{-3}$ Pa). The valves are controlled manually or PID regulated from a Labview PC using the detected MAS rate. The control PC regulates the boiler heater and the needle valves and continuously monitors at the same time the LHe level in the dewar, inner dewar pressure, the spinner gas pressures after the needle valves, and the probe temperature. After the valves, the drive and bearing gas streams are independently transferred to the NMR probe using a vacuum-insulated flexible tube and joined to the rigid gas-transfer path in the probe via a cryogenic vacuum-insulated bayonet. The above-described 3.2 mm HC probe was operated for these open-circuit He MAS experiments as well; the only difference to the N_2 MAS experiment is that the spent GHe exiting from the probe is not discarded but collected into a dedicated pipeline connected to a large-scale liquefaction center on campus ($\sim$500 m away from the NMR lab) for recycling.

To initiate the sample cooling, the He dewar was pressurized to about 150 kPa on gauge using the boiler heater. By adjusting the needle valves, the pressure is adjusted mostly on the bearing line (e.g., 130 kPa bearing line and 20 kPa drive line) to obtain a slow ($\sim$1 kHz) sample rotation below the mechanical resonance frequency characteristic to the rotor geometry and the kinematic viscosity of GHe, and at the same time allowing a substantial amount of gas flow ($\sim$30 l min^{-1}) for cooling. The kinematic viscosity of GHe dramatically changes over the temperature drop, and a number of characteristic behaviors were observed during the cooling process. For example, at $T \sim 200$ K, a gradual increase of the bearing gas flow rate is observed due to the transition of the gas flow through the bearing clearance from the turbulent to laminar regime, accelerating the cooling rate. When the temperature reaches $T \sim 120$ K, the drive gas pressure was manually increased to >40 kPa to spin up the sample rotor to $\nu_R > 2$ kHz by going swiftly across the mechanical resonance condition of the rotor located at $\nu_R = 2$–2.2 kHz. An attempt to spin up the sample at higher temperature should be avoided as the density of GHe is too low (i.e., insufficient bearing stiffness), and the rotor easily collides against the ceramics bearing stator. After the increase of the MAS rate at $\sim$120 K, the sample temperature rapidly falls to $T \sim 35$ K in $\sim$30 min. The whole cooling process from room temperature down to 35 K typically takes $\sim$2 h. A similar amount of time is required to warm up the probe to ambient temperature for sample exchange.

9.3.1.2 Performance of LHe Boil-off MAS Probe System

The minimum sample temperature obtained with the LHe boil-off MAS probe system was $T \sim 20$ K, with the input spinner gas temperature to the probe of $\sim$10 K. The maximum MAS rate at 20 K was $\sim$3 kHz for a loaded 3.2 mm Si_3N_4 rotor, with the total gas flow (on bearing and driving) of $\sim$100 l min^{-1} (at the standard condition). The sample temperature could be regulated very precisely ($\pm$0.5 K) over $\sim$10 h using the internal probe heater feedback regulated from the sample chamber temperature continuously monitored using a RuO_2 sensor. (The temperature in the sample rotor measured from the ^{79}Br T_1 calibration was the same as the sample chamber temperature within $\pm$2 K.) It was on the other hand difficult to completely avoid the fluctuation of the MAS rate due to the intrinsic instability of the boiled-off GHe streams, as well as the limited agility and the highly nonlinear response of the cryogenic needle valves. Liquid He, which has very low heat of vaporization ($\sim$70 times less than for LN_2), is highly sensitive to the heat convection in the dewar and to the finite heat transfer through the dewar wall that changes with the liquid level, affecting the evaporation rate. This contributes to the difficulty in long-term stabilization of the boiled-off stream even with a precise heater controller. Thus, the best stability for the MAS rate was about $\pm$30 Hz with the above setup. The maximum MAS rate was also relatively low due to the suboptimal rotor design at that time.

Typical LHe consumption was $\sim$6 l h^{-1}; thus, our 100 l dewar lasts $\sim$10 h, which is good enough for recording several 2D data, or a few 3D datasets in one time. Because the dewar is pressurized, it is impossible to refill it during the NMR measurement. A longer measurement requires one to completely depressurize the dewar to refill, changing the sample temperature considerably. Thus, it is usually recommended to freshly start the measurement on the next day. The high operational cost is an issue as well: for example, performing the low-temperature experiments repeatedly over the days on one week easily consumes $\sim$300 l of LHe, which is clearly not suitable to a routine application. Although in our system the spent GHe exhaust from the probe is directly collected and recycled off-site, the recycling cost entails maintenance cost for the large-scale purifier and liquefaction facilities, as well as that for the required electricity and for the professional workforces. This off-site recycling strategy also requires one to plan purchasing the recycled LHe back from the facility and transporting it in advance and on time for every measurement, disrupting the flexibility of experiments.

9.3.2 Closed-cycle He MAS Probe System

Despite the number of motivations for implementing ultralow temperature ($T \ll 100$ K) MAS NMR, as outlined in Section 9.1, establishing a long-term stable and cost-efficient He MAS NMR probe system has long been an overwhelming challenge; thus, He MAS NMR has not been widely used in applications. As we have seen in Section 9.3.1, the issues associated with the conventional LHe boil-off strategy includes (i) the difficulty to obtain a stable MAS rate and sample temperature throughout the measurement, (ii) the difficulty to maintain the condition for a long time (>10 h) due to the difficulty of refilling the pressurized boiler dewar, and (iii) the high operational cost due to the high commercial price of LHe. Moreover, general issues unrelated to the sample spinning involve (iv) the high propensity of a probe for arcing from the ionized helium at high-voltage parts of the RF circuit, introducing unrecoverable noise in the NMR data and (v) the inhomogeneous signal broadening due to distributed molecular conformation frozen at low temperature, making an analysis in a 1D spectrum very difficult.

An elegant and highly efficient approach overcoming these issues has recently been proposed by the authors' group, reporting the first example of a LHe-free closed-cycle MAS NMR probe system[23] (patents pending: G20150053, 2015-0168082). Cooling a sufficient amount of GHe 'on the fly' for a cryogenic fast MAS in a closed-loop structure leads to engineering challenges, but was realized as described in the following sections. As a result, the above-mentioned difficulties (i)–(iv) were clearly resolved, and the issue (v) was significantly improved by facilitating high-dimensional data acquisition at ultralow temperatures.

9.3.2.1 Design and Operation of Closed-Cycle He MAS Probe System

The closed-cycle He MAS probe system[23] consists of (i) the cryogenic helium circulation (CHC) unit that involves a heat-exchanger vessel encasing Gifford-McMahon (GM) gas chillers, a gas compressor rack, regulators, and buffer tanks and (ii) a dedicated DNP MAS NMR probe that guarantees a complete He-hermiticity for the entire gas-transfer lines and in the sample chamber (Figure 9.3). The GHe introduced to the system from a gas cylinder is independently compressed for the driving and bearing gas streams, buffered, regulated, then cooled with the GM cooler, and finally sent to the NMR probe for sample cooling and spinning. The exhaust GHe being still very cold will be recycled in the initial cooling stage of the heat exchanger, and then returns to the original compressor completing a full cycle. This innovative in situ gas circulation mechanism enables a stable cryogenic spinner gas flow, and has realized exquisitely stable MAS virtually indefinitely without consuming GHe. Details follow.

First of all, the system including the CHC unit and the connected NMR probe need to be purged using low-flow ($\sim 2 \, l \, min^{-1}$), high-purity (99.99%) GHe until the He purity measured at the vent valve reaches '100%' (which takes ~ 10 min). This is followed by 10 repeated pressurize-and-vent cycles to completely expel any remaining impurity gas out of the system. From a fear of disturbing the rotor caps and boiling the liquid sample, the usual 'pump-down' gas replacement procedure was avoided. Finally, roughly $25 \, l$ of 'persistent' GHe is introduced to the system to start the cooling process. This 'persistent' GHe will not be consumed at all during the low-temperature experiment that may last for weeks or even longer.

A metal bellows compressor (Senior Aerospace, MB-602), installed on the driving and the bearing gas line, produces on each line $\sim 100 \, l \, min^{-1}$ of a compressed (<0.3 MPa) GHe stream. The pump requires no oil lubricant and minimizes the contamination of the spinner gas in the long term. The gas is then independently buffered with a $5 \, l$ front buffer tank, regulated with a mass flow controller (MFC, Horiba/STEC Inc., SEC-N132MGM), and sent to the heat-exchanging vessel. The main vacuum vessel encases two 10 K GM coolers (Sumitomo Heavy Industries, RDK-408S), one each on the driving and bearing gas line, being able to cool $\sim 250 \, l \, min^{-1}$ of GHe in total from room temperature down to ~ 25 K with a 15 kW power input for the GM coolers. More recently, we have extended the system with a booster heat-exchanger vessel encasing three additional GM coolers (two on the drive and one on the bearing line) that is installed right after the main heat-exchanger vessel in series (Figure 9.3). Further, a bellows compressor was added on the driving gas line to double the maximum driving gas flow to $\sim 200 \, l \, min^{-1}$. With this

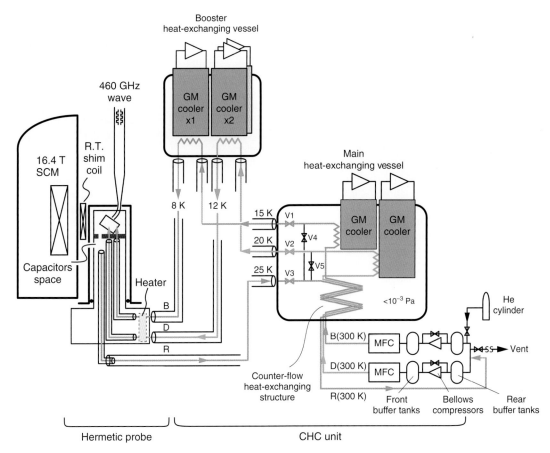

Figure 9.3.   A diagram of the closed-cycle He MAS system dedicated to a 16.4 T DNP-NMR spectrometer. The system consists of the cryogenic helium circulation (CHC) unit and the dedicated NMR probe. The path for the bearing, drive, and return gas streams are indicated by blue, green, and orange lines, respectively. Vacuum-insulated gas transfer lines are indicated by a cylinder enclosing the arrow of the gas flow. The booster heat-exchanger vessel can be bypassed for experiments at $T > 35$ K. By closing valves V1–V3 and opening V4 and V5 in the main heat-exchanger vessel, an independent closed-cycle cooling of the CHC unit is possible while leaving the probe at ambient temperature

boosted setup, totally ~350 l min^{-1} of GHe is cooled to ~9 K with a 38 kW power input.

The cold spinner gases are then independently transferred to the NMR probe using a vacuum-insulated flexible tube and joined to the rigid transfer line internal to the probe using a vacuum-insulated bayonet. The probe design is almost identical to the abovedescribed 3.2 mm HC probe used for the N$_2$-MAS (see Section 9.2) and LHe boil-off MAS systems (see Section 9.3.1), except that the RF circuit is tuned for the NMR at 16.4 T. For the moment, a regular Kel-F MAS module with standard bearing and driving stator ceramics are used without modification, but the

bearing clearance was low-temperature optimized by fine-tuning the Si$_3$N$_4$ rotor outer diameter. The vespel rotor cap used at both ends of the rotor is specially designed so that it does not get loose at ultralow temperature (patent pending). No dielectric sheath or silicon coating was applied to the RF coil.

To prevent probe arcing under the He atmosphere, the GHe exit from the MAS module after the sample cooling and -spinning was directed through a 15 mm diameter feedthrough to a space just below the sample chamber, where the variable/chip capacitors for RF circuit tuning and matching are located. The 6 mm diameter inlet to the gas exhaust pipe is located just below the capacitors' space. The large amount of gas flow ($\sim$350 l min^{-1} at maximum) and a slightly positive pressure in the capacitors' room due to the pressure loss at the exhaust pipe inlet seem to contribute to suppressing the discharge arcing by reducing the mean free path of ionized helium atoms.

The vacuum-jacketed exhaust pipe returns the spent GHe down the probe and back toward the heat-exchanger vessel via an insulated flexible transfer tube by preserving its low temperature (25–180 K depending on the set sample temperature). This cold exhaust GHe is reused in the initial cooling stage with a tube-in-tube-type counter flow heat-exchanging structure,[23] and as a result is warmed to ambient temperature, fully exploiting its cooling capacity. This cooling power recycling mechanism was crucial for reducing the required number of coolers and for containing the small size of the entire system. The ambient return gas is also important for operating the compressors, filters, buffer tanks, and the high-precision MFCs all at room temperature without requiring more complex, expensive, and sometimes precision-compromising cryogenic specifications. The cooling power recycling strategy was adopted in a recently reported open-circuit N$_2$ MAS NMR probe system, where it played a major role in reducing the LN$_2$ consumption at the heat exchanger.[48] Finally, the gas is returned to the original bellows compressor via a 5 l rear buffer tank, where it is recompressed and sent out for a fresh cooling cycle.

9.3.2.2 Performance of Closed-cycle He MAS Probe System

Using only two GM coolers on the main heat-exchanger vessel, a minimum sample temperature of $T = 35$ K was obtained at $v_R = 4$ kHz. With the boosted version using five GM coolers in total, the minimum temperature reached $T = 15$ K at $v_R = 4$ kHz. The maximum spinning rate at higher temperature was: $v_{R,max} = 4.8$ kHz at 20 K, 7 kHz at 35 K, 11 kHz at 60 K, 14.5 kHz at 85 K, and 18 kHz at 200 K. It is remarkable that a single MAS probe system seamlessly covers such a wide temperature range. The spinning rate was found to scale almost linearly with the temperature, suggesting that the temperature-dependent volume of the driving gas flow is a limiting factor. For instance, a typical driving gas flow of $\sim$180 l min^{-1} under standard conditions corresponds only to $\sim$20 l min^{-1} at $T = 35$ K, assuming an ideal gas. Bouleau *et al.* previously reported

$v_R \sim 13$ kHz at $T \sim 35$ K with their LHe-cooled MAS probe system[55] using $\sim$250 l min^{-1} of GHe on the driving gas line (D. Lee, CEA, personal communication). Simple expansion of the system with more compressors as well as the coolers makes the system massive and less desirable when space is limited. Thus, modifications of the bearing stator, driving turbine, and the MAS module housing are in order, and an extensive study using fluid dynamics simulations is underway.

Low operational cost is one of the primary advantages of the present system. The total amount of required GHe for a single run including that used for the purging is only $\sim$700 l, corresponding to <1 l of LHe, occupying a very small fraction of the total operational cost. In fact, a regular 14 m^3 He gas cylinder typically lasts for $\sim$3 months. The GM cooler is robust and needs infrequent maintenance. Also, only a small shack (5 m long × 1 m wide × 2 m high) is required outside the NMR lab for installing five air-cooled compressors. The electricity required for the gas chiller is 192 kWh per GM cooler per day, and the number of required GM coolers depends on the desired temperature range: all five GM coolers (960 kWh day^{-1}) were needed for the ultralow temperature regime, $T = 15$–35 K, while only two were sufficient (384 kWh day^{-1}) for low-temperature regime, $T = 35$–100 K, and one on the bearing line for the high-temperature regime, $T = 100$–200 K. As compared with the conventional LHe boil-off system, the closed-cycle He MAS system thus suppresses the operational cost by a factor of >15–30 for measurement at low- to ultralow temperatures. On the basis of the fact that even the state-of-the-art N$_2$-MAS system consumes $\sim$90 l day^{-1} of LN$_2$ at the heat exchanger in addition to the electricity for the precooling step,[48] the closed-cycle He MAS system runs at a factor of 3–4 lower operational cost in the high-temperature regime.

Another outstanding feature of the closed-cycle system is the long-term stability of the measurement condition. For both the sample temperature and the MAS rate, an exquisite stability of $\pm$0.5 K and $\pm$2 Hz, respectively, was obtained at $T = 40$ K, with the longest uninterrupted run of over two weeks confirmed so far.[23] During the weeks of operation, no loss of the gas flow/pressure was observed, and no condensation developed anywhere in the system/probe, thus apparently even longer experiments are possible. Similar stability was observed at all temperatures tested between $T = 20$ and 200 K. Unlike the conventional LHe boil-off system, the gas production step is strikingly steady, making the intrinsic MAS rate very stable without any active feedback control. For example, the MAS rate only exhibited a small ($\pm15\,$Hz) and slow change with a time period of a day, presumably coming from the change in the cooling capacity of the GM coolers that weakly depend on the environment temperature that changed over the day and night. This MAS rate fluctuation was easily removed by setting the maximum driving gas flow at the MFC. Thus, the active feedback control was unnecessary although possible to implement. On the basis of the stability of the gas flow and cooling capacity, the reproducibility of the conditions is very high over repeated experiments, e.g., obtaining $T = 35.0\,$K at $v_R = 5.0\,$kHz repeatedly on different samples and on different days is easy, which should improve the quality of data in a series measurement and comparison.

A limitation of the current system is the slow cooling process. With both the main and the booster heat-exchanger vessels in use, it takes roughly 8 h to cool the sample from room temperature down to $T \sim 90\,$K due to the large total mass and the heat capacity of the system, although no human attendance is required during this first-stage cooling process. As also described in Section 9.3.1.1, once the temperature reaches $T \sim 120\,$K, one can safely spin up the sample rotor by increasing the spinner gas flow, and this enables much faster cooling later on; it takes only another $\sim 1\,$h to reach the ultralow temperature regime, $T = 15$–$35\,$K. Taking advantage of the ability to cool the CHC unit independently to the NMR probe by closing the valves V1–V3 and opening V4 and V5 (Figure 9.3), the startup procedure can be streamlined so that one can cool only the CHC unit overnight unattended and start cooling the probe and the sample the next morning. The probe cools fast due to its small heat capacity in comparison to the exchanger vessels. In any case, this long cooling process required at every sample exchange considerably reduces the throughput of the measurement. A cryogenic sample ejection/insertion system[22] is an obvious solution to this problem, and an original system is currently under construction.

It is naturally possible to operate the closed-cycle He MAS system with all the GM coolers turned off for room-temperature NMR experiments. Considering the ~ 3 times higher speed of sound of GHe at room temperature ($\sim 970\,$m s^{-1}) vs that of GN$_2$ ($\sim 330\,$m s^{-1}), the maximum MAS can exceed the current maximum achieved with the conventional air-driven sample

spinning. A limiting issue will be the strength of the rotor material to stand the centrifugal force, and a study to overcome this issue is underway in the authors' lab.

A version of the closed-cycle He MAS system dedicated only to the high-temperature regime, $T = 70$–$200\,$K, is also under investigation. The system should require only one GM cooler mainly cooling the bearing gas; thus, the size, the initial installation cost, and the operational cost will be significantly reduced. In the near future, the closed-cycle MAS probe system can become a 'staple' equipment not only for DNP MAS NMR but also for regular MAS NMR measurements in the low to ultralow sample temperature regimes.

9.4 CRYOGENICALLY OPTIMIZED HIGH-FIELD DNP AND SENSITIVITY

This section looks at DNP results obtained with the abovedescribed N$_2$-cooling MAS system at $B_0 = 14.1\,$T (see Section 9.4.1), and that based on the closed-cycle He MAS at $B_0 = 16.4\,$T (see Section 9.4.2), illustrating their performance in practical applications. We also discuss the effective sensitivity gain at ultralow sample temperatures as compared with the conventional DNP performed at $T \sim 100\,$K.

9.4.1 DNP with N$_2$-MAS Probe System

A stable and cost-effective DNP NMR at $T \sim 100\,$K is possible with a N$_2$-cooling and -spinning MAS NMR probe system such as that described in Section 9.2.1. Figure 9.4 shows a set of representative DNP-enhanced 2D ^{13}C–^{13}C correlation NMR spectra of uniformly ^{13}C-labeled glucose recorded with our 3.2 mm HC probe and the custom-built DNP spectrometer operating at 14.1 T. The entire DNP system is described in a previous chapter reporting the first DNP enhancement achieved at a field $>10\,$T.[46] The spectra in Figure 9.4(b) and (c) were taken with glucose *dissolved* in the standard DNP juice (20 mM TOTAPOL in d$_8$-glycerol/D$_2$O/H$_2$O; 6/3/1 w/w/w). Thus, the observed linewidths of $\sim 4\,$ppm (Figure 9.4c) are mainly due to the inhomogeneous broadening caused by the structural heterogeneity of the glucose molecules frozen in the glass matrix.

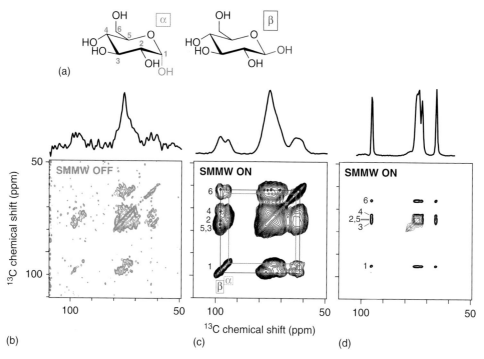

Figure 9.4. Two-dimensional ^{13}C–^{13}C DARR correlation spectra of ^{13}C-labeled glucose taken at $B_0 = 14.1$ T, $T = 100$ K, and $\nu_R = 7$ kHz. The mixing time was 50 ms. Shown in (a) are the structures of α and β anomers of glucose. (b,c) Spectra taken for the glucose dissolved in the DNP juice without (b) and with (c) the 395 GHz SMMW irradiation. The recycle delay was 15 s for maximizing the signal intensity. (d) A DNP-enhanced spectrum of crystalline glucose wetted with the DNP juice. The recycle delay was 60 s. The low-temperature nuclear polarization enhancement $\varepsilon_{DNP} \times \varepsilon_{Curie}$ was ~30 in both (c) and (d). The acquisition time with two scans per transient was ~1 h (b and c) and ~3 h (d). (Reproduced with permission from Ref. 30. © Elsevier, 2016)

Still, the signals from the α and β anomers are clearly distinguishable in the SMMW-on spectrum. The observed DNP enhancement factor was $\varepsilon_{DNP} \sim 10$. The low-temperature nuclear polarization enhancement, taking account of the Curie effect, $\varepsilon_{Curie} \sim 3$, was thus $\varepsilon_{DNP} \times \varepsilon_{Curie} \sim 30$. At the same field condition ($B_0 = 14.1$ T), ε_{DNP} of up to ~150 has been reported.[25] The relatively low ε_{DNP} we obtained is attributed to the poor transmission of the 395 GHz SMMW (only ~100 mW delivered to the sample) as well as a poorly defined $\mathbf{B}_1$ field direction in the SMMW beam radiated in the TE_{06} mode and directly transmitted using an overmoded smooth-wall waveguide in our spectrometer.[30,46] A 395 GHz CW gyrotron equipped with an internal quasi-optical mode convertor for a Gaussian beam output is currently under test. A corrugated waveguide transmission system is also under construction.

One can polarize with DNP not only monodispersed molecules but also suspended/wetted molecular assemblies such as crystalline powder samples. Figure 9.4(d) shows the same correlation spectrum obtained for a crystalline glucose sample. About 20 mg of ground crystalline powder was center-packed into a 3.2 mm rotor, and then wetted with a few drops of the DNP juice before quickly freezing in the probe. This relatively large active sample volume is suited for measurements on bulky samples, such as membrane proteins, whole cells, and material surfaces. The signals for the crystalline glucose were all much sharper (full-width at half maximum, FWHM ~1 ppm) than those of the dissolved glucose. Although glucose is soluble to water, sharp crystalline glucose signals are clearly preserved with the above preparation procedure (Figure 9.4d). It is intriguing to see that crystalline glucose only exhibits the α

anomer signals, clearly seen for the C1 resonance, while the dissolved glucose exhibits an equal mixture of anomers. Obviously, the anomerization reaction proceeds only in solution. Also, an almost identical enhancement of ε_{DNP} ~10 was observed. A similar enhancement factor obtained for the crystalline and noncrystalline samples indicates that the enhanced ^{1}H polarization reaches the core of the crystals. Assuming the ^{1}H–^{1}H spin diffusion constant of 0.8 nm^2 ms^{-1}, the polarization should propagate over a distance of 0.25 μm during the polarization buildup-time constant of ~100 s for the crystalline glucose sample. Thus, the crystals are deduced to have a largest dimension of less than about 0.5 μm in our sample. A similar test should be useful with a polymer blend for quantifying the domain size.[57]

As this crystal-wetting procedure allows the active sample volume to be predominantly occupied by the sample of interest (with only ~30 vol% for the DNP juice), while obtaining a similar DNP enhancement to a dissolved molecule in favorable cases, it achieves high *absolute* sensitivity. Although the longer buildup-time constant observed for the crystalline (~100 s) versus dissolved (~5 s) glucose causes a factor of ~4 loss of the unit-time sensitivity, the gains from the DNP enhancement and the Curie factor $\varepsilon_{DNP} \times \varepsilon_{Curie}$ usually overwhelm the loss. The temperature-dependent variation of the buildup time (e.g., 100 K vs room temperature) is generally much smaller, especially for a doped sample, and has an even smaller impact on the overall sensitivity gain. It is of note that the DNP MAS NMR is suited to analyze extremely small, e.g., submicrometer (protein) crystal samples with excellent sensitivity and resolution. This is in contrast to X-ray crystallography that often struggles with crystals in such an immature size. Applications of the sample-wetting procedure are wide: there are a number of reports on DNP with wetted crystalline molecules and pharmaceutical compounds at around T ~ 100 K.[26,28] A similar sample preparation protocol is also valid for large molecular assemblies such as secretion needle proteins[32] as well as membrane-embedded proteins. The suspended membrane pellet involving the protein of interest is often mixed with the DNP juice, and then thoroughly spun down to squeeze out the excess matrix fraction.[29] In one case, the membrane sample was loaded into a rotor, then wet with a few drops of the DNP juice from the top, and equilibrated.[43]

9.4.2 DNP with Closed-cycle He MAS Probe System

The rather steep decrease of the CE-based DNP enhancement with increasing field strength evokes a concern to the cost-efficiency of high-field DNP spectroscopy. However, as discussed in the following paragraphs, the loss of efficiency seems to be largely recovered using the ultralow sample temperature (T ~ 30 K) MAS DNP setting.[37,56]

Figure 9.5 shows a representative 460 GHz-700 MHz DNP-enhanced NMR spectrum of 1M ^{13}C-urea taken at $T = 40$ K, $v_R = 4$ kHz using AMUPol as the polarizing agent. The closed-cycle He MAS probe system described in the Section 9.3.2 was employed. The spectrometer setup including the 460 GHz gyrotron and its transmission system has been described elsewhere.[30] The simplistic DNP enhancement factor calculated comparing the SMMW on- and off-signal intensities, ε_{DNP}, of ~100 was obtained with ~3 W of the SMMW power input to the probe. The enhancement was still in the regime where it linearly increases with the input SMMW power (data now shown), indicating the importance of the high-power

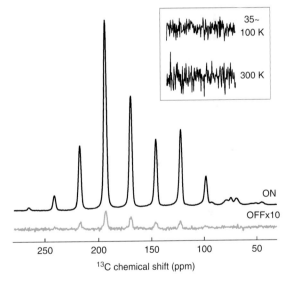

Figure 9.5. A representative 460 GHz-700 MHz DNP-enhanced ^{13}C-CP MAS NMR spectrum of ^{13}C-urea taken with (black) and without (gray) the SMMW irradiation. Data were recorded at $T = 40$ K, $v_R = 4$ kHz, and $B_0 = 16.4$ T. The sample was 1 M ^{13}C-urea and 20 mM AMUPol dissolved in the standard DNP juice. Inset: comparison of the thermal noise measured at $T = 100$ and 300 K

SMMW capability for an optimal sensitivity gain even at ultralow sample temperature. As expected, the thermal noise decreased with lowering temperature: the standard deviation noise intensity was a factor of 1.5 smaller at ~100 K or lower temperatures, as compared with that at room temperature as the noise sources including the sample, coil, leads, and capacitors are kept cold in the probe (Figure 9.5, inset).

Figure 9.6(a) plots the temperature dependence of the DNP enhancement observed for the two most popular biradical polarizing agents, TOTAPOL[13] (black data) and AMUPol[58] (gray data). First of all, it is generally stated that the CE-based DNP enhancement exponentially increases with lowering sample temperature. And for both biradicals, the enhancement factor of ε_{DNP} ~100 was obtained at $T = 30\text{--}40$ K (Figure 9.6a). Almost identical temperature dependence was previously observed for TOTAPOL using the open-circuit He MAS system described in the Section 9.3.1 at $B_0 = 14.1$ T.[56] It is thus suggested that the electron oversaturation or the MAS-induced EPR spectral diffusion[24,59] is still not a serious limiting factor in the tested temperature range. Within the temperature range tested, a factor of 1.5–2 larger values of ε_{DNP} were obtained for AMUPol than for TOTAPOL. On the other hand, the quenching factor ε_θ ($= \chi_{depo} \times \chi_{paramag}$, where $\chi_{paramag}$ is the paramagnetic quenching factor and χ_{depo} is a factor due to the depolarization/Thurber effect[60]) observed at $T = 40\text{--}90$ K was considerably poorer for AMUPol ($\varepsilon_\theta = 0.25\text{--}0.3$, i.e., the SMMW off-signal intensity is only <30% of that for the undoped sample) than for TOTAPOL ($\varepsilon_\theta = 0.35\text{--}0.5$) for the dissolved urea sample. Thus, the enhancement factor accounting for the quenching effect $\varepsilon_{DNP} \times \varepsilon_\theta$ turned out to be similar for both polarizing agents (Figure 9.6a, dashed lines). Note, however, that the quenching effect is not a universal factor, but strongly depends on the sample setup, for example, it affects supramolecular assemblies or crystalline/membrane-embedded samples wetted with the DNP juice much less than the dissolved molecule due to less paramagnetic broadening effect.

Figure 9.6(b) takes a closer look at the temperature-dependence of the quenching factor ε_θ measured for both biradicals. For AMUPol, the contributions from the depolarization effect and the paramagnetic quenching are separately measured (gray broken lines): the depolarization factor was found to decrease with lowering the sample temperature from χ_{depo} ~0.6 at 90 K to ~0.4 at 40 K, while the paramagnetic quenching stayed almost unchanged

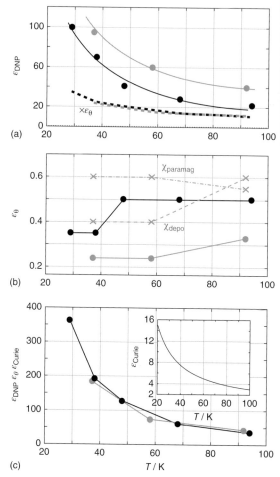

Figure 9.6. Temperature dependence of the DNP enhancement factor ε_{DNP} (a), the quenching factor ε_θ (b), and the factor taking account of the quenching as well as the Curie factor $\varepsilon_{DNP} \times \varepsilon_\theta \times \varepsilon_{Curie}$, (c). Black and gray symbols show data for TOTAPOL and AMUPol, respectively. In (a), dashed lines include the effect of the quenching factor ε_θ. In (b), two contributions to the quenching factor, i.e., the depolarization and paramagnetic quenching factors, are separately measured and plotted for AMUPol (dashed lines). The inset in (c) plots the temperature-dependent enhancement ε_{Curie} of the static nuclear polarization relative to that at room temperature

over the tested temperature range (at $\chi_{paramag}$ ~0.5). Note also that the value of the depolarization factor (χ_{depo} ~0.6 at T ~100 K) we obtained at $B_0 = 16.4$ T is substantially larger than that reported at $B_0 = 9.4$ T (χ_{depo} ~0.4 at T ~100 K) in accordance with previous observations.[61] Thus, this is generally suggesting a decreasing impact of the depolarization effect with increasing external field condition.

Figure 9.6(c) plots the low-temperature polarization enhancement including the Curie factor, $\varepsilon_{DNP} \times \varepsilon_\theta \times \varepsilon_{Curie}$. At $T \sim 30$ K, the enhancement factor is reaching ~400, which is a factor of >10 greater than that obtained with DNP at conventional temperature $T \sim 100$ K using TOTAPOL or AMUPol. In fact, the impact of using ultralow temperature MAS as a tool to recover the lost DNP efficiency at high fields is outstanding among other approaches such as the improvement of the polarizing agents and the SMMW irradiation efficiency. The increase obtained from the development of polarizing agents (TOTAPOL vs AMUPol) was a factor of 1.5–2 (Figure 9.6a), and the use of smaller diameter rotor for a better SMMW irradiation (3.2 vs 2.5/1.3 mm rotor) gave a factor of ~2.[13,62] Note also that the sensitivity enhancement observed here is subject to a further increase in the future by combining the lower temperature MAS with low-temperature-optimized polarizing agent, a strategy to increase the SMMW power distribution, a more sophisticated SMMW irradiation scheme, and the introduction of cold preamps and duplexer.

For an elaborate discussion on the sensitivity gain, even more factors can be taken into account such as the gains from the decreased thermal noise of the RF circuit at low temperatures (ε_{noise}) and the increasing inductive NMR signal at higher fields (ε_{B0}), and on the other hand, the losses from the slower polarization buildup at low temperatures (ε_{time}) and lower sample filling factor due to the volume occupied by the DNP juice (ε_{fill}), etc. to give a more adapted sensitivity factor $\varepsilon^+ = \varepsilon_{DNP} \times \varepsilon_\theta \times \varepsilon_{Curie} \times \varepsilon_{noise} \times \varepsilon_{B0} \times \varepsilon_{time} \times \varepsilon_{fill}$. However, as partly mentioned earlier, many of these factors are strongly probe design- or sample-dependent and makes the discussion more specific to each case. For example, the factors ε_{time}, ε_{fill}, and ε_θ depend on the form of the sample (monodispersed vs aggregated/assembled/crystalized), the EPR parameters (radical species, its concentration, the electron T_{1e}, and the existence of the direct/indirect association of unpaired electrons with the molecule of interest), and the number of methyl groups in the molecule and are often put aside for a general and basic assessment of the DNP efficiency.

To illustrate the historical development toward high-field DNP, Figure 9.7 plots in chronological order the sensitivity factor that takes account of the gain from the increasing field strength $\varepsilon_{DNP} \times \varepsilon_{Curie} \times \varepsilon_{B0}$. Each data point was calculated as looked up in the first

DNP report for a given field strength. Note that both ε_{Curie} and ε_{B0} are sample-independent factors. The gain from the field strength shall be referenced here to $B_0 = 1$ T so that $\varepsilon_{B0} = B_0^{3/2}$, where B_0 is the field strength (in Tesla) used for the DNP measurement. The 'power-of-3/2 trend' was experimentally verified previously (Figure S4 in Ref. 23). In Figure 9.7, open and filled circles show data for the Air/N_2-spinning SE- and CE-DNP, respectively, while open and filled stars show data obtained with our N_2-MAS and He-MAS DNP setups, respectively. The rise of the high-field CE-DNP in the 1990s and on (filled circles), as well as that of ultralow temperature MAS DNP at very high fields $B_0 > 10$ T in the 2010s (filled stars), is clearly seen in the plot. In 2017, the closed-cycle He MAS-based DNP spectrometer marked at $B_0 = 16.4$ T the sensitivity factor of ~6.6×10^4 (=$\varepsilon_{DNP} \times \varepsilon_{Curie} \times \varepsilon_{B0} = 100 \times 10 \times (16.4)^{3/2}$) (Figure 9.7, symbol labeled with 'p'), which is sixfold greater than that reported for the same field condition, but at conventional temperature (~90 K), ~1.2×10^4 (=$60 \times 3 \times (16.4)^{3/2}$) (labeled with 'j'). The efficiency of the emerging overhauser (OE) DNP (Figure 9.7, open triangles),

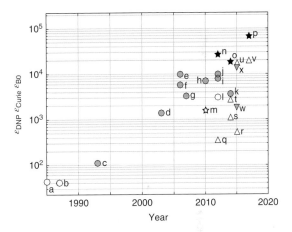

Figure 9.7. Recent history of the sensitivity factor taking into account the effect of the dynamic and static nuclear polarization and increasing static field, $\varepsilon_{DNP} \times \varepsilon_{Curie} \times \varepsilon_{B0}$. Data obtained at $B_0 > 10$ T with the N_2-MAS and He-MAS DNP setups described in this chapter are shown with open (m = 46) and filled stars (n = 56/o = 30/p = 37), respectively. Open and filled circles are for the air/N_2-spinning SE- and CE-DNP data, respectively (a = 66/b/g = 14/c = 9/e,f = 13/d = 67/h = 68/i = 19/j = 45/k = 69/l = 70). Recent advances observed for the OE DNP and the heterobiradical are shown with open (q = 71/s,t = 63/r,u = 72/v = 73) and filled triangles (w,x = 74), respectively

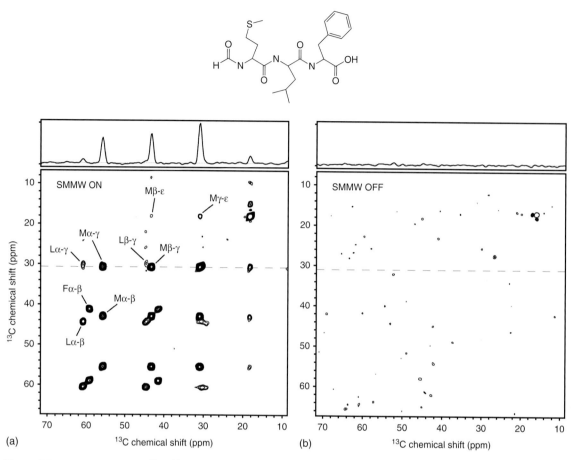

Figure 9.8. Two-dimensional ^{13}C–^{13}C DARR correlation spectra of crystalline powder sample of MLF (the chemical structure shown on top) wetted with TOTAPOL-containing TCE solution, taken with the SMMW on at $T = 60$ K (a), and off at room temperature (b). The recycle delay was 15 s (a) or 2 s (b). $B_0 = 16.4$ T, $v_R = 6$ kHz. The spectral region for the aliphatic carbon is shown. For both data sets, a 7°-pulse was employed for the initial ^{1}H excitation to emulate a mass-limited sample. Absolutely no probe arcing was observed during the data acquisition, even with high-power ^{1}H RF fields during the CP (50 kHz) and the data acquisition for decoupling (70 kHz) over a total irradiation time of >15 ms, as well as the low-power DARR ^{1}H field (8 kHz) during the 100 ms mixing time

although known to behave favorably with increasing field strength,[63] does not yet match the efficiency of the ultralow temperature CE-DNP. The efficiency of the OE-DNP under ultralow temperature MAS is to be investigated. A recent chapter reports on the static OE DNP enhancement at $T = 1.2$ K.[64]

As an example of the unique ability of the ultralow temperature MAS-DNP, Figure 9.8 shows a DNP-enhanced 2D ^{13}C–^{13}C correlation spectrum taken with a crystalline tripeptide sample, N-formyl-Met-Leu-Phe-OH (MLF) at 60 K. The crystalline powder of MLF has long been one of the

most widely used standard test samples for MAS NMR, but no DNP enhancement has been reported so far at conventional temperatures, $T \sim 100$ K, presumably because of the unusually high methyl group density (three methyl groups in a stretch of three amino-acids, and its dense accumulation) in the crystal that impedes the accumulation of the dynamic nuclear polarization. The fine powder sample of the [U-^{13}C,^{15}N]-MLF crystal as grown was wetted with 20 mM TOTAPOL-containing tetrachloroethane (TCE) solution, loaded to a 3.2 mm Si$_3$N$_4$ rotor then cooled in the NMR probe. A signal enhancement of $\varepsilon_{DNP} > 4$ was observed only at $T \sim 60\,K$ or lower temperatures using the closed-cycle He MAS probe system. The ^1H spin lattice relaxation time T_{1H} that steeply increased below $100\,K$ ($T_{1H} \sim 1\,s$ at $300\,K$, $\sim 3\,s$ at $100\,K$, and $\sim 15\,s$ at $60\,K$) seems to correlate with the observability of the DNP enhancement. Still, the relatively low enhancement observed may further be attributed to the impedance for the spin diffusion existing at the crystal surface, but could not be investigated here in detail as the matrix signal from the natural abundance TCE was not clearly detected even in the SMMW-on spectrum, masked by the signals from MLF and their spinning sidebands.

Although the sensitivity gain from DNP together with the Curie factor of $\varepsilon_{DNP} \times \varepsilon_{Curie} \sim 20$ ($= 4 \times 300/60$) at $60\,K$ is small, it can be substantial in many applications. To illustrate this point, the data shown in Figure 9.8 was recorded with only $620\,\mu g$ of MLF loaded to the rotor, and a small angle ($\sim 7°$) excitation was employed to emulate an effective sample quantity of $\sim 75\,\mu g$ ($= \sin(7°) \times 620$), or $\sim 180\,nmol$ of MLF. As shown in Figure 9.8(a), the DNP-enhanced data showed all the expected signals clearly above the noise level with only two scans, while no signal was detected at room temperature without DNP (Figure 9.8b). This suggests that the ultralow temperature MAS DNP can be a general tool to increase the DNP efficiency for methyl-abundant chemical/biological systems including membrane proteins by elongating T_{1H}, allowing the enhancement factor to approach that obtained with the standard DNP samples such as urea and proline bearing no methyl groups. It was also shown previously that resolving signals into $>1D$ space greatly helps extracting information from a low-temperature-broadened spectrum.[23] Ultralow-temperature high-dimensional ($>2D$) spectroscopy preserving high DNP efficiency thus represents an important breakthrough for the analysis of big and complex chemical systems at unprecedented signal receptivity and resolution.

9.5 SUMMARY AND OUTLOOK

In this chapter, we described three cryogenic MAS platforms: one N_2-based and two He-based MAS NMR probe systems in comparison with other existing instruments where possible. The primary concern with the cryogenic MAS facility includes the stability of the measurement condition, installation and operational costs, and the ease of operation. All the described systems incorporated various modifications and novel features to meet these challenges, such as the GN_2 separation from the atmospheric air, precooling with an electric gas chiller, the use of a LAr bath for heat exchange for N_2-based MAS, and advanced vacuum heat-insulation structures, and on-campus or in situ He recycling techniques for the He-based MAS platforms.

The closed-cycle MAS system cools the spinner gas streams 'on the fly' in a completely LHe-free manner and runs virtually indefinitely without losing the spinner GHe, establishing, for the first time, a stable, economically run, and easy-to-use platform for low- to ultralow temperature $T = 20–200\,K$, for either DNP-enhanced or regular MAS NMR spectroscopy. Further technical developments are currently underway for faster MAS, a sample-ejection/insertion mechanism compatible with the closed-cycle feature, and a cold preamplifier and duplexer to gain from the reduced thermal noise combined with DNP.

The DNP enhancement factor was found to improve exponentially with lowering the sample temperature, reaching $\varepsilon_{DNP} \sim 100$ at $B_0 = 16.4\,T$ and $T = 30–40\,K$ with TOTAPOL/AMUPol biradical (Figure 9.6a). The gain accounting for the quenching effect and the Curie factor $\varepsilon_{DNP} \times \varepsilon_\theta \times \varepsilon_{Curie}$ is reaching 400 at $T \sim 30\,K$, which is by a factor of >10 greater than that obtained at conventional DNP temperatures, $T \sim 100\,K$. This observation proves the prospect of the currently popular CE-based DNP for even higher field conditions in the coming years.

In combination with more advanced SMMW irradiation schemes,[30] low-temperature-optimized polarizing agents, and the actively reduced thermal noise, the ultimate sensitivity gain should exceed $10\,000$ in the near future, and this should open up avenues to, e.g., in-cell structural biology at physiological concentration; studies on supramolecular assemblies; isotope labeling-free structural biology; and detection of trace amounts of natural compounds, material surfaces, pharmaceutical contamination, and minor metabolites. DNP based on (transient) nonthermal electron polarization should also benefit from the low-temperature MAS NMR platforms[65] by longer triplet lifetimes and slower nuclear/electron relaxation. Although ultralow temperature high-dimensional spectroscopy for DNP facilitates analysis of complex systems, development of new sample preparation protocols for improving the low-temperature spectral resolution, and/or numerical approaches for extracting information from overlapped signals, should also be an important part of the future innovation.

RELATED ARTICLES IN EMAGRES

High-Frequency Dynamic Nuclear Polarization

Sensitivity Enhancement Utilizing Parahydrogen

Sensitivity of the NMR Experiment

Dynamic Nuclear Polarization and High-Resolution NMR of Solids

Chemically Induced Dynamic Nuclear Polarization

Overhauser, Albert W.: Dynamic Nuclear Polarization

REFERENCES

1. M. R. Hansen, R. Graf, and H. W. Spiess, *Chem. Rev.*, 2016, **116**, 1272.

2. S. Kazmierski, T. Pawlak, A. Jeziorna, and M. J. Potrzebowski, *Polym. Adv. Technol.*, 2016, **27**, 1143.

3. A. J. Howarth, A. W. Peters, N. A. Vermeulen, T. C. Wang, J. T. Hupp, and O. K. Farha, *Chem. Mater.*, 2017, **29**, 26.

4. T. Polenova, R. Gupta, and A. Goldbourt, *Anal. Chem.*, 2015, **87**, 5458.

5. T. R. Carver and C. P. Slichter, *Phys. Rev.*, 1953, **92**, 212.

6. A. W. Overhauser, *Phys. Rev.*, 1953, **92**, 411.

7. C. D. Jefferies, *Phys. Rev.*, 1957, **106**, 164.

8. C. D. Jefferies, *Phys. Rev.*, 1960, **117**, 1056.

9. L. R. Becerra, G. J. Gerfen, R. J. Temkin, D. J. Singel, and R. G. Griffin, *Phys. Rev. Lett.*, 1993, **71**, 3561.

10. M. Rosay, V. Weis, K. E. Kreischer, R. J. Temkin, and R. G. Griffin, *J. Am. Chem. Soc.*, 2002, **124**, 3214.

11. K. N. Hu, *Solid State Nucl. Magn. Reson.*, 2011, **40**, 31.

12. Y. Matsuki, T. Maly, O. Ouari, H. Karoui, F. Le Moigne, E. Rizzato, S. Lyubenova, J. Herzfeld, T. Prisner, P. Tordo, and R. G. Griffin, *Angew. Chem. Int. Ed.*, 2009, **48**, 4996.

13. C. S. Song, K. N. Hu, C. G. Joo, T. M. Swager, and R. G. Griffin, *J. Am. Chem. Soc.*, 2006, **128**, 11385.

14. V. S. Bajaj, M. K. Hornstein, K. E. Kreischer, J. R. Sirigiri, P. P. Woskov, M. L. Mak-Jurkauskas, J. Herzfeld, R. J. Temkin, and R. G. Griffin, *J. Magn. Reson.*, 2007, **189**, 251.

15. M. K. Hornstein, V. S. Bajaj, R. G. Griffin, and R. J. Temkin, *IEEE Trans. Plasma Sci. IEEE Nucl. Plasma. Sci. Soc.*, 2006, **34**, 524.

16. M. K. Hornstein, V. S. Bajaj, R. G. Griffin, and R. J. Temkin, *IEEE Trans. Plasma Sci. IEEE Nucl. Plasma Sci. Soc.*, 2007, **35**, 27.

17. T. Idehara, J. C. Mudiganti, L. Agusu, T. Kanemaki, I. Ogawa, T. Fujiwara, Y. Matsuki, and K. Ueda, *J. Infrar. Millim. Teraherz. Waves*, 2012, **33**, 724.

18. T. Idehara, Y. Tatematsu, Y. Yamaguchi, E. M. Khutoryan, A. N. Kuleshov, K. Ueda, Y. Matsuki, and T. Fujiwara, *J. Infrar. Millim. Teraherz. Waves*, 2015, **36**, 613.

19. A. B. Barnes, E. A. Nanni, J. Herzfeld, R. G. Griffin, and R. J. Temkin, *J. Magn. Reson.*, 2012, **221**, 147.

20. T. H. Chang, T. Idehara, I. Ogawa, L. Agusu, and S. Kobayashi, *J. Appl. Phys.*, 2009, **105**, 063304.

21. M. K. Hornstein, V. S. Bajaj, R. G. Griffin, K. E. Kreischer, I. Mastovsky, M. A. Shapiro, J. R. Sirigiri, and R. J. Temkin, *IEEE Trans Electron Devices*, 2005, **52**, 798.

22. A. B. Barnes, M. L. Mak-Jurkauskas, Y. Matsuki, V. S. Bajaj, P. C. A. Van Der Wel, R. Derocher, J. Bryant, J. R. Sirigiri, R. J. Temkin, J. Lugtenburg, J. Herzfeld, and R. G. Griffin, *J. Magn. Reson.*, 2009, **198**, 261.

23. Y. Matsuki, S. Nakamura, S. Fukui, H. Suematsu, and T. Fujiwara, *J. Magn. Reson.*, 2015, **259**, 76.

24. K. Thurber and R. Tycko, *J. Magn. Reson.*, 2016, **264**, 99.

25. M. Rosay, M. Blank, and F. Engelke, *J. Magn. Reson.*, 2016, **264**, 88.

26. K. Marker, M. Pingret, J. M. Mouesca, D. Gasparutto, S. Hediger, and G. De Paepe, *J. Am. Chem. Soc.*, 2015, **137**, 13796.

27. T. C. Ong, M. L. Mak-Jurkauskas, J. J. Walish, V. K. Michaelis, B. Corzilius, A. A. Smith, A. M. Clausen, J. C. Cheetham, T. M. Swager, and R. G. Griffin, *J. Phys. Chem. B*, 2013, **117**, 3040.

28. A. C. Pinon, A. J. Rossini, C. M. Widdifield, D. Gajan, and L. Emsley, *Mol. Pharm.*, 2015, **12**, 4146.

29. V. S. Bajaj, M. L. Mak-Jurkauskas, M. Belenky, J. Herzfeld, and R. G. Griffin, *Proc. Natl. Acad. Sci. U. S. A.*, 2009, **106**, 9244.

30. Y. Matsuki, T. Idehara, J. Fukazawa, and T. Fujiwara, *J. Magn. Reson.*, 2016, **264**, 107.

31. M. Valla, A. J. Rossini, M. Caillot, C. Chizallet, P. Raybaud, M. Digne, A. Chaumonnot, A. Lesage, L. Emsley, J. A. Van Bokhoven, and C. Coperet, *J. Am. Chem. Soc.*, 2015, **137**, 10710.

32. P. Fricke, D. Mance, V. Chevelkov, K. Giller, S. Becker, M. Baldus, and A. Lange, *J. Biomol. NMR*, 2016, **65**, 121.

33. D. Le, F. Ziarelli, T. N. Phan, G. Mollica, P. Thureau, F. Aussenac, O. Ouari, D. Gigmes, P. Tordo, and S. Viel, *Macromol. Rapid Commun.*, 2015, **36**, 1416.

34. W. R. Gunther, V. K. Michaelis, M. A. Caporini, R. G. Griffin, and Y. Roman-Leshkov, *J. Am. Chem. Soc.*, 2014, **136**, 6219.

35. I. V. Sergeyev, B. Itin, R. Rogawski, L. A. Day, and A. E. Mcdermott, *Proc. Natl. Acad. Sci. U. S. A.*, 2017, **114**, 5171.

36. D. Mance, P. Gast, M. Huber, M. Baldus, and K. L. Ivanov, *J. Chem. Phys.*, 2015, **142**, 234201.

37. Y. Matsuki and T. Fujiwara, in Experimental Approaches of NMR Spectroscopy, ed. The Nuclear Magnetic Resonance Society Of Japan, Springer: Tokyo, 2017, 9811059659 (I.S.B.N.).

38. R. Tycko, S. E. Barrett, G. Dabbagh, L. N. Pfeiffer, and K. W. West, *Science*, 1995, **268**, 1460.

39. P. Beckett, M. S. Denning, I. Heinmaa, M. C. Dimri, E. A. Young, R. Stern, and M. Carravetta, *J. Chem. Phys.*, 2012, **137**, 114201.

40. V. S. Bajaj, P. C. A. Van Der Wel, and R. G. Griffin, *J. Am. Chem. Soc.*, 2009, **131**, 118.

41. C. Beduz, M. Carravetta, J. Y. Chen, M. Concistre, M. Denning, M. Frunzi, A. J. Horsewill, O. G. Johannessen, R. Lawler, X. Lei, M. H. Levitt, Y. Li, S. Mamone, Y. Murata, U. Nagel, T. Nishida, J. Ollivier, S. Rols, T. Room, R. Sarkar, N. J. Turro, and Y. Yang, *Proc. Natl. Acad. Sci. U. S. A.*, 2012, **109**, 12894.

42. Q. Z. Ni, E. Markhasin, T. V. Can, B. Corzilius, K. O. Tan, A. B. Barnes, E. Daviso, Y. Su, J. Herzfeld, and R. G. Griffin, *J. Phys. Chem. B*, 2017, **121**, 4997.

43. J. Becker-Baldus, C. Bamann, K. Saxena, H. Gustmann, L. J. Brown, R. C. Brown, C. Reiter, E. Bamberg, J. Wachtveitl, H. Schwalbe, and C. Glaubitz, *Proc. Natl. Acad. Sci. U. S. A.*, 2015, **112**, 9896.

44. M. L. Mak-Jurkauskas, V. S. Bajaj, M. K. Hornstein, M. Belenky, R. G. Griffin, and J. Herzfeld, *Proc. Natl. Acad. Sci. U. S. A.*, 2008, **105**, 883.

45. A. B. Barnes, E. Markhasin, E. Daviso, V. K. Michaelis, E. A. Nanni, S. K. Jawla, E. L. Mena, R. Derocher, A. Thakkar, P. P. Woskov, J. Herzfeld, R. J. Temkin, and R. G. Griffin, *J. Magn. Reson.*, 2012, **224**, 1.

46. Y. Matsuki, H. Takahashi, K. Ueda, T. Idehara, I. Ogawa, M. Toda, H. Akutsu, and T. Fujiwara, *Phys. Chem. Chem. Phys.*, 2010, **12**, 5799.

47. P. J. Allen, F. Creuzet, H. J. M. Degroot, and R. G. Griffin, *J. Magn. Reson.*, 1991, **92**, 614.

48. B. J. Albert, S. H. Pahng, N. Alaniva, E. L. Sesti, P. W. Rand, E. P. Saliba, F. J. Scott, E. J. Choi, and A. B. Barnes, *J. Magn. Reson.*, 2017, **283**, 71.

49. K. R. Thurber, A. Potapov, W. M. Yau, and R. Tycko, *J. Magn. Reson.*, 2013, **226**, 100.

50. M. Concistre, O. G. Johannessen, E. Carignani, M. Geppi, and M. H. Levitt, *Acc. Chem. Res.*, 2013, **46**, 1914.

51. A. Hackmann, H. Seidel, R. D. Kendrick, P. C. Myhre, and C. S. Yannoni, *J. Magn. Reson.*, 1988, **79**, 148.

52. D. A. Hall, D. C. Maus, G. J. Gerfen, S. J. Inati, L. R. Becerra, F. W. Dahlquist, and R. G. Griffin, *Science*, 1997, **276**, 930.

53. V. Macho, R. Kendrick, and C. S. Yannoni, *J. Magn. Reson.*, 1983, **52**, 450.

54. A. Samoson, T. Tuherm, J. Past, and A. Reinhold, *Top. Curr. Chem.*, 2004, **246**, 15.

55. E. Bouleau, P. Saint-Bonnet, F. Mentink-Vigier, H. Takahashi, J.-F. Jacuot, M. Bardet, F. Aussenac, A. Purea, F. Engelke, S. Hediger, D. Lee, and G. Depaepe, *Chem. Sci.*, 2015, **6**, 6806.

56. Y. Matsuki, K. Ueda, T. Idehara, R. Ikeda, I. Ogawa, S. Nakamura, M. Toda, T. Anai, and T. Fujiwara, *J. Magn. Reson.*, 2012, **225**, 1.

57. J. Clauss, K. Schmidt-Rohr, and H. W. Spiess, *Acta Polym.*, 1993, **44**, 1.

58. C. Sauvee, M. Rosay, G. Casano, F. Aussenac, R. T. Weber, O. Ouari, and P. Tordo, *Angew. Chem. Int. Ed.*, 2013, **52**, 10858.

59. K. R. Thurber and R. Tycko, *J. Chem. Phys.*, 2012, **137**, 084508.

60. K. R. Thurber and R. Tycko, *J. Chem. Phys.*, 2014, **140**, 184201.

61. F. Mentink-Vigier, S. Paul, D. Lee, A. Feintuch, S. Hediger, S. Vega, and G. De Paepe, *Phys. Chem. Chem. Phys.*, 2015, **17**, 21824.

62. D. J. Kubicki, A. J. Rossini, A. Purea, A. Zagdoun, O. Ouari, P. Tordo, F. Engelke, A. Lesage, and L. Emsley, *J. Am. Chem. Soc.*, 2014, **136**, 15711.

63. T. V. Can, M. A. Caporini, F. Mentink-Vigier, B. Corzilius, J. J. Walish, M. Rosay, W. E. Maas,

M. Baldus, S. Vega, T. M. Swager, and R. G. Griffin, *J. Chem. Phys.*, 2014, **141**, 064202.

64. X. Ji, T. V. Can, F. Mentink-Vigier, A. Bornet, J. Milani, B. Vuichoud, M. A. Caporini, R. G. Griffin, S. Jannin, M. Goldman, and G. Bodenhausen, *J. Magn. Reson.*, 2018, **286**, 138.

65. K. Tateishi, M. Negoro, A. Kagawa, and M. Kitagawa, *Angew. Chem. Int. Ed.*, 2013, **52**, 13307.

66. R. A. Wind, M. J. Duijvestijn, C. Van Der Lugt, A. Manenschijn, and J. Vriend, *Prog. Nucl. Magn. Reson. Spectrosc.*, 1985, **17**, 33.

67. V. S. Bajaj, C. T. Farrar, M. K. Hornstein, I. Mastovsky, J. Vieregg, J. Bryant, B. Elena, K. E. Kreischer, R. J. Temkin, and R. G. Griffin, *J. Magn. Reson.*, 2003, **160**, 85.

68. M. Rosay, L. Tometich, S. Pawsey, R. Bader, R. Schauwecker, M. Blank, P. M. Borchard, S. R. Cauffman, K. L. Felch, R. T. Weber, R. J. Temkin, R. G. Griffin, and W. E. Maas, *Phys. Chem. Chem. Phys.*, 2010, **12**, 5850.

69. E. J. Koers, E. A. W. Van Der Cruijsen, M. Rosay, M. Weingarth, A. Prokofyev, C. Sauvee, O. Ouari, J. Van Der Zwan, O. Pongs, P. Tordo, W. E. Maas, and M. Baldus, *J. Biomol. NMR*, 2014, **60**, 157.

70. B. Corzilius, A. A. Smith, and R. G. Griffin, *J. Chem. Phys.*, 2012, **137**, 054201.

71. O. Haze, B. Corzilius, A. A. Smith, R. G. Griffin, and T. M. Swager, *J. Am. Chem. Soc.*, 2012, **134**, 14287.

72. M. Lelli, S. R. Chaudhari, D. Gajan, G. Casano, A. J. Rossini, O. Ouari, P. Tordo, A. Lesage, and L. Emsley, *J. Am. Chem. Soc.*, 2015, **137**, 14558.

73. S. R. Chaudhari, P. Berruyer, D. Gajan, C. Reiter, F. Engelke, D. L. Silverio, C. Coperet, M. Lelli, A. Lesage, and L. Emsley, *Phys. Chem. Chem. Phys.*, 2016, **18**, 10616.

74. G. Mathies, M. A. Caporini, V. K. Michaelis, Y. Liu, K. N. Hu, D. Mance, J. L. Zweier, M. Rosay, M. Baldus, and R. G. Griffin, *Angew. Chem. Int. Ed.*, 2015, **54**, 11770.

Chapter 10

Versatile Dynamic Nuclear Polarization Hardware with Integrated Electron Paramagnetic Resonance Capabilities

Alisa Leavesley[1], Ilia Kaminker[1], and Songi Han[1,2]

[1]*Department of Chemistry and Biochemistry, University of California Santa Barbara, Santa Barbara, CA, USA*

[2]*Department of Chemical Engineering, University of California Santa Barbara, Santa Barbara, CA, USA*

10.1 INTRODUCTION

Dynamic nuclear polarization (DNP) is an increasingly popular technique, especially with the advent of commercial DNP systems. However, the design for reliable DNP experiments is still limited primarily to a proven set of experimental conditions and sample types that produce effective signal enhancements and consistent DNP performance. A rational design of DNP experiments will benefit from the analysis

Handbook of High Field Dynamic Nuclear Polarization.
Edited by Vladimir K. Michaelis, Robert G. Griffin, Björn Corzilius and Shimon Vega
© 2020 John Wiley & Sons, Ltd. ISBN: 978-1-119-44164-9
Also published in eMagRes (online edition)
DOI: 10.1002/9780470034590.emrstm1564

of the DNP mechanisms under a variety of sample formulations, requiring the study of both the nuclear and electron spin dynamics in the system – ideally under identical experimental conditions. The study of the electron (de)polarization profile has helped to identify the DNP mechanism(s) in a system,[1] such as differentiating between a truncated cross effect (CE) that has the appearance of the Overhauser effect,[2] evaluating the contributions of CE and solid-effect (SE) DNP,[3,4] and contributed to the development of new DNP theory, such as the heteronuclear-CE.[5] Careful analysis of the DNP polarization profile and the electron spin relaxation properties revealed that glass polymorphism generated by specific solvent mixtures induced radical clustering, as identified by altered electron spin phase memory time (T_m).[6] The measurement of T_m also helped determine that the incoherent term of the electron–nuclear hyperfine coupling is suppressed when the electron spins are saturated under DNP conditions, which reduces the NMR linewidth.[7] These studies exemplify the benefits of a dual DNP/EPR spectrometer that yield mechanistic insight and unravel experimental misunderstandings. Thus, a dual DNP/EPR spectrometer needs to be intelligently designed to effectively detect and analyze the electron and nuclear spins in the system. This understanding is not new.[1,8–11] However, high-field instrumentation with dual DNP/EPR capabilities is scarce, with only a few setups available to date, including custom instrumentation at 95 GHz at the Weizmann Institute and Nottingham University,[12–15] 140 GHz at MIT,[16,17] and 200 GHz at UCSB.[18,19] A versatile dual DNP/EPR instrument requires broad-band microwave (μw) capabilities to select for multiple paramagnetic species, allow for μw manipulation through an arbitrary waveform generator (AWG), and be capable of pump-probe, multi-frequency EPR experiments. Solid-state (ss) μw sources are an essential feature to achieve this type of versatility. However, only in recent years have ss-μw sources become powerful enough for DNP and pulsed EPR experiments. A number of ss-μw source-based high-field EPR spectrometers operational at >95 GHz have led the way, including that of van Tol and coworkers at the National High Magnetic Field Lab (110–395 GHz),[20,21] Takahashi at USC (115 and 230 GHz),[22,23] Griffin at MIT (140 GHz),[24] Britt at Davis (263 GHz), Prisner at the Goethe University in Frankfurt (180 GHz),[25] Sherwin at UCSB (240 GHz),[26,27] Lubitz at the Max Planck Institute (122 and 244 GHz),[28] Freed at Cornell University (250 GHz),[29] not to forget the commercial Bruker pulsed EPR system, the E780 (263 GHz),[30] Schmidt at Leiden University (275 GHz),[31,32] and Möbius at the Free University Berlin (360 GHz).[33,34] Tycko and Thurber at the National Institute of Health (264 GHz) have demonstrated the use of a ss-μw source for magic-angle spinning (MAS) DNP operation,[35–37] while Zilm and coworkers at Yale University (200 GHz) have debuted at the 2017 ISMAR, a dual MAS DNP/EPR instrument operational at room temperature using diamond P1 centers for signal.[38] We forecast for MAS and static experiments, that dual DNP/EPR will become a critical capability for advancing the field of DNP, by rationally improving DNP performance and increasing the application scope of DNP. This motivates our description of the hardware, operation, and experimental parameters for the Han lab UCSB high-field (6.9 T) dual DNP/EPR instrument, which can operate between 188–201 GHz and 3.5 K to room temperature. This overview of our dual DNP/EPR system and select diagnostic DNP and EPR studies are presented, so that other groups can build and improve upon this platform and the presented results.

10.2 HARDWARE

A versatile dual DNP/EPR spectrometer benefits from a modular design, so that modifications for specific experiments and hardware developments can be easily incorporated. Designing for modularity requires a particular eye for the type of experiments one wants to achieve with the instrument. The type of DNP to be performed will alter the required components for the spectrometer, where dissolution dynamic nuclear polarization (d-DNP) will need a rapid melting apparatus and generally cryogenic handling capabilities down to ~1 K, MAS DNP requires a stator and other associated hardware for MAS, while static DNP may require wider μw bandwidths or a cavity for easy conversion between DNP and EPR. However, all dual DNP/EPR instruments require at least an NMR console and a superconducting magnet corresponding to the magnetic field of choice, sample temperature control, μw source and manipulation stage for transmission, a probe to direct the μw to the sample and support the sample, and detection capabilities for the EPR signal. In this section we will overview some of the options available for each of these essential components for dual DNP/EPR instruments, while describing specific examples for the 194 GHz static dual DNP/EPR system installed in the Han lab at UCSB, where the general schematics of the instrument, as previously described by Siaw *et al.*[18] are shown in Figure 10.1(a) with a picture of the actual instrument in Figure 10.1(b). In this study, detailed descriptions of recent advances to integrate AWGs and 2-μw source capabilities will be.[39] In the following sections we detail the components and design of the different modules in the dual DNP/EPR instrument with shaped pulse capabilities for μw transmission, pump-probe electron spin experiments, and synchronized radio frequency (rf)- and μw-pulsing capabilities.

10.2.1 High-field Magnet

A dual DNP/EPR instrument is capable of acquiring both NMR and EPR data. For MAS DNP operation, particularly biological samples, where achieving sub-ppm resolution is important, a static-field NMR magnet with 10^{-7}–10^{-8} field homogeneity is a necessary prerequisite. An alternative for static NMR/DNP and d-DNP is a variable-field EPR magnet, where a ~10 ppm homogeneity can be achieved, and the magnetic field is easily adjusted and swept by changing

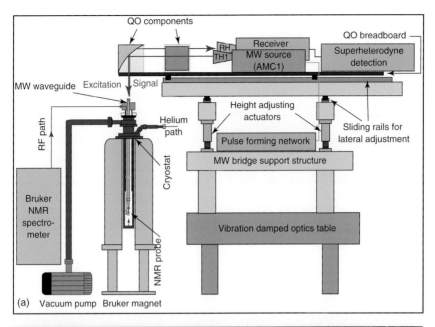

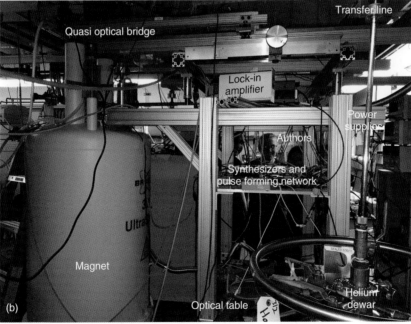

Figure 10.1. Instrument overview schematic (a) and picture (b) of the dual 194 GHz DNP/EPR spectrometer at UCSB

the current in the superconducting coil. There are other important parameters to consider when choosing the best magnet for a given system – cryogen-free versus a wet design, bore size, type and number of cryogenic shims, and the presence of a dedicated superconducting and/or room temperature sweep coil. All these choices will affect the cost of the magnet itself but have added benefits; a cryogen-free system will decrease helium consumption, while the majority of the magnet operating costs will be electricity. Variable field and sweep coils in the magnet allow for traditional field-swept EPR experiments, and so avoid some of the problems associated with frequency-swept EPR experiments, such as standing waves in the quasi-optics and frequency-dependent source output power. A variable-field magnet also increases the scope of paramagnetic species that can be employed as polarizing agents in DNP by meeting the resonance conditions for species with $g \neq 2$. In our example system a standard, commercial, 7 Tesla, wide-bore (89 mm) NMR magnet from Bruker (Avance D300WB) was used, which was not equipped with a sweep coil.

10.2.2 NMR Console

An NMR console can be homebuilt or purchased commercially. Hilty and Takeda provide two different bases for building-your-own NMR consoles.[40,41] In the example systems that we discuss here, a commercial Bruker Avance DSX console with 1 kW ^{1}H and X channel amplifiers was used. Due to the large difference in NMR signal intensity between DNP enhanced and unenhanced signals, the dynamic range of the commercial analog-to-digital convertor (ADC) was insufficient. If the receiver gain of the spectrometer was adjusted for the unenhanced NMR signal intensity, the enhanced signal under DNP conditions saturated the ADC. Alternatively, if the receiver gain was adjusted for the intensity of the DNP-enhanced signal, the signal-to-noise ratio (SNR) of the unenhanced signal became unsatisfactory. Varying receiver gain throughout the experiment was precluded by nonlinear variations in the gain, which resulted in large errors in the absolute NMR signal intensity, even after a calibration procedure. Implementation of a set of calibrated variable attenuators (Pasternack) between the pre-amplifier and the console allowed for operation with a high receiver gain setting and prevented saturation of the ADC for DNP-enhanced signals, which has been the standard operating procedure for the Han lab UCSB instrument. The unattenuated signal is then calculated according to $V_{in} = V_{out}10^{0.05A}$, where A is the applied attenuation in dB, V_{in} is the voltage input, and V_{out} is the voltage output after the attenuator(s), which is a standard dB to voltage calculation. The integrals of the NMR peaks scale linearly with the signal voltage.

10.2.3 Cryostat and Temperature Control

DNP at temperatures above 200 K has been successfully demonstrated at high magnetic fields (> 5 T) for BDPA and TEKPol radicals dissolved in *ortho*-terphenyl;[42] however, other examples of DNP performed at high temperatures and high magnetic fields are limited.[43,44] Freezing a sample to cryogenic temperatures with liquid helium or nitrogen significantly improves the electron spin Boltzmann statistics and prolongs the electron spin relaxation times, subsequently increasing the achievable NMR signal under DNP. Thus, operation at a low ($\leq$100 K) and stable temperatures is typically desired for most DNP experiments. For MAS-DNP, there are three current designs for spinning and cooling the sample: nitrogen only (>90 K),[45] a hybrid nitrogen (spinning) and helium (cooling) system (>20 K),[35,46] and a helium-only design (>6 K).[47,48] For static DNP/NMR, d-DNP, and EPR systems, a cryostat is used to cool the sample to low temperatures – typically either using a continuous-flow or a cryogen-free design. A decision must be made if the cryostat will be top- or bottom-loaded, whether an optical window is needed, and if a cryogen-free cryostat will be used or not – the latter option can significantly reduce the operational costs. The Han lab UCSB system uses a custom Janis STVP-200-NMR continuous-flow cryostat (Janis Research Co. LLC) designed for operation from <3 to 325 K, with a 50 W voltage controlled resistive heater for temperature control. A Cernox temperature sensor (LakeShore Cryogenics) near the cryogen inlet for the cryostat acts as the detector for a Labview proportional-integral-derivative (PID) algorithm to control the voltage for the heater, maintaining stable temperatures ($\pm$0.05 K) over long experimental times. Liquid helium from a slightly pressurized, 3 psi, liquid helium dewar (Praxair) is supplied to the cryostat with a flow-controlled liquid helium transfer line (Janis Research Co LLC). The cryostat outlet is maintained at atmospheric pressures for operation at temperatures >5 K and reaches a stable temperature in less than an hour. For operations at below <5 K, a Sogevac SV65B rotary cane pump (Oerlikon Leybold Vacuum) is used to evacuate the sample chamber of the cryostat, while still flowing the liquid helium to depress the boiling point of the helium, allowing for operation down to ~3 K. The cryostat itself is top-loaded into the 89 mm bore of the NMR magnet (room temperature shims are removed) with a hollow 52 mm bore for a top-loaded dual DNP/EPR probe (Figure 10.1a).

10.2.4 Microwave Source and Bridge Control

The μw source is the heart of a dual DNP/EPR spectrometer. The requirements for μw power vary significantly between different DNP experiments, with few millliWatts being sufficient for d-DNP at very low ~1.2 K temperatures to several Watts or tens of Watts needed for MAS DNP performed at ≥100 K. The required frequency range depends on the choice of the operating magnetic field, and typically varies between 95 GHz (3.3 T) for d-DNP and DNP/EPR instruments to 527 GHz (17 T, 800 MHz ^{1}H) for high-resolution MAS DNP. Another important consideration is whether continuous wave (cw) μw are sufficient as used in d-DNP and MAS DNP systems, or whether a more sophisticated manipulation of the μw is required for pulsed operation: frequency, phase, and/or amplitude modulation (AM). If shaping of the μw is desired, the conventional approach utilized in pulsed EPR at ~9–35 GHz, is to preform and shape the pulses at X-band (8–12 GHz) and low power (mW), potentially multiply to higher frequencies, and amplify the power output to hundreds of Watts or a few kiloWatts, before guiding the μw beam to the sample. This approach works well up to ~95 GHz,[13,49–52] but has limited utility at 150–300 GHz, with current state-of-the-art extended interaction klystron (EIK) amplifiers providing only a few watts of power.[30,53,54] Above 300 GHz, this design becomes obsolete, since high-power amplifiers are not available for these frequencies. The commercially available DNP spectrometer made by Bruker relies on a different design concept: a gyrotron source is used,[45] which inherently limits the operation to cw, but allows for tens of Watts of power output up to 527 GHz (800 MHz ^{1}H).[55] However, if low μw power is sufficient, typically when operating at liquid helium temperatures below 50 K, as found in ultralow temperature MAS-, static-, and d-DNP, a ss-μw source can be used.[14,18,37] Beyond being the cheapest option, a ss-μw source is significantly more agile than either a klystron or a gyrotron, and thus it is an ideal μw source for a versatile dual DNP/EPR instrument. Besides simply considering the output power of the μw source, the ss-μw source is easier and faster to manipulate, where modulation of the μw, beyond changing the frequency in cw operation, significantly increases the scope of potential DNP and EPR experiments that the instrument can perform. Such examples will be provided with case studies in later sections.

10.2.4.1 Solid-state Microwave Sources

ss-MW sources can have large operational bandwidths, but suffer from relatively low-power outputs of ≤500 mW at high fields (>150 GHz). Only in recent years have ss-μw sources in the 100–200 GHz range matured enough to be effectively utilized for EPR and DNP experiments without further amplification, with generally ~150 mW produced at 200 GHz and tens of milliwatts at higher frequencies. High-field ss-μw synthesizers come in a variety of forms, as gunn diodes, metal-oxide semiconductor (CMOS) chips, or low-frequency synthesizers that are amplified and multiplied to the desired high-field frequency. The latter type of sources generally rely on a crystal that will produce oscillatory irradiation in the tens of gigahertz range when a voltage is applied, which is then multiplied to the desired frequency through a series of doublers and amplifiers collectively referred to as amplification-multiplication chains (AMCs). Virginia Diodes Inc (VDI) and Elva-1 produce sources at frequencies ranging from 40 GHz to 3.2 THz, and from 26 to 180 GHz, respectively.[56,57] In our example system, ~ 12 GHz cw-irradiation is produced with a yttrium iron garnet (YIG) crystal (Microlambda and VDI) and is multiplied x16 by a VDI AMC to reach ~140 mW at 200 GHz, with a bandwidth of 190–201 GHz. The resultant linearly polarized 200 GHz μw are transmitted through a corrugated WR4.3UG to a 12.7 mm i.d. transmission horn (Thomas Keating Ltd) to generate a Gaussian beam suitable for quasi-optical transmission to the sample (see Section 10.2.5). In the last year we have acquired from VDI a 500 mW, 194 GHz source that combines a W-band (~95 GHz) amplifier with a custom AMC to boost the output power to >400 mW over 188–196 GHz. Similar developments were initially incepted by A. I. Smirnov and A. A. Nevzorov at NC State.[58] All of the synthesizers have a 1 Hz frequency resolution and a ~10 ms response time to changing frequencies. μw AM is integrated into the AMCs through an analog voltage (0–5 V) control, which has a response time of ≤0.5 ms. The frequency and AM are paramount features for agile μw manipulation that are necessary for accurate DNP and EPR experiments, where elevating the μw irradiation beyond the cw mode will improve the versatility of the system by allowing for pulsed-DNP and pulsed-EPR experiments, especially for coherent control of the electron spin systems. However, the slow response times of the YIG synthesizer (~10 ms frequency switching time) and the integrated AM control for the AMCs limit the instrument's capabilities for electron spin manipulation that require manipulations on a faster (ns–μs) timescale. To allow for greater flexibility in system performance and to overcome the above limitations, a two-source system with AWG capabilities was implemented.

10.2.4.2   Synthesizer Selection

Our strategy to reduce the frequency switching time in two-frequency experiments, e.g., for ELectron-electron DOuble Resonance (ELDOR) measurements, is to employ two separate YIG-based synthesizers. The selection between the two cw-μw synthesizers is done with ss-switches (SWM-DJV-1DT-2ATT, American Microwave Corp.) that have a short, < 10 ns, response time (shown in Figure 10.2a,b – orange block), followed by a power combiner (Narda). Selecting which switch is open or closed will determine which synthesizer is used to direct the ~12 GHz μw signal to the AMC(s), and so acts as the first cut for generating, and defining the length of, μw pulses. If cw-irradiation is desired, the switches are simply opened for the entire duration of the experiment. If two different synthesizers are used in the spectrometer, it is important to ensure that the input power to the pulse-forming network module (teal block in Figure 10.2b) is the same, regardless of which synthesizer is selected. This avoids saturating some of the components in the pulse-forming network module when using one synthesizer versus the other and is typically achieved by adding attenuators after the synthesizer with the higher output power.

10.2.4.3   Pulse-forming Network with AWG Capabilities

A pulse-forming network further develops the pulse from the initial cut provided by the source selection module. This is where phase, frequency, and amplitude manipulations can be performed to shape the pulse to the desired outcome. Various components can be incorporated to manipulate the 12 GHz signal in the pulse-forming network prior to passing through the AMCs to reach 194 GHz, which is a significant advantage of the ss-μw source-based system compared to a gyrotron-based system. The schematics for the example system's pulse-forming unit are shown in Figure 10.2(b) – in the teal block. The signal from the synthesizer selection module is sent to a four-way SP4T switch (F9140AH, General Microwave) with

a 10 ns response time to select the type of pulse manipulation. The first 'bypass' channel is used if no other manipulations are necessary, such as in straightforward cw-DNP or cw-EPR experiments. The second channel is equipped with a voltage-controlled phase shifter (SLPS-122-25V, Spacek Labs) that shifts the phase of the μw pulse by an arbitrary value relative to the 'bypass' channel, as needed for two-step phase cycled pulsed EPR. As the response of the voltage-controlled phase shifters is frequency dependent, a calibration of the phase shift per applied voltage as a function of irradiation frequency is necessary to optimize the pulse phases to ensure ideal 0°/180° phase cycling. Since the AMC will multiple the phase by 16x, the phase is shifted by 180°/16 = 11.25° at 12 GHz for 0°/180° two-step phase cycling at 200 GHz.

Channels 3 and 4 are used for arbitrary shaping of the μw pulse by mixing, via an IQ mixer (IQ0618LXP, Marki microwaves), the 12 GHz cw-μw with an arbitrary waveform produced by the AWG. The IQ mixer generates a waveform at the carrier frequency – here 12 GHz, with an arbitrary phase and amplitude limited by the temporal resolution of the AWG. This IQ mixer-based design eliminates the need for subsequent filtering of unwanted sidebands as would be required for a single-sided mixer-based design. Additionally, specifying the I and Q inputs doubles the bandwidth of the resultant waveform, which can extend from $\omega_{LO} - \omega_{clock}$ to $\omega_{LO} + \omega_{clock}$, where ω_{LO} is the local oscillator (LO) frequency of the IQ mixer and ω_{clock} is the clock frequency of the AWG. In the Han lab UCSB system, two AWGs are installed: one is a homebuilt AWG developed with the Martinis group at UCSB,[59] with 16 kB of memory and a clock frequency of 1 GHz, which allows for 1 ns temporal resolution at a maximum pulse length of 15 μs.[39] The second AWG is commercial (DAx22000-8M, Chase Wavepond) with 8 MB of memory and a 2.5 GHz clock frequency; this increases the length of the shaped pulses to ~3.3 ms with 400 ps time resolution. Two AWGs may be required for pulse sequences where the total length of the sequence is longer than 3.3 ms, and thus exceeds the memory of the AWG – this is typical for DNP experiments. In many cases, the seconds-long irradiation for DNP experiments consists of a repeating unit of <3 ms. In this case one AWG is programmed to generate the repeating unit and produces the waveform upon receiving an external trigger, which is supplied repeatedly for the length of the experiment. The second AWG can be used for a different set of shaped pulses, for example, detection pulses in an ELDOR experiment.

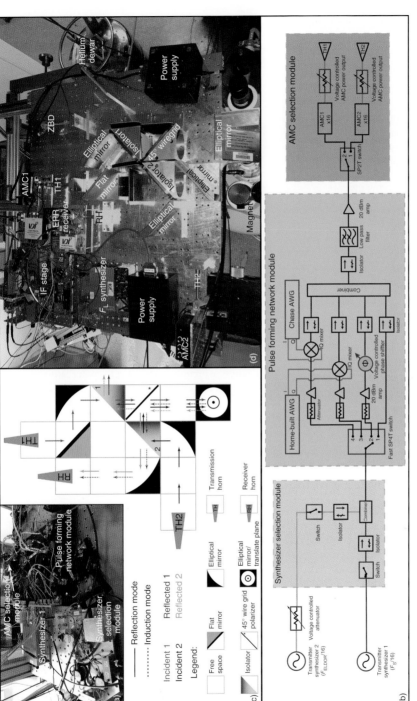

Figure 10.2. Picture (a) and schematic (b) of the μw manipulation capabilities at 12 GHz for the dual DNP/EPR spectrometer. Orange block denotes synthesizer selection module; teal block details the components of the homemade pulse-forming unit with phase cycling, bypass, homebuilt AWG, and Chase AWG channels available (bottom to top); purple block denotes the AMC selection module. Components shown in red are computer controlled by Specman4EPR software. Blue line denotes the path of 12 GHz signal at main (F_S) frequency generated by transmitter source 1. Orange line denotes the path of the 12 GHz signal at the second (F_{ELDOR}) frequency generated by transmitter source 2. Purple line denotes the path of signal mixed with the AWGs green indicate the 194 GHz μw. Black lines denote the paths shared by both F_S and F_{ELDOR} signals. (c) Quasi-optical design for a two-source system. Each tile represents a distance of 125 mm or f/2. The dark part of the isolator indicates where the Faraday rotator is located. TH1 (transmission horn 1) results in induction mode EPR detection. TH2 results in reflection mode detection, and RH (receiver horn) captures the EPR signal for detection. Arrows indicate the μw pathway in free space through the quasi-optics as defined by color and solid vs dashed in the figure. (d) Picture of the μw bridge supporting the quasi optics, AMC1, AMC2, and the EPR detector (EPR receiver system and intermediate frequency (IF) stage).

Another notable advantage of AWGs is that to the first order the phases generated by mixing the pulse with the AWG waveform are independent of the pulse frequency. This allows for frequency-independent phase cycling and no calibration table that is inherent to the traditional phase shifter-based approach. Since the AMCs will multiply all phase and frequency manipulations by 16x, this should be accommodated in the programming of the AWG waveform.

Under normal operating conditions AMCs are operated under saturating conditions to minimize the AM noise. If AM is required, it is possible to reduce the output power of the AMC by reducing the amplitude of the input signal to the AMC. To achieve these types of AMs, a voltage-controlled attenuator can be applied to the synthesizer(s) to reduce the amplitude of the 12 GHz signal, or AM can be applied to the pulse using the AWG(s), such that the AMCs run under non-saturating conditions. There is a very narrow dynamic range over which AMs at 12 GHz are carried over to 194 GHz, <0.5 dB, but more significantly hysteresis effects and increased AM noise makes reproducible operation of the AMC(s) under these conditions challenging. The increased AM noise of the AMC under non-saturating conditions is a direct consequence of the very steep and nonlinear dependence of the output μw power at 200 GHz on the input power at 12 GHz. This amplifies the AM noise present in the input signal. However, under saturating conditions the AM noise of the AMC is suppressed because the output μw power is not dependent on the input power. The successful implementation, as well as the challenges and pitfalls, of amplitude-modulated AWG pulses for pulsed EPR at 200 GHz have been recently reported by Kaminker *et al.*[39]

Following the four channels: bypass, phase shifter, and two AWGs, a combiner (Narda) is used to allow all four channels to be directed to either of our two AMCs. Note that special care was taken to ensure that the power is the same (within 1 dB) after the combiner regardless of the channel used. Note that the AWG channels are calibrated for waveforms at maximum intensity. After the combiner, the manipulated 12 GHz signal is sent through an isolator (Quest) and a low-pass filter (Marki microwave) to remove leaked second harmonics of the carrier frequency from the IQ mixers for AWG operations. The resultant signal is then amplified (ZVA-183-S+, Mini-circuits) and sent to the AMC selection module.

10.2.4.4 AMC Selection

As described later, we find that in order to minimize the hysteresis effects of switching amplitude and/or frequency within a single AMC, pump-probe type experiments are best performed with two AMCs. The selection module that controls which AMC will receive the manipulated 12 GHz signal from the pulse-forming network is depicted in the purple block of Figure 10.2(b). This configuration utilizes a SP2T switch (F9120AH, General Microwaves) with a 10 ns time resolution after the pulse-forming module to direct the manipulated μw to the desired AMC. After the signal has been selected for either of the two AMCs, attenuators are used to set the input power for each AMC to ensure operation is under saturating, but safe, conditions when a full amplitude input pulse is generated. We will refer to the two AMCs by AMC1/TH1 and AMC2/TH2 in the following sections, where TH is the transmission horn.

10.2.4.5 Spectrometer Control

The control of all μw switches in the pulse-forming unit and the superheterodyne receiver system (see Section 10.2.6), triggering of AWGs, and synchronization with the NMR spectrometer are performed by a 24-channel pulse generator with 3.3 ns time resolution (PulseBlaster ESR-PRO SpinCore). The voltages to the voltage-controlled attenuator and voltage-controlled phase shifter are provided by a DAC board (USB-6001 National Instruments). All the spectrometer control electronics are controlled by a SpecMan4EPR software as described in the Section 10.2.8.

10.2.5 Quasi-optics Design for Transmission and EPR Detection

After the AMCs have produced the 194 GHz waveform, quasi-optics are used to direct the μw through free space on the μw bridge to the sample in the magnet and for EPR experiments to guide the EPR signal back from the sample to the EPR detector located on the μw bridge. At lower frequencies, waveguides are generally used for low-loss transmission from source to sample; however, the losses associated with the passive components of waveguide transmission such as circulators and isolators become prohibitively high at sub-THz frequencies. At frequencies ≥ 95 GHz, quasi-optics is an effective solution to manipulate the μw and minimize power losses. The schematic representation of the quasi-optical design for two-source one-receiver operation is shown in Figure 10.2(c) with the picture of the actual quasi-optical bridge in Figure 10.2(d). In Figure 10.2(c) each square is 125×125 mm, representing half of the 250 mm focal length, f, of the elliptical mirrors used in the system. All quasi-optical components were purchased from Thomas Keating Ltd. A corrugated WR4.3UG to 12.7 mm i.d. horn translates the 194 GHz μw irradiation from the AMC to free space, producing a linearly polarized Gaussian beam with a beam diameter of 8 mm ($=12.7 \times 0.64$). Once the μw beam leaves the horn it will begin to diverge. Thus, a combination of two elliptical mirrors is used to refocus the beam and form an image on the top of the corrugate waveguide leading to the sample. A *f-2f-f* configuration between the horn, mirrors, and waveguide is used to prevent image effects such as parity inversion, frequency dependence, and phase distortions due to Gaussian beam propagation,[60-62] and to maximize the quality of the beam waist at the top of the waveguide. Isolators are used to protect the AMCs from reflected power and reduce the standing waves in the system. An isolator is a combination of a wiregrid polarizer, a Faraday rotator that rotates the beam polarization by 45°, and an absorber mounted on a half cube in a triangular configuration. Two isolators are used in the current Han lab UCSB quasi-optical μw bridge as shown in Figure 10.2(c) and are labeled as 1 and 2.

The transmission pathways through the quasi-optics from both AMCs will now be described for the example UCSB system. Following the μw beam with vertical polarization from AMC1/TH1, a flat mirror 125 mm away from the horn is used to change the direction of the beam without altering the divergence of the beam due to limited space on the quasi-optical breadboard. An elliptical mirror placed 250 mm away from the horn (125 mm from the flat mirror) is then used to refocus the beam through isolator 1, which contains a horizontal wiregrid. In a wiregrid polarizer, the electric field polarization is transmitted if it is perpendicular to the wires of the wiregrid polarizer and is reflected if it is parallel. The Faraday rotator in the beam rotates the electric field polarization by 45° counter-clockwise to 135°. The now 135° polarized beam is then transmitted through the 45° wiregrid (E-field is perpendicular) to the second elliptical mirror spaced *2f* from the first elliptical mirror, which refocuses and translates the beam onto the waveguide 250 mm below the second elliptical

mirror (the corrugated waveguide will be discussed in Section 10.2.6). After the μw beam has interacted with the sample, the polarization is no longer linear, but elliptical. Thus, the returning beam will be split by the 45° wiregrid; the component orthogonal to the incident beam is reflected by the 45° wiregrid (commonly referred to as the induction-mode EPR signal). The other component with the same polarization as the incident beam (commonly referred to as reflection mode EPR signal) is transmitted through the 45° wiregrid and rotated a further 45° counter-clockwise by the Faraday rotator, resulting in a horizontal E-field polarization, which is reflected off the horizontal wiregrid in isolator 1 and sent to the absorber. The induction-mode EPR signal from AMC1/TH1 passes through a Faraday rotator on isolator 2 and is converted to vertical polarization that will be transmitted through the horizontal wiregrid in the isolator 2. The beam will then be refocused with an elliptical mirror *2f* from the first elliptical mirror after the waveguide to the receiver horn for EPR detection positioned 250 mm after the last elliptical mirror.

Next, we consider the path of the μw beams produced from AMC2/TH2 that have horizontal E-field polarization. The first elliptical mirror is placed 250 mm from the horn and refocuses the beam towards isolator 2, which has a horizontal wiregrid. As the polarization is horizontal, the beam will be completely reflected by the wiregrid and sent through the Faraday rotator. This beam has 45° polarization after isolator 2 that will be parallel to the 45° wiregrid, and thus will also be reflected towards the second elliptical mirror, placed *2f* from the first elliptical mirror on the AMC2/TH2 beam path and refocused on the waveguide below. After interacting with the sample, the returning signal will similarly be elliptical and split by the 45° wiregrid; however, in the opposite fashion compared to the reflected beam originating from AMC1/TH1, since the polarization of the incident beam from AMC2/TH2 is perpendicular to that of AMC1/TH1. Thus, the induction-mode EPR signal will be transmitted through the 45° wiregrid and sent into the absorber of isolator 1, while the reflection-mode EPR signal will be reflected by the 45° wiregrid and sent to the receive horn for EPR detection. In other words, the receiver detects the induction-mode EPR signal from AMC1/TH1, and the reflection-mode EPR signal from AMC2/TH2. In both cases, the vertically polarized EPR signal at the receiver position is orthogonal to B_0 at the sample position. It is important to note that if the absorber in isolator 1 was removed and another elliptical mirror was placed to allow for the *f-2f-f* configuration to a second EPR detector, then induction- and reflection-mode EPR signals could be read for both AMCs, depending on which EPR receiver was used. An important consideration is that the reflection-mode EPR signal has significantly more power (reflected incident beam plus EPR signal) than induction-mode EPR signal (only EPR signal). The higher power of the reflection-mode EPR signal could damage the EPR detection scheme. As such, precautions to prevent EPR detector damage should be taken when operating in the reflection mode. In our system, the detector can only safely withstand reflected pulses, if pulse lengths do not exceed 2 µs and a 0.5% duty cycle; this is ensured by the spectrometer control software (see Section 10.2.8). If cw-irradiation from AMC2/TH2 is required, such experiments cannot be combined with EPR detection as an absorber must be placed before the receiver horn to protect the EPR detector (EPR detection schemes will be discussed in Section 10.2.7).

10.2.6 Dual DNP/EPR Probe

A dual DNP/EPR probe needs two primary components for effective dual operation: NMR circuitry (i.e., NMR coil/tuning/matching circuit) and EPR circuitry (i.e., waveguide and optionally a resonant cavity). A commercial low-temperature NMR probe can be coupled with a waveguide to direct the µw to the sample. However, the example system uses a homebuilt probe that is top-loaded into the cryostat with inductively coupled NMR circuitry and a waveguide with no cavity to direct the µw for DNP and EPR operation. The waveguide is a 0.9 m long corrugated waveguide supplied by Thomas Keating Ltd with an i.d. of 12.7 mm that is tapered down to a 5.3 mm i.d. to concentrate the µw onto the sample. A 42 mm long, smooth-walled, copper extension with a 5.3 mm i.d. is used to reach the sweet spot of the magnetic field and helps support the NMR circuitry. The corrugations of the main waveguide and taper allow for low-loss power transmission of the µw from free space in the quasi-optics to the sample, as a 1 m long waveguide causes <0.5 dB of loss. The waveguide is capped at the top with a transparent polymethylpentene (TPX®) window to allow for a complete vacuum seal. For complete transmission of the µw beam, the window must be a multiple of $\lambda/2$ thick – here 3.55 mm.[60] The corrugated waveguide

transmits the HE_{11} mode of the µw beam, which has a 98% coupling efficiency from free space (TEM_{00}) into the waveguide, for which the beam waist needs to be 0.64 times that of the corrugated waveguide inner radius for ideal coupling. A potential alternative to an expensive corrugated waveguide that has yet to be attempted in a DNP or high-field EPR instrument is a smooth-walled overmoded waveguide, which can be implemented with three stipulations: there is only a 91% ideal coupling efficiency from free space to the waveguide, the waveguide will transmit the TE_{11} mode, and the waveguide i.d. needs to be 0.76 times of the beam waist for optimal transmission. The loss of µw power through the coupling efficiency for both smooth-walled and corrugated waveguides is due to higher order mode conversions. Additional loss of ~1% will result in smooth-walled waveguides due to resistive wall losses, while this effect is negligible for corrugated waveguides. Thus, a smooth-walled waveguide may be a low-cost alternative that should be considered if µw power is not a limiting factor. However, smooth-walled waveguides have not been systematically studied for quasi-optical DNP and EPR systems at high magnetic fields, and thus further investigations are necessary before a final judgment can be made.

The NMR circuitry in the example system utilizes inductive coupling to improve the circuit filling factor.[63,64] Each individual nuclei has its own NMR coil tuned to its resonance frequency, where resonances >200 MHz are generated with an Alderman–Grant type coil design with a ±7 MHz tuning range, while nuclei with resonances below <200 MHz are made with saddle coils that have a ±2 MHz tuning range at room temperature. The NMR coils have a copper skirt extending beyond the basic coil for tuning of the circuit at cryogenic temperatures. The copper skirt in conjunction with the copper extension of the waveguide and the tuning ring (Figure 10.3a) with sapphire between each layer acts as a capacitor, where the degree of overlap between the tuning ring and the NMR coil skirt will change the resonant frequency of the NMR coil, allowing the NMR coil to have tuning capabilities from room temperature to 4 K.[63] The NMR signal and rf-pulses are transmitted and received between the NMR coil and the console via a pick-up loop. The pick-up loop is inductively coupled to the NMR coil; this was done to minimize the number of grounding loops in the system, increase the coil's filling factor, and allow for greater experimental versatility, since different

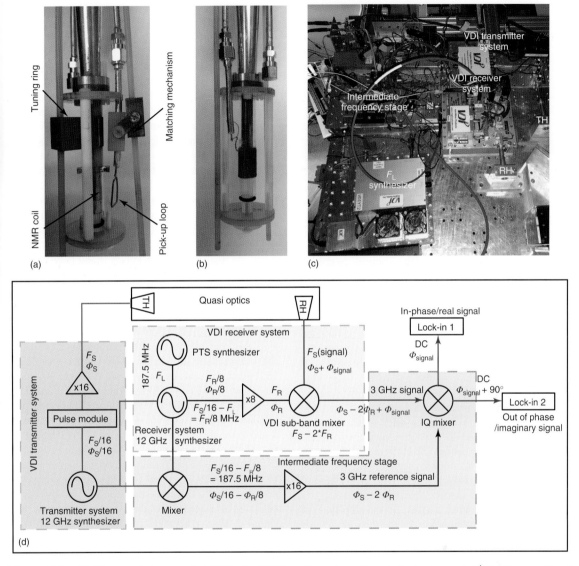

Figure 10.3. (a) Close up of the probe in NMR/pulsed EPR configuration with an inductively coupled ^{1}H Alderman–Grant coil. (b) The probe in the cw-EPR configuration with the modulation coil in place. Picture (c) and schematic (d) of the superheterodyne EPR detection used in the UCSB dual 194 GHz instrument. The orange block denotes the primary VDI transmitter, where all μw manipulations are abbreviated to pulse module. The yellow block denotes the VDI receiver system and dedicated synthesizer. The green block denotes an abbreviated IF stage. Incident μw are shown as blue lines, while μw carrying EPR signal are shown in red. Reference frequencies are shown as black

nuclei simply need the NMR coil to be exchanged for operation (private communication, Toby Zens). Matching is accomplished in these inductively coupled NMR coils by adjusting the distance between the window in the NMR coil and the pick-up loop, where larger distances result in poorer couplings and a reduction in the Ohmic match. Adjusting the distance between the pick-up loop and the coil can be done either manually at room temperature or with a matching mechanism as depicted in Figure 10.3(a) for cryogenic manipulations. More detailed and complete description of the NMR circuitry and DNP probe will be forthcoming in a future publication. For pulsed-EPR operations, no modifications to the probe are necessary, while cw-EPR may need a modulation coil to modulate the B_0 for lock-in based acquisition, this configuration is depicted in Figure 10.3(b).

10.2.7 EPR Detection Hardware and Methods

To enable EPR detection the spectrometer should include a sensitive detector for the relevant frequency range, e.g., 190–200 GHz for the Han lab UCSB system, where the vertically polarized EPR signal at the receive horn (see Section 10.2.5) originates from the plane orthogonal to B_0 at the sample. A couple of detector options are available: direct magnitude detection of the signal amplitude is possible with a zero-biased Schottky diode (ZBD) (Virginia Diodes Inc). If phase-sensitive detection is required, a homodyne or heterodyne detection scheme should be implemented. A ZBD is a good starter detector for initial tests of the system due to its simplicity, as it converts the amplitude of the µw to a voltage that can be directly detected; however, a ZBD has limited sensitivity that in practice makes it only suitable for lock-in amplifier-based cw-EPR detection, but is difficult for pulsed EPR detection. Lock-in-based cw-EPR detection using ZBD requires modulation of the magnetic field B_0, usually in the form $B_0 = B_0^0 + \sin(\omega_{mod}t) \times B_{mod}$, where B_0^0 is the central magnetic field of the superconducting magnet, ω_{mod} is the modulation frequency, and B_{mod} is the amplitude of the magnetic field modulation. In the Han lab UCSB system the field modulation is achieved by a separate modulation coil around the sample (Figure 10.3b) mounted on the EPR/NMR probe, and an AC current at frequency ω_{mod} is passed through the wire and provides the required B_0 modulation. The resulting cw-EPR signal is thus modulated by $\sin(\omega_{mod}t)$ and is demodulated with a lock-in amplifier (Stanford Research Systems – SR830). A lock-in amplifier acquires quadrature signal, where a zero-order phase correction can be used to convert all of the signal into a single trace. The modulation–demodulation process results in the acquisition of the derivative EPR signal. If the system is equipped with a variable field magnet or a sweep coil, then the magnetic field can be swept to acquire the EPR spectrum while keeping the µw frequency constant. However, if changing the magnetic field is not feasible such as when a conventional NMR magnet is used, then the EPR spectra can be acquired by sweeping the µw frequency, as demonstrated with the Han lab UCSB DNP/EPR system.[18]

For high sensitivity EPR detection and/or if phase information is of interest a heterodyne detection method is recommended. While a homodyne detection is also possible (reference and signal at the same carrier frequency are mixed to DC), this method is not recommended because of the large $1/f$ noise associated with a homodyne detection scheme. Therefore, a heterodyne detection scheme with an intermediate frequency is the method of choice. In this scheme, the sub-THz EPR signal is initially down-converted to an intermediate frequency using a sub-band mixer. Thus, when the intermediate frequency signal is mixed with the LO, the $1/f$ noise is significantly reduced for the final down-conversion to DC with an IQ mixer. The Han lab UCSB DNP/EPR system utilizes a heterodyne detection scheme with a 3 GHz intermediate frequency, which was chosen given the readily commercially available and affordable µw components needed for the intermediate frequency stage of the heterodyne EPR detection scheme, as depicted in Figure 10.3(c,d). To detect the 194 GHz EPR signal (F_S), the Gaussian beam is collected by a receive horn connected to a sub-band mixer (Rx-143, Virginia Diode Inc.) with a LO reference frequency (F_R) of 95.5 GHz, which results in the signal being down-converted to 3 GHz ($3\,\text{GHz} = F_S - 2F_R$). To accommodate changes in the F_S (for frequency stepped echo detected or frequency swept cw-EPR experiments), the F_R should be set such that after the sub-band mixer the down-converted signal is always at 3 GHz. The LO of the sub-band mixer is produced by an AMC (Rx-143, Virginia Diodes Inc.) multiplying the signal from a dedicated synthesizer operating at $F_R/8$ by eightfold (receiver synthesizer in Figure 10.3c). This dedicated synthesizer from VDI is a special YIG synthesizer designed to output the frequency difference between two analog inputs (i) the transmitter synthesizer frequency ($F_S/16$) and (ii) an offset frequency, F_L, set to 187.5 MHz ($3\,\text{GHz}/16 = 187.5\,\text{MHz}$) generated by a rf-source (Programmed Test Sources Inc.). The output frequency of the receiver synthesizer thus becomes $F_R/8 = F_S/16 - F_L$. This enables F_S and $2F_R$ to always be 3 GHz apart and ensures that the EPR signal after the sub-band mixer is exactly at 3 GHz, independent of the F_S. The 3 GHz EPR signal then passes through a low-noise amplifier (S020040M4601, Lucix), an isolator (L3 Narda-ATM), and a filter (K&L microwave) before being mixed with a 3 GHz reference signal in an IQ mixer (IQ0255LMP, Marki microwaves) to produce the EPR signal at DC. The 3 GHz reference signal is mixed (M10616NA, Marki microwaves) down from the primary synthesizers to 187.5 MHz. and subsequently multiplied 16x (WFM-T-187.5-3000, Wilmanco) to 3 GHz. The IQ mixer produces two quadrature EPR signals at DC: absorption and dispersion mode. For cw-EPR, each quadrature from the IQ mixer would need to be sent to a synchronized lock-in amplifier to demodulate the cw-EPR signal. For pulsed EPR echo-detected experiments the signal is directly digitized with a 1 GHz dual-channel digital-to-analog converter (AP240, Keysight, formerly ACQIRIS). A full detailed description of the heterodyne EPR detection scheme and intermediate frequency stage are presented by Siaw *et al.*[18]

10.2.8 Integrated Software Control

An integrated software system is useful so that one program can be used to control all the components of the instrument. In practice, it is convenient to retain control of the NMR part of the instrument with a commercial software that was supplied with the NMR spectrometer (e.g., TOPSPIN 1.3 in the Han lab UCSB system). In this case one only needs to take care of the synchronization between the NMR rf-pulses and DNP/EPR MW pulses. In the Han lab UCSB system, this has been achieved with Specman4EPR (Femi Instruments LLC) software package.[65] Specman4EPR directly controls the μw components, either via the DAC (USB-6001, National Instruments) for voltage-controlled components or by programming the 24-channel digital pulse generator (PulseBlaster ESR-PRO SpinCore) that controls all of the four μw switches in the system. Specman4EPR software also controls the frequency of the two transmitter synthesizers (synthesizer 1 and synthesizer 2 in Figure 10.2b) and programs the waveforms into the AWGs (Homebuilt and Chase Wavepond). In addition, the EPR signal is recorded from either ADCs in pulsed EPR experiments, or from the lock-in amplifiers in cw-EPR or power calibration experiments. The synchronization with the NMR software (TOPSPIN 1.3) is achieved by providing a TTL trigger(s) to the NMR spectrometer, in synchrony with the NMR pulse sequence that is designed to advance to the next step upon receiving the trigger. This ensures that all the timings in the experiment are set in the SpecMan4EPR software.

10.2.9 Performance Diagnostics

Power loss analyses of the different components in the instrument should be performed to accurately measure the μw power at the sample position, and to aid in diagnosing instrument performance issues that arise. A combination frequency counter and power meter (EIP 548A, EIP Microwave Inc) can be used to characterize the various components at 12 GHz from the synthesizer to many parts of the AMC selection module, as well as the 3 GHz components of the heterodyne detection scheme. The insertion loss of each component was measured and tabulated, so that if performance declines, then measurement of the power after key components such as amplifiers and mixers can be used to determine if a component has become faulty. This streamlines the process of diagnosing poor μw performance. An oscilloscope (Agilent MS071048) is helpful for the analysis of the AWG outputs at 12 GHz. A photoacoustic absolute power meter (Thomas Keating) was used to determine the power losses through the quasi-optical components at ~200 GHz, of which the Faraday rotators cause the largest power loss of 2 dB per pass through a rotator. A detailed description of how quasi-optical power loss analyses can be performed is provided in the work by Siaw *et al.*[18] A pyroelectric (ELTEC instruments Inc.) detector mounted at the sample position inside the magnet is used for aligning the μw beam path from the bridge, with the waveguide in the probe insert – such aligning is done regularly. Once calibrated in reference to the absolute power meter, the pyroelectric detector can act as a pseudo-power meter for the complete system. Another way to analyze the system's μw performance is by monitoring the EPR echo intensity of a standard sample – diamond is a good standard to choose because it does not degrade with time and has a large room temperature EPR signal of its P1 centers. A network analyzer (Hewlett Packard – 8753A) is used to check the performance of the various homebuilt NMR coils for each nuclei regarding their Q-factor, resonance frequency tuning range, and matching to the 50 Ohm rf-output from the NMR console.

10.3   EXPERIMENTAL RESULTS

10.3.1   Hole-burning ELDOR and ED-NMR

The basic ELDOR experiment carried out with the pulse sequence shown in Figure 10.4(a) (inset) can be delineated into two distinct regimes characterized by the length of the excitation pulse, t_{excite}: (i) hole-burning ELDOR – if t_{excite} is longer than the time required for the whole electron spin system

to reach steady state, typically on the order of T_{1e}, then the effects observed in the ELDOR spectra are characteristic of electron spectral diffusion (eSD) and (ii) ELDOR-detected NMR (ED-NMR) – if t_{excite} is short ($\ll T_{1e}$) the ELDOR spectra are dominated by signals originating from the excitation of forbidden electron–nuclear transitions of proximal hyperfine coupled nuclei (ED-NMR). The hole-burning ELDOR is relevant for the study of DNP mechanisms and probing the state of the electron spins under DNP

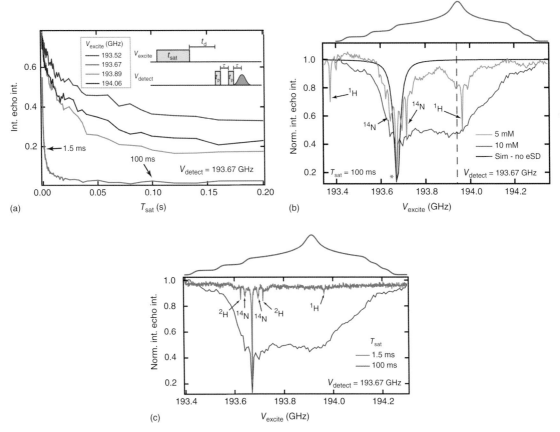

Figure 10.4.   (a) Electron excitation rates for 10 mM TOTAPOL in 6 : 3 : 1 d_8-glycerol:D$_2$O:H$_2$O at 4 K at $v_{detect} = 193.67$ GHz for a variety v_{excite} as denoted in the figure. The inset is the ELDOR pulse sequence applied to obtain the experimental data presented here. (b) Experimental ELDOR spectrum of 5 mM (green) and 10 mM (red) TOTAPOL in a 6 : 3 : 1 d_8-glycerol:D$_2$O:H$_2$O glass at 4 K overlaid with an ELDOR simulation without eSD (black), with $t_{sat} = 100$ ms. The nitroxide EPR spectrum is shown as reference in gray above the ELDOR spectrum. The hyperfine forbidden transitions are annotated in the figure, and the allowed EPR transition is annotated by '*'. (c) ELDOR spectra taken with 1.5 ms (green) and 100 ms (red) excitation pulse lengths for 10 mM TOTAPOL in 6 : 3 : 1 d_8-glycerol:D$_2$O:H$_2$O at 4 K at $v_{detect} = 193.67$ GHz. The simulated EPR spectrum for TOTAPOL at 7 T and 4 K is shown above, based on an experimental EPR spectrum acquired at 8.56 T and 4.5 K with traditional field-sweeping capabilities.[27] Experimental parameters for all spectra: repetition time = 500 ms, $t_p = 750$ ns, $\tau = 500$ ns, and $t_d = 10$ μs.

conditions, since the long t_{excite} pulse mimics the prolonged irradiation found in DNP experiments.

10.3.1.1 Hole-burning ELDOR

In hole-burning ELDOR experiments, strong μw irradiation (t_{excite}) is applied for sufficient time to saturate the electron spins resonant with the irradiation frequency. A saturated or perturbed electron spin system will have a reduced or eliminated EPR signal. In the absence of spin–spin interactions, the shape of the hole 'burnt' into the EPR spectrum will have a Lorentzian lineshape with the width depending on the strength of the μw irradiation, the electron spin lattice relaxation time (T_{1e}), and the electron phase memory time (T_m) – a measure of transverse relaxation. The presence of eSD resulting from electron spin–spin dipolar interactions will manifest itself as a deviation from the Lorentzian lineshape, where the strength of eSD can be extracted through analysis of ELDOR spectra, as has been previously demonstrated.[1,66]

To acquire a 1D ELDOR/ED-NMR spectrum a constant field technique is used, where the magnetic field and detection frequency (v_{detect}) are held constant while the excitation frequency (v_{excite}) is sampled across the EPR spectrum, as shown by the pulse sequence in the inset of Figure 10.4(a). A complete (2D) set of ELDOR/ED-NMR spectra can be acquired by either stepping v_{detect} across the whole EPR spectrum, or if the spectrometer is equipped with a variable field magnet or a sweep coil, by changing the magnetic field.[4,67] In a DNP experiment the length of μw irradiation (typically minutes at low temperatures) is long enough for the electron spin system to reach a steady state. In ELDOR experiments minutes-long irradiation, typical for DNP experiments, is impractical because the acquisition of a 2D ELDOR dataset will become unrealistically long. Instead t_{excite} is set for the shortest time required for the electron spin system to reach the steady state. This can be done either by taking complete ELDOR profiles at several t_{excite}, or by acquiring ELDOR experiments as a function of t_{excite} at only a few v_{excite} values. The advantage of the latter approach is that it is a less time-consuming method to determine a t_{excite} that is representative of DNP conditions, compared to acquiring the complete ELDOR dataset for each t_{excite} value. Typically, we find that the longest time to reach a steady state occurs near the edges of the EPR spectrum, and thus the v_{excite} values should be extended to cover the whole EPR spectrum. Four of such ELDOR vs t_{excite}

measurements are shown in Figure 10.4(a) for 10 mM TOTAPOL. These rates show that if $v_{excite} = v_{detect}$, then the electron spins will become saturated at approximately 10 ms, while if $v_{detect} \neq v_{excite}$, then the process takes about an order of magnitude longer. The reduction of EPR signal intensity at $v_{detect} \neq v_{excite}$ is a result of polarization transferred between two electron spins with different resonance frequencies via eSD. The efficiency of polarization transfer decreases with the increase in the frequency separation between the electron spins and is proportional to the relative spin populations of the two frequencies. Therefore, excitation in the middle of the EPR spectrum will take the longest time to reach a steady state on the sides of the spectra, where frequency separation between v_{excite} and v_{detect} is large, and spin population is low. Besides determining representative DNP t_{excite}'s, the rate of electron depolarization between v_{excite} and v_{detect} can be extracted by fitting the ELDOR vs t_{excite} data to $y = Ae^{-t/t_{eSD}} + C$, where t_{eSD} is the rate of spectral diffusion between electrons at v_{excite} and v_{detect}. The extracted t_{eSD} were found to be 1 ms for $v_{detect} = 193.67$ GHz ($v_{excite} = v_{detect}$, $\Delta v = 0$ MHz), 17 ms for $v_{detect} = 193.887$ GHz (center of the nitroxide EPR spectrum, $_{\Delta v = 217 MHz}$), 25 ms for $v_{detect} = 193.52$ GHz (edge of the EPR spectrum $\Delta v = -150$ MHz), and 18 ms for $v_{detect} = 194.06$ GHz (intermediate position on EPR spectrum, $\Delta v = 390$ MHz). It is clear that the extracted t_{eSD} values correlate better with the relative spin populations between the two electron spin frequencies than with the frequency separation, Δv.

After determining t_{excite}, hole-burning ELDOR spectra can be acquired, where the hole-burning ELDOR profiles for 5 mM (green) and 10 mM (red) TOTAPOL in 6:3:1 d_8-glycerol:D_2O:H_2O at 4 K are shown in Figure 10.4(b). Both spectra have a sharp peak at $v_{detect} = v_{excite}$, corresponding to the allowed EPR transition, which is denoted by '*'. The weaker, sharp features correspond to forbidden transitions between the electron and its hyperfine coupled nuclei, where the ^{14}N and ^{1}H forbidden electron–nuclear transitions are annotated in Figure 10.4(b). These forbidden transitions are the basis for ED-NMR and will be discussed further in Section 10.3.1.2. A simulation of the hole-burning experiment without any eSD effects is shown in black in Figure 10.4(b). The inability of this simulation to capture the broad depolarization at the center of the EPR spectrum at 197.85 GHz that is especially pronounced in the 40 mM sample suggests that this feature is caused by eSD. The eSD process is driven by electron–electron dipolar interactions that propagate the μw-induced electron spin depolarization to the off-resonance electron spins. Therefore, we expect the eSD-induced depolarization to be more pronounced in the samples with higher electron spin concentrations. In addition, as was discussed previously, eSD is proportional to the relative spin population and is thus most pronounced in the ELDOR spectra at the frequency corresponding to the center of the EPR spectrum.[1] The 40 mM 4-amino TEMPO sample has significantly larger eSD effects compared to the 10 mM sample as seen by the larger depolarization feature at the center of the EPR spectrum at 193.92 GHz (marked with dashed line in Figure 10.4b).

10.3.1.2 ELDOR Detected NMR

The hole burnt into the electron spin system is dominated by the allowed EPR transitions excited with the saturation pulse. However, if electron–nuclear hyperfine interactions are present in the system, additional holes due to excitation of forbidden electron–nuclear transitions can be detected as well – this is the basis for the ED-NMR experiment. These holes originate from electron spin depolarization due to simultaneous electron–nuclear spin flip-flops when the μw irradiation matches the difference in energy between the electron–nuclear mixed states from the forbidden transitions.[68] ED-NMR can be thought of as detecting the SE for *all* the different NMR active nuclei that are weakly hyperfine coupled to the unpaired electrons in the system, with symmetric peaks centered around the allowed EPR transition. The frequency difference between the symmetric peaks and the allowed EPR transition correspond to the nuclear Larmor frequency of the hyperfine coupled nuclei. ED-NMR to detect hyperfine coupling fingerprints was first proposed by Schossler, Wacker, and Schweiger in the mid-1990s.[69]

1D and 2D ED-NMR spectra are acquired in the same manner as those for hole-burning ELDOR; however, the length and/or power of the excitation pulse needs to be adjusted to prevent broadening of the allowed EPR transition, which can mask the forbidden hyperfine transitions of low γ-nuclei. This sharp depolarization due to the allowed EPR transition when $\nu_{excite} = \nu_{detect}$ is termed the central peak blindness. For ED-NMR, the detection pulses to form an echo can be used to selectively enhance the observation of specific types of forbidden transitions, such as selecting for differences in the microenvironment (via detection of multiple A_{zz}) versus detecting double quantum forbidden transitions.[67,70] The echo detection method shown in Figure 10.4(a) can be substituted with FID detection; however, recent research has significantly favored the echo over FID detection for ELDOR-based experiments.[69,71] Finally, the delay between the excitation and the detection pulses for both DNP-relevant hole burning and the ED-NMR experiments, t_d, needs to be sufficiently long to eliminate unwanted electron spin transverse coherences – i.e., longer than T_m.[71]

ED-NMR with short excitation pulses is of particular interest because it can detect the nuclei that are hyperfine coupled to a paramagnetic spin, which is extremely beneficial in determining the identity, structure, and spatial relations of nuclei immediately surrounding an unpaired electron. A comparison of ED-NMR data (green trace) and hole-burning ELDOR data (red trace) of the same 10 mM TOTAPOL sample in 6 : 3 : 1 d_8-glycerol:D_2O:H_2O at 4 K, 7 T, and with $\nu_{detect} = 193.67$ GHz are shown in Figure 10.4(c). As forecasted, the only difference between the ED-NMR and hole-burning ELDOR acquisitions lies in t_{excite}, with $t_{excite} = 1.5$ ms for the former and $t_{excite} = 100$ ms for the latter. The nuclei that are hyperfine coupled to the TOTAPOL are denoted for the resolved ED-NMR resonance in Figure 10.4(c), where it should be noted no eSD is observed. In the hole-burning ELDOR spectrum representing DNP conditions, it is difficult to identify which nuclei are hyperfine coupled to the electron given the extensive depolarization across the EPR spectrum due to eSD.[1]

10.3.2 Two-source ELDOR

As described in Section 10.3.1.1 in order to acquire ELDOR profiles, ν_{excite} is stepped across the entire EPR spectrum – ideally with the same μw power across the whole frequency range. However, the power output from the AMCs *is* frequency dependent; thus, to get (DNP and EPR) frequency-dependent data with the same nominal power across the frequency range, a power calibration for the relevant frequency range (for the UCSB DNP system: 192 to 195 GHz) is required to acquire DNP and EPR spectra for commonly used DNP radicals. For AMCs operating under saturating conditions, the power variation across the typical frequency range for nitroxides is approximately 10–15%, which is acceptable, but this variation becomes significantly worse for AMCs operating under non-saturating conditions. As was discussed in Section 10.3.1.2, to prevent central peak blindness, a two-frequency ELDOR experiment may require the pulses at each frequency to be produced with different output powers. The slow, on the EPR timescale, response time of the built-in AMC amplitude control (see Section 10.2.4.1) of ~0.5 ms prohibit directly manipulating the AMCs' amplitude in the course of single EPR pulse sequence. This makes acquiring ELDOR electron depolarization profiles with a single AMC challenging when different power levels are required for the excitation and detection pulses.

In the following, we provide an illustrative example on the importance of power calibration and a benefit of using the 2-AMC setup for acquiring ELDOR spectra with power-dependent t_{excite} pulses. When power-dependent electron depolarization profiles are obtained for a 40 mM 4-hydroxy TEMPO (4HT) sample in 1 : 1 DMSO:H_2O at 4 K with a single AMC and without power calibration, clear distortions of the background signal are observed, as shown in Figure 5(a). For this dataset, a single AMC was used, and thus the μw power level was varied for the entire ELDOR pulse sequence, i.e., both excitation and detection pulses. Lowering the intensity of the detection pulses results in a lower echo intensity and thus requires more signal averaging to acquire data with a similar SNR as the ELDOR spectra acquired with higher μw powers. The distortions in the background of the ELDOR profiles are due to variations in the power output as a function of frequency, and are most noticeable for intermediate irradiation powers, as the distortion effects are minimized at the maximum AMC output where the performance is optimal, and at low powers where frequency-dependent power variation is minimal.

In an effort to obtain reliable power-dependent ELDOR data without sacrificing SNR for the lower power level experiments and to overcome the slow response time of the AMC amplitude control we attempted to modulate the output power of the AMC by changing the power of the input (~12 GHz) signal to the AMC. This was achieved by adding a voltage-controlled attenuator between synthesizer 2 (F_{ELDOR}) and the pulse-forming network. This allows the synthesizer used for the excitation pulse to be attenuated at each v_{excite} in an ELDOR experiment and leaves the detection pulses, generated from synthesizer 1 ($F_S/16$), to be at full power. As the AMC is designed to operate under saturating conditions, the AMC will be forced to run

under the less-ideal non-saturating conditions, which results in higher AM noise for the excitation pulses. The response of the AMC to changes in the power level of the input signal are highly nonlinear under non-saturation conditions. Therefore, a calibration of the output at high frequency (~194 GHz) as a function of the input μw power at low frequency (~12 GHz) is required for this method to work. However, the very narrow dynamic range where the AMC performs under non-saturating conditions limits the precision that can be achieved with this methodology. Nevertheless, when this power calibration was carried out the resulting power-dependent ELDOR spectra were significantly improved, yet some artifacts were still clearly visible in the ELDOR spectra. These artifacts could be traced to the 'heating (memory) effects' of the AMC, where the detection pulses were dependent on the frequency and power of the preceding excitation pulse.

The use of a second AMC to completely separate the excitation and detection channels eliminates these 'memory' effects, since the detection and excitation pulses are now generated by separate AMCs. The more robust built-in AM can now also be used to control the power level of the excitation pulses, as each AMC only requires a single power level in a given ELDOR experiment. Note that as detailed in Section 10.2.7 our sub-band mixer EPR detector detects induction-mode EPR signal from AMC1/TH1, and reflection-mode EPR signal from AMC2/TH2, as shown in Figure 10.2(c). In the reflection-mode EPR configuration, the signal and the reflected excitation pulses are both transmitted into the detector; thus, the excitation pulses must be short (2 μs with 5% duty cycle) in order to prevent damage to the sub-band mixer. Therefore, AMC1/TH1 was used for the long excitation pulses and AMC2/TH2 for the shorter detection pulses. Power-dependent ELDOR profiles of the 40 mM 4-hydroxy TEMPO (4HT) sample in 1:1 DMSO:H_2O at 4 K were acquired using the just-introduced two-source configuration by solely varying the power of the excitation pulses and are shown in Figure 10.5(b). Notably, the previously observed distortions to the background signal in the ELDOR acquired with a single AMC (Figure 10.5a) are eliminated. Thus, the implementation of a second source for two-source ELDOR improves the quality of power-dependent electron depolarization profiles and allows for acquisitions of highly reproducible electron depolarization profiles with clean baselines.

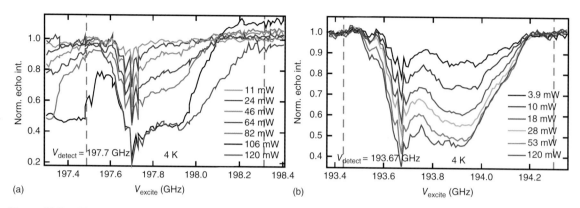

Figure 10.5. Electron depolarization profiles of 40 mM 4-hydroxy TEMPO in 1:1 DMSO:H$_2$O at 4 K and at 7 T with attenuation of AMC power output for the whole pulse sequence (a) and with two-source ELDOR (b). For each the applied μw power during the saturation pulse was varied and is depicted inside the figure. The dashed gray lines correspond to the width of the EPR line. Experimental parameters: (a) repetition time = 400 ms, t_{excite} = 100 ms, ν_{excite} = 197.7 GHz, t_p = 500 ns, τ = 500 ns, and t_d = 10 μs. (b) repetition time = 500 ms, t_{excite} = 100 ms, ν_{excite} = 193.67 GHz, t_p = 750 ns, τ = 500 ns, and t_d = 10 μs.

10.3.3 Arbitrary Waveform-shaped Pulses for EPR

Integration of an AWG allows for enhanced control over the pulse amplitude, frequency, and phases for more elaborate spin manipulations than is possible with rectangular pulses. AWG has been utilized by NMR for decades,[72–74] but only in the last few years has the technology became mature enough, reaching sub-ns time resolution, for AWGs to shape μw pulses in EPR experiments, mainly at X-band (9.5 GHz) and Q-band (35 GHz) frequencies.[59,75–80] We have recently integrated AWG capabilities in our 200 GHz DNP/EPR spectrometer.[39] An AWG can be used as a convenient way to provide phase cycling capabilities for rectangular pulse experiments, beyond what is possible with a traditional two-channel spectrometer design, as the AWG phase manipulation is frequency independent to the first order. But more importantly, phase- and amplitude-modulated pulses can be used to enhance the performance of the system relative to a rectangular pulse, for example, by extending the bandwidth of a pulse to much beyond $\gg\omega_1$ of a rectangular pulse. This is beneficial for systems where the available source technology is power limited (see Section 10.2.4), yielding low ω_1 of order 0.5 MHz for EPR and DNP experiments. The solution to enhance ω_1 through the use of a resonant cavity as is commonly employed for EPR is not compatible with high-frequency DNP experiments, given the relatively large sample volumes (30–50 μl) frequently used for DNP experiments. Therefore, AWG-enhanced shaped pulses become a very attractive approach to address the ω_1 limitations, as they allow for fine control of the spin dynamics and spin manipulation over the bandwidth that can exceed the available ω_1 by orders of magnitude.

Due to the nonlinear nature of the AMC and small dynamic range of the input signal when operating under non-saturating conditions, AM becomes challenging.[39] As such we will limit our discussion of AWG functionality to frequency and phase modulations. Fortunately, under power limiting conditions, phase-shaped full amplitude pulses typically provide the best performance.[81–83] One common frequency-modulated pulse to increase the bandwidth of a pulse is the chirp pulse, where the frequency is changed in a linear fashion over the duration of the pulse. The chirp pulse is defined by the following parameters: (i) the chirp pulse bandwidth $\Delta\omega_{chirp}$, (ii) the rate at which the instantaneous frequency is swept across this bandwidth, k, and (iii) its amplitude. Since the instantaneous frequency of the chirp pulse is changing linearly from $\omega_{mw} - \frac{\Delta\omega_{chirp}}{2}$ to $\omega_{mw} + \frac{\Delta\omega_{chirp}}{2}$ during the chirp pulse duration, t_{chirp}, the chirp rate is given by $k = \frac{\Delta\omega_{chirp}}{t_{chirp}}$. The phase profile of the chirp pulse, $\Theta_{pulse}(t)$, is: $\Theta_{pulse}(t) = \left(\left(-\frac{\Delta\omega_{chirp}}{2}\right)t + \frac{kt^2}{2}\right)$.

The pulse sequence for measuring the inversion profile of a saturation pulse is shown in Figure 10.6(a).

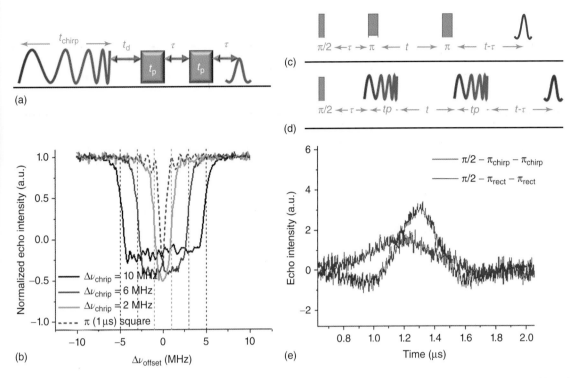

Figure 10.6. (a) Pulse sequence used for measuring inversion profile of a chirp pulse. (b) Experimental inversion profiles of a 10 μs chirp pulse for BDPA in o-terphenyl sample. Refocused echo pulse sequence with (c) all rectangular pulses and (d) chirp-refocusing pulses. (e) Refocused echoes as measured on the 3.4 nm Gd^{3+} ruler molecule using pulse sequences in (c) blue trace and pulse sequence in (d) red trace. Experimental parameters of BDPA sample: $t_p = 500$ ns; $\tau = 800$ ns; $t_d = 2.4$ μs; $k = 1$ MHz μs^{-1}; Gd^{3+} sample: $\pi/2 = 187.5$ ns; $\pi = 375$ ns; $\pi_{chirp} = 1$ μs; $\Delta\omega_{chirp} = 4$ MHz; $\tau = 1$ μs; $t = 6$ μs; $t_p = 1$ μs; repetition time 100 μs. (Reproduced with permission from Ref. 39. © Elsevier, 2017)

Here, the intensity of the echo formed by the two t_p pulses was monitored as a function of the offset frequency of the t_{chirp} inversion pulse. The advantage of using the phase-modulated chirp pulses is illustrated in Figure 10.6(b). An inversion efficiency of three different chirp pulses with $\Delta\omega_{chirp}$ of 2, 6, and 10 MHz clearly exceed that of a rectangular-π pulse (dashed blue line). Notably for this sample of 1% BDPA in o-terphenyl, the inversion bandwidth of the shaped pulses exceeds that of the rectangular pulse by more than a factor of ten.

The gain in performance by using chirp pulses was also demonstrated for coherent EPR experiments with a refocused echo pulse sequence, where the rectangular pulses (Figure 10.6c) are replaced with a pair of refocusing chirp pulses (Figure 10.6d). Figure 10.6(e) clearly shows the enhanced performance for a 3.4 nm Gd^{3+} ruler molecule, as the echo intensity is two times higher when the chirp-refocusing pulses were used.

These examples illustrate the improved electron spin control that is enabled by using AWG phase-shaped pulses.

10.3.4 Using EPR to Understand DNP Results

As DNP is the transfer of polarization from the electrons to the hyperfine coupled nuclei in a system, it is imperative to understand the electron and nuclear behaviors individually and how they interact with each other, under conditions relevant for DNP. Thus, EPR can help elucidate surprising DNP results and refine our understanding of the mechanisms that drive DNP processes. In the following section, we provide a few examples of how the EPR results improved our understanding of DNP or enhanced the DNP performance. In the modern era of DNP-revival, such experiments were initially carried out by the Nottingham group[15] using longitudinal EPR detection in their 95 GHz/3.3 T DNP polarizer, including double-resonance pump-probe experiments.[84] This was followed by work from the Weizmann Institute,[4,66] MIT,[11] and UCSB[1] using conventional EPR detection at 3.3, 5, and 7 T, respectively. The following section provides a detailed description of the kind of new understanding that was enabled by EPR and especially double-resonance ELDOR experiments.

10.3.4.1 Electron Spectral Diffusion and Indirect Cross Effect

At very low temperatures (<1.5 K) typical for d-DNP experiments, DNP is assumed to proceed via the multi-electron–multi-nuclear thermal mixing (TM) mechanism.[85–88] In contrast, at higher temperatures the traditional SE and CE DNP mechanisms only involve one and two electron spins, respectively. In this case, only the electrons on-resonance with the µw irradiation can contribute to the DNP process in static DNP experiments.[89–94] This view was challenged by the measurement of the electron spin polarizations across the EPR spectrum in the course of a DNP experiment by ELDOR and by observation of the reduction of DNP efficiency with an increase of µw power – oversaturation.[4,95] The ELDOR experiments were initially acquired at 3.3 T on a dual DNP/EPR instrument and revealed significant depolarization of the electron spins occurring across the EPR spectrum for samples with radical concentrations (10–40 mM) typical of DNP experiments, as shown in Figure 10.8(a).[12] The observation of extensive electron spin depolarization led to the development of a theoretical model that successfully described the experimental results by incorporating eSD effects. To do this the model accounted for the exchange of polarization between electron spins with different resonance frequencies.[4] The realization of the importance of eSD in the DNP process under conditions where a common spin-temperature does not apply led to the development of the indirect cross effect (iCE) model.[66] In the traditional CE DNP mechanism, the extent of nuclear polarization is proportional to the polarization differential between the two electrons forming the CE pair, which in the absence of eSD is generated by the µw irradiation on-resonance with one of the two electron spins. The iCE model extends this idea by allowing nuclear polarization to result from the electron spin polarization differential in CE pairs, where neither of the spins is on-resonance with the µw irradiation. Instead the polarization differential is created by the unequal influence of eSD on the two electron spins. This model correctly simulated the shapes of experimentally observed DNP profiles based *solely* on ELDOR data (see example in Figure 10.7a) for electron depolarization under the same conditions, deeming the model successful. Remarkably, the model accurately predicted changes in the shape of the DNP profiles observed with changes in temperature as shown in Figure 10.7(b–d). Since the iCE model allows for a significant portion of the electron spins to take part in DNP, it explains why DNP becomes so efficient at high radical concentrations. This is especially true for inhomogeneously broadened EPR spectra, >500 MHz wide which in the absence of eSD would have only a small fraction of participating electron spins to drive CE DNP. In addition, the iCE model correctly predicts that at very high radical concentrations and low temperatures, where eSD is very efficient, the saturation of the electron spectrum can exceed the optimal for DNP. When this occurs, most of the electron spins are saturated, and thus, the electron spins pairwise polarization differential is reduced, which decreases the DNP efficiency – the effect is also known as 'oversaturation'.[95]

10.3.4.2 Polymorphism-induced Radical Clustering

A robust technique requires consistent data acquisition; however, many DNP experimentalists observe at times inconsistent results from day-to-day with nominally the same sample. A dual DNP/EPR spectrometer is an essential tool to determine the cause behind such an observation. A sample of 40 mM 4-amino TEMPO in two solvent compositions: $5:4:1$ (541) and $6:3:1$ (631) ratios of d_8-glycerol:D_2O:H_2O at 4 K and at 7 T were studied with two freezing methods: fast cool (FC) and thermally controlled (TC) to help understand potential causes of inconsistent DNP results with the 541-based samples.[6] The freezing protocol–temperature as a function of time are depicted in Figure 10.8(a), where the FC is frozen at a rate of 10 K min^{-1}, while the TC is frozen at a rate of 1 K min^{-1}, held at an isotherm 10 K above the glass transition temperature (T_g) for 1 h, and is cooled to 100 K at a rate of 1 K min^{-1}, below which a 10 K min^{-1} freezing rate is adopted. The DNP profiles for the two solvent compositions and freezing methods taken in triplicate are shown in Figure 10.8(b,c), where the average data is shown and the standard deviation

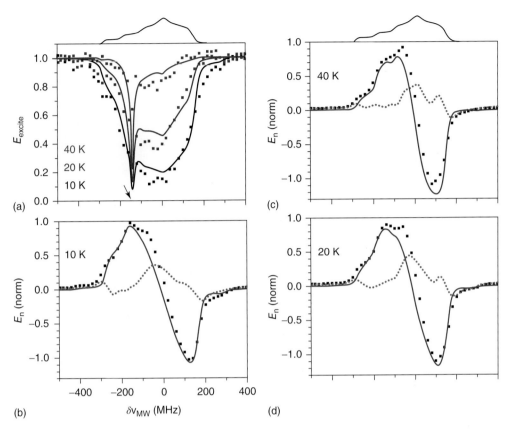

Figure 10.7. (a) Measured (square symbols) and simulated (lines) steady-state E_{excite} profiles, detected at δv_{detect} equal to -140 MHz. (b–d) Measured (square symbols) and simulated iCE (red lines) DNP profiles. The measured spectra were obtained at 40 K (c), 20 K (d), and 10 K (b), using a t_{excite} value of about $5T_{1n}$. The simulated spectra were generated using the iCE model based on the ELDOR data in (a–c). The EPR-line shape of TEMPOL is given by the gray line at the top of the figure. (Reproduced from Ref. 66 with permission from the PCCP Owner Societies)

is denoted by error bars. The 631 systems had more consistent DNP enhancements and DNP profiles than the 541 systems. It is clear from the DNP profiles that the 541-TC had the least consistent DNP profile line-shape and DNP enhancement values (the enhancement values for each sample are shown in the corresponding figure legend). The question is whether electron spin dynamics can resolve the mechanism behind this observation of inconsistency with the 541-TC system.

The electron phase memory time (T_m) is a measure of the electron decoherence of the spin system, and can be directly correlated with the local radical concentration at high radical concentrations and liquid helium temperatures.[26] This is because stronger electron–electron dipolar interactions will cause faster decoherence, and the strength of electron–electron interactions is enhanced by an increase in the local electron spin concentration, if electron–electron dipolar coupling dominates the decoherence mechanism. T_m was acquired for all four systems at 4 K and 7 T and is shown in Figure 10.8(d). 541-TC has a reduced T_m compared to the other systems, indicating that 541-TC has a higher local radical concentration compared to the other systems, while the nominal global radical concentration between all four systems is the same. This means the combination of the 541-solvent system and the TC freezing method induces radical clustering. The cause of radical clustering was determined to be glass polymorphism.[6] Thus, variations in glass polymorphs formed from day-to-day alter the radical distribution, as indicated by variations in T_m, where glass polymorphism-induced radical clustering is the

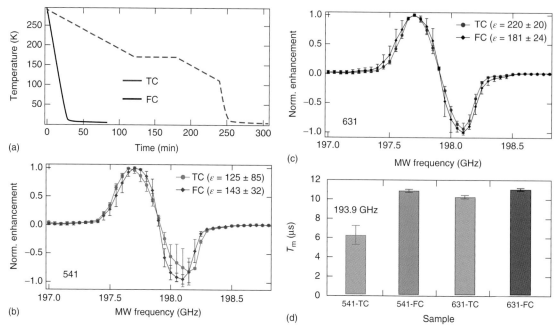

Figure 10.8. (a) Temperature cooling curves for the two freezing methods: fast cool (FC) and thermally controlled (TC). Normalized DNP profiles of the two freezing methods taken in triplicate for a 40 mM 4AT sample in 5 : 4 : 1 (b) and 6 : 3 : 1 (c) of d_8-gylcerol:D_2O:H_2O. The average data is plotted, and the standard deviation is shown as error bars, with the maximum positive enhancements and their error for each system noted in parenthesis in the figure legend. DNP signal was recorded after 60 s of μw irradiation. The enhancements for 541-TC and 541-FC were 125 ± 85 and 143 ± 32, respectively, while the enhancements for 631-TC and 631-FC were 220 ± 20 and 181 ± 24, respectively. (d) T_m for each system is shown when irradiated at 193.7 GHz (the center of the EPR spectrum). Error bars represent the error from fitting the raw experimental data with $y = A\exp^{-2t/T_m}$. Experimental parameters were $t_p = 750$ ns and repetition time $= 500$ ms. (Reproduced from Ref. 6 with permission from the PCCP Owner Societies)

leading cause of the inconsistent DNP results for the 541-TC system.[6]

10.3.4.3 Repression of Electron–Nuclear Coupling

The inherent presence of paramagnetic species in DNP processes can render determining the precise NMR signals near the unpaired electron challenging in terms of shifts in the peak position and line broadening. These effects come from the hyperfine interactions (dipolar and scalar) between the electron and nuclear spins, where both coherent (strength of the electron–nuclear interaction) and incoherent (fluctuations in the local field due to spin flip-flops – i.e., paramagnetic relaxation enhancement (PRE)) electron–nuclear interactions affect the NMR linewidth.[96] The shifting of the peak position

is influenced by the contact shift and pseudo-contact shift – i.e., isotropic and anisotropic dipolar interactions, respectively.[97] To acquire undistorted NMR spectra, a reduction of these paramagnetic effects is desirable via electron–nuclear decoupling – similarly to [1]H-decoupling of heteronuclei.[98–100] However, full electron–nuclear decoupling is difficult to achieve due to the wide electron spectral bandwidth at high fields (>5 T) that is challenging to completely saturate, as a nitroxide radical is ~1 GHz broad at 7 T due to the g-factor anisotropy. We demonstrated that large portions of a nitroxide EPR spectrum can be saturated by using high radical concentrations at low temperatures.[1,39] Thus, a partial reversal of the PRE by electron spin saturation (REPRESSION) can be used to narrow the NMR linewidth, $\Delta\nu_{FWHM}$, as shown in a recent publication by Jain *et al.*[7] This study shows for a Li + ion electrolyte solution doped with a high concentration of nitroxide radicals (> 20 mM), that hyperfine-decoupling (hfDC) can reduce the NMR linewidth broadening and peak shifting induced by paramagnetic effects, as shown in Figure 10.9(a).

To elucidate the mechanism behind the narrowing of Δv_{FWHM}, the DNP profiles were taken with and without a hfDC pulse, where the hfDC pulse frequency, v_{hfDC}, does not have to be the same frequency as the DNP μw irradiation, v_{DNP}, as depicted by the pulse sequence shown in Figure 10.9(b). The DNP profiles in Figure 10.9(c) illustrate that a hfDC pulse immediately prior to acquisition does not alter the DNP enhancement or profile lineshape.[7] However, if the v_{hfDC} is chosen to equal the center of the EPR spectrum for the paramagnetic species, then the Δv_{FWHM} will be the most narrow as shown by Figure 10.9(d). This is because the μw irradiation saturates the largest number of electron spins at the center of the EPR spectrum, where the large field fluctuations caused by electron saturation will suppress the incoherent term of the hyperfine interaction – the PRE effect. This was confirmed via Δv_{FWHM} and T_m measurements as a function of applied μw power and gating time, τ_{gate} (the time between the DNP μw pulse and rf-pulse for NMR acquisition where no μw are applied to the system) as shown in Figure 10.9(e,f). As the μw power and τ_{gate} are varied, T_m (a measure of the field fluctuations) and Δv_{FWHM} (a measure of hfDC strength) follow the same trends, confirming that hfDC results from RE-PRESSION, which suppresses the incoherent term of electron–nuclear hyperfine interactions.[7]

10.3.4.4 *AWG-enhanced DNP*

The shaped pulses made possible by integration of AWG capabilities for EPR can also be applied to DNP experiments to either improve the efficiency of the polarization transfer[101–104] or to include more electron spins in the DNP process by means of ultra-broadband excitation.[105] In a chirp-DNP experiment, the monochromatic irradiation typical of conventional DNP experiments is substituted by a train of chirp pulses of duration, t_{chirp}, and bandwidth, $\Delta \omega_{chirp}$, that are repeated continuously for the total length of the DNP experiment (Figure 10.10a). Chirp-DNP consistently outperforms conventional cw-DNP under multiple experimental conditions and with different radicals as the polarizing agents (Figure 10.10b). The increase in DNP efficiency is caused by the increased bandwidth of the shaped, chirp pulses compared to cw-excitation, where more

electrons over the well-defined bandwidth can partake in the CE DNP process.[105] The ability to precisely control the extent of EPR line depolarization was confirmed by chirp ELDOR experiments. Here, the monochromatic pump pulse in conventional pulsed ELDOR experiments was replaced with a train of chirp pulses identical to the one used in the chirp-DNP experiment (Figure 10.10c). A comparison of chirp ELDOR spectra acquired with $\Delta \omega_{chirp} = 300$ MHz and regular pulsed ELDOR spectra relying on a monochromatic pump pulse is presented in Figure 10.10(d) for 2.5 mM of the monoradical 4-amino-TEMPO (4AT). The chirp ELDOR spectra (red and neon green traces) confirms that a complete saturation of the EPR spectrum over the full $\Delta \omega_{chirp} = 300$ MHz bandwidth (indicated by black double-sided arrows) is achieved. This exceeds the electron spin ω_1 of 0.4–0.5 MHz by a factor >600, and is in stark contrast to the narrow <10 MHz full width half maximum (FWHM) saturation achieved with cw-irradiation (magenta and olive traces, Figure 10.10d). Thus, chirp ELDOR confirms that chirp-train excitation allows for electron spins over the whole $\Delta \omega_{chirp}$ bandwidth to participate in DNP, where these electron spins were previously not utilized by cw-DNP.[105]

10.4 CONCLUSIONS

Versatile and agile dual DNP/EPR spectrometers are necessary to broaden the scope of DNP experiments for a broader range of applications, and to understand the underlying physics and mechanisms for DNP processes, especially when new sample formulations or odd results are to be explored. The instrument and experimental designs outlined here enables the reader to build a modular two ss-μw source design for an AWG-capable dual DNP/EPR instrument. The advantage of a modular design allows the instrument to be easily modified to accommodate exactly what is needed for the desired experiments or available budget, while also providing easy access for upgrading one module at a time or adding new modules. Although alternative μw sources were mentioned, the ss-μw source is the heart of a versatile dual capability instrument owing to its wide bandwidths (~10 GHz) and tunability that allows the user to access a range of g-factors found with atypical radicals or paramagnetic transition metals for DNP and EPR experiments. The AWG capability provides the agility necessary for precise μw pulse shaping and broadening

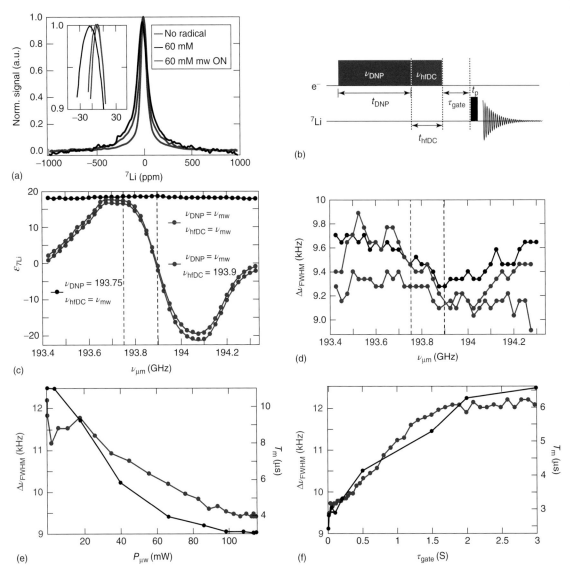

Figure 10.9. The normalized ^{7}Li NMR spectra for a sample of 1 M LiPF$_6$ dissolved in 1:1 ethylene carbonate: dimethyl carbonate with no radical (blue), with 60 mM TEMPO methacrylate with (red) and without (black) μw irradiation at 4 K and at 7 T (a). (b) The pulse sequence used for DNP experiments. (c) The DNP profile for $\nu_{DNP} = \nu_{hfDC}$ (red); $\nu_{DNP} \neq \nu_{hfDC}$, $\nu_{hfDC} = 193.9$ GHz (blue); and $\nu_{DNP} \neq \nu_{hfDC}$, $\nu_{DNP} = 193.75$ GHz (black), where $t_{DNP} + t_{hfDC} = 60$ s, and $t_{hfDC} = 0.5$ s, $\tau_{gate} = 0$ s. (d) ^{7}Li $\Delta\nu_{FWHM}$ for the same experimental conditions as (c). For both (c) and (d) the dashed lines indicate the position of maximum positive DNP enhancement (193.75 GHz) and the maximum hfDC (193.9 GHz). Electron T_m and ^{7}Li DNP-enhanced $\Delta\nu_{FWHM}$ as a function of μw power (e) and τ_{gate} (f). For (e) and (f), $t_{hfDC} = 0$ s, $\nu_{DNP} = 193.75$ GHz, and $t_{DNP} = 60$ s.[7] (Reprinted with permission from S.K. Jain, T.A. Siaw, A. Equbal, C.B. Wilson, I. Kaminker, S. Han, Reversal of Paramagnetic Effects by Electron Spin Saturation, J. Phys. Chem. C. 122 (2018) 5578–5589. doi:10.1021/acs.jpcc.8b00312. Copyright (2018) American Chemical Society)

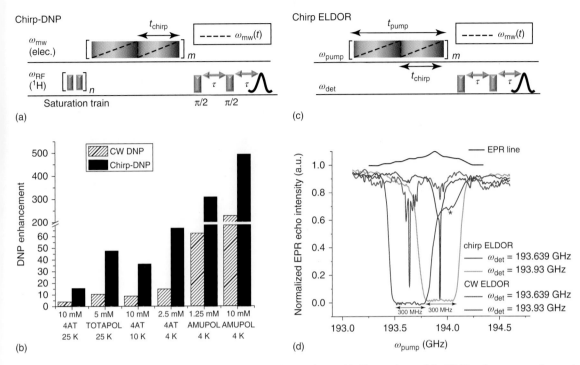

Figure 10.10. Pulse sequence schematic for the chirp-DNP experiment (a). Comparison of the NMR enhancement observed for cw-DNP and chirp-DNP under a range of conditions and with different radicals used as polarizing agents with a total μw irradiation of 60 s (b). Pulse sequence schematic for the chirp-ELDOR experiment (c). Comparison of the experimental ELDOR and chirp-ELDOR spectra measured on 2.5 mM 4AT at 4 K (d).[105] (Reprinted with permission from I. Kaminker, S. Han, Amplification of Dynamic Nuclear Polarization at 200 GHz by Arbitrary Pulse Shaping of the Electron Spin Saturation Profile., J. Phys. Chem. Lett. In Press (2018). doi:10.1021/acs.jpclett.8b01413. Copyright (2018) American Chemical Society)

the excitation bandwidth for DNP and EPR, while the addition of a second μw source increases the performance and eliminates artifacts for pump-probe type ELDOR experiments.

ACKNOWLEDGMENTS

This material is based upon work supported by the National Science Foundation (CHE 1505038, MCB #1617025, and MCB #1244651), the National Institute of Health (R21EB022731 and R21GM103477), and the Binational Science Foundation (Grant #2014149). The content of this work is solely the responsibility of the authors and does not necessarily represent the official views of the NSF, NIH, or BSF. IK acknowledges the support of the long-term postdoctoral fellowship by the Human Frontier Science Foundation.

RELATED ARTICLES IN EMAGRES

Spin Dynamics

EPR Spectroscopy of Nitroxide Spin Probes

Electron Paramagnetic Resonance Instrumentation

Very-high-frequency EPR

ELDOR-detected NMR

Shaped Pulses in EPR

Radiofrequency Coils for NMR: A Peripatetic History of Their Twists and Turns

High-Frequency Dynamic Nuclear Polarization

Dynamic Nuclear Polarization: Applications to Liquid-State NMR Spectroscopy

Shaped Pulses

Electron–Nuclear Hyperfine Interactions

Dynamic Nuclear Polarization and High-Resolution NMR of Solids

REFERENCES

1. A. Leavesley, D. Shimon, T. A. Siaw, A. Feintuch, D. Goldfarb, S. Vega, I. Kaminker, and S. Han, *Phys. Chem. Chem. Phys.*, 2017, **19**, 3596. DOI: 10.1039/C6CP06893F.

2. A. Equbal, Y. Li, A. Leavesley, S. Huang, A. Rajca, S. Rajca, and S. Han, *J. Phys. Chem. Lett.*, 2018, **9**, 2175. DOI: 10.1021/acs.jpclett.8b00751.

3. Y. Hovav, A. Feintuch, and S. Vega, *J. Chem. Phys.*, 2011, **134**, 74509. DOI: 10.1063/1.3526486.

4. Y. Hovav, I. Kaminker, D. Shimon, A. Feintuch, D. Goldfarb, and S. Vega, *Phys. Chem. Chem. Phys.*, 2015, **17**, 226. DOI: 10.1039/C4CP03825H.

5. I. Kaminker, D. Shimon, Y. Hovav, A. Feintuch, and S. Vega, *Phys. Chem. Chem. Phys.*, 2016, **18**, 11017. DOI: 10.1039/c5cp06689a.

6. A. Leavesley, C. B. Wilson, M. S. Sherwin, and S. Han, *Phys. Chem. Chem. Phys.*, 2018, **20**, 9897. DOI: 10.1039/c8cp00358k.

7. S. K. Jain, T. A. Siaw, A. Equbal, C. B. Wilson, I. Kaminker, and S. Han, *J. Phys. Chem. C.*, 2018, **122**, 5578. DOI: 10.1021/acs.jpcc.8b00312.

8. K. N. Hu, C. Song, H. H. Yu, T. M. Swager, and R. G. Griffin, *J. Chem. Phys.*, 2008, **128**, 52302. DOI: 10.1063/1.2816783.

9. H. Takahashi, C. Fernández-De-Alba, D. Lee, V. Maurel, S. Gambarelli, M. Bardet, S. Hediger, A. L. Barra, and G. De Paëpe, *J. Magn. Reson.*, 2014, **239**, 91. DOI: 10.1016/j.jmr.2013.12.005.

10. E. Ravera, D. Shimon, A. Feintuch, D. Goldfarb, S. Vega, A. Flori, C. Luchinat, L. Menichetti, and G. Parigi, *Phys. Chem. Chem. Phys.*, 2015, **17**, 26969. DOI: 10.1039/c5cp04138d.

11. A. A. Smith, B. Corzilius, O. Haze, T. M. Swager, and R. G. Griffin, *J. Chem. Phys.*, 2013, **139**, 214201. DOI: 10.1063/1.4832323.

12. A. Feintuch, D. Shimon, Y. Hovav, D. Banerjee, I. Kaminker, Y. Lipkin, K. Zibzener, B. Epel, S. Vega, and D. Goldfarb, *J. Magn. Reson.*, 2011, **209**, 136. DOI: 10.1016/j.jmr.2010.12.010.

13. P. A. S. Cruickshank, D. R. Bolton, D. A. Robertson, R. I. Hunter, R. J. Wylde, and G. M. Smith, *Rev. Sci. Instrum.*, 2009, **80**, 103012. DOI: 10.1063/1.3239402.

14. J. Leggett, R. Hunter, J. Granwehr, R. Panek, A. J. Perez-Linde, A. J. Horsewill, J. McMaster, G. Smith, and W. Kockenberger, *Phys. Chem. Chem. Phys.*, 2010, **12**, 5883. DOI: 10.1039/b926434e.

15. J. Granwehr, J. Leggett, and W. Köckenberger, *J. Magn. Reson.*, 2007, **187**, 266. DOI: 10.1016/j.jmr. 2007.05.011.

16. A. A. Smith, B. Corzilius, J. A. Bryant, R. Derocher, P. P. Woskov, R. J. Temkin, and R. G. Griffin, *J. Magn. Reson.*, 2012, **223**, 170. DOI: 10.1016/j.jmr.2012.07.008.

17. L. R. Becerra, G. J. Gerfen, B. F. Bellew, J. A. Bryant, D. A. Hall, S. J. Inati, R. T. Weber, S. Un, T. F. Prisner, A. E. Mcdermott, K. W. Fishbein, K. E. Kreischer, R. J. Temkin, D. J. Singel, and R. G. Griffin, *J. Magn. Reson. Ser. A.*, 1995, **117**, 28. DOI: 10.1006/jmra.1995.9975.

18. T. A. Siaw, A. Leavesley, A. Lund, I. Kaminker, and S. Han, *J. Magn. Reson.*, 2016, **264**, 131. DOI: 10.1016/j.jmr.2015.12.012.

19. B. D. Armstrong, D. T. Edwards, R. J. Wylde, S. A. Walker, and S. Han, *Phys. Chem. Chem. Phys.*, 2010, **12**, 5920. DOI: 10.1039/c002146f.

20. G. W. Morley, L. C. Brunel, and J. van Tol, *Rev. Sci. Instrum.*, 2008, **79**, 64703. DOI: 10.1063/1.2937630.

21. T. Dubroca, J. Mckay, X. Wang, and J. van Tol, *IEEE Int. Microw. Symp.*, 2017, 1400. DOI: 10.1109/MWSYM.2017.8058878.

22. F. H. Cho, V. Stepanov, C. Abeywardana, and S. Takahashi, 'Methods Enzymology', 1st edn, Elsevier Inc., 2015, p 95. DOI: 10.1016/bs.mie.2015.07.001

23. F. H. Cho, V. Stepanov, and S. Takahashi, *Rev. Sci. Instrum.*, 2014, **85**, 75110. DOI: 10.1063/1.4889873.

24. M. Bennati, C. T. Farrar, J. A. Bryant, S. J. Inati, V. Weis, G. J. Gerfen, P. Riggs-Gelasco, J. Stubbe, and R. G. Griffin, *J. Magn. Reson.*, 1999, **138**, 232. DOI: 10.1006/jmre.1999.1727.

25. M. M. Hertel, V. P. Denysenkov, M. Bennati, and T. F. Prisner, *Magn. Reson. Chem.*, 2005, **43**, 248. DOI: 10.1002/mrc.1681.

26. D. T. Edwards, S. Takahashi, M. S. Sherwin, and S. Han, *J. Magn. Reson.*, 2012, **223**, 198. DOI: 10.1016/j.jmr.2012.07.004.

27. D. T. Edwards, Z. Ma, T. J. Meade, D. Goldfarb, S. Han, and M. S. Sherwin, *Phys. Chem. Chem. Phys.*, 2013, **15**, 11313. DOI: 10.1039/c3cp43787f.

28. E. Reijerse, P. P. Schmidt, G. Klihm, and W. Lubitz, *Appl. Magn. Reson.*, 2007, **31**, 611. DOI: 10.1007/BF03166606.

29. K. A. Earle and J. H. Freed, *Appl. Magn. Reson.*, 1999, **16**, 247. DOI: 10.1007/BF03161937.

30. Bruker, ELEXSYS E780, 2018. https://www.bruker.com/products/mr/epr/elexsys/e780/overview.html (accessed 28 January 2018).

31. H. Blok, J. A. J. M. Disselhorst, S. B. Orlinskii, and J. Schmidt, *J. Magn. Reson.*, 2004, **166**, 92. DOI: 10.1016/j.jmr.2003.10.011.

32. H. Blok, J. A. J. M. Disselhorst, H. Van Der Meer, S. B. Orlinskii, and J. Schmidt, *J. Magn. Reson.*, 2005, **173**, 49. DOI: 10.1016/j.jmr.2004.11.019.

33. Y. A. Grishin, M. R. Fuchs, A. Schnegg, A. A. Dubinskii, B. S. Dumesh, F. S. Rusin, V. L. Bratman, and K. Möbius, *Rev. Sci. Instrum.*, 2004, **75**, 2926. DOI: 10.1063/1.1778071.

34. K. Möbius, A. Savitsky, A. Schnegg, M. Plato, and M. Fuchst, *Phys. Chem. Chem. Phys.*, 2005, **7**, 19. DOI: 10.1039/b412180e.

35. K. Thurber and R. Tycko, *J. Magn. Reson.*, 2016, **264**, 99. DOI: 10.1016/j.jmr.2016.01.011.

36. K. R. Thurber, A. Potapov, W. M. Yau, and R. Tycko, *J. Magn. Reson.*, 2013, **226**, 100. DOI: 10.1016/j.jmr.2012.11.009.

37. K. R. Thurber, W. M. Yau, and R. Tycko, *J. Magn. Reson.*, 2010, **204**, 303. DOI: 10.1016/j.jmr.2010.03.016.

38. A. Carroll and K. Zilm, Dual DNP/EPR MAS Spectrometer and Studies of P1 Center of Diamond, Private Communication, July 2017.

39. I. Kaminker, R. Barnes, and S. Han, *J. Magn. Reson.*, 2017, **279**, 81. DOI: 10.1016/j.jmr.2017.04.016.

40. H. Y. Chen, Y. Kim, P. Nath, and C. Hilty, *J. Magn. Reson.*, 2015, **255**, 100. DOI: 10.1016/j.jmr.2015.02.011.

41. K. Takeda, *J. Magn. Reson.*, 2008, **192**, 218. DOI: 10.1016/j.jmr.2008.02.019.

42. M. Lelli, S. R. Chaudhari, D. Gajan, G. Casano, A. J. Rossini, O. Ouari, P. Tordo, A. Lesage, and L. Emsley, *J. Am. Chem. Soc.*, 2015, **137**, 14558. DOI: 10.1021/jacs.5b08423.

43. Ü. Akbey, A. H. Linden, and H. Oschkinat, *Appl. Magn. Reson.*, 2012, **43**, 81. DOI: 10.1007/s00723-012-0357-2.

44. M.-A. Geiger, M. Orwick-Rydmark, K. Märker, W. T. Franks, D. Akhmetzyanov, D. Stöppler, M. Zinke, E. Specker, M. Nazaré, A. Diehl, B.-J. van Rossum, F. Aussenac, T. Prisner, Ü. Akbey, and H. Oschkinat, *Phys. Chem. Chem. Phys.*, 2016, **18**, 30696. DOI: 10.1039/C6CP06154K.

45. M. Rosay, L. Tometich, S. Pawsey, R. Bader, R. Schauwecker, M. Blank, P. M. Borchard, S. R. Cauffman, K. L. Felch, R. T. Weber, R. J. Temkin, R. G. Griffin, and W. E. Maas, *Phys. Chem. Chem. Phys.*, 2010, **12**, 5850. DOI: 10.1039/c003685b.

46. A. Lund, M. F. Hsieh, T. A. Siaw, and S. Han, *Phys. Chem. Chem. Phys.*, 2015, **17**, 25449. DOI: 10.1039/c5cp03396a.

47. D. Lee, E. Bouleau, P. Saint-Bonnet, S. Hediger, and G. De Paëpe, *J. Magn. Reson.*, 2016, **264**, 116. DOI: 10.1016/j.jmr.2015.12.010.

48. Y. Matsuki, T. Idehara, J. Fukazawa, and T. Fujiwara, *J. Magn. Reson.*, 2016, **264**, 107. DOI: 10.1016/j.jmr.2016.01.022.

49. B. E. Sturgeon and R. D. Britt, *Rev. Sci. Instrum.*, 1992, **63**, 2187. DOI: 10.1063/1.1143136.

50. C. E. Davoust, P. E. Doan, and B. M. Hoffman, *J. Magn. Reson. - Ser. A.*, 1996, **119**, 38. DOI: 10.1006/jmra.1996.0049.

51. Bruker Biospin, Welcome to Electron Paramagnetic Resonance (EPR), 2018. https://www.bruker.com/products/mr/epr.html (accessed 2 May 2018).

52. A. V. Astashkin, J. H. Enemark, and A. Raitsimring, *Concepts Magn. Reson. Part B Magn. Reson. Eng.*, 2006, **35**, 125. DOI: 10.1002/cmr.b.

53. C. and P. Industries, Extended Interaction Klystrons (EIKs), 2018. http://www.cpii.com/product.cfm/4/40/158 (accessed 28 January 2018).

54. A.E.I. Systems, Pulse TWT Amplifiers, 2018. http://www.applsys.com/pulse_twt.html (accessed 28 January 2018).

55. C. and P. Industries, Gyrotrons, 2018. http://www.cpii.com/product.cfm/1/18/30 (accessed 28 January 2018).

56. Viringia Diodes Inc, 2018. http://vadiodes.com/en/ (accessed 28 January 2018).

57. Elva-1, 2018. http://www.elva-1.com/ (accessed 28 January 2018).

58. A. I. Smirnov and A. Nezerov, High Powered W-Band Sources to Boost High Frequency Power Output, 2017.

59. T. Kaufmann, T. J. Keller, J. M. Franck, R. P. Barnes, S. J. Glaser, J. M. Martinis, and S. Han, *J. Magn. Reson.*, 2013, **235**, 95. DOI: 10.1016/j.jmr.2013.07.015.

60. P. F. Goldsmith, Quasioptical Systems: Gaussian Beam Quasioptical Propagation and Applications, IEEE Press: New York, 1998.

61. J. A. Murphy and C. O. Sullivan, in 'Terahertz Spectrosc. Imaging', eds K. E. Peiponen, A. Zeitler, and M. Kuwata-Gonokami, Springer: Berlin, 2013, p 29. DOI: 10.1007/978-3-642-29564-5

62. I. Ogawa, T. Idehara, M. L. Pereyaslavets, and W. Kasparek, *Int. J. Electron.*, 2000, **87**, 865. DOI: 10.1080/00207210050028 /97.

63. B. Marsden, V. Lim, B. Taber, and A. Zens, *J. Magn. Reson.*, 2016, **268**, 25. DOI: 10.1016/j.jmr.2016.03.007.

64. B. Taber and A. Zens, *J. Magn. Reson.*, 2015, **259**, 114. DOI: 10.1016/j.jmr.2015.07.011.

65. B. Epel, specman4epr, 2018. http://specman4epr.com (accessed 4 February 2018).

66. Y. Hovav, D. Shimon, I. Kaminker, A. Feintuch, D. Goldfarb, and S. Vega, *Phys. Chem. Chem. Phys.*, 2015, **17**, 6053. DOI: 10.1039/C4CP05625F.

67. A. Nalepa, K. Möbius, W. Lubitz, and A. Savitsky, *J. Magn. Reson.*, 2014, **242**, 203. DOI: 10.1016/j.jmr.2014.02.026.

68. A. Potapov, B. Epel, and D. Goldfarb, *J. Chem. Phys.*, 2008, **128**, 52320. DOI: 10.1063/1.2833584.

69. P. Schosseler, T. Wacker, and A. Schweiger, *Chem. Phys. Lett.*, 1994, **224**, 319. DOI: 10.1016/ 0009-2614(94)00548-6.

70. M. Florent, I. Kaminker, V. Nagarajan, and D. Goldfarb, *J. Magn. Reson.*, 2011, **210**, 192. DOI: 10.1016/j.jmr.2011.03.005.

71. N. Cox, W. Lubitz, and A. Savitsky, *Mol. Phys.*, 2013, **111**, 2788. DOI: 10.1080/00268976.2013.830783.

72. J. Baum, R. Tycko, and A. Pines, *Phys. Rev. A.*, 1985, **32**, 3435. DOI: 10.1103/PhysRevA.32.3435.

73. J. M. Bohlen, M. Rey, and G. Bodenhausen, *J. Magn. Reson.*, 1989, **84**, 191. DOI: 10.1016/ 0022-2364(89)90018-8.

74. D. Kunz, *Magn. Reson. Med.*, 1986, **3**, 377. DOI: 10.1002/mrm.1910030303.

75. A. Doll, S. Pribitzer, R. Tschaggelar, and G. Jeschke, *J. Magn. Reson.*, 2013, **230**, 27. DOI: 10.1016/j.jmr.2013.01.002.

76. A. Doll, M. Qi, S. Pribitzer, N. Wili, M. Yulikov, A. Godt, and G. Jeschke, *Phys. Chem. Chem. Phys.*, 2015, **17**, 7334. DOI: 10.1039/C4CP05893C.

77. P. E. Spindler, S. J. Glaser, T. E. Skinner, and T. F. Prisner, *Angew. Chemie - Int. Ed.*, 2013, **52**, 3425. DOI: 10.1002/anie.201207777.

78. P. E. Spindler, I. Waclawska, B. Endeward, J. Plackmeyer, C. Ziegler, and T. F. Prisner, *J. Phys. Chem. Lett.*, 2015, **6**, 4331. DOI: 10.1021/acs.jpclett. 5b01933.

79. P. Schöps, P. E. Spindler, A. Marko, and T. F. Prisner, *J. Magn. Reson.*, 2015, **250**, 55. DOI: 10.1016/j.jmr.2014.10.017.

80. M. Tseitlin, R. W. Quine, G. A. Rinard, S. S. Eaton, and G. R. Eaton, *J. Magn. Reson.*, 2011, **213**, 119. DOI: 10.1016/j.jmr.2011.09.024.

81. M. A. Smith, H. Hu, and A. J. Shaka, *J. Magn. Reson.*, 2001, **151**, 269. DOI: 10.1006/jmre.2001.2364.

82. T. E. Skinner, T. O. Reiss, B. Luy, N. Khaneja, and S. J. Glaser, *J. Magn. Reson.*, 2003, **163**, 8. DOI: 10.1016/S1090-7807(03)00153-8.

83. K. Kobzar, T. E. Skinner, N. Khaneja, S. J. Glaser, and B. Luy, *J. Magn. Reson.*, 2008, **194**, 58. DOI: 10.1016/j.jmr.2008.05.023.

84. J. Granwehr and W. Köckenberger, *Appl. Magn. Reson.*, 2008, **34**, 355. DOI: 10.1007/s00723-008-0133-5.

85. S. C. Serra, A. Rosso, and F. Tedoldi, *Phys. Chem. Chem. Phys.*, 2013, **15**, 8416. DOI: 10.1039/ C3CP44667K.

86. W. T. Wenckebach, *J. Magn. Reson.*, 2017, **277**, 68. DOI: 10.1016/j.jmr.2017.01.020.

87. S. Jannin, A. Comment, and J. J. van der Klink, *Appl. Magn. Reson.*, 2012, **43**, 59. DOI: 10.1007/ s00723-012-0363-4.

88. A. Abragam and W. G. Proctor, *Phys. Rev.*, 1958, **109**, 1441. DOI: 10.1103/PhysRev.109.1441.

89. Y. Hovav, O. Levinkron, A. Feintuch, and S. Vega, *Appl. Magn. Reson.*, 2012, **43**, 21. DOI: 10.1007/s00723-012-0359-0.

90. Y. Hovav, A. Feintuch, and S. Vega, *J. Magn. Reson.*, 2012, **214**, 29. DOI: 10.1016/j.jmr.2011.09.047.

91. D. Banerjee, D. Shimon, A. Feintuch, S. Vega, and D. Goldfarb, *J. Magn. Reson.*, 2013, **230**, 212. DOI: 10.1016/j.jmr.2013.02.010.

92. A. A. Smith, B. Corzilius, A. B. Barnes, T. Maly, and R. G. Griffin, *J. Chem. Phys.*, 2012, **136**, 15101. DOI: 10.1063/1.3670019.

93. C. T. Farrar, D. A. Hall, G. J. Gerfen, S. J. Inati, and R. G. Griffin, *J. Chem. Phys.*, 2001, **114**, 4922. DOI: 10.1063/1.1346640.

94. A. Karabanov, G. Kwiatkowski, and W. Köckenberger, *Mol. Phys.*, 2014, **112**, 1838. DOI: 10.1080/00268976.2014.884287.

95. T. A. Siaw, M. Fehr, A. Lund, A. Latimer, S. A. Walker, D. T. Edwards, and S. Han, *Phys. Chem. Chem. Phys.*, 2014, **16**, 18694. DOI: 10.1039/c4cp02013h.

96. J. Koehler and J. Meiler, *Prog. Nucl. Magn. Reson. Spectrosc.*, 2011, **59**, 360. DOI: 10.1016/j.pnmrs.2011.05.001.

97. T. O. Pennanen and J. Vaara, *Phys. Rev. Lett.*, 2008, **100**, 4. DOI: 10.1103/PhysRevLett.100.133002.

98. M. Ernst, H. Zimmermann, and B. H. Meier, *Chem. Phys. Lett.*, 2000, **317**, 581. DOI: 10.1016/S0009-2614(99)01423-2.

99. G. De Paëpe, B. Eléna, and L. Emsley, *J. Chem. Phys.*, 2004, **121**, 3165. DOI: 10.1063/1.1773155.

100. M. Leskes, R. S. Thakur, P. K. Madhu, N. D. Kurur, and S. Vega, *J. Chem. Phys.*, 2007, **127**, 24501. DOI: 10.1063/1.2746039.

101. A. Henstra, P. Dirksen, and J. Schmidt, *J. Magn. Reson.*, 1988, **77**, 389. DOI: 10.1016/0022-2364(88)90190-4.

102. A. Henstra, P. Dirksen, and W. T. Wenckebach, *Phys. Lett. A.*, 1988, **134**, 134. DOI: 10.1016/0375-9601(88)90950-4.

103. S. K. Jain, G. Mathies, and R. G. Griffin, *J. Chem. Phys.*, 2017, **147**, 164201. DOI: 10.1063/1.5000528.

104. T. V. Can, J. J. Walish, T. M. Swager, and R. G. Griffin, *J. Chem. Phys.*, 2015, **143**, 54201. DOI: 10.1063/1.4927087.

105. I. Kaminker and S. Han, *J. Phys. Chem. Lett.*, 2018, **9**, 3110. DOI: 10.1021/acs.jpclett.8b01413.

Chapter 11

Dissolution Dynamic Nuclear Polarization Methodology and Instrumentation

Dennis Kurzbach[1] and Sami Jannin[2]

[1] Laboratoire des biomolécules, LBM, Département de chimie, École normale supérieure, PSL University, Sorbonne Université, CNRS, Paris, France
[2] Université de Lyon, CNRS, Université Claude Bernard Lyon 1, ENS de Lyon, Institut des Sciences Analytiques, UMR 5280, Villeurbanne, France

11.1 INTRODUCTION

Dissolution dynamic nuclear polarization (d-DNP) aims at hyperpolarizing nuclear spins of samples in the solid state before dissolution and detection in the liquid state by means of a conventional magnetic resonance or imaging (NMR or MRI) machine. It consists in (i) performing DNP at cryogenic temperatures to improve the nuclear spin polarization and to establish a strong polarization, (ii) subsequently dissolving and rapidly transferring the sample into the liquid state, and (iii) performing magnetic resonance measurements at ambient temperatures. The method was first published by Ardenkjaer-Larsen et al. in 2003.[1] The dramatic overall signal enhancements that sometimes exceed 10 000-fold arise from both the DNP enhancement and the temperature jump that confers an extra polarization factor.

d-DNP renders a plethora of applications possible, such as real-time monitoring of in vitro and in vivo metabolomics,[2–6] chemical reaction monitoring,[7] investigation of nuclear spin dynamics,[8–11] hyperpolarized protein NMR,[12–15] or real-time imaging.[16,17] The very intense signals provided by d-DNP applications generally outdate the need for signal averaging opening new avenues for time-resolved studies on the millisecond to second time scale and render many materials accessible that could not be studied before due to a lack of NMR sensitivity.

The principle setup of a d-DNP experiment is displayed in Figure 11.1. One of the major challenges of d-DNP is that temperature jumps, dissolutions, transfers, and injections need to be performed in ways

Handbook of High Field Dynamic Nuclear Polarization.
Edited by Vladimir K. Michaelis, Robert G. Griffin, Björn Corzilius and Shimon Vega
© 2020 John Wiley & Sons, Ltd. ISBN: 978-1-119-44164-9
Also published in eMagRes (online edition)
DOI: 10.1002/9780470034590.emrstm1563

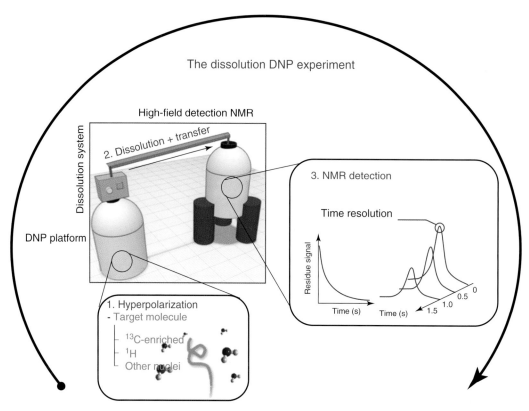

Figure 11.1. Workflow of a dissolution DNP experiment. 1. ^1H nuclei or other nuclear spins present in a sample are hyperpolarized together. 2. The hyperpolarized sample is rapidly dissolved and transferred to the detection NMR spectrometer. 3. After injection of the sample into the detection spectrometer an NMR spectrum will be detected, typically with a sub-Hertz repetition rate by using, e.g., small-angle detection pulses

that preserve most of the hyperpolarization. Therefore, very rapid procedures and highly specialized experimental setups are necessary.

Furthermore, sophisticated experimental protocols are required to achieve such a peculiar experiment. The purpose of this chapter is to review the general methodology behind d-DNP, and instrumentational aspects going from basic tasks, such as sample preparation, over operational aspects to future perspectives.

11.2 FROM SAMPLE PREPARATION AT ROOM TEMPERATURE TO HIGH POLARIZATION AT LOW TEMPERATURES

In this section, we focus on the experimental conditions for the first step of a d-DNP experiment, i.e.,

the hyperpolarization of a sample at low temperature by incident microwave irradiation that excites paramagnetic polarizing agents (PAs), which are mixed with and dipolarly coupled to the nuclear spins under study.

Before d-DNP can be performed with a molecule of interest, one needs to go through an optimized sample preparation step. Routinely, a solution of molecules is doped with a paramagnetic PA, i.e., a stable radical in the tens of millimolar concentration range. Different PAs are currently used and their nature and performances are discussed in Section 11.2.1.1. As for all DNP applications, PAs are necessary for the DNP process since one exploits the high electron spin polarization by transferring it to nuclear spins. However, in d-DNP applications the presence of PAs is also often disturbing NMR detection after dissolution rendering them ambivalent entities.

Once prepared, a sample is frozen within the DNP apparatus. One important feature is that DNP generally requires that samples form glasses upon freezing and do not crystallize. This ensures that the sample remains largely homogeneous as it is at its glass transition temperature. Details, examples, and counterexample are given in Sections 11.2.1.2 and 11.2.2.1.

After vitrification, the d-DNP sample is placed in the d-DNP instrument (often called a polarizer), which is primarily composed of a cryostat and a magnet to cool down the sample in a moderate magnetic field. The combination of temperature and magnetic field T^{DNP} and B_0^{DNP} needs to be carefully chosen to yield a maximum electron spin polarization of the PAs, even under microwave saturation, since it is this quantity, which is ultimately transferred to the surrounding nuclear spins to overcome their low polarization at thermal equilibrium. The experimental DNP conditions and corresponding optimization procedures are discussed in detail in Section 11.2.2.2.

The transfer of polarization from the highly polarized electron spins to the surrounding nuclear spins is mediated by microwave irradiation close to the electron spin resonance frequency as described in Section 11.2.2.3. Depending on the DNP conditions and PAs employed, different DNP effects can take place as described in Section 11.2.2.5. Recently, new DNP methods have been developed that significantly improve the DNP performances (both rates and absolute polarizations) for broad temperature and magnetic field regimes, namely *Microwave Modulation*, *Cross-Polarization*, as described in the corresponding sections.

11.2.1 Sample Preparation

In the following several aspects concerning the sample preparation for static low-temperature DNP prior to dissolution will be reviewed. Without aiming at covering all possible aspects, the section will focus on the radicals (denoted as polarization agents) necessary for DNP and the matrices that dissolve these molecules. We assume that sample preparations that consider these two factors are applicable to a large variety of target molecules that are to be hyperpolarized, however, it should be kept in mind that depending on the target under study optimal sample conditions might significantly vary (see, e.g., Refs 18–21).

11.2.1.1 Polarizing Agents

The largest share of d-DNP experiments aims at NMR observation of ^{1}H or ^{13}C nuclei, although other nuclei, such as ^{2}H, ^{15}N, or ^{31}P, have been reported to be feasible to d-DNP.[8,22–25] For ^{1}H and ^{13}C spins, currently two prominent classes of PAs are used for hyperpolarization in d-DNP applications. The first comprises tri-aryl methyl (TAM, also frequently called 'trityl') radicals that exhibit narrow electron paramagnetic resonance (EPR) lines and that are very efficient for hyperpolarization of ^{13}C nuclei, and the second class comprises nitroxide radicals that exhibit broad EPR lines and that are most often used for ^{1}H hyperpolarization. Their DNP efficiencies are particularly dependent on the strength of the external magnetic field and the experimental temperature.

The use of trityl radicals at typical concentrations of around 15–20 mM can be very efficient and regularly leads to polarization levels >50% for ^{13}C.[26,27] However, DNP build-up times can become very long (often the proton Larmor frequency is significantly larger than the trityl line width enabling only solid effect (SE) processes, which proceed slowly) such that these radicals are often not suitable for proton hyperpolarization, in contrast to nitroxides, which can lead to very fast hyperpolarization build-up processes, yet, inefficient for nuclei with low gyromagnetic ratios γ. The addition of other paramagnetic agents such as Gd^{3+} to trityl preparations can sometimes further boost the obtainable nuclear polarization.[28,29] Recently, 1,3-bisdiphenylene-2-phenylallyl (BDPA) as a narrow-band alternative to trityl has found first applications in d-DNP.[29] Chemical modifications of the basic BDPA molecule can yield comparable or even faster and more efficient polarization build-up than current state-of-the-art trityl-based PAs.

The second strategy employs nitroxide radicals at typical radical concentrations around 30–50 mM. Nitroxide radicals are less efficient than trityl radicals for direct ^{13}C polarization under similar conditions. However, they have the advantage that they are very efficient, with respect to build-up rates and polarization levels, for hyperpolarizing nuclear spins with high gyromagnetic ratios and relatively short relaxation times such as ^{1}H.

The use of cross-polarization (CP) techniques[30,31] as described in Section 11.2.2.6 and as commonly used in solid-state NMR can be combined with the use of nitroxides to indirectly hyperpolarize low-γ nuclei. Frequently, such procedures significantly boost the performances of ^{13}C DNP, often beyond the polarization that can be achieved with trityl radicals. The principle idea is that the high and rapidly building ^{1}H polarization can be efficiently transferred during the DNP process to the ^{13}C spins by radiofrequency (rf) pulses at the ^{1}H and ^{13}C resonances. Polarization levels >50% can be reached on a regular basis for ^{13}C. In principle, the use of bi-nitroxide radicals such as AMUPol or TOTAPol that are popular in low-temperature magic-angle spinning (LT-MAS)-DNP applications can further boost the DNP performance, e.g., if low PA concentrations are desired and (thermal mixing) TM-based DNP is not effective. However, bi-radicals have so far only found limited application in the realm of d-DNP as the feature only limited DNP efficiency in static solids.[32–34] On the contrary, tri-radicals might be advantageous in this case (see Ref. 35).

11.2.1.2 DNP Matrices

As DNP for dissolution applications is typically performed at very low temperatures, samples usually need to be vitrified for the hyperpolarization process, often by direct immersion of a sample holder in liquid helium (lHe). In this regard, several points need to be considered when aiming at the optimization of polarization levels:

1. The sample matrix, i.e., typically the solvent that dissolves the PA and the molecule under study, must not crystallize upon cooling. Crystallization has spurious effects on DNP performances, since it entails a separation of radical-rich phases and crystalline solvent phases causing inhomogeneities and variations in local radical concentrations that reduce the efficiency of DNP.

2. As spin diffusion is transporting the hyperpolarization from regions nearby PAs (within the diffusion barrier)[36] to other parts of a sample, the solvent matrix should contain a sufficiently high concentration of active nuclear spins to allow for effective diffusion of the polarization. In principle, ^{13}C spins in the case of direct carbon DNP or ^{1}H in the case of proton-based experiments.

   As the dipolar moment of protons is relatively large, spin diffusion is typically quite efficient for these nuclei even when counteracted by relaxation processes.

   In contrast, in the case of direct ^{13}C polarization, it has been found that when the concentration

of ^{13}C-enriched molecules is too low to ensure efficient spin diffusion, DNP becomes inefficient. One way to tackle this problem is to add more ^{13}C-enriched species to the mixture to compensate for this frequently encountered problem.[37]

Another solution is to assist carbon-based spin diffusion through the addition of a small concentration of protons, which can substantially increase the rate of diffusion by facilitating flip–flop processes and thus also assist DNP (so-called proton-driven ^{13}C spin diffusion, for details, see Refs 37, 38). However, too high proton concentrations can also hamper DNP, such that partially (80–90%) deuterated mixtures are usually preferred for efficient DNP.

3. Empirically it was found that maximum polarization is often reached with partially deuterated water–glycerol mixtures with a particular composition of H_2O:D_2O:glycerol-d_8/1:3:6, which is often coined 'DNP-juice'. This mixture (and slight variations) is typically also used in d-DNP applications.[32] This sample matrix prevents on the one hand crystallization through the presence of glycerol as cryo-protectant and on the other hand contains sufficient protons to efficiently assist ^{13}C spin diffusion without causing additional relaxation processes. Note that 'DNP-juice' was originally optimized for MAS-DNP and is not necessarily performant for all applications of static DNP. The advantage of DNP-juice is yet, applicability to a large variety of samples. Other sample preparations that aim at increasing ^{13}C concentrations can, e.g., be mixtures of urea and glycerol[1] or pyruvic acid without further additions, except PAs.[39]

4. In principle, every (partially deuterated) solvent that dissolves the desired PAs and target molecules can be employed as DNP matrix. However, when choosing the appropriate matrix for d-DNP, one should keep in mind that the sample will ultimately be dissolved and transferred to the liquid-state. (This is discussed in more detail in Section 11.3.1.) The hydrophobicity of the solvent used for dissolution should be carefully considered. For example, the largest share of dissolution experiments is performed by use of D_2O as dissolving liquid, which may entail a phase separation if the solvent used to produce the glassy DNP matrix is too hydrophobic. Likewise, if hydrophobic solvents are used as dissolution liquid, organic solvents as DNP matrix

(e.g., partially deuterated ethanol or DMSO) can be beneficial for d-DNP experiments as these are miscible with a large share of solvents and do not entail a separation of phases upon mixing with the dissolution liquid. Interestingly though, phase-separation upon dissolution has even been put to use by Harris *et al.*[40] to scavenge radicals that dissolve in a different phase than the hyperpolarized substrate of interest. This aspect of removal of PAs from the DNP mixture will be treated in more detail in Section 11.3.2.

11.2.2 Experimental DNP

In this section, we will review – without aiming at completeness – experimental aspects and recent developments that need to be considered when targeting DNP efficiency (prior to dissolution of the sample) with respect to polarization times as well as achievable steady-state polarization levels. The discussed features are of general nature and applicable to a broad variety of systems.

11.2.2.1 Sample Vitrification

After sample preparation at room temperature, the next step in a d-DNP experiment is typically sample vitrification. As mentioned above, crystallization of the sample that entails a spatial separation of PAs from crystalline parts of the sample should be avoided for efficient DNP. The simplest technique to rapidly vitrify a sample and to achieve a glassy non-crystalline state is plunging it into lHe, which is typically present in the cryostat of a low-temperature DNP apparatus. Additionally, several other techniques have been developed to optimize the speed of vitrification in cases where plunging in lHe is not feasible. For example, isopentane (at ca 173 K) super-cooled in liquid nitrogen (lN$_2$) can rapidly vitrify a sample as the cold liquid does not evaporate, like lHe or lN$_2$, when in contact with the warm sample body. Alternatively, small droplets of ca 10–50 μl can be dripped into lN$_2$ for vitrification and subsequently transferred to the polarizer, which allows for visual confirmation of the transparency (as an indication of successful prevention of crystallization) of the sample before DNP.

Interestingly, it was found that the most often (*vide supra*) used water–glycerol mixtures can undergo ripening processes prior to vitrification, which

might either boost DNP or reduce its efficiency depending on the concentrations and nature of the radicals used.[41] After preparation of the sample, water–glycerol mixtures undergo nano-phase separations (NPS) that entail the transient formation of water-rich and water-depleted phases, which can be trapped if the sample is rapidly vitrified during NPS. Thus, despite a homogeneous, glassy macroscopic appearance, water–glycerol mixtures are often nano-heterogeneous. Most importantly, the different nanophases have different affinities for different radicals. For example, for the TEMPOL radical – as frequently used in combination with water–glycerol matrices – significant variations in DNP performance were observed for different ripening periods T_{ripe} between sample preparation and vitrification. Exemplarily, it was observed that for the most-often used 50 mM TEMPOL in H$_2$O:D$_2$O:glycerol-d_8/1:4:5 mixtures that an optimum in DNP was reached for a delay of 45 min between sample preparation and vitrification (a plus of 20%, see Figure 11.2). For other radicals, comparable behavior has been reported, e.g., for 25 mM AMUPol, variations of 17% and for 15 mM Trityl even 40% have been observed, depending on the ripening delay T_{ripe}. Similar behavior has been found for other water/glycerol mixtures. Details can be found in Ref. 41.

Such behavior can be of importance in many fields of research, such as dissolution DNP of biomolecules,[4] drug screening,[42] in vivo imaging, and cancer monitoring,[30] since many of these studies employ water–glycerol mixtures.

11.2.2.2 Temperature and Magnetic Field

The polarization, P, of an ensemble of electron (or nuclear) spins $1/2$ with gyromagnetic ratios γ_e (or γ_n) precessing at a temperature T^{DNP} in the magnetic field of the polarizer B_0^{DNP} is given by

$$P = \tanh\left(\frac{\hbar\gamma B_0^{DNP}}{2k_B T^{DNP}}\right) \qquad (11.1)$$

with $\hbar$ and k_B the Planck and Boltzmann constants, respectively.

Therefore, to reach high electron spin polarizations, samples (once vitrified) typically need to be cooled to very low temperatures, often below 4.2 K. This is achieved by immersing the sample in a lHe bath and by subsequently applying a vacuum evaporating the cryogenic fluid resulting in a cool down of the lHe

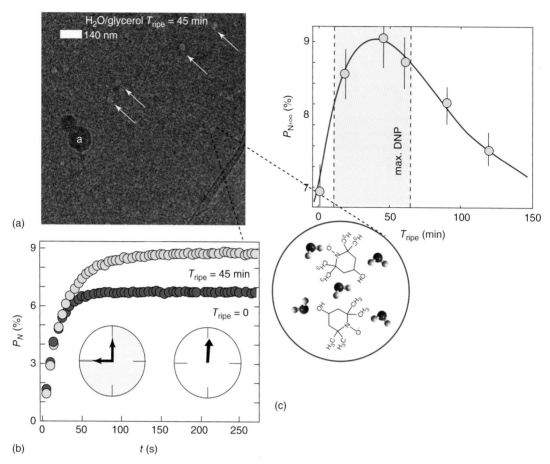

Figure 11.2. (a) NPS observed via cryo-transmission electron microscopy (TEM) in water–glycerol mixtures 45 min after sample preparation. (b) Build-up curves of proton polarization (P_N) induced by microwave saturation (187.9 GHz with a 2 kHz frequency modulation over a bandwidth of 100 MHz, 50 mM TEMPOL) of the EPR transitions of TEMPOL at 4 K and 6.7 T. Yellow: Sample vitrified immediately after preparation ($T_{ripe} < 1$ min). Blue: Sample vitrified after a ripening interval $T_{ripe} = 45$ min after sample preparation. (c) Steady-state proton polarization ($P_{N,\infty}$) at 4 K as a function of the interval T_{ripe} between preparation and vitrification of the 'DNP-juice'. The solid line serves to guide for the eye

bath below its superfluid transition, to temperatures between 0.8 and 1.5 K. As lHe is expensive, reduced consumption rates and even cryogen-free systems are becoming economically desirable.

Temperatures down to ca 1.5 K can be reached on a regular basis in lHe bath cryostats. At such temperatures, external magnetic fields over 3 T are typically required to achieve high electron spin polarizations. In practice, state-of-the-art d-DNP polarizers typically operate at 3.3, 6.7 T, or 9.4 T.[27,43,44] While theoretically higher nuclear polarization levels are often expected at higher fields, it is practically not always straightforward to correlate maximum

hyperpolarization levels with magnetic field strength. Transitions between different DNP mechanisms (as discussed in Section 11.2.2.5) in dependence of the magnetic field, temperature, nature of the radical, and sample composition can cause deviations from the expected monotonic field-dependence.

11.2.2.3 Microwave Irradiation

To perform DNP, the samples must be irradiated with microwaves that are slightly off-resonance with respect to the employed PAs' central transition frequency. The optimal frequency and power necessary to achieve optimal DNP depends strongly on the sample and apparatus used. Therefore, it is desirable to employ a microwave source that allows for variation of the frequency in order to adapt the experimental setup to different PAs. The magnetic fields used in d-DNP translate into microwave frequencies between W-band (94 GHz, 3.3 T, e.g., the HyperSense system), 2×W-band (188 GHz, 6.7 T, e.g., the Bruker prototype system) to 263 GHz (9.7 T, e.g., Cryogenics systems), for which suitable microwave sources are commercially available, which deliver sufficient microwave power. As the source is typically an external device an appropriate waveguide is necessary to apply the microwave to the sample situated inside the cryostat within the superconducting magnet. The output power needed to efficiently drive DNP in typical d-DNP conditions (1–4 K, 3.3–9.4 T) does rarely exceed 100 mW.

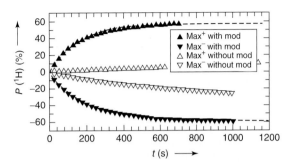

Figure 11.3. ^{1}H DNP build-up curves (positive and negative) measured at $T = 1.2$ K and $B_0 = 6.7$ T, with and without frequency modulation (sample: 10 : 40 : 50 (v/v/v) H_2O:D_2O:glycerol-d_8 mixture with 25 mM TEMPOL). The optimal frequencies $f_{\mu w} = 187.85$ and 188.3 GHz were set for positive or negative DNP, respectively, with a microwave power $P_{\mu w} = 87.5$ mW. An amplitude of $\Delta f_{\mu w} = 100$ MHz was used for frequency modulation. (Reproduced with permission from Ref. 49. © Elsevier, 2014)

11.2.2.4 *Microwave Frequency Modulation*

To achieve maximum DNP at low temperatures with static samples as used in d-DNP applications, it is desirable to saturate only a part of the EPR transition of the used PAs. Too narrow or too extensive excitation may lead to low performances.[45] An important aspect is the optimization of the concentration of PAs in order to achieve appropriate rates of electronic spectral diffusion (eSD). eSD distributes saturation from irradiated spin packets in heterogeneously broadened EPR lines to other spins via dipolar coupling between the unpaired electrons.[36] If the electron concentration is too high and eSD too fast, the bandwidth of saturation induced by the microwave becomes too wide. If the electron concentration is too low, the eSD efficiency is low and the saturation bandwidth narrow. Hence, one can indirectly optimize the polarization of the electrons via the PA concentration, which dictates how the microwave saturation propagates across the EPR spectrum.

Another more practical way to 'tune' the saturation bandwidth is to apply a microwave frequency modulation over a well-controlled range that precisely defines the range of excited frequencies, a technique discovered already in 1996,[46] followed by works by Tycko and co-workers[47] and Goldfarb and co-workers[48] but which recently has been extended to DNP at 1.2 K.[49] To this end, the microwave irradiation, centered around a frequency $f_{\mu w}$, is modulated over a range $\Delta f_{\mu w}$ that typically spans 50–100 MHz for nitroxide radicals. The modulation frequency

typically needs to be of at least $f_{\text{mod}} > 1$ kHz to span a sufficiently large frequency range and to compete with electron spin-lattice relaxation, which is normally quite fast (the order of magnitude is typically $T_{1e} \sim 0.1–1$ s under d-DNP conditions).[49] Modulation of the microwave in frequency causes a wider range of EPR transitions centered around the microwave carrier frequency to be excited leading to faster and higher DNP, which in return allows one to decrease the PA concentration while still achieving substantial polarization enhancements (see Figure 11.3).

11.2.2.5 *DNP Effects*

The principle DNP mechanisms in dissolution applications are not different from those encountered in other DNP uses, i.e., the solid effect (SE), the cross effect (CE), thermal mixing (TM), and Overhauser effect (OE)[50] are the principle mechanisms identified so far. In practice, one can often not discern a single dominant mechanism at temperatures between 1.2 and 4 K and magnetic fields between 3.3 and 9.4 T. Sometimes, a superposition of mechanisms is observed with relative contributions that depend on a multitude of factors such as PA type, concentration, solvent, type of target nucleus, and of course temperature and magnetic field.

For the narrow-band PAs trityl and BDPA, the well-resolved SE as established by Abragam and Goldman[51] in 1978 dominates ^{1}H hyperpolarization as the nuclear Larmor frequency is much larger than the EPR linewidth. Counterintuitively, it was recently found that the BDPA radical can additionally lead to contributions of OE DNP and the SE in a single sample even at temperatures as low as 1.2 K, although OE requires the presence of efficient cross-relaxation pathways.[50]

TM is often the major mechanism for nitroxides[52] and even for narrow-band radicals such as trityls TM can dominate hyperpolarization of nuclei with low gyromagnetic ratios.[53] However, it has been shown recently that at lower concentrations of nitroxide radicals (<30 mM), contributions of SE, CE, and TM cannot easily be disentangled.[24] Han, Vega, and co-workers[34,48,54–62] found that at higher temperatures (>4 K) and at intermediate (10–40 mM) nitroxide radical concentrations mixtures of differential SE and CE can account for some important experimental features.

A crucial aspect concerning the contribution of a particular mechanism to DNP is eSD. The eSD rate depends on a multitude of factors, including temperature, the total width of the EPR line, and the homogeneous width (directly related to the concentration of the PAs) and it is not yet completely clear how to theoretically describe eSD.[63] In cases of heterogeneous line broadening (as typical for nitroxides), eSD is necessary to spread the microwave saturation from the irradiation frequency across a broad portion of the EPR spectrum. Consider, for example, an EPR line whose spectral width is larger than the nuclear Larmor frequency at a given magnetic field. Only eSD that establishes a flow of magnetization over a bandwidth larger than the nuclear Larmor frequency can enable efficient TM as the latter prerequisites triple spin flips embracing two electrons with a frequency difference corresponding to the nuclear Larmor frequency.[36] Additionally, eSD needs to be fast, i.e., effective on time scales shorter than the electronic T_1 to facilitate TM. When eSD becomes ineffective, CE and SE are likely to become dominant over TM. Therefore, the PA concentration often needs to be carefully optimized to achieve maximum efficiency of a particular mechanism.

In practice, for highly concentrated (>50 mM) nitroxide radicals at 6.7 T, one can observe a transition from a fast and complete eSD regime over the whole EPR line (the common approximation for theoretical treatment of TM[64]), to a slow eSD regime (leading to a dominance of CE and SE), if the temperature is lowered from 4.2 to 1.2 K (see, for example, Ref. 65).

11.2.2.6 *Cross-Polarization*

Using commercially available and reasonably priced nitroxide radicals and their derivatives as PAs, quite high polarization levels can rapidly be reached for ^{1}H spins. Currently, the highest polarization levels are obtained at a field of 6.7 T and temperatures around 1.2 K reaching >90%[66] (e.g., with 40 mM TEMPOL in water–glycerol mixtures containing ^{13}C-enriched molecules[24]). Yet, the direct polarization of heteronuclei such as ^{13}C is less efficient using this type of PA. Among many, one reason is that pure TM as DNP mechanism leads to a single spin temperature common among all nuclear spin species present in a sample. Therefore, the ^{13}C polarization typically is restricted to a quarter (corresponding to $\gamma(^{13}\mathrm{C})/\gamma(^1\mathrm{H})$ in the high-temperature approximation) of the ^{1}H polarization as has been shown by Goldman, de Boer, and others.[67,68]

A promising compromise to boost the ^{13}C polarization with nitroxide radicals as PAs is CP.[69–71] In a CP experiment, we take advantage of the high and fast-building ^{1}H polarization, and transfer it with rf pulses to the ^{13}C spins. This transfer is performed with a dedicated NMR probe located in the polarizer, and only takes a few milliseconds. The probe must be designed such that it allows on the one hand for double resonance NMR experiments and on the other hand dissolution after CP-based hyperpolarization.

Unfortunately, the $^1\mathrm{H} \rightarrow {}^{13}\mathrm{C}$ polarization transfer via CP is often limited by the excitation bandwidths of the employed rf pulses and not fully efficient in a static solid at temperatures close to 1 K. Thus, it is frequently necessary to repeat the CP sequence a few times to yield an optimal ^{13}C polarization. Recently it was shown that Hartmann–Hahn type CP is performing better in cases where the electron polarization is in thermal equilibrium (close to unity at 1.2 K). Hence, CP is more efficient when the microwave irradiation is interrupted (so-called gated microwave irradiation) ca 0.5 s before the Hartmann–Hahn CP radio-pulse sequence to let the electron spins fully relax to their thermal equilibrium. Through this, better CP performances can be achieved, especially at temperatures below 4.2 K (see Figure 11.4a for the gated pulse sequence).[72] The underlying reason is a reduced proton relaxation rate in the rotating frame $R_{1,\rho}$ during the Hartmann–Hahn spin-lock, which increases the reservoir of polarization that can be transferred. Figure 11.4(b) shows a prototypical ^{13}C CP-DNP build-up in the case of CP with continuous

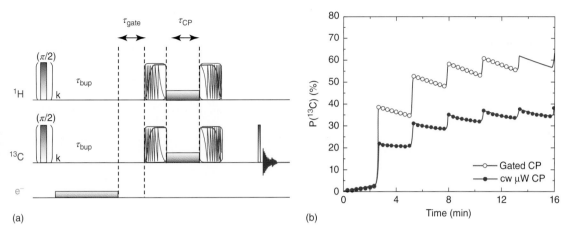

(a)   (b)

Figure 11.4. (a) Pulse sequence for multiple-contact cross-polarization enhanced by DNP. The adiabatic half passage pulses sweep over 100 kHz in 175 µs. The rf amplitude for ^{13}C is ramped from 50% to 100% of the average rf amplitude, which was $\gamma B_1(^1H)/(2\pi) = 20$ kHz on both channels. Microwave irradiation is gated off during the intervals $\tau_{gate} = 500$ ms and τ_{CP}. (b) Build-up of polarization $P(^{13}C)$ during a multiple CP pulse sequence applied every 2.5 min with continuous (red) or gated (blue) microwave irradiation ($P_{\mu w} = 87.5$ mW, $f_{\mu w} = 188.3$ GHz, and frequency modulation with an amplitude $\Delta f_{\mu w} = 50$ MHz and a frequency $f_{mod} = 10$ kHz) in 3 M sodium [1-^{13}C]acetate with 40 mM TEMPOL at 1.2 K and 6.7 T. (Reproduced from Ref. 72 with permission from the PCCP Owner Societies)

microwave irradiation (red) in comparison to the case of microwave-gated CP (blue). The ^{13}C polarization initially builds-up by direct DNP (0–2.6 min), in several subsequent iterations (obvious through steep increases in $P(^{13}C)$ in Figure 11.4b) the application of a CP pulse sequence transfers part of the ^{1}H polarization to the ^{13}C nuclei. Many details on the instrumentation and methods can be found in Ref. 66.

The iterative multiple-contact (MC) CP approach features quite short build-up times as compared to conventional DNP applications (a few minutes versus hours), which renders it an appealing technique for application of the dissolution DNP methodology that can systematically lead to a boost in obtainable polarization levels as compared to direct DNP techniques.[69,73] In the example of Figure 11.4, >60% polarization for ^{13}C in acetate is obtained in ca 15 min. An analysis of the parameters influencing MC-CP efficiency can be found in Refs 74, 75. With respect to static DNP at low temperatures, the long ^{1}H build-up times and broad NMR lines that often exceed the excitation pulse width of the applied pulses render the MC-CP approach more efficient compared to single-contact techniques.

Note that the indirect CP-DNP approach can readily be combined with the use of partially deuterated solvent mixtures as discussed above in Section 11.2.1.2. One of the most popular solvent mixtures is 90%

deuterated water–glycerol as it allows for dissolution of a plethora of hydrophilic target molecules and tends to form at the same time a glassy matrix upon freezing. The 10% protons present in the water–glycerol matrix are used as polarization source, while the target of interest is dispersed in the matrix and ^{13}C hyperpolarized via CP.[2,4,22,73,76]

11.3 FROM POLARIZED SOLIDS TO THE LIQUID-STATE NMR OR MRI MEASUREMENTS

In this section, we focus on the experimental conditions for the second step of a d-DNP experiment, i.e., the dissolution and transfer of the hyperpolarized sample. Once a high nuclear spin polarization is obtained, the frozen d-DNP sample needs to be dissolved to ambient temperatures and simultaneously transferred to the liquid-state. As discussed in Section 11.3.1, this process needs to be performed rapidly in a high-field and low-temperature environment within the polarizer. Optionally, the PAs can be removed from the hyperpolarized solution by different means as described in Section 11.3.2. In Section 11.3.3, we discuss strategies to rapidly transfer the hyperpolarized solution to the detection NMR spectrometer, with the possible use of a *magnetic tunnel* to prevent exacerbated hyperpolarization losses as described in the corresponding section.

11.3.1 Dissolution

For a dissolution DNP experiment, the sample needs to be dissolved after hyperpolarization at low temperatures. This is typically achieved using pressurized, superheated solvents. For hydrophilic samples, heavy water (typically at 10–11 bar and 150–180 °C) is often used. For more hydrophobic samples, methanol and toluene are popular solvents that allow for efficient dissolution. Typically, a burst of 2–5 mL of the superheated dissolution liquid is squirted onto the frozen sample dissolving it, diluting it, and at the same time propelling it towards the transfer system.

The dissolution procedure normally takes place in the polarizer at lHe temperatures. This is an important feature of d-DNP. Indeed, if the sample is removed from the polarizer in the solid state (without a phase transition to the liquid state), and transported in a room temperature and/or low magnetic-field environment, one observes a quasi-instantaneous loss of the polarization by fast nuclear spin relaxation (in part due to the presence of paramagnetic PAs). This can be overcome, in some particular cases, by a very fast pneumatic sample transfer[77] or by heterogeneous sample formulations as described in Section 11.5.1. Yet, as the largest share of current experiments employs dissolution within the polarizer, we want to focus on the fact that the dissolution needs to be fast to prevent excessive nuclear spin relaxation during the warm-up phase, and more practically to avoid freezing of the dissolution liquid. Dilution is also an important factor because paramagnetic PAs used for DNP tend to erode hyperpolarization in the liquid state, especially at low magnetic fields.[78] This effect (see Ref. 79 for a brief theoretical treatment) is to a first approximation linearly dependent on the PAs' concentration for the small molecules typically used in d-DNP experiments, therefore dilution can be highly beneficial.

To achieve a rapid dissolution, many systems make use of a 'dissolution stick system'. The dissolution stick is inserted into the magnet (after pressurization) bringing an outlet for the superheated dissolution liquid close to the hyperpolarized sample. Through this the superheated liquid is guided towards the vitrified sample to dissolve it. At the same time, the dissolution stick allows guiding the dissolved sample

back out of the magnet ready to be transferred to the detection NMR system. It must be considered that the hyperpolarized sample, which normally amounts to ca 50–500 μL is significantly diluted during the dissolution process. A minimization of the volume of dissolution liquid could, hence, be desirable for many applications to augment the concentration of the target molecule, given that paramagnetic relaxation does not become too efficient.

11.3.2 PAs Removal

Even after dilution, the presence of paramagnetic PAs after dissolution often remains a problem of d-DNP, which can cause paramagnetic relaxation enhancements (PRE) leading to a fast loss of hyperpolarization. Additionally, a PA's presence can be disadvantageous for *in vivo* applications because of its toxicity or because of interactions with the molecules under study. It is therefore desirable to remove the paramagnetic substances during the dissolution process. To this end, several methods have been published to date.

1. Nitroxide-based PAs can be reduced into a diamagnetic form using reactive substances, such as ascorbic acid.[80] The method includes the production of two different types of vitrified samples, one (sample 1) housing the molecule of interest and the radicals for hyperpolarization, the other (sample 2) houses the reduction agent. During DNP, only sample 1 is hyperpolarized, while sample 2 remains inert. Yet, both samples are dissolved simultaneously and rapidly mixed leading to a reduction of the radicals (e.g., $R_2NO\cdot$ to R_2NOH) suppressing PRE effects. However, this method can only be applied in combination with target molecules that are not reacting with the reduction agent.

2. A special type of trityl-based PA (the GE Healthcare product OX063) can be removed via filtration by rapid column chromatography after precipitation by a pH jump. This method has been used for d-DNP-based imaging for prostate cancer monitoring.[16] As almost complete PA removal can be achieved, hyperpolarized liquids can thus be produced that contain practically no paramagnetic impurities and are as such feasible for injection into a living organism, even into humans.

3. Alternatively, PAs can be generated optically with high-power UV irradiation in some specific sample formulations and used for DNP at

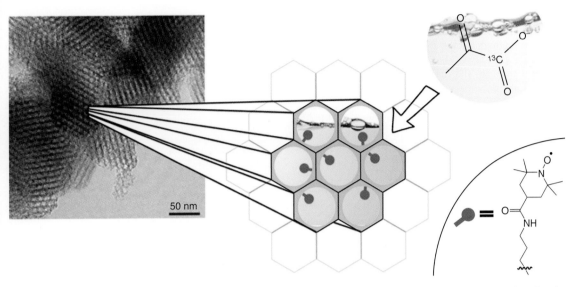

Figure 11.5. HYPSO 1.0 impregnated with a solution of the analyte to be polarized without addition of any glass-forming agents. The red dots schematically represent the PAs. (Reproduced with permission from Gajan, D.; Bornet, A.; Vuichoud, B.; Milani, J.; Melzi, R.; van Kalkeren, H. A.; Veyre, L.; Thieuleux, C.; Conley, M. P.; Grüning, W. R.; Schwarzwälder, M.; Lesage, A.; Copéret, C.; Bodenhausen, G.; Emsley, L.; S. Jannin, Proc. Natl. Acad. Sci. 2014, 111, 14693–14697, Copyright (2014) National Academy of Sciences)

low-temperature. The UV irradiation causes the formation of radical pairs by splitting chemical bonds, which can be used as PAs and are, under some circumstances, not stable at room temperature and are therefore instantaneously eliminated through recombination during the dissolution step.[81]

4. Finally, another recent approach to solve the problem of PA contamination is the use of spin-labeled thermoresponsive polymers[82] that precipitate upon dissolution or silica-based microporous materials with covalently attached radical that can be removed during the sample transfer. The advantage of these PAs is that they can be filtered in-line after dissolution without the need for further steps such as pH jumps of mixing with secondary products.[83,84]

For example, terminally nitroxide-labeled poly-N-isopropyl acryl amide (pNiPAam) can be dissolved in water–glycerol mixtures as used for DNP at low temperatures. Upon dissolution with superheated D_2O at 180–200 °C, pNiPAam will precipitate as it features a lowest critical solution temperature (LCST) between ca 30 and 60 °C depending on the solution pH, molecular

weight, and polymer concentration.[85] The formed precipitates are typically of macroscopic size allowing for complete filtration.

Another example are mesoporous silica matrices that are often used for heterogeneous catalysis. These structures can be surface-functionalized with nitroxide radicals and used for DNP after impregnation with a solution that contains the molecules of interest. Since the PAs are covalently attached to the meso-structures, simple in-line filtration after dissolution allows one to produce hyperpolarized solution free of any contaminants. These new polarizing matrices (see Figure 11.5) were coined hybrid polarizing solids (HYPSOs). Either nitroxides or trityl PAs can be used to coat the mesoporous materials.[83,86,87]

11.3.3 Sample Transfer

After the sample has been hyperpolarized and dissolved (and optionally the PA removed), it must be transferred to a detection magnet and therefore leave the magnetic field of the polarizing apparatus. The transfer must proceed rapidly as spin relaxation will cause the loss of hyperpolarization during the transfer delay. This is typically achieved by pressing the liquid bolus through a tube of ca 1 mm diameter with gaseous helium at 6–10 bar directly into an NMR tube already waiting in the detection NMR spectrometer. Several high-pressure systems that can operate at pressures up to 50 bar have been devised to increase the speed of sample transfer, e.g., by Bowen and Hilty[88]. Additionally, the sample can rest in a phase separation compartment for a short time (<1 s) before being injected into the NMR tube to allow for degassing.

Yet, one problem cannot be solved by improved speed; often the sample makes a passage through a magnetic field of varying strength, possibly even through a zero-field zone, before entering the detection magnet. This peculiarity of d-DNP has several consequences with respect to nuclear spin relaxation after dissolution. While contributions from chemical shift anisotropy (CSA)-based relaxation are dampened when the sample passes through a low-field zone, contributions due to dipolar relaxation become typically stronger. For nuclei, whose relaxation behavior is dominated by the latter mechanism, such as ^{1}H nuclei, zero-field passages need thus to be circumvented. In general, the dependence of the relaxation behavior in varying fields on the presence of PAs is quite complicated such that precise predictions are challenging, yet for protons in small molecules a tendency of increased PRE at low-field can be anticipated (see Figure 11.6).

11.3.4 Magnetic Tunnel

An approved strategy to reduce relaxation due to zero-field passages features the use of a magnetic tunnel between DNP apparatus and the detection spectrometer in which a static magnetic field is maintained by a Halbach array of permanent magnets (see Figure 11.7).[89] From theoretical considerations and practical feasibility, a field of 0.9 T turned out to be optimally suited to prevent strong paramagnetic relaxation. Such, even quite long distances up to 5–10 m can be bridged without critical loss of polarization due to low-field passage. For high-γ nuclei, such as protons, e.g., in water, the use of a magnetic tunnel can significantly prolong relaxation times rendering access to materials possible that could not be studied before.

For low-γ nuclei one must differentiate different cases. The impact of a magnetic tunnel is only very

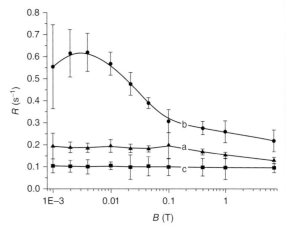

Figure 11.6. Proton longitudinal ('spin-lattice') relaxation rates of bromothiophene carboxylate (BTC) determined with a home-built relaxometer[78] as a function of the static field B_0 expressed on a logarithmic scale. (a) Triangles for sample 1 containing 50 mM BTC in D_2O with naturally dissolved O_2; (b) dots for sample 2 after addition of 0.25 mM TEM-POL; and (c) squares for sample 3 after addition of 30 mM sodium ascorbate to scavenge radicals and paramagnetic oxygen

small if CSA contributions to relaxation are negligible, e.g., for ^{13}C. In contrast, for cases of fast CSA-driven relaxation, such as ^{31}P,[90] the optimal field during transfer needs to equilibrated between suppression of CSA relaxation and impact of dipolar contributions such that the relaxation times that are typically quite short at higher field can be extended to render d-DNP feasible.

11.3.5 Sample Injection

A crucial step in every d-DNP experiment is the injection of the hyperpolarized sample into an appropriate cell or NMR tube that fits the detection NMR spectrometer. Different sophisticated sample injection mechanisms have been developed so far to avoid key problems like macro- and microbubbles that perturb shims and field homogeneity as well as convection and turbulences. Especially, when using pulsed field gradients (PFGs) for NMR detection, the latter point should be considered.

One of the most sophisticated systems currently available has been developed by Frydman and

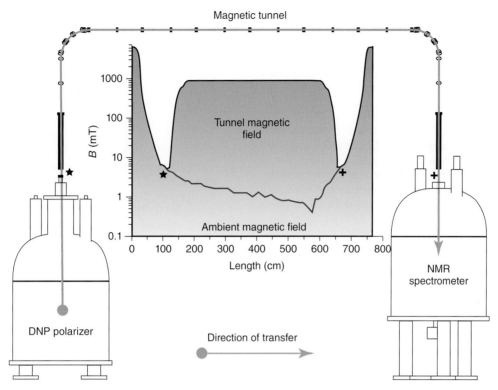

Figure 11.7. Sketch of the difference between the magnetic field passed during the transfer with and without a magnetic tunnel. The magnetic field strength during the transfer of the hyperpolarized fluid from the DNP polarizer to the unshielded 300 MHz NMR spectrometer through a magnetic tunnel (black line) or without tunnel (red line). The star and cross indicate the polarizer–tunnel and tunnel–spectrometer junctions. (Source: Milani *et al.* [89], http://pubs.rsc.org/en/content/articlehtml/2016/cp/c6cp00839a. Licensed under CC BY 3.0)

co-workers[13,91]. It makes use of fast-switching MPLC valves to guide the liquid after dissolution and of pressure gradients to push the sample towards the detection spectrometer and stabilize it after injection into an NMR tube. It allows to transfer the sample from the polarizer to the spectrometer in less than 3 s and at the same time for non-violent injections causing a minimum of turbulences by using back-pressure that is applied to a high-pressure tube waiting in the detection spectrometer ready to be filled with a hyperpolarized liquid.

Another very advanced system has been developed by Hilty and co-workers, which combines high-pressure transfer systems with a flow through-cell that can be closed after it has been filled with a hyperpolarized liquid. This system allows for dissolution and injection in less than one second without any significant turbulences.[88]

11.3.6   NMR Detection

After hyperpolarization of a substrate, dissolution, transfer, PA removal, and injection into a detection spectrometer the main goal of d-DNP is detection of hyperpolarized spins with enhancements that can exceed four orders of magnitude. Naturally, the main advantage of the tremendous signal enhancements is that they outdate the need for NMR signal averaging. A plethora of different applications has emerged from this potential, such as (among many other excellent examples) hyperpolarized imaging[17,30] and metabolomics,[9] monitoring of kinetics of chemical reactions or of protein–ligand binding,[2,4,92] creation and observation of non-equilibrium quantum states such as long-lived states or singlet–triplet imbalances,[8,9,11] hyperpolarized multidimensional spectroscopy,[12,13] or drug screening.[93,94]

For all these applications, d-DNP provides sufficiently intense signals after dissolution and transfer of a hyperpolarized substrate that allow for the detection of time series of sequential NMR spectra since a single scan is normally sufficient for a spectrum with adequate signal-to-noise ratio (SNR), even for nuclei with low gyromagnetic ratios and intrinsically low sensitivity. Typical repetition rates are on the order of one second depending on the length of the accumulated free induction decay (FID) and the employed radiopulse sequence. The time window amenable to signal observation is thereby determined by the relaxation properties of the hyperpolarized target molecules (cf. Section 11.3.2), which after some time do not provide sufficiently intense signals anymore as to allow for single-scan detection. This limitation of the d-DNP methodology, the inevitable return of the spins to their thermal equilibrium polarization, poses a significant experimental challenge, especially for fast-relaxing nuclei such as ^{1}H or ^{31}P. Thus, d-DNP typically demands for rapid detection schemes, typically on the order of a minute.

Yet, in general, the entire wealth of NMR pulse sequences is applicable to d-DNP hyperpolarized substrates after dissolution and injection, limited only by the relaxation time of the hyperpolarized spins. Such, also rapid multidimensional detection, schemes[13] are possible and even detection of single-scan 2D spectra is possible in a time-resolved manner by spatial encoding (SPEN) of the indirectly detected dimension. Such a 2D spectrum can be detected in less than 1 s enabling the measurement of a series of multidimensional spectra with a sub-Hertz repetition rate.[76]

11.4 COMMERCIAL DESIGNS AND INSTRUMENTATIONAL ASPECTS

Several systems are currently (2018) available that enable hyperpolarization at low temperatures and subsequent dissolution of the sample (Figure 11.8). Among the most prominent systems are:

1. The Oxford Instruments product Hypersense® operating at temperatures between 1.4 and 4.2 K, which is currently operational in over a dozen laboratories. Being the first commercial apparatus for d-DNP, the Hypersense® is a product that aims at complete automation of the polarization and dissolution processes, such that only a minimum of

manual operations is necessary. It thus enables basic applications for many users on a routine basis. The lHe used to cool the variable temperature insert (VTI) in which the sample is hyperpolarized is automatically provided from the reservoir, which cools the superconducting magnet surrounding the VTI. An external pumping stage is necessary to cool the lHe bath below 4.2 K by applying high vacuum. Dissolution of the sample is achieved via a 'dissolution stick' system and is fully automatic.

2. Another commercially available cryogen-free apparatus from Cryogenics that can be adapted for DNP can achieve likewise temperatures as low as 1.4 K. The apparatus functions via a closed He cycle that uses pressure gradients to cool down the magnet as well as the cryostat that contains the probe. The helium consumption is reduced to a minimum by using specially designed helium recycling systems. An appropriate dissolution system has recently been made available by the start-up 'Polarize'. Additionally CP-DNP techniques have been developed for the system allowing for substantial heteronuclear polarization.[43]

3. A Bruker prototype with a special VTI insert embedded in a wide-bore bore magnet that can reach temperatures down to 1.2 K using commercial two-stage pumping stations. In this polarizer, the lHe is provided by an external reservoir, which needs to be additionally connected to the system (see Figure 11.8). The system allows for daily routine applications achieving high polarization for ^{1}H, ^{2}H, ^{13}C, and ^{31}P in less than 1 h.[24] This allows several d-DNP experiments a day for various nuclei. The sample dissolution and transfer is enabled by a dissolution stick that is rapidly inserted into the VTI after hyperpolarization of a sample. The dissolution and transfer process takes in total 1−4 s depending on the size of the NMR tube (5 or 10 mm). For ^{13}C, signal enhancements >10 000 are routinely achieved.

4. The GE Healthcare product SPINlab, which operates at 5 T and 0.9 K. This product is optimized for use in biomedical applications and like the Hypersense® aims at complete automation and is licensed for use in human *in vivo* applications.

The Hypersense® operates at W-band (94 GHz Larmor frequency for electrons), i.e., 3.3 T, while

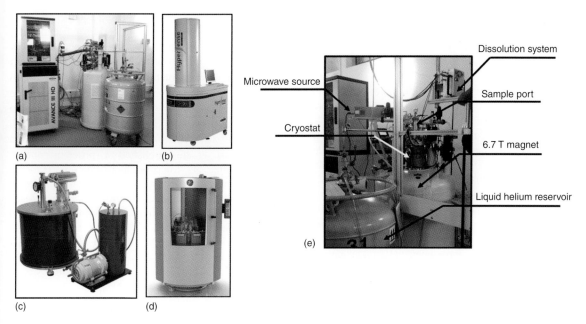

Figure 11.8. (a) The Bruker prototype DNP system at Ecole Normale Supérieure in Paris. (b) The Oxford Instruments Hypersense DNP system. It operates at magnetic fields of 3.3 T and includes an automatized dissolution system. (c) The Cryogenics cryogen-free DNP system for which a dissolution system is not yet commercially available. (d) The GE Healthcare SPINlab system aiming at clinical applications and full automation. (e) Detailed view of the system in (a). Major components of the system are indicted. The cryostat is inserted into a 6.7 T NMR magnet

the Bruker system works at 6.7 T (188 GHz Larmor frequency for electrons). The Cryogenics magnets are available at fields up to 11.7 T.

11.4.1 Sample Holder and Dissolution System

The samples for d-DNP applications must be placed in a multifunctional compartment in all the above-mentioned designs. These need not only to allow for efficient DNP, i.e., need to be immersed in a bath of lHe, but also for a rapid dissolution.

For example, in the Bruker prototype system, samples are placed in 6 mm outer diameter sample cups made of PEEK. This material allows for microwaves (μw) and rf irradiation to pass and can at the same time withstand the harsh pressure and temperature jumps during the dissolution procedures (*vide supra*). The sample cup is mounted to a hollow rod (to guide the dissolution system after hyperpolarization) to be inserted in the magnet and the lHe bath.

PEEK has the advantage that it is a very robust material. However, it features the disadvantage that it is

not proton-free. That is, for measurements of absolute polarization levels within the polarizer different sample holders need to be employed.

For the SPINlab from GE Healthcare a flow-through system (fluid path) has been developed that allows to inject and dissolve a sample with a minimum number of removable parts and simultaneously preventing contamination by external impurities, which makes this system particularly appealing for medicinal *in vivo* applications and amenable to produce hyperpolarized liquids that are injected into human propends. The setup is detailed in Ref. 95.

11.5 FUTURE PERSPECTIVES AND CONCLUDING REMARKS

Concluding, d-DNP has undergone substantial developments in the past 15 years and has emerged as a powerful supplement for nuclear magnetic resonance spectroscopy. Especially, instrumentational developments have brought the methodology to a point where a daily routine use in many NMR laboratories is possible. Presently, two distinct routes seem to lead the way to the future of dissolution DNP; one comprises medicinal applications, the other analytical uses in fundamental research.

In view of the first way, d-DNP is a promising candidate for noninvasive monitoring of metabolomic activity of cancer tissue *in vivo*. From the instrumentational point of view, the culprit of this methodology presents itself as the combination of a very complex methodology situated in the domain of nuclear physics that still undergoes substantial developments and optimization processes with actual applications to alive patients that must not be harmed due to the experimental procedure and that demand for results that significant complement other existing and established clinical methods. However, great steps have been undertaken and are currently commenced to develop d-DNP into a reliable method for metabolomic imaging.

In view of the second way, d-DNP presents itself currently as a promising tool for a plethora of applications in analytical chemistry, molecular biology, and physics, especially with potential to monitor non-equilibrium dynamics. Yet, routine use on a daily basis is still rare as the methodology is quite complex, form the theoretical as well as from the instrumentational point of view. A 'black box'-type usage of

d-DNP is furthermore complicated by the fact that current commercially available and user-friendly designs are very expensive. Two interesting alternatives are the use of (i) transportable hyperpolarization and (ii) hyperpolarization centers for external users.

11.5.1 Transporting Hyperpolarization

Recently, Ji *et al.* published a proof-of-concept study aiming at 'transportable hyperpolarization'. They hyperpolarize ^{13}C nuclei in microcrystals of ^{13}C-enriched glycine and alanine using the CP approach. Having a very long longitudinal relaxation time, these spins could survive the transfer from a fix-installed polarization device to a mobile transportation Dewar that contained lHe at a static magnetic field of 1 T. The concept is depicted in Figure 11.9.

Within the Dewar the long-lived ^{13}C hyperpolarization of the microcrystalline amino acids could be transported over long distances and subsequently dissolved (after 16 h storage) into a detection NMR spectrometer by use of a specially designed dissolution system. This development opens perspectives towards hyperpolarization centers, where hyperpolarized samples of interest are prepared and subsequently shipped to the laboratory in need of the substance.

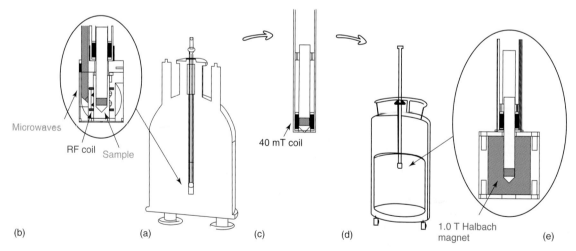

Figure 11.9. (a) The DNP polarizer consists of a 6.7-T wide bore magnet and a 1.2-K cryostat equipped with (b) a DNP probe where the sample is inserted and irradiated with microwaves for ^{1}H DNP and radiofrequency fields for ^{1}H–^{13}C cross-polarization. The sample can then be manually removed from the polarizer in (c) a transfer stick comprising a coil sustaining a magnetic field of ca 40 mT (100 turns, current of 4 A) and subsequently inserted in (d) a conventional liquid helium transport Dewar with (e) a magnetic insert providing a 1.0 T static magnetic field for storage or transport. (Source: Ji *et al.* [96], https://www.nature.com/articles/ncomms13975. Licensed under CC BY 4.0)

Another recent development enables very long-lived hyperpolarization is based on the use of UV-irradiation and transient production of PAs.[97] Annihilation of PAs after suspension of the UV-irradiation yields prolongations of hyperpolarization lifetimes of over 18× likewise enabling dissolution experiments remote to the polarization site.

11.5.2 Hyperpolarization Centers

As d-DNP prerequisites expert operators that are well-trained in the operation of the experimental setup, centers in which external users can perform their experiments are an interesting option that enables more applications without the need for an apparatus for every user. Positive examples are the EPFL polarizer (Lausanne, Switzerland) and the Bruker prototype that has been operational in a high-field NMR facility in Paris (France) since beginning 2016 and is available for external users via a national infrastructure network. In 2018, a second version of this Bruker prototype polarizer was made available in Lyon (France) and for external users. In the last few years, a substantial number of international users has applied and successfully planned and performed experiments with local experts embracing studies going from basic methods development (e.g., SPEN-DOSY[76]) over fundamental understanding of DNP (e.g., TM[24]) to applications (e.g., intrinsically disordered proteins, IDPs[12]).

ACKNOWLEDGMENTS

The authors are deeply indebted to Prof. Geoffrey Bodenhausen for his continuous support and dedication for the advances of the d-DNP methodology and the international d-DNP community. The authors thank a seemingly endless number of students, co-workers, and colleagues for their dedication and energy, without which none of the here reviewed advances would have been possible. The authors thank the European Research Council under the European Union's Horizon 2020 research and innovation programme (ERC Grant Agreement n 714519/HP4all), the EPFL, the Swiss National Science Foundation, Bruker BioSpin, the Ecole Normale Supérieure Paris, the ENS-Lyon, the French CNRS, and Lyon 1 University.

REFERENCES

1. J. H. Ardenkjaer-Larsen, B. Fridlund, A. Gram, G. Hansson, L. Hansson, M. H. Lerche, R. Servin, M. Thaning, and K. Golman, *Proc. Natl. Acad. Sci. U. S. A.*, 2003, **100**, 10158.

2. A. Sadet, M. Emmanuelle, M. Weber, A. Jhajharia, D. Kurzbach, G. Bodenhausen, E. Miclet, and D. Abergel, *Chem. Eur. J.*, 2018, **24**, 5456.

3. J. N. Dumez, J. Milani, B. Vuichoud, A. Bornet, J. Lalande-Martin, I. Tea, M. Yon, M. Maucourt, C. Deborde, A. Moing, L. Frydman, G. Bodenhausen, S. Jannin, and P. Giraudeau, *Analyst*, 2015, **140**, 5860.

4. E. Miclet, D. Abergel, A. Bornet, J. Milani, S. Jannin, and G. Bodenhausen, *J. Phys. Chem. Lett.*, 2014, **5**, 3290.

5. P. R. Jensen, M. Karlsson, M. H. Lerche, and S. Meier, *Chem. Eur. J.*, 2013, **19**, 13288.

6. T. Harris, G. Eliyahu, L. Frydman, and H. Degani, *Proc. Natl. Acad. Sci. U. S. A.*, 2009, **106**, 18131.

7. H. Zeng, Y. Lee, and C. Hilty, *Anal. Chem.*, 2010, **82**, 8897.

8. A. Jhajharia, E. M. Weber, J. G. Kempf, D. Abergel, G. Bodenhausen, and D. Kurzbach, *J. Chem. Phys.*, 2017, **146**, 041101.

9. D. Mammoli, B. Vuichoud, A. Bornet, J. Milani, J. N. Dumez, S. Jannin, and G. Bodenhausen, *J. Phys. Chem. B*, 2015, **119**, 4048.

10. P. R. Vasos, A. Comment, R. Sarkar, P. Ahuja, S. Jannin, J. P. Ansermet, J. A. Konter, P. Hautle, B. van den Brandt, and G. Bodenhausen, *Proc. Natl. Acad. Sci. U. S. A.*, 2009, **106**, 18469.

11. M. C. Tayler, I. Marco-Rius, M. I. Kettunen, K. M. Brindle, M. H. Levitt, and G. Pileio, *J. Am. Chem. Soc.*, 2012, **134**, 7668.

12. D. Kurzbach, E. Canet, A. G. Flamm, A. Jhajharia, E. M. Weber, R. Konrat, and G. Bodenhausen, *Angew. Chem. Int. Ed. Engl.*, 2017, **56**, 389.

13. G. Olsen, E. Markhasin, O. Szekely, C. Bretschneider, and L. Frydman, *J. Magn. Reson.*, 2016, **264**, 49.

14. M. Ragavan, L. I. Iconaru, C. G. Park, R. W. Kriwacki, and C. Hilty, *Angew. Chem. Int. Ed. Engl.*, 2017, **56**, 7070.

15. J. Kim, M. Liu, and C. Hilty, *J. Phys. Chem. B*, 2017, **121**, 6492.

16. S. J. Nelson, J. Kurhanewicz, D. B. Vigneron, P. E. Larson, A. L. Harzstark, M. Ferrone, M. van Criekinge,

J. W. Chang, R. Bok, I. Park, G. Reed, L. Carvajal, E. J. Small, P. Munster, V. K. Weinberg, J. H. Ardenkjaer-Larsen, A. P. Chen, R. E. Hurd, L. I. Odegardstuen, F. J. Robb, J. Tropp, and J. A. Murray, *Sci. Transl. Med.*, 2013, **5**, 198ra108.

17. K. Golman, R. in't Zandt, and M. Thaning, *Proc. Natl. Acad. Sci. U. S. A.*, 2006, **103**, 11270.

18. M. Karlsson, P. R. Jensen, J. Ø. Duus, S. Meier, and M. H. Lerche, *Appl. Magn. Reson.*, 2012, **43**, 223.

19. P. Ahuja, R. Sarkar, S. Jannin, P. R. Vasos, and G. Bodenhausen, *Chem. Commun.*, 2010, **46**, 8192.

20. B. Lama, J. H. Collins, D. Downes, A. N. Smith, and J. R. Long, *NMR Biomed.*, 2016, **29**, 226.

21. D. Mammoli, N. Salvi, J. Milani, R. Buratto, A. Bornet, A. A. Sehgal, E. Canet, P. Pelupessy, D. Carnevale, S. Jannin, and G. Bodenhausen, *Phys. Chem. Chem. Phys.*, 2015, **17**, 26819.

22. D. Kurzbach, E. M. M. Weber, A. Jhajharia, S. F. Cousin, A. Sadet, S. Marhabaie, E. Canet, N. Birlirakis, J. Milani, S. Jannin, D. Eshchenko, A. Hassan, R. Melzi, S. Luetolf, M. Sacher, M. Rossire, J. Kempf, J. A. B. Lohman, M. Weller, G. Bodenhausen, and D. Abergel, *J. Chem. Phys.*, 2016, **145**, 194203.

23. A. Nardi-Schreiber, A. Gamliel, T. Harris, G. Sapir, J. Sosna, J. M. Gomori, and R. Katz-Brull, *Nat. Commun.*, 2017, **8**, 341.

24. D. Guarin, S. Marhabaie, A. Rosso, D. Abergel, G. Bodenhausen, K. Ivanov, and D. Kurzbach, *J. Phys. Chem. Lett.*, 2017, **8**, 5531.

25. J. Milani, B. Vuichoud, A. Bornet, R. Melzi, S. Jannin, and G. Bodenhausen, *Rev. Sci. Instrum.*, 2017, **88**, 015109.

26. P. Niedbalski, C. Parish, A. Kiswandhi, and L. Lumata, *Magn. Reson. Chem,*, 2016, **54**, 962.

27. F. Jahnig, G. Kwiatkowski, A. Dapp, A. Hunkeler, B. H. Meier, S. Kozerke, and M. Ernst, *Phys. Chem. Chem. Phys.*, 2017, **19**, 19196.

28. L. Lumata, Z. Kovacs, A. D. Sherry, C. Malloy, S. Hill, v. Tol, L. Yu, L. Song, and M. E. Merritt, *Phys. Chem. Chem. Phys.*, 2013, **15**, 9800.

29. L. Lumata, M. E. Merritt, C. R. Malloy, A. D. Sherry, and Z. Kovacs, *J. Phys. Chem. A*, 2012, **116**, 5129.

30. S. J. Nelson, D. Vigneron, J. Kurhanewicz, A. Chen, R. Bok, and R. Hurd, *Appl. Magn. Reson.*, 2008, **34**, 533.

31. D. M. Wilson and J. Kurhanewicz, *J. Nucl. Med.*, 2014, **55**, 1567.

32. C. Sauvee, G. Casano, S. Abel, A. Rockenbauer, D. Akhmetzyanov, H. Karoui, D. Siri, F. Aussenac, W. Maas, R. T. Weber, T. Prisner, M. Rosay, P. Tordo, and O. Ouari, *Chemistry*, 2016, **22**, 5598.

33. M. Rosay, L. Tometich, S. Pawsey, R. Bader, R. Schauwecker, M. Blank, P. M. Borchard, S. R. Cauffman, K. L. Felch, R. T. Weber, R. J. Temkin, R. G. Griffin, and W. E. Maas, *Phys. Chem. Chem. Phys.*, 2010, **12**, 5850.

34. D. Shimon, A. Feintuch, D. Goldfarb, and S. Vega, *Phys. Chem. Chem. Phys.*, 2014, **16**, 6687.

35. W. M. Yau, K. R. Thurber, and R. Tycko, *J. Magn. Reson.*, 2014, **244**, 98.

36. T. Wenckebach, Essentials of Dynamic Nuclear Polarization. Spindrift Publications: The Netherlands, 2016.

37. K. Schmidt-Rohr and H. W. Spiess, Multidimensional Solid-Sate NMR and Polymers.

38. M. Veshtort and R. G. Griffin, *J. Chem. Phys.*, 2011, **135**, 134509.

39. H. A. Yoshihara, E. Can, M. Karlsson, M. H. Lerche, J. Schwitter, and A. Comment, *Phys. Chem. Chem. Phys.*, 2016, **18**, 12409.

40. T. Harris, C. Bretschneider, and L. Frydman, *J. Magn. Reson.*, 2011, **211**, 96.

41. E. M. M. Weber, G. Sicoli, H. Vezin, G. Frébourg, D. Abergel, G. Bodenhausen, and D. Kurzbach, *Angew. Chem. Int. Ed.*, 2018, **57**, 5171.

42. R. Buratto, A. Bornet, J. Milani, D. Mammoli, B. Vuichoud, N. Salvi, M. Singh, A. Laguerre, S. Passemard, S. Gerber-Lemaire, S. Jannin, and G. Bodenhausen, *Chem. Rev.*, 2014, **9**, 2509.

43. M. Baudin, B. Vuichoud, A. Bornet, J. Milani, G. Bodenhausen, and S. Jannin, *J. Magn. Reson.*, 2018, **294**, 115

44. O. Szekely, G. L. Olsen, I. C. Felli, and L. Frydman, *Anal. Chem.*, 2018.

45. T. A. Siaw, M. Fehr, A. Lund, A. Latimer, S. A. Walker, D. T. Edwards, and S. I. Han, *Phys. Chem. Chem. Phys.*, 2014, **16**, 18694.

46. Spin Muon Collaboration (SMC), B. Adeva, E. Arik, S. Ahmad, A. Arvidson, B. Badelek, M. K. Ballintijn, G. Bardin, G. Baum, P. Berglund, L. Betev, I. G. Bird, R. Birsa, P. Björkholm, B. E. Bonner, N.de Botton, M. Boutemeur, F. Bradamante, A. Bressan, A. Brüll, J. Buchanan, S. Bültmann, E. Burtin, C. Cavata, J. P. Chen, J. Clement, M. Clocchiatti, M. D. Corcoran, D. Crabb, J. Cranshaw, J. Çuhadar, S. Dalla Torre, A. Deshpande, R.van Dantzig, D. Day, S. Dhawan,

C. Dulya, A. Dyring, S. Eichblatt, J. C. Faivre, D. Fasching, F. Feinstein, C. Fernandez, B. Frois, C. Garabatos, J. A. Garzon, T. Gaussiran, M. Giorgi, E.von Goeler, I. A. Golutvin, A. Gomez, G. Gracia, N.de Groot, M. Grosse Perdekamp, E. Gülmez, D.von Harrach, T. Hasegawa, P. Hautle, N. Hayashi, C. A. Heusch, N. Horikawa, V. W. Hughes, G. Igo, S. Ishimoto, T. Iwata, M.de Jong, E. M. Kabuß, T. Kageya, R. Kaiser, A. Karev, H. J. Kessler, T. J. Ketel, I. Kiryushin, A. Kishi, Y. Kisselev, L. Klostermann, D. Krämer, V. Krivokhijine, V. Kukhtin, J. Kyynäräinen, M. Lamanna, U. Landgraf, K. Lau, T. Layda, J. M.Le Goff, F. Lehar, A.de Lesquen, J. Lichtenstadt, T. Lindqvist, M. Litmaath, S. Lopez-Ponte, M. Lowe, A. Magnon, G. K. Mallot, F. Marie, A. Martin, J. Martino, T. Matsuda, B. Mayes, J. S. McCarthy, K. Medved, G.van Middelkoop, D. Miller, J. Mitchell, K. Mori, J. Moromisato, G. S. Mutchler, A. Nagaitsev, J. Nassalski, L. Naumann, B. Neganov, T. O. Niinikoski, J. E. J. Oberski, A. Ogawa, S. Okumi, C. S. Özben, A. Penzo, C. A. Perez, F. Perrot-Kunne, D. Peshekhonov, R. Piegaia, L. Pinsky, S. Platchkov, M. Plo, D. Pose, H. Postma, J. Pretz, T. Pussieux, J. Pyrlik, I. Reyhancan, J. M. Rieubland, A. Rijllart, J. B. Roberts, S. E. Rock, M. Rodriguez, E. Rondio, O. Rondon, L. Ropelewski, A. Rosado, I. Sabo, J. Saborido, G. Salvato, A. Sandacz, D. Sanders, I. Savin, P. Schiavon, K. P. Schüler, R. Segel, R. Seitz, Y. Semertzidis, S. Sergeev, F. Sever, P. Shanahan, E. Sichtermann, G. Smirnov, A. Staude, A. Steinmetz, H. Stuhrmann, K. M. Teichert, F. Tessarotto, W. Thiel, M. Velasco, J. Vogt, R. Voss, R. Weinstein, C. Whitten, R. Willumeit, R. Windmolders, W. Wislicki, A. Witzmann, A. Yañez, N. I. Zamiatin, A. M. Zanetti, and J. Zhao, *Nucl. Instrum. Methods Phys. Res. Sect. A*, 1993, **372**, 339.

47. K. R. Thurber, W. M. Yau, and R. Tycko, *J. Magn. Reson.*, 2010, **204**, 303.

48. Y. Hovav, A. Feintuch, S. Vega, and D. Goldfarb, *J. Magn. Reson.*, 2014, **238**, 94.

49. A. Bornet, J. Milani, B. Vuichoud, A. J. P. Linde, G. Bodenhausen, and S. Jannin, *Chem. Phys. Lett.*, 2014, **602**, 63.

50. X. Ji, T. V. Can, F. Mentink-Vigier, A. Bornet, J. Milani, B. Vuichoud, M. A. Caporini, R. G. Griffin, S. Jannin, M. Goldman, and G. Bodenhausen, *J. Magn. Reson.*, 2017, **286**, 138.

51. A. Abragam and M. Goldman, *Rep. Prog. Phys.*, 1978, **41**, 395.

52. S. Jannin, A. Comment, and J. J. van der Klink, *Appl. Magn. Reson.*, 2012, **43**, 59.

53. L. Lumata, A. K. Jindal, M. E. Merritt, C. R. Malloy, A. D. Sherry, and Z. Kovacs, *J. Am. Chem. Soc.*, 2011, **133**, 8673.

54. A. Leavesley, D. Shimon, T. A. Siaw, A. Feintuch, D. Goldfarb, S. Vega, I. Kaminker, and S. Han, *Phys. Chem. Chem. Phys.*, 2017, **19**, 3596.

55. Y. Hovav, D. Shimon, I. Kaminker, A. Feintuch, D. Goldfarb, and S. Vega, *Phys. Chem. Chem. Phys.*, 2015, **17**, 6053.

56. Y. Hovav, I. Kaminker, D. Shimon, A. Feintuch, D. Goldfarb, and S. Vega, *Phys. Chem. Chem. Phys.*, 2015, **17**, 226.

57. Y. Hovav, A. Feintuch, and S. Vega, *Phys. Chem. Chem. Phys.*, 2013, **15**, 188.

58. Y. Hovav, O. Levinkron, A. Feintuch, and S. Vega, *Appl. Magn. Reson.*, 2012, **43**, 21.

59. Y. Hovav, A. Feintuch, and S. Vega, *J. Magn. Reson.*, 2012, **214**, 29.

60. D. Shimon, Y. Hovav, A. Feintuch, D. Goldfarb, and S. Vega, *Phys. Chem. Chem. Phys.*, 2012, **14**, 5729.

61. A. Feintuch, D. Shimon, Y. Hovav, D. Banerjee, I. Kaminker, Y. Lipkin, K. Zibzener, B. Epel, S. Vega, and D. Goldfarb, *J. Magn. Reson.*, 2011, **209**, 136.

62. Y. Hovav, A. Feintuch, and S. Vega, *J. Magn. Reson.*, 2010, **207**, 176.

63. W. T. Wenckebach, *J. Magn. Reson.*, 2017, **284**, 104.

64. E. M. M. Weber, H. Vezin, J. G. Kempf, G. Bodenhausen, D. Abergel, and D. Kurzbach, *Phys. Chem. Chem. Phys.*, 2017, **19**, 16087.

65. A. Bornet, De l'usage des protons hyperpolarisés pour augmenter la sensibilité de la RMN, Lausanne, EPFL.

66. A. Bornet and S. Jannin, *J. Magn. Reson.*, 2016, **264**, 13.

67. S. F. J. Cox, V. Bouffard, and M. Goldman, *J. Phys. C Solid State Phys.*, 1973, **6**, L100.

68. W. de Boer, M. Borghini, K. Morimoto, T. O. Niinikoski, and F. Udo, *J. Low Temp. Phys.*, 1974, **15**, 249.

69. S. Jannin, A. Bornet, S. Colombo, and G. Bodenhausen, *Chem. Phys. Lett.*, 2011, **517**, 234.

70. A. Bornet, J. Milani, S. T. Wang, D. Mammoli, R. Buratto, N. Salvi, T. F. Segawa, V. Vitzthum, P. Mieville, S. Chinthalapalli, A. J. Perez-Linde, D. Carnevale, S. Jannin, M. Caporini, S. Ulzega, M. Rey, and G. Bodenhausen, *Chimia*, 2012, **66**, 734.

71. A. Potapov, K. R. Thurber, W. M. Yau, and R. Tycko, *J. Magn. Reson.*, 2012, **221**, 32.

72. A. Bornet, A. Pinon, A. Jhajharia, M. Baudin, X. Ji, L. Emsley, G. Bodenhausen, J. H. Ardenkjaer-Larsen, and S. Jannin, *Phys. Chem. Chem. Phys.*, 2016, **18**, 30530.

73. B. Vuichoud, J. Milani, A. Bornet, R. Melzi, S. Jannin, and G. Bodenhausen, *J. Phys. Chem. B*, 2014, **118**, 1411.

74. F. Saidi, F. Taulelle, and C. Martineau, *J. Pharm. Sci.*, 2016, **105**, 2397.

75. J. Raya and J. Hirschinger, *J. Magn. Reson.*, 2017, **281**, 253.

76. L. Guduff, D. Kurzbach, C. V. Heijenoort, D. Abergel, and J.-N. Dumez, *Chem. Eur. J.*, 2017, **23**, 16722.

77. B. Meier, Conference Poster, EUROMAR, 2017, Warsaw, Poland.

78. P. Mieville, S. Jannin, and G. Bodenhausen, *J. Magn. Reson.*, 2011, **210**, 137.

79. G. M. Clore and J. Iwahara, *Chem. Rev.*, 2009, **109**, 4108.

80. P. Mieville, P. Ahuja, R. Sarkar, S. Jannin, P. R. Vasos, S. Gerber-Lemaire, M. Mishkovsky, A. Comment, R. Gruetter, O. Ouari, P. Tordo, and G. Bodenhausen, *Angew. Chem. Int. Ed. Engl.*, 2010, **49**, 6182.

81. T. R. Eichhorn, Y. Takado, N. Salameh, A. Capozzi, T. Cheng, J. N. Hyacinthe, M. Mishkovsky, C. Roussel, and A. Comment, *Proc. Natl. Acad. Sci. U. S. A.*, 2013, **110**, 18064.

82. T. Cheng, M. Mishkovsky, M. J. Junk, K. Munnemann, and A. Comment, *Macromol. Rapid Commun.*, 2016, **37**, 1074.

83. D. Gajan, A. Bornet, B. Vuichoud, J. Milani, R. Melzi, H. A. van Kalkeren, L. Veyre, C. Thieuleux, M. P. Conley, W. R. Grüning, M. Schwarzwälder, A. Lesage, C. Copéret, G. Bodenhausen, L. Emsley, and S. Jannin, *Proc. Natl. Acad. Sci.*, 2014, **111**, 14693.

84. B. Vuichoud, E. Canet, J. Milani, A. Bornet, D. Baudouin, L. Veyre, D. Gajan, L. Emsley, A. Lesage, C. Coperet, C. Thieuleux, G. Bodenhausen, I. Koptyug, and S. Jannin, *J. Phys. Chem. Lett.*, 2016, **7**, 3235.

85. D. Kurzbach, M. J. N. Junk, and D. Hinderberger, *Macromol. Rapid Commun.*, 2013, **34**, 119.

86. D. Baudouin, H. A. van Kalkeren, A. Bornet, B. Vuichoud, L. Veyre, M. Cavailles, M. Schwarzwalder, W. C. Liao, D. Gajan, G. Bodenhausen, L. Emsley, A. Lesage, S. Jannin, C. Coperet, and C. Thieuleux, *Chem. Sci.*, 2016, **7**, 6846.

87. W. R. Gruning, H. Bieringer, M. Schwarzwalder, D. Gajan, A. Bornet, B. Vuichoud, J. Milani, D. Baudouin, L. Veyre, A. Lesage, S. Jannin, G. Bodenhausen, C. Thieuleux, and C. Coperet, *Helvetis Chim. Acta*, 2017, **100**, e1600122.

88. S. Bowen and C. Hilty, *Phys. Chem. Chem. Phys.*, 2010, **12**, 5766.

89. J. Milani, B. Vuichoud, A. Bornet, P. Mieville, R. Mottier, S. Jannin, and G. Bodenhausen, *Rev. Sci. Instrum.*, 2015, **86**, 024101.

90. M. F. Roberts, Q. Cui, C. J. Turner, D. A. Case, and A. G. Redfield, *Biochemistry*, 2004, **43**, 3637.

91. T. Harris, O. Szekely, and L. Frydman, *J. Phys. Chem. B*, 2014, **118**, 3281.

92. Q. Chappuis, J. Milani, B. Vuichoud, A. Bornet, A. D. Gossert, G. Bodenhausen, and S. Jannin, *J. Phys. Chem. Lett.*, 2015, **6**, 1674.

93. R. Buratto, D. Mammoli, E. Chiarparin, G. Williams, and G. Bodenhausen, *Angew Chem. Int. Ed.*, 2014, **53**, 11376.

94. H. Min, G. Sekar, and C. Hilty, *Chem. Rev.*, 2015, **10**, 1559.

95. R. M. Malinowski, K. W. Lipso, M. H. Lerche, and J. H. Ardenkjaer-Larsen, *J. Magn. Reson.*, 2016, **272**, 141.

96. X. Ji, A. Bornet, B. Vuichoud, J. Milani, D. Gajan, A. J. Rossini, L. Emsley, G. Bodenhausen, and S. Jannin, *Nat. Commun.*, 2017, **8**, 13975.

97. A. Capozzi, T. Cheng, G. Boero, C. Roussel, and A. Comment, *Nat. Commun.*, 2017, **8**, 15757.

Chapter 12

Introduction to Dissolution DNP: Overview, Instrumentation, and Human Applications

Jan H. Ardenkjaer-Larsen[1,2]

[1]*Technical University of Denmark, Lyngby, Denmark*
[2]*GE Healthcare, Brøndby, Denmark*

12.1 INTRODUCTION

The aim of this chapter is to introduce the basic principles and progress toward the clinical application of hyperpolarization using dissolution dynamic nuclear polarization (dDNP). Less than a decade has passed

Handbook of High Field Dynamic Nuclear Polarization.
Edited by Vladimir K. Michaelis, Robert G. Griffin, Björn Corzilius and Shimon Vega
© 2020 John Wiley & Sons, Ltd. ISBN: 978-1-119-44164-9
Also published in eMagRes (online edition)
DOI: 10.1002/9780470034590.emrstm1549

from first published[1] to first in man.[2–4] During this time a research field has emerged studying the basic physics of DNP, developing advanced dDNP instrumentation, optimizing acquisition hardware and pulse sequences, and applying the method to study a range of bioprobes in biological systems.

Hyperpolarization indicates that the polarization is no longer determined by the static magnetic field of the scanner. A polarizer enhances the polarization of the nuclear spins outside the imaging system. The hyperpolarization method can be based on several principles,[5] of which, two have successfully been applied to molecules in solution: para-hydrogen induced polarization (PHIP) and dDNP. The dDNP method has been particularly successful in making solutions of biologically interesting molecules with highly polarized nuclear spins. The method takes advantage of DNP in the solid state followed by rapid dissolution in a suitable solvent. The dissolution step retains almost completely the nuclear spin polarization, creating a solution with a non-thermal nuclear polarization approaching unity.

Hyperpolarized metabolic magnetic resonance (MR) grew out of vision that MR as modality has excellent properties for studying the biochemical changes associated with disease, and that early diagnosis, staging, and response monitoring required characterization of the disease at the cellular and biochemical level. The MR spectrum provides both identification and quantification of the metabolites involved. However, magnetic resonance spectroscopy (MRS) suffers from several limitations: poor sensitivity, leading to long scan-times and poor spatial resolution, and limited spectral resolution due to a crowded spectrum. The long scan time means that only steady-state concentrations or slow dynamic changes can be measured. To overcome the tremendous challenge with sensitivity, hyperpolarization seemed the most powerful vista to take. The potential for 10 000-fold enhancement of nuclear spin polarization would compensate the low metabolite concentrations and enable otherwise impossible studies.

12.2 DYNAMIC NUCLEAR POLARIZATION

DNP was first described theoretically by Overhauser in 1953,[6] and a few months later demonstrated by Carver and Slichter[7] in metallic lithium. Overhauser predicted that saturating the conduction electrons of a metal would lead to a dynamic polarization of the nuclear spins. This was a fundamental discovery causing disbelief at the time: that heating of one spin system could lead to the cooling of another. Abragam soon extended the prediction by Overhauser for metals to electron spins in solution,[8] and most NMR spectroscopists are today familiar with the nuclear and electronic Overhauser effect. Relaxation processes that couple the spins via molecular motions drive this effect. Soon after, the Solid Effect was described for spins in the solid state coupled by dipolar interactions.[9] Later, DNP in the solid state was extended mechanistically to processes involving several electron spins.[10] The theory of DNP in the solid state, however, has failed to provide a quantitative description of the general case. However, the theoretical description of DNP has made significant progress in recent years. For a recent comprehensive description of DNP based on spin temperature concepts see Refs 11–13. In the solid state, microwave irradiation is close to the resonance frequency of the electron spin transfer, in part, the high electron spin polarization to the nuclear spins. The efficiency of this process depends on several parameters characterizing the various spin systems, but also on technical factors such as microwave frequency and power.

Polarization is the difference in population between the two possible energy eigenstates for a spin $1/2$, and is given by

$$P = \frac{N^+ - N^-}{N^+ + N^-} = \tan h \left(\frac{\gamma \hbar B_0}{2 k_B T} \right) \quad (12.1)$$

where N^+ and N^- denote the number of spins parallel (spin up) and anti-parallel (spin down) to the external magnetic field, respectively. According to equation (12.1) a temperature can be assigned to the spins for any polarization. For example, $P = 0$, complete saturation, corresponds to an infinite spin temperature. Depending on the sign of the polarization, the spin temperature can be either negative or positive, and will approach zero (cooling) as the polarization approaches unity. As a point of caution, it is often stated that spins are either pointing up or down, i.e., is in one of the two eigenstates. However, this is not correct. Quoting Slichter[14]: 'We emphasize that an arbitrary orientation can be specified, since sometimes the belief is erroneously held that spins may only be found pointing either parallel or antiparallel to the quantizing field. One of the beauties of quantum theory is that it contains features of both discreteness and continuity. In terms of the two quantum states with $m = \pm 1/2$ we can describe an expectation value of the magnetization which goes all the way from parallel to antiparallel, including all values in between. … '. The spin is therefore in a superposition of the eigenstates, and it is perfectly valid to speak of populations (probabilities) of the eigenstates. The proton polarization in a 3 T MR scanner becomes 10×10^{-6}, while the polarization for ^{13}C is only 2.5×10^{-6}. A theoretical enhancement of the polarization of 100 000 and 400 000 times, respectively, is therefore possible, if unity polarization could be achieved.

12.3 ELECTRON PARAMAGNETIC AGENT (EPA)

DNP requires the presence of unpaired electrons, typically in the form of an organic radical, but a few metal ions have high efficiency for DNP at low temperature, Cr(V) in particular. The magnetic moment of the electron is 658 times higher than that of the proton. This means that the electron spin will reach unity polarization at a moderate magnetic field strength and liquid helium temperature, Figure 12.1. At, e.g., 3.35 T and 1 K the electron spin polarization is already 98%. The choice of Electron Paramagnetic Agent (EPA) will depend on a number of factors. Firstly, the EPA needs to be chemically stable and dissolve readily in the matrix of interest. Secondly, the electron paramagnetic resonance (EPR) spectrum of the radical should have a width that allows DNP to be effective for the nucleus of interest, i.e., a linewidth that exceeds the

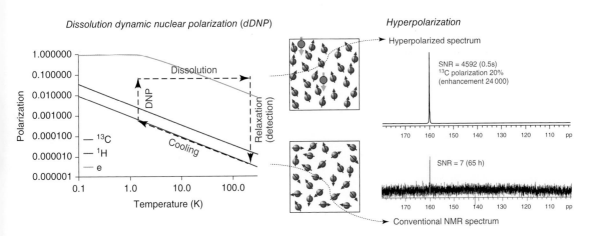

Figure 12.1. The principle of dDNP is illustrated. At room temperature and 3 T the ^{13}C nuclear spins are only weakly polarized to about 2.5 ppm, left graph. However, electron spins have a 2700 times stronger magnetic moment and are easily polarized. When the sample is cooled the electron spin polarization reaches almost unity (97.8% at 3.35 T and 1 K). By irradiation with microwaves close to the resonance frequency of the electron spins, electron–electron-nuclear transitions are induced, and the nuclear spin polarization will be enhanced 100-fold by DNP to several tens of percent. This process is slow at low temperatures, and takes typically 15–60 min. When the sample is polarized, superheated water or buffer dissolves the sample within seconds, and a room temperature solution of the hyperpolarized molecule is obtained. The hyperpolarized nuclear spins relax to thermal equilibrium (e.g., 2.5 ppm) with the longitudinal relaxation constant, T_1, of typically 40–80 s for carboxylic acids. A hyperpolarized ^{13}C spectrum of urea obtained in a single transient is shown to the right along with the thermal spectrum after 65 h of averaging

Larmor frequency of the nuclear spin. Thirdly, the EPA should have low toxicity and, ideally, be removable from the hyperpolarized solution. In practice, the above criteria mean that two classes of EPA are available, namely nitroxides and trityls.[15,16] The nitroxides belong to a class of molecules that have been studied extensively by EPR, and which have been used for DNP for many samples. Nitroxides are characterized by having a broad EPR spectrum. The EPR linewidth is approx. 4.0 promille (‰) of the EPR frequency, compared to the ^1H resonance frequency, which is 1.5‰ of the EPR frequency. Some of them have reasonable chemical stability and come with different degrees of hydrophilicity. The trityl is another class of EPA with superior properties for direct polarization of low gamma nuclei such as ^{13}C, ^{15}N, and ^2H. These radicals have a linewidth that is only approx. 0.80‰ of the resonance frequency, much less than the proton resonance frequency, but perfectly matched for ^{13}C, which has a resonance frequency, which is 0.37‰ of the EPR frequency. The trityls exist with a range of hydrophilicities, and they are typically very stable chemically (Figure 12.2).

It has been shown that a small amount of gadolinium can positively affect the DNP enhancement in the solid.[18] Other paramagnetic ions and molecules (Mn^{2+} and O_2) can in part have the same effect. A shortening of the EPA relaxation times can explain the effect. Adding $1–2\,mmol\,l^{-1}$ chelated Gd^{3+} leads to a 50–100% improvement of the DNP enhancement factor for pyruvic acid. The effect seems to be general to most samples, but each sample has to be optimized similarly to the concentration of the EPA. There is no direct DNP effect of the Gd^{3+} by itself under the conditions typically used (almost unity polarization of the EPA). Finally, Gd^{3+} may enhance the solid-state polarization by DNP, but care should be taken in avoiding accelerated relaxation in the liquid state. Free Gd^{3+} ions would cause detrimental liquid-state relaxation and pose an in vivo safety risk. After dissolution, the low concentration of radical and chelated Gd will have a negligible effect on T_1 in most cases.

A new class of EPA, with wider prospects, is UV-generated radicals.[19] A precursor molecule, e.g., pyruvic acid itself, can be excited with UV light to produce a radical species with good ESR properties for DNP. The radical is labile and has to be generated at low temperature, e.g., 77 K. In the dissolution process the radical annihilates and decomposes into acetate and bicarbonate. However, the most

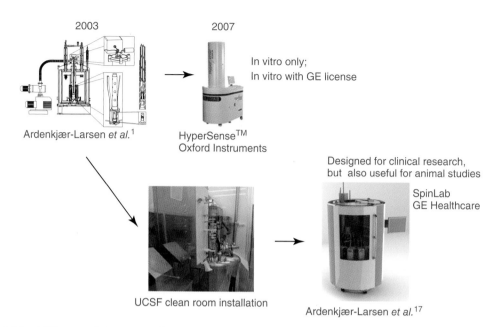

Figure 12.2. Different polarizer generations. The original polarizer design was described in Ref. 1. This polarizer was installed in a clean room at UCSF for the patient study published in Ref. 2. Oxford Instruments commercialized the Hypersense under license from General Electric. Later, General Electric introduced the SPINlab for clinical studies[17]

exciting prospect for these radicals is the potential for generating radical-free, hyperpolarized solids that can be transported over longer distances (see Section 12.7).

12.4 IMAGING AGENTS

Of the growing list of agents investigated in vivo using this method (Table 12.1, for a more comprehensive list see Ref. 43), the most studied is [1-^{13}C]pyruvate. This agent has shown great utility in oncology, as exemplified by the many studies summarized later in this chapter. [1-^{13}C]pyruvate was also the first agent to be used in a human study of hyperpolarized metabolic imaging.[2] Many of the most promising agents are reviewed in relation to the different applications of dDNP.

12.5 DNP SAMPLE PREPARATION

The first step in hyperpolarizing a new imaging agent by dDNP is to find a formulation with high concentration of the agent, and good solubility of the EPA. In order for the DNP process to be effective, the EPA must be homogeneously distributed within the sample. Many molecules will be crystalline, or have a tendency to crystallize as saturated aqueous solutions. This will cause the EPA to concentrate in domains and lead to poor a DNP effect. To prevent this, the sample should stay amorphous when frozen to ensure homogenous distribution of the EPA. Three examples of molecules that are liquids at room temperature and stay amorphous when frozen without additives are [1-^{13}C]pyruvic acid (or any other isotopic labeling), 2-keto-[1-^{13}C]isocaproic acid, and *bis*-1,1-(hydroxymethyl)-[1-^{13}C]cyclopropane-d$_8$ (HP001). All three molecules are liquids at room temperature and dissolve well a hydrophilic EPA.

For other compounds, it is necessary to prevent crystallization by mixing or dissolving the compound in a suitable solvent such as glycerol or dimethylsulfoxide (DMSO) can be used as solvent for the molecule and the EPA. For in vivo studies, it is necessary to formulate the molecule in a concentrated form in order to achieve a high concentration of the molecule after dissolution. To give an estimate of the requirements, a patient dose of 0.1 mmol kg^{-1} body weight can be assumed for an imaging agent with low toxicity, requiring approximately 10 mmol of compound. This means approximately 1 g of imaging agent, assuming

Table 12.1. Imaging agents that have been polarized with dDNP and imaged in vivo

Agent	Products	Refs
[1-^{13}C]pyruvate	[1-^{13}C]lactate, [1-^{13}C]alanine, [^{13}C]bicarbonate, ^{13}CO$_2$	20
[2-^{13}C]pyruvate	[2-^{13}C]lactate, [2-^{13}C]alanine, [1-^{13}C]acetyl-carnitine, [1-^{13}C]citrate, [5-^{13}C]glutamate	21
[1,2-^{13}C$_2$]pyruvate	[1,2-^{13}C$_2$]lactate, [1,2-^{13}C$_2$]alanine, [1-^{13}C]acetyl-carnitine, [1-^{13}C]citrate, [5-^{13}C]glutamate, [^{13}C]bicarbonate, ^{13}CO$_2$	22
[1-^{13}C]lactate	[1-^{13}C]pyruvate, [1-^{13}C]alanine, [^{13}C]bicarbonate, ^{13}CO$_2$	23
^{13}C-bicarbonate	^{13}CO$_2$	24
[1,4-^{13}C$_2$]fumarate	[1,4-^{13}C$_2$]malate	25
[2-^{13}C]dihydroxyacetone	[2-^{13}C]glycerol-3-phosphate, [2-^{13}C]phosphoenolpyruvate	26
[2-^{13}C]fructose	[1-^{13}C]fructose-6-phosphate	27
[5-^{13}C]glutamine	[5-^{13}C]glutamate	28
[1-^{13}C]ethylpyruvate	[1-^{13}C]pyruvate, [1-^{13}C]lactate, [1-^{13}C]alanine, [^{13}C]bicarbonate, ^{13}CO$_2$	29
[U-^{13}C$_2$,U-^{2}H]glucose	Full glycolysis, lactate, alanine, bicarbonate	30
[1-^{13}C]acetate	[1-^{13}C]acetylcarnitine	31, 32
[^{13}C]urea/[^{13}C,^{15}N$_2$]urea	None	33, 34
bis-1,1-(hydroxymethyl)-[1-^{13}C]cyclopropane-d8 (HP001)	None	35
α-keto-[1-^{13}C]isocaproate	[1-^{13}C]leucine	36
[1-^{13}C]dehydroascorbic acid	[1-^{13}C]ascorbic acid	37, 38
[1-^{13}C]alanine	[1-^{13}C]lactate, [1-^{13}C]pyruvate, [^{13}C]bicarbonate	39
[1-^{13}C]glycerate	[1-^{13}C]pyruvate, [1-^{13}C]lactate	40
[1,3-^{13}C$_2$]acetoacetate	[1,3-^{13}C$_2$]β-hydroxybutyrate, [1-^{13}C]acetyl-carnitine, [5-^{13}C]glutamate, [1-^{13}C]citrate	41, 42
[1-^{13}C] β-hydroxybutyrate	[1,3-^{13}C$_2$]acetoacetate, [1-^{13}C]acetyl-carnitine	41

a molecular weight of 100 g mol^{-1}. To keep the sample size reasonable requires a solubility of 30–50%. A solvent mixture with high solubility for the molecule and EPA, preventing sample crystallization and with good in vivo tolerance, therefore has to be chosen. An example of a biologically compatible formulation is fumaric acid (e.g., [1,4-^{13}C$_2$, 2,3-D$_2$]fumaric acid) in DMSO. DMSO is a widely used solvent for pharmaceuticals and has a good safety profile. As a saturated solution of fumaric acid with a molarity of 3.6 mol l^{-1} or 1 : 1.8 (w/w), the solution forms an amorphous solid when frozen, if the cooling rate is not too slow. Dry DMSO should be used as small amounts of water will decrease solubility and increase supersaturation.

Another means of improving the solubility involves changing the counter ion of salts. Solubility typically increases with increasing size of the counter ion, and two examples of this can be mentioned: The cesium salt of bicarbonate, CsH^{13}CO$_3$, and the TRIS salt of acetate.[44] Both of these salts have higher solubility than their sodium counterpart. Finally, for amino acids (zwitter ions at neutral pH) it has shown that either high or low pH preparations increase the aqueous solubility by reducing the charge of the molecule to a point[44] that no or little glycerol is needed to form an amorphous sample.

Finally, Ji *et al.*[45] showed that microcrystals of solids can be polarized by impregnation with a water/glycerol solution of a nitroxide. The proton polarization of the microcrystals is enhanced by spin diffusion from the impregnation. However, reduced polarization and longer polarization time is observed.

12.6 DISSOLUTION AND RELAXATION IN THE LIQUID STATE

To make the polarized solid sample useful for in vivo imaging, we dissolve the sample in a suitable buffer. Depending on the DNP sample, the dissolution may involve neutralization of the agent with acid or base. Buffering of the solution may be required to maintain control of pH within the physiologic range of 6.8–8.1. Physiological buffers such as tris(hydroxymethyl)aminomethane (TRIS) or 4-(2-hydroxyethyl)piperazine-1-ethanesulfonate (HEPES) are commonly used. An isotonic formulation is desired. This may mean lowering the concentration of the imaging agent, if it is hypertonic, or adding sodium chloride if hypotonic. The dissolution has to be efficient and fast compared to the nuclear T_1 in order to preserve the nuclear polarization during the phase transition. Formulating the solid sample as beads or powder may improve the dissolution in terms of polarization loss and recovery of the imaging agent. Optimizing the fluid dynamics as well as providing the necessary heat is essential for optimal performance of more difficult agents.[46] To minimize relaxation, dissolution is performed inside the cryostat in the high field of the polarizer (e.g., ~3 T in the case of a 3.35 T polarizer), but above the liquid helium surface. As an example, the T_1 of the C-1 of [1-^{13}C]pyruvic acid at 9.4 T is ~1.6 s at 0 °C (unpublished data), and is attributed to dipolar relaxation by the solvent and methyl protons. According to theory, the minimum T_1 scales with B_0, which means that a minimum T_1 of 0.7 s should be expected during the dissolution in the 3 T polarizer field. Dissolution should happen on a faster time scale to avoid a loss of polarization. The severity of the problem will depend on the target spin and sample properties, but several parameters can be controlled, e.g., distances to other spins (labeling position), the abundance of other spins (full or partial deuteration, as well of the solvent), and the concentration of the EPA.

Any paramagnetic impurities that could increase the relaxation rate can be chelated by adding, for example, ethylenedinitrotetraacetic acid (EDTA) to the dissolution solvent. In most cases, the EPA or Gd chelate do not cause significant relaxation after dissolution and may be safe to inject into animals. For preclinical imaging it is not required to remove the EPA. The same applies to the Gd chelate in case it is used in the formulation. However, the solution may undergo a filtration or chromatography step to remove the EPA involved in the DNP process. In case a Gd chelate has been added, this agent may be removed as well. The filtration can either be in-line with the dissolution process or a subsequent step. In either case, the filtration is completed in a matter of a few seconds with insignificant loss of polarization or target molecule. As an example, in the case of clinical studies with pyruvate, the solid sample is neat pyruvic acid (^{13}C-labeled) with 15 mM trityl radical (AH111501). The solid sample is dissolved in water-for-injection, which causes the trityl to

precipitate under the acidic conditions. The solid EPA is then removed by filtration, and the pyruvic acid is neutralized post dissolution.

12.7 dDNP INSTRUMENTATION

It is important to choose magnetic field and temperature conditions for DNP that lead to high nuclear polarization, but at the same time are easily achievable. Temperatures of ~1 K can be achieved by pumping on liquid helium. In the original dDNP polarizer design, liquid helium was supplied to the sample space through a needle valve from the magnet cryostat, but an alternative design has used a separate helium dewar.[47] In both cases, the low temperature is achieved with large mechanical pumps (200–500 m^3 h^{-1}) that vent the gas to atmosphere. This makes the helium consumption and running costs high. The first polarizer had a magnetic field strength of 3.35 T, since microwave sources are readily available at 94 GHz for irradiation of the electron spin. The microwave power is guided to the sample through oversized waveguide in stainless steel to reduce thermal heat loads. Similarly, the NMR probe is constructed with a stainless steel coaxial cable. To ensure that the sample is effectively cooled, it is typically immersed in liquid helium. The gas phase does not provide sufficient thermal conductivity (vapor pressure of 1 mbar at 1.2 K). The sample is loaded in a cup that is immersed in the helium bath. At dissolution the sample is raised out of the helium bath, but stays within the magnetic field, and a dissolution stick is introduced quickly and seals with the sample cup before the superheated solvent is injected at a typical pressure of 10 bar (water vapor pressure at 180 °C). The latent heat of an organic sample is typically 500–1000 J g^{-1}, and this energy is easily provided by the superheated solvent (4.2 J K^{-1} g^{-1}). By proper design of the dissolution system, low dilution factors can be obtained.[48]

At a temperature of approximately 1 K, the electron T_1, T_{1e}, is in the range of 0.1–1 s. The EPR line is effectively homogeneous at the concentrations used for DNP, and the resonant absorption for complete saturation can be estimated as $P = \gamma \hbar N / 2 T_{1e}$, which for 15 mM electron spins and T_{1e} of 1 s is equal to 0.5 mW cm^{-3}. For shorter relaxation time and higher spin concentration, more MW power is needed, but the required power is relatively modest for low-temperature DNP. Nitroxides are inhomogeneously broadened and microwave modulation is beneficial to efficiently saturate the EPR line.[49]

A DNP polarizer designed with sterile use intent[17] was published with these main features:

1. To provide a sterile barrier to the product through a single-use fluid path (FP).
2. To eliminate consumption of liquid cryogens.
3. To increase throughput by having four independent parallel samples.
4. To increase the size of the individual DNP samples up to 2.0 ml.
5. To automate the operation and remove operator variability and interventions.
6. To add quality control (QC).

The polarizer (SPINlab, GE Healthcare) operates at 5 T and 0.9 K in a closed cycle operation. There is no loss of cryogens in the operation. The cryostat consists of a superconducting magnet with liquid helium, but the cooling power is provided by a cryocooler. The sample cooling is provided by a sorption pump, which is a canister filled with 7 kg of charcoal that will adsorb helium when cooled and releases the helium gas again when heated. The charcoal is cooled by closing a thermal switch connecting the sorption pump to the cryocooler driving the temperature of the charcoal to 5–6 K. The helium gas adsorbs, the pressure drops, and the helium bath temperature drops. This can continue until the charcoal saturates and the helium bath runs dry. The adsorbed helium can then be regenerated by opening the thermal switch, heating the charcoal electrically, and condensing the released helium gas back into the helium vessel surrounding the sample space. The sample space contains up to four samples simultaneously and has four independent loading systems. On the top side of the gate valves are four air locks that together with the sliding seal on the FP allow loading and unloading of samples without breaking the vacuum of the sample space. This improves throughput and reduces dramatically heat load to helium bath. By gradually inserting the sample, more than 95% of the sample heat is rejected to the cryocooler and only 30 J ends up in the bath for a 1 g sample. With this design a base temperature of less than 0.9 K can be sustained for more than 3 days. When the system is heavily used, the heat loads increase, but full-day operation is achieved. The system is regenerated overnight to be ready for next day use.

The polarizer uses a closed FP to provide the sterile barrier. It is a sealed plastic component with all the pharmaceuticals pre-loaded. It can be single use and disposable for clinical use, but can be reused for non-sterile applications (in vitro and animal research).

It has a vial that can hold up to 2 ml of sample. The vial is glued or welded to a long tube that allows it to reach the low-temperature region. The other end of the tube is connected to a syringe with the solvent that sits outside the cryostat in a heater. Inside this outer tube is an inner tube, which allows solvent to be ejected onto the sample during dissolution. The syringe allows the solvent to be heated to a desired temperature and released at a controlled flow rate. The dissolved sample returns along the outer tubing, past the syringe, and into the receiver where all the QC is performed. A sliding seal on the outer tubing allows the sample to be moved through the cryostat without breaking the vacuum. In contrast to other published dissolution systems, where the solvent vapor pressure is the drive pressure for the solvent, this design allows independent control of solvent temperature and flow rate, which is critical in ensuring an optimal dissolution and full preservation of polarization for more complicated samples.[50] The FP has the advantage that the all-plastic component does not leach any paramagnetic metals into the sample that can reduce T_1 of the hyperpolarized agent.

QC is essential for clinical use of hyperpolarization. dDNP does not involve any chemistry, but is a complex compounding process. In the case of pyruvate, the most used formulation is based on pyruvic acid with trityl, and it involves a neutralization step. The pyruvic acid formulation is advantageous since neat pyruvic acid is 14 M, and a human dose can be achieved with approximately 1.5 g of pyruvic acid (or 1.2 ml). Other formulations with lower molarity require correspondingly larger sample volume. This is a requirement that needs to be considered for any agent formulation that will be translated to the clinic. In the case of pyruvate, the QC involves measurement of pH, temperature, volume, pyruvate concentration, radical concentration (to verify removal), and polarization. These measurements are performed optically and by NMR without product contact. The time from start of dissolution to completion of all QC measurements is approximately 30 s. Even with 30 s, plus the time to inject, significant polarization is lost for agents with relatively short T_1 (e.g., glucose with a T_1 of ca 15 s). Hyperpolarization, in general, pushes the boundaries of 'bed side' pharmacy, and safety needs to be demonstrated by validation or QC. For small animal studies, automated transfer and injection of the hyperpolarized agent has been demonstrated, with a total time from dissolution to injection of approximately 5 s.[51] It is unlikely that this will be acceptable clinically, but it may be possible to reduce the time for QC, transfer, and injection.

It has recently been shown that higher nuclear polarization can be obtained for both nitroxides and trityls by increasing the magnetic field strength[52–54] or lowering the temperature.[17] For [1-^{13}C]pyruvic acid, the ^{13}C polarization improved from 27% at 3.35 T to 60% at 4.64 T in the solid state and 60% at 7 T, all at 1–1.2 K.

Cross-polarization from protons to low gamma nuclei has proven to be very efficient for dilute proton systems and small samples.[55] At 6.7 T and 1.2 K, a ^{13}C polarization of 78% with a time constant of 470 s was achieved for [^{13}C]urea. However, efficient cross-polarization has yet to be demonstrated for large samples with abundant protons. This is due to the strong dipolar broadening of the proton resonance (linewidth of 60–80 kHz) in fully protonated samples and the size of the sample, as well as the tendency of probe arcing in the helium atmosphere of the polarizer.

A centralized production site (within a city or region) for hyperpolarized solid samples could be imagined. Two approaches have been proposed.[19,45] In both cases long relaxation times (16–20 h) have been demonstrated in a permanent magnetic field at liquid helium temperature. This could open up for during-the-day delivery of hyperpolarized solid samples, leaving only the dissolution and QC at the local site. This is obviously, not only interesting for medical applications, but could open up for wider access to hyperpolarization.

12.8 IMAGING

12.8.1 Hardware Requirements

The sensitivity of the MR experiment is proportional to the polarization, which is created by the polarizer. For an MR scanner operating at a few megahertz or higher, the dominating source of noise is the imaged object (i.e., the patient), which scales linearly with frequency in the same way as the induced signal voltage. This means that a scanner field strength of only approximately 0.5 T should be sufficient to reach the crossover point between electronic and sample noise. However, for small coil elements (surface coils or array coils) or small animals, the required field strength is higher, and even at 3 T (32 MHz), it can be challenging to provide sufficient loading of the coil(s).

Standard clinical MR systems and coils are designed to transmit and receive radio frequency (RF) signals at ^{1}H resonance frequency only. However, multi-nuclear

spectroscopy is available from most manufacturers of whole-body MR scanners. This option allows the system to perform MR experiments on non-proton nuclei of interest such as ^{13}C and ^{15}N. Multi-nuclear spectroscopy requires a broadband RF power amplifier, in addition to the standard ^{1}H narrowband amplifier. Secondly, the system will not have a built-in coil for transmission at the ^{13}C frequency. The ^{13}C coil can be designed for both transmission and reception, e.g., as a birdcage volume coil or surface coil, or reception can be by single or multiple coil elements in close proximity to the object. Narrowband low-noise-preamplifiers for ^{13}C are required that are built into the MR scanner or into the coil, with one preamplifier generally required for each receive channel. Since it is desirable to perform both ^{1}H anatomical imaging and hyperpolarized ^{13}C metabolic imaging during the same exam without repositioning the subject, the ^{13}C RF coil design and setup need to preserve the ability to perform ^{1}H imaging with minimal compromise of image quality. The coil configuration and design can be further optimized for imaging a particular organ/anatomy. For example, in the first proof-of-concept clinical trial of hyperpolarized ^{13}C metabolic imaging in prostate cancer patients, a ^{13}C transmit-only volume coil built into a custom patient table was used in conjunction with a receive-only endo-rectal coil containing both a ^{13}C and a ^{1}H element for signal reception. The scanner body coil was used for ^{1}H RF transmission during ^{1}H imaging.[2] A multichannel ^{13}C receive-only array coil suited for other human applications has also been demonstrated recently.[56] Receive-array coils may be particularly advantageous for hyperpolarized imaging in order to improve the encoding efficiency to cover the spatial, spectral, and kinetic dimensions within the time window of the hyperpolarized signal. The challenge is, however, to eliminate electronic noise for small coil elements at the relatively low ^{13}C resonance frequency ($\sim$32 MHz). Regardless of the coil design and combination, the MR system needs to be configured such that the correct coils/channels are active or disabled during specific periods of the scans to avoid signal degradation due to coupling between coils.

Spatial encoding of the MR signal in imaging takes place by time-varying magnetic field gradients. It is important to note that for a given magnetic field gradient, the spatial variation in resonance frequency is proportional to the gyromagnetic ratio, and, thus, the maximum spatial resolution achievable for ^{13}C imaging is approximately a quarter of that of the proton under the same imaging conditions. Pulse sequences for ^{13}C imaging need to take this limitation into consideration.

It may be possible to circumvent the low gyromagnetic ratio limitation by transferring the ^{13}C or ^{15}N magnetization to neighboring ^{1}H nuclei for detection, see indirect detection further down. However, simultaneous RF transmission at both ^{1}H and the low γ nucleus frequencies is required for the polarization transfer pulse sequence, and this capability may not be available on some clinical MR scanners. Proton decoupling during ^{13}C acquisition may be advantageous to remove long-range couplings and increase T_2^*.[57]

12.8.2 Magnetic Resonance Spectroscopic Imaging (MRSI)

Hyperpolarized ^{13}C MR imaging requires acquisition of both the spatial distribution and kinetics of the metabolite signals. Early work in hyperpolarized ^{13}C imaging employed concentric phase encoding and variable flip angle[20] to acquire chemical shift images (CSI) in two dimensions within a time window that coincides with the maximum signal of metabolic products. The optimum acquisition depends on the bolus injection, the organ of interest, and perfusion. Typical CSI scan parameters would be a 16×16 matrix and 80 ms repetition time, i.e., a total scan time of 20 s. More efficient encoding can be achieved with echo planar spectroscopic imaging (EPSI). EPSI with flyback or symmetric gradient waveform traverses time and one spatial frequency domain in a single readout period, shortening the acquisition, and allowing either single-time-point 3D MRSI or time-resolved multi-slice 2D MRSI[57,58] on a standard clinical 3 T system (with a maximum gradient strength of $40\,\text{mT}\,\text{m}^{-1}$ and slew rate of $150\,\text{mT}\,\text{m}^{-1}\,\text{ms}^{-1}$). There is a trade-off between spectral bandwidth and spatial resolution in the design of these gradient trajectories. Typically, a 5 mm resolution is achievable with 500 Hz spectral bandwidth without spectral aliasing of [1-^{13}C]pyruvate and its metabolic products (except ^{13}C-bicarbonate, which can be folded into a spectral region with no signal). On a preclinical MR scanner, 3D EPSI with high temporal resolution (2 s) has been demonstrated due to the better RF and gradient performance of the system.[59] A similar trade-off also exists for spiralCSI, which employs spiral readout gradients to sample a full plane in a single readout, and concatenates the spiral gradients multiple times for chemical shift encoding. However, even with multiple interleaves to minimize the impact of gradient slew rate, the 2D spiral readout time can result in a spectral bandwidth that is insufficient to fully cover the metabolite chemical shift range. On a clinical system, spiralCSI completes a 2D MRSI of a single slice in 375 ms, a 50-fold reduction in scan time compared to the conventional CSI method. However, for clinical applications that require a large field-of-view (FOV), spiralCSI acquisition time may increase drastically due to the increase of interleaves required to maintain the same spatial resolution and spectral bandwidth. In addition, spiralCSI encodes a circular FOV and can become inefficient for a region of interest with an asymmetric FOV. On the other hand, EPSI allows asymmetric FOV and the FOV in the one direction encoded by the EPSI readout is virtually unlimited due to the high sampling rate available on all clinical systems. The efficiency of the various acquisition schemes has been analyzed by Durst *et al.*[60]

Dynamic metabolic imaging can be used to determine rate constants, and signal averaging over the time course for each voxel can regain most of the signal-to-noise-ratio (SNR) observed in optimized single-time-point methods. In preclinical studies, five dimensional (three spatial, the spectral and the kinetics), MRSI has been demonstrated by using spiralCSI and compressed sensing, both giving high-quality images and dynamic curves. Taking advantage of the considerable sparsity in hyperpolarized ^{13}C spectra, compressed sensing pseudo-randomly under-sampled the spectral and two spatial dimensions during EPSI flyback readout, yields up to a factor of 7.53 in acceleration[61] relative to the conventional 3D EPSI sequence.[62] The acceleration can be used to improve spatial resolution and decrease acquisition time, or to cover a larger FOV, which will be useful for clinical applications. The trade-off of this technique is the loss of metabolite peaks with low SNR.

Another approach is IDEAL spiralCSI.[63] IDEAL requires a priori information of the resonance frequencies of the ^{13}C metabolites to minimize the number of excitations for spectral decoding. Spectral decoding is accomplished by varying the echo time (delay between excitation pulse and spiral magnetic field gradient) from excitation to excitation in an optimal way based on the number and specific spectral frequencies of the metabolites. There is no trade-off between spatial resolution and spectral bandwidth and therefore the spatial resolution can be as high as SNR permits. This technique has been demonstrated in time-resolved 2D imaging, with a potential of combining with a pulse-and-acquire FID acquisition to obtain a spectrum to define the a priori information.

12.8.3 RF Designs to Optimize SNR

The signal from the injected, relatively concentrated, hyperpolarized ^{13}C substrate is initially often 5–10 times larger than the signals of its metabolic products. Multiband spectral–spatial RF excitation pulses[64] use spectral selectivity to excite the injected hyperpolarized ^{13}C substrate with a lower flip angle than the metabolic products. This way substrate polarization is preserved and product SNR maximized. The metabolic products are observable for a longer time than a uniformly constant flip angle strategy. A recent development combining multiband RF pulse design and compressed sensing random sampling created a sequence for time-resolved 3D MRSI acquisition[61,65] with good SNR. The flip-angle of the injected ^{13}C substrate and that of the products can be optimized for optimal contrast-to-noise-ratio (CNR) ratio for a particular organ or for disease characterization. An alternative to spectroscopically resolving multiple metabolites is to excite each metabolite selectively by a spectral–spatial pulse.[62] A multi-slice cardiac-gated sequence consisting of a large flip-angle spectral–spatial excitation pulse with a single-shot spiral trajectory was developed for ^{13}C imaging of cardiac metabolism.[66] The sequence alternates between the frequencies corresponding to each metabolite. Likewise, Gordon *et al.*[67] demonstrated a symmetric EPI with spectral–spatial excitation based on the product sequence available on a clinical 3 T scanner. The limitation of (spectral) selective excitation is that the length of the RF pulse often becomes long and leads to large signal loss during the pulse.

12.8.4 T_2-based Sequences

Long T_2 relaxation times of ^{13}C metabolites were reported in a rat hepatocellular carcinoma (HCC) study[58] using a single-voxel preparation pulse followed by a train of spin-echoes to measure the T_2 decay of the signal within the voxel. T_2s of [1-^{13}C]alanine and [1-^{13}C]lactate were found to be longer in HCC tumors (1.2 and 0.9 s, respectively) than in normal liver (0.4 and 0.5 s, respectively). A large SNR gain is expected by using T_2-based sequences as compared to T_2^*-based

sequences. Most of the sequences discussed earlier are limited by signal decay given by T_2^* that can be as long as 100–200 ms for [1-^{13}C]pyruvate, but is reduced to about 25 ms for [1-^{13}C]lactate and [1-^{13}C]alanine due to stronger J_{CH} coupling. T_2-based sequences, such as multi-echo balanced-steady-state-free-precession (bSSFP)[68,69] and fast spin echo echo planar spectroscopic imaging (FSE-EPSI),[70] have significant signal gain and are excellent for single-time-point and dynamic MRSI. The challenge to utilize this strategy for time-resolved MRSI lies in the strong RF depletion during the echo train. In multi-echo bSSFP, a series of gradient echoes are acquired between each RF pulse. The gradient echoes are used to spectrally decode the involved metabolites. More recently, the spectral selectivity of the RF pulse was exploited in a 3D bSSFP sequence to separate the spectral components of pyruvate and its metabolites as well as co-polarized urea.[71] The sequence was able to acquire a $32 \times 16 \times 16$ spatial matrix in less than a second.

The long-range coupling from the C-1 carbon of lactate to the methyl protons enables indirect detection of the carbon signal. Wang *et al.*[72] have recently demonstrated this in vivo on a clinical 3 T scanner in a murine tumor model using a reverse INEPT (insensitive nuclei enhanced by polarization transfer) sequence. However, at this point it is still unclear whether the theoretical sensitivity gain can be realized given the relaxation losses during transfer and the imperfections of pulses.

12.9 KINETIC MODELING

Hyperpolarized pyruvate-to-lactate signal–time curves have been described by a two-site exchange[73] model. The conversion of [1-^{13}C]pyruvate to [1-^{13}C]lactate, as observed in hyperpolarized metabolic imaging, is a combination of flux (net creation of lactate) and exchange (^{13}C enrichment of the endogenous lactate pool). The impact of lactate pool size and proof of exchange was shown in cells preconditioned with unlabeled lactate. With increasing lactate pool size, an increase in the hyperpolarized [1-^{13}C]lactate was observed. The importance of exchange has been demonstrated in a lymphoma model by magnetization transfer between pyruvate and lactate after inversion of the magnetization for one of the species.[74] The authors concluded that the steady-state lactate pool size is likely the limiting factor of the observed hyperpolarized [1-^{13}C]lactate signal.

The availability of the reduced form of nicotinamide adenine dinucleotide, NADH, from sources beyond lactate dehydrogenase (LDH) catalyzed exchange, also impacts the conversion of hyperpolarized [1-^{13}C]pyruvate to [1-^{13}C]lactate. For example, added NADH from aldolase processing of ethanol in liver[75] has been shown to increase the flux of hyperpolarized [1-^{13}C]pyruvate to [1-^{13}C]lactate.

Under saturating conditions, the apparent rate constant, k_{pl}, increases as the pyruvate dose decreases.[76] The small-tip angle, pulse-and-acquire dynamic curves are biased by the substrate dose, bolus shape, and accumulated in flow of [1-^{13}C]lactate. These factors can be eliminated by using a saturation recovery method,[68] resulting in dynamic curves that describe the instantaneous metabolic conversion at the local tissue level during the passage of hyperpolarized [1-^{13}C]pyruvate.

For quantification of hyperpolarized spectra, a commonly used software is jMRUI.[77] For the estimation of kinetic rate constants, mathematical modeling is typically performed[63,78] to analyze the signal–time curves. It is not always required, since it was shown that the ratio of observed metabolites scales linearly with the kinetic rate constants estimated using mathematical modeling approaches.[79,80] In specific cases, it is possible to derive a mathematical framework, which calculates the kinetic rate constants directly based on the ratio of observed metabolites, and their longitudinal relaxation time, obliterating the need for using complex mathematical modeling.[63] The slowly varying kinetics (sparsity) of the metabolism can be exploited to accelerate the dynamic acquisition with methods such as k-t-PCA and k-t-SPARSE,[81] using Principal Component Analysis and compressed sensing, respectively.

12.10 APPLICATIONS

12.10.1 Oncology

The first clinical study of hyperpolarized ^{13}C-pyruvate metabolic imaging of prostate cancer patients was a Phase 1/2a Ascending-Dose Study to Assess the Safety and Tolerability and Imaging Potential of Hyperpolarized Pyruvate (^{13}C) Injection in Subjects with Prostate Cancer.[2] In this 33 patient study the primary objective was to assess the safety of the hyperpolarized pyruvate (^{13}C) injection. The secondary objectives were to

determine: (i) The kinetics of hyperpolarized pyruvate injection delivery and metabolism in the prostate and (ii) to determine the SNR for ^{13}C pyruvate metabolites in regions of cancer as a function of the dose of the hyperpolarized pyruvate (^{13}C) injection. The highest dose level was well tolerated, and excellent contrast-to-noise for [1-^{13}C]lactate was observed in the tumor. The dynamic spectroscopy showed rapid bolus arrival of the hyperpolarized [1-^{13}C]pyruvate to the prostate, within 30 s; no detectable lactate in the normal prostate, but a large lactate signal from the cancer. When knowing the pharmacokinetics, the imaging could be initiated at the optimal time point, to generate metabolic maps from an EPSI sequence. Figure 12.3 shows an example of a patient that had a known abnormality in the right gland on the T_2 proton image, but nothing noticeable on the left side. However, the lactate/pyruvate image showed bilateral disease, which was later confirmed by pathology.

Using hyperpolarized [1-^{13}C]pyruvate, Hu *et al.*[83] studied liver metabolism in fasted rats and found higher lactate to alanine signal ratios and lower alanine signal level in the fasted rats than in free-fed rats. The low alanine signal is most likely due to a reduction of alanine aminotransferase (ALT) activity in fasted rat liver during gluconeogenesis. Alanine is also a good biomarker for HCC detection. Using hyperpolarized [1-^{13}C]pyruvate, Darpolor *et al.*[84] found elevated alanine and lactate levels, consistent with enzyme expression analysis on rat HCC tissue extract. Interestingly, ^{13}C MRSI showed high alanine signals specifically in HCC tumors whereas it showed high lactate signals in the HCC tumors and in blood vessels. Low ^{13}C-alanine signals in vessels may be due to the much slower transport of alanine than lactate from cells to blood. Therefore, within the time window of hyperpolarization, less alanine signal was observed in vessels and normal liver (Figures 12.4 and 12.5).

Hyperpolarized pyruvate studies in glioma rat models[85,86] have shown that the lactate signal correlates with response to treatment. The pyruvate uptake was higher in the tumor than in the normal brain due to the disruption of the blood brain barrier (BBB) in gliomas. The first human brain tumor case reports demonstrate both baseline lactate and bicarbonate signal in the normal brain, and differentiation of the lactate signal within the tumor.[4,87]

Day *et al.*[73] reported decreased flux between pyruvate and lactate in lymphoma tumors when treated with etoposide and interrogated with hyperpolarized [1-^{13}C]pyruvate. Etoposide induces

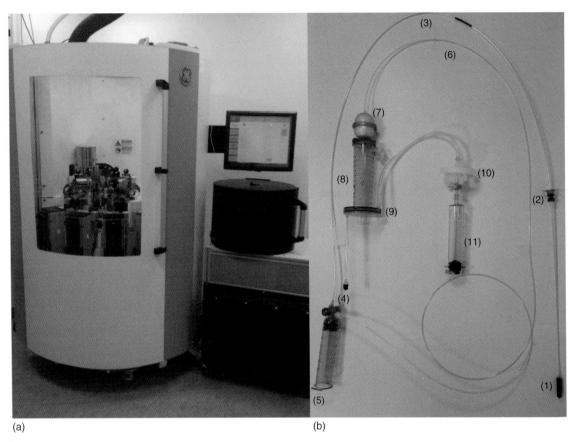

(a) (b)

Figure 12.3. (a) SPINlab polarizer. The polarizer operates at 5 T and 0.9 K. The polarizer has four independent channels that allow simultaneous polarization of up to four samples (fluid paths). The QC module (cylindrical black unit) is seen to the right of the photo below the touch screen. (b) Fluid path (FP). (1) Vial, (2) dynamic seal, (3) coaxial tube, (4) valve, (5) dissolution syringe, (6) transfer tube, (7) EPA filter, (8) receiver vessel, (9) QC appendage, (10) sterile assurance filter (11), and Medrad 65 ml MR syringe. The vial loads into the polarizer (behind the sliding door). The vial is inserted into one of the four airlocks and the dissolution syringe is inserted into the corresponding heater-pressure module. After some pump–flush cycles, the airlock gate-valve will open and the vial can be pushed through the dynamic seal to the 0.9 K sample space. The receiver vessel is initially loaded into the corresponding warmer module seen to the lower right in (a). Shortly before dissolution the receiver vessel is moved from the warmer to the QC module. (Reproduced with permission from Ref. 82. © Elsevier, 2016)

apoptosis and loss of NADH due to activation of poly-ADP-ribose-polymerase (PARP). Loss of the coenzyme NADH leads to reduced LDH activity. Another hyperpolarized ^{13}C substrate that has been studied in lymphoma is 2-keto-[1-^{13}C]isocaproate (KIC).[88] KIC is metabolized to leucine by branched chain amino acid transferase (BCAT), a biomarker for metastasis in some tumors and a target of proto-oncogene c-myc. Following injection of hyperpolarized KIC a sevenfold higher signal of ^{13}C-leucine in murine lymphoma than in healthy

tissue was found. In the same study, no ^{13}C-leucine was observed in rat mammary adenocarcinoma. Ex vivo BCAT expression analysis yielded a high BCAT level in murine lymphoma and very low in rat mammary tumor, consistent with the hyperpolarized ^{13}C metabolic imaging findings. In a study of treatment response in lymphoma tumors with hyperpolarized [1,4-^{13}C$_2$]fumarate, Gallagher *et al.*[25] found that [1,4-^{13}C$_2$]malate from the labeled fumarate is a sensitive marker of necrosis. Fumarate uptake is slow, and malate is only observed in tissue with

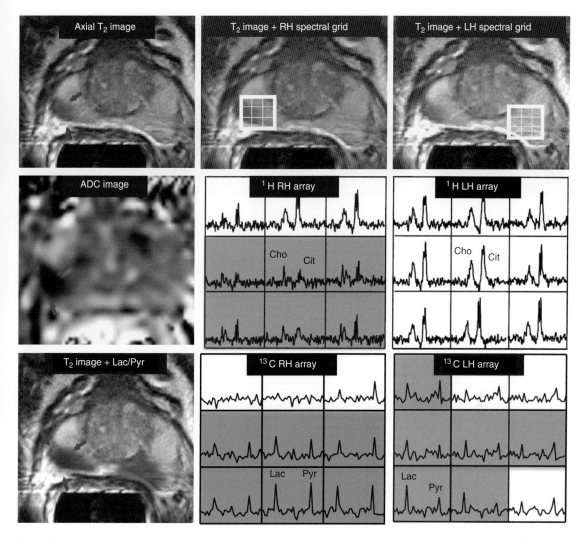

Figure 12.4. 2D single-time-point MRSI data. Images were obtained from a patient with serum PSA of 9.5 ng ml^{-1}, who was diagnosed with bilateral biopsy-proven Gleason grade $3 + 3$ prostate cancer and received the highest dose of hyperpolarized [1-^{13}C]pyruvate (0.43 ml kg^{-1}). The axial T_2-weighted image shows a unilateral region of reduced signal intensity (red arrows), which is consistent with a reduction in the corresponding ADC. The ^{1}H spectral arrays supported these findings, with voxels with reduced citrate and elevated choline/citrate (highlighted in pink) on the right side of the gland and voxels with normal metabolite ratios on the left side. The ^{13}C spectral arrays show voxels with elevated levels of hyperpolarized [1-^{13}C]lactate/[1-^{13}C]pyruvate (highlighted in pink) on both the right and left sides of the prostate. The location of colored regions in the metabolite image overlay had a ratio of [1-^{13}C]lactate/[1-^{13}C]pyruvate greater than or equal to 0.6. (Reproduced with permission from Ref. 2. © American Association for the Advancement of Science, 2013)

disintegrating cell membrane allowing fumarase to leak into the interstitial space. The conversion was 2.4-fold higher in etoposide-treated lymphoma tumors, and the malate signal correlated strongly with necrosis.

12.10.2 Cardiology

Generation and utilization of adenosine triphosphate, ATP, in the heart are tightly regulated by physiological conditions and energetic needs. The heart uses fatty

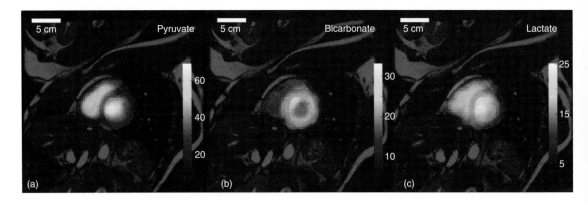

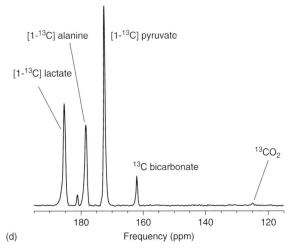

Figure 12.5.   Cardiac ^{13}C images from human subject displayed as color overlays on top of grayscale anatomical images. The [1-^{13}C]pyruvate substrate (a) is seen mainly in the blood pool within the cardiac chambers. Flux of pyruvate through the pyruvate dehydrogenase complex is reflected in the ^{13}C-bicarbonate image (b). The [1-^{13}C]lactate product (c) appeared with a diffuse distribution covering the muscle and chambers. Representative ^{13}C spectrum (d) of substrate and products from the heart of the subject. (Reproduced with permission from C.H. Cunningham, J.Y. Lau, A.P. Chen, B.J. Geraghty, W.J. Perks, I. Roifman, G.A. Wright, K.A. Connelly, Hyperpolarized ^{13}C Metabolic MRI of the Human Heart: Initial Experience., Circ. Res. (2016) CIRCRESAHA.116.309769. © American Heart Association, 2016)

acids, carbohydrates, and ketones as the substrates for energy production depending on availability. Altered myocardial substrate utilization is associated with diseases such as cardiomyopathy, hypertension, and diabetes; it also occurs during ischemia and reperfusion. Since all substrates are converted to acetyl-CoA prior to entering the Krebs cycle, measurement of the metabolic fluxes of acetyl-CoA production from various substrates can be used to monitor the changes in substrate selection and utilization. Pyruvate dehydrogenase (PDH) is the enzyme that decarboxylates the carbohydrate-derived pyruvate to acetyl-CoA and

CO_2, and the control of this enzyme's expression and activity is closely tied to myocardial substrate selection, thus the ability of using hyperpolarized ^{13}C pyruvate to noninvasively probe PDH flux is potentially a powerful diagnostic tool in cardiology.

Indeed, a number of recent reports in small and large animal models have demonstrated the ability of hyperpolarized ^{13}C MR imaging and spectroscopy to characterize the PDH flux noninvasively in normal hearts, and hearts during ischemia-reperfusion and cardiac disease.[89–91] In normal hearts, the ^{13}CO$_2$ derived from [1-^{13}C]pyruvate due to cardiac PDH flux is observed mostly as ^{13}C-bicarbonate (in equilibrium with ^{13}CO$_2$) signal in MRS data, and some [1-^{13}C]lactate and [1-^{13}C]alanine signals can also be observed. In spatially resolved ^{13}C MR imaging data obtained from large animal models and human subjects, the substrate signal was found to be localized mostly in the cardiac chambers while ^{13}C-bicarbonate was localized in the myocardium; [1-^{13}C]lactate signal was more diffuse and observed in both the blood and cardiac muscle.[3,92,93] Recently, it was demonstrated that lactate produced in other organs, the liver predominantly, can bias estimation of the metabolism.[94]

In models of ischemia and reperfusion, impaired PDH flux can be observed as decreased ^{13}C-bicarbonate signal shortly following reperfusion.[89,95] Potentially, the viability of the affected tissue may be probed by following the recovery of the PDH flux (or the lack of it) post reperfusion and assessment of interventions targeting this metabolic pathway may also benefit from this technique. Change in PDH flux due to diabetes has been investigated in small animal models.[91] It has also been reported in a porcine pacing model of dilated cardiomyopathy (DCM) that disease progression can be followed noninvasively with ^{13}C metabolic imaging using hyperpolarized [1-^{13}C]pyruvate. Altered cardiac PDH flux was found to be strongly associated with onset of decompensated DCM.[93] Monitoring cardiac substrate utilization in patients may provide valuable information regarding progression of these diseases and aid clinical management.

Monitoring of cardiac Krebs cycle flux in real time using hyperpolarized [2-^{13}C]pyruvate is also feasible since the C2 position on pyruvate is carried into the cycle through acetyl-CoA (instead of being released as ^{13}CO$_2$), and changes of Krebs cycle flux can be assessed by measuring changes in the [5-^{13}C]glutamate signal.[22] Along with PDH flux, these additional parameters obtainable by hyperpolarized ^{13}C MR provide insights into cardiac energetics and cellular environment that were not previously accessible noninvasively by other imaging modalities and may become valuable clinical tools in cardiology.

Although most of the efforts so far in utilizing hyperpolarized ^{13}C MR metabolic imaging in cardiology have been focused on probing substrate utilization using [1-^{13}C]pyruvate, ketone body utilization was recently studied in the rat heart.[41] The authors show that downstream metabolites (acetyl-carnitine and Krebs cycle intermediates) can be detected from both β-hydroxybutyrate and acetoacetate.

12.10.3 Neurology

Measuring BBB transport, inflammation, and redox in the brain has the potential to address unmet clinical needs in neurodegenerative disease, traumatic brain injury (TBI), and stroke. A substantial amount of pyruvate passes the normal BBB, and is converted into lactate and bicarbonate in brain tissue.[21,96,97] Thus, it should be possible to measure quantitatively BBB transport abnormalities and metabolism. However, a strategy to overcome the BBB transport limitation explored the use of the more lipophilic ethyl ester of pyruvate.[29]

TBI is known to cause perturbations in the energy metabolism of the brain. The ^{13}C-bicarbonate signal was found to be 25% lower in the injured hemisphere compared with the non-injured hemisphere, while the hyperpolarized bicarbonate-to-lactate ratio was 34% lower in the injured hemisphere in rats with moderate TBI after injection of hyperpolarized [1-^{13}C]pyruvate.[98]

Ethylpyruvate passes the BBB more efficiently than pyruvate due to the lipophilicity of the compound, and it hydrolyzes quickly to pyruvate and ethanol.[29] Ethylpyruvate has also been used therapeutically, and it may become a preferred substrate clinically for brain applications in the future. Several other dDNP agents have also been studied in the brain including ketoisocaproate (KIC)[36] and glucose.[30] KIC has higher uptake than pyruvate, and the leucine reported on branched-chain-amino-acid activity. With [2,3,4,6,6-^{2}H$_5$, 3,4-^{13}C$_2$]-D-glucose, the C-1 lactate signal was detected from the whole brain. All studies to date have been on the normal brain in rodents, and more studies in disease models will be required. Secondly, it is likely that clinical studies with pyruvate in patients with neurological disease will appear in the coming years. Some indication of the prospects for brain imaging is found in the initial brain tumor studies in patients (see Section 12.10.1).

12.10.4 pH Mapping

Many pathologies are associated with changes in tissue acid–base balance, including inflammation and ischemia.[99,100] For instance, most tumors have an acidic extracellular pH compared to normal tissue, and this can be correlated with prognosis and response to treatment.[101,102] Many chemotherapeutics are believed to be ineffective due to the acidic extracellular environment in tumors. Despite the importance of pH and its relationship to disease, there is currently no clinical tool available to image the spatial distribution of pH in humans. Gallagher *et al.* have shown that hyperpolarized ^{13}C-bicarbonate can be used to map pH in tumors[24] by using the Henderson–Hasselbalch equation and the relative signals of ^{13}C-bicarbonate and ^{13}CO$_2$. The tumor showed lower pH than the surrounding healthy tissues. Bicarbonate is well tolerated and has been used clinically already. However, it has two limitations: (i) the T_1 is relatively short in vivo and (ii) a significant fraction of the CO$_2$ exchanges on first pass in the lungs. Therefore, exogenous pH probes have been developed to overcome these limitations. Zymonic acid[103] and carbonates of glycerol[104] have longer T_1, and they have two intramolecular ^{13}C, one of which shifts with pH. It is also possible to determine pH from the CO$_2$ and bicarbonate produced when [1-^{13}C]pyruvate enters the Krebs cycle. In the heart, pH may also be assessed from the ^{13}CO$_2$ and ^{13}C-bicarbonate produced from [1-^{13}C]pyruvate,[22,90,105] if sufficient SNR is obtained for the ^{13}CO$_2$ signal.

12.10.5 Redox Imaging

An interesting imaging agent for the study of neurodegenerative disease may be [1-^{13}C]dehydroascorbic acid (DHA).[37,38] This molecule is transported across the BBB by the GLUT1 transporter, and the conversion rate of DHA to ascorbate is a biomarker for oxidative stress. Von Morze *et al.* investigated the ketone body ^{13}C-acetoacetate and its conversion to ^{13}C-β-hydroxybutyrate in vivo, catalyzed by β-hydroxybutyrate dehydrogenase as a marker of mitochondrial redox state in the kidney of rats treated with metformin[42] (notice that the same substrate is also discussed in Section 12.10.2). Likewise, both the lactate-to-pyruvate[106] and the alanine-to-pyruvate[39] ratio is a biomarker of redox state being proportional to free cytosolic [NADH]/[NAD$^+$].

12.11 OUTLOOK

dDNP is a research field in the infancy and the science is flourishing. This review presents state-of-the-art in field, as well as outlines some of the areas where further progress is needed to translate the technology to the clinic.

For hyperpolarized metabolic MR to have a role in medical imaging relies on finding indications with a real unmet medical need. With currently more than 20 SPINlab installed worldwide and seven sites with approval for human use, the coming years open up for multisite clinical trials and more extensive clinical research. At www.clinicaltrials.gov 15 recruiting clinical studies can be found, involving hyperpolarized pyruvate. The outlook for hyperpolarized metabolic imaging is still very promising. The coming years will hopefully allow multisite clinical trials to identify the patient populations that could form basis for full clinical development of hyperpolarized [1-^{13}C]pyruvate.

ACKNOWLEDGEMENT

The author gratefully acknowledges the Danish National Research Foundation (DNRF124).

RELATED ARTICLES IN EMAGRES

Hyperpolarization Methods for MRS

Integration of ^{13}C Isotopomer Methods and Hyperpolarization Provides a Comprehensive Picture of Metabolism

Maciel, Gary E.: A Little Polarization Can be Good for You

Overhauser, Albert W.: Dynamic Nuclear Polarization

REFERENCES

1. J. H. Ardenkjaer-Larsen, B. Fridlund, A. Gram, G. Hansson, L. Hansson, M. H. Lerche, R. Servin, M. Thaning, and K. Golman, *Proc. Natl. Acad. Sci. U. S. A.*, 2003, **100**, 10158. DOI: 10.1073/pnas.1733835100.

2. S. J. Nelson, J. Kurhanewicz, D. B. Vigneron, P. E. Z. Larson, A. L. Harzstark, M. Ferrone, M. van Criekinge, J. W. Chang, R. Bok, I. Park, G. Reed, L. Carvajal, E. J. Small, P. Munster, V. K. Weinberg, J. H. Ardenkjaer-Larsen, A. P. Chen, R. E. Hurd, L.-I. Odegardstuen, F. J. Robb, J. Tropp, and J. A. Murray, *Sci. Transl. Med.*, 2013, **5**, 198ra108. DOI: 10.1126/scitranslmed.3006070.

3. C. H. Cunningham, J. Y. Lau, A. P. Chen, B. J. Ger-
    aghty, W. J. Perks, I. Roifman, G. A. Wright, and
    K. A. Connelly, *Circ. Res.*, 2016, **119**, 1177. DOI:
    10.1161/CIRCRESAHA.116.309769.

4. V. Z. Miloushev, K. L. Granlund, R. Boltyanskiy, S. K.
    Lyashchenko, L. M. DeAngelis, I. K. Mellinghoff,
    C. W. Brennan, V. Tabar, T. J. Yang, A. I. Holodny,
    R. E. Sosa, Y. W. Guo, A. P. Chen, J. Tropp, F. Robb,
    and K. R. Keshari, *Cancer Res.*, 2018, **78**, 3755. DOI:
    10.1158/0008-5472.CAN-18-0221.

5. B. M. Goodson, N. Whiting, A. M. Coffey, P. Niko-
    laou, F. Shi, B. M. Gust, M. E. Gemeinhardt,
    R. V. Shchepin, J. G. Skinner, J. R. Birchall, M.
    J. Barlow, E. Y. Chekmenev, Hyperpolarization meth-
    ods for MRS, eMagRes, John Wiley & Sons Ltd,
    Chichester, UK, 2015: 797–810. DOI: 10.1002/
    9780470034590.emrstm1457.

6. A. W. A. W. Overhauser, *Phys. Rev.*, 1953, **92**, 411.
    DOI: 10.1103/PhysRev.92.411.

7. T. R. Carver and C. P. Slichter, *Phys. Rev.*, 1953, **92**,
    212.

8. A. Abragam and W. G. Proctor, *Phys. Rev.*, 1957, **106**,
    160. DOI: 10.1103/PhysRev.106.160.

9. C. D. Jeffries, *Phys. Rev.*, 1957, **106**, 164. DOI:
    10.1103/PhysRev.106.164.

10. A. Abragam, M. Goldman, Nuclear Magnetism: Or-
    der and Disorder, Clarendon Press, 1982. http://findit.
    dtu.dk/en/catalog/2300286932 (accessed January 16,
    2018).

11. W. T. Wenckebach, Essentials of dynamic nu-
    clear polarization, n.d. http://www.wenckebach.net/
    html/dnp-book.html (accessed February 5, 2018).

12. W. T. Wenckebach, *J. Magn. Reson.*, 2017, **284**, 104.
    DOI: 10.1016/J.JMR.2017.10.001.

13. W. T. Wenckebach, *J. Magn. Reson.*, 2017, **277**, 68.
    DOI: 10.1016/J.JMR.2017.01.020.

14. C. P. Slichter, Principles of Magnetic Resonance,
    Berlin, Heidelberg: Springer Berlin Heidelberg, 1990.
    DOI: 10.1007/978-3-662-09441-9.

15. K.-N. Hu, *Solid State Nucl. Magn. Reson.*, 2011, **40**,
    31. DOI: 10.1016/j.ssnmr.2011.08.001.

16. A. S. Lilly Thankamony, J. J. Wittmann, M. Kaushik,
    and B. Corzilius, *Prog. Nucl. Magn. Reson. Spec-
    trosc.*, 2017, **102–103**, 120. DOI: 10.1016/j.pnmrs.
    2017.06.002.

17. J. H. Ardenkjaer-Larsen, A. M. Leach, N. Clarke,
    J. Urbahn, D. Anderson, and T. W. Skloss, *NMR
    Biomed.*, 2011, **24**, 927. DOI: 10.1002/nbm.1682.

18. J. H. Ardenkjaer-Larsen, S. MacHoll, and H. Jóhan-
    nesson, *Appl. Magn. Reson.*, 2008, **34**, 509. DOI:
    10.1007/s00723-008-0134-4.

19. A. Capozzi, T. Cheng, G. Bocro, C. Roussel, and
    A. Comment, *Nat. Commun.*, 2017, **8**, 15757. DOI:
    10.1038/ncomms15757.

20. K. Golman, R. I. Zandt, M. Lerche, R. Pehrson,
    and J. H. Ardenkjaer-Larsen, *Cancer Res.*, 2006, **66**,
    10855. DOI: 10.1158/0008-5472.CAN-06-2564.

21. M. Marjańska, I. Iltis, A. A. Shestov, D. K. Deelc-
    hand, C. Nelson, K. Uğurbil, and P.-G. Henry,
    J. Magn. Reson., 2010, **206**, 210. DOI: 10.1016/j.jmr.
    2010.07.006.

22. A. P. Chen, R. E. Hurd, M. A. Schroeder, A. Z. Lau,
    Y. Gu, W. W. Lam, J. Barry, J. Tropp, and
    C. H. Cunningham, *NMR Biomed.*, 2012, **25**, 305.
    DOI: 10.1002/nbm.1749.

23. A. P. Chen, J. Kurhanewicz, R. Bok, D. Xu, D. Joun,
    V. Zhang, S. J. Nelson, R. E. Hurd, and D. B. Vi-
    gneron, *Magn. Reson. Imaging*, 2008, **26**, 721. DOI:
    10.1016/j.mri.2008.01.002.

24. F. A. Gallagher, M. I. Kettunen, S. E. Day, D.-E. Hu,
    J. H. Ardenkjaer-Larsen, R. in't Zandt, P. R. Jensen,
    M. Karlsson, K. Golman, M. H. Lerche, and
    K. M. Brindle, *Nature*, 2008, **453**, 940. DOI:
    10.1038/nature07017.

25. F. A. Gallagher, M. I. Kettunen, D.-E. Hu, P. R. Jensen,
    R. in't Zandt, M. Karlsson, A. Gisselsson, S. K. Nel-
    son, T. H. Witney, S. E. Bohndiek, G. Hansson,
    T. Peitersen, M. H. Lerche, and K. M. Brindle, *Proc.
    Natl. Acad. Sci. U. S. A.*, 2009, **106**, 19801. DOI:
    10.1073/pnas.0911447106.

26. I. Marco-Rius, P. Cao, C. von Morze, M. Merritt,
    K. X. Moreno, G.-Y. Chang, M. A. Ohliger, D. Pearce,
    J. Kurhanewicz, P. E. Z. Larson, and D. B. Vi-
    gneron, *Magn. Reson. Med.*, 2017, **77**, 1419. DOI:
    10.1002/mrm.26226.

27. K. R. Keshari, D. M. Wilson, A. P. Chen, R. Bok,
    P. E. Z. Larson, S. Hu, M. Van Criekinge, J. M. Mac-
    donald, D. B. Vigneron, and J. Kurhanewicz, *J. Am.
    Chem. Soc.*, 2009, **131**, 17591. DOI: 10.1021/
    ja9049355.

28. C. Canapè, G. Catanzaro, E. Terreno, M. Karlsson,
    M. H. Lerche, and P. R. Jensen, *Magn. Reson. Med.*,
    2015, **73**, 2296. DOI: 10.1002/mrm.25360.

29. R. E. Hurd, Y.-F. Yen, D. Mayer, A. Chen, D. Wil-
    son, S. Kohler, R. Bok, D. Vigneron, J. Kurhanewicz,
    J. Tropp, D. Spielman, and A. Pfefferbaum, *Magn. Re-
    son. Med.*, 2010, **63**, 1137. DOI: 10.1002/mrm.22364.

30. M. Mishkovsky, B. Anderson, M. Karlsson, M. H. Lerche, A. D. Sherry, R. Gruetter, Z. Kovacs, and A. Comment, *Sci. Rep.*, 2017, **7**, 11719. DOI: 10.1038/s41598-017-12086-z.

31. A. Flori, M. Liserani, F. Frijia, G. Giovannetti, V. Lionetti, V. Casieri, V. Positano, G. D. Aquaro, F. A. Recchia, M. F. Santarelli, L. Landini, J. H. Ardenkjaer-Larsen, and L. Menichetti, *Contrast Media Mol. Imaging*, 2014, **10**, 194. DOI: 10.1002/cmmi.1618.

32. U. Koellisch, C. Laustsen, T. S. Nørlinger, J. A. Østergaard, A. Flyvbjerg, C. V. Gringeri, M. I. Menzel, R. F. Schulte, A. Haase, and H. Stødkilde-Jørgensen, *Physiol. Rep.*, 2015, **3**, e12474. DOI: 10.14814/phy2.12474.

33. C. von Morze, R. A. Bok, G. D. Reed, J. H. Ardenkjaer-Larsen, J. Kurhanewicz, and D. B. Vigneron, *Magn. Reson. Med.*, 2014, **72**, 1599. DOI: 10.1002/mrm.25071.

34. E. S. S. Hansen, N. J. Stewart, J. M. Wild, H. Stødkilde-Jørgensen, and C. Laustsen, *Magn. Reson. Med.*, 2016. DOI: 10.1002/mrm.26483.

35. C. von Morze, R. A. Bok, G. D. Reed, J. H. Ardenkjaer-Larsen, J. Kurhanewicz, and D. B. Vigneron, *Magn. Reson. Med.*, 2013. DOI: 10.1002/mrm.25071.

36. S. A. Butt, L. V. Søgaard, P. O. Magnusson, M. H. Lauritzen, C. Laustsen, P. Åkeson, J. H. Ardenkjær-Larsen, P. Keson, and J. H. Ardenkjær-Larsen, *J. Cereb. Blood Flow Metab.*, 2012, **32**, 1508. DOI: 10.1038/jcbfm.2012.34.

37. S. E. Bohndiek, M. I. Kettunen, D. Hu, B. W. C. Kennedy, J. Boren, F. A. Gallagher, and K. M. Brindle, *J. Am. Chem. Soc.*, 2011, **133**, 11795. DOI: 10.1021/ja2045925.

38. K. R. Keshari, J. Kurhanewicz, R. Bok, P. E. Z. Larson, D. B. Vigneron, and D. M. Wilson, *Proc. Natl. Acad. Sci. U. S. A.*, 2011, **108**, 18606. DOI: 10.1073/pnas.1106920108.

39. J. M. Park, C. Khemtong, S.-C. Liu, R. E. Hurd, and D. M. Spielman, *Magn. Reson. Med.*, 2017, **77**, 1741. DOI: 10.1002/mrm.26662.

40. J. M. Park, M. Wu, K. Datta, S.-C. Liu, A. Castillo, H. Lough, D. M. Spielman, and K. L. Billingsley, *J. Am. Chem. Soc.*, 2017, **139**, 6629. DOI: 10.1021/jacs.7b00708.

41. J. J. Miller, D. R. Ball, A. Z. Lau, and D. J. Tyler, *NMR Biomed.*, 2018, e3912. DOI: 10.1002/nbm.3912.

42. C. von Morze, M. A. Ohliger, I. Marco-Rius, D. M. Wilson, R. R. Flavell, D. Pearce, D. B. Vigneron, J. Kurhanewicz, and Z. J. Wang, *Magn. Reson. Med.*, 2018. DOI: 10.1002/mrm.27054.

43. K. R. Keshari and D. M. Wilson, *Chem. Soc. Rev.*, 2014, **43**, 1627. DOI: 10.1039/c3cs60124b.

44. M. Karlsson, P. R. Jensen, J. Ø. Duus, S. Meier, and M. H. Lerche, *Appl. Magn. Reson.*, 2012, **43**, 223. DOI: 10.1007/s00723-012-0336-7.

45. X. Ji, A. Bornet, B. Vuichoud, J. Milani, D. Gajan, A. J. Rossini, L. Emsley, G. Bodenhausen, and S. Jannin, *Nat. Commun.*, 2017, **8**, 13975. DOI: 10.1038/ncomms13975.

46. S. Bowen and J. H. Ardenkjaer-Larsen, *J. Magn. Reson.*, 2013, **236**, 26. DOI: 10.1016/j.jmr.2013.08.007.

47. A. Comment, B. van den Brandt, K. Uffmann, F. Kurdzesau, S. Jannin, J. A. Konter, P. Hautle, W. T. Wenckebach, R. Gruetter, and J. J. van der Klink, *Appl. Magn. Reson.*, 2008, **34**, 313. DOI: 10.1007/s00723-008-0119-3.

48. S. Bowen and J. H. Ardenkjaer-Larsen, *J. Magn. Reson.*, 2014, **240**, 90. DOI: 10.1016/j.jmr.2014.01.009.

49. B. Adeva, E. Arik, S. Ahmad, A. Arvidson, B. Badelek, M. K. Ballintijn, G. Bardin, G. Baum, P. Berglund, L. Betev, I. G. Bird, R. Birsa, P. Björkholm, B. E. Bonner, N. de Botton, M. Boutemeur, F. Bradamante, A. Bressan, A. Brüll, J. Buchanan, S. Bültmann, E. Burtin, C. Cavata, J. P. Chen, J. Clement, M. Clocchiatti, M. D. Corcoran, D. Crabb, J. Cranshaw, J. Çuhadar, S. Dalla Torre, A. Deshpande, R. van Dantzig, D. Day, S. Dhawan, C. Dulya, A. Dyring, S. Eichblatt, J. C. Faivre, D. Fasching, F. Feinstein, C. Fernandez, B. Frois, C. Garabatos, J. A. Garzon, T. Gaussiran, M. Giorgi, E. von Goeler, I. A. Golutvin, A. Gomez, G. Gracia, N. de Groot, M. Grosse Perdekamp, E. Gülmez, D. von Harrach, T. Hasegawa, P. Hautle, N. Hayashi, C. A. Heusch, N. Horikawa, V. W. Hughes, G. Igo, S. Ishimoto, T. Iwata, M. de Jong, E. M. Kabuß, T. Kageya, R. Kaiser, A. Karev, H. J. Kessler, T. J. Ketel, I. Kiryushin, A. Kishi, Y. Kisselev, L. Klostermann, D. Krämer, V. Krivokhijine, V. Kukhtin, J. Kyynäräinen, M. Lamanna, U. Landgraf, K. Lau, T. Layda, J. M. Le Goff, F. Lehar, A. de Lesquen, J. Lichtenstadt, T. Lindqvist, M. Litmaath, S. Lopez-Ponte, M. Lowe, A. Magnon, G. K. Mallot, F. Marie, A. Martin, J. Martino, T. Matsuda, B. Mayes, J. S. McCarthy, K. Medved, G. van Middelkoop, D. Miller, J. Mitchell, K. Mori, J. Moromisato, G. S. Mutchler, A. Nagaitsev,

J. Nassalski, L. Naumann, B. Neganov, T. O. Ninikoski, J. E. J. Oberski, A. Ogawa, S. Okumi, C. S. Özben, A. Penzo, C. A. Perez, F. Perrot-Kunne, D. Peshekhonov, R. Piegaia, L. Pinsky, S. Platchkov, M. Plo, D. Pose, H. Postma, J. Pretz, T. Pussieux, J. Pyrlik, I. Reyhancan, J. M. Rieubland, A. Rijllart, J. B. Roberts, S. E. Rock, M. Rodriguez, E. Rondio, O. Rondon, L. Ropelewski, A. Rosado, I. Sabo, J. Saborido, G. Salvato, A. Sandacz, D. Sanders, I. Savin, P. Schiavon, K. P. Schüler, R. Segel, R. Seitz, Y. Semertzidis, S. Sergeev, F. Sever, P. Shanahan, E. Sichtermann, G. Smirnov, A. Staude, A. Steinmetz, H. Stuhrmann, K. M. Teichert, F. Tessarotto, W. Thiel, M. Velasco, J. Vogt, R. Voss, R. Weinstein, C. Whitten, R. Willumeit, R. Windmolders, W. Wislicki, A. Witzmann, A. Yañez, N. I. Zamiatin, A. M. Zanetti, and J. Zhao, *Nucl. Instrum. Methods Phys. Res. Sect. A Accel. Spectrometers Detect. Assoc. Equip.*, 1996, **372**, 339. DOI: 10.1016/0168-9002(95)01376-8.

50. J. Jam, S. Dey, L. Muralidharan, A. M. A. M. Leach, and J. H. Ardenkjaer-Larsen, Jet impingment melting with vaporization: A numerical study, In 2008 Proceedings of the ASME Summer Heat Transfer Conference, HT 2008, 2009, 559–567, 2014, http://www.scopus.com/inward/record.url?eid=2-s2.0-70349120429&partnerID=tZOtx3y1.

51. T. Cheng, M. Mishkovsky, J. A. M. Bastiaansen, O. Ouari, P. Hautle, P. Tordo, B. van den Brandt, and A. Comment, *NMR Biomed.*, 2013, **26**, 1582. DOI: 10.1002/nbm.2993.

52. H. Jóhannesson, S. Macholl, and J. H. Ardenkjaer-Larsen, *J. Magn. Reson.*, 2009, **197**, 167. DOI: 10.1016/j.jmr.2008.12.016.

53. T. Cheng, A. Capozzi, Y. Takado, R. Balzan, and A. Comment, *Phys. Chem. Chem. Phys.*, 2013, **15**, 20819. DOI: 10.1039/C3CP53022A.

54. H. A. I. Yoshihara, E. Can, M. Karlsson, M. H. Lerche, J. Schwitter, and A. Comment, *Phys. Chem. Chem. Phys.*, 2016, **18**, 12409. DOI: 10.1039/C6CP00589F.

55. A. Bornet, A. Pinon, A. Jhajharia, M. Baudin, X. Ji, L. Emsley, G. Bodenhausen, J. H. Ardenkjaer-Larsen, and S. Jannin, *Phys. Chem. Chem. Phys.*, 2016, **18**, 30530. DOI: 10.1039/C6CP05587G.

56. S. Ringgaard, R. F. Schulte, J. Tropp, C. Kögler, T. Lanz, M. A. Navarro, J. H. Ardenkjaer-Larsen, F. Robb, H. Stødkilde-Jørgensen, and C. Laustsen, *Proc. Int. Soc. Mag. Reson. Med.*, 2016, 2152.

57. A. P. Chen, J. Tropp, R. E. Hurd, M. Van Criekinge, L. G. Carvajal, D. Xu, J. Kurhanewicz, and D. B. Vigneron, *J. Magn. Reson.*, 2009, **197**, 100. DOI: 10.1016/j.jmr.2008.12.004.

58. Y.-F. Yen, P. Le Roux, D. Mayer, R. King, D. Spielman, J. Tropp, K. Butts Pauly, A. Pfefferbaum, S. Vasanawala, and R. Hurd, *NMR Biomed.*, 2010, **23**. DOI: 10.1002/nbm.1481.

59. J. Wang, A. J. Wright, D. Hu, R. Hesketh, and K. M. Brindle, *Magn. Reson. Med.*, 2017, **77**, 740. DOI: 10.1002/mrm.26168.

60. M. Durst, U. Koellisch, A. Frank, G. Rancan, C. V. Gringeri, V. Karas, F. Wiesinger, M. I. Menzel, M. Schwaiger, A. Haase, and R. F. Schulte, *NMR Biomed.*, 2015, **28**, 715. DOI: 10.1002/nbm.3301.

61. P. E. Z. Larson, S. Hu, M. Lustig, A. B. Kerr, S. J. Nelson, J. Kurhanewicz, J. M. Pauly, and D. B. Vigneron, *Magn. Reson. Med.*, 2011, **65**, 610. DOI: 10.1002/mrm.22650.

62. C. H. Cunningham, A. P. Chen, M. Lustig, B. A. Hargreaves, J. Lupo, D. Xu, J. Kurhanewicz, R. E. Hurd, J. M. Pauly, S. J. Nelson, and D. B. Vigneron, *J. Magn. Reson.*, 2008, **193**, 139. DOI: 10.1016/j.jmr.2008.03.012.

63. O. Khegai, R. F. Schulte, M. A. Janich, M. I. Menzel, E. Farrell, A. M. Otto, J. H. Ardenkjaer-Larsen, S. J. Glaser, A. Haase, M. Schwaiger, and F. Wiesinger, *NMR Biomed.*, 2014, **27**, 1256. DOI: 10.1002/nbm.3174.

64. P. E. Z. Larson, A. B. Kerr, A. P. Chen, M. S. Lustig, M. L. Zierhut, S. Hu, C. H. Cunningham, J. M. Pauly, J. Kurhanewicz, and D. B. Vigneron, *J. Magn. Reson.*, 2008, **194**, 121. DOI: 10.1016/j.jmr.2008.06.010.

65. P. E. Z. Larson, R. Bok, A. B. Kerr, M. Lustig, S. Hu, A. P. Chen, S. J. Nelson, J. M. Pauly, J. Kurhanewicz, and D. B. Vigneron, *Magn. Reson. Med.*, 2010, **63**, 582. DOI: 10.1002/mrm.22264.

66. A. Z. Lau, A. P. Chen, N. R. Ghugre, V. Ramanan, W. W. Lam, K. A. Connelly, G. A. Wright, and C. H. Cunningham, *Magn. Reson. Med.*, 2010, **64**, 1323. DOI: 10.1002/mrm.22525.

67. J. W. Gordon, D. B. Vigneron, and P. E. Z. Larson, *Magn. Reson. Med.*, 2017, **77**, 826. DOI: 10.1002/mrm.26123.

68. J. Leupold, S. Månsson, J. S. Petersson, J. Hennig, and O. Wieben, *MAGMA*, 2009, **22**, 251. DOI: 10.1007/s10334-009-0169-z.

69. J. Leupold, O. Wieben, S. Månsson, O. Speck, K. Scheffler, J. S. Petersson, and J. Hennig, *Magn. Reson. Mater. Phys. Biol. Med.*, 2006, **19**, 267. DOI: 10.1007/s10334-006-0056-9.

70. S. Månsson, J. S. Petersson, and K. Scheffler, *Magn. Reson. Med.*, 2012, **68**, 1894. DOI: 10.1002/mrm.24183.

71. H. Shang, S. Sukumar, C. von Morze, R. A. Bok, I. Marco-Rius, A. Kerr, G. D. Reed, E. Milshteyn, M. A. Ohliger, J. Kurhanewicz, P. E. Z. Larson, J. M. Pauly, and D. B. Vigneron, *Magn. Reson. Med.*, 2017, **78**, 963. DOI: 10.1002/mrm.26480.

72. J. Wang, F. Kreis, A. J. Wright, R. L. Hesketh, M. H. Levitt, and K. M. Brindle, *Magn. Reson. Med.*, 2018, **79**, 741. DOI: 10.1002/mrm.26725.

73. S. E. Day, M. I. Kettunen, F. A. Gallagher, D.-E. Hu, M. Lerche, J. Wolber, K. Golman, J. H. Ardenkjaer-Larsen, and K. M. Brindle, *Nat. Med.*, 2007, **13**, 1382. DOI: 10.1038/nm1650.

74. M. I. Kettunen, D. Hu, T. H. Witney, R. McLaughlin, F. A. Gallagher, S. E. Bohndiek, S. E. Day, and K. M. Brindle, *Magn. Reson. Med.*, 2010, **63**, 872. DOI: 10.1002/mrm.22276.

75. D. M. Spielman, D. Mayer, Y.-F. Yen, J. Tropp, R. E. Hurd, and A. Pfefferbaum, *Magn. Reson. Med.*, 2009, **62**, 307. DOI: 10.1002/mrm.21998.

76. M. L. Zierhut, Y.-F. Yen, A. P. Chen, R. Bok, M. J. Albers, V. Zhang, J. Tropp, I. Park, D. B. Vigneron, J. Kurhanewicz, R. E. Hurd, and S. J. Nelson, *J. Magn. Reson.*, 2010, **202**, 85. DOI: 10.1016/j.jmr.2009.10.003.

77. A. Naressi, C. Couturier, J. M. Devos, M. Janssen, C. Mangeat, R. de Beer, and D. Graveron-Demilly, *Magma Magn. Reson. Mater. Phys. Biol. Med.*, 2001, **12**, 141. DOI: 10.1007/BF02668096.

78. L. Menichetti, F. Frijia, A. Flori, F. Wiesinger, V. Lionetti, G. Giovannetti, G. D. Aquaro, F. A. Recchia, J. H. Ardenkjaer-Larsen, M. F. Santarelli, and M. Lombardi, *Contrast Media Mol. Imaging*, 2012, **7**, 85. DOI: 10.1002/cmmi.480.

79. D. K. Hill, Y. Jamin, M. R. Orton, N. Tardif, H. G. Parkes, S. P. Robinson, M. O. Leach, Y.-L. Chung, and T. R. Eykyn, *NMR Biomed.*, 2013, **26**, 1321. DOI: 10.1002/nbm.2957.

80. C. J. Daniels, M. A. McLean, R. F. Schulte, F. J. Robb, A. B. Gill, N. McGlashan, M. J. Graves, M. Schwaiger, D. J. Lomas, K. M. Brindle, and F. A. Gallagher, *NMR Biomed.*, 2016, **29**, 387. DOI: 10.1002/nbm.3468.

81. P. Wespi, J. Steinhauser, G. Kwiatkowski, and S. Kozerke, *NMR Biomed.*, 2017, e3876. DOI: 10.1002/nbm.3876.

82. R. M. Malinowski, K. W. Lipsø, M. H. Lerche, and J. H. Ardenkjær-Larsen, *J. Magn. Reson.*, 2016, **272**. DOI: 10.1016/j.jmr.2016.09.015.

83. S. Hu, A. P. Chen, M. L. Zierhut, R. Bok, Y.-F. Yen, M. A. Schroeder, R. E. Hurd, S. J. Nelson, J. Kurhanewicz, and D. B. Vigneron, *Mol. Imaging Biol.*, 2009, **11**, 399. DOI: 10.1007/s11307-009-0218-z.

84. M. M. Darpolor, Y.-F. Yen, M.-S. Chua, L. Xing, R. H. Clarke-Katzenberg, W. Shi, D. Mayer, S. Josan, R. E. Hurd, A. Pfefferbaum, L. Senadheera, S. So, L. V. Hofmann, G. M. Glazer, and D. M. Spielman, *NMR Biomed.*, 2011, **24**, 506. DOI: 10.1002/nbm.1616.

85. I. Park, R. Bok, T. Ozawa, J. J. Phillips, C. D. James, D. B. Vigneron, S. M. Ronen, and S. J. Nelson, *J. Magn. Reson. Imaging*, 2011, **33**, 1284. DOI: 10.1002/jmri.22563.

86. S. E. Day, M. I. Kettunen, M. K. Cherukuri, J. B. Mitchell, M. J. Lizak, H. D. Morris, S. Matsumoto, A. P. Koretsky, and K. M. Brindle, *Magn. Reson. Med.*, 2011, **65**, 557. DOI: 10.1002/mrm.22698.

87. I. Park, P. E. Z. Larson, J. W. Gordon, L. Carvajal, H.-Y. Chen, R. Bok, M. Van Criekinge, M. Ferrone, J. B. Slater, D. Xu, J. Kurhanewicz, D. B. Vigneron, S. Chang, and S. J. Nelson, *Magn. Reson. Med.*, 2018. DOI: 10.1002/mrm.27077.

88. M. Karlsson, P. R. Jensen, R. in't Zandt, A. Gisselsson, G. Hansson, J. Ø. Duus, S. Meier, and M. H. Lerche, *Int. J. Cancer*, 2010, **127**, 729. DOI: 10.1002/ijc.25072.

89. K. Golman, J. S. Petersson, P. Magnusson, E. Johansson, P. Akeson, C.-M. Chai, G. Hansson, and S. Månsson, *Magn. Reson. Med.*, 2008, **59**, 1005. DOI: 10.1002/mrm.21460.

90. M. E. Merritt, C. Harrison, C. Storey, F. M. Jeffrey, A. D. Sherry, and C. R. Malloy, *Proc. Natl. Acad. Sci. U. S. A.*, 2007, **104**, 19773. DOI: 10.1073/pnas.0706235104.

91. M. A. Schroeder, L. E. Cochlin, L. C. Heather, K. Clarke, G. K. Radda, and D. J. Tyler, *Proc. Natl. Acad. Sci. U. S. A.*, 2008, **105**, 12051. DOI: 10.1073/pnas.0805953105.

92. F. Frijia, M. F. Santarelli, U. Koellisch, G. Giovannetti, T. Lanz, A. Flori, M. Durst, G. D. Aquaro, R. F. Schulte, D. De Marchi, V. Lionetti, J. H. Ardenkjaer-Larsen, L. Landini, L. Menichetti, and V. Positano, *J. Med. Biol. Eng.*, 2016, **36**. DOI: 10.1007/s40846-016-0113-4.

93. M. A. Schroeder, A. Z. Lau, A. P. Chen, Y. Gu, J. Nagendran, J. Barry, X. Hu, J. R. B. Dyck, D. J. Tyler, K. Clarke, K. A. Connelly, G. A. Wright, and C. H. Cunningham, *Eur. J. Heart Fail.*, 2013, **15**, 130. DOI: 10.1093/eurjhf/hfs192.

94. P. Wespi, J. Steinhauser, G. Kwiatkowski, and S. Kozerke, *Magn. Reson. Med.*, 2018. DOI: 10.1002/mrm.27197.

95. M. E. Merritt, C. Harrison, C. Storey, A. D. Sherry, and C. R. Malloy, *Magn. Reson. Med.*, 2008, **60**, 1029. DOI: 10.1002/mrm.21760.

96. D. Mayer, Y.-F. Yen, A. Takahashi, S. Josan, J. Tropp, B. K. Rutt, R. E. Hurd, D. M. Spielman, and A. Pfefferbaum, *Magn. Reson. Med.*, 2011, **65**, 1228. DOI: 10.1002/mrm.22707.

97. J. M. Park, S. Josan, T. Grafendorfer, Y.-F. Yen, R. E. Hurd, D. M. Spielman, and D. Mayer, *NMR Biomed.*, 2013, **26**, 1197. DOI: 10.1002/nbm.2935.

98. S. J. DeVience, X. Lu, J. Proctor, P. Rangghran, E. R. Melhem, R. Gullapalli, G. M. Fiskum, and D. Mayer, *Sci. Rep.*, 2017, **7**, 1907. DOI: 10.1038/s41598-017-01736-x.

99. H. J. Adrogué, F. J. Gennari, J. H. Galla, and N. E. Madias, *Kidney Int.*, 2009, **76**, 1239. DOI: 10.1038/ki.2009.359.

100. J. R. Casey, S. Grinstein, and J. Orlowski, *Nat. Rev. Mol. Cell Biol.*, 2010, **11**, 50. DOI: 10.1038/nrm2820.

101. N. Raghunand, X. He, R. van Sluis, B. Mahoney, B. Baggett, C. W. Taylor, G. Paine-Murrieta, D. Roe, Z. M. Bhujwalla, and R. J. Gillies, *Br. J. Cancer*, 1999, **80**, 1005. DOI: 10.1038/sj.bjc.6690455.

102. I. F. Robey, B. K. Baggett, N. D. Kirkpatrick, D. J. Roe, J. Dosescu, B. F. Sloane, A. I. Hashim, D. L. Morse, N. Raghunand, R. A. Gatenby, and R. J. Gillies, *Cancer Res.*, 2009, **69**, 2260. DOI: 10.1158/0008-5472.CAN-07-5575.

103. S. Düwel, C. Hundshammer, M. Gersch, B. Feuerecker, K. Steiger, A. Buck, A. Walch, A. Haase, S. J. Glaser, M. Schwaiger, and F. Schilling, *Nat. Commun.*, 2017, **8**, 15126. DOI: 10.1038/ncomms15126.

104. D. E. Korenchan, R. R. Flavell, C. Baligand, R. Sriram, K. Neumann, S. Sukumar, H. VanBrocklin, D. B. Vigneron, D. M. Wilson, and J. Kurhanewicz, *Chem. Commun. (Camb.)*, 2016, **52**, 3030. DOI: 10.1039/c5cc09724j.

105. A. Z. Lau, J. J. Miller, and D. J. Tyler, *Magn. Reson. Med.*, 2017, **77**, 1810. DOI: 10.1002/mrm.26260.

106. C. E. Christensen, M. Karlsson, J. R. Winther, R. Jensen, and M. H. Lerche, *J. Biol. Chem.*, 2013. DOI: 10.1074/jbc.M113.498626.

Chapter 13

Liquid-state Overhauser DNP at High Magnetic Fields

Vasyl P. Denysenkov and Thomas F. Prisner

Goethe University, Frankfurt-am-Main, Germany

13.1 INTRODUCTION

Dynamic nuclear polarization (DNP) belongs to one of the earliest methodological discoveries in the field of magnetic resonance spectroscopy. Shortly after the theoretical prediction by Overhauser in 1953,[1] it was experimentally proven by Carver and Slichter[2,3] on metallic Li. Soon afterward, it was also shown that DNP works with radicals dissolved in organic liquids.[4–6] Subsequently, theoretical models based on the statistical modulation of the hyperfine (hf) interaction between the unpaired electron spin S of the radical and the nuclear spin I of the solvent molecules were developed.[6,7] This allowed relating the observed DNP enhancements to the translational and rotational dynamics of the polarizing agent molecule (carrying the unpaired electron spin S) and the polarizing target molecule (carrying the observed nuclear spin I). The spectral densities $J(\omega)$, describing the dynamics of the hf interaction, were derived by describing both molecules bearing the two spins as hard spheres, with diffusion and rotation rates described by the Stokes–Einstein model.[8,9] In this way, DNP experiments allowed determining molecular diffusion coefficients, or vice versa, prediction of the enhancement factors when the correlation times τ_c are known. All Overhauser DNP experiments in the 1960–1980s were performed at magnetic fields below 1.5 T.[10] Dynamic nuclear polarization was almost completely abandoned, as nuclear magnetic resonance (NMR) moved to higher magnetic fields to achieve higher chemical shift dispersion and higher Boltzmann polarization because the theoretical models predicted a steep decrease of the enhancements at higher resonance frequencies. At this time, liquid-state Overhauser DNP experiments at higher magnetic field have solely been performed by Dorn *et al.*[11] by executing the polarization transfer at lower magnetic field values (0.3 T) and then transferring the liquid sample into a larger magnetic field for NMR detection. Of course, this method reduces the overall DNP efficiency, owing to the polarization transfer being performed at a lower magnetic field (lower Boltzmann polarization). Only after the pioneering solid-state DNP work performed at high magnetic fields by the Griffin and co-workers[12] and the Ardenkjaer-Larsen

Handbook of High Field Dynamic Nuclear Polarization.
Edited by Vladimir K. Michaelis, Robert G. Griffin, Björn Corzilius and Shimon Vega
© 2020 John Wiley & Sons, Ltd. ISBN: 978-1-119-44164-9
Also published in eMagRes (online edition)
DOI: 10.1002/9780470034590.emrstm1557

et al.,[13] investigations of Overhauser DNP at high magnetic fields were started.[14–17]

Similar to solid-state DNP experiments at high magnetic fields, very strong signal enhancements up to a factor of 100 could be observed in liquids as well. In this chapter, we will describe experimental requirements to perform Overhauser DNP at high magnetic fields. Excitation of the electron paramagnetic resonance (EPR) transitions at high magnetic fields requires MW in the 100 GHz to 1 THz range. Resonant structures with spatial separation of the electrical and magnetic components of the MW are necessary to avoid the absorption of the terahertz radiation by liquids. Such double-resonance (NMR and EPR) structures, working at high magnetic fields, will be described in this chapter. The MW resonant structure limits the sample dimension much below the MW wavelength. This also strongly reduces the overall sample amount accessible for liquid DNP experiments at high magnetic fields. The field dependence of all factors influencing the Overhauser DNP efficiency and the overall NMR signal-to-noise improvement will be discussed in this chapter.

The experimental procedures to disentangle the different factors influencing the Overhauser DNP enhancement will be explained in detail. This enables the quantitative determination of the coupling factor ξ, which describes the dynamic modulation of the coupling between the electron spin S and the nuclear spin I, and a critical comparison of the results with the theoretical models. In this chapter, we will concentrate on the dipolar coupling between the electron and nuclear spins, which is the typical case for nitroxide radicals in organic solvents, while the proton spins are observed. Scalar coupling, as observed for carbon nuclear spins,[18] will be discussed in another contribution to this book. The recently discovered Overhauser DNP enhancements in insulating solids[19] will not be discussed in this chapter.

13.2 THEORY OF OVERHAUSER DNP

Overhauser DNP as well as the nuclear Overhauser effect (NOE) can be described by the Solomon equations.[20] In the case of Overhauser DNP, the change of the electron spin magnetization S_z is determined by its fast intrinsic relaxation rate R_{1S} and its interaction with the MW excitation, described by the MW field strength in radian frequency units ω_1 (Figure 13.1).

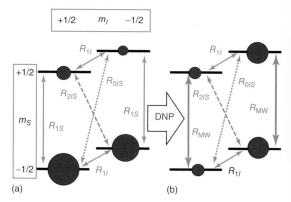

Figure 13.1. Energy-level diagram for a coupled electron spin $S = 1/2$ and nuclear spin $I = 1/2$. (a) Populations at thermal equilibrium with relaxation rates between the levels. (b) Changed populations of the levels if the allowed electron spin transitions are saturated by the MW excitation, resulting in an enhanced nuclear spin polarization. Solid arrows depict single-quantum transitions, while dashed and dotted arrows depict double- and zero-quantum transitions, respectively. Arrows in green indicate saturation of EPR transitions

If the electron spin system can be described by a single homogeneous Lorentzian line, the saturation factor s of the electron spin system is given by:[6]

$$s = \frac{S_0 - \langle S_z \rangle}{S_0} = \frac{\omega_1^2 T_2}{R_1 + \omega_1^2 T_2} \qquad (13.1)$$

where S_0 is the value of $\langle S_z \rangle$ in thermal equilibrium.

The saturation factor s describes how efficiently the electron spin transitions are saturated by the MW radiation. It ranges from 0 for no saturation, i.e., Boltzmann magnetization S_0, to 1 for fully saturated electron spin transitions with equalized populations ($\langle S_z \rangle = 0$). In the case of nitroxide radicals with three hf lines ($I(^{14}N) = 1$), nitrogen spin relaxation and Heisenberg spin exchange lead to more complicated equations.[21,22] Unfortunately, at high magnetic fields a direct determination of the electron spin relaxation times T_{1S} and T_{2S} is difficult or even impossible.[23,24] Thus, experimental procedures to determine the saturation factor s are required, which will be described below.

The steady-state solution of the Solomon equation for the nuclear spin magnetization I_z results in the Overhauser formula for the DNP enhancement factor ε:[6]

$$\varepsilon = \frac{\langle I_z \rangle - I_0}{I_0} = f \cdot s \cdot \xi \cdot \frac{\gamma_e}{\gamma_n} \qquad (13.2)$$

where γ_e and γ_n are the gyromagnetic ratios of the electron and the nucleus, respectively, i.e., $\gamma_e/\gamma_n \approx -658$ for protons. The leakage factor f describes the degree of nuclear spin relaxation caused by the presence of the unpaired electron spin S. The leakage factor equals to 1 when the nuclear spin relaxation is solely determined by the electron spin and 0 when there is no contribution from the electron spin (which usually is only the case if the radical concentration is 0). This factor can easily be determined experimentally by measuring the nuclear spin relaxation times with (T_{1I}^S) and without (T_{1I}^0) radicals:

$$f = 1 - \frac{T_{1I}^S}{T_{1I}^0} = \frac{2R_I + R_{0IS} + R_{2IS}}{2R_I + R_{0IS} + R_{2IS} + R_I^0} \qquad (13.3)$$

where R_I^0 is the nuclear spin relaxation rate independent of the electron spin interaction, and R_{0IS} and R_{2IS} describe the zero- and double-quantum transition rates, involving coupled electron-nuclear spin flips (Figure 13.1).

The third, most complicated factor on the right-hand side of equation (13.2) is the coupling factor ξ. It accounts for the dynamic modulation of the hf interaction between the DNP agent (free radical) and the target molecule (solvent).

The Overhauser effect benefits from the dipolar (anisotropic) and scalar (isotropic) hf interactions between the electron and nuclear spins. In the case of protons, which is our focus here, only dipolar interaction has to be taken into consideration.[7] The coupling factor ξ can be described by spectral density functions $J(\omega)$ at the respective Larmor frequencies of the nuclear and electron spins:[6,25]

$$\xi = \frac{R_{2SI} - R_{0SI}}{2R_I + R_{2SI} - R_{0SI}} = \frac{5J(\omega_S)}{3J(\omega_I) + 7J(\omega_S)} \approx \frac{5J(\omega_S)}{3J(\omega_I)} \qquad (13.4)$$

Here, ω_S and ω_I are the Larmor frequencies of the electron and nuclear spins, respectively, at the magnetic field of the experiment. Equation (13.4) assumes an isotropic environment, like simple liquid, and $\omega_S \gg \omega_I$. The second approximation step applies at high magnetic fields where $J(\omega_I) \gg J(\omega_S)$. $J(\omega)$ can be calculated from the hard-sphere model assuming Stokes–Einstein diffusional and rotational motion of the two molecules[8,9] or using MD simulations.[26,27] This enables the quantitative comparison of the predictions for the coupling factor with experimentally determined values, when all the other factors in equation (13.2) are known.

While optimization of f and s is rather a technical issue, the coupling factor reflects the nature of the polarization transfer between the electron and nuclear spins, which cannot easily be controlled. The coupling factor ranges from 0 to 1/2 if the magnetic dipole–dipole coupling is the dominant mechanism, a common case for protons of target molecules in liquids. This yields a maximal theoretical enhancement ($\xi = 0.5, s = 1, f = 1$) to upto −330. The coupling factor and its temperature dependence can also be extracted from the field dependence of the solvent water proton relaxivity (often called nuclear magnetic relaxation dispersion, NMRD) profiles[28–30] assuming the hard-sphere model. Thus, the coupling factors determined from MD simulation, NMR dispersion measurements, and DNP experiments can be quantitatively compared.

13.2.1   Instrumental Aspects of Overhauser DNP at High Magnetic Fields

MW absorption at frequencies above 94 GHz (corresponding to an external magnetic field of higher than 3.4 T) limits the penetration depth into liquid aqueous samples to micrometer-length scales, much smaller than standard NMR sample sizes (Figure 13.2). However, MW magnetic field strengths of 1 mT (corresponding to $\omega_1 = 1.76 \times 10^8$ s^{-1}) are needed to sufficiently saturate the electron spin transitions, because of the fast relaxation times of radicals in liquid solutions at room temperature. It would correspond to approximately 500 W of MW power at 260 GHz for a freely propagating MW beam focused on a 2×1 mm^2 area.

Irradiating such high MW power onto the liquid sample would lead to unacceptably large heating and immediate evaporation of the sample. Thus, an MW resonant structure is necessary to perform DNP experiments in liquid solutions at high magnetic fields. Placing the sample in a node of the electrical field component within the MW cavity allows to reduce the absorption and thus heating strongly, while the MW magnetic field component, necessary to excite the electron spin transitions, is maximal at the sample position and in addition enhanced by the resonator Q value. The resonance structure is resonant also at the nuclear Larmor frequency to perform NMR excitation/detection.

ENDOR (electron-nuclear double-resonance) probes have both RF and MW resonance circuits and could be used for liquid-state DNP.[28] However, commercial ENDOR probes are usually not compatible with NMR magnets and are not optimized

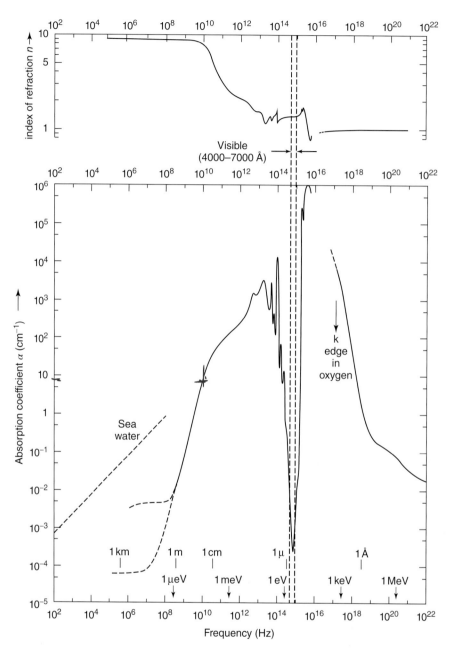

Figure 13.2.   Frequency dependence of the refraction index (a) and absorption coefficient (b) for liquid water.[31] Blue ink marking at 260 GHz. (Reproduced with permission from Ref. 31. © John Wiley & Sons, 1962)

for NMR detection. For this reason, several research groups have fabricated homemade probes consisting of double resonance to accomplish DNP experiments in liquids.

A combination of a dielectric resonator (for MW excitation at 95 GHz) with a U-shaped wire (for RF excitation at 144 MHz) was described by the van Bentum et al.[17] A modified Bruker W-band ENDOR

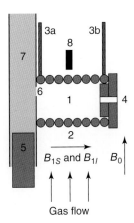

Figure 13.3. Schematic diagram of the Warwick DNP/EPR probe working at 3.4 T: (1) W-band cavity, (2) RF coil, (3a, b) RF leads, (4) sample access, (5) variable tuner, (6) coupling iris, (7) waveguide to spectrometer, and (8) thermocouple. (For reproduction of material from PCCP[16])

double-resonance structure was designed by the Warwick group for DNP at 3.4 T (Figure 13.3).[16]

We developed double-resonance structures for DNP working at 260 GHz EPR and 400 MHz proton NMR frequencies. Our first approach was a fundamental-mode double-resonance structure, similar to the ones used for ENDOR[32] and solid-state DNP[33] at lower frequencies (Figure 13.4a). It consists of a cylindrical TE_{011} MW cavity, where the solenoidal slotted cylinder of the cavity serves as the RF coil for NMR excitation/detection. The six-turn helical coil of 1.48 mm inner diameter was fabricated

of a copper tape and connected to an RF circuit tuned to the 400 MHz NMR frequency.

The MW is confined inside the solenoidal coil and two plungers made of MACOR with flat caps coated with a 1 μm thick silver film. Both plungers have 0.15 mm holes in the center to align the liquid sample (3–5 nl) inside a quartz capillary along the cylinder axis. One plunger can be moved remotely via a gearbox for tuning the MW frequency (to be in resonance with the electron spin Larmor frequency at the given magnetic field). The MW coupling is achieved by an elliptical centered iris and via a fundamental-mode (WR-4) waveguide that touches the solenoidal helix in the middle, grounding the coil at this position with respect to the RF. The angular electric field distribution of TE_{01n} modes is maintained since the gaps between the helical turns are almost parallel to the surface currents. Moreover, these gaps serve as filter against unwanted MW modes, resulting in a clean frequency swept tuning picture showing only the TE_{01n} modes. In case of a perfectly symmetric solenoid, the RF current has a maximum in the center turn corresponding to the virtual ground.[35] Therefore, the RF field distortion due to the electrical contact between the grounded waveguide and the center turn of the solenoid is minimized. In addition, the probe is further equipped with a field-modulation coil placed near the helix for cw-EPR experiments, which are performed in a first step to find the on-resonance condition for MW pumping.[34] This fundamental-mode helical–cylindrical double-resonance structure used for our first demonstrations of the applicability of DNP at a magnetic field of 9.2 T gains a rather large B_1 field values for a given MW power. This is due to the fact

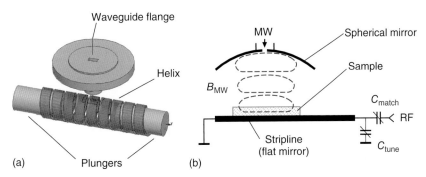

Figure 13.4. Two double-resonance structures used for DNP of liquids at 9.4 T magnetic field. (a) Fundamental-mode helix-cylindrical double-resonance structure[34] with a maximum aqueous sample volume of 5 nl (Reproduced with permission from Ref. 34. © Springer-Verlag, 2008). (b) Higher mode stripline-Fabry–Pérot resonance structure with an aqueous sample volume of 90 nl

that the MW is confined in a very small volume. For the same reason, the active sample size is restricted in all three dimensions to values small compared to the MW wavelength, resulting in a very small aqueous sample volume of only 3–5 nl. Together with a poor spectral resolution and a low NMR filling factor, this resulted in small overall signal amplitudes, which hamper the observation of NMR signals of biomolecules.

To overcome this limitation we developed a new type of the double-resonance structure for liquid-state DNP consisting of a Fabry–Pérot resonator for the MW and a stripline resonator for the NMR frequencies (Figure 13.4b). This new double-resonance structure restricts the sample size to a much smaller value than the MW wavelength in only one dimension, offering a 30-fold increase in aqueous sample volume (90 nl) and strongly improved NMR sensitivity and linewidth.[36] On the other hand, higher MW power is necessary to obtain B_1 fields large enough to saturate the electron spin transitions for such an overmoded cavity. Here, a copper stripe (5 μm thickness, 24×5 mm^2 area), deposited on a quartz substrate (0.5 mm thickness), is placed inside an RF-shielded box ($30 \times 10 \times 5$ mm^3) between two parallel copper plates, which serve as ground planes of the stripline. The structure contains RF tune and match capacitors as circular structures attached to the box. The spherical mirror, forming the Fabry–Pérot resonator, is positioned above the stripe and irradiates the structure with MW through an iris

hole with 0.4 mm diameter and 0.05 mm thickness. Near critical coupling to the loaded resonator could be obtained with this configuration, as predicted by finite element calculations. The stripe is used as an NMR resonant structure[37] as well as a plain mirror of the MW Fabry–Pérot resonator, which is adjusted to the TE$_{005}$ mode. A droplet of the liquid sample is placed inside the sample holder (of 2.5 mm diameter and 0.02 high) made of PTFE film, which is placed on the plain mirror surface. The droplet is squeezed with a 0.12 mm quartz cover plate keeping the sample at the node of the MW electrical field component. Proton-free Krytox lubricant (DuPont) seals the sample. The cover plate is gently pressed against the holder with an additional spacer ring placed between the quartz cover plate and the upper removable ground plate of the copper shield box.[36]

The MW resonator also allows for a sensitive detection of the EPR signal, which is important to optimize and adjust the EPR condition. EPR detection is accomplished in reflection mode by a home-build MW bridge (Figure 13.5b). Our MW bridge is based on metal-dielectric waveguide components, square cross-section metal tubes with lateral dimensions larger than the wavelength, and with a dielectric coating of the inner surface (Institute of Radiophysics and Electronics in Kharkov, Ukraine). The MW bridge can operate in three different modes: tuning, EPR detection, and DNP excitation. In the first case, the

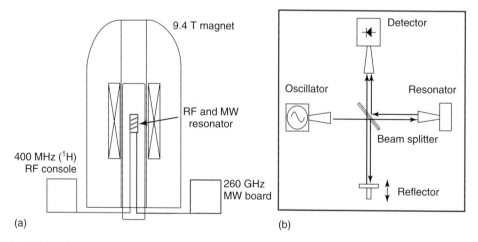

Figure 13.5. Schematic representation of the in situ 9.4 T DNP setup for liquid samples (a)[38] and the microwave bridge with Michelson interferometer configuration[34] used for tuning, EPR detection, and DNP operation (b). The two microwave sources connected (low-power tunable solid-state source and high-power gyrotron) are schematically depictured as oscillator in the diagram. ((a) Reproduced with permission from Ref. 38. © Elsevier, 2016; (b) Reproduced with permission from Ref. 34. © Springer-Verlag, 2008)

bridge allows to monitor the MW resonator dip of the double-resonance DNP probe. In the second mode, detection of the cw-EPR signal with field modulation and lock-in detection is possible. In the cw-EPR configuration, the z-shim coils of the magnet can be used to sweep the magnetic field. For these two modes, a solid-state MW source (VDI-S019b, Virginia Diodes Inc., USA) of 50 mW MW power and a frequency sweep range between 258 and 264 GHz is used. For EPR detection, the MW bridge works in a Michelson interferometer configuration with a suppression of 34 dB of the MW reflection from the cavity, allowing high-sensitive detection using a zero-bias Schottky diode detector (Virginia Diodes Inc., USA). The third mode of the bridge is used for DNP experiments, with a high-power gyrotron (GYCOM, Russia, 16 W CW MW power) for excitation. The NMR spectrometer consists of a 400 MHz NMR Avance console (Bruker, Germany) and a wide bore cryomagnet (9.4 T, sweep range ±80 mT). The sketch of the overall setup is depictured in Figure 13.5. More details about the experimental setup could be found elsewhere.[39]

The most important and critical part of a liquid-state high-field in situ DNP spectrometer is the double-resonance probe as mentioned earlier, which ultimately defines the efficiency of the microwave excitation, the heat dissipation of the sample, the DNP enhancement, as well as the NMR performance.

Another option to achieve DNP of liquids at high magnetic fields is the 'shuttle' approach, where the polarization transfer occurs at lower magnetic fields (typically 0.3 T/X-band), and the NMR signal is detected at a much higher magnetic field.

The group of researchers in Göttingen uses a shuttling setup, where the NMR tube is shifted quickly between the main field of the superconducting magnet (7 T) and a fringe field position (0.34 T, DNP position) homogenized by ferroshim tubes[40] (Figure 13.6).

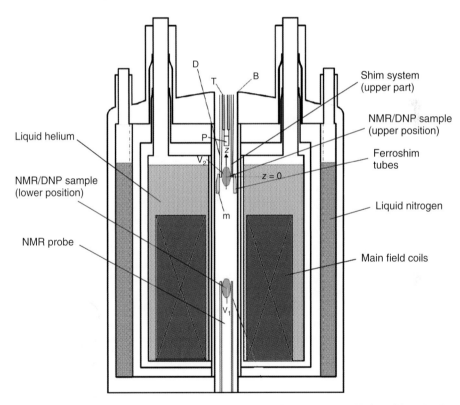

Figure 13.6. Two-center shuttle DNP magnet. Cross section of the NMR cryomagnet with ferroshim tubes integrated in the room-temperature bore generating a second homogeneous field region of 0.32 T (upper position of the NMR/DNP sample) 468 mm above the main magnetic field center of 14 T (lower position of the NMR/DNP sample). (Reproduced with permission from Ref. 41. © Elsevier, 2012)

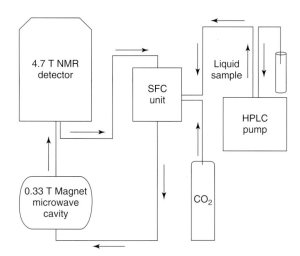

Figure 13.7. Apparatus for continuous-flow HPLC shuttle DNP setup coupled with supercritical CO_2.[42] (Wang et al.[42] https://pubs.rsc.org/en/content/articlehtml/2015/sc/c5sc02499d. Licensed under CC BY 3.0)

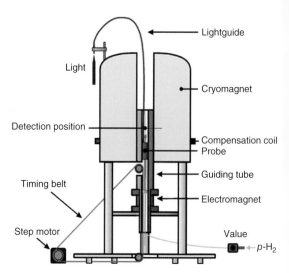

Figure 13.8. Scheme of the Berlin/Novosibirsk field-cycling DNP/CIDNP/PHIP setup. Here the magnetic field is switched by moving the whole NMR probe into the fringe field of the NMR cryomagnet or into an auxiliary electromagnet. The probe moves inside a guiding tube driven by a step motor connected to the probe by a timing belt. (Reproduced with permission from Ref. 44. © Elsevier, 2016)

The Dorn group developed a shuttle DNP apparatus utilizing supercritical fluids like CO_2, which has very short correlation times that makes it attractive to achieve high DNP efficiencies at high magnetic fields. In this case, the polarization transfer is performed at 0.33 T and NMR detection at 4.7 T[42] (Figure 13.7). In the case of solutions and fluids, the ^1H DNP enhancement is usually dominated by the modulation of the dipolar coupling between the radical and the target molecule. At high magnetic fields (>4 T), fluctuations in the subpicosecond regime are necessary to achieve spectral densities at the respective electron spin Larmor frequency to obtain efficient Overhauser DNP enhancements [see Formula (13.4)]. Translational and rotational motion of molecules, which account largely for the DNP enhancements at lower magnetic fields, are for many solvents too slow to significantly contribute to spectral densities in the 100 GHz to 1 THz region.

This method was further developed to do in situ DNP at 3.4 T by the Kentgens group[43] where the polarization transfer step and the detection are both performed in the same place.

In the Vieth group, the whole NMR probe head can be shuttled between 100 µT (corresponding to as low as 72 MHz EPR frequency) and 7 T field[44] (see Figure 13.8). In this setup, hyperpolarization can be achieved at variable magnetic fields by irradiating light on the sample via a flexible light guide

chemically induced dynamic nuclear polarization (CIDNP), bubbling the sample with *para*-hydrogen gas parahydrogen induced polarization (PHIP), or pumping electronic spin transitions (DNP).

13.3   EXPERIMENTAL RESULTS

13.3.1   Overhauser DNP with Sample Shuttling

Dynamic nuclear polarization experiments on the sample shuttling setup from Göttingen were performed on small molecules such as water, ethanol, glucose, and 2,2-dimethyl-2-silapentane-5-sulfonic acid sodium salt (DSS) in D_2O, which can be dissolved in sufficiently high concentration to obtain a sizeable Boltzmann polarized reference signal to estimate DNP enhancements.[40] Ethanol is only slightly larger than water and showed only modest polarization losses arising during the shuttling process by relaxation. The multiplet structure of the NMR spectrum gave insight into the obtained resolution taking into account the shuttling process and the radical-induced line broadening. Glucose is a larger molecule and showed larger relaxation losses during the shuttling process in comparison to water, ethanol, or DSS. This demonstrates one of the problems of the shuttling DNP approach for proton NMR at high magnetic fields. The other limitation is the DNP enhancement reduction, by factor of the ratio between the polarizing and detection magnetic field strength. Additionally, the sample size is limited due to MW penetration depth at X-band frequencies.

A 5 mM deuterated ^{15}N TEMPONE solution (80/20 H_2O/D_2O) with 10 mM DSS and 0.1 M ethanol was used for the experiments. This concentration is a compromise to achieve a fast and efficient polarization transfer at the low magnetic field and not too fast proton relaxation during the shuttling process. In order to minimize the relaxation losses during the shuttling transfer, the shuttle time between the low-field and the high-field positions was achieved to be as short as 37 ms. Mechanical oscillations after the arrival of the sample at the 14 T NMR detection field settled after 70 ms. The low-field EPR line was excited by MW for saturation of the electron spin system. The maximum enhancement factor of $\varepsilon = -1.7$ for ethanol was estimated as the ratio between the integral of the respective signal with and without DNP (Figure 13.9).

The proton signals from water, ethanol, and the DSS reference all show a negative enhancement. The shift and broadening of the water peak is caused by microwave heating. Enhancement factors in the range of $\varepsilon = -1.4$ to -2.8 for the different protons of D-glucose were achieved. The relaxation losses for the glucose protons were estimated to be between 10% and 30% due to T_1 relaxation within the shuttling process.

13.3.2 Experimental Results of In Situ DNP at 9.2 T Magnetic Field

Figure 13.10 shows a DNP experiment performed on a water sample doped with 1 M 4-hydroxy-2,2,6,6-tetramethylpiperidine-1-oxyl (TEMPOL). The strong negatively enhanced signal belongs to the sample part inside the MW resonator. It yields a surprising large enhancement factor of $\varepsilon = -83$. Notably, while the NMR signal of water outside of the MW cavity remains unchanged at the same position of approximately -7 ppm, the DNP-enhanced signal significantly changes its spectral position with respect to the reference signal, as

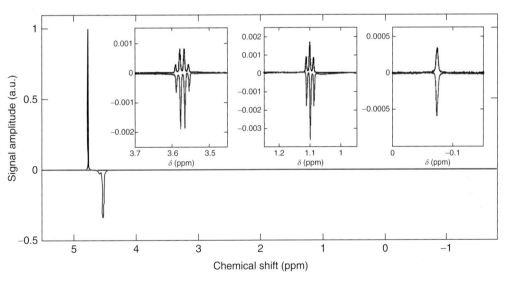

Figure 13.9. Proton DNP for sample shuttle of ethanol: 600 MHz proton spectrum of ethanol at room temperature (black) vs the DNP-enhanced spectrum (red) of 0.1 M ethanol 80/20 H_2O/D_2O with 5 mM TEMPONE-D, ^{15}N and 10 mM DSS solution. The reference signal was measured with a standard experiment consisting of a short $\pi/2$ pulse with duration of 7.1 μs. MW irradiation time at the low-field position was 12 s with an MW power of 16 W. Direct after the microwave irradiation, the sample was shuttled back to the high-field position in $t_{shuttle} = 40$ ms, and after a settling time $t_{set} = 70$ ms, the detection pulse was applied. (Reproduced with permission from Ref. 41. © Springer-Verlag, 2012)

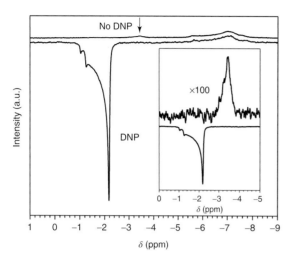

Figure 13.10. ^{1}H NMR/DNP signals of water protons with 1 M of ^{14}N-TEMPOL in a 30 μm ID capillary, without (positive signal) and with (negative signal) MW. NMR shift of the DNP signal arises from a combination of sample heating (60 °C) and from the suppression of the paramagnetic shift, caused by electrical and magnetic components of the MW field. The MW power was approximately 0.6 W, corresponding to a B_1 field of 3.5 G in both the experiments.[45] (Reproduced from Ref. 45 with permission from the PCCP Owner Societies)

indicated by the arrow in the figure. This shift is caused by a combination of MW heating and the suppression of the paramagnetic shift[46] by the saturating MW excitation. Both contributions to the chemical shift are important for a proper understanding of the DNP experiments at high magnetic fields and are required for quantitative determination of the saturation factor s as explained later.

The build-up of the DNP polarization for the in situ setup is defined by the nuclear spin relaxation time T_{1l}^{S} in equation (13.3). This has been demonstrated by progressive ^{1}H NMR saturation measurements under DNP conditions.[47] The DNP build-up time of $T_{\mathrm{DNP}} = 0.65$ s fits very well to the paramagnetic relaxation time of the sample without MW irradiation of $T_{1l}^{S} = 0.7$ s (Figure 13.11). This is an advantage of the in situ DNP method at high fields compared to the shuttle-DNP approach, where the shuttle time has to be taken into account. Therefore, NMR experiments under in situ DNP conditions can be performed even more rapidly, compared to samples without radicals in the solution.

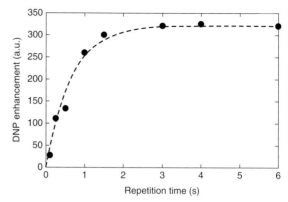

Figure 13.11. DNP build-up time of 50 mM TEMPOL in water measured under continuous MW excitation by progressive saturation of the proton NMR signal intensity as a function of the RF repetition time at 9.4 T

Besides the experimental determination of the DNP enhancement factor ε, an independent experimental determination of the leakage factor f and the saturation factor s is necessary to get access to the coupling factor ξ [equation (13.2)] as discussed earlier. Only then, a quantitative comparison with the theoretical models describing Overhauser DNP is possible [equation (13.4)]. The leakage factor can be easily determined by measuring the proton relaxation time with and without the radicals [equation (13.3)]. More complicated is the experimental determination of the saturation factor s.

Simple cw-EPR saturation measurements, as described by equation (13.1), are difficult to perform at high magnetic fields. This is due to the very fast relaxation times of most radicals in liquid solution at room temperature[23] and the more complicated lineshape at high magnetic fields due to g-tensor anisotropy[48] and the hf-splitting.[22] Pulsed ELDOR (electron–electron double-resonance) experiments allow obtaining very reliable saturation factors at lower microwave frequencies,[22,24] but are difficult to perform at high microwave frequencies, owing to limited MW power and the very short electron spin relaxation times of nitroxide radicals in liquid water solutions (<300 ns at 260 GHz).

Therefore, another method is used often in Overhauser DNP experiments to get rid of the saturation factor s in equation (13.2).[47] It is based on the linear dependence of $1/\varepsilon$ on the inverse MW power $1/P_{\mathrm{MW}}$ [from equations (13.1) and (13.2)]. This can be most easily achieved by plotting $1/\varepsilon$ against the inverse

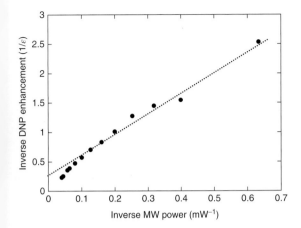

Figure 13.12. Plot of the inverse DNP enhancement against the inverse MW power to extract the maximum achievable DNP enhancement ε_{max} for $s = 1$ from the abscissa. Experimental data obtained at 260 GHz for a 50 mM TEMPOL aqueous solution

MW power ($1/P_{MW}$), as shown in Figure 13.12 for a 50 mM TEMPOL in water solution at 260 GHz. Linear fitting of the experimental data allows to extract the maximum DNP enhancement ε_{max} (for $s = 1$) from the abscissa of the graph. Careful inspection of the experimental data revealed one problem related to this method at high fields: applying higher MW power increases not only the saturation factor but also the sample temperature due to MW heating. This of course changes all parameters involved in the theoretical description of Overhauser DNP. The viscosity of the solution depends on the temperature and thus the dynamic modulation of the hf interactions, which influences the nuclear spin relaxation times as well as the DNP efficiency. This effect can be seen in Figure 13.12, where the DNP enhancements obtained at high MW power systematically deviate from the expected linear dependence. This is caused by heating of the sample up to the boiling point, which also increases the mobility and therefore the DNP enhancement. Additionally, the electron spin relaxation times are a function of temperature, thereby further influencing the saturation factor and the nuclear spin relaxation times.

The methods described above can hardly be used to extract reliable quantitative saturation factors because MW heating is more severe at higher MW frequencies (and thus at higher magnetic fields).

Fortunately, there is another possibility to extract the electron spin saturation value at high magnetic fields.

In a significantly large static magnetic field, the nuclei experience a sizable paramagnetic shift due to the unpaired electron spins of the radicals in the solution, The shift depends on the radical concentration arising from the difference of electron spins in the up and down states, e.g., the electron spin polarization $<S_z>$.[46] If the electron spin system is saturated, the contribution of electron spin polarization to the chemical shift diminishes. After full saturation, i.e., equalized electron spin populations, the paramagnetic shift disappears and the chemical shift reaches an asymptotic constant value.

The idea to determine the saturation of the EPR line has been used in one-dimensional organic conductors (fluoranthenyl radical anion salts) to suppress the Knight shift.[49] For radicals in solution at high magnetic fields and high radical concentrations, this effect can also be observed.[45] There is also a temperature effect on the chemical shift. For this reason, experiments with and without radicals have to be performed to separate the paramagnetic shift from the temperature effect (Figure 13.13). This allows determining the saturation factor s from the observation of the chemical shift of the water proton NMR line. Additionally, this experiment allows to define the sample temperature for a given MW irradiation power.

From Figure 13.13 it can also be determined that for a 20 μm ID capillary, an MW power of roughly 0.1 mW is sufficient to achieve a saturation factor $s > 0.9$. For larger MW powers, the saturation factor s does not improve anymore, but the sample is heated further, leading to still increasing DNP enhancements.[45] This method allowed us to also obtain reliable quantitative values of the saturation parameter s at high magnetic fields.

13.3.3 Comparison of High-field Overhauser DNP Results with Theoretical Models

The leakage factors f, the saturation factors s, as well as the ^{1}H DNP enhancement factors ε could be quantitatively determined for solutions of TEMPOL in water, toluene, acetone, and DMSO at a magnetic field of 9.4 T by the experimental methods described before. The coupling parameters ξ at this high magnetic field could be calculated[50] from these experimental results. Substantial ^{1}H DNP enhancements were found for all of these solvents, as well as for other dissolved metabolites, such as pyruvate, lactate, and alanine.[51] The large DNP enhancements observed for Fremy's salt and TEMPOL dissolved in water at 9.4 T, with

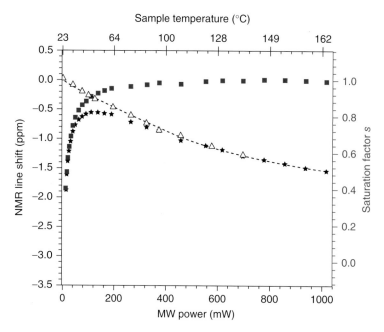

Figure 13.13. NMR line shifts of the peak of a pure water sample (open triangles) and a 1 M ^{14}N-TEMPOL in water sample (open stars) plotted against the incident MW power. Through the microwave heating calibration, each irradiation power can be assigned to a sample temperature (temperature scale is given on the top). By subtracting the temperature shift from the total NMR shift, the pure paramagnetic shift is obtained, which can be scaled to yield the saturation factor *s* (solid squares, scale on the right). (Reproduced from Ref. 45 with permission from the PCCP Owner Societies)

DNP enhancement factors of $\varepsilon = -83$ and $\varepsilon = -80$ respectively, as well as the surprising low dependence of the DNP enhancement with respect to the viscosity of the different solvents used, were rather surprising and unexpected. Large proton Overhauser DNP enhancements have also been reported by other groups at 3.4 T magnetic field for nitroxide radicals in water[17] and toluene.[16] Figure 13.14 summarizes the largest ^{1}H DNP enhancements reported for nitroxide radicals in water as a function of the magnetic field (respectively, MW frequency). Despite the fact that all these data points are obtained under rather different experimental conditions (MW power and B_1 field strength, sample temperature, and radical concentration), it is rather striking that the field/frequency dependence seems not to follow a steep decrease at higher magnetic fields as predicted from the hard-sphere model. Molecular dynamic simulations on the other side predict rather well the experimentally observed values[27] and exhibit a less steep decrease of the coupling factor as a function of the MW frequency. At a magnetic field of 9.4 T, dynamics of the hf coupling with correlations times shorter than 3 ps have to be present

to explain such large Overhauser DNP enhancements. However, translational diffusion correlation times between nitroxide radicals and solvent molecules extracted from DNP experiments at lower magnetic fields (<1 T) assuming the classical model with both spins in the center of two spherical molecules are much larger than 3 ps. On the other hand, local dynamics between the unpaired electron spin of the nitroxide and the proton of the solvent can have much shorter correlation times. For example, weak hydrogen bond vibrations between the radical and the solvent molecule in the range of several wavenumbers will start to contribute to spectral density in the terahertz region. Indeed, low-frequency vibrational modes might even have an inverse frequency dependence to spectral densities with an increasing contribution at higher magnetic fields. The classical models do not include such high-frequency local dynamics. MD simulations demonstrated that the parameters extracted from the hard-sphere models (such as the distance of the closest approach between both the spins or the correlation times) are indeed frequency dependent and thus should be treated with some caution or

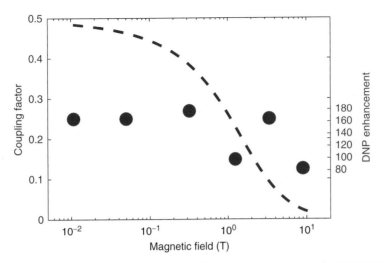

Figure 13.14. Maximum experimental observed Overhauser DNP enhancements (red dots) of TEMPOL in water at different magnetic field values. The experiment data are from: <0.3 T,[52] 0.34 T,[24] 1.5 T,[53] 3.4 T,[17] and 9.2 T.[54] The blue dash line depicts the frequency dependence of the coupling factor predicted by the classical hard-sphere model for pure translational motion with a correlation time of 20 ps

even taken more as empirical parameters.[38] At least they cannot be taken for a quantitative prediction of Overhauser DNP enhancements at high magnetic fields.

Of course, increased heating of the sample at higher frequencies (and therefore faster dynamics) could obscure the expected field profile. Nevertheless, MD simulations also predict very nicely and quantitatively the weak viscosity dependence of the high-field coupling factors. At 9.4 T, the DNP enhancements of TEMPOL in solvents with very different viscosities, such as toluene, acetone, and DMSO, were only slightly different.[26,27,55] Thus, MD simulations predict reliable coupling factors at different MW frequencies, temperatures, and solvent viscosities and can even explain small enhancement differences of different solvent protons.[26] In addition, MD simulations were also successful for explaining scalar hf couplings and DNP enhancements, like for ^{13}C nucleus,[56] which is out-of-reach for the hard-sphere model. The dependence of the coupling factor on the closest distance of approach between both spins showed to be crucial to achieve correct coupling factors. Atomistic MD simulations naturally account for such geometric (e.g., off-centered spin positions in the molecules) and dynamic (e.g., local librational modes or weak hydrogen bonding) effects and are therefore very well suited to predict Overhauser DNP efficiencies at high magnetic

fields, which rely much more on such details when compared to low magnetic fields.

13.3.4 Applications of High-field DNP

Our Fabry–Pérot resonator for MW excitation at 260 GHz described above can accommodate 90 nl of a disc-shaped sample within about 3 mm diameter. This is an ideal geometry for partially aligned multilayers of lipid bilayers as depicted in Figure 13.15(A) and (B). Several layers of lipid bilayers 1,2-dioleoyl-*sn*-glycero-3-phosphocholine (DOPC) doped with nitroxide radicals and exposed to water were deposited directly onto the surface of the flat mirror with a total thickness of approximately 20 μm (approximately 2000 bilayers, 160 μg). This assures that the wet lipid layers are situated in the maximum of the MW magnetic field and at a minimum of the MW electrical field component. The low main phase transition temperature of DOPC of −17 °C ensures that the sample is in the fluid phase at room temperature. For NMR detection of the lipid proton signal the flat mirror of the Fabry–Pérot resonator serves as a stripline RF probe tuned to 400 MHz (proton spin Larmor frequency).

The stripline NMR probe together with the Fabry–Pérot resonator for MW excitation provides a

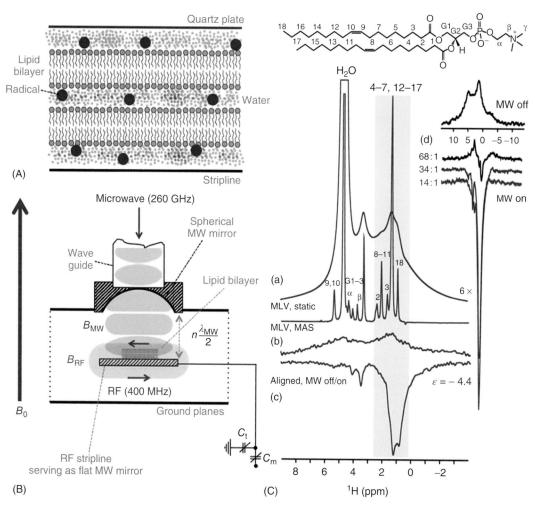

Figure 13.15. (A) Schematic sketch of the ordered lipid bilayers in a water/radical mixture at ambient temperature. (B) Schematic diagram of the stripline (RF)/Fabry–Pérot (MW) double-resonance probe structure used for ^{1}H DNP experiments of the partially ordered fluid lipid membrane layers that are deposited directly onto the flat metal stripline, which creates the RF field at the surface for NMR excitation and which is used for sensitive FID detection. The stripline together with an opposite spherical mirror forms the microwave Fabry–Pérot resonator.[36] (C) (a) ^{1}H NMR spectra of DOPC vesicles (static and MAS at 600 MHz). (b) ^{1}H NMR spectrum of DOPC doped with TEMPOL (DOPC : TEMPOL 34 : 1) and hydrated with D_2O (DOPC : D_2O 1 : 27). (c) Upon microwave irradiation (5.6 W), an enhancement of −4.4 was observed for the acyl chain proton resonances. (d) The enhancement increases with the amount of TEMPOL present.[57] (Reproduced with permission from Ref. 57. © American Chemical Society, 2014)

very sensitive double-resonance structure for such flat samples. With the inserted sample the quality factor Q of the MW resonator was about 100, leading to a factor of 10 enhanced MW B_1 field strength at the sample. The RF $\pi/2$-pulse length for protons of the stripline resonance structure was 20 µs at 1 W of applied RF power, resulting in RF conversion factor of 10 kHz $W^{-1/2}$.

Under MAS (magic angle spinning) conditions with a commercial solid-state NMR probe, many proton resonances of such a lipid sample are well resolved as shown in Figure 13.15(C) (MLV MAS spectrum, in red).[57] In our static double-resonance probe with the sample aligned with the membrane normal parallel to B_0, the NMR lines are significantly broadened, but the resonances of the acyl chain and choline protons of the lipids can still be distinguished from the water signal. This is because the fast anisotropic rotational diffusion of the lipids in the fluid phase partially averages the homonuclear proton dipole–dipole coupling network. Thus, the separation of the lipid and water resonances allowed a proof-of-concept study to see whether DNP enhancement occurs only for water protons or also for the lipid protons.

When DOPC was doped with the monoradical TEMPOL at a molar ratio of 34 : 1 and hydrated with 670 D_2O molecules per lipid, the HDO water proton resonance strongly dominates the NMR spectrum. Upon irradiation with MW, we observed a negative Overhauser DNP enhancement of the water protons, as reported also at lower magnetic fields.[58]

If the hydration level was further reduced to a level of 27 : 1 D_2O : DOPC, while keeping the bilayer in the fluid phase, the proton NMR signal from the lipid could also be detected. Under MW irradiation, a strong negative enhancement of the acyl chain and also of the γ-choline protons was observed.[57] The DNP enhancement increased with increasing amount of radicals in the membrane and by increasing the MW power. The radical to lipid ratio was varied between 1 : 14 and 1 : 68, and a MW power of up to 5.6 W was applied to the sample. The best enhancement for the acyl chain protons was obtained at the highest tested radical amount (TEMPOL/DOPC of 1/14), while the DNP effect for the γ-choline resonance is only detected for lower TEMPOL concentrations. One possible explanation of this difference could be that the hydrophilic TEMPOL radical quenches the NMR signals from protons at the membrane surface. It was also observed that the overall shape of the NMR spectra changed under DNP conditions and became slightly narrower compared to the Boltzmann polarized NMR spectra. This might indicate different DNP enhancement factors for the individual protons along the lipid molecule.

The same type of experiments were also performed using bilayers consisting of 1,2-dimyristoyl-*sn*-glycero-3-phosphocholine (DMPC). In contrast to DOPC, DMPC has shorter acyl chains without double bonds and a higher phase transition temperature of 24 °C. Using TEMPOL radicals, a DNP enhancement similar to the results obtained with DOPC was observed. Surprisingly, replacing the TEMPOL monoradical with the biradicals bTbK[59] or TOTAPOL,[60] which are typically used for cross-effect

DNP in the solid state,[61] even larger Overhauser DNP enhancements ($\varepsilon = -10$) were achieved.[57]

Thus, the Overhauser DNP experiments performed on partially ordered lipid layers in the fluid phase further highlight and emphasize our findings and conclusions obtained on TEMPOL dissolved in organic solvents. Quantitative comparison of the coupling factors derived from 9.4 T DNP experiments and MD simulations suggests that the translational diffusion cannot explain the observed high-field Overhauser DNP enhancements. Instead, local fast dynamics of the lipids have to account for the observed DNP enhancements at such high magnetic fields. This is in line with the experimental findings that the larger (slower diffusing) biradicals produce higher DNP enhancements compared to the smaller (faster diffusing) TEMPOL radical. In this specific case the different location of TEMPOL and bTbK at the lipid bilayer can be responsible for the observed differences between them.

13.4 SUMMARY AND OUTLOOK

Similar to solid-state DNP, as shown by several groups over the past few years, large DNP enhancements can also be achieved in the liquid-state at high magnetic fields. Fast dynamics in the subpicosecond timescale between the radical and the target molecule exist and supply enough spectral density to obtain substantial NMR signal enhancements. A detailed understanding of these dynamics might help to optimize the DNP agent–target system for specific applications. Because both spins have to come very close to achieve efficient polarization transfer at high magnetic fields, high-field Overhauser DNP experiments might also be interesting for the investigation of specific noncovalent binding sites between molecules or transient complexes formed in catalytic reactions or signal transduction in biology.

Presently, the main limitation of in situ high-field Overhauser DNP applications in the liquid-state results from the MW heating of the sample. One way to reduce the sample heating is by inserting the sample in the node of the standing electrical field component inside an MW resonator. Unfortunately, this limits the dimension of the sample at least in one dimension to a size much smaller than the MW wavelength and thus much smaller than the standard NMR samples. Thus, no in situ Overhauser DNP enhancement in aqueous solutions at high magnetic fields for typical NMR sample sizes has been published so far. Another way to avoid this drawback is the Shuttle-DNP approach, in which the polarization transfer is accomplished at lower MW frequencies, where MW absorption is less problematic due to the wavelength of the MW in the centimeter range. The disadvantage of this approach is that the overall achievable DNP enhancement is reduced by the ratio between the polarizing and NMR detection field strengths. In addition, especially for proton spins, polarization might get lost within the shuttle time. Therefore, at the moment the achieved Overhauser DNP enhancements are not so high, although carbon nuclear spins show more promising results.

Moreover, applications of the method seem potentially interesting only if very small sample volumes are acceptable. This might be in the field of analytical NMR spectroscopy, as for example, a 400 MHz NMR spectrometer directly coupled to an HPLC output, allowing NMR characterization of the different fractions. A new stripline-Fabry–Pérot DNP probe, allowing flow-through operation, is currently under construction in our laboratory. Such a probe and setup would also be interesting for applications in metabolomics, where again small sample volumes have to be investigated. In overmoded MW resonance structures, as for example the Fabry–Pérot resonator, the sample size has to be reduced only in its height. Such a configuration seems ideally suited for investigations of surface phenomena, as for example heterogeneous catalysis. In principle, such a configuration could also be directly coupled to an NMR microscopy probe, providing spatially resolved images of two-dimensional objects with higher sensitivity. In addition, higher contrast could be envisioned for heterogeneous samples with a liquid-accessible and nonaccessible phase. Furthermore, such a setup could be used to investigate flow dynamics in chemical microreactors and flow systems. Microreactors can be used to follow chemical reactions kinetics. Dynamic nuclear polarization-enhanced educts could be used to follow and unravel the chemical mechanism of the reaction steps in detail.[62] So far, dissolution DNP has been used for such applications; however, direct Overhauser DNP in the liquid phase could be advantageous to allow observation of fast reactions and to observe proton nuclear spins directly. Our first [1]H Overhauser DNP experiments performed on lipid bilayers at 9.4 T and ambient temperature[57] indicate that such experiments may also be successful on membrane proteins embedded in ordered lipid films. Conformational heterogeneity of such proteins, sometimes observed in solid-state DNP work at low temperatures,

might be reduced at ambient temperatures. Additional information might be accessible using different kinds of radicals (hydrophobic vs hydrophilic or covalently attached). The recently observed Overhauser DNP effect in solids might be closely related to the fast dynamic modes of the hf coupling observed in high-field Overhauser DNP in liquids.

CONFLICTS OF INTEREST

There are no conflicts to declare.

ACKNOWLEDGMENTS

This research was supported by the German Research Foundation (DFG), the German–Israel Research Program (DIP), and the European Commission (Bio-DNP Design Study). We acknowledge Clemens Glaubitz, Johanna Becker-Baldus, Orawan Jakdetchai, Deniz Sezer, Petr Neugebauer, Mark Prandolini, Alex Krahn, Frank Engelke, Marat Gafurov, and Jan Krummenacker for their experimental and theoretical contributions to the 9.4 T DNP studies reviewed here.

RELATED ARTICLES IN EMAGRES

Griffin, Robert G.: Perspectives of Magnetic Resonance

Overhauser, Albert W.: Dynamic Nuclear Polarization

Dynamic Nuclear Polarization: Applications to Liquid-State NMR Spectroscopy

High-Frequency Dynamic Nuclear Polarization

REFERENCES

1. W. Overhauser, *Phys. Rev.*, 1953, **92**, 411.

2. T. R. Carver and C. P. Slichter, *Phys. Rev.*, 1953, **92**, 212.

3. C. P. Slichter, *Phys. Chem. Chem. Phys.*, 2010, **12**, 5741.

4. H. G. Beljers, L. van der Kint, and J. S. van Wieringen, *Phys. Rev.*, 1954, **95**, 1683.

5. A. V. Kessenikh, V. I. Lushchikov, A. A. Manenkov, and Y. V. Taran, *Sov. Phys. Solid State*, 1963, **5**, 321.

6. K. H. Hausser and D. Stehlik, *Adv. Magn. Reson.*, 1968, **3**, 79.

7. W. Müller-Warmuth and K. Meise-Gresch, *Adv. Magn. Reson.*, 1983, **11**, 1.

8. Y. Ayant, E. Belorizky, J. Alizon, and J. Gallice, *J. Phys.*, 1975, **36**, 991.

9. L.-P. Hwang and J. H. Freed, *J. Chem. Phys.*, 1975, **63**, 4017.

10. R. A. Dwek, R. E. Richards, and D. Taylor, *Ann. Rep. NMR Spectrosc.*, 1969, **2**, 293.

11. H. C. Dorn, R. Gitti, K. H. Tsai, and T. E. Glass, *Chem. Phys. Lett.*, 1989, **155**, 227.

12. L. R. Becerra, G. J. Gerfen, R. J. Temkin, D. J. Singel, and R. G. Griffin, *Phys. Rev. Lett.*, 1993, **71**, 3561.

13. J. H. Ardenkjaer-Larsen, B. Fridlund, A. Gram, G. Hansson, L. Hansson, M. H. Lerche, R. Servin, M. Thaning, and K. Golman, *PNAS*, 2003, **100**, 10158.

14. N. M. Loening, M. Rosay, V. Weis, and R. G. Griffin, *J. Am. Chem. Soc.*, 2002, **124**, 8808.

15. M. J. Prandolini, V. P. Denysenkov, M. Gafurov, B. Endeward, and T. F. Prisner, *J. Am. Chem. Soc.*, 2009, **131**, 6090.

16. E. V. Kryukov, M. E. Newton, K. J. Pike, D. R. Bolton, R. M. Kovalczyk, A. P. Howes, M. E. Smith, and R. Dupree, *Phys. Chem. Chem. Phys.*, 2010, **12**, 5757.

17. P. J. M. van Bentum, G. H. A. van der Heijden, J. A. Villanueva-Garibay, and A. P. M. Kentgens, *Phys. Chem. Chem. Phys.*, 2011, **13**, 17831.

18. G. Liu, M. Levien, N. Karschin, G. Parigi, C. Lucinat, and M. Bennati, *Nat. Chem.*, 2017, **9**, 676.

19. O. Haze, B. Corzilius, A. A. Smith, and R. G. Griffin, *J. Am. Chem. Soc.*, 2012, **134**, 14287.

20. I. Solomon, *Phys. Rev.*, 1955, **99**, 559.

21. R. D. Bates and W. S. Drozdoski, *J. Chem. Phys.*, 1977, **67**, 4038.

22. B. H. Robinson, D. A. Haas, and C. Mailer, *Science*, 1994, **263**, 490.

23. W. Froncisz, T. G. Camenisch, J. J. Ratke, J. R. Anderson, W. K. Subczyski, R. A. Stangeway, J. H. Sibradas, and J. S. Hyde, *J. Magn. Reson.*, 2008, **193**, 297.

24. M.-T. Türke and M. Bennati, *Phys. Chem. Chem. Phys.*, 2011, **13**, 3630.

25. D. Sezer, M. J. Prandolini, and T. F. Prisner, *Phys. Chem. Chem. Phys.*, 2009, **11**, 6626.

26. D. Sezer, *Phys. Chem. Chem. Phys.*, 2013, **15**, 526.

27. D. Sezer, *Phys. Chem. Chem. Phys.*, 2014, **16**, 1022.

28. P. Höfer, G. Parigi, C. Luchinat, P. Carl, G. Guthausen, M. Reese, T. Carlomagno, C. Griesinger, and M. Bennati, *J. Am. Chem. Soc.*, 2008, **130**, 3254.

29. M. Bennati, C. Luchinat, G. Parigi, and M.-T. Turke, *Phys. Chem. Chem. Phys.*, 2010, **12**, 5902.

30. E. Ravera, C. Luchinat, and G. Parigi, *J. Magn. Reson.*, 2016, **254**, 78.

31. J. D. Jackson, Classical Electrodynamics, John Wiley & Sons Inc: New York, 1962.

32. K. Gruber, J. Forrer, A. Schweiger, and H. H. Gunthard, *J. Phys. E: Sci. Instrum.*, 1973, **7**, 569.

33. V. Weis, M. Bennati, M. Rosay, J. A. Bryant, and R. G. Griffin, *J. Magn. Reson.*, 1999, **140**, 293.

34. V. P. Denysenkov, M. J. Prandolini, A. Krahn, M. Gafurov, B. Endeward, and T. F. Prisner, *Appl. Magn. Reson.*, 2008, **34**, 289.

35. F. Engelke, *Concept Magn. Reson.*, 2002, **15**, 129.

36. V. Denysenkov and T. F. Prisner, *J. Magn. Reson.*, 2012, **217**, 1.

37. P. J. M. van Bentum, J. W. G. Janssen, A. P. M. Kentgens, J. Bart, and J. G. E. Gardeniers, *J. Magn. Reson.*, 2007, **189**, 104.

38. T. Prisner, V. Denysenkov, and D. Sezer, *J. Magn. Reson.*, 2016, **264**, 68.

39. V. P. Denysenkov, M. J. Prandolini, M. Gafurov, D. Sezer, B. Endeward, and T. F. Prisner, *Phys. Chem. Chem. Phys.*, 2010, **12**, 5786.

40. A. Krahn, P. Lottmann, T. Marquardsen, A. Tavernier, M. T. Turke, M. Reese, A. Leonov, M. Bennati, P. Hoefer, F. Engelke, and C. Griesinger, *Phys. Chem. Chem. Phys.*, 2010, **12**, 5830.

41. C. Griesinger, M. Bennati, H.-M. Vieth, C. Luchinat, G. Parigi, P. Höfer, F. Engelke, S. J. Glaser, V. Denysenkov, and T. F. Prisner, *Prog. Nucl. Magn. Res. Spectrosc.*, 2012, **64**, 4.

42. X. Wang, W. C. Isley, S. I. Salido, Z. Sun, L. Song, K. H. Tsai, C. J. Cramerb, and H. C. Dorn, *Chem. Sci.*, 2015, **6**, 6482.

43. S. G. J. van Meerten, M. C. D. Tayler, A. P. M. Kentgens, and P. J. M. van Bentum, *J. Magn. Reson.*, 2016, **267**, 30.

44. A. S. Kiryutin, A. N. Pravdivtsev, K. L. Ivanov, Y. A. Grishin, H.-M. Vieth, and A. V. Yurkovskaya, *J. Magn. Reson.*, 2016, **263**, 79.

45. P. Neugebauer, J. G. Krummenacker, V. P. Denysenkov, G. Parigi, C. Luchinat, and T. F. Prisner, *Phys. Chem. Chem. Phys.*, 2013, **15**, 6049.

46. I. Bertini, C. Luchinat, and G. Parigi, 'Solution NMR of Paramagnetic Molecules', Elsevier, 2001, Vol. **2**.

47. T. Prisner and M. Prandolini, in 'Multifrequency Electron Paramagnetic Resonance: Theory and Applications', ed S. K. Misra, John Wiley & Sons, Inc: New York, 2011, Chapter 24.

48. K. A. Earle, D. E. Budil, and J. H. Freed, *J. Phys. Chem.*, 1993, **97**, 13289.

49. R. A. Wind, H. Lock, and M. Mehring, *Chem. Phys. Lett.*, 1987, **141**, 283.

50. P. Neugebauer, J. G. Krummenacker, V. P. Denysenkov, C. Helmling, C. Luchinat, G. Parigi, and T. F. Prisner, *Phys. Chem. Chem. Phys.*, 2014, **16**, 18781.

51. J. G. Krummenacker, V. P. Denysenkov, and T. F. Prisner, *Appl. Magn. Reson.*, 2012, **43**, 139.

52. S. E. Korchak, A. S. Kiryutin, K. L. Ivanov, A. V. Yurkovskaya, A. A. Grishin, H. Zimmermann, and H.-M. Vieth, *Appl. Magn. Reson.*, 2010, **37**, 515.

53. J. Krummenacker, V. Denysenkov, M. Terekhov, L. Schreiber, and T. F. Prisner, *J. Magn. Reson.*, 2012, **215**, 94.

54. M. Gafurov, V. Denysenkov, M. Prandolini, and T. Prisner, *Appl. Magn. Reson.*, 2012, **43**, 119.

55. S. E. Küçük, P. Neugebauer, T. F. Prisner, and D. Sezer, *Phys. Chem. Chem. Phys.*, 2015, **17**, 6618.

56. S. E. Kücük and D. Sezer, *Phys. Chem. Chem. Phys.*, 2016, **18**, 9353.

57. O. Jakdetchai, V. Denysenkov, J. Becker-Baldus, B. Dutagaci, T. F. Prisner, and C. Glaubitz, *J. Am. Chem. Soc.*, 2014, **136**, 15533.

58. R. Kausik and S. Han, *Phys. Chem. Chem. Phys.*, 2011, **13**, 7732.

59. Y. Matsuki, T. Maly, O. Ouari, H. Karoui, F. Le Moigne, E. Rizzato, S. Lyubenova, J. Herzfeld, T. Prisner, P. Tordo, and R. G. Griffin, *Angew. Chem. Int. Ed.*, 2009, **48**, 4996.

60. C. S. Song, K. N. Hu, C. G. Joo, T. M. Swager, and R. G. Griffin, *J. Am. Chem. Soc.*, 2006, **35**, 11385.

61. A. S. L. Thankamony, J. J. Wittmann, M. Kaushik, and B. Corzilius, *Prog. Nucl. Magn. Res. Spectrosc.*, 2017, **102**, 120.

62. S. Bowen and C. Hilty, *Angew. Chem. Int. Ed.*, 2008, **47**, 5235.

Chapter 14

Overhauser DNP in Liquids on ^{13}C Nuclei

Marina Bennati[1,2] and Tomas Orlando[1]

[1]*Max Planck Institute for Biophysical Chemistry, Göttingen, Germany*
[2]*University of Göttingen, Göttingen, Germany*

14.1 INTRODUCTION

Overhauser type of dynamic nuclear polarization (DNP) was historically the first predicted mechanism.[1] It is based on electron–nuclear cross-relaxation and acts when pumping one or several allowed electron spin transitions coupled to nuclear spin transitions.[2,3] Most importantly, its efficiency requires that the hyperfine (hf) interaction between electron and nuclear spins is modulated at the electron spin resonance frequency,

Handbook of High Field Dynamic Nuclear Polarization.
Edited by Vladimir K. Michaelis, Robert G. Griffin, Björn Corzilius and Shimon Vega
© 2020 John Wiley & Sons, Ltd. ISBN: 978-1-119-44164-9
Also published in eMagRes (online edition)
DOI: 10.1002/9780470034590.emrstm1581

thus at correlation times of tens of picoseconds to subpicoseconds. Such short correlation times are provided by fast dynamics such as the motion of conduction electrons in metals[4] or molecular diffusion in the liquid state.[5] In the latter case, a polarizer molecule – usually a nitroxide radical – is mixed in solution with the target molecule, from which the NMR spectrum is recorded.

Similar to all DNP experiments described in this chapter, until the 1990s, DNP in liquids was developed at low and medium magnetic fields, i.e., $0.3-1$ T, owing to the lack of microwave sources to explore high-field performance.[6] In the past two decades, the availability of high-frequency EPR instrumentation has boosted the method up to magnetic fields of 9 T,[7–10] with two very recent reports also at 14 T.[11,12] This revival has led to fundamental progress in understanding the liquid DNP mechanism.[13–16] For ^{1}H nuclei at magnetic fields >1 T, negative signal enhancements up to 10^2 (at room temperature) in magnitude have been observed.[16,17] These conditions result in an emissive enhanced NMR signal due to the magnetic electron–nuclear dipolar interaction modulated by molecular diffusion. Moreover, application of liquid DNP at high fields in aqueous solutions at ambient temperatures is aggravated by dielectric losses. This problem can only be solved using mw resonant structures with the sample placed in a minimum of the electric field of the standing wave.[18,19]

To date, investigation of liquid DNP for nuclei with low gyromagnetic ratio, with ^{13}C being the prototype one, has been mainly restricted to magnetic fields around 1 T.[6] As first, the Boltzmann signal of these nuclei is difficult to detect at very low fields (e.g., 0.34 T/X-band frequencies). Furthermore, both old and recent literature on ^{13}C in the low to medium field range has revealed an intriguing dependence of the NMR signal enhancements on the ^{13}C chemical environment, reporting enhancement values sometimes positive or negative.[20–23] This effect is inherent to the subtle hf interaction mechanism to the ^{13}C nucleus, which usually displays a mixture of large scalar and dipolar contributions, depending on the physical and chemical properties of the polarizer/target system. Specifically, modulation of a scalar hf interaction between a polarizer molecule and a ^{13}C nucleus, residing on a target molecule, requires formation of a transient polarizer/target complex. Within this complex, a suited orbital overlap has to occur that 'transfers' unpaired spin density via a spin polarization mechanism[24] to the ^{13}C nucleus. This mechanism is substantially different than in ^1H dipolar Overhauser DNP, as reported in water[7,17,25] or toluene,[26] where dipolar interaction is modulated through space via molecular diffusion.

Our recent ^{13}C DNP experiments[12,27] at high magnetic field (3.4 T) have started shining some light in the complex but fascinating mechanism of scalar Overhauser DNP. The huge achievable enhancements (up to 10^3 at room temperature) might be potentially useful for a general application in high-field solution NMR. The complex atomistic details of scalar hf interaction prevent us – at the current stage – to generalize the ^{13}C results to other low-gamma nuclei. Therefore, we restrict this review chapter to ^{13}C DNP, describing our understanding on DNP of representative small organic molecules with their unexpected field dependence. A possible generalization of this mechanism to ^{13}C in other chemical environment is discussed in the last section.

14.2 GENERAL THEORY OF OVERHAUSER DNP IN LIQUIDS

In a liquid DNP experiment, thermal polarization of electron spins is transferred to coupled nuclear spins by saturating the electron spin resonance line. The process is best described by considering the four energy levels of an electron spin $S = \frac{1}{2}$ coupled to a nuclear spin $I = \frac{1}{2}$ (e.g., a proton, ^1H) (Figure 14.1).

Continuous mw pumping drives the electron spin system into saturation, creating a deviation of the

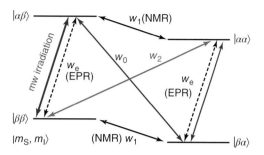

Figure 14.1. Energy-level scheme of an electron spin $S = \frac{1}{2}$ coupled to a proton; α and β denote the quantum numbers $+1/2$ and $-1/2$, respectively. Transition probabilities among possible pairs of levels are indicated. w_e and w_1 are probabilities that flip only the electron or the nuclear quantum number, respectively, and w_0 and w_2 are zero- and double-quantum transition probabilities, respectively

electron spin expectation value $\langle S_z \rangle$ from Boltzmann equilibrium S_0. This is expressed through the effective saturation factor s_{eff} of the radical, which takes into account the saturation of the pumped EPR transition plus the related ELDOR (electron–electron double resonance) effect on the other EPR hf lines.[28] Electron spin lattice and electron–nuclear cross-relaxation redistribute populations in this spin system and build up nuclear polarization (proportional to $\langle I_z \rangle$). If transition probabilities $w_{0,1,2}$ between a pair of energy states are labeled according to Figure 14.1, the steady-state solution of the coupled differential equation for the individual populations leads to the well-known Overhauser equation, which relates the generated nuclear spin polarization to the involved transition probabilities as:[2,5]

$$\varepsilon = \frac{\langle I_z \rangle}{I_0} = 1 - \frac{w_2 - w_0}{w_0 + 2w_1 + w_2}$$
$$\times \frac{w_0 + 2w_1 + w_2}{w_0 + 2w_1 + w_2 + R_{1dia}} \cdot s_{eff} \cdot \frac{|\gamma_e|}{\gamma_I}$$
$$= 1 - \xi \cdot f \cdot s_{eff} \cdot \frac{|\gamma_e|}{\gamma_I} \tag{14.1}$$

where ε is the NMR signal enhancement, I_0 is the nuclear Boltzmann polarization, γ_e and γ_I are the gyromagnetic ratios of electron and nuclear spins, respectively, R_{1dia} is the diamagnetic contribution to the total nuclear spin relaxation rate, and the terms ξ and f are called coupling and leakage factors, respectively. The most important parameter in DNP is the coupling factor ξ and its dependence on the resonance frequency, i.e., the external static magnetic field.

However, this parameter can be properly evaluated only if f and – particularly – s_{eff} are determined in separate experiments, as was pointed out in early[6] and more recent literature.[13] f is easily obtained from T_{1I} measurements of the target nucleus in solution with (T_{1I}) and without $(T_{1I,dia})$ the paramagnetic polarizer, as $f = 1 - T_{1I}/T_{1I,dia}$. Determination of s_{eff} can occur from continuous wave (CW) EPR saturation experiments,[29] but it becomes challenging when the EPR absorption is split into several hf lines, as in nitroxide radicals. In some recent chapters, we have reported extensive discussion on the saturation factor for nitroxide radicals by means of ELDOR spectroscopy,[27,28,30] and other authors reported analysis at high fields (9.2 T) by considering the paramagnetic shifts.[31] Overall, when using high polarizer concentrations ($c \geq 30$ mM) and irradiation in mw cavities, leakage and saturation factors for nitroxide radicals might approach unity. In this case, the observed enhancement directly reflects the coupling factor ξ,

$$\varepsilon_{max} \approx \xi \cdot \frac{|\gamma_e|}{\gamma_I} \qquad (14.2)$$

14.3 FIELD DEPENDENCE OF COUPLING FACTORS FOR MIXED DIPOLAR AND SCALAR RELAXATION

Equation (14.1) defines the coupling factor as proportional to the difference between double- and zero-quantum transition probabilities; however, this definition alone does not give much insight into the physical picture, and relaxation theory has to be recalled. We follow here the semiclassical description of hf relaxation as described by Solomon[2], Abragam[32], and Hausser and Stehlik[5]. In the liquid state, the contribution of dipolar and scalar hf relaxation can be expressed through additive terms in the coupling factor:[5]

$$\xi = \frac{w_2^{dip} - w_0^{dip} - w_0^{scalar}}{w_0^{dip} + 2w_1^{dip} + w_2^{dip} + w_0^{scalar}} \qquad (14.3)$$

where the superscripts refer to the different scalar and dipolar mechanisms, respectively. The relations between probabilities $w_{0,1,2}$ and spectral densities for a fast, randomly fluctuating hf Hamiltonian were derived

by Solomon[2] and can be expressed as:[3]

$$w_0 = w_0^{dip} + w_0^{scalar} = k\, J(\omega_I - \omega_e, \tau_{dip})$$
$$+ k_a\, J(\omega_I - \omega_e, \tau_{scalar})$$
$$w_1 = \frac{3}{2}k\, J(\omega_I, \tau_{dip})$$
$$w_2 = 6k\, J(\omega_I + \omega_e, \tau_{dip}) \qquad (14.4)$$

where $J(\omega, \tau)$ is the spectral density function, τ_{dip} and τ_{scalar} are the correlation times for dipolar and scalar relaxation, and k and k_a are prefactors that represent the mean squared fluctuation of dipolar and scalar interactions, respectively. We note that scalar interaction contributes only to zero and not to single- and double-quantum transitions. With this consideration, equation (14.3) for a scalar-only mechanism leads to a negative coupling factor $\xi_{scalar} = -1$, which is completely field independent. It was suggested[5] to introduce an additional term to include fast electron spin relaxation as a possible source for the time dependence of hf scalar coupling (scalar relaxation of the second kind). This effect would induce pure nuclear relaxation transitions, generating a w_1^{scalar} [not included in equation (14.3)] weighted by a factor $\beta : w_0^{scalar} + \beta \cdot w_1^{scalar}$. However, this term can be neglected when the polarizer is an organic radical with $T_{1e} \gtrsim 1$ ns, such as a nitroxide.

In the other limiting case of a dipolar-only mechanism, the coupling factor was discussed in several previous chapters and leads to a field dependence on the order $\propto 1/\omega_e^2$, as readily seen from the expression for a Lorentzian spectral density $J(\omega, \tau) = \tau/(1 + \omega^2\tau^2)$ in the regime $\omega^2\tau^2 \gg 1$. However, most of the time scalar and dipolar relaxation occur concomitantly and, for a further discussion, it is convenient to introduce the experimentally measurable nuclear relaxation rates $R_1 = 1/T_1$:

$$R_1 = R_{1,para} + R_{1,dia}$$
$$R_{1,para} = R_{1,scalar} + R_{1,dip}$$
$$R_{1,dip} = k(7J(\omega_e, \tau_{dip}) + 3J(\omega_I, \tau_{dip}))$$
$$R_{1,scalar} = k_a J(\omega_e, \tau_{scalar}) \qquad (14.5)$$

With these definitions, the coupling factor in equation (14.3) can be rewritten as:[33]

$$\xi = \frac{5}{7}\left[1 - \frac{3k\, J(\omega_I, \tau_{dip})}{R_{1,scalar} + R_{1,dip}}\right] - \frac{12}{7}\frac{R_{1,scalar}}{R_{1,scalar} + R_{1,dip}} \qquad (14.6)$$

In a first simple example, we illustrate the frequency dependence of the coupling factor from

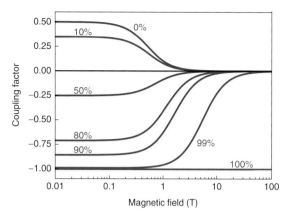

Figure 14.2. Coupling factors calculated according to equation (14.6) for mixed scalar and dipolar relaxation mechanisms using one single correlation time for both $\tau_{dip} = \tau_{scalar} = 20$ ps as well as a spectral density of Lorentz type. The percentage is defined as: $100 \times (R_{1,scalar}/R_{1,para})$

equation (14.6) for mixed scalar and dipolar contributions calculated assuming a Lorentzian spectral density function and identical correlation times for scalar and dipolar relaxation. The plots in Figure 14.2 were obtained for several ratios of $R_{1,scalar}/R_{1,para}$. The deleterious effect of counteracting dipolar and scalar mechanisms becomes visible from this plot. For almost equal contributions, the resulting coupling factor becomes very small, and DNP enhancements are difficult to observe. Moreover, even for the case of a dominant scalar mechanism, the presence of dipolar contribution leads to a decay of the coupling factor at high fields, as discussed in the following paragraphs.

However, in a real sample the situation is more complicated because:

1. correlation times for dipolar and scalar mechanisms are different;
2. multiple scalar and dipolar relaxation pathways might act concomitantly; and
3. precise form of the spectral density might be unknown.

Concerning the first point, we noted[27] that, differently than in Figure 14.2, the correlation times for scalar and dipolar relaxation are likely substantially different. Indeed, correlation times for molecular diffusions are on the order of tens or hundreds of picoseconds (e.g., for nitroxide radical in water $\tau_D \approx 15$–20 ps[7] and the trityl radical in water $\tau_D \approx 140$ ps[29]). Instead, correlation times

that modulate isotropic hf interaction might be on the subpicosecond time scale,[6] or the time scale of fast molecular collisions. Indeed, for the isotropic hf interaction, it is expected that small changes in bond lengths and orbital overlap are sufficient for causing appreciable changes in the spin polarization mechanism. Because of the different correlation times and the fact that w_0 and w_2 both depend on ω_e via $J(\omega_e)$, the dipolar contributions in equation (14.3), w_0^{dip} and w_2^{dip}, decay faster with increasing magnetic field than the scalar part w_0^{scalar}. The resulting behavior is a field dependence of the coupling factor, which can be distinguished into two regimes: (i) the low-field regime, where ξ is dominated by the effective ratio $R_{1,scalar}/R_{1,para}$; (ii) the high-field regime, in which the dipolar contribution has mainly decayed and the coupling factor reduces to the form:[27]

$$\xi \approx \frac{-w_0^{scalar}}{2w_1^{dip} + w_0^{scalar}} \tag{14.7}$$

We note that the dipolar contribution is still present in the denominator of equation (14.7) and drives ξ to zero in the high-field limit. To illustrate the more realistic behavior of equation (14.6), we first introduce established expressions for spectral densities, which were found to properly describe DNP data, particularly at high fields (>1 T). For the dipolar contribution, we employ the spectral density derived from the hard-sphere force-free (FF) model:[33–35]

$$J_{diff}(\omega, \tau_d)$$
$$= \frac{1 + 5z/8 + z^2/8}{1 + z + z^2/2 + z^3/6 + 4z^4/81 + z^5/81 + z^6/648} \tag{14.8}$$

with $z = \sqrt{2\omega\tau_d}$ and where $\tau_d = r_d^2/D$ describes the correlation time for molecular diffusion, r_d is the distance of closest approach, and D is the sum of the diffusion coefficients of radical and solvent molecules. For this function, the prefactor k in equation (14.5) has the form:

$$k = \frac{32\,000\,\pi}{405} \left(\frac{\mu_0}{4\pi}\right)^2 \frac{N_A C \gamma_I^2 g_e^2 \mu_B^2 S(S+1)}{r_d D} \tag{14.9}$$

where N_A is the Avogadro's constant, C is the molar concentration of the paramagnet (in mol dm^{-3}), μ_B is the Bohr magneton, and g_e is the electron g factor. To describe the temporal modulation of the scalar hf interaction in liquids, we adopted the so-called pulse model proposed by Müller-Warmuth *et al.*,[21] by which spin density is transferred from the radical to the target

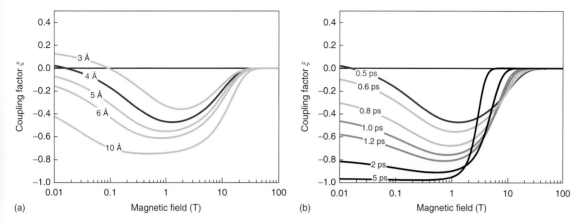

Figure 14.3. Coupling factors as a function of magnetic field calculated in the presence of scalar interaction (with random encounters of duration τ_i) and translational diffusion with a correlation time τ_D, according to equations (14.6), (14.8), and (14.9). Curves were computed for different distances r_d of closest approach (left, a) and contact times τ_i (b). Only one type of encounter was considered, thus setting x_i in equation (14.10) equal to one. For the red curve, the ratio of $R_{1,scalar}/R_{1,para}$ was 28% in the low-field limit. This ratio changes with (i) field due to the different field dependencies of the two spectral densities in equations (14.8) and (14.10) and (ii) with r_d, as it is included in the prefactor k, and τ_i, since $J_{pulsed} \propto \tau_i$ in equation (14.10). The diffusion constant D in equation (14.9) was set at 1.7×10^{-9} m² s⁻¹ (value for CCl₄ at room temperature)

nucleus through random collisions of duration τ_i at an average rate $1/\tau_p$. The spectral density within this model was expressed as:

$$J_{pulsed}(\omega_e) = \frac{A^2}{\hbar^2 \tau_p} \left[\sum_i x_i \tau_i \exp\left(-\omega_e \tau_i\right) \right]^2 \quad (14.10)$$

where x_i are the relative fractions of encounters with duration τ_i and $A^2 \equiv \langle A^2 \rangle$ is the mean squared amplitude of the scalar hf in energy units. The corresponding prefactor k_a [equation (14.5)] equals $2/3 S(S+1)$.[21] Inserting equations (14.8)–(14.10) in equation (14.6), the field dependence of ξ is calculated as displayed in Figure 14.3. Depending on the initial ratio $R_{1,scalar}/R_{1,para}$ and the respective correlation times in equations (14.8) and (14.10), the coupling factor – in absolute value – might decrease toward high fields or first increase up to a maximum and then decays to zero. These general thoughts provided us the starting point to rationalize the recent unexpected observation that ^{13}C Overhauser DNP enhancements of certain model systems, such as ^{13}CCl₄ doped with TEMPONE (4-oxo-2,2,6,6-tetramethylpiperidine), do not decrease but rather increase with magnetic field. Experimental observations from the various reports in the literature are summarized in the following sections.

14.4 EXPERIMENTAL RESULTS ON ORGANIC MOLECULES

Initial systematic studies of ^{13}C DNP enhancements up to 1 T were reported by Müller-Warmuth and Meise-Gresch.[6] They investigated different ^{13}C-labeled organic solvents such as benzene and CCl₄, using various aromatic polarizers such as BDPA (bis-diphenylene-phenyl-allyl). Scalar ^{13}C enhancements on ^{13}CCl₄ up to two orders of magnitude were reported,[36] which were interpreted in terms of the pulsed model [equation (14.10)] with a contact time τ_i on the order of 1 ps. Later on, low-field (0.3 T) Overhauser ^{13}C-DNP was introduced by Dorn and coworkers to enhance sensitivity in continuous-flow NMR with detection at 4.7 T.[37,38] The authors employed nitroxide radicals as polarizers, either directly mixed in the liquid phase or immobilized in a silica phase. Although the reported enhancements were extrapolated for $s_{eff} = 1$ and $f = 1$, some very interesting features emerged. The maximal enhancements ε_{max} [equation (14.2)] were found for the system nitroxide/^{13}CHCl₃, with a ^{13}C coupling factor reaching unity at 0.34 T. Instead, for the slightly different system nitroxide/^{13}CCl₄, the coupling factor was a factor of 5 smaller, which was a striking observation. A more recent study by Wang et al.[23] and a report by Lingwood and Han,[22] both at low field (0.34 T/9 GHz

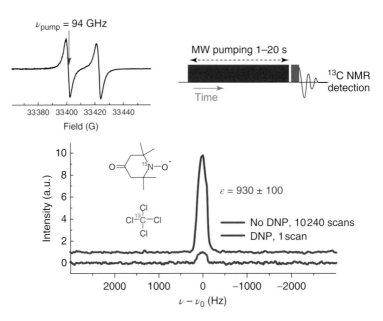

Figure 14.4. ^{13}C DNP NMR on ^{13}CCl$_4$ with 30 mM ^{15}N-TEMPONE-d$_{16}$ at 3.4 T (ν_e = 94 GHz, ν_{NMR} (^{13}C) = 36 MHz, RT). Black: Boltzmann signal with 10 240 scans and recycle delay of 15 s, and the spectrum scaled down by a factor of 10 240$^{1/2}$ for comparison at the same noise level to the DNP spectrum; red: DNP with 1 scan and 15 s of mw irradiation ($P_{mw} \leq 250$ mW, $B_1 \leq 2$ G). Sample volume is ~500 nl. Inset left: 94 GHz EPR spectrum of ^{15}N-TEMPONE at RT and the irradiation position of the pump pulse in the EPR line. Inset right: schematic representation of the DNP sequence, showing a MW pulse lasting several seconds followed by ^{13}C NMR FID signal detection. (Reproduced with permission from Ref. 27. © Springer Nature, 2017)

EPR), established that the chemical environment of the ^{13}C nucleus has a dramatic impact on the magnitude and sign of the ^{13}C enhancements. Han *et al.* pointed out the role of ^{1}H DNP in the so-called three spin effect of ^{13}C DNP, also recognized in earlier studies,[5,20] and they showed that it becomes negligible at high ($\geq$20–30 mM) polarizer concentrations. The first high-field (5 T) Overhauser DNP study of ^{13}C was performed by Griffin and coworkers.[39] They investigated the ^{13}CCl$_4$/BDPA system and reported a ^{13}C enhancement of ≈40. However, the experimental setup was different than in the previous literature, since they employed an NMR-optimized probe head without microwave cavity and a gyrotron source for mw pumping.

Motivated by our previous studies on ^{1}H-Overhauser DNP and by large enhancements ($\varepsilon \approx 600$) that we had observed in ^{13}CHCl$_3$/TEMPONE in a shuttle DNP experiment,[40] we addressed the question whether ^{13}C signal enhancements are observable at high magnetic fields. Taking advantage of an optimized DNP instrument operating at 3.4 T/94 GHz equipped with a low-power microwave source ($P_{mw} \leq 250$ mW)

and an ENDOR resonator,[41] we examined the room-temperature ^{13}C signal enhancements of ^{13}CCl$_4$, ^{13}CHCl$_3$, and other organic compounds doped with TEMPONE.[27] The positive (scalar) enhancement for ^{13}CCl$_4$ is displayed in Figure 14.4 and amounted to 930 ± 100, whereas the positive enhancement of ^{13}CHCl$_3$ was 550 ± 60. A control experiment with ^{13}CDCl$_3$ led to a similar enhancement (680 ± 70), indicating that the role of the three-spin effect was negligible. Saturation and leakage factors were measured independently. With the available information on s_{eff} and f, the DNP coupling factors could be calculated, resulting in -0.47 and -0.37 (within an estimated error of ± 15 %) for ^{13}CCl$_4$ and ^{13}CHCl$_3$, respectively.

Besides the large unexpected enhancements, these data did not confirm the large difference (factor of 5) in coupling factors reported by Dorn and coworkers at 0.34 T for these two compounds, suggesting a complex field dependence.[23]

To examine a more general performance of ^{13}C DNP at 3.4 T, we also investigated organic molecules with ^{13}C chemical environments different from chlorinated

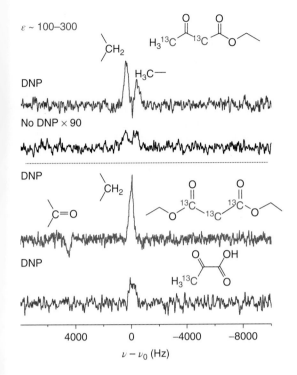

Figure 14.5. ^{13}C DNP NMR (3.4 T) spectra of ethyl ace-toacetate (red), diethyl malonate (olive), and pyruvic acid (blue) doped with ~30–40 mM ^{15}N-TEMPONE. Black trace is the Boltzmann spectrum of an ethyl acetoacetate sample containing three times more ^{13}C of the corresponding DNP sample. The number of scans for the Boltzmann spectrum is 900 times that of the DNP spectrum, and the spectrum is scaled down by a factor of $900^{1/2}$ for comparison at the same noise level to the DNP spectrum. In the DNP spectrum of di-ethyl malonate (olive), the negative NMR signal results from the carbonyl group dominated by dipolar DNP. (Reproduced with permission from Ref. 27. © Springer Nature, 2017)

carbons.[27] We selected ethyl acetoacetate and pyruvate as they are involved in several important metabolic pathways.[42,43] ^{13}C DNP enhancements on the order of 100 up to 300 were observed in the α-^{13}C carbons of ethyl acetoacetate, diethyl malonate, and pyruvic acid (Figure 14.5). The Boltzmann signal of these samples was not always detectable as was far below the sensitivity limit (~2–5 × 10^{19} ^{13}C spins) of our EPR/DNP spectrometer. Also, the low resolution of the NMR spectra is due to the low homogeneity of the 3.4 T EPR magnet (no shim coils are typically used in EPR experiments). Nevertheless, the experiments

provided the important information that the scalar ^{13}C DNP mechanism is also efficient in ^{13}CH$_2$ and ^{13}CH$_3$ groups.

14.5 MECHANISTIC ANALYSIS

A more detailed understanding of the ^{13}C DNP mechanism was achieved by nuclear-magnetic relaxation dispersion (NMRD), which measures the field dependence of the relaxation rate R_1 of the target nucleus. Relaxation rates normalized over the concentration of paramagnet give the so-called relaxivities, which are usually plotted vs the magnetic field or vs the nuclear Larmor frequency.[5,33] From simulations of the relaxivity field profiles, the DNP coupling factor can be extracted according to equations (14.5) and (14.6). The ^{13}C relaxivity profiles of ^{13}CCl$_4$ and ^{13}CHCl$_3$ recorded at room temperature are displayed in Figure 14.6. The high radical concentration ensured that the diamagnetic contribution [defined in equation (14.5)] could be neglected. One can readily observe that the two profiles are quite different at low fields, but converge to a similar value at around 1 T. The difference at low fields immediately reports on the contribution of the H atom in CHCl$_3$.

To disentangle the different relaxation contributions, a simple relaxation model was formulated, which was based on the consideration that scalar relaxation requires an intermolecular transfer of unpaired spin density from the nitroxide to the target ^{13}C nucleus. However, in both ^{13}CCl$_4$ and ^{13}CHCl$_3$ compounds, no direct orbital overlap can occur between the SOMO (singly occupied molecular orbital) of the radical and the ^{13}C atomic orbitals due to the carbon sp^3 hybridization and the four atoms (either Cl or H) bound to it. Therefore, spin density transfer can only occur through a spin-polarization mechanism that involves the orbitals of the Cl and/or H atoms. This transfer is then 'mediated' by statistic encounters between the polarizer and the target molecule either through the four Cl atoms (^{13}CCl$_4$) or through three Cl and one proton (^{13}CHCl$_3$). The observed relaxivity can be formulated as the sum of scalar and dipolar relaxation contributions mediated by these four types of encounters:

$$R_{1,\text{para}}(^{13}\text{CCl}_4) = R_{1,\text{scalar,Cl}} + R_{1,\text{dip, Cl}}$$

$$R_{1,\text{para}}(^{13}\text{CHCl}_3) = \frac{3}{4}R_{1,\text{para}}(^{13}\text{CCl}_4) + R_{1,\text{dip, H}}$$

$$+ R_{1,\text{scalar, H}} \qquad (14.11)$$

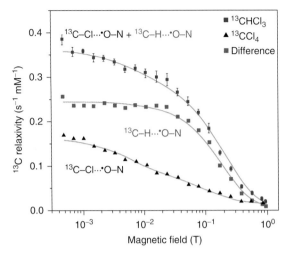

Figure 14.6. Comparison of ^{13}C relaxivity for ^{13}CCl$_4$ (black triangles) and ^{13}CHCl$_3$ (blue squares). At fields around 1 T, relaxivities of both compounds reach similar values, and the difference at lower fields is due to the presence of the proton in CHCl$_3$. Pink squares: Curve is obtained after subtraction of 3/4 the relaxivity of ^{13}CCl$_4$ from ^{13}CHCl$_3$; the latter represents approximately the contribution from the H atom in CHCl$_3$. (Reproduced with permission from Ref. 27. © Springer Nature, 2017)

Following this model, the H contribution can be obtained by subtracting 3/4 of the ^{13}CCl$_4$ relaxivity from the CHCl$_3$ relaxivity. The resulting difference curve (Figure 14.6) does not show a typical behavior for a pure dipolar contribution, where the high-field value amounts to 3/10 of the low-field value.[5,33] Instead, all the three profiles in Figure 14.6 could be reproduced using mixtures of scalar and dipolar relaxation according to equation (14.11) with the spectral densities defined in equations (14.8)–(14.10).

The relaxivity profile of ^{13}CCl$_4$ could be reproduced by a scalar relaxation mediated by one predominant type of polarizer/target encounter at a correlation time on the order of $\tau_1 \sim 1$ ps [equation (14.10)]. This short correlation time was consistent with the value proposed earlier by Müller-Warmuth *et al.*[21] from the analysis of data at 1 T and is related to the average duration of a fast, elastic intermolecular collision, as observed in infrared and light-scattering spectra.[44] The profile of ^{13}CHCl$_3$ could be fitted with parameters shared with ^{13}CCl$_4$ plus the additional contributions through the H atom [equation (14.11)]. To reproduce the H-mediated relaxivity profile, which decays at a field around 1 T (Figure 14.6), the spectral density for scalar relaxation required a correlation time on the order of $\sim$24 ps, i.e., longer than for a Cl-mediated encounter. Although the NMRD curves were available only up to 1 T, the relaxation rates extracted from these NMRD fits allowed for an estimate of the coupling factors at 3.4 T to $\xi = -0.47$ (CCl$_4$) and $\xi = -0.37$ (CHCl$_3$), in good agreement with the experimental coupling factors reported in Section 14.5. We note that a more precise analysis of the DNP mechanism in the high-field range ($\geq$1 T) requires a complimentary determination of ^{13}C DNP coupling factors at different magnetic fields to accurately determine the residual dipolar contribution in equation (14.7). A more systematic investigation in this direction will be reported in the near future.[12]

A rationale for the lower ^{13}C coupling factor of CHCl$_3$ as compared to ^{13}CCl$_4$ at 3.4 T was obtained by considering the DFT-optimized structures of the complexes with TEMPONE through the Cl and H atoms, as illustrated in Figure 14.7. According to these structures, the distance of the closest approach r_d, which governs the dipolar relaxation via equations (14.8) and (14.9), becomes substantially shorter when the encounter is mediated by the H atom as compared

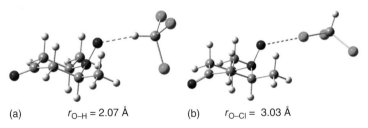

(a) $r_{O-H} = 2.07$ Å   (b) $r_{O-Cl} = 3.03$ Å

Figure 14.7. DFT-optimized structure of possible encounters between TEMPONE and CHCl$_3$, leading to different distances of closest approach. Color code: white, H; gray, C; blue, N; red, O; green, Cl. (Reproduced with permission from Ref. 27. © Springer Nature, 2017)

to the Cl atom, corresponding to a stronger dipolar interaction. Therefore, the residual dipolar contribution at high fields, see equation (14.7), is overall larger in CHCl$_3$ as compared to CCl$_4$, consistent with the smaller coupling factor.

14.6 PERSPECTIVES

The DNP mechanistic analysis on the two model systems ^{13}CCl$_4$ and ^{13}CHCl$_3$ doped with TEMPONE gave first insight into the basic relaxation mechanisms underlying the observed ^{13}C DNP enhancements at 3.4 T. Nevertheless, these results raise the question whether substantial ^{13}C DNP signal enhancements can also be achieved at higher magnetic fields and for other ^{13}C chemical environments. An answer to these questions requires more experimental investigations, but trends might be indicated by the data available so far at 0.34 and 3.4 T. ^{13}C enhancements reported at 0.34 T for several organic compounds[23] were about one to two orders of magnitude and either positive or negative. This strongly depends on the specific ^{13}C chemical environment that plays a key role in defining the two counteracting mechanisms, i.e., dipolar and scalar. However, since the field dependence of these two could be different, it was possible to observe that at 3.4 T on various organic molecules, such as pyruvate and dimethyl-malonate, ^{13}C in CH$_2$ and CH$_3$ groups display a sizable scalar DNP enhancement up to two orders of magnitude.

We conclude that both chlorine as well as H-atoms can mediate intermolecular scalar interaction – via a spin polarization mechanism – between the nitroxide and ^{13}C nuclei, although the mechanism seems more efficient in the halogen case. According to our model calculations in Figure 14.3, the decreased ^{13}C enhancements in the CH$_2$ and CH$_3$ groups as compared to the chlorinated samples might be due to either the fact that H atoms mediate scalar relaxation at a slower correlation time than Cl, leading to a faster decay of the scalar spectral density at high fields, or that the counteractive dipolar contribution is larger in an encounter complex formed through the H atom, as illustrated in Figure 14.7. More experimental data combined with a field-dependent analysis will be required to distinguish more precisely between these two mechanisms.

Initial data directed to a more systematic investigation on the role of the halogen atom indicate that the large ^{13}C enhancements are also observable with bromine instead of chlorine.[12] Therefore, a route for liquid ^{13}C DNP applications might be the specific labeling of a target ^{13}C nucleus with halogen atoms. Owing to the large occurrence of halogens in pharmaceutical products and their well-assessed chemistry, one might be able to use liquid DNP to study specific interactions of these products with biological substrates. As the Overhauser DNP experiment can be repeated for signal averaging and allows for multidimensional NMR spectroscopy, it might have important application in studies of natural products or samples available in very small amounts or concentration.

For general application of ^{13}C DNP in solution NMR, we expect that enhancements up to two orders of magnitudes should be attainable in several compounds in the medium magnetic field region (3–10 T). An essential step toward this direction will be the development of suitable liquid DNP probe heads capable of storing NMR-size sample volumes (i.e., >1 µl) at this field/frequency range. Overall, the availability of an efficient DNP mechanism in liquid at high fields will offer several, new opportunities, with more ideas and perspectives expected in the future.

RELATED ARTICLES IN EMAGRES

Overhauser, Albert W.: Dynamic Nuclear Polarization

REFERENCES

1. A. W. Overhauser, *Phys. Rev.*, 1953, **92**, 411.

2. I. Solomon, *Phys. Rev.*, 1955, **99**, 559.

3. A. Abragam, Principles of Nuclear Magnetism, Oxford University Press Inc.: New York, 1962.

4. T. R. Carver and C. P. Slichter, *Phys. Rev.*, 1953, **92**, 212.

5. D. Hausser and D. Stehlik, *Adv. Magn. Reson.*, 1968, **3**, 79.

6. W. Müller-Warmuth and K. Meise-Gresch, *Adv. Magn. Reson.*, 1983, **11**, 1.

7. P. Höfer, G. Parigi, C. Luchinat, P. Carl, G. Guthausen, M. Reese, T. Carlomagno, C. Griesinger, and M. Bennati, *J. Am. Chem. Soc.*, 2008, **130**, 3254.

8. E. V. Kryukov, M. E. Newton, K. J. Pike, D. R. Bolton, R. M. Kowalczyk, A. P. Howes, M. E. Smith, and R. Dupree, *Phys. Chem. Chem. Phys.*, 2010, **12**, 5757.

9. P. J. M. van Bentum, G. H. A. van der Heijden, J. A. Villanueva-Garibay, and A. P. M. Kentgens, *Phys. Chem. Chem. Phys.*, 2011, **13**, 17831.

10. M. J. Prandolini, V. P. Denysenkov, M. Gafurov, B. Endeward, and T. F. Prisner, *J. Am. Chem. Soc.*, 2009, **131**, 6090.

11. T. Dubroca, A. N. Smith, K. J. Pike, S. Froud, R. Wylde, B. Trociewitz, J. McKay, F. Mentink-Vigier, J. van Tol, S. Wi, W. Brey, J. R. Long, L. Frydman, and S. Hill, *J. Magn. Reson.*, 2018, **289**, 35.

12. T. Orlando, R. Dervisoglu, M. Levien, I. Tkach, L. Andreas, T. Prisner, V. Denysenkov, and M. Bennati, *Angew. Chem. Int. Ed.*, 2018. DOI: 10.1002/anie.201811892.

13. M. Bennati, I. Tkach, and M. T. Türke, in 'Electron Paramagnetic Resonance', ed B. Gilbert, Royal Society of Chemistry: Cambridge, 2011, p 155.

14. C. Griesinger, M. Bennati, H. M. Vieth, C. Luchinat, G. Parigi, P. Hofer, F. Engelke, S. J. Glaser, V. Denysenkov, and T. F. Prisner, *Prog. Nucl. Magn. Reson. Spectrosc.*, 2012, **64**, 4.

15. J. M. Franck, A. Pavlova, J. A. Scott, and S. Han, *Prog. Nucl. Magn. Reson. Spectrosc.*, 2013, **74**, 33.

16. T. F. Prisner, V. Denysenkov, and D. Sezer, *J. Magn. Reson.*, 2016, **264**, 68.

17. M. T. Türke, I. Tkach, M. Reese, P. Hofer, and M. Bennati, *Phys. Chem. Chem. Phys.*, 2010, **12**, 5893.

18. V. Denysenkov and T. Prisner, *J. Magn. Reson.*, 2012, **217**, 1.

19. D. Yoon, A. I. Dimitriadis, M. Soundararajan, C. Caspers, J. Genoud, S. Alberti, E. de Rijk, and J.-P. Ansermet, *Anal. Chem.*, 2018, **90**, 5320.

20. H. Brunner and K. H. Hausser, *J. Magn. Reson.*, 1972, **6**, 605.

21. W. Müller-Warmuth, R. Vilhjalmsson, P. A. M. Gerlof, J. Smidt, and J. Trommel, *Mol. Phys.*, 1976, **31**, 1055.

22. M. D. Lingwood and S. G. Han, *J. Magn. Reson.*, 2009, **201**, 137.

23. X. Wang, W. C. Isley, S. I. Salido, Z. Sun, L. Song, K. H. Tsai, C. J. Cramer, and H. C. Dorn, *Chem. Sci.*, 2015, **6**, 6482.

24. M. Bennati, *eMagRes*, 2017, **6**, 271.

25. M. T. Türke, G. Parigi, C. Luchinat, and M. Bennati, *Phys. Chem. Chem. Phys.*, 2012, **14**, 502.

26. N. Enkin, G. Liu, I. Tkach, and M. Bennati, *Phys. Chem. Chem. Phys.*, 2014, **16**, 8795.

27. G. Liu, M. Levien, N. Karschin, G. Parigi, C. Luchinat, and M. Bennati, *Nat. Chem.*, 2017, **9**, 676.

28. M. T. Türke and M. Bennati, *Phys. Chem. Chem. Phys.*, 2011, **13**, 3630.

29. J. H. Ardenkjaer-Larsen, I. Laursen, I. Leunbach, G. Ehnholm, L. G. Wistrand, J. S. Petersson, and K. Golman, *J. Magn. Reson.*, 1998, **133**, 1.

30. N. Enkin, G. Q. Liu, M. D. Gimenez-Lopez, K. Porfyrakis, I. Tkach, and M. Bennati, *Phys. Chem. Chem. Phys.*, 2015, **17**, 11144.

31. P. Neugebauer, J. G. Krummenacker, V. Denysenkov, G. Parigi, C. Luchinat, and T. F. Prisner, *Phys. Chem. Chem. Phys.*, 2013, **15**, 6049.

32. A. Abragam, *Phys. Rev.*, 1955, **98**, 1729.

33. M. Bennati, C. Luchinat, G. Parigi, and M. T. Türke, *Phys. Chem. Chem. Phys.*, 2010, **12**, 5902.

34. J. Freed, *J. Chem. Phys.*, 1978, **68**, 4034.

35. C. F. Polnaszek and R. G. Bryant, *J. Chem. Phys.*, 1984, **81**, 4038.

36. J. Trommel, PhD thesis, University of Leiden, 1978.

37. S. Stevenson and H. C. Dorn, ^{13}C Dynamic Nuclear Polarization: A Detector for Continuous-Flow, Online Chromatographya. *Anal. Chem.*, 1994, **66**, 2993.

38. S. Stevenson, T. Glass, and H. C. Dorn, *Anal. Chem.*, 1998, **70**, 2623.

39. N. M. Loening, M. Rosay, V. Weis, and R. G. Griffin, *J. Am. Chem. Soc.*, 2002, **124**, 8808.

40. M. Reese, M. T. Türke, I. Tkach, G. Parigi, C. Luchinat, T. Marquardsen, A. Tavernier, P. Hofer, F. Engelke, C. Griesinger, and M. Bennati, *J. Am. Chem. Soc.*, 2009, **131**, 15086.

41. M. T. Türke and M. Bennati, *Appl. Magn. Reson.*, 2012, **43**, 129.

42. P. R. Jensen, S. C. Serra, L. Miragoli, M. Karlsson, C. Cabella, L. Poggi, L. Venturi, F. Tedoldi, and M. H. Lerche, *Int. J. Cancer*, 2015, **136**, E117.

43. S. J. Nelson, J. Kurhanewicz, D. B. Vigneron, P. E. Z. Larson, A. L. Harzstark, M. Ferrone, M. van Criekinge, J. W. Chang, R. Bok, I. Park, G. Reed, L. Carvajal, E. J. Small, P. Munster, V. K. Weinberg, J. H. Ardenkjaer-Larsen, A. P. Chen, R. E. Hurd, L. I. Odegardstuen, F. J. Robb, J. Tropp, and J. A. Murray, *Sci. Transl. Med.*, 2013, **5**, 198ra108.

44. J. Bucaro and T. Litovitz, *J. Chem. Phys.*, 1971, **55**, 3585.

PART B
Applications

Chapter 15

DNP and Cellular Solid-state NMR

Alessandra Lucini Paioni, Marie A.M. Renault, and Marc Baldus

Utrecht University, Utrecht, The Netherlands

15.1 INTRODUCTION

In recent years, in-cell NMR (nuclear magnetic resonance) has become a powerful method to study molecules in living bacterial or mammalian cells.[1–4] Yet, such studies have been limited to molecules that rapidly tumble in their natural environment, which puts significant constraints regarding molecular size and complexity that can be investigated. On the other hand, solid-state nuclear magnetic resonance (ssNMR) is a method that can be applied to gain atomic-level insight into heterogeneous molecular systems, largely irrespective of molecular size,[5] starting with applications to collagen,[6] membrane proteins,[7] and even entire bacterial cells.[8] For example, ssNMR represents one of the few structural methods that can probe membrane protein assemblies at an atomic level.[7,9–15]

In addition to the well-established use of synthetic bilayer settings, cellular ssNMR concepts have been developed in recent years that permit extension of such studies to cellular preparations.[16–24] Such studies have, for example, been motivated by observations that the natural cell membrane environment is critically involved in membrane protein function.[25,26] At the same time, ssNMR has profited from revolutionary enhancements in sensitivity, mainly due to the advent of dynamic nuclear polarization (DNP, Ref. 27 and related chapters in this handbook), and it has seen significant progress in the field of ^1H detection,[28,29] which increases spectral resolution and can further enhance spectroscopic sensitivity. First reports have appeared recently that make combined use of such methods to examine biomolecules in situ, for example, in the natural bacterial cellular environment such as the cytoplasm,[30] embedded in membranes,[16,20,31–34] or spanning the entire cell envelope (CE).[18] In the following sections, we describe these and other in situ ssNMR studies and discuss specific aspects related to the use of DNP on such systems.

15.2 METHODOLOGICAL ASPECTS OF DNP-SUPPORTED CELLULAR ssNMR

15.2.1 Sample Preparation for Cellular Solid-state NMR

Virtually all NMR studies conducted today on proteins or other biomolecules use isotope labeling to enhance

Handbook of High Field Dynamic Nuclear Polarization.
Edited by Vladimir K. Michaelis, Robert G. Griffin, Björn Corzilius and Shimon Vega
© 2020 John Wiley & Sons, Ltd. ISBN: 978-1-119-44164-9
Also published in eMagRes (online edition)
DOI: 10.1002/9780470034590.emrstm1561

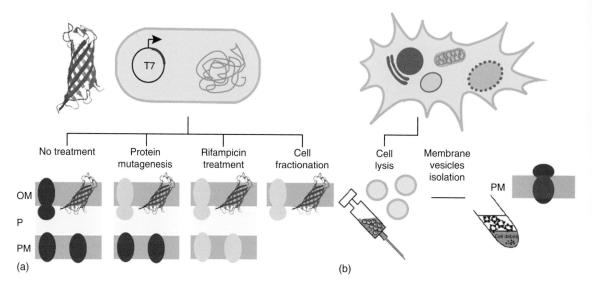

Figure 15.1. Example of prokaryotic and eukaryotic cellular ssNMR systems. (a) Accumulation of isotope-labeled membrane protein in *Escherichia coli* using the bacteriophage T7 promoter-driven overexpression. The quality of cellular solid-state NMR spectra of cell-envelope (CE) and whole-cell (WC) preparations can be significantly improved by the deletion of abundant *E. coli* proteins (OmpA and OmpF), rifampicin treatment during protein production, and the isolation of the subcellular compartment of interest after cell fractionation. (b) Preparation of EGFR-rich membrane vesicles from A431 cells cultured in [^{13}C, ^{15}N]-labeled DMEM medium. Cells were scraped from the plates and vesiculated by passing them through a syringe 10 times. After removal of the unbroken cells and cell nuclei by spinning at low speed, the membrane vesicles were spun down at high speed and loaded into an ssNMR rotor. Note that all methods can also be used to study whole cells

NMR signals and to allow for multidimensional ssNMR correlation spectroscopy. In the case of proteins, tailored isotope-labeling approaches in prokaryotes and eukaryotes are usually required. In practice, cellular NMR applications furthermore involve dedicated procedures to control the concentration of the protein target and limit background labeling artifacts[3] (Figure 15.1). Yet, with increasing size and molecular complexity, or whenever target proteins are inserted into the membrane, isotope labeling and protein overexpression may not be sufficient to detect ssNMR signals of a particular protein in its natural biological environment.

An example of such a situation refers to the outer membrane of Gram-negative bacteria where endogenous protein expression levels of naturally highly abundant outer membrane proteins, i.e., OmpA and OmpF, may be comparable to the membrane protein of interest, even under the conditions of overexpression. In this case, studies using CEs or whole cells (WC) of Gram-negative bacteria are facilitated by the use of an *Escherichia coli* deletion strain to remove signals from naturally highly abundant membrane proteins.[16,17] In such preparations, signals from other cellular components including lipids, nucleotides, peptidoglycan (PG), and lipopolysaccharides (LPS) remain visible at intensities similar to the (overexpressed) protein of interest.

Another strategy to minimize background labeling relates to inhibiting the expression of endogenous proteins by adding antibiotics such as rifampicin to the growth medium.[21,22] Most common and popular inducible recombinant expression systems used to produce heterologous and homologous proteins in bacteria often rely on nonbacterial RNA polymerases, such as the bacteriophage T7 system, that are not inhibited by rifampicin. If one induces expression of the recombinant protein after rifampicin treatment, the biosynthesis of endogenous proteins rapidly declines, without affecting the production of the protein of interest. Therefore, by coordinating the inhibition of endogenous gene expression with exposure to isotopically labeled media and induction of recombinant protein expression, the rifampicin treatment provides a means to enhance the signal-to-noise ratio of ssNMR spectra of the protein of interest in cellular environments. Obviously, nonproteinaceous cell components such as nucleic acids, lipids, and polysaccharides remain labeled. To further suppress such contributions, specific editing methods such as those discussed in Refs 16, 22 can be used (Figure 15.1a).

Owing to their easy handling and well-established genetic tools, bacteria are broadly used to express homologous genes in minimal media. However, proteins from higher organisms often require specific eukaryotic machineries to ensure proper protein folding, the addition of posttranslational modifications, and the correct export in native cell compartments. For NMR purposes, posttranslational modifications can be achieved in bacteria by the coexpression of specific enzymes during the production of the protein target. For example, the O-linked β-*N*-acetylglucosamine transferase (OGT) transfers *N*-acetylglucosamine (GlcNAc) from uridine 5′-diphospho-*N*-acetylglucosamine (UDP-GlcNAc) to serine and threonine residues of numerous target proteins, including those produced in *E. coli*. We previously utilized the latter aspect to produce ^{13}C,^{15}N-labeled nucleoporins (Nups) that are natively heavily modified by O-linked β-*N*-acetylglucosamines.[35] Concomitantly, significant progress has been made in the past few years to express proteins from eukaryotic cells. Fully and specifically ^{13}C,^{15}N-labeled proteins suitable for NMR studies were produced from different eukaryotic cell types (see, e.g., Ref. 13 for a recent review). Yet, the costs for producing such samples can still be prohibitive.

When NMR on intact (bacterial or eukaryotic) cells is critical, isolating specific cell compartments or organelles offers a means to study molecular components such as membrane proteins in their native cell environment with enhanced spectroscopy sensitivity (see, e.g., Ref. 36). In our laboratory, we could recently produce fully and specifically isotope-labeled plasma membrane vesicles derived from the human cancer cells A431.[23] These cells are known to express high levels of the epidermal growth factor receptor (EGFR). Solid-state NMR analysis of isotope-labeled vesicles allowed us to study functional EGF receptors in the natural membrane environment. For isotope labeling, we adapted published procedures using a combination of dialyzed fetal calf serum and amino acid mixtures obtained from algae extracts to produce an [^{13}C,^{15}N]-enriched medium. Next, we produced plasma membrane vesicles from lysed A431 cells that contain an increased level of functional receptor. Those

vesicles are about an order of magnitude smaller than eukaryotic cells, thereby enhancing signal-to-noise for ssNMR studies significantly (Figure 15.1b).

15.2.2   Dynamic Nuclear Polarization

Spectroscopic sensitivity is of outermost importance for studying complex molecular systems, and the advent of commercial DNP spectrometers has been critical to research described in this chapter. In the past decade, we have seen the development of high-frequency gyrotron microwave sources, low-temperature MAS probes that permit in situ microwave irradiation of the samples, and tailored biradical polarizing agents. As a result, DNP has become the most widely applicable hyperpolarization technique to enhance the sensitivity of NMR experiments. Collectively, these advances have made it possible to apply DNP on a routine basis, creating novel opportunities for NMR studies on complex and large molecular assemblies in life and material sciences. Briefly, DNP takes advantage of the higher equilibrium polarization of electrons and enhances NMR signals by transferring the large polarization of electrons to nearby nuclei via microwave (μw) irradiation. Usually, DNP studies involve the addition of soluble paramagnetic compounds to a sample dissolved in a glassy matrix at low temperatures, and under magic-angle spinning conditions nuclear polarization is most efficiently generated by the so-called cross-effect, involving a pair of electrons in the form of a biradical and a nuclear spin (see Ref. 27 and related chapters).

In 2012, we investigated the use of DNP to conduct ssNMR studies on ^{13}C,^{15}N-labeled cellular preparations of *E. coli* containing a recombinant membrane protein.[17] Bacterial cells and CE extracts were typically suspended in a matrix consisting of a well-defined aqueous buffer containing d_8-glycerol as a cryoprotectant and TOTAPOL[37] as polarizing agent. The biradicals TOTAPOL and AMUpol[38] are compatible with cell viability when used at a concentration of 5–40 mM and provide significant sensitivity enhancements at low temperature, i.e., 100 K. Figure 15.2 compares ^{13}C and ^{15}N cross-polarization (CP) spectra of uniformly ^{13}C,^{15}N-labeled WC with the CE isolated from *E. coli* cells, recorded on a DNP NMR instrument operating at 400 MHz ^{1}H Larmor frequency and a MAS rate of 8 kHz, in the presence and absence of microwave irradiation. For optimal DNP analysis, the

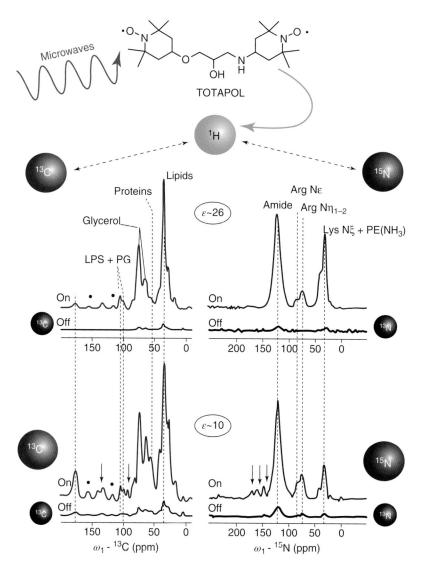

Figure 15.2. Comparison of ^{13}C (left) and ^{15}N (right) CP MAS spectra of (U-^{13}C,^{15}N)-labeled cell envelope (CE, top) and whole cell (WC, bottom) with (top trace, 'on') and without (bottom trace, 'off') DNP, using continuous-wave microwave irradiation and a recycling delay of 10 s. Significant differences in the spectra of the two preparations are denoted by arrows. Asterisks indicate MAS side bands. Assignments of major *E. coli* molecular components are, if available, annotated, and signal enhancements are given. Natural abundance ^{13}C signals from glycerol ($\delta^{13}C = 73$ and 63 ppm) present in the glassy matrix and from the silicon plug ($\delta^{13}C = 4$ ppm) are visible in the ^{13}C CP spectra of the WC and CE preparations

concentration of the polarizing agent has to be adjusted to obtain a compromise between the optimal DNP enhancement and the degradation of the spectrum sensitivity through line broadening, signal quenching, and enhanced relaxation caused by paramagnetic effects.[39] At low temperature, line broadening caused by conformational disorder becomes also challenging, in particular for components such as PG, which may exhibit a high degree of dynamics at room temperature. In most critical cases, this line broadening can be important enough to cancel the gain acquired by DNP. In addition, specific interactions of the radical

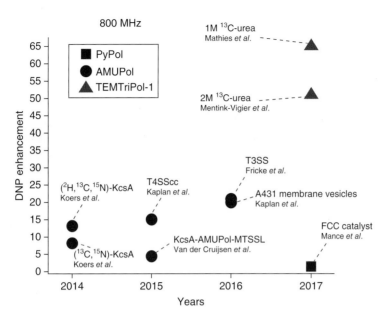

Figure 15.3. **DNP enhancement factors seen at the 800 MHz DNP conditions.** Data are shown for the case of proteoliposomal preparations of KcsA and the type 4 secretion system core complex (T4SScc), as well as EGFR-rich plasma membrane vesicles of A431 cells. In addition, this figure contains DNP enhancement factors measured on type 3 secretion system (T3SS) needles and a FCC catalyst that exhibits significantly lower proton densities. Finally, we have included recently published literature[41,44] on DNP enhancements obtained with DNP polarizing agents formed by a nitroxide (TEMPO) tethered to a trityl radical

molecule with the surface of proteins, lipids, and polysaccharides may lead to a nonuniform distribution of active radicals inside the sample. This property can be exploited to enhance signals associated with a particular compartment or a protein target by linking the radical to lipids, ligands, or the protein itself.

Currently, DNP NMR instruments can operate at 400, 600, 700, and up to 800 MHz ^{1}H Larmor frequency. Regarding the maximum achievable DNP enhancement, recent studies have shown that they can significantly vary, depending on sample preparation[19,40] or choice of the polarizing agent.[13,41] However, the enhancement generally reduces with increasing magnetic field in the case of biradicals that utilize the cross-effect. Mance *et al.* have examined[42] the magnetic field dependence of biradicals including AMUPol.[38] These studies suggested an approximately $1/B_0^2$ dependence of cross-effect DNP that is governed by a relative weak effective hyperfine coupling that describes the initial electron-nuclear polarization transfer step. The corresponding distance amounts to approximately 10 Å, which is in line with experimentally observed paramagnetic relaxation enhancement (PRE)

effects under low-temperature DNP conditions.[40,43] In Figure 15.3, a summary of some of the DNP enhancement factors obtained in our laboratory using a 800 MHz DNP setup is given. Data are shown for the case of proteoliposomal preparations of KcsA[40,43] and the type 4 secretion system core complex[18] (T4SScc), as well as EGFR-rich plasma membrane vesicles of A431 cells.[23] In addition, Figure 15.3 contains DNP enhancement factors measured on type 3 secretion needles[45] and a FCC catalyst (using PyPol[38] as DNP polarizing agent) that exhibits significantly lower proton densities.[46] Finally, we have included recently published literature[41,44] on – currently – the highest DNP enhancements (even when compared to obtained with DNP polarizing agents formed by a nitroxide (TEMPO) tethered to a trityl radical). These compounds exhibit a stronger exchange interaction and an improved electron–electron coupling that changes the magnetic field profile of the DNP enhancement. Moreover, recent work showed[44] that these compounds exhibit lower depolarization effects, which adds to an increased DNP performance. Thus far, these compounds have been mostly tested on model systems such as ^{13}C urea (Figure 15.3), which tends to exhibit stronger DNP enhancements than seen for applications to (membrane) proteins. Ongoing work is devoted to make these compounds broadly suitable for life and material science applications.

Next to sensitivity, spectral resolution is another critical factor for DNP applications on complex (bio)molecules. In our previous work, we have shown that high-field DNP can significantly enhance the prospects to conduct in-depth structural investigations of complex molecules such as membrane proteins. For example, we compared the spectral resolution observed in two-dimensional ($^{13}C,^{13}C$) and ($^{15}N,^{13}C$) correlation datasets on fully $^{13}C,^{15}N$-labeled variants of the KcsA potassium channel at 400 and 800 MHz DNP conditions using soluble[40] and tagged[43] AMUPol variants to data obtained at ambient temperatures at 700 MHz. In addition, we examined the benefits of probing membrane protein structure by directly attaching mono- or biradicals to induce local paramagnetic relaxation (PRE) effects.[43] In general, we observed that high-field DNP conditions can significantly enhance spectral resolution,[23,40,45] and we also showed that the intrinsic paramagnetic properties of the polarizing agents can be used as direct structural probes during the spectroscopic analysis. For instance, in the case of the membrane-embedded KcsA channel,[40] our results help to pinpoint the water-accessible pore of the channel in membranes. The most solvent-exposed surface shell exhibits strong PREs, at least under the conditions used in our experiments.

15.2.3 Complementary Approaches and Methods

A precise characterization of cellular preparations by complementary methods is crucial for proper sample optimization, the subsequent acquisition of high-quality ssNMR spectra, and their interpretation. As discussed earlier, the concentration of labeled target protein must be high enough with respect to other proteins to provide well-resolved resonances and unambiguous identification by ssNMR. The protein composition of cellular preparations can be primarily investigated in a semiquantitative manner by polyacrylamide gel electrophoresis (PAGE). Gel glycine–SDS-PAGE are the commonly used SDS electrophoretic techniques for separating proteins in a mass range of 1–500 kDa (Figure 15.4a). When protein identification becomes challenging

in one dimension, 2D electrophoretic system in conjunction with protein identification techniques can be used to establish proteome maps and yielding information about the occurrence and abundance of proteins in cellular preparations. To probe the level of EGFR expression, we, for example, conducted mass spectrometry (MS) experiments[23] on A431 cells and the isolated membrane vesicles (Figure 15.4b). Using a well-established semiquantitative analysis based on summed ions intensities over all detected peptides, we found actin to be the most abundant protein in whole A431 cells, followed by other abundant soluble molecules including heat shock and histone proteins. While the EGFR expression level was lower than these proteins, EGFR still represented the most abundant membrane protein in the cells, in line with the previous findings.[47] Interestingly, when moving to isolated membrane vesicles, we found EGFR enriched by a factor 5.5, making EGFR, together with actin, the most abundant protein in the membrane vesicles. As membrane proteins are typically less detectable by MS than soluble proteins,[48] we concluded that the MS-based estimation of EGFR level is at the lower limit. When analyzing WC or cellular preparations encompassing several cell compartments, the homogeneous targeting of the protein of interest is also a critical factor for obtaining high-quality ssNMR spectra and physiologically relevant information. Fluorescence microscopy in combination with fluorescent active ligands is an extremely powerful method to localize the protein of interest. An example is shown in Figure 15.4(c), where the use of confocal microscopy on A431 cells and EGFR negative cells confirmed that EGFR is fully targeted to the membrane. For further characterization, cryoelectron microscopy[18,23] and tomography[49] can provide global information about intermolecular interactions and supramolecular organization of protein assemblies with respect to the membrane or other cell compartments. As an example, Figure 15.4(d) displays the single-molecular cryo-EM map of T4SScc in cellular ssNMR preparations and after subsequent solubilization.[18]

On a spectroscopic level, electron paramagnetic resonance (EPR[49–51]) and solution NMR[1–4] can provide insights on the conformation and dynamics of soluble proteins and nucleic acids at atomic-level resolution in various cellular systems including bacteria, yeast, insect cells, oocytes, and mammalian cells.[52] While in-cell solution-state NMR experiments can be carried out at physiological temperatures, the fast relaxation of nitroxide spin labels at room temperature requires EPR

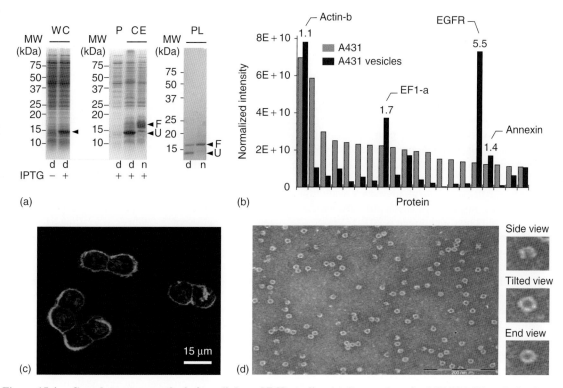

Figure 15.4. Complementary methods for cellular ssNMR studies. (a) Coomassie-stained SDS/PAGE analysis of whole cells (WC), cell envelope (CE), and protoplasm (P) fractions obtained from exponentially growing *E. coli* OmpA-OmpF-cells in presence (+) and in absence (−) of IPTG inducer. Molecular mass markers (MM) are indicated next to the gels. Samples were denatured by boiling in SDS (d) or left on ice (n) before electrophoresis. F and U denote the positions of folded and heat-denatured forms of PagL, respectively. As a reference, proteoliposomes (PL) containing purified PagL protein were analyzed concomitantly. (b) Relative abundance of EGFR in A431 cells (light blue) and membrane vesicles (dark blue) assessed by mass spectrometry. Normalized intensities of the 20 most-abundant proteins were calculated by summing the intensities over all peptides detected in the tryptic digests of the cells and vesicles for the annotated proteins. Enrichment factors (labels) were calculated from the normalized intensities, clearly revealing that only EGFR is highly significantly enriched in the membrane vesicles. (c) Confocal microscopy of A431 cells incubated with Alexa488-tagged EGF (in green). Blue represents DAPI staining of nuclei. (d) Single-molecule Cyro-EM data on solubilized T4SScc preparations used for cellular ssNMR studies

experiments to be conducted at cryogenic temperatures similar to the DNP ssNMR conditions. Finally, the use of UV–vis spectroscopy provides extremely valuable information for small molecules that may be of interest in the context of in situ life or material-science applications.[53]

15.3 APPLICATIONS

'On-cell' ssNMR approaches have been used for decades to study the composition, structure, dynamics, and function of bacterial cell walls. In more recent years, a number of cellular ssNMR applications aimed to provide a more detailed understanding of molecular mechanisms associated with bacterial pathogenicity and virulence in a functional-relevant environment. Indeed, host–pathogen recognition, cell adhesion, or regulation of cell activity involves interactions with many different CE components. The ability to characterize (membrane) proteins and associated biomolecules in a nonperturbative manner is crucial to understanding the structure and function of the bacterial cell walls, determining drug modes of action and developing new-generation therapeutics.

The bacterial CE is a complex multilayered structure that usually falls into one of the two major groups. Gram-negative bacteria are surrounded by a thin PG cell wall, which itself is surrounded by an outer membrane (OM) containing proteins (OMPs), phospholipids, and LPS. Gram-positive bacteria lack an OM but are surrounded by layers of PG many times thicker than is found in the Gram-negatives. Threading through these layers of PG are long anionic polymers, called teichoic acids. Reliable protocols and a number of successful applications have emerged in the past few years, allowing bacterial membrane proteins to be investigated in situ with atomic-level resolution. These protocols involved β-barrel proteins inserted into the bacterial outer membrane (OMPs),[17] proteins embedded in the inner membrane of bacteria such as ion and proton channels,[31,33] retinal proteins,[32] electron-transport proteins,[24] or periplasmic proteins associated with the PG.[54]

In 2012, we conducted the first structural study of a recombinant membrane protein in intact *E. coli* CE settings. As our model system, we selected the 150 residue integral membrane-protein PagL from *Pseudomonas aeruginosa*, an OM enzyme that removes a fatty acyl chain from LPS and helps the bacterium to evade the host immune system. The protocols involved general expression and purification procedures that lead to uniformly $^{13}C,^{15}N$-labeled preparations of WC and CEs containing (or not) the membrane protein target (Figure 15.4a). PagL was expressed under control of the bacteriophage T7 promoter, which is inducible with IPTG, in a mutant *E. coli* BL21Star(DE3) strain lacking OmpF and OmpA (BL21dm). The suppression of these major OMPs prevented to a large extent the accumulation of the unprocessed signal-peptide-bearing precursor of PagL and led to significant amounts of mature protein in the host membrane. For optimal analysis of major cell-associated molecular components, *E. coli* cultures were switched from unlabeled to $^{13}C,^{15}N$-isotope-labeled growth conditions at the beginning of the exponential growth phase, when recombinant protein production was induced, leading to the incorporation of isotopes in PagL and coexpressed endogenous molecular components. WC and CE samples were prepared from the same exponentially growing culture. To characterize the presumably rigid membrane-associated—molecular components in WC and CE, we performed a set of 2D $^{13}C-^{13}C$ correlation experiments employing dipolar-based magnetization transfer steps. Both preparations yielded NMR

spectra of astonishing quality considering sample complexity and noncrystallinity, with well-dispersed cross-peaks characteristic for protein and lipid signals. As anticipated, we observed an improvement in both sensitivity and spectral resolution for the CE preparation, potentially due to the single contribution of CE-associated components. These results were corroborated by SDS/PAGE analysis, which showed a significant decrease of the amount of proteins after removal of the protoplasm by cell lysis and ultracentrifugation (Figure 15.4a). The stability of WC and CE preparations was monitored by 1D ^{13}C ssNMR using either dipolar- or scalar-based magnetization transfers steps and did not reveal any marked spectroscopic changes over time. To examine in further detail the conformation of PagL in CE, we performed 2D ^{15}N-edited ^{13}C correlation experiments in which signals arising from nonproteinaceous molecular components are largely reduced (Figure 15.5a). Because of the favorable spectroscopic dispersion among Thr, Ser, and Gly residues in 2D and 3D CC/NC correlation experiments, we could subsequently perform a residue-specific analysis based on sequential resonance assignments of PagL residues located in different topological regions of the protein. Complementary protocols were further established to characterize major endogenous macromolecular components of *E. coli* CE that surround the target protein PagL. The lipoprotein lpp or Braun's protein belongs to the most abundant CE proteins in exponentially growing *E. coli* cells. Lpp is found in both 'free' and 'bound' forms, the latter being covalently attached to the PG network. In solution, the 56 residue polypeptide moiety, called Lpp-56, associates to form a hydrophilic homotrimer composed of a three-stranded coiled-coil domain and two helix-capping motifs, but a model for a lipophilic superhelical assembly containing six subunits has also been proposed. We performed a series of 2D $^{15}N-^{13}C$ and $^{13}C-^{13}C$ correlation experiments on CE isolated from noninduced WC and compared the results with predictions based on the available high-resolution 3D structure of Lpp-56 (Figure 15.5b). We found good agreement between our data and predicted intraresidue backbone C–C as well as N–Cα correlation patterns, suggesting the predominance of well-folded Lpp in the CE preparations. Finally, highly flexible nonproteinaceous CE-associated components such as PG and LPS that play crucial roles in bacterial physiology, virulence, and antibiotic resistance were readily identified using through-bond $^{1}H-^{13}C$ and

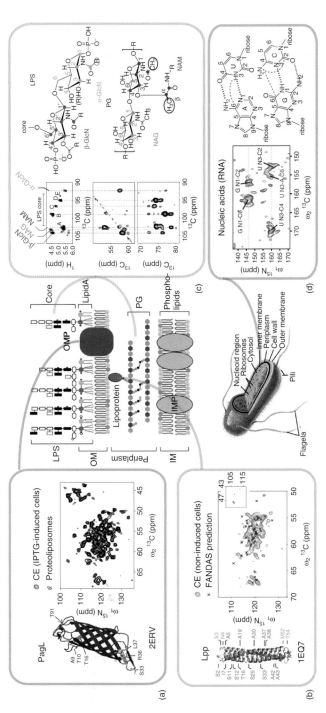

Figure 15.5. **Structural characterization of complex macromolecules in their native bacterial cellular setting.** (a) Conformational analysis of the recombinant outer membrane protein PagL. Overlay of 2D NCA correlation spectra of CE isolated from IPTG-induced WC (black) and PL (red). Major structural changes found between liposome and CE preparations are highlighted in orange on the 3D model of PagL with the corresponding residue numbers. (b) Characterization of the Braun lipoprotein Lpp. 2D NCA correlation spectrum of CE isolated from noninduced WC overlaid with N–Cα correlations (red crosses) predicted from the crystal structure of Lpp-56. Spectral region characteristic of Gly residues (Inset, upper right corner). Residues for which backbone correlations significantly deviate from predictions are labeled and highlighted in orange, while others are colored in green. (c) Resonance assignment of flexible LPS and PG using through-bond 2D (^{1}H–^{13}C) INEPT and 2D (^{13}C–^{13}C) INEPT-TOBSY correlation spectroscopy. The CC and HC correlations that belong to the same spin system are connected by dashed lines. Assigned resonances from LPS and PG glucosamine units are highlighted onto chemical structures. β/α-GlcN, β/α-N-acetylglucosamine; PG NAG, peptidoglycan N-acetyl glucosamine; NAM, N-acetyl muramic acid; LP, lipopolysaccharide. (d) Identification of nucleic acids. Selected spectral regions from DNP-enhanced 2D NCO spectra obtained on (U-^{13}C, ^{15}N)-labeled WC showing characteristic correlations of RNAs. Indicated assignments are based on BMRB average chemical shifts from common RNA nucleotides. Chemical structure of standard RNA bases with Watson–Crick base pairing throughout. Intra- and intermolecular contacts between cytosine N4 or uracil N3 and other C atoms that support our assignment are indicated

^{13}C–^{13}C correlation spectra obtained on CE and WC preparations (Figure 15.5c). Using this strategy, we obtained *de novo* ssNMR assignments of LPS and PG glucosamine units and some of their substituents. Solid-state NMR chemical shifts were compatible with the presence of polysubstituted glucosamine units within the lipid A of LPS and backbone PG moieties. However, carbohydrates from the core of the LPS could not be identified unambiguously due to the large overlap of ^{13}C and ^1H resonances and the inherent structural heterogeneity of the LPS core region.

With increasing levels of molecular complexity, spectroscopic sensitivity becomes a critical factor, and sensitivity-enhanced ssNMR methods such as DNP are mandatory to perform structural analysis in whole-cell settings. Initial studies suggest that low temperatures and cryo-protection with 15–40% glycerol are beneficial for WC sample stability and viability. When analyzing PagL-containing cellular preparations, i.e., WC and CE under low-temperature DNP conditions, we observed significant DNP enhancement factors in spectral regions characteristic for recombinant protein signals and major nonproteinaceous cellular components (Figure 15.2). Because of a favorable spectral resolution at low temperature, multidimensional ^{13}C–^{13}C and ^{13}C–^{15}N DNP-enhanced correlation spectra allowed the transposition of NMR assignments from major CE-associated macromolecules, including PagL, PG, and LPS. Furthermore, (U-^{13}C,^{15}N)-whole-cell preparations revealed signals from nucleic acids, notably RNA that are particularly abundant during bacterial exponential growth phase (Figure 15.5d). Taken together, these results showed that DNP not only increases spectroscopic sensitivity by at least an order of magnitude but also permits the detection of molecular components that are not associated with specific compartments.

Subsequently, we adapted cellular ssNMR protocols to investigate larger proteins and CE-associated complexes (Figure 15.6a) as well as eukaryotic membrane proteins (Figure 15.6b). Firstly, we studied the T4SScc, a 1 MDa protein machine consisting of 14 copies of 3 proteins (VirB7, VirB9, and VirB10). T4SScc is part of a larger machine (T4SS) that spans the periplasm and is embedded in both the inner and outer membranes of Gram-negative bacteria. To obtain structural information about T4SScc in its cellular setting, we coexpressed all the three subunits in wild-type *E. coli* BL21 (DE3) as well as in cells deficient in OmpA and OmpF (BL21dm)

as described earlier. Either uniformly labeled with ^{13}C and ^{15}N or selectively labeled T4SScc preparations with [^{13}C,^{15}N]-(Gly,Ser,Leu,Val) or with ([^{13}C,^{15}N]-Thr, [^{15}N]-Val) in the CE of BL21dm cells were used for in situ structural studies. We exploited the well-established correlation between NMR resonance frequencies and protein secondary structure to obtain structural information about the backbone fold of cell-embedded T4SScc using 3D and 2D ssNMR data sets (Figure 15.6a, left) obtained under DNP conditions. Taken together, our data identified unique sequential correlations that help to describe the fold of the T4SScc in the CE (Figure 15.6a, right), in particular protein regions that have been elusive for X-ray or single-particle cryo-EM studies.

Using a similar DNP-supported approach and dedicated sample preparation methods (see above), we more recently studied structural and dynamical aspects of the EGF receptor in A431 vesicles. In particular, we monitored specific sequential protein correlations before (Figure 15.6b, left, orange) and after (Figure 15.6b, left, cyan) binding of the natural EGFR ligand EGF by 3D DNP ssNMR. The corresponding sequential correlations that result from our amino-acid-specific ^{13}C and ^{15}N labeling are indicated in Figure 15.6 (b, right), and, together with ssNMR data obtained at ambient temperature, led to a model of the activation step of EGFR that, according to our data, involves changes in overall as well as local receptor dynamics.

15.4 CONCLUSIONS

In recent years, significant progress has been made in developing ssNMR methods and protocols that make use of DNP for the study of cellular (mostly membrane-embedded) protein preparations. In our chapter, we described the conceptual details of such methods and its first applications in the context of the outer membrane protein PagL. We also briefly discussed structural findings on the T4SScc system as well as the EGF receptor in a eukaryotic membrane setting. For further information on these studies, we refer the interested reader to our original publications.

Thus far, most DNP-supported studies of complex biomolecules utilized chemical-shift information or relied on the determination of specific distance or torsional restraints. In the future, further efforts will be devoted to increase the number of structural constraints that can be obtained from such

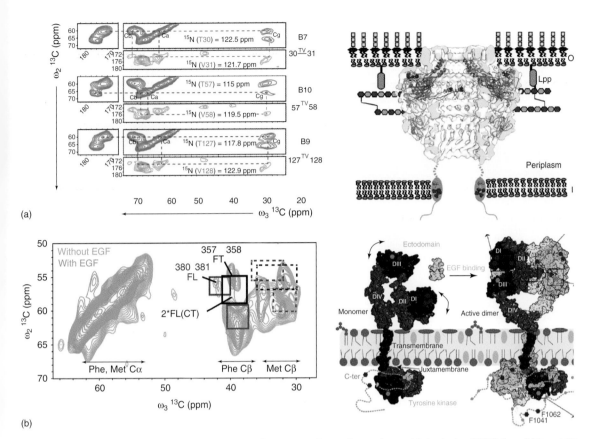

Figure 15.6. **DNP-enhanced ssNMR on large membrane proteins and complexes.** (a) Analysis of DNP-based 2D and 3D ssNMR data sets of tailor-labeled cell-embedded T4SScc. Left, 2D (^{13}C,^{13}C) planes taken out of 3D NCACX (green) and 3D NCOCX (light blue) of TV-T4SScc acquired at 400 MHz DNP. Expected correlations were identified in the spectra. Right, summary of residue-specific ssNMR probes and their location in reference to the cell envelope including the T4SScc electron microscopy map with the outer membrane complex fitted inside (PDB 3JQO). The N terminus of VirB9 is also docked into the map. Identified residues in ssNMR spectra are shown as orange balls for the TV-T4SScc and red balls for the GSLV-T4SScc. (b) Analysis of DNP-enhanced 2D and 3D ssNMR datasets of EGFR in A341 cell vesicles. (Left) 2D planes of 3D N-edited CC correlation spectra before (orange) and after (cyan) addition of EGF for the indicated ^{15}N chemical shifts. Red, blue, and black boxes represent the chemical-shift ranges expected for Phe (solid lines) and Met (dotted lines) Cα and Cβ correlations in α-helical, random-coil, and β-strand conformations. (Right) Generic model of EGFR activation via conformational selection in the ECD. At high temperatures, the unbound receptor exhibits dynamics in both the ECD and CT. Upon binding to the ligand EGF (shown in yellow), the receptor dimerizes and exhibits less dynamics, both on a global and local scale. Residues probed by ssNMR in the MFTL sample are highlighted in orange (MT and FT residue pairs) and magenta (ML and FL pairs)

studies – for example, by tailoring the rich arsenal of ^{13}C/^{15}N-detected experiments that can give 3D structural information for cellular ssNMR conditions. Likewise, dedicated methods that probe the supramolecular setting[55] will be of high interest. Such studies may, at least for bacterial systems where deuteration is straightforward, also go hand-in-hand with cellular ssNMR ^{1}H detection approaches.[26,33]

While the emphasis in many ssNMR studies so far has largely been on membrane proteins, we[17] and others[30] have already shown that ssNMR studies can also be conducted on WC. As discussed in Section 15.1, it seems likely that future cellular ssNMR studies may be possible on large cell-embedded complexes that are difficult to study by in-cell solution NMR. With these aspects in mind, the combination of DNP and cellular solid-state NMR is likely to provide detailed structural insight into biomolecules in their natural setting that are difficult to obtain by any other method at present and that is complementary to modalities such as cryoelectron tomography or high-resolution fluorescence spectroscopy.

ACKNOWLEDGMENTS

We gratefully acknowledge our collaborators and colleagues for their invaluable contributions to cited publications from our own research group. These studies were supported through grants from NWO, the EU and NIH as well as the DFG, the Max-Planck-Society and the Volkswagen foundation.

RELATED ARTICLES IN EMAGRES

Long-Lived States and Coherences for Line Narrowing, DNP, and Study of Interactions

Tissue and Cell Samples by HRMAS NMR

In-cell Nuclear Magnetic Resonance Spectroscopy

Cohen, Jack S.: Recollections of Early NMR Applications to Proteins and Cells

REFERENCES

1. Z. Serber, A. T. Keatinge-Clay, R. Ledwidge, A. E. Kelly, S. M. Miller, and V. Dotsch, *J. Am. Chem. Soc.*, 2001, **123**, 2446.

2. E. Luchinat and L. Banci, *IUCrJ.*, 2017, **4**, 108.

3. D. I. Freedberg and P. Selenko, *Live Cell NMR. Annu. Rev. Biophys.*, 2014, **43**, 171.

4. C. Li, J. Zhao, K. Cheng, Y. Ge, Q. Wu, Y. Ye, G. Xu, Z. Zhang, W. Zheng, X. Zhang, X. Zhou, G. Pielak, and M. Liu, *Annu. Rev. Anal. Chem.*, 2017, **10**, 157.

5. M. Renault, A. Cukkemane, and M. Baldus, *Angew. Chem. Int. Ed.*, 2010, **49**, 8346.

6. D. A. Torchia and D. L. Vanderhart, *J. Mol. Biol.*, 1976, **104**, 315.

7. F. M. Marassi and S. J. Opella, *Curr. Opin. Struct. Biol.*, 1998, **8**, 640.

8. G. Tong, Y. Pan, H. Dong, R. Pryor, G. E. Wilson, and J. Schaefer, *Biochemistry*, 1997, **36**, 9859.

9. J. Herzfeld and J. C. Lansing, *Annu. Rev. Biophys. Biomol. Struct.*, 2002, **31**, 73.

10. H.-X. Zhou and T. A. Cross, *Annu. Rev. Biophys.*, 2013, **42**, 361.

11. S. Wang and V. Ladizhansky, *Prog. Nucl. Magn. Reson. Spectrosc.*, 2014, **82**, 1.

12. L. A. Baker and M. Baldus, *Curr. Opin. Struct. Biol.*, 2014, **27**, 48.

13. M. Kaplan, C. Pinto, K. Houben, and M. Baldus, *Q. Rev. Biophys.*, 2016, **49**, e15.

14. M. Hong, Y. Zhang, and F. Hu, *Annu. Rev. Phys. Chem.*, 2012, **63**, 1.

15. S. J. Ullrich and C. Glaubitz, *Acc. Chem. Res.*, 2013, **46**, 2164.

16. M. Renault, R. Tommassen-van Boxtel, M. P. Bos, J. A. Post, J. Tommassen, and M. Baldus, *Proc. Natl. Acad. Sci. U. S. A.*, 2012, **109**, 4863.

17. M. Renault, S. Pawsey, M. P. Bos, E. J. Koers, D. Nand, R. Tommassen-van Boxtel, M. Rosay, J. Tommassen, W. E. Maas, and M. Baldus, *Angew. Chem. Int. Ed.*, 2012, **51**, 2998.

18. M. Kaplan, A. Cukkemane, G. C. P. van Zundert, S. Narasimhan, M. Daniëls, D. Mance, G. Waksman, A. M. J. J. Bonvin, R. Fronzes, G. E. Folkers, and M. Baldus, *Nat. Methods*, 2015, **12**, 649.

19. T. Jacso, W. T. Franks, H. Rose, U. Fink, J. Broecker, S. Keller, H. Oschkinat, and B. Reif, *Angew. Chem.*, 2012, **124**, 447.

20. R. Fu, X. Wang, C. Li, A. N. Santiago-Miranda, G. J. Pielak, and F. Tian, *J. Am. Chem. Soc.*, 2011, **133**, 12370.

21. L. A. Baker, G. E. Folkers, T. Sinnige, K. Houben, M. Kaplan, E. A. W. van der Cruijsen, and M. Baldus, *Methods Enzymol.*, 2015, **557**, 307.

22. L. A. Baker, M. Daniëls, E. A. W. van der Cruijsen, G. E. Folkers, and M. Baldus, *J. Biomol. NMR*, 2015, **62**, 199.

23. M. Kaplan, S. Narasimhan, C. de Heus, D. Mance, S. Van Doorn, K. Houben, D. Popov-Celeketic, R. Damman, E. A. Katrukha, P. Jain, W. J. C. Geerts, A. J. R. Heck, G. E. Folkers, L. C. Kapitein, S. Lemeer, P. M. P. van Bergen en Henegouwen, and M. Baldus, *Cell*, 2016, **167**, 1241.

24. K. Yamamoto, M. A. Caporini, S.-C. Im, L. Waskell, and A. Ramamoorthy, *Biochim. Biophys. Acta*, 2015, **1848**, 342.

25. E. A. W. van der Cruijsen, A. V. Prokofyev, O. Pongs, and M. Baldus, *Biophys. J.*, 2017, **112**, 99.

26. L. A. Baker, T. Sinnige, P. Schellenberger, J. de Keyzer, C. A. Siebert, A. J. M. Driessen, M. Baldus, and K. Grünewald, *Structure*, 2018, **26**, 161.

27. Q. Z. Ni, E. Daviso, T. V. Can, E. Markhasin, S. K. Jawla, T. M. Swager, R. J. Temkin, J. Herzfeld, and R. G. Griffin, *Acc. Chem. Res.*, 2013, **46**, 1933.

28. Y. Ishii and R. Tycko, *J. Magn. Reson.*, 2000, **142**, 199.

29. S. Asami and B. Reif, *Acc. Chem. Res.*, 2013, **46**, 2089.

30. S. Reckel, J. J. Lopez, F. Löhr, C. Glaubitz, and V. Dötsch, *Chembiochem*, 2012, **13**, 534.

31. Y. Miao, H. Qin, R. Fu, M. Sharma, T. V. Can, I. Hung, S. Luca, P. L. Gor'kov, W. W. Brey, and T. A. Cross, *Angew. Chem. Int. Ed.*, 2012, **51**, 8383.

32. M. E. Ward, S. Wang, R. Munro, E. Ritz, I. Hung, P. L. Gor'kov, Y. Jiang, H. Liang, L. S. Brown, and V. Ladizhansky, *Biophys. J.*, 2015, **108**, 1683.

33. J. Medeiros Silva, D. Mance, M. Daniëls, S. Jekhmane, K. Houben, M. Baldus, and M. Weingarth, *Angew. Chem. Int. Ed.*, 2016, **55**, 13606.

34. C. Pinto, D. Mance, M. Julien, N. Daniels, M. Weingarth, and M. Baldus, *J. Struct. Biol.*, 2018. DOI: 10.1016/j.jsb.2017.11.015.

35. A. A. Labokha, S. Gradmann, S. Frey, B. B. Hülsmann, H. Urlaub, M. Baldus, and D. Görlich, *EMBO J.*, 2012, **32**, 204.

36. L. Barbieri, I. Bertini, E. Luchinat, E. Secci, Y. Zhao, L. Banci, and A. R. Aricescu, *Nat. Chem. Biol.*, 2013, **9**, 297.

37. C. Song, K.-N. Hu, C.-G. Joo, T. M. Swager, and R. G. Griffin, *J. Am. Chem. Soc.*, 2006, **128**, 11385.

38. C. Sauvee, M. Rosay, G. Casano, F. Aussenac, R. T. Weber, O. Ouari, and P. Tordo, *Angew. Chem. Int. Ed.*, 2013, **52**, 10858.

39. S. Lange, A. H. Linden, Ü. Akbey, W. T. Franks, N. M. Loening, B.-J. van Rossum, and H. Oschkinat, *J. Magn. Reson.*, 2012, **216**, 209.

40. E. J. Koers, E. A. W. van der Cruijsen, M. Rosay, M. Weingarth, A. Prokofyev, C. Sauvee, O. Ouari, J. van der Zwan, O. Pongs, P. Tordo, W. E. Maas, and M. Baldus, *J. Biomol. NMR*, 2014, **60**, 157.

41. G. Mathies, M. A. Caporini, V. K. Michaelis, Y. Liu, K.-N. Hu, D. Mance, J. L. Zweier, M. Rosay, M. Baldus, and R. G. Griffin, *Angew. Chem. Int. Ed.*, 2015, **54**, 11770.

42. D. Mance, P. Gast, M. Huber, M. Baldus, and K. L. T. Ivanov, *J. Chem. Phys.*, 2015, **142**, 234201.

43. E. A. W. van der Cruijsen, E. J. Koers, C. Sauvee, R. E. Hulse, M. Weingarth, O. Ouari, E. Perozo, P. Tordo, and M. Baldus, *Chem. Eur. J.*, 2015, **21**, 12971.

44. F. Mentink-Vigier, G. Mathies, Y. Liu, A.-L. Barra, M. A. Caporini, D. Lee, S. Hediger, R. G. Griffin, and G. De Paëpe, *Chem. Sci.*, 2017, **8**, 8150.

45. P. Fricke, D. Mance, V. Chevelkov, K. Giller, S. Becker, M. Baldus, and A. Lange, *J. Biomol. NMR*, 2016, **65**, 121.

46. D. Mance, J. van der Zwan, M. E. Z. Velthoen, F. Meirer, B. M. Weckhuysen, M. Baldus, and E. T. C. Vogt, *Chem. Commun.*, 2017, **53**, 3933.

47. H. Haigler, J. F. Ash, S. J. Singer, and S. Cohen, *Proc. Natl. Acad. Sci. U. S. A.*, 1978, **75**, 3317.

48. V. Santoni, M. Molloy, and T. Rabilloud, *Electrophoresis*, 2000, **21**, 1054.

49. F.-X. Theillet, A. Binolfi, B. Bekei, A. Martorana, H. M. Rose, M. Stuiver, S. Verzini, D. Lorenz, M. van Rossum, D. Goldfarb, and P. Selenko, *Nature*, 2016, **530**, 45.

50. A. Martorana, G. Bellapadrona, A. Feintuch, E. Di Gregorio, S. Aime, and D. Goldfarb, *J. Am. Chem. Soc.*, 2014, **136**, 13458.

51. I. Krstić, R. Hänsel, O. Romainczyk, J. W. Engels, V. Dötsch, and T. F. Prisner, *Angew. Chem. Int. Ed.*, 2011, **50**, 5070.

52. R. Hänsel, L. M. Luh, I. Corbeski, L. Trantirek, and V. Dötsch, *Angew. Chem. Int. Ed.*, 2014, **53**, 10300.

53. A. D. Chowdhury, K. Houben, G. T. Whiting, S.-H. Chung, M. Baldus, and B. M. Weckhuysen, *Nat. Catal.*, 2018, **1**, 23.

54. P. Schanda, S. Triboulet, C. Laguri, C. M. Bougault, I. Ayala, M. Callon, M. Arthur, and J.-P. Simorre, *J. Am. Chem. Soc.*, 2014, **136**, 17852.

55. M. Weingarth and M. Baldus, *Acc. Chem. Res.*, 2013, **46**, 2037.

Chapter 16

Cryo-trapped Intermediates of Retinal Proteins Studied by DNP-enhanced MAS NMR Spectroscopy

Johanna Becker-Baldus and Clemens Glaubitz

Biophysical Chemistry & Centre for Biomolecular Magnetic Resonance, Goethe University Frankfurt, Frankfurt, Germany

16.1 INTRODUCTION

Understanding membrane proteins and their functional mechanism requires information at atomic resolution, ideally as a 'molecular movie' along the reaction coordinate of the processes they are involved in. Techniques that in principle provide such information, either as snapshots or in a time-resolved fashion, are X-ray crystallography, cryo-electron microscopy, and NMR spectroscopy. Owing to their importance for the biological processes at the cell membrane and as drug targets, membrane proteins are of special interest but impose constraints on these techniques. Crystallization sometimes requires protein modifications and the crystal lattice can introduce artificial restrains on the system. Cryo-electron microscopy suffers much less from these restrictions but can in many cases only be applied to relatively large systems and the resolution is often still limited despite tremendous progress during the last couple of years. When considering NMR spectroscopy, solid-state NMR is the method of choice because membrane proteins can be studied in a native-like lipid bilayer. In favorable cases, global high resolution structural information can be obtained by this technique.[1] More often, specific questions have been addressed, which can then be analyzed with high precision.[2]

Major challenges of solid-state NMR are sensitivity and resolution. Both can be improved by utilizing high magnetic fields. However, especially for membrane proteins, this is often not sufficient because the amount of sample that can be investigated is limited and unfavorable protein dynamics can compromise resolution and sensitivity. A fundamental solution for improving sensitivity is offered, in principle, by DNP-enhanced MAS NMR spectroscopy. So far, this approach has been demonstrated to work best at lower fields (e.g., 400 MHz) and lower temperatures (100 K).

Handbook of High Field Dynamic Nuclear Polarization.
Edited by Vladimir K. Michaelis, Robert G. Griffin, Björn Corzilius and Shimon Vega
© 2020 John Wiley & Sons, Ltd. ISBN: 978-1-119-44164-9
Also published in eMagRes (online edition)
DOI: 10.1002/9780470034590.emrstm1552

In order to understand the mechanisms underlying the functions provided by membrane proteins, intermediate states have to be identified, prepared, and structurally characterized. In many cases, these intermediates cannot be prepared as a clean state, which increases the demand for sensitivity further.

How to generate intermediate states depends very much on the system under investigation. Trapping of states can be done chemically by using reagents that are modified and prevent the reaction from completion.[3] The lipid bilayer composition in combination with pH alterations was used to shift open and closed state equilibriums in the potassium channel KcsA.[4] Alternatively, a reaction can be stopped at a certain time point by fast freezing (freeze-quenching). Low temperatures can also favorably be used to stabilize chemically trapped states or proteins with low stability in general (e.g., GPCRs).[5] More details with respect to cryo-trapping are given in Section 16.2.

Both cryo-trapping and cryo-stabilization can be ideally combined with solid-state NMR methods, which perform well under cryogenic conditions such as cross effect DNP, see Section 16.3. This method is well established and due to the large signal-enhancement cryo-trapped intermediates including minor species or cryo-stabilized samples can be investigated. In the main part of this chapter, Section 16.4, we will focus on retinal proteins for which the reaction can be triggered by light. Many of the resulting intermediate states can be trapped by choosing defined illumination conditions. In addition, specific mutations can be utilized for increasing the population of certain intermediates.

16.2 CRYO-TRAPPING OF INTERMEDIATE STATES

The underlying mechanisms of many biochemical processes involve the formation of one or several distinct intermediates, which determine the pathway during the course of the reaction from the initial to the final state. Disentangling these mechanisms requires collecting as much structural knowledge as possible about these intermediate states. However, the rapidity of processes such as enzymatic catalysis, transport through membranes, signaling, and sensing, as well as the short lifetimes of intermediate states, prevent their accumulation at concentrations and timescale needed for high-resolution structural methods.

A solution to this problem is the accumulation and stabilization of intermediates at low temperature ('cryoenzymology').[6] Such stabilization is often referred to as cryo-trapping. In this procedure, a reaction is triggered but then has not enough thermal energy to complete. This can be achieved by either starting the reaction at low temperature or by quickly freezing the sample after the reaction has started. The latter procedure is referred to as rapid freeze-quench. It has mostly been used in combination with EPR spectroscopy to follow enzymatic reactions with a paramagnetic center.[7] Applications in solid-state NMR have also been reported: Enzyme intermediate complexes of uridine diphosphate-*N*-acetylglucosamine enolpyruvyl transferase were studied after freeze-quench,[8] and a partially folded state, a so-called folding intermediate, of the chicken villin headpiece domain (HP35) could be observed and analyzed.[9]

In the abovementioned examples, the reactions were triggered either by rapid mixing or a temperature jump followed by rapid freeze-quench. In case of photoreceptors, reactions are started by light and can be investigated in a similar way. In addition, it is also possible to precool the sample and then start the reaction by light removing the need for a rapid freeze-quench step. This enlarges the experimental portfolio for such systems. For NMR applications, the low temperature used during the recording of cryo-trapped samples can be ideally combined with DNP-enhancement methods.

We will show the case of retinal proteins, which are archetypal, membrane-bound photoreceptors, that the combination of cryo-trapping and DNP enables the study of several reaction intermediates.

16.3 DNP-ENHANCED MAS NMR OF MEMBRANE PROTEINS

In solid-state NMR, membrane proteins are usually studied within a native-like lipid bilayer. Two approaches have been established to deal with the large anisotropic NMR interactions in these immobilized samples.

The first method, 'oriented solid-state NMR', utilizes magnetic or mechanical alignment of membranes, which results in the ideal case in anisotropic but well-resolved spectra. It has been used mainly for peptides, and high resolution depends directly on a high degree of alignment, which can be difficult to achieve. Its combination with DNP has been demonstrated.[10] The second method is magic-angle spinning, which effectively suppresses all unwanted anisotropic interactions. After the removal of excess water, proteoliposomes are transferred into the containers used for sample spinning. As proteoliposomes can be directly used in this method and no alignment is needed, MAS NMR is usually the preferred approach when studying membrane proteins by NMR. For DNP experiments, special equipment is needed, which enables MAS at low temperatures and irradiation by high power microwaves. Such instrumentation is now commercially available.

So far, cross-effect DNP is the only enhancement mechanism that has already left the 'proof-of-concept' stage, which has standard protocols established and that is widely used in a broad range of applications.[11] It is therefore the method of choice for membrane protein samples. Its main drawback is that the experiment has to be conducted at low temperature, typically around 100 K. In addition, it works better at lower rather than higher magnetic field.[12] Therefore, most successful applications so far were based on data acquired on 400 MHz/263 GHz spectrometers. Given the rapid progress in the field with respect to alternative enhancement mechanisms, novel radicals, and improved hardware, more experimental flexibility with respect to higher fields and variable temperatures could be envisaged in the future.

At room temperature, proteins can adopt structurally different states, which are in equilibrium with each other. Often, suitable experimental conditions can be found to shift the equilibrium toward one or the other form. But even if exchange between such substates can be excluded, all proteins have some conformational degree of freedom and at room temperature different conformations are sampled. On the NMR timescale, these motions are averaged and for most resonances a single NMR frequency or a relatively narrow distribution is observed. Under cryogenic conditions, e.g. at a temperature of 100 K, the whole conformational ensemble is therefore 'cryo-trapped' (Figure 16.1a and b).

This situation results often in severe line broadening, which is illustrated for 2D $^{13}C-^{13}C$ correlation spectra of an intensively labeled sample of green proteorhodopsin (PR) (Figure 16.1a). Nonfrozen samples at high field (850 MHz) provide well-structured lineshapes in 2D spectra (Figure 16.1c), and most of the overlapped signals can be fully resolved and assigned in three dimensions. When recording such spectra at 400 MHz, a field currently highly suitable for cross-effect DNP, resolution is compromised (Figure 16.1d). In addition, under cryogenic

conditions (100 K), the same spectrum becomes dominated by broad lines (Figure 16.1e). The broadening is heterogeneous in nature.[15] This situation is more pronounced for proteins, which sample a large conformational space and therefore increasing the magnetic field would not always improve resolution but also results in complex lineshapes. However, these spectra contain valuable site-specific information about conformational sampling.[16]

Therefore, very sparse labeling schemes are the key for the successful design of DNP-enhanced MAS NMR experiments. An example is shown in Figure 16.1f, where ^{13}C-labeled retinal was incorporated into green PR. The spin system can be identified and the chemical shift of each single atom can be extracted. In a similar way, selective labeling of amino acid pairs has been used to reduce signal overlap and to answer specific questions in great detail.[17] Selective labeling has also been used to extract information from DNP-enhancement measurements in a cellular context looking at the bacterial type IV secretion system,[18] the epidermal growth factor receptor[16] and in the cytochrome-P450-cytochrome-b$_5$ complex.[19] In a similar way, selective labels have been introduced by labeling the ligand in the human ABC transporter TAP and the human G-Protein coupled bradykinin-1 receptor.[5]

16.3.1   Sample Preparation

A prerequisite for cross-effect DNP-enhancement is sample doping with suitable polarizing agents, usually biradicals in a deuterium-enriched, glass-forming matrix. The current gold standard biradical most compatible with biochemical requirements for use in aqueous environment is the commercially available AMUPOL.[20] Standard protocols for soluble samples require to dissolve the sample in 60% D_8-glycerol, 30% D_2O, and 10% H_2O (v/v/v) with 10 mM AMUPOL. Similarly, solid samples can be impregnated with this solution. However, resuspension of a membrane protein pellet in this solvent leads to disruption of the lipid bilayer and in addition, the high deuterium content of the solvent prevents re-pelleting of the proteoliposomes. An alternative method was developed in our lab. Briefly, 100 μl of a solution of 30% D_8-glycerol, 60% D_2O, and 10% H_2O (v/v/v) with 20 mM AMUPOL is added on top of the pellet and the sample is then incubated for 24 h at 4 °C. Then the radical solution is removed and the pellet is

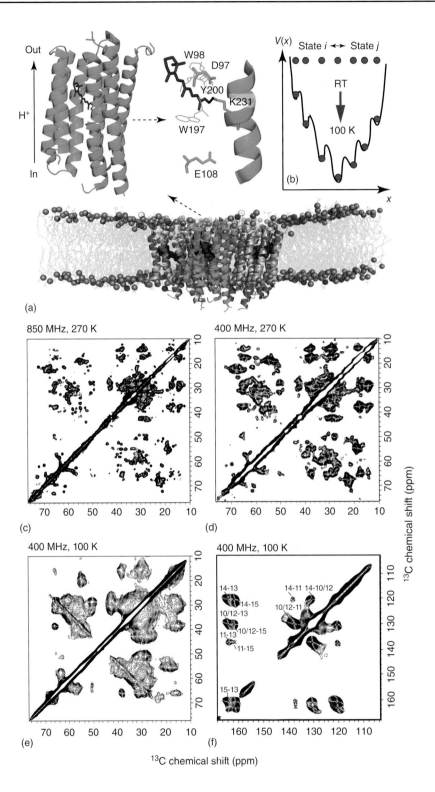

(a)

(b)

(c) 850 MHz, 270 K

(d) 400 MHz, 270 K

(e) 400 MHz, 100 K

(f) 400 MHz, 100 K

^{13}C chemical shift (ppm)

packed into a MAS rotor and stored at −80 °C until the measurement (Figure 16.2). For very tight pellets or large sample volumes, the procedure can be optimized by dividing the pellet into smaller parts and incubating them separately before recombining them in the MAS rotor. Using these procedures, enhancements between 20 and 60 have been routinely achieved for different types of membrane proteins.

16.4 RETINAL PROTEINS

16.4.1 Occurrence

Retinal proteins are found in all domains of life. They are characterized by a molecular architecture with seven transmembrane helices and a retinal cofactor covalently bound via a Schiff base linkage to a lysine in the middle of transmembrane helix 7. In animals, rhodopsins carry a 11-*cis* retinal cofactor and couple to G-proteins. In contrast, microbial rhodopsins carry all-*trans*-retinal and act as ion pumps, ion channels, sensors, or kinases. Their functional mechanism is cyclic and referred to as photocycle.

In this chapter, we will only focus on microbial rhodopsins. The first discovered and most studied member of this group is bacteriorhodopsin (BR) from *Halobacterium salinarum*, a halophilic archaea. BR is a trimeric proton pump and occurs in a 2D crystalline arrangement in the native purple membrane. For biophysical studies, this membrane is used directly without any further purification. In the native organism, the proton pumping generates a pH gradient, which is then used for ATP synthesis. This direct form of photosynthesis has attracted great attention and BR is, therefore, one of the most thoroughly studied membrane proteins. However, it was always considered as evolutionary less important due to its rare occurrence and because of the extreme conditions needed by *H. salinarum*. This view has changed with the availability of extensive metagenomic screens in the early twenty-first century, which provided evidence for a widespread distribution of microbial rhodopsins, which triggered extensive research efforts (Figure 16.3).[21] An intriguing question guiding this field of research is how the same fold with the same chromophore can carry out such a diverse array of functions. Another reason for the strong interest in these proteins is their application in optogenetics.[22]

16.4.2 Photocycle and Characterization of the Intermediates

16.4.2.1 Photocycle

Microbial retinal proteins undergo a cyclic reaction upon illumination. The photocycle starts from the ground state with the chromophore in the all-*trans*,15-*anti* conformation. For most systems, this is also the thermodynamically stable state in the dark. There are a few exceptions. For example, in BR after several hours in the dark, an equilibrium between the all-*trans*,15-*anti* conformation, and the 13-*cis*,15-*syn* conformation is established.[23] This process is called dark adaptation. Illumination of the ground state starts the photocycle. As an example, the photocycle of BR is shown in Figure 16.4.[24] The cycle is started by light, which leads to an all-*trans*,15-*anti* to 13-*cis*,15-*anti* conformational change of the chromophore. In BR, this first intermediate was termed K-state and a K-like state is observed for all microbial retinal proteins. Its absorption maximum is red-shifted compared to the ground state. After the K-state, a number of different intermediates follow but not all of them can be detected in all retinal proteins. The reaction

Figure 16.1. Effect of field and temperature on MAS NMR spectra of the extensively labeled membrane protein green proteorhodopsin (PR) in proteoliposomes. (a) Visualization of the green PR pentamer in the lipid bilayer with one protomer highlighted in green and the retinal chromophore in blue (the image was created based on the blue PR X-ray structure PDB 4XTO). Proton donor E108, proton acceptor D97, Schiff base K231, and some key residues in the binding pocket are highlighted. (b) Cartoon representation of the conformational space sampled by a protein. The observed NMR signal in a nonfrozen sample is a time average of the sampled conformations. By lowering the temperature, the different conformations do not have enough thermal energy to interconvert anymore and the observed NMR signal is the sum of the different conformations present in the sample. (c) ^{13}C-^{13}C PDSD spectrum of extensively labeled PR (uniform $^{13}C^{15}N$ except labeling of WYFIV) in DMPC/DMPA membranes at 270 K and 850 MHz, (d) at 270 K and 400 MHz, (e) at 100 K and 400 MHz, (f) PR containing retinal labeled at positions C-10 to C-15 in DMPC/DMPA at 100 K. Experimental conditions of the data shown here are described in Refs 13, 14

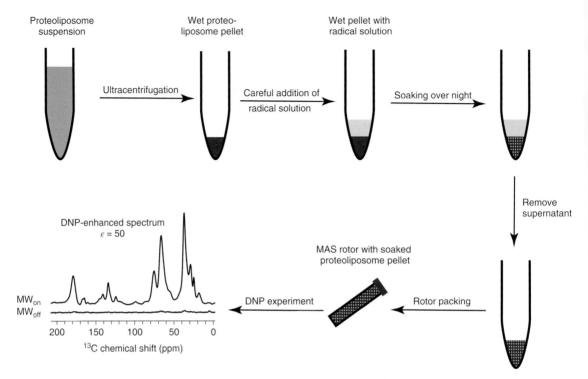

Figure 16.2. Scheme for the preparation of proteoliposomes for DNP-measurements. Proteoliposomes are concentrated by ultracentrifugation. The obtained wet protein pellet is incubated with a solution containing the polarizing agent and glycerol (typically 20 mM Amupol, 30% D_8-glycerol, 60% D_2O, and 10% H_2O). The radical solution can then be removed. The volume of the resulting pellet is not increased significantly by this procedure and the pellet can be transferred to an MAS rotor for the DNP experiment. The spectra shown here were recorded on PR-containing proteoliposomes

proceeds by forming the blue-shifted L-state. It has the same chromophore conformation as the K-state. It is assumed that the protein prepares for deprotonation of the Schiff base in the L-intermediate. Schiff base deprotonation then results in the strongly blue-shifted M-intermediate. Two different M-intermediates are observed. They differ in the H-bond pattern of the protein. The N-state is formed by reprotonation of the Schiff base and again different substrates have been described. Reconversion of the 13-*cis* chromophore to the all-*trans* conformation generates the red-shifted O-state, which then relaxes to the ground state. The described changes in the chromophore are accompanied by de- and reprotonation reactions of sites in the opsin and changes in the pK_a values of the involved functional groups.

To understand all steps of the photocycle, ideally, we would need to know the conformation and protonation states of the opsin and the retinal-Schiff base chromophore in all intermediates. The first change is

caused by the photoconversion of the chromophore and is then transferred to the protein, which vice versa influences the chromophore leading to deprotonation and reconversion of the involved bonds. Thus a crucial first step in understanding the photocycle reactions is to elucidate the chromophore structure. This has been done for BR using Resonance Raman and solid-state NMR spectroscopy. Ideally, also the opsin structure, side chain protonation states and position of water molecules have to be known. For BR, many crystal structures of the ground and some of the intermediate states are available providing such information.[24b] For Anabaena sensory rhodopsin (ASR), blue proteorhodopsin (BPR), channelrhodopsin-2 (ChR2), and Krokinobacter eikastus rhodopsin 2 (KR2) ground state structures have been determined.[25] Unfortunately, it is generally difficult to obtain crystals of the photointermediates and such structure so far could only be determined for BR. In addition, the chromophore conformation and the protonation state

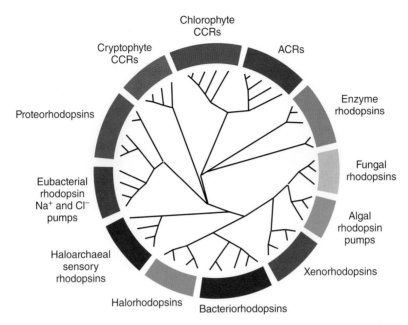

Figure 16.3. Cladogram of the microbial rhodopsin superfamily, which contains many thousands of members, is extremely widespread and is found in prokaryotes and eukaryotes throughout very diverse ecosystems. They are based on a common molecular architecture with 7 transmembrane helices and a retinal chromophore bound in a Schiff base linkage to a lysine in helix 7 (Figure 16.1a). Based on subtle evolutionary modification of this protein scaffold, diverse functions are produced including ion pumps, ion channels, sensors, or kinases. (Reproduced with permission from E. G. Govorunova, O. A. Sineshchekov, H. Li, J. L. Spudich, Annu. Rev. Biochem., 2017, 86, 845–872. © Annual Reviews, 2017)

of the involved residues on the opsin usually cannot be derived from X-ray structures. In the absence of crystal structures, information on the opsin is mainly obtained from infrared spectroscopy, which, however, suffers from a lack of atomic resolution.

Optical- and IR- and to some extent Raman-spectroscopy are great tools to follow the photocycle in real time, but site-resolved structural data are difficult to extract. For an insight at atomic resolution with X-ray or solid-state NMR trapping of the intermediate states is required. With Resonance Raman spectroscopy signals from the chromophore are enhanced selectively. Several Resonance Raman studies are available but interpretation is often ambiguous due to the difficulties in assigning the bands. In principle, such an assignment can be done using a large selection of differently isotope-labeled samples and has been done for BR.[26] The assignment established for BR has been transferred to other retinal proteins but this is an error-prone procedure.

Assignment of the chromophore signals by NMR is much more straightforward and thus the spectra yield

atomic resolution information. Combining selective labeling of the chromophore with DNP signal enhancement and trapping protocols yielded detailed information on a range of retinal proteins and their intermediates, see subsequent sections.

Ideally, also the changes of the protein are monitored during all photocycle steps by solid-state NMR. However, due to the large size of the systems highly selective labeling schemes have to be used due to the line broadening observed at the low temperatures required for trapping the photointermediates.

16.4.2.2 Schiff Base Protonation State and Conformation

There are several ways to experimentally extract information on the Schiff base in retinal proteins. In all intermediates but the M-like states the Schiff base in protonated. Deprotonation results in a large blue shift of the absorption maximum to 360–410 nm. It can be readily detected by UV–Vis spectroscopy also in a time-resolved manner. In solid-state NMR

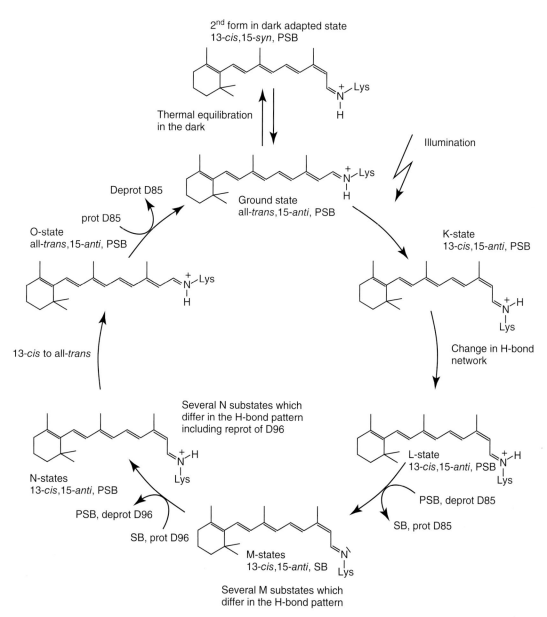

Figure 16.4. Evolution of the chromophore in BR during the photocycle. Exemplarily, the carbon atom-numbering scheme of the Schiff base retinal is given for the ground state structure. Conformational changes of the chromophore as well as the protonation state of the Schiff base (SB for Schiff base, PSB for protonated Schiff base) are given. In addition, protonation changes of the proton acceptor (D85) and the proton donor (D96) are indicated

deprotonation causes a huge change in chemical shift to more than 300 ppm and thus a deprotonated Schiff base is easily identified. In addition, the ^{13}C chemical shift of the C-13 in retinal undergoes a large chemical shift change, which is characteristic for the deprotonated Schiff base due to the change in charge delocalization in the retinal polyene chain. The chemical shift of the protonated Schiff base usually correlates well with the distance to the counter ion and the wavelength of the absorption maximum. An empirical relationship has been established for model compounds and can be used to analyze if there is a deviation of the retinal protein under investigation from the expected behavior.[27]

More challenging is the detection of the Schiff base conformation. It can either be in the 15-*anti* or the 15-*syn* conformation. The ^{13}C-14 chemical shift can be regarded as a marker for this conformation as it shows a lower chemical shift when moving from the 15-*anti* to the 15-*syn* conformation due to the gamma-gauche effect with the Schiff base lysine Cε. Interpretation of this shift is most straightforward in the ground state. In the photointermediates, changes in the retinal-binding pocket and of the chromophore itself might also alter the retinal chemical shift. A method not directly involving the chemical shifts for determining the Schiff base conformation was introduced by Thompson *et al.*[28] Using rotational resonance, the distance between C-14-retinal and Cε-lysine was probed and found to be much shorter in the 15-*syn* case compared to 15-*anti*. The Schiff base conformation can also be analyzed by Resonance Raman spectroscopy. It has been shown that the C-14/C-15 stretch vibration is very sensitive to deuteration in the 15-*syn* but not in the 15-*anti*-form. Ideally, the ^{2}H exchange experiment is combined with isotope labeling to clearly identify the C-14/C-15 stretch vibration.

16.4.2.3 Retinal Conformation

In microbial rhodopsins, the retinal conformation changes at the C-13 double bond. This change is characterized by a large chemical shift change on the C-20 and the C-12 sides due to the altered contact with C-15, also a gamma-gauche effect. When 13-*trans* switches to 13-*cis*, the ^{13}C-12 signal moves to higher chemical shifts and the ^{13}C-20 signal moves to lower chemical shifts. This effect has been used to deduce the C-13 bond conformation in BR photointermediates.[29]

In Resonance Raman spectroscopy, a marker band for the C-13-conformation is the C-12D/C-14D rock frequency in 12,14-D$_2$-retinal.[26]

16.4.3 Photointermediates by (DNP-enhanced) NMR

16.4.3.1 Chromophore Isotope Labeling

Investigation of large biomolecules by (DNP-enhanced) solid-state NMR usually requires enrichment of the ^{13}C and ^{15}N isotopes. When studying the chromophore in retinal proteins ideally the retinal cofactor is ^{13}C labeled at the atomic sites of interest.[30] The lysine side chains of the protein are ^{15}N-labeled resulting in a ^{15}N-labeled Schiff base linker.

Preparation of the sample depends on the protein of interest. Lysine labeling is usually possible if the host organism can grow in defined medium and scrambling is not too severe. Alternatively, the whole protein can be ^{15}N-labeled as the Schiff base is often found in an isolated region of the spectrum.

Different synthetic routes have been chosen for incorporating labeled retinal. In the simplest case, isotope-labeled retinal is just added to the growth medium using an expression system in which retinal is not produced.[31] However, this approach requires large quantities of labeled material. Another possibility is to grow the protein with natural abundance retinal and then to remove the retinal by chemically bleaching the protein. Then the sample is reconstituted with the isotope-labeled retinal.[32] A third approach is to grow the protein in question in the absence of retinal to obtain the opsin. Then isotope-labeled retinal is added to the opsin-containing membranes from which the retinal protein can then be purified.[33] In cases when the host system produces retinal itself, isotope labeling of the medium will result in labeled retinal.[34]

16.4.3.2 Illumination Setup

The generation of photointermediates requires illumination of the sample. The samples typically consist of proteoliposomes, which have been treated with radical to enable DNP-enhancements, see above. The proteoliposome pellet is transferred to an optically transparent sapphire rotor. In most situations, it is best to illuminate the sample directly in the NMR spectrometer to exclude temperature changes during sample transfer.

Several setups, which allow illumination of the sample inside an NMR spectrometer have been described and used in the past.

To our knowledge, only two of them have been combined with a DNP setup. The first example is from the Griffin lab.[35] A multimode fiber was put into a 4 mm MAS stator in a way that the resulting diffuse light beam was perpendicular to the rotor. Our lab used a similar approach within a commercial Bruker 3.2 mm DNP probe.[36]

An interesting illumination approach in a non-DNP probe was presented by the Naito lab.[37] They used a rotor spinning system that employs the drive tip on the bottom of the rotor. The top cap of the zirconia rotor, which only has a sealing function was replaced by a conical glass rod penetrating inside the sample. The sample was then illuminated from inside by shining light onto the glass top cap of the spinning rotor resulting in an illumination of the sample from the inside.

16.4.3.3 *Trapping Protocols*

NMR experiments of the photointermediates require stable trapping of the intermediates in question. Therefore, trapping protocols have to be established. Figure 16.5(a) shows the energy landscape of the different intermediates schematically.

The K-state is formed by photoconversion of the C13 bond and can be generated by illuminating the sample at low temperatures, which prevent decay of the K-state. For DNP experiments this is usually done at temperatures around 100 K. As the absorption maxima of K-state and ground state overlap, the K-state is also converted back to the ground state during the illumination procedure. In addition, often the samples are not transparent enough to ensure illumination of the whole sample. Therefore, generally, a mixture between ground and K-state is obtained.

Starting from a ground state/K-state mixture obtained by illumination at 100 K, the sample temperature can be increased in the dark which enables the system to populate further photocycle intermediates due to the higher thermal energy available. Such a process is called *thermal relaxation* (Figure 16.5b). For DNP measurements, the sample is then cooled again for detection. The type of intermediate trapped this way depends on the height of the energy barrier between the different intermediates and the temperature used for relaxation. In some cases, several intermediates can be detected using different temperatures, in others, only the decay of K to the ground state is

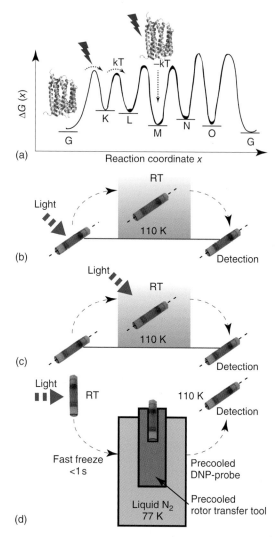

Figure 16.5. Catching photointermediates for DNP-enhanced MAS NMR. (a) Energy landscape sketch for BR. The ground state is the lowest energy state and illumination creates the first photointermediate, the K-state. When enough thermal energy is present, the next photointermediate can be populated and so on until the ground state is reached again. Using suitable protocols, certain photointermediates can be enriched and trapped. (b) Thermal relaxation: By illumination at very low temperatures only the K-state is produced. This can then be converted to other states by raising the temperature. (c) Thermal trapping: The illumination is done at higher temperatures to directly obtain later states. (d) Freeze-quench: States with a long lifetime can be trapped by producing the state under continuous illumination followed by a freeze-quench step, which then preserves this state without light at low temperatures

observed, see subsequent sections. Photointermediates generated this way are always obtained in a mixture with the ground state because this is already the case in the initial K-state preparation.

Another approach is to directly illuminate the sample at the temperature required for generating the desired intermediate. This is referred to as *thermal trapping* (Figure 16.5c). After illumination, the sample is then cooled for the measurement. The advantage of this technique is that in principle it is possible to completely eliminate the ground state as not all photointermediates can be directly converted back to the ground state by light. However, also the opposite is possible, some intermediates cannot be obtained because light converts them back to the ground state (e.g., blue light quenching of the M-state). As another complication, it is possible that the illumination results in a photocycle branching by causing the generated photointermediate to undergo a photoconversion resulting in a side product usually not observed using the thermal relaxation approach.

Also, mixtures of the above-described protocols can be envisioned, whereby a photointermediate generated by illuminating at a certain temperature then undergoes thermal relaxation at an even higher temperature. In addition, different photointermediates can have different absorption maxima and the result of the trapping can depend on the illumination wavelength.

A third approach is to use fast-freezing protocols (*freeze-quenching*; Figure 16.5d). Under continuous illumination at room temperature the slowest intermediate, often the very last step of the photocycle, is accumulated. The sample can then be quickly frozen to preserve this enrichment during the measurement, which is then carried out in the dark. However, a fast temperature jump is not possible inside the NMR-probe. Therefore, the rotor has to be illuminated outside, quickly frozen, and then the cold rotor has to be transferred into the precooled NMR probe.

16.4.3.4 Bacteriorhodopsin

As detailed previously, BR has been studied intensively by NMR and other methods. In the dark, two different forms exist. Solid-state NMR played a crucial role in establishing the chromophore conformation of these states and showed that the dark state consists of a mixture of all-*trans*,15-*anti* and 13-*cis*,15-*syn* retinal Schiff bases.[23,38] Exposure of BR to light leads to light adaption resulting in a pure all-*trans*,15-*anti*-state, which is the starting point of

the photocycle. Reconversion to the dark-adapted state takes several hours. There are numerous early solid-state NMR studies on BR photointermediates, but they generally suffered from low signal-to-noise ratios.[39] This changed when the first studies using DNP-enhanced solid-state NMR were carried out on the system.[34,35] A large number of chemical shifts for the retinal Schiff base as well as for the retinal itself could be collected for several photointermediates using thermal relaxation and thermal trapping. An example of the obtained spectra is shown in Figure 16.6.

Here, we briefly summarize the findings on the different photointermediates. In the K intermediate, a strong shielding of the Schiff base chemical shift is observed, which was explained by the loss of counter ion interaction. In contrast, in the subsequent L intermediate, the Schiff base nitrogen is deshielded indicating a reestablishment of a strong counter ion interaction. Interestingly, L is not a pure state but 3–4 distinct L species could be observed but only one of these can then be relaxed further to give the deprotonated M intermediate. Two different M intermediates with different chemical shifts for the deprotonated Schiff base could be observed indicating a different hydrogen-bonding interactions in the two states. In the N intermediate, the Schiff base is reprotonated but the chromophore is still in the 13-*cis*, 15-*anti* conformation.[40]

The main findings on BR were obtained before DNP-enhanced MAS NMR methods became available. This was possible because the protein naturally occurs in native purple membranes with a protein to lipid ration of just 3 : 1, which can directly be used for solid-state NMR spectroscopy but also for many other biophysical methods. Other retinal proteins can only be studied at much lower density and only DNP-based methods have enabled the detailed study of the chromophore in these systems, see the following section.

16.4.3.5 Channelrhodopsin-2

Light-gated cation channels are found in small green algae. The most prominent representative is ChR2 from *Chlamydomonas reinhardtii* and serves as a light-gated cation channel, which is responsible for phototaxis in the organism. The discovery of channelrhodopsins led to the new field of optogenetics.[22] Briefly, channelrhodopsins can be introduced in neuronal cells, which can then be depolarized upon

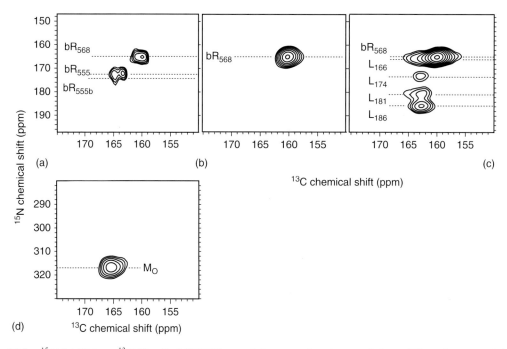

Figure 16.6. ^{15}N-Schiff base ^{13}C-15-retinal TEDOR correlation experiments recorded on differently prepared bacteri-orhodopsin (BR) samples. (a) Dark-adapted BR. (b) Light-adapted BR. (c) L-state trapping. (d) M-state trapping. (Reproduced with permission from V. S. Bajaj, M. L. Mak-Jurkauskas, M. Belenky, J. Herzfeld, R. G. Griffin, Proc Natl Acad Sci U S A, 2009, 106 (23), 9244–9249, Copyright (2009) National Academy of Sciences)

illumination with spatiotemporal resolution. The photocycle is coupled to the channel activity and the molecular basis of this coupling is an intensive area of research. The photocycle starts with formation of a K-like state (P_1^{500}), which then deprotonates to yield the M-like state P_2^{410}. Channel opening occurs in the deprotonated state but even after reprotonation, which leads to the P_3^{520} state, the channel remains open. Channel closing coincides with the formation of a long-lived late state P_4^{480}. However, channel closing kinetics are biexponential indicating a branching of the photocycle (Figure 16.7a).

Structural studies of ChR2 are complicated by the fact that it has to be expressed in eukaryotic systems. Expression of the opsin and regeneration with labeled retinal is possible and also ^{15}N-labeled ChR2 could be obtained from *Pichia pastoris*. In early studies, Channelrhodopsin was proposed to have a chromophore mixture in the ground state similar to BR.[41] However, DNP-enhanced solid-state NMR data clearly showed that in the dark solely the all-*trans*,15-*anti* form is present.[36,42]

Several states could be trapped and analyzed using DNP-enhanced MAS NMR.[36] Figure 16.7(b) shows an example of a thermal relaxation experiment using 14,15-^{13}C$_2$-Retinal-ChR2. The K-state can be generated as in the proton pump retinal proteins by illumination at low temperature, which was confirmed by low-temperature optical spectroscopy (Figure 16.7c). So far no trapping protocol for the deprotonated P_2^{410} and the reprotonated P_3^{520} could be established. The late state P_4^{480} can be trapped using thermal relaxation from K or freeze-quenching. Illumination of this state leads to another photoproduct not observable otherwise (Figure 16.7d). Interestingly, using the thermal trapping method the ground state can be completely depopulated. The chromophore structures of the ground state and the states that can be trapped have been characterized by chemical shift, C-14/C-15 distance, and C-14/C-15 dihedral angle measurements. DNP-enhancement was mandatory for the described measurements, as illumination requires optically less dense samples, which imposes restrictions on sample size.

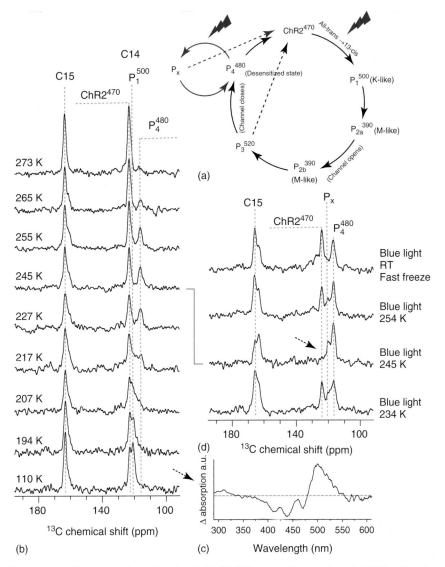

Figure 16.7. Photointermediates of channelrhodopsin-2 (ChR2). (a) Photocycle of ChR2. The late desensitized state P_4^{480} can also react to light and triggers a secondary photocycle. (b) Thermal relaxation experiment on 14,15-^{13}C$_2$-Retinal-^{15}N-ChR2. (c) UV–Vis spectrum after illumination at 150 K showing that the ground state population is reduced and the K-like population is increased. (d) Thermal trapping and freeze-quench (top row) experiments on 14,15-^{13}C$_2$-Retinal-^{15}N-ChR2. In contrast to the thermal relaxation, experiment conditions can be found for which the ground state can be completely depopulated, see arrow in the third spectrum in (d). (Reproduced with permission from J. Becker-Baldus, C. Bamann, K. Saxena, H. Gustmann, L. J. Brown, R. C. Brown, C. Reiter, E. Bamberg, J. Wachtveitl, H. Schwalbe, C. Glaubitz, Proc Natl Acad Sci U S A, 2015, 112 (32), 9896–9901, Copyright (2015) National Academy of Sciences)

16.4.3.6 *Proteorhodopsin*

Proteorhodopsins are proton pumps but differ from the archeal BR (see Figure 16.1a and review Bamann *et al.*[43]). They occur in many different organisms and are found in two forms: blue or green light absorbing. They seem to support their host organism under starvation conditions but additional sensing functions have been discussed as well. The observable photointermediates are similar to BR although the kinetics of the photocycle is different. Green PR has been intensively investigated by standard solid-state

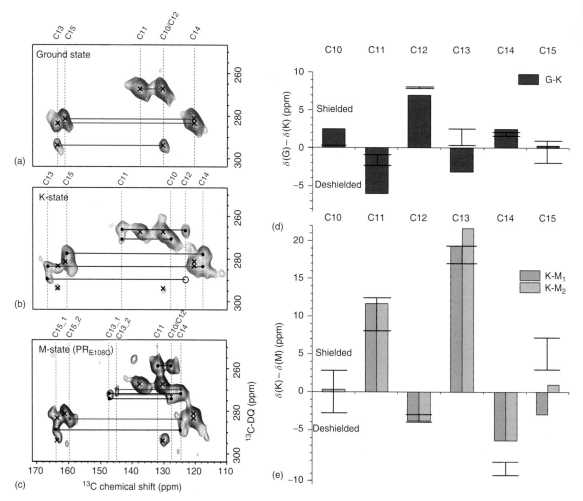

Figure 16.8. Photointermediates of proteorhodopsin. Double-quantum single-quantum spectra of PR containing retinal labeled at positions C-10–C-15 in the ground state (a), K-state (b) and M-state (c). The K-state spectra contain a residual ground state population. The M-state was trapped using the proton donor mutation E108Q, which increases the M-state population. Two different M-states, as well as a residual ground state population, were detected. (d) Chemical shift differences between ground state and K-state. Black bars denote the range of chemical shift changed expected due to the conformational change of the chromophore. (e) Chemical shift differences between K-state and M-states. Black bars denote the range of chemical shift change expected due to the deprotonation of the Schiff base. Deviations from the expected changes indicate protein-specific retinal interactions, which are of functional importance. (Reprinted with permission from M. Mehler, C. E. Eckert, A. J. Leeder, J. Kaur, T. Fischer, N. Kubatova, L. J. Brown, R. C. D. Brown, J. Becker-Baldus, J. Wachtveitl, C. Glaubitz, J. Am. Chem. Soc., 2017, 139 (45), 16143–16153. Copyright (2017) American Chemical Society)

NMR methods in its ground state. In an effort to explain the color tuning mechanisms in the L105Q and the A178R variants, the chemical shifts, as well as the geometry of the chromophore, were studied in detail using DNP-enhanced MAS NMR studies on the 14,15-^{13}C$_2$-Retinal-^{15}N-Schiff base.[13,15] In addition, photointermediates of PR could be investigated using this technique yielding information on the K-state and two different M-states (Figure 16.8).[14]

Proteorhodopsin forms pentamers. The protomer interface has been probed by DNP-enhanced MAS-NMR,[44] which revealed critical contacts in the protomer interface using mixed-labeled PR complexes. Owing to the random mixing of the samples, the number of 'right' contacts in the sample is low. In addition, the involved distances are long resulting in low signal-to-noise ratios due to the small dipole–dipole couplings. These experiments benefited not only from the provided DNP signal-enhancement but from the low temperatures, which reduce dynamics and thus result in longer T_2' times enabling dipolar-based long distance measurements.

16.5 OUTLOOK

Most studies on retinal protein photointermediates have so far concentrated on the retinal Schiff base chromophore, but studying the opsin in these different states is equally important. Resolution especially at the low temperatures required for DNP-enhanced experiments, and trapping is a major challenge for such studies. The observed line broadening is heterogeneous in nature and moving to higher magnetic fields will not improve the situation drastically but comes at the cost of reduced DNP-enhancement. Many states can still be trapped at higher temperatures (180–260 K, depending on the state), which already provide improved line width. Therefore, further technical improvements, which enable DNP experiments at higher fields and under variable temperature conditions, would be highly desirable. In addition, faster MAS probes, which at present also become available under DNP conditions, might open up new perspectives, since DNP efficiency improves[45], decoupling becomes more efficient and ^{1}H detection experiments might become possible.

At present, the best approach is to study highly selectively labeled samples under established DNP-enhancement conditions. Such an approach has already been used, for example, in studies of the ABC exporter MsbA.[17,46]

Advances in fast freezing methods will extend the range of photointermediates in DNP-enhanced studies. In addition, other reaction intermediates or folding intermediates might become accessible by ultra-fast freezing methods. Such frozen samples can then ideally be studied by DNP-enhanced MAS NMR spectroscopy.

ACKNOWLEDGMENTS

Research in the Glaubitz lab (www.glaubitz-lab.de) is supported by various grants provided by the Deutsche Forschungsgemeinschaft and by the State of Hesse through infrastructure support of the Biomolecular Magnetic Resonance Centre (www.bmrz.de). Lenica Reggie is acknowledged for help with recording the spectra in Figure 16.1.

RELATED ARTICLES IN EMAGRES

Lipid Dynamics and Protein–Lipid Interactions in Integral Membrane Proteins: Insights from Solid-State NMR

Dynamic Nuclear Polarization and High-Resolution NMR of Solids

High-Frequency Dynamic Nuclear Polarization

Membrane Associated Systems: Structural Studies by MAS NMR

Structure Determination of Solid Proteins Using MAS and Isotopic Enrichment

REFERENCES

1. S. Wang, R. A. Munro, L. Shi, I. Kawamura, T. Okitsu, A. Wada, S. Y. Kim, K. H. Jung, L. S. Brown, and V. Ladizhansky, *Nat. Methods*, 2013, **10**, 1007.

2. V. Ladizhansky, *Biochim. Biophys. Acta*, 2017, **1865**, 1577.

3. (a) B. G. Caulkins, R. P. Young, R. A. Kudla, C. Yang, T. J. Bittbauer, B. Bastin, E. Hilario, L. Fan, M. J. Marsella, M. F. Dunn, and L. J. Mueller, *J. Am. Chem. Soc.*, 2016, **138**, 15214; (b) H. Kaur, A. Lakatos-Karoly, R. Vogel, A. Noll, R. Tampe, and C. Glaubitz, *Nat. Commun.*, 2016, **7**, 13864.

4. E. A. van der Cruijsen, A. V. Prokofyev, O. Pongs, and M. Baldus, *Biophys. J.*, 2017, **112**, 99.

5. (a) E. Lehnert, J. Mao, A. R. Mehdipour, G. Hummer, R. Abele, C. Glaubitz, and R. Tampe, *J. Am. Chem. Soc.*, 2016, **138**, 13967; (b) L. Joedicke, J. Mao, G. Kuenze, C. Reinhart, T. Kalavacherla, H. R. A. Jonker, C. Richter, H. Schwalbe, J. Meiler, J. Preu, H. Michel, and C. Glaubitz, *Nat. Chem. Biol.*, 2018, **14**, 284.

6. A. L. Fink, *Acc. Chem. Res.*, 1977, **10**, 233.

7. (a) J. Kim, D. J. Darley, W. Buckel, and A. J. Pierik, *Nature*, 2008, **452**, 239; (b) J. M. Kuchenreuther, W. K. Myers, T. A. Stich, S. J. George, Y. Nejatyjahromy, J. R. Swartz, and R. D. Britt, *Science*, 2013, **342**, 472; (c) I. Kaminker, A. Sushenko, A. Potapov, S. Daube, B. Akabayov, I. Sagi, and D. Goldfarb, *J. Am. Chem. Soc.*, 2011, **133**, 15514.

8. Y. Li, F. Krekel, C. A. Ramilo, N. Amrhein, and J. N. S. Evans, *FEBS Lett.*, 1995, **377**, 208.

9. K. N. Hu, W. M. Yau, and R. Tycko, *J. Am. Chem. Soc.*, 2010, **132**, 24.

10. E. S. Salnikov, H. Sarrouj, C. Reiter, C. Aisenbrey, A. Purea, F. Aussenac, O. Ouari, P. Tordo, I. Fedotenko, F. Engelke, and B. Bechinger, *J. Phys. Chem. B*, 2015, **119**, 14574.

11. A. S. Lilly Thankamony, J. J. Wittmann, M. Kaushik, and B. Corzilius, *Prog. Nucl. Magn. Reson. Spectrosc.*, 2017, **102–103**, 120.

12. D. Mance, P. Gast, M. Huber, M. Baldus, and K. L. Ivanov, *J. Chem. Phys.*, 2015, **142**, 234201.

13. M. Mehler, F. Scholz, S. J. Ullrich, J. Mao, M. Braun, L. J. Brown, R. C. Brown, S. A. Fiedler, J. Becker-Baldus, J. Wachtveitl, and C. Glaubitz, *Biophys. J.*, 2013, **105**, 385.

14. M. Mehler, C. E. Eckert, A. J. Leeder, J. Kaur, T. Fischer, N. Kubatova, L. J. Brown, R. C. D. Brown, J. Becker-Baldus, J. Wachtveitl, and C. Glaubitz, *J. Am. Chem. Soc.*, 2017, **139**, 16143.

15. J. Mao, N. N. Do, F. Scholz, L. Reggie, M. Mehler, A. Lakatos, Y. S. Ong, S. J. Ullrich, L. J. Brown, R. C. Brown, J. Becker-Baldus, J. Wachtveitl, and C. Glaubitz, *J. Am. Chem. Soc.*, 2014, **136**, 17578.

16. M. Kaplan, S. Narasimhan, C. de Heus, D. Mance, S. van Doorn, K. Houben, D. Popov-Celeketic, R. Damman, E. A. Katrukha, P. Jain, W. J. C. Geerts, A. J. R. Heck, G. E. Folkers, L. C. Kapitein, S. Lemeer, P. M. P. V. E. Henegouwen, and M. Baldus, *Cell*, 2016, **167**, 1241.

17. R. Spadaccini, H. Kaur, J. Becker-Baldus, and C. Glaubitz, *Biochim. Biophys. Acta*, 2018, **1860**, 833.

18. M. Kaplan, A. Cukkemane, G. C. van Zundert, S. Narasimhan, M. Daniels, D. Mance, G. Waksman, A. M. Bonvin, R. Fronzes, G. E. Folkers, and M. Baldus, *Nat. Methods*, 2015, **12**, 649.

19. K. Yamamoto, M. A. Caporini, S. C. Im, L. Waskell, and A. Ramamoorthy, *Sci. Rep.*, 2017, **7**, 4116.

20. C. Sauvee, M. Rosay, G. Casano, F. Aussenac, R. T. Weber, O. Ouari, and P. Tordo, *Angew. Chem. Int. Ed. Engl.*, 2013, **52**, 10858.

21. (a) E. G. Govorunova, O. A. Sineshchekov, H. Li, and J. L. Spudich, *Annu. Rev. Biochem.*, 2017, **86**, 845; (b) O. Beja, E. N. Spudich, J. L. Spudich, M. Leclerc, and E. F. DeLong, *Nature*, 2001, **411**, 786.

22. E. G. Govorunova and L. A. Koppel, *Biochem. Mosc.*, 2016, **81**, 928.

23. G. S. Harbison, S. O. Smith, J. A. Pardoen, C. Winkel, J. Lugtenburg, J. Herzfeld, R. Mathies, and R. G. Griffin, *Proc. Natl. Acad. Sci. U. S. A.*, 1984, **81**, 1706.

24. (a) J. K. Lanyi, *Annu. Rev. Physiol.*, 2004, **66**, 665; (b) C. Wickstrand, R. Dods, A. Royant, and R. Neutze, *Biochim. Biophys. Acta*, 2015, **1850**, 536.

25. (a) L. Vogeley, O. A. Sineshchekov, V. D. Trivedi, J. Sasaki, J. L. Spudich, and H. Luecke, *Science*, 2004, **306**, 1390; (b) T. Ran, G. Ozorowski, Y. Gao, O. A. Sineshchekov, W. Wang, J. L. Spudich, and H. Luecke, *Acta Crystallogr. D Biol. Crystallogr.*, 2013, **69**, 1965; (c) O. Volkov, K. Kovalev, V. Polovinkin, V. Borshchevskiy, C. Bamann, R. Astashkin, E. Marin, A. Popov, T. Balandin, D. Willbold, G. Buldt, E. Bamberg, and V. Gordeliy, *Science*, 2017, **358**; (d) H. E. Kato, K. Inoue, R. Abe-Yoshizumi, Y. Kato, H. Ono, M. Konno, S. Hososhima, T. Ishizuka, M. R. Hoque, H. Kunitomo, J. Ito, S. Yoshizawa, K. Yamashita, M. Takemoto, T. Nishizawa, R. Taniguchi, K. Kogure, A. D. Maturana, Y. Iino, H. Yawo, R. Ishitani, H. Kandori, and O. Nureki, *Nature*, 2015, **521**, 48; (e) I. Gushchin, V. Shevchenko, V. Polovinkin, K. Kovalev, A. Alekseev, E. Round, V. Borshchevskiy, T. Balandin, A. Popov, T. Gensch, C. Fahlke, C. Bamann, D. Willbold, G. Buldt, E. Bamberg, and V. Gordeliy, *Nat. Struct. Mol. Biol.*, 2015, **22**, 390.

26. S. O. Smith, J. Lugtenburg, and R. A. Mathies, *J. Membr. Biol.*, 1985, **85**, 95.

27. J. Hu, R. G. Griffin, and J. Herzfeld, *Proc. Natl. Acad. Sci. U. S. A.*, 1994, **91**, 8880.

28. L. K. Thompson, A. E. McDermott, J. Raap, C. M. van der Wielen, J. Lugtenburg, J. Herzfeld, and R. G. Griffin, *Biochemistry*, 1992, **31**, 7931.

29. M. E. Hatcher, J. G. Hu, M. Belenky, P. Verdegem, J. Lugtenburg, R. G. Griffin, and J. Herzfeld, *Biophys. J.*, 2002, **82**, 1017.

30. J. Lugtenburg, *Pure Appl. Chem.*, 1985, **57**, 753.

31. N. Pfleger, M. Lorch, A. C. Woerner, S. Shastri, and C. Glaubitz, *J. Biomol. NMR*, 2008, **40**, 15.

32. D. Oesterhelt and L. Schuhmann, *FEBS Lett.*, 1974, **44**, 262.

33. S. O. Smith, J. A. Pardoen, P. P. J. Mulder, B. Curry, J. Lugtenburg, and R. Mathies, *Biochemistry*, 1983, **22**, 6141.

34. V. S. Bajaj, M. L. Mak-Jurkauskas, M. Belenky, J. Herzfeld, and R. G. Griffin, *Proc. Natl. Acad. Sci. U. S. A.*, 2009, **106**, 9244.

35. M. L. Mak-Jurkauskas, V. S. Bajaj, M. K. Hornstein, M. Belenky, R. G. Griffin, and J. Herzfeld, *Proc. Natl. Acad. Sci. U. S. A.*, 2008, **105**, 883.

36. J. Becker-Baldus, C. Bamann, K. Saxena, H. Gustmann, L. J. Brown, R. C. Brown, C. Reiter, E. Bamberg, J. Wachtveitl, H. Schwalbe, and C. Glaubitz, *Proc. Natl. Acad. Sci. U. S. A.*, 2015, **112**, 9896.

37. Y. Tomonaga, T. Hidaka, I. Kawamura, T. Nishio, K. Ohsawa, T. Okitsu, A. Wada, Y. Sudo, N. Kamo, A. Ramamoorthy, and A. Naito, *Biophys. J.*, 2011, **101**, L50.

38. G. S. Harbison, S. O. Smith, J. A. Pardoen, P. P. J. Mulder, J. Lugtenburg, J. Herzfeld, R. Mathies, and R. G. Griffin, *Biochemistry*, 2002, **23**, 2662.

39. J. G. Hu, B. Q. Sun, A. T. Petkova, R. G. Griffin, and J. Herzfeld, *Biochemistry*, 1997, **36**, 9316.

40. K. V. Lakshmi, M. R. Farrar, J. Raap, J. Lugtenburg, R. G. Griffin, and J. Herzfeld, *Biochemistry*, 1994, **33**, 8853.

41. M. Nack, I. Radu, C. Bamann, E. Bamberg, and J. Heberle, *FEBS Lett.*, 2009, **583**, 3676.

42. S. Bruun, D. Stoeppler, A. Keidel, U. Kuhlmann, M. Luck, A. Diehl, M. A. Geiger, D. Woodmansee, D. Trauner, P. Hegemann, H. Oschkinat, P. Hildebrandt, and K. Stehfest, *Biochemistry*, 2015, **54**, 5389.

43. C. Bamann, E. Bamberg, J. Wachtveitl, and C. Glaubitz, *Biochim. Biophys. Acta*, 2014, **1837**, 614.

44. J. Maciejko, M. Mehler, J. Kaur, T. Lieblein, N. Morgner, O. Ouari, P. Tordo, J. Becker-Baldus, and C. Glaubitz, *J. Am. Chem. Soc.*, 2015, **137**, 9032.

45. S. R. Chaudhari, D. Wisser, A. C. Pinon, P. Berruyer, D. Gajan, P. Tordo, O. Ouari, C. Reiter, F. Engelke, C. Coperet, M. Lelli, A. Lesage, and L. Emsley, *J. Am. Chem. Soc.*, 2017, **139**, 10609.

46. H. Kaur, B. Abreu, D. Akhmetzyanov, A. Lakatos-Karoly, C. M. Soares, T. Prisner, and C. Glaubitz, *J. Am. Chem. Soc.*, 2018, in press, DOI: 10.1021/jacs.8b06739.

Chapter 17

DNP Solid-state NMR of Biological Membranes

Burkhard Bechinger

Institut de Chimie, Université de Strasbourg/CNRS, UMR7177, Strasbourg, France

17.1 INTRODUCTION

Among spectroscopic approaches, nuclear magnetic resonance (NMR) is uniquely able to provide information about the structure, dynamics, and molecular interactions at atomistic resolution, when at the same time, it suffers from the low sensitivity inherent to this noninvasive technology. Despite this limitation, NMR has greatly expanded in only a few decades and has become a routine technique in structural biology with multidimensional solution NMR being established for the study of biomolecular complexes of up to several hundred Daltons in size.[1] Furthermore, solid-state NMR has been applied to study large supramolecular assemblies such as amyloid

fibers, protein aggregates, membrane polypeptides, and lipids or other biomolecular complexes.[2–5] In order to overcome the existing limitations in sensitivity, different dynamic nuclear polarization (DNP) approaches have been developed for solution- and solid-state NMR spectroscopy. In this chapter, special focus is given to the investigation of biological membranes and the polypeptides associated with them by two complementary solid-state NMR approaches. In this context, DNP has been applied to different rhodopsins, peptides bound to membrane transporters, membrane-associated peptides, fibrils and phages as well as whole cells, cell walls, and membranes.[6,7] The technique has thus allowed the detection of otherwise invisible conformers,[8] of dilute components,[9] or to work with much smaller samples on shorter time frames.[10]

Lipids, polypeptides, and lipopolysaccharides are all important components of biological membranes, and because of their biomedical importance, their structure and interactions are studied by a multitude of biophysical approaches. However, the supramolecular complexes formed by these macromolecules are beyond the limitations of routine multidimensional solution NMR spectroscopy. Therefore, both membrane protein structure and the interplay between polypeptide and lipid supramolecular structure are important research topics of solid-state NMR, whereas studies of other membrane components, which also play important roles in many biological processes, remain sparse.

Ideally, a biomacromolecule should be studied in an environment that mimics as closely as possible its

Handbook of High Field Dynamic Nuclear Polarization.
Edited by Vladimir K. Michaelis, Robert G. Griffin, Björn Corzilius and Shimon Vega
© 2020 John Wiley & Sons, Ltd. ISBN: 978-1-119-44164-9
Also published in eMagRes (online edition)
DOI: 10.1002/9780470034590.emrstm1558

physiological environment, but often it is necessary to work with crystals or powders of microcrystals that, respectively, allow application of diffraction techniques or that result in well-resolved spectral lines during solid-state NMR spectroscopic analysis. Furthermore, the pH and salt concentrations of solutions are adjusted in such a manner to prevent interference with radio frequency (RF) performance or to slow chemical exchange processes that would, e.g., lead to the suppression of the important amide resonances. For membrane protein studies, mimetics of liquid crystalline bilayers have been developed such as small micelles, isotropic bicelles, or nanodiscs (Figure 17.1a–c). These can be investigated by solution-state NMR spectroscopy and possibly by alternative DNP approaches.

Even a small unilamellar vesicle reaches a size of several hundred megadaltons, much too big for multidimensional solution NMR spectroscopy. Therefore, extended lipid bilayers can only be investigated by this approach in cases where local motions and fast exchange between free and membrane-associated states result in exploitable resonances. As an

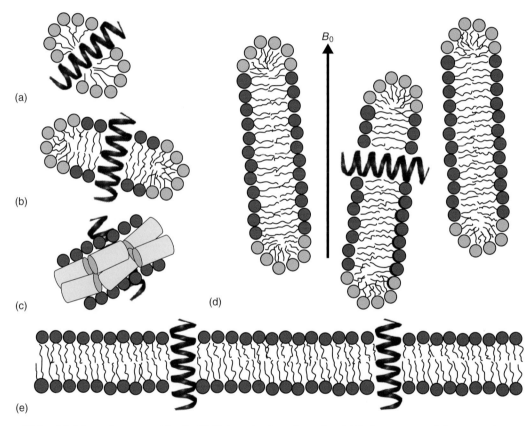

(a)

(b)

(c)

(d)

(e)

Figure 17.1. Model membranes used for the NMR investigation of proteins. (a) In the presence of detergents, fast tumbling micelles keep the membrane polypeptides solubilized. (b) Bicelles form when short- and long-chain phospholipids are mixed. In the presence of a large proportion of short-chain lipids, these tumble isotropically and are used for solution-state NMR investigations. (c) Nanodiscs form in the presence of the predominantly helical membrane scaffolding protein, a derivative of apolipoprotein A, that wraps around the lipid bilayer of typically around 100 lipids. (d) Large bicelles form when the ratio of short- to long-chain phospholipids is lowered. At the correct temperature, hydration, and composition, these magnetically align relative to B_0. Bicelle-associated polypeptides, such as the transmembrane helix shown, have been investigated by solid-state NMR spectroscopy. Partially oriented macromolecules that are soluble in the surrounding aqueous buffer are studied by solution NMR approaches (not shown). (e) Solid-state NMR experiments are performed with extended lipid bilayers in the form of either vesicles or stacks of oriented bilayers, which are prepared on mechanical supports and aligned relative to the magnetic field direction

alternative, solid-state NMR methods have been developed to investigate membrane-associated lipids and polypeptides.[7,11,12]

17.2 SOLID-STATE NMR OF MEMBRANES

17.2.1 Solid-state NMR Approaches to Investigate Membrane-associated Peptides, Proteins, and Lipids

Two fundamentally different concepts are used to investigate proteins and lipids in extended bilayer environments by solid-state NMR spectroscopy (Figure 17.1d,e). On the one hand, magic angle spinning (MAS) is used to average the anisotropies of NMR interactions resulting in one- and multidimensional spectra that represent isotropic chemical shift resonances (Figure 17.2a). The spectra can be used to reveal distances and torsion angles for structural analysis. In the past, the MAS solid-state NMR structural investigations of biomolecules relied strongly on ^{15}N and ^{13}C heteronuclei. Because ^{1}H are abundant in biomolecules, they form a network of strong [1]H–[1]H dipolar interactions. Sample spinning speeds >60 kHz, which allow decoupling of the ^{1}H dipolar network,

have become available only recently,[13] but not yet at the cryotemperatures generally used to exploit DNP in MAS solid-state NMR spectroscopy.[14]

On the other hand, the investigation of static, uniaxially oriented samples is an alternative and highly complementary approach for the structure determination of membrane proteins because it provides angular information from anisotropic chemical shifts (e.g., ^{15}N and ^{13}C), quadrupolar (^{2}H) and dipolar interactions (e.g., [1]H–[15]N) from labeled sites relative to the external magnetic field direction.[15–17] It can also be used to investigate the orientational order of phospholipids (^{31}P at 100% natural abundance or [2]H-labeled sites.[18] The concept has been transferred to solution NMR spectroscopy by measuring orientational restraints through residual dipolar couplings, albeit these are orders of magnitude smaller than the dipolar couplings measured in the solid state.

Uniaxially oriented samples have been prepared mechanically by depositing peptide–lipid mixtures or proteoliposomes onto glass (Figure 17.1e) or polymer surfaces (e.g., the polymer sheet wrapped into a spiral shown in Figure 17.2b).[19] While proteins are reconstituted in aqueous environments and the resulting proteoliposomes spread on the solid surface, peptides are usually dissolved in the presence of organic solvents, mixed with lipid solutions, dried onto the surface,

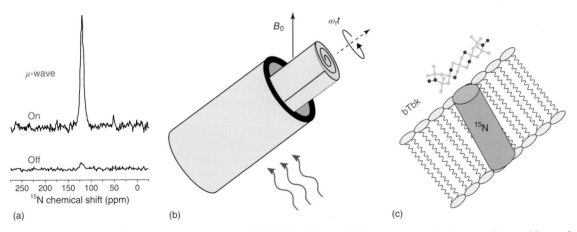

(a)   (b)   (c)

Figure 17.2. Magic angle-oriented sample spinning (MAOSS) solid-state NMR spectroscopy has been used to provide proof of concept that lipid bilayers remain oriented at 100 K and can be used for DNP experiments. (a) The ^{15}N spectrum of a transmembrane polypeptide labeled with ^{15}N at a single peptide bond is shown in the absence and presence of microwave irradiation. (b) The lipid bilayers are oriented along polymer surfaces that are rolled into a spiral (yellow) and introduced into the MAS rotor (blue). The microwave irradiation is indicated schematically by red arrows. (c) The biradical bTbk used in this proof-of-concept study and a lipid bilayer with the membrane normal oriented perpendicular to the rotor axis are shown to the right of the rotor arrangement presented in (b). (Reprinted with permission from E. Salnikov, M. Rosay, S. Pawsey, O. Ouari, P. Tordo, B. Bechinger, J Am Chem Soc 2010, 132 (17), 5940. Copyright 2017 American Chemical Society)

and carefully rehydrated. The sample can then be inserted into the magnetic field (B_0) at any angle. The alignment of the normal parallel to B_0 is best suited to obtain structural constraints, whereas a perpendicular arrangement provides valuable additional data on the rotational diffusion and thereby the oligomerization size within a membrane.[15]

Alternatively, mixtures of short- and long-chain phospholipids exhibit a complex phase diagram including the formation of bicelles (Figure 17.1b,d). The phospholipid bilayer that forms the core of the bicellar assemblies spontaneously orients with the normal perpendicular to the magnetic field of the NMR spectrometer within a narrow range of temperatures, a well-chosen ratio of long-/short-chain lipids and a suitable hydration (Figure 17.1d).[20] Notably, magnetically oriented bicelles occur under full hydration, albeit care has to be taken to adjust the water content relative to the lipid. Rotational diffusion of the bicelle and the protein within the membrane is sufficiently fast for high-resolution solid-state NMR spectra that can thus be analyzed in terms of angular constraints. Phase diagrams in which magnetically oriented bicelles occur have also been observed in the presence of detergents, styrene maleic acid polymers, or amphipathic polypeptides instead of the short-chain phospholipids.[21,22]

Notably, for both MAS and oriented solid-state NMR spectroscopy, the preparation and isotopic labeling of the polypeptide are extremely important steps before spectral acquisition, and tightly connected to the structural information that is obtained. Uniformly and selectively (one or a few types of amino acids throughout the protein) labeled proteins and peptides have been prepared by molecular genetics and biochemical approaches.[12] The polypeptides are reconstituted into oriented membranes and typically investigated by two-dimensional separated local field spectroscopy such as polarization inversion spin exchange at magic angle (PISEMA). Assignment of the resonances remains a challenge, but, once achieved, provides a comprehensive data set for structural analysis through orientational restraints.[16]

Specifically, labeled peptides (i.e., at single sites) are more easily prepared by solid-phase peptide synthesis and investigated in a similar manner. Assignment of a singly labeled site is straightforward, and the gramicidin A structure has been determined from 120 such angular restraints. While the ^{15}N chemical shift of a single labeled peptide bond is sufficient to provide qualitative information if a helical domain is

aligned parallel to the surface or transmembrane, the combination of ^{15}N chemical shifts and quadrupolar splittings from the alanine methyl group ($CH-C^2H_3$) has been shown to result in detailed information about the topology of a protein domain relative to the membrane normal.[15] Notably, the approach has allowed to refine the high-resolution structural information of a polypeptide domain obtained from micellar environments to fit NMR data recorded from phospholipid bilayers.[23]

Topological/angular information can also be obtained from nonoriented membranes (vesicles) under the condition that fast rotational averaging occurs around the bilayer normal, which is the case for peptides and small proteins in a liquid crystalline membrane (Figure 17.1d,e). In this case, the averaged powder pattern line shapes can be analyzed for angular information.[24] While this observation has for many decades been used to characterize the phospholipid macromolecular phases (using ^{31}P NMR), it has been introduced much later to also work for membrane-associated polypeptides (using ^{15}N or ^{2}H NMR). Additional information about the lipid fatty acyl chain order parameters or changes in lipid head group conformation upon electrostatic interactions are obtained using ^{2}H solid-state NMR spectroscopy from static nonoriented samples.[11]

Notably, acquisition of powder pattern line shapes can be sped up by slow spinning around the magic angle; thus; the total spectral intensity is focused in a pattern of spinning side bands from which at the same time the anisotropies and corresponding powder pattern line shape can be calculated. Along this line, magic angle-oriented sample spinning (MAOSS) has been developed to obtain angular information from slowly spinning-oriented samples (Figure 17.2b).[25,26] These can be prepared in large MAS rotors (7 mm diameter) by inserting circular glass plates with the sample normal parallel to the rotor axis,[25] or by preparing membranes aligned along the surface of polymer films that can be wrapped into a cylindrical arrangement (rotors as small as 3.2 mm have been used; in Figure 17.2b, a rotor is schematically illustrated in blue, and the polymer sheet in yellow).[27] In the latter case, the membrane normal is oriented perpendicular to the rotor axis. Therefore, a more elaborate analysis of the spinning side band pattern is required to obtain tilt angle and mosaicity.[26]

Indeed, both MAS and oriented samples have been used for structure determination of membrane polypeptides where considerably more structures deposited at the protein database have been determined from the latter when compared to MAS solid-state NMR spectroscopy, some using mixed information.[7] When lipids are studied, the information obtained from static oriented or nonoriented samples dominates over publications using MAS solid-state NMR spectroscopy.

17.2.2 Sensitivity Concerns for Solid-state NMR Spectroscopy of Membranes

NMR is an inherently insensitive technique, and structural analysis typically requires quantitative amounts of isotopically labeled biomacromolecules. Nevertheless, its many advantages such as the accessibility of all backbone and side chain atoms, including hydrogens that form the outermost 'interaction layer', the suitability to measure dynamics, and conformational exchange have resulted in a fast development of multidimensional solution NMR approaches and the development of NMR spectrometers working at high magnetic fields. In solution, the signals usually derive from the sensitive ^1H nucleus, and the line widths of biomacromolecules cover a few Hertz. Sensitivity is even more of an issue for solid-state NMR spectroscopy because the spectra are commonly recorded from low-γ heteronuclei, and the line widths are usually considerably broader even under MAS conditions. Ultrafast MAS, which makes accessible well-resolved ^1H spectra also in the solid state, has become available only recently. To achieve ultrafast spinning, the rotor outer diameter is reduced to ≤ 1.3 mm and the active volume of such high-speed samples consequently only a few microliters.[28] The reduced sample size imposes limitations in itself.

The line widths are even more pronounced for static uniaxially oriented membrane samples where it can happen that due to slow or intermediate conformational exchange, even in a macroscopically well-aligned sample, the integrated resonance intensity is spread over the full chemical shift anisotropy.[5] Even a well-oriented polypeptide can still be characterized by a line width at half height (LWHH) of 100 Hz.[16,23] Therefore, obtaining good signal-to-noise is most difficult for oriented samples where the acquisition of a one-dimensional ^{15}N solid-state NMR spectrum usually takes hours, even days, particularly if motions typical to liquid crystalline environment hamper the efficiency of cross-polarization. Thus, acquiring multidimensional NMR spectra is possible only for samples with relatively sharp lines and even in favorable cases requires long acquisition periods. While DNP has been shown to enhance the possibilities of solid-state NMR from microcrystalline samples, protein fibers, rhodopsins/purple membranes, or of surfaces in material sciences, it should be of particular value for oriented samples where signal-to-noise issues are even more pronounced, and DNP can much extend the scope of such studies.[6,7,29]

17.2.3 DNP/Solid-state NMR Spectroscopy of Membranes

A variety of different mechanisms have been proposed to explain the DNP effects, depending on the dynamics in the sample (liquid or solid, as well as local dynamics), the properties of the radical, the magnetic field strength, etc.[30] The cross-effect is the most efficient mechanism for electron-nucleus polarization transfer in the solid state, although Overhauser DNP, driven by electron-nuclear cross-relaxation and thought to be limited to solution and electrically conductive solids, has also been found effective under some conditions in insulating solids.[30-33] The cross-effect requires that the samples are prepared in the presence of suitable biradicals where two electrons are coupled at appropriate strength and relative orientation and that electron relaxation is slowed by cooling to temperatures usually obtained with liquid nitrogen or liquid helium. Furthermore, the preparation protocol and sample composition should ensure a homogeneous dispersion of the biradicals as well as spin diffusion of the polarization within the frozen sample. From an instrumental point of view, a strong microwave source, a microwave guide, and a probe head that efficiently transfers the incoming microwaves to the sample active volume, and which is tuned to ^1H and the heteronuclei of interest, are required.[6,34] The sample has to be prepared and packed in such a manner to allow microwave penetration ideally throughout all of the active volume of the NMR coil and the efficient dissipation of the heat from such irradiation.[30,35]

When model compounds that have been embedded into glasses made from partially deuterated glycerol/water solutions are investigated under MAS conditions, enhancement factors that range from very low to up to 515 have been obtained, the latter coming close to the theoretical limit of 660.[6] For proteins in such glassy matrices, enhancement factors are somewhat lower but still impressive. These enhancement factors are measured by comparing the ^{13}C or ^{15}N signal intensities obtained from cross-polarization spectra in the presence and absence of microwave irradiation (Figure 17.2a). It should be noted that the 'real gain in sensitivity' is lower because the latter should take into consideration a comparison of signal intensities that are obtained under optimal conditions each, without and with DNP. Notably, some paramagnetic centers have been shown to cause considerable depolarization under MAS and in the absence of microwave irradiation.[36,37] Comparison to this value correspondingly increases the apparent enhancement when the microwaves are switched on.[6] The radicals also cause paramagnetic relaxation enhancement (bleaching) in their direct neighborhood, but they also allow for faster experimental repetition times; thus, it is important to find the optimal radical concentration.[38,39] It should also be taken into consideration that changes regarding instrumentation are necessary to allow for microwave penetration of the sample and DNP,[35] and that the samples are often diluted by the cryoprotectant solution carrying the radicals. A number of suggestions have been made to come up with a more realistic value for the still significant merit of DNP where the above, the altered relaxation behavior, and more are taken into consideration.[39,40] Furthermore, to date, the choice of DNP instruments remains limited; thus, for example, the advantages in sensitivity and resolution of the very highest magnetic field cannot be exploited. Notably, cooling the sample to 100 K provides additional benefits by increasing the signal intensity through the Boltzmann factor (approximately threefold at 100 K when compared to ambient temperatures) and by freezing motions that interfere with dipolar interactions and thus cross-polarization, an effect that can be pronounced for molecules within liquid crystalline bilayers.

On the other hand, the lines tend to broaden at cryogenic temperatures when fast motion, averaging different conformational or topological states, is frozen. Previous investigations indicated that the line broadening is inhomogeneous in nature and correlated to conformational and topological heterogeneity [10] and solvent exposure.[6] The line broadening effects are smallest for residues that are in protein–protein contact when compared to those exposed to solvent.[6] In addition, residues of membrane polypeptides in contact with the lipids show more line broadening

when compared to those exposed to the aqueous phase.[34] Problems of assignment and resolution that are associated with line broadening can be overcome by the use of specific and selective labeling schemes. It should be noted that problems in resolution are highly system and sample dependent. For example, excellent resolution has been obtained for fibers,[41] and resolution has even been improved in frozen oriented samples under DNP conditions in cases where motions at ambient temperature can excessively average the chemical shift anisotropy.[10]

17.2.4 Membrane Samples for DNP Solid-state NMR Measurements

When solid-state NMR/DNP samples are typically prepared, the radicals are introduced by preparing a homogeneous solution that forms a glassy matrix when frozen. It is assumed that the electron polarization is transferred to the nuclei of the solvent via the hyperfine couplings and then relayed by spin diffusion to the macromolecules of interest.[6] A high degree of solvent deuteration (e.g., 60/30/10 vol% d_8-glycerol/D_2O/H_2O, also called *DNP juice*) assures an optimized ^{1}H relaxation rate and that the polarization is channeled efficiently to the ^{1}H of the molecules of interest by sufficiently fast spin diffusion. Thus, the radical concentration itself is optimized empirically to allow high polarization when keeping paramagnetic bleaching and other detrimental effects low.[38,39]

When compared to the two orders of magnitude signal enhancements that are obtained for the model compounds or proteins in glassy matrices, the DNP efficiency is considerably reduced in the presence of membranes. Thus, after soaking a membrane sample in 'DNP juice', enhancement factors of typically 15–30 have been obtained under MAS conditions for membrane samples.[34] The situation is even worse for matrix-free, oriented membrane samples where enhancement factors around or below 10 have initially been obtained (Figure 17.2a),[27] although new polarizing agents promise that better values are possible also in routine applications of DNP to membranes (Figure 17.3a,e).[38] It should be noted that a detailed comparison should take into account the biradicals used, the field strength, the sample temperature as well as other details of the sample preparation protocol, and the NMR measurement.

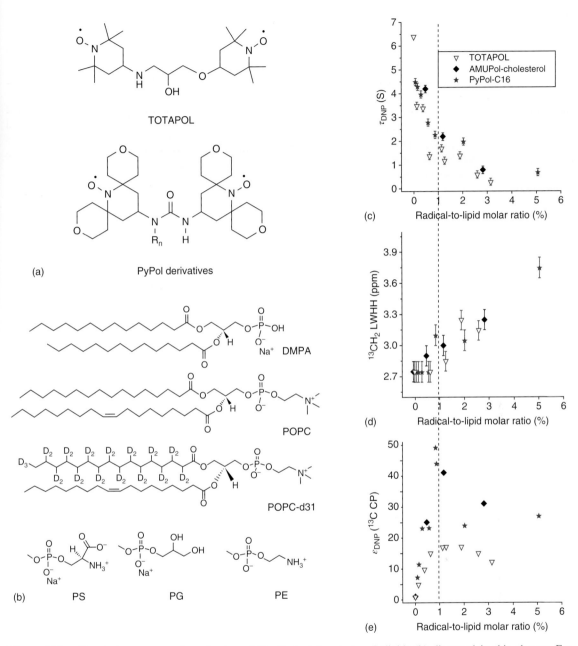

Figure 17.3. Chemical structures of a selection of biradicals (a) and phospholipids (b) discussed in this chapter. For PS, PG, and PE, only the lipid head group is shown. The following PyPol derivatives have been investigated earlier[38]: R_1 = H, PyPol; R_2 = $(CH_2CH_2O)_4CH_3$, AMUPOL; R_3 = $(CH_2CH_2O)_2(CH_2)_2NHC(O)(CH_2)_{14}CH_3$, PyPol-C16; R_4 = $(CH_2CH_2O)_2(CH_2)_2NHC(O)O$-cholesteryl, AMUPOL-cholesterol. The DNP buildup rates (c), the ^{13}C line width of the $^{13}CH_2$ signal (d), and the DNP enhancement (e) are shown as a function of biradical-to-lipid molar ratio in the presence of TOTAPOL, AMUPOL-cholesterol, or PyPol-C16 when investigated in POPC membranes and in the absence of bulk water. ((a, c–e) Reproduced with permission from Ref. 38. © John Wiley and Sons, 2017. (b) Reprinted with permission from Avanti Polar Lipids. © 2018 Avanti Polar Lipids)

17.2.5 Considerations to Improve the Efficiency of DNP/Solid-state NMR of Membranes

One may ask what makes it so much more difficult to obtain a good DNP effect in the presence of membranes, and why is the situation even more difficult for mechanically oriented settings? Observations made with these, other interface-rich and/or matrix-free systems suggest that the inhomogeneous distribution of radicals and other sample-dependent factors have so far prevented higher DNP efficiencies.[34,42]

First, it should be noted that during NMR investigations membranes are usually highly concentrated. Mechanically oriented samples are typically hydrated in 93% relative humidity and lack bulk solvent.[19] Therefore, in such a highly anisotropic environment, the biradical partitions between the membrane and the solvent phase, accumulating, e.g., at the bilayer interface (Figure 17.2c).[34,42] As a consequence, the local concentrations are highly variable and difficult

to optimize. Because in the proximity of the radical the nuclear signals are quenched, there is an optimal concentration of biradical, enough for DNP to reach a maximum of nuclei by spin-diffusion but not too much for paramagnetic quenching to take over. Interestingly, the optimal overall concentration in the presence of membranes comes close to that also found for glassy samples.[38] The radical-nuclei interactions also become apparent by increasing ^{1}H T_1 relaxation rates (Figure 17.3c) and the NMR line width (Figure 17.3d). Notably, upon addition of bulk DNP juice to a membrane paste, a considerable increase in the enhancement factor is obtained when at the same time the sample is diluted.[42] Biradicals that are well water soluble or membrane anchored have ameliorated some of these problems and result in considerable improvements of the DNP enhancement (Figures 17.3a,e and 17.4a).[38]

Notably, the enhancement factors exhibit a strong dependence on the membrane lipid composition.[38] Model membranes made of 1-palmitoyl-2-oleoyl-*sn*-glycero-3-phosphocholine(POPC) showed the best results for a large range of biradicals (Figure 17.3b

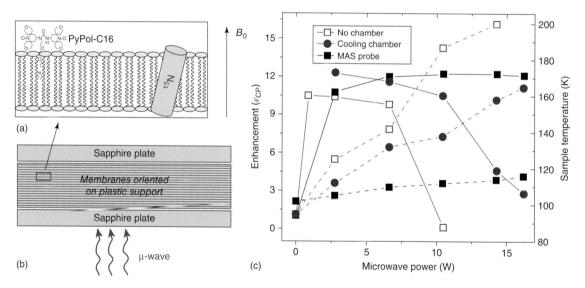

Figure 17.4. (a) Schematic illustration of a ^{15}N-labeled transmembrane helical polypeptide reconstituted into phospholipid bilayers in the presence of the PyPol-C16 biradical. The alignment of the membrane normal is parallel to the magnetic field direction of the NMR spectrometer (B_0). (b) Suitable supports to prepare oriented membranes for DNP investigations are made of thin polymer sheets stabilized by sapphire plates. (c) The enhancement factors of 1.5 M ^{15}NH$_4$Cl in glycerol/water in the presence of AMUPOL are shown as a function of microwave irradiation (solid lines, left ordinate) for a commercial MAS probe (black), a newly designed static flat-coil probe with (blue), and without a dedicated cooling chamber (red). The corresponding sample temperatures are shown as hatched lines (right ordinate). (Reprinted with permission from E. Salnikov, H. Sarrouj, C. Reiter, C. Aisenbrey, A. Purea, F. Aussenac, O. Ouari, P. Tordo, I. Fedoenko, F. Engelke, B. Bechinger, J Phys Chem B 2015, 119 (46), 14574. Copyright 2015 American Chemical Society)

shows the chemical structures of some of the biradicals and phospholipids mentioned in this chapter). In contrast, the presence of phosphatidylethanolamine or of negatively charged lipids had detrimental effects on the DNP enhancements. Interestingly, there is no correlation with the presence of methyl groups, which are most abundant for the phosphatidylcholine head group and which in other context have been shown to accelerate electron relaxation in a manner that is unfavorable for DNP (Figure 17.3b).[43,44] Along this line, partial deuteration of the lipids did not show beneficial effects during first investigations (Figure 17.3b).[34,38] Although the reasons for this lipid dependence have not been investigated in detail, the data are suggestive that the lipid packing and curvature and the resulting partitioning, aggregation, and penetration depth of the radicals into the membrane interface are important.

It should be noted that the design of biradicals is a complex topic where the relative angle between the g-tensors, the distance, structural flexibility, and accessibility of conformational space are all important issues. Therefore, not all biradicals have the same performance, and this is also the case in a membrane environment.[38] Furthermore, the performance of biradicals may vary with the magnetic field and/or the temperature. Notably, AMUPOL, which has an excellent performance in glassy matrices, also turned out to be one of the best when anchored to the membrane and when investigated in the matrix-free environment of mechanically oriented membranes (Figures 17.3 and 17.4).[38]

Because reconstitution of polypeptides into membranes, preparation of oriented membranes, and equilibration in defined relative humidity can take several days and involve the exposure to different buffers/solvents and air, the stability of the biradicals remains a concern. Care should be taken to avoid exposure to acidic conditions at ambient or elevated temperature,[42] and spin counting by EPR can be a valuable control to optimize such protocols. Taking an example where only half the radical centers survive sample preparation and storage, the statistical probability of finding a biradical is reduced to $1/4$, which is a concentration well below the optimum.

A second reason for the lower DNP efficiencies observed for oriented membranes when compared to glassy samples under MAS conditions lies in the fact that sample spinning at moderate frequencies results in much increased enhancement factors by stimulating the quantum mechanical polarization transfer,[38,45,46] and also nuclear depolarization in the absence of microwave irradiation.[36] Continuously reorienting the sample relative to the static parts of the NMR probe head also has positive effects on the cooling and microwave exposure of the sample (cf. the following discussion).

Third, sample heating due to microwave irradiation can be quite pronounced, which becomes apparent when an increase in microwave power reduces the enhancement (Figure 17.4c).[38] Therefore, it is important to efficiently cool the samples, for example, by constructing a cooling chamber around the coil,[35] using sapphire rotors instead of zirconium,[30] or in the case of mechanically oriented samples using sapphire plates in combination with polymer sheets instead of glass plates (Figure 17.4b).[38] Notably, turning or spinning the sample also helps in heat dissipation because the sample is more uniformly exposed not only to the incoming microwave irradiation but also the cooling gas. Even slow turning at a few Hertz as it occurs involuntary from incoming bearing gas (i.e., even in the absence of drive) significantly improves the cooling of the sample and microwave exposure, and thereby the enhancement factors.[38]

For oriented membranes, the DNP performance was much improved by designing an NMR probe with four channels (^{1}H, ^{13}C, ^{15}N, and microwaves) and a flattened coil where the filling factor and geometry are optimized for stacks of oriented membranes.[35] Furthermore, the geometry and materials used for orienting, microwave penetration, and cooling the samples were optimized (Figure 17.4b). In static oriented samples, the air flow and the microwaves always arrive at the same side of the sample. Therefore, to avoid excessive sample heating, a dedicated cooling chamber was designed around the static NMR coil (Figure 17.4c).[35] In order to further improve sample cooling and microwave penetration, replacing the commonly used glass plates by polymer sheets and sapphire supports was necessary and much improved DNP performance.[35] Notably, the microwave power has to be adjusted to promote strong DNP enhancement without too much sample heating (Figure 17.4c).

Furthermore, as discussed earlier, the optimization of the biradical type, concentration, and reconstitution protocol have considerably helped to improve DNP performance.[38] The ^{1}H buildup times have served as a valuable indicator if the biradicals distribute uniformly in the sample or if they aggregate and thereby lose efficiency for DNP.[38]

Another concern is the alignment of the lipid bilayer and the topology of peptide domains when investigated at the cryo temperatures. Therefore, the membrane topology of the PGLa antimicrobial peptide was investigated as a function of sample conditions.[10] This helical sequence is known to respond in a sensitive manner to membrane lipid composition, peptide-to-lipid ratio, and the presence of other peptides. While cooling the sample down to 100 K did not change the alignment of this helical peptide parallel to the 1,2-dimyristoyl-*sn*-glycero-3-phosphocholine/1,2-dimyristoyl-*sn*-glycero-3-phospho-(1'-*rac*-glycerol) DMPC/DMPG 3/1 mole/mole bilayer surface, the hydration of the membrane had a pronounced effect on the tilt angle.[10] Furthermore, when a transmembrane helical sequence was investigated, the tilt angle changed into a more upright position, from 22° to 10°, to compensate the thickening of the membrane upon passage from the liquid crystalline to the gel state.[10]

17.3 APPLICATIONS

Despite the comparatively modest enhancements that are obtained in the presence of membranes when compared to molecules embedded in glassy matrices made from aqueous solutions, a number of problems have already been tackled successfully using DNP solid-state NMR; yielding scientific data would have been difficult if not impossible to obtain without signal enhancement techniques. A number of publications are reviewed in references[6,7,34,47]; therefore, here we will focus only on a few examples. Recently, several publications describe the use of DNP to investigate the protein–protein contact sites of membrane proteins. This includes NHHC experiments on clusters of K[+] channels that have been labeled with either ^{15}N or ^{13}C and reconstituted in the same membrane at low peptide-to-lipid molar ratio (1/400) in combination with molecular dynamics simulations. DNP enhancements of 20–55 at 263 GHz/400 MHz revealed that the number of protein–protein interfaces increased upon channel opening.[48]

Furthermore, interactions between transmembrane domains and between lipids and protein were monitored for the bitopic cytochrome P450 – cytochrome b$_5$ complex when associated with bilayers. The signal enhancements were around 10–20 with AMUPOL and ≤4 with TOTAPOL at 395 GHz/600 MHz.[49] Distances were measured between [13]C- and ^{15}N-labeled protomers at the oligomeric

interface of heptahelical green proteorhodopsin in 1,2-dimyristoyl-*sn*-glycero-3-phosphocholine/1,2-dimyristoyl-*sn*-glycero-3-phosphate (DMPC/DMPA) 9/1 mole/mole membranes with enhancements of about 60 at 263 GHz/400 MHz.[50] Indeed, DNP has served during the investigation of several different rhodopsins, among which, early work by Griffin and coworkers revealed the early K-intermediate and four so far undetected L-intermediates of the bacteriorhodopsin photocycle.[7] Recently, the photocycle of this protein has been further analyzed.[3]

Furthermore, DNP allowed to investigate a peptide bound to the TAP ABC transporter by MAS solid-state NMR[9] or of the PGLa antimicrobial peptide that has been reconstituted into oriented phospholipid bilayers under static matrix-free conditions by two-dimensional solid-state NMR with considerable time savings.[10] Paramagnetic functional groups have also been tagged to proteins and lipids to study interfaces, protein oligomerization, or protein–lipid interactions.[6]

While the reduction of nitroxide spin tags inside the cell requires that the samples are frozen within a short time,[51] trityl radicals are not susceptible to reduction, and (endogenous) paramagnetic metal ions are stable in cellular environments. Thus, the latter have been used successfully for in-cell EPR [52] and DNP/MAS solid-state NMR studies.[53,54]

Furthermore, DNP/solid-state NMR has been applied to bacterial outer membrane proteins, a bacterial secretion machinery, or native cellular membranes.[7] Finally, it should be noted that dissolution DNP/NMR has been used to study pyruvate membrane crossing and metabolic conversion in the cell as a function of different levels of expression of the MCT4 membrane monocarboxylate transporters.[55]

17.4 PERSPECTIVES

DNP/solid-state NMR has only started to be more widely accessible a few years ago, and its application to membranes is only at its beginning. When compared to the very first experiments, considerable progress has been made in improving the signal enhancement by better understanding the requirements for sample preparation protocols,[34] biradicals,[38] experimental setup, and other parameters.[35] Because a particular strength of solid-state NMR is its capacity to investigate proteins and lipids in a near physiological environment, i.e., liquid crystalline bilayers or whole cells,[7,18] it is desirable to expand DNP experiments also to higher temperatures. Indeed, significant signal increase for lipid bilayers under room-temperature conditions has been achieved by utilizing the Overhauser effect. Experiments were carried out on aligned bilayers at 263 GHz/400 MHz using a stripline structure combined with a Fabry–Perot microwave resonator (sample size 160 µg). A signal enhancement of protons of up to −10 was observed.[32] Furthermore, by choosing an adequate solvent containing orthoterphenyl, ^{1}H cross-effect DNP enhancements of over 80 have been obtained at 240 K below the glass transition of orthoterphenyl and in the presence of the biradical TEKPol. At ambient temperatures, ^{1}H enhancements around 15–20 remain, a comparatively good result.[56] Additional benefits could arise from preparing liquid crystalline membranes from lipids that remain in the liquid crystalline state at low temperature and/or the incorporation of cholesterol. On the other end of the temperature range, MAS DNP experiments have been performed in the 25–30 K range (cooling with liquid He), which bears the advantage that good DNP enhancements are obtained with a 800 mW extended interaction oscillator source (rather than a more powerful but also more expensive gyrotron).[57]

ACKNOWLEDGMENTS

The author expresses his gratitude to Evgeniy Salnikov, Olivier Ouari, Paul Tordo, Ilya Fedotenko, Hartmut Oschkinat, Trent Franks, Fabien Aussenac, Frank Engelke, Christian Reiter, Priyanga Bandara, Armin Purea, Melanie Rosay, Shane Pawsey, and Eline Koers for fruitful collaborations and discussion. The continuous support by Werner Maas, Alain Belguise, and Bruker Biopsin is greatly appreciated. The financial contributions of the Agence Nationale de la Recherche (projects membraneDNP 12-BSV5-0012, MemPepSyn 14-CE34-0001-01, In-Membrane 15-CE11-0017-01, Biosupramol 17-CE18-0033-3 and the LabEx Chemistry of Complex Systems 10-LABX-0026_CSC), the University of Strasbourg, the CNRS, the Région Alsace, and the RTRA International Center of Frontier Research in Chemistry are gratefully acknowledged. The author also wishes to thank the *Institut Universitaire de France* for providing additional time to be dedicated to research.

RELATED ARTICLES IN EMAGRES

Aligned Membrane Proteins: Structural Studies

Exploring Transporters within the Small Multidrug Resistance Family Using Solid-State NMR Spectroscopy

Fast Magic Angle Spinning for Protein Solid-State NMR Spectroscopy

Gramicidin Channels: Orientational Constraints for Defining High-Resolution Structures

Lipid Polymorphism

Membrane Associated Systems: Structural Studies by MAS NMR

Lipid Dynamics and Protein–Lipid Interactions in Integral Membrane Proteins: Insights from Solid-State NMR

Structure and Function Studies of Energy and Signal Transducing Proteins by Solid-State NMR

Bilayer Membranes: Deuterium and Carbon-13 NMR

Residual Dipolar Couplings

Glycolipids

Sonicated Membrane Vesicles

Conditional Membrane Proteins: Solution NMR Studies of Structure, Dynamics, and Function

Multiple-Resonance Multi-Dimensional Solid-State NMR of Proteins

REFERENCES

1. R. Rosenzweig and L. E. Kay, *J. Am. Chem. Soc.*, 2016, **138**, 1466.

2. S. E. D. Nelson, K. N. Ha, T. Gopinath, M. H. Exline, A. Mascioni, D. D. Thomas, and G. Veglia, *Biochim. Biophys. Acta*, 2018, **1860**, 1335.

3. Q. Z. Ni, T. V. Can, E. Daviso, M. Belenky, R. G. Griffin, and J. Herzfeld, *J. Am. Chem. Soc.*, 2018, **140**, 4085.

4. W. Qiang, W. M. Yau, J. X. Lu, J. Collinge, and R. Tycko, *Nature*, 2017, **541**, 217.

5. A. Itkin, E. S. Salnikov, C. Aisenbrey, J. Raya, V. Raussens, J. M. Ruysschaert, and B. Bechinger, *ACS Omega*, 2017, **2**, 6525.

6. R. Rogawski and A. E. McDermott, *Arch. Biochem. Biophys.*, 2017, **628**, 102.

7. V. Ladizhansky, *Biochim. Biophys. Acta*, 2017, **1865**, 1577.

8. M. L. Mak-Jurkauskas, V. S. Bajaj, M. K. Hornstein, M. Belenky, R. G. Griffin, and J. Herzfeld, *Proc. Natl. Acad. Sci. U. S. A.*, 2008, **105**, 883.

9. E. Lehnert, J. Mao, A. R. Mehdipour, G. Hummer, R. Abele, C. Glaubitz, and R. Tampe, *J. Am. Chem. Soc.*, 2016, **138**, 13967.

10. E. S. Salnikov, C. Aisenbrey, F. Aussenac, O. Ouari, H. Sarrouj, C. Reiter, P. Tordo, F. Engelke, and B. Bechinger, *Sci. Rep.*, 2016, **6**, 20895.

11. T. R. Molugu, S. Lee, and M. F. Brown, *Chem. Rev.*, 2017, **117**, 12087.

12. M. Kaplan, C. Pinto, K. Houben, and M. Baldus, *Q. Rev. Biophys.*, 2016, **49**, e15.

13. O. Saurel, I. Iordanov, G. Nars, P. Demange, T. Le Marchand, L. B. Andreas, G. Pintacuda, and A. Milon, *J. Am. Chem. Soc.*, 2017, **139**, 1590.

14. S. Chaudhari, P. Berruyer, D. Gajan, C. Reiter, F. Engelke, D. Silverio, C. Coperet, M. Lelli, A. Lesage, and L. Emsley, *Phys. Chem. Chem. Phys.*, 2016, **18**, 10616.

15. B. Bechinger, J. M. Resende, and C. Aisenbrey, *Biophys. Chem.*, 2011, **153**, 115.

16. T. Gopinath, K. R. Mote, and G. Veglia, *J. Biomol. NMR*, 2015, **62**, 53.

17. A. Ramamoorthy, Y. Wei, and D. Lee, *Ann. Rep. NMR Spectrosc.*, 2004, **52**, 1.

18. B. Bechinger and E. S. Salnikov, *Chem. Phys. Lipids*, 2012, **165**, 282.

19. C. Aisenbrey, P. Bertani, and B. Bechinger, in *Antimicrobial Peptides*, eds A. Guiliani and A. C. Rinaldi, Humana Press/Springer: New York, 2010, p 209.

20. E. J. Dufourc, N. Harmouche, C. Loudet-Courrèges, R. Oda, A. Diller, B. Odaert, A. Grélard, and S. Buchoux, in *New Developments in NMR No. 3, Advances in Biological Solid-State NMR: Proteins and Membrane-Active Peptides*, eds F. Separovic and A. Naito, Royal Society of Chemistry: London, 2014.

21. J. Wolf, C. Aisenbrey, N. Harmouche, J. Raya, P. Bertani, N. Voievoda, R. Süss, and B. Bechinger, *Biophys. J.*, 2017, **113**, 1290.

22. T. Ravula, N. Z. Hardin, S. K. Ramadugu, S. J. Cox, and A. Ramamoorthy, *Angew. Chem. Int. Ed. Engl.*, 2018, **57**, 1342.

23. M. Michalek, E. Salnikov, S. Werten, and B. Bechinger, *Biophys. J.*, 2013, **105**, 699.

24. J. Raya, B. Perrone, B. Bechinger, and J. Hirschinger, *Chem. Phys. Lett.*, 2011, **508**, 155.

25. C. Glaubitz and A. Watts, *J. Magn. Reson.*, 1998, **130**, 305.

26. C. Sizun and B. Bechinger, *J. Am. Chem. Soc.*, 2002, **124**, 1146.

27. E. Salnikov, M. Rosay, S. Pawsey, O. Ouari, P. Tordo, and B. Bechinger, *J. Am. Chem. Soc.*, 2010, **132**, 5940.

28. D. Lalli, M. N. Idso, L. B. Andreas, S. Hussain, N. Baxter, S. Han, B. F. Chmelka, and G. Pintacuda, *J. Am. Chem. Soc.*, 2017, **139**, 13006.

29. Y. Su, L. Andreas, and R. G. Griffin, *Annu. Rev. Biochem.*, 2015, **84**, 465.

30. T. V. Can, Q. Z. Ni, and R. G. Griffin, *J. Magn. Reson.*, 2015, **253**, 23.

31. S. Pylaeva, K. L. Ivanov, M. Baldus, D. Sebastiani, and H. Elgabarty, *J. Phys. Chem. Lett.*, 2017, **8**, 2137.

32. O. Jakdetchai, V. Denysenkov, J. Becker-Baldus, B. Dutagaci, T. F. Prisner, and C. Glaubitz, *J. Am. Chem. Soc.*, 2014, **136**, 15533.

33. T. V. Can, M. A. Caporini, F. Mentink-Vigier, B. Corzilius, J. J. Walish, M. Rosay, W. E. Maas, M. Baldus, S. Vega, T. M. Swager, and R. G. Griffin, *J. Chem. Phys.*, 2014, **141**, 064202.

34. S. Y. Liao, M. Lee, T. Wang, I. V. Sergeyev, and M. Hong, *J. Biomol. NMR*, 2016, **64**, 223.

35. E. Salnikov, H. Sarrouj, C. Reiter, C. Aisenbrey, A. Purea, F. Aussenac, O. Ouari, P. Tordo, I. Fedoenko, F. Engelke, and B. Bechinger, *J. Phys. Chem. B*, 2015, **119**, 14574.

36. F. Mentink-Vigier, S. Paul, D. Lee, A. Feintuch, S. Hediger, S. Vega, and G. De Paepe, *Phys. Chem. Chem. Phys.*, 2015, **17**, 21824.

37. K. R. Thurber and R. Tycko, *J. Chem. Phys.*, 2014, **140**, 184201.

38. E. S. Salnikov, S. Abel, G. Karthikeyan, H. Karoui, F. Aussenac, P. Tordo, O. Ouari, and B. Bechinger, *ChemPhysChem*, 2017, **18**, 2103.

39. H. Takahashi, C. Fernandez-De-Alba, D. Lee, V. Maurel, S. Gambarelli, M. Bardet, S. Hediger, A. L. Barra, and G. De Paepe, *J. Magn. Reson.*, 2014, **239**, 91.

40. V. Vitzthum, F. Borcard, S. Jannin, M. Morin, P. Mieville, M. A. Caporini, A. Sienkiewicz, S. Gerber-Lemaire, and G. Bodenhausen, *ChemPhysChem*, 2011, **12**, 2929.

41. P. Fricke, D. Mance, V. Chevelkov, K. Giller, S. Becker, M. Baldus, and A. Lange, *J. Biomol. NMR*, 2016, **65**, 121.

42. E. S. Salnikov, O. Ouari, E. Koers, H. Sarrouj, T. Franks, M. Rosay, S. Pawsey, C. Reiter, P. Bandara, H. Oschkinat, P. Tordo, F. Engelke, and B. Bechinger, *Appl. Magn. Reson.*, 2012, **43**, 91.

43. H. Sato, V. Kathirvelu, A. Fielding, J. P. Blinco, A. S. Micallef, S. E. Bottle, S. S. Eaton, and G. R. Eaton, *Mol. Phys.*, 2007, **105**, 2137.

44. D. J. Kubicki, G. Casano, M. Schwarzwalder, S. Abel, C. Sauvee, K. Ganesan, M. Yulikov, A. J. Rossini, G. Jeschke, C. Coperet, A. Lesage, P. Tordo, O. Ouari, and L. Emsley, *Chem. Sci.*, 2016, **7**, 550.

45. F. Mentink-Vigier, U. Akbey, Y. Hovav, S. Vega, H. Oschkinat, and A. Feintuch, *J. Magn. Reson.*, 2012, **224**, 13.

46. K. R. Thurber and R. Tycko, *J. Chem. Phys.*, 2012, **137**, 084508.

47. A. S. Lilly Thankamony, J. J. Wittmann, M. Kaushik, and B. Corzilius, *Prog. Nucl. Magn. Reson. Spectrosc.*, 2017, **102–103**, 120.

48. K. M. Visscher, J. Medeiros-Silva, D. Mance, J. Rodrigues, M. Daniels, A. Bonvin, M. Baldus, and M. Weingarth, *Angew. Chem. Int. Ed. Engl.*, 2017, **56**, 13222.

49. K. Yamamoto, M. A. Caporini, S. C. Im, L. Waskell, and A. Ramamoorthy, *Sci. Rep.*, 2017, **7**, 4116.

50. J. Maciejko, M. Mehler, J. Kaur, T. Lieblein, N. Morgner, O. Ouari, P. Tordo, J. Becker-Baldus, and C. Glaubitz, *J. Am. Chem. Soc.*, 2015, **137**, 9032.

51. K. K. Frederick, V. K. Michaelis, B. Corzilius, T. C. Ong, A. C. Jacavone, R. G. Griffin, and S. Lindquist, *Cell*, 2015, **163**, 620.

52. J. J. Jassoy, A. Berndhauser, F. Duthie, S. P. Kuhn, G. Hagelueken, and O. Schiemann, *Angew. Chem. Int. Ed. Engl.*, 2017, **56**, 177.

53. M. Kaushik, T. Bahrenberg, T. V. Can, M. A. Caporini, R. Silvers, J. Heiliger, A. A. Smith, H. Schwalbe, R. G. Griffin, and B. Corzilius, *Phys. Chem. Chem. Phys.*, 2016, **18**, 27205.

54. B. Corzilius, A. A. Smith, and R. G. Griffin, *J. Chem. Phys.*, 2012, **137**, 054201.

55. R. Balzan, L. Fernandes, L. Pidial, A. Comment, B. Tavitian, and P. R. Vasos, *Magn. Reson. Chem.*, 2017, **55**, 579.

56. M. Lelli, S. R. Chaudhari, D. Gajan, G. Casano, A. J. Rossini, O. Ouari, P. Tordo, A. Lesage, and L. Emsley, *J. Am. Chem. Soc.*, 2015, **137**, 14558.

57. W. M. Yau, K. R. Thurber, and R. Tycko, *J. Mag. Reson.*, 2014, **244**, 98.

Chapter 18

DNP in Materials Science: Touching the Surface

Pierrick Berruyer[1], Lyndon Emsley[2], and Anne Lesage[1]

[1]*Institut des Sciences Analytiques, UMR 5280, University of Lyon, CNRS, Université Claude Bernard Lyon 1, ENS Lyon, Villeurbanne, France*
[2]*Institut des Sciences et Ingénierie Chimiques, Ecole Polytechnique Fédérale de Lausanne (EPFL), Lausanne, Switzerland*

18.1 INTRODUCTION

Chemists have always demonstrated a strong and continuously renewed interest in the design of materials having specific functions. This is usually done

Handbook of High Field Dynamic Nuclear Polarization.
Edited by Vladimir K. Michaelis, Robert G. Griffin, Björn Corzilius and Shimon Vega
© 2020 John Wiley & Sons, Ltd. ISBN: 978-1-119-44164-9
Also published in eMagRes (online edition)
DOI: 10.1002/9780470034590.emrstm1554

by exploiting and/or tuning their surface properties. To name a few examples, metal–organic frameworks (MOFs) with large gas storage capacity are obtained by functionalizing the surface sites of their porous network;[1] the efficiency of heterogeneous catalysts like alumina- or silica-based industrial catalysts is related to the nature and number of their catalytically active surface entities that can be optimized to increase their performance;[2] nanoparticles with tailor designed surface functionalities are engineered for theranostic purposes, energy-based or environmental applications.[3,4] In all these applications, establishing surface structure–activity relationships is essential in order to rationally design materials with improved properties. This, in turn, depends on the chemists being able to observe, characterize and determine the structure of active sites at surfaces, i.e., having appropriate and effective analytical tools.

A range of analytical methods can be used to characterize surfaces at the molecular level, such as scanning tunneling microscopy (STM), noncontact atomic force microscopy (nc-AFM), or X-ray-based approaches like low-energy electron diffraction (LEED) and extended X-ray absorption fine structure (EXAFS) spectroscopy. However, these techniques are not broadly applicable and/or only yield a partial description of surface sites. Over the years, solid-state nuclear magnetic resonance (NMR) spectroscopy was developed as a method of choice to probe surfaces with atomic resolution. In particular, NMR is an extremely versatile technique as it does not require any long-range atomic order and can be applied to almost any sample formulation, and as such often represents a valuable complementary analytical tool to other surface sensitive techniques.

There are many reports in literature where conventional solid-state NMR spectroscopy under magic-angle spinning (MAS) has been successfully applied to characterize surfaces. For example, in 2007, Basset and coworkers used NMR in combination with EXAFS to reveal the dissociation mechanism of N_2 on tantalum surface sites supported onto silica MCM-41.[5] In 2010, Pruski and coworkers obtained detailed insights into the conformation of fluorinated organic groups anchored on silica surfaces by NMR and theoretical calculations.[6] However, despite a few successful studies, NMR on surfaces remains very challenging, often requiring materials with high-surface areas, isotopic labeling of the sites of interest and/or prohibitively long acquisition times. This is mainly because NMR spectroscopy is an intrinsically low-sensitivity technique, and this limitation becomes critical when the substrates to be described are located (diluted) at a surface.

Dynamic nuclear polarization (DNP) is an elegant and efficient way to circumvent the lack of sensitivity of NMR. The technique was exploited for materials in the late 1980s and early 1990s, to investigate dehydrogenated amorphous silicon,[7] chemical vapor deposited diamond films[8,9] or interfaces between polymers.[10] These pioneering experiments were recorded at low magnetic fields and under static or low MAS conditions, and thus at that time, the approach did not further develop and find wide application in the field of materials.

Thanks to the ongoing research work of Griffin and coworkers at MIT and notably the introduction of high-frequency and high-power microwave sources, DNP experienced a tremendous resurgence in the 2000s, establishing itself as the leading hyperpolarization technique to boost the sensitivity of solid-state NMR experiments, and opening previously unconceivable perspectives.[11] In 2010, the first application of DNP-enhanced solid-state NMR spectroscopy on surfaces at high magnetic field (9.4 T) was reported.[12] This demonstration experiment was done on functionalized mesoporous amorphous silica SBA-15, to characterize organic ligands covalently incorporated at the surface of the framework. Signal enhancement

factors of up to 50 were obtained by impregnating the material with a polarizing solution containing TOTAPOL as the electron source. Over the last 7 years, this approach coined surface enhanced NMR spectroscopy (SENS) has been successfully applied to a wide range of materials, often revealing novel structural features at surfaces.[13–15]

Numerous developments and applications demonstrate today unequivocally that DNP SENS is a versatile and unique approach to *touch* surfaces in materials science. In this chapter, we will describe recent advances in this field. In particular, we will present experimental aspects related to sample preparation and then review various applications of DNP SENS. We will finally discuss current and future perspectives of DNP-enhanced NMR for surface description.

18.2 PRINCIPLES OF DNP SURFACE ENHANCED NMR SPECTROSCOPY

18.2.1 The Indirect DNP Approach

DNP SENS was first demonstrated on organic–inorganic mesoporous silica materials having a pore diameter of 6 nm and obtained by a sol–gel process using a templating route. The surface was functionalized by covalently bound phenol or imidazolium fragments. As the pristine materials did not contain any unpaired electrons, incipient wetness impregnation was used to infuse the pores of the materials with a solution containing free radicals – the polarizing agents – namely a solution of 30 mM TEMPO or 25 mM TOTAPOL in water.[12] After impregnation, the sample was packed in a 3.2 mm sapphire rotor, transferred to a DNP low-temperature MAS probe cooled at approximately 100 K and spun at the magic angle at 8 kHz spinning frequency. As represented in Figure 18.1(a), the continuous irradiation of the sample with microwaves at or near the electron EPR frequency can result in the (partial) transfer of the electron spin polarization to the adjacent protons (the different transfer mechanisms are not discussed in this chapter). ^1H–^1H spin diffusion, mediated by the protonated molecules of the frozen solvent matrix, then distributes the proton hyperpolarization through the whole sample toward the surface where it is finally transferred to other nuclei (carbon-13 spins in this case) via cross-polarization (CP). We note here the solvent plays two roles. First, it allows to introduce the source of unpaired electrons into the sample. Second,

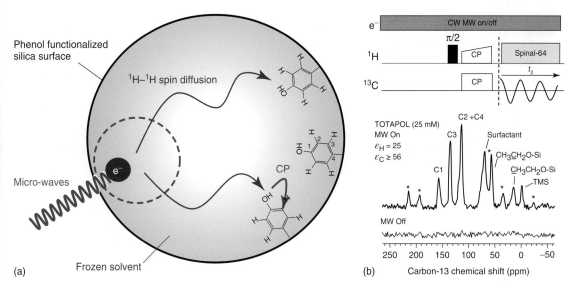

Figure 18.1. (a) Schematic representation of a DNP SENS experiment as first described in Ref. 12. The figure shows the surface of the phenol-functionalized silica framework. The material is filled with the frozen solution in which the polarizing agent (typically a binitroxide) has been dissolved before impregnation. The EPR resonance of the latter is irradiated via the application of continuous microwaves, which can generate a ^{1}H enhanced polarization that is subsequently distributed to the whole sample via ^{1}H–^{1}H spin diffusion. After reaching the surface phenol moieties, the ^{1}H hyperpolarization is transferred by CP to ^{13}C nuclei. (b) Pulse sequence used for 1D DNP-enhanced CPMAS NMR spectroscopy. Continuous microwave (MW) irradiation can be either switched on or off. Carbon-13 CPMAS spectra of mesoporous silica materials functionalized with phenols with (top) and without (bottom) microwave irradiation to induce DNP. The experiments were recorded at 9.4 T and 105 K sample temperature. The signals arising from the phenol moieties are clearly observed (atom labels are given in (a)). The aliphatic signals were assigned to ethyl surface moieties remaining from the material synthesis. (Reprinted with permission from Lesage, A.; Lelli, M.; Gajan, D.; Caporini, M. A.; Vitzthum, V.; Mieville, P.; Alauzun, J.; Roussey, A.; Thieuleux, C.; Mehdi, A.; Bodenhausen, G.; Coperet, C.; Emsley, L., Surface enhanced NMR spectroscopy by dynamic nuclear polarization. *J. Am. Chem. Soc.* 2010, **132** (44), 15459–61. Copyright (2010) American Chemical Society)

once frozen, it provides a rigid protonated matrix that diffuses the nuclear hyperpolarization across the sample. The importance of ^{1}H–^{1}H spin diffusion in DNP MAS experiments has been demonstrated and discussed in several recent papers.[16–19] This approach allowed the authors to record one-dimensional (1D) ^{13}C{^{1}H} CPMAS spectra of the surface functionalities at natural abundance within minutes (Figure 18.1b), and two-dimensional (2D) ^{1}H–^{13}C HETCOR (HETeronuclear CORrelation) spectra within hours. Later, the expeditious acquisition of ^{29}Si{^{1}H} CPMAS and HETCOR DNP SENS spectra on similar materials has been reported using the same experimental methodology.[20] This approach is commonly described as indirect DNP as the observed nuclei are not directly polarized from the electrons, but from neighboring hyperpolarized protons receiving their polarization from adjacent solvent molecules. Along

similar lines, in 2013, DNP NMR experiments on mesoporous silica nanoparticles loaded with surfactant were performed.[21] In these experiments, the proton-enhanced magnetization was transported by proton–proton spin diffusion by the surfactant molecules filling the pores before reaching the surface. The indirect DNP approach usually yields extremely large signal amplification factors. For example, enhancements as high as 213 have been reported on impregnated porous silica materials, at a magnetic field of 9.4 T and a sample temperature of about 100 K.[22]

18.2.2 The Direct DNP Approach

In the direct DNP method, the hyperpolarization is directly transferred from the electrons to the observed (low-gamma) nuclei. In 2011, Lafon and coworkers reported a direct DNP-enhanced NMR experiment on porous silicas impregnated with a 15 mM TOTAPOL solution, and for which a ^{29}Si enhancement of 30 could be achieved.[23] The authors showed that while indirect DNP (based on $\{^{1}H\}^{29}$Si CP) reveals only the silica surface or the first layer of ^{29}Si nuclei, the direct DNP approach (based on ^{29}Si single pulse) allows to also detect the sub-layers of the silica framework. We note here that indirect DNP will indeed only reveal the surface when the material does not contain any ^{1}H in the bulk. If not, ^{1}H$-^{1}$H spin diffusion will possibly propagate the hyperpolarization inside the core of the substrate.[16,18] This assumption is, however, usually correct in many materials of interest. A similar strategy was applied for the investigation of the structure of disk-shaped nanoparticles, where the comparison between direct and indirect DNP-enhanced ^{29}Si NMR allowed the identification of Q^2 defect sites close to the surface.[24] Direct DNP has also been demonstrated for quadrupolar nuclei such as ^{51}V, ^{27}Al, and ^{17}O.[25–29]

In all the examples reported so far where indirect and direct DNP were compared, the two approaches appear to be complementary. Indirect DNP is very selective of the first atomic layer whereas direct DNP probes the surface sub-layers. In the following sections, unless otherwise indicated, we will refer to indirect DNP approaches as these are the most widely used strategies and as they provide much higher signal amplification factors.

18.3 DNP FORMULATION FOR SURFACES

Sample formulation is an essential step to ensure the success of a DNP SENS experiment. While the very first applications of DNP SENS were extremely promising, they were still far from being optimized and thus from providing the highest possible sensitivity gains. The introduction of even more efficient polarizing agents has led to a continuous and significant increase of the signal enhancements (see Chapter 5). Several other factors and experimental parameters playing a key role have been carefully considered and optimized over the years.

18.3.1 Polarizing Solutions

As the materials studied by MAS DNP do not usually contain unpaired electrons, the sample needs to be doped with stable free radicals. These are typically stable organic radicals (usually binitroxides) dissolved in aqueous media (e.g., the so-called DNP juice, namely glycerol-d$_8$/D$_2$O/H$_2$O: 60/30/10: v/v/v) or in organic solvents (e.g., 1,1,2,2-tetrachloroethane (TCE)). MAS DNP experiments are usually performed at low temperature (LT), i.e., at around 100 K on commercial instruments, and the polarizing matrix is thus frozen. Experiments are conducted under cryogenic conditions as the electron relaxation times increase with decreasing temperatures, which facilitates the saturation of the electron spins and leads to higher DNP enhancements.[30] Avoiding radical aggregation upon sample cooling is key to ensure the success of the DNP experiment. Radical aggregation, which typically results from the formation of crystalline domains, leads to a locally high concentration of electrons, detrimental to DNP efficiency.[31] Hence, glassy matrices are usually used like a mixture of water with DMSO,[32] or with glycerol.[11,30,33,34] In 2012, Zagdoun *et al.* compared different glassy organic solvents for applications to hydrophobic surfaces or water-sensitive materials.[35] They reported that the most efficient nonaqueous solvents are halogenated ones, such as TCE or 1,2-dichloroethane.[36] Later, ortho-terphenyl (OTP) was introduced as an alternative efficient DNP matrix.[35]

The proton content of the DNP matrix, i.e., the ratio between deuterated and protonated solvents, is another important parameter that impacts the enhancement factors. While protons in the matrix are necessary to convey the enhanced polarization from the source to the surface, a high concentration of proton spins will also lead to a dispersion of the hyperpolarization. Generally, the enhanced polarization will be transported through ^{1}H$-^{1}$H spin diffusion by both the protons of the solvent and of the surface. The optimal solvent deuteration is thus dependent on the proton content in the material itself and needs to be optimized on a case by case basis. In this chapter, as already pointed out briefly, we note that spin diffusion is a highly relevant process in most MAS DNP experiments and has been investigated with various diffusion numerical models in Refs. 16, 18, 19, and 21.

For surface studies, the choice of the DNP matrix usually relies on its capacity to impregnate the material or wet its surface. As-synthetized silicas exhibit surface silanols, which gives a hydrophilic character to the surface as they enable the formation of hydrogen bonds between Si–OH groups and water molecules. In this case, an aqueous polarizing medium (e.g., water

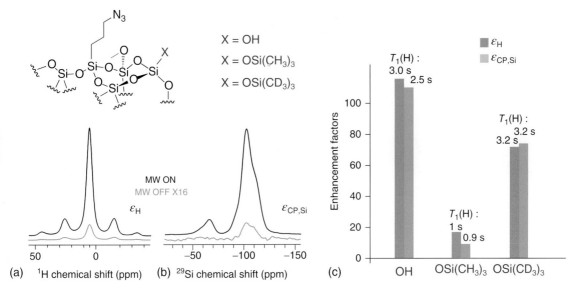

Figure 18.2. (a) ^{1}H and (b) ^{29}Si one-dimensional DNP-enhanced spectra of azide-functionalized SBA-15 silica. (c) ^{1}H enhancement (ε_H), ^{29}Si enhancement (via CP, $\varepsilon_{Si,CP}$) for the material with different passivating groups. (Reproduced with permission from Ref. 38. © John Wiley and Sons, 2013)

and TOTAPOL or AMUPOL) will easily impregnate the silica surface. Passivating the surface, i.e., replacing silanols by hydrophobic organic moieties is often a necessary step before grafting any reactive complex. For instance, Conley *et al.*[37] explained that the passivation of surface silanols was mandatory in order to synthesize supported Ir metal carbenes. This passivation step is usually achieved by replacing of surface silanols by chlorosilane, or tetramethylsilyl chloride, and thus modifing the hydrophilic nature of the surface. Impregnation will then require an organic solvent, as water will not penetrate the porous passivated material and wet the passivated surface. In 2013, the influence of various passivating groups on the DNP enhancement has been studied, as described in Figure 18.2.[38] In particular, they showed that the presence of surface protonated methyl groups had a detrimental impact on the DNP performance: an enhancement $\varepsilon_{Si,CP} = 9$ was measured when the surface was passivated with protonated trimethylsilyl groups (TMS) while $\varepsilon_{Si,CP}$ reached 74 when deuterated TMS were incorporated.

This drastic increase was attributed to changes in the relaxation properties of the sample, methyl groups acting as relaxation sinks that reduce the proton longitudinal relaxation time and lead to a rapid dissipation of the enhanced magnetization. This

was a key observation in the context of the surface characterization by DNP as many materials have functional groups containing methyls.

18.3.2 Radical Concentration

The biradical concentration was one of the first and most straightforward parameter to be optimized. To do so, the concept of a surface overall sensitivity enhancement factor was first introduced. We will refer here to the definition of the overall sensitivity gain proposed in 2012 in Ref. 39. In a DNP-enhanced NMR experiment, the common way to measure the sensitivity gain is to evaluate the DNP enhancement factor ε, calculated as the ratio of the integrated signal intensities microwaves on and off. This is, however, not sufficient to estimate the real sensitivity gain with respect to a conventional NMR experiment recorded at room temperature and without the presence of a paramagnetic polarizing agent. The introduction of radical species in the material leads to a partial loss of potential NMR signals due to paramagnetic effects. In addition, the Boltzmann polarization and the longitudinal relaxation times will be different at room and cryogenic temperatures. In order to take into account all the effects, the

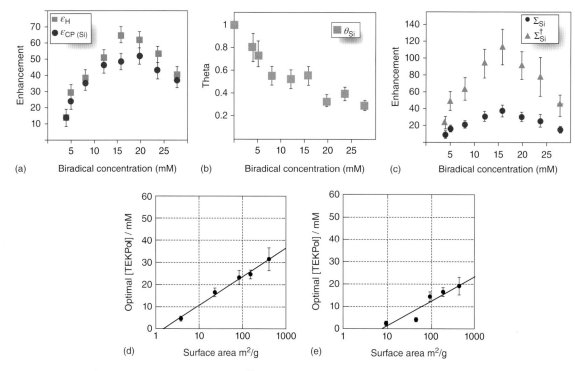

Figure 18.3. (a) ^{1}H enhancement (ε_H) and indirect ^{29}Si enhancement ($\varepsilon_{Si,CP}$) recorded on azide-functionalized SBA-15 silica at different concentrations of bCTbK in TCE. (b) ^{29}Si quenching factor (θ_{Si}) as a function of bCTbK concentration. (c) Overall sensitivity enhancements as a function of bCTbK concentration. $\Sigma^\dagger_{Si}$ takes the Boltzmann temperature effect into account, whereas Σ_{Si} does not. (d and e) Optimal TEKPOL concentration as a function of the surface area for, respectively, Siral-5 and silica nanoparticles as reported recently by Perras *et al.* (Adapted with permission from Ref. 35. © Royal Society of Chemistry, 2012. Adapted with permission from Ref. 43. © Elsevier, 2018)

overall sensitivity enhancement can be thus defined as:

$$\Sigma^\dagger = \varepsilon \times \theta \times \sqrt{\frac{T_1}{T_{DNP}}} \times \frac{298\,K}{105\,K} \qquad (18.1)$$

where ε is the observed DNP enhancement, T_1 the spin–lattice relaxation time measured on a radical-free material at room temperature, and T_{DNP} the buildup time measured on a doped sample under microwave irradiation. θ is the contribution factor calculated as the ratio of the integrated signal intensity of the materials impregnated with the polarizing solution and with a pure solvent, at LT, at a particular MAS rate, and in the absence of microwaves. While the definition of $\Sigma^\dagger$ reflects the overall sensitivity gain of MAS DNP experiments, it does not provide insights into the mechanism that leads to the reduction of the NMR signal upon radical addition. In particular, in 2014, the concept of depolarization that occurs when nitroxide

biradicals are used in combination with magic-angle spinning has emerged,[40–42] and we note here that the contribution factor θ encompasses both paramagnetic bleaching and depolarization effects. This aspect and a detailed discussion of the parameters contributing to the overall sensitivity gain of solid-state DNP is out of the scope of this chapter. The reader is asked to refer to the Chapter 4.

In 2012, Zagdoun *et al.*[35] measured the overall sensitivity gain $\Sigma^\dagger$ for DNP-enhanced ^{29}Si NMR experiments on surfaces of hybrid silica-based mesostructured materials as a function of the radical concentration, in this case, the bCTbk polarizing agent dissolved in TCE. As shown in Figure 18.3(a), (b), and (c), an increase in biradical concentration led to a significant reduction of the contribution factor θ and an optimal radical concentration of 16 mM was found.

Importantly, this optimized biradical concentration – deduced here experimentally for a high-surface area silica matrix – cannot be generalized to all classes of materials. Indeed, the distribution of the polarizing agent and its affinity with the surface plays a key role in the overall DNP efficiency. Recently, Pruski and coworkers have reported a systematic analysis of the optimal radical concentration for DNP SENS experiments done in two types of materials, the silica-aluminas Siral-5, and silica nanoparticles.[43] They examined in detail the dependence of the DNP overall sensitivity with the surface area of the materials and with the radical concentration (TEKPOL dissolved in TCE). They reported for both materials that the optimal radical concentration increased linearly with the logarithm of their surface area (Figure 18.3d and e). While the trends for silica-aluminas and silica nanoparticles were similar, the optimal concentration was, however, different because of different surface–radical interactions, favorable interactions leading to an increase of the radical concentration at the surface.

18.3.3 Specific Formulations

The investigation of the materials of interest are not always compatible with the experimental conditions imposed by the DNP experiment. For example, some surface species react with the radical and degrade, and some nanoparticles tend to aggregate at LT. Several strategies have been proposed to make DNP suitable with the characterization of highly reactive surface entities. They will be described in Section 18.7. In 2015, it has been shown that colloidal semiconductors nanocrystals, e.g., quantum dots (QDs), suspended in a radical solution, aggregate when the sample is frozen at 100 K, and it has been proposed to disperse the nanoparticles into mesoporous silica prior to the DNP experiment.[44] This approach prevents aggregation of the particles upon sample cooling. It yielded ^{31}P CP enhancements at the surface of 56 for InP QDs, while much lower enhancement factors were obtained using conventional incipient wetness impregnation procedures. This strategy has been successfully applied on oleate capped CdSe (zinc-blende polymorph) QDs, allowing the rapid acquisition of a DNP-enhanced ^{111}Cd–^{13}C dipolar HMQC spectrum, highlighting the coordination of the oleate carboxylate to the QDs surface.

A formulation for DNP MAS experiments coined *DNP Jelly* based on an acrylamide gel has been proposed.[45] Remarkably, this DNP matrix does not require the presence of a glassy agent, such as glycerol, and can be applied in cases where conventional impregnation protocols fail. With this new formulation, the authors were able to characterize CdTe-COOH nanoparticles by recording with high-sensitivity a ^{113}Cd DNP-enhanced CP-CPMG (cross-polarization Carr-Purcell-Meil-Gibson) spectrum of the surface while protecting the sample from aggregation.

18.3.4 Suppression of Solvent NMR Signals

As described in the previous sections, the sample is made suitable for the DNP SENS experiment by the addition of a polarizing solution containing radical species. The solvent often gives rise to resonances that mask the NMR peaks of interest. This is the case for organic solvents or glycerol that typically yield ^{13}C resonances of strong intensity. The transverse coherence lifetimes of the organic solvent spins are usually substantially shorter than those of the surface sites.[46] Thus it has been proposed to insert a spin-echo block prior to signal acquisition or a proton spin-lock period prior to the CP step of the CP experiment to exploit this differential relaxation behavior and filter out the solvent resonances.[46] In particular, the CPMAS echo pulse sequence was found to lead to a strong and selective attenuation of the solvent resonances in functionalized mesoporous silica impregnated with TEKPOL in TCE. A similar observation had been made previously on periodic mesoporous organosilicates.[47] A different strategy to remove the undesired solvent peaks has been proposed in parallel that consists in using deuterated solvents in combination with ^{2}H–^{13}C dipolar recoupling sequences that yield a rapid and selective dephasing of the solvent carbon-13 resonances.[48] Several experimental approaches are thus now available to obtain solvent resonance-free DNP-enhanced NMR spectra on surfaces.

In the following section, we will review applications where the tremendous sensitivity gains and time-saving provided by DNP SENS have been successfully applied to reveal the surface structure of various types of materials. This section is not intended to be exhaustive, in particular as the application fields of DNP SENS are continuously growing. Porous materials with high surface areas were among the first materials to be investigated by DNP SENS. The latest developments concern more demanding systems, including very low surface area materials or materials with highly reactive surface sites.

18.4 DNP SENS ON POROUS MATERIALS

18.4.1 Porous Silica Frameworks Incorporating Organometallic Complexes

Porous silica matrices are attractive materials as inert surfaces to covalently attach organic or organometallic ligands and have been largely studied by DNP SENS. After the first proof-of-principle on phenol- and imidazolium-functionalized mesoporous silica,[12,20] DNP SENS has been rapidly applied to more and more complex silica-supported fragments. For example, a dirhodium catalyst anchored to a bifunctional (NH_2 and COOH) SBA-15 mesoporous silica by DNP-enhanced ^{15}N NMR spectroscopy at natural abundance was characterized.[49] ^{15}N DNP SENS was used to establish the presence of nitrogen-containing zirconium sites supported at the surface of catalytically active mesoporous silica nanoparticles.[50] Meanwhile, it was soon realized that the sensitivity gains allowed by DNP could be exploited to unveil structural features that were previously unobservable and escaped any other analytical techniques. In 2013, the presence of secondary interactions between the metal center and the silica surface in ruthenium complexes immobilized on SBA-15 silica has been reported. They were observed indirectly in 2D $^1H-^{29}Si$ DNP-enhanced correlation spectra, in cases where the catalyst precursor was anchored to the surface via a flexible tether. From this observation, the authors postulated that the presence of surface interactions enabled the stabilization of reactive intermediates, in turn leading to higher catalytic performances with flexible Ru complexes.[51] The same consortium also observed similar interactions in a well-defined Pd-N heterocyclic carbene anchored on a hybrid material.[52] In 2015, metal–surface interactions could be qualitatively observed by 2D $^1H-^{13}C$ DNP SENS on an Ir(I)-NHC catalyst immobilized on SBA-15, providing a rational explanation for the outstanding catalytic activity of these supported complexes in alkene hydrogenation.[53] In 2017, Berruyer *et al.*[22] went a step further with the full three-dimensional structure determination of a platinum complex supported onto mesoporous silica. Their approach combined the measurement of inter-nuclear pairwise distances by DNP-enhanced $^{13}C-\{^{15}N\}$ and $^{29}Si-\{^{15}N\}$ rotational-echo double-resonance (REDOR) spectroscopy, with sophisticated structure calculation protocols. The procedure is schematically represented in Figure 18.4(a). The REDOR measurements are reported in Figure 18.4(b) together with the calculated dephasing curves obtained for the best fit structure shown in Figure 18.4(c). This study points to the presence of a single well-defined conformation folded toward the surface with an interaction between the platinum metal center and the surface oxygen atoms of silica. DNP SENS can also be used to study the spatial distribution of the grafted moieties on the porous silica surface obtained through different synthetic routes, namely co-condensation or post-grafting. It has been found that post-grafting leads to a more homogenous distribution of the moieties than co-condensation.[54]

18.4.2 Metal–Organic Frameworks (MOFs) and Mesoporous Organic Polymers (MOPs)

Several DNP SENS studies were reported on MOFs, which are an important class of crystalline porous hybrid materials, attracting much attention for their potential in gas storage and separation, molecular sieving processes, and heterogeneous catalysis.[55] In 2012, DNP SENS experiments on MOFs were performed to characterize parent and proline-functionalized frameworks by ^{13}C and ^{15}N NMR spectroscopy at natural abundance.[56] Interestingly, it was observed that the DNP enhancement could be increased by allowing the nonfunctionalized MOFs – impregnated with a bTbK solution in TCE – to *rest* at room temperature for 25 h over the bench. This effect was not observed on the functionalized MOFs, suggesting that the polarizing agent was partially blocked outside of the porous network by the functional groups, while slowly diffusing inside the channels of the parent framework. For the functionalized MOFs, extremely low enhancements were observed, which was explained by the fact that the radical molecules could not penetrate the framework and resided on or near the surface of the crystallites. Later, Pourpoint *et al.*[57] investigated MIL-100(Al) by DNP MAS. They showed that, contrary to what had been observed previously with bTbK, TOTAPOL could diffuse in the porous network of MIL-100(Al). They obtained enhancement factors of about eight that allowed them to record with high-sensitivity $^{27}Al-^{13}C$ HMQC and

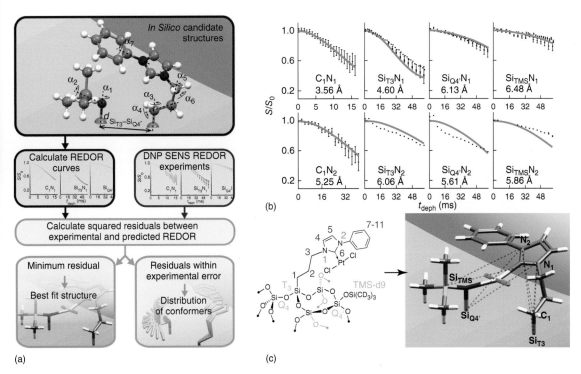

Figure 18.4. (a) Surface Sites 3D structure determination process based on REDOR DNP SENS experiments. (b) $^{13}C-\{^{15}N\}$ and $^{29}Si-\{^{15}N\}$ DNP SENS REDOR recorded on platinum surface sites anchored on mesoporous SBA-15 silica surface. (c) Structure of the studied anchored platinum sites on mesoporous SBA-15 silica surface and the resulting 3D structure obtained with DNP SENS. (Reprinted with permission from Berruyer, P.; Lelli, M.; Conley, M. P.; Silverio, D. L.; Widdifield, C. M.; Siddiqi, G.; Gajan, D.; Lesage, A.; Coperet, C.; Emsley, L., Three-Dimensional Structure Determination of Surface Sites. *J. Am. Chem. Soc.* 2017, **139** (2), 849–855. Copyright (2017) American Chemical Society)

S-REASPDOR correlation experiments, and thus probe aluminum–carbon proximities in this material. Along similar lines, the coordination of Pt^{2+} in a zirconium MOF UiO-66-NH was investigated using DNP-enhanced $^{15}N\{^1H\}$ CPMAS experiments in combination with EXAFS and density functional theory (DFT) calculations.[58] More recently, ^{195}Pt NMR of coordinated Pt^{2+} species supported on the UiO-66-NH2 MOF has been reported in Ref. 60. The challenge here laid in (i) the low concentration of Pt(II) due to its dispersion in the MOF, and (ii) the extreme broadening of ^{195}Pt signal due to a large CSA. The sensitivity gain provided by DNP allowed them to record a static ^{195}Pt BRAIN-CP/WURST-CPMG (broadband adiabatic inversion – CP/wideband, uniform rate, and smooth truncation – Carr-Purcell-Meil-Gibson)[59] spectrum (Figure 18.5) within a relatively short experimental time (7.5 h). The authors reported

an enhancement of 30, achieved by mixing the MOF material with sodium chloride particles whereas only 6 was obtained without the sodium chloride particles.[60] This DNP amplification effect due to NaCl particles has been described and attributed previously to a better penetration of the microwaves in the sample in Ref. 61.

In 2013, Blanc *et al.*[62] demonstrated that high sensitivity ^{15}N and ^{13}C CPMAS spectra could be recorded at natural isotopic abundance on mesoporous organic polymers (MOPs) by DNP SENS, opening new perspectives for the high-throughput characterization of porous polymer libraries. Relatively low enhancements (< 20) were reported likely because the radical molecules cannot penetrate the small pore size network of MOPs. In this chapter, we note that the DNP enhancement factors reported so far for MOFs and MOPs are in general much lower than

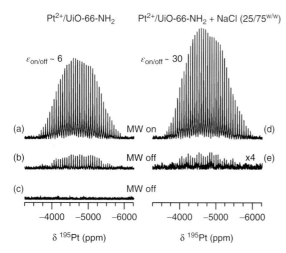

Figure 18.5. DNP-enhanced ^{195}Pt$\{^1$H$\}$ BRAIN-CP subspectra of Pt^{2+}/UiO-66-NH$_2$ (a,b) and the Pt^{2+}/UiO-66-NH$_2$/NaCl (25/75: w/w) mixture (d,e) taken with (a,d) and without (b,e) microwave irradiation. Subspectrum of Pt^{2+}/UiO-66-NH$_2$ (c) was acquired using direct excitation by WURST-CPMG without microwave irradiation. (Reprinted with permission from Kobayashi, T.; Perras, F.A.; Goh, T.W.; Metz, T.L.; Huang, W.; Pruski, M., DNP-enhanced Ultrawideline Solid-State NMR Spectroscopy: Studies of Platinum in Metal-Organic Framework. *J. Phys. Chem. Lett.* 2016, 7, 2322–2327. Copyright (2016) American Chemical Society)

those usually obtained with porous silica materials, and that much remains in terms of optimizing their formulation.

18.4.3 Zeolites

Zeolites were also successfully investigated by DNP SENS. Sn-ß zeolites were investigated by DNP-enhanced ^{119}Sn NMR spectroscopy at natural abundance.[63,64] In particular, Copéret and coworkers combined ^{119}Sn magic-angle turning (MAT) experiments recorded under DNP conditions with computational chemistry and Mössbauer spectroscopy to unambiguously demonstrate the presence of Sn(IV) active sites in an octahedral environment involving water molecules at the surface of the ß-zeolite framework.[63] More recently, Chmelka and coworkers implemented DNP-enhanced 2D *J*-based and dipolar-based ^{29}Si–^{29}Si correlation experiments to

gain new insights into the structural rearrangements occurring during the crystallization of zeolites.[65]

18.4.4 Porous Aluminas

Porous aluminas, used as catalysts or catalyst supports, are another important class of materials that benefited a lot from the recent advent of DNP SENS. In 2014, by combining both direct and indirect DNP approaches with ^{27}Al MQMAS experiments, it was demonstrated that pentacoordinated Al^{3+} ions were only located at the surface of hydrated γ-alumina.[26] Highly dispersed silanol sites on γ-alumina have been recently investigated by Stair and coworkers.[66,67] The authors demonstrated the existence of these isolated surface sites in materials obtained using two different synthetic approaches: (i) atomic layer deposition and (ii) chemical liquid deposition of tetraethylorthosilicate (TEOS) under anhydrous conditions. Here, DNP enabled the implementation of 2D ^{29}Si double quantum/single quantum correlation experiments, and in combination with other analytical techniques, yielded new insights into the formation mechanism of the isolated silanol sites in SiO$_x$/Al$_2$O$_3$. In 2017, DNP SENS was used to characterize substrate–surface interactions on alumina-supported metal catalysts and the important gain in sensitivity was here crucial to perform long-range ^{13}C$\{^{27}$Al$\}$ RESPDOR (rotational-echo saturation pulse double-resonance) measurements.[68]

18.5 DNP SENS ON LOW SURFACE AREA MATERIALS

The very first applications of DNP SENS have been reported on porous systems with high surface specific areas, but the approach has rapidly spread to more challenging materials having extremely low surface areas, and/or highly diluted surface sites, and for which the low sensitivity of NMR has for years been an insurmountable barrier. A few examples of these landmark studies are reviewed in the following section.

In Ref. 69, DNP SENS was used to get new insights into the adsorption of dilute quantities (~0.1% by weight of solids) of the disaccharide sucrose on low-surface-area (~1 m^2 g^{-1}) cementitious Ca$_3$SiO$_5$ particles and explained the relationships between

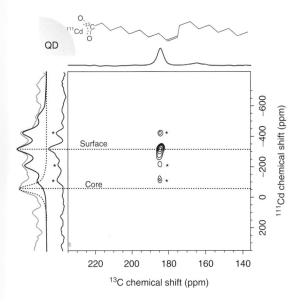

Figure 18.6. DNP-enhanced dipolar-HMQC spectrum correlating ^{13}C-1-oleate ligands and ^{111}Cd of CdSe. The dark blue curves are the projections of the two axes. The light grey curve is a ^{111}Cd{^{1}H} CPMAS spectrum. The solid black line shows the isotropic signal ^{111}Cd for the surface and the dashed black line shows the ^{111}Cd signal of the core. First order rotary resonance recoupling (R3) was employed. (Reprinted with permission from Piveteau, L.; Ong, T.-C.; Rossini, A. J.; Emsley, L.; Copéret, C.; Kovalenko, M. V., Structure of Colloidal Quantum Dots from Dynamic Nuclear Polarization Surface Enhanced NMR Spectroscopy. *J. Am. Chem. Soc.* 2015, **137**, 13964–13971. Copyright (2015) American Chemical Society)

the adsorption modes – hydrogen-bond-mediated or electrostatic adsorption – and the hydration properties of the silicate particles. Three-dimensional structures of cementitious calcium silicate hydrate were later proposed by combining DNP-enhanced solid-state NMR with atomistic calculations, including structure relaxation by DFT and classical molecular dynamics simulations to test structure stability.[70] DNP SENS has also emerged as a method of choice to investigate the surface structure nanoparticles (NPs). In 2012, Oschkinat and coworkers reported the use of DNP to observe spherical silica nanoparticles of different sizes using indirect ^{13}C and ^{29}Si DNP along with direct ^{29}Si DNP. In particular, they found that the DNP enhancement significantly increases when decreasing the size of the particle for both direct and indirect DNP.[71] In Ref. 72, the network of

organosiloxanes on silica nanoparticles is unraveled from through-bond and through-space ^{29}Si–^{29}Si DNP SENS. Protesescu *et al.*[73] used DNP SENS, in combination with Mössbauer and X-ray absorption spectroscopies, to reveal the layered structure of colloidal Sn/SnO/SnO$_2$ nanoparticles. DNP SENS demonstrated that the surface is made of amorphous SnO$_2$ while a crystalline β-Sn core was observed by X-ray measurements. Mössbauer spectroscopy revealed the existence of interlayered amorphous SnO. In 2015, the analysis of oleate-capped 4 nm CdSe QDs has been reported using a specific formulation, i.e., the dispersion of the nanoparticles in mesoporous SiO$_2$ as described in Section 18.3.3.[44] The authors applied a DNP-enhanced ^{13}C–^{111}Cd dipolar HMQC experiment to directly observe interactions between the oleate ligand and the QDs surface as reported in Figure 18.6. We note that in this study ^{13}C-isotopic enrichment has been necessary to record 2D spectra with sufficient sensitivity.

In Ref. 74, isotopic labeling was also combined with DNP SENS to obtain unprecedented insights into the structure of trigonal bipyramidal geometry (TBP) and square pyramidal geometry (SP) metallacyclobutane intermediates in supported metathesis catalysts. The molecular structures of the TBP and SP isomers, formed after reaction of ethylene with tungsten alkylidene sites, are shown schematically in Figure 18.7(a). Quantitative measurements of C_α–C_β distances were performed using dipolar-recoupling experiments that were found to be in close agreement with the distances in the DFT optimized structures. DNP-enhanced 2D refocused INADEQUATE experiments also revealed the presence of metallacyclopentane complexes as a minor surface species, by a secondary reaction resulting from the presence of a large excess of ethylene (Figure 18.7b). This observation provided new clues to understand the deactivation processes in these heterogeneous metathesis catalysts. More recently, DNP-enhanced ^{13}C–^{13}C 2D spectroscopy elucidated the degradation of methionine on Pd nanoparticles supported on γ-Al$_2$O$_3$.[75]

DNP SENS has been also applied to study the composition of the solid-electrolyte interphase (SEI), which is the thin layer that usually forms on the surface of anodes due to the reduction of the electrolyte, on reduced graphene oxide anode of Li-ion batteries and to rationalize the effect of additives in the stability of the SEI of Li-ion batteries silicon nanowires anodes.[76,77]

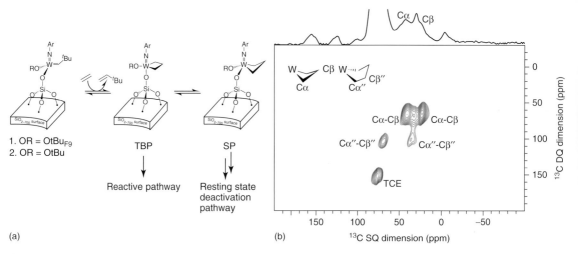

Figure 18.7. (a) Structure of tungsten metathesis catalyst immobilized on $SiO_{2-(700)}$ particles and its metathesis reaction intermediates TBP and SP metallacyclobutane. (b) Refocused $^{13}C-^{13}C$ INADEQUATE obtained by reacting **2** with large excess of ^{13}C-ethylene yielding **SP-2** and hypothesized metallacyclopentane. (Reproduced with permission from Ref. 74. © John Wiley and Sons, 2016)

18.6 DNP SENS FOR THE OBSERVATION OF INSENSITIVE SURFACE NUCLEI

Many of the nuclei of interest in crystalline or amorphous inorganic or hybrid materials at the heart of modern chemistry have low gyromagnetic ratio nuclear spin and/or are present at very low natural abundance. These are highly relevant targets for DNP, as conventional NMR spectroscopy on these nuclei is extremely insensitive. This is, for example the case of oxygen, which is present in many materials, but whose NMR active isotope ^{17}O is a quadrupolar nucleus with a natural abundance of only 0.04%. The ability to perform DNP-enhanced $^{17}O\{^1H\}$ CP-MAS NMR spectroscopy was first demonstrated on $H_2{}^{17}O$[78] and later on $Mg(OH)_2$ particles using both the direct and indirect polarization transfer methods.[79] The implementation of more sophisticated transfer schemes allowed to measure one-bond O–H distances at natural ^{17}O isotopic abundance on $Mg(OH)_2$, $Ca(OH)_2$ particles.[80] This was permitted by the implementation of a series of PRESTO-QCPMG (phase-shifted recoupling effects a smooth transfer of order – quadrupolar Carr-Purcell-Meil-Gibson) experiments with variable recoupling periods. Using a similar approach, the authors could then probe the detailed structure of hydrogen-bonded silanol groups at the surface of mesoporous silica nanoparticles and aluminosilicates.[80,81] In Ref. 29,

the first surface-sensitive DNP-enhanced ^{17}O NMR experiments on nanoparticles, namely CeO_2 nanoparticles isotopically enriched in ^{17}O, were performed. While indirect $^1H-^{17}O$ DNP revealed signals corresponding to Ce–OH terminations and H_2O molecules adsorbed at the surface, and as such was not extremely informative, direct ^{17}O DNP led to the observation of the first three layers of the CeO_2 nanoparticles. Different polarization buildup times were measured for the three layers, correlated with their surface proximity. These experiments were performed at a magnetic field of 14.1 T. In 2017, application of DNP-enhanced ^{17}O NMR at 18.8 T on $Mg(OH)_2$ microcrystalline particles has been reported, comparing different formulations (namely TEKPOL or BDPA in TCE or OTP) and different polarization transfer schemes (cross-effect and Overhauser effect DNP) suitable for high magnetic field.[82] Surface ^{17}O enhancements as high as 17 were reported. Despite OE DNP with BDPA in OTP at 18.8 T on $Mg(OH)_2$ microcrystalline particles gave higher $\varepsilon(^{17}O$ CP), CE DNP with TEKPOL in OTP or TCE was demonstrated to be more efficient in terms of overall sensitivity gain as it allows faster recycling. Other applications of DNP SENS to insensitive nuclei concern ^{89}Y NMR on hydrated yttrium-doped barium zirconates,[83] and ^{43}Ca NMR on nanocrystalline apatites.[84]

18.7 HIGHLY REACTIVE AND SENSITIVE MATERIALS

The many examples spanned in the previous sections demonstrate the outstanding potential of MAS DNP-enhanced solid-state NMR to probe atomic-scale surface structures in a vast range of materials. In principle, DNP SENS is, however, limited to surfaces compatible with the presence of free radicals. Notably, the approach fails when the polarizing agent is reduced by reactive surface sites and/or when it modifies their structure. In Ref. 51, the authors reported that no signal enhancement factor was observed for silica-supported ruthenium complexes, likely because of the interaction/reaction between TOTAPOL and the Ru centers.

Two different strategies have been recently proposed to prevent the degradation of both the polarizing agent and the surface sites upon impregnation with the polarizing solution. The first strategy consists in investigating reactive complexes confined in porous materials of small pore diameter (2.5–3.0 nm) to avoid the penetration of bulky radicals like TEKPOL inside the material.[85] In this chapter, surface enhancements as high as 30 on tungsten-organometallic complexes immobilized on MCM-41 has been reported. The second more versatile approach relies on the design of polarizing agents in which the presence of free radical center is encapsulated and thus protected inside an organic shell. Dendritic polarizing agents (Figure 18.8), in which the biradical is protected by bulky organosilicon moieties has also been proposed.[86] With this strategy, the authors could record an enhanced $^{13}C\{^1H\}$ CPMAS spectrum of reactive tungsten organometallic complexes supported onto $SiO_{2-(700)}$ particles with a DNP amplification factor of about 10, while no enhancement was obtained when the material was impregnated with a solution of TEKPOL in TCE. These two landmark studies open undoubtedly new horizons, yet the characterization of sensitive materials by DNP SENS still remains one of the main challenges of hyperpolarized NMR for surfaces.

18.8 CONCLUSIONS

In less than a decade, DNP-enhanced solid-state NMR spectroscopy has demonstrated its capacity to provide unique structural information in materials science. While a large panel of analytical techniques are available to probe surfaces (IR spectroscopy, EXAFS, tunneling scanning microscopy, etc.), DNP SENS opens up new complementary routes, inconceivable just a few years ago, to characterize the atomic structure of surfaces with unprecedented resolution. Key achievements have been reported, such as, to

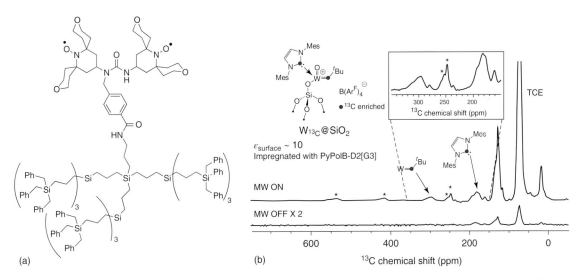

Figure 18.8. Structure of PyPolB-D2[G3]. DNP $^{13}C\{^1H\}$ CPMAS spectra of a tungsten complex anchored on silica impregnated with 16 mM PyPolB-D2[G3] in TCE (inset shows the enlarged region between 250 and 350 ppm). (Source: Liao[86], http://pubs.rsc.org/en/content/articlehtml/2016/sc/c6sc03139k. Licensed under CC BY 3.0)

name but a few, the full three-dimensional structure determination of organometallic surface sites, the detailed investigation of the adsorption of chemicals on low-surface-area materials, and the surface characterization of QDs. This new ability to touch surfaces by DNP SENS will continue to revolutionize our molecular understanding of materials properties in substrates as diverse as faceted surfaces, metal nanoparticle-based systems, industrial heterogeneous catalysts, and gas or energy storage materials. New instrumental and methodological developments, notably at very high magnetic field (18.8 T and above) and very fast MAS, will continue to push forward the current frontiers so as to tackle increasingly complex surfaces.

REFERENCES

1. T. Islamoglu, S. Goswami, Z. Li, A. J. Howarth, O. K. Farha, and J. T. Hupp, *Acc. Chem. Res.*, 2017, **50**, 805.

2. J. D. Pelletier and J. M. Basset, *Acc. Chem. Res.*, 2016, **49**, 664.

3. L. M. Martinez-Prieto and B. Chaudret, *Acc. Chem. Res.*, 2018, **51**, 376.

4. Y. Braeken, S. Cheruku, A. Ethirajan, and W. Maes, *Materials (Basel)*, 2017, **10**, E1420.

5. P. Avenier, M. Taoufik, A. Lesage, X. Solans-Monfort, A. Baudouin, A. de Mallmann, L. Veyre, J.-M. Basset, O. Eisenstein, L. Emsley, and E. A. Quadrelli, *Science*, 2007, **317**, 1056.

6. K. Mao, T. Kobayashi, J. W. Wiench, H. T. Chen, C. H. Tsai, V. S. Lin, and M. Pruski, *J. Am. Chem. Soc.*, 2010, **132**, 12452.

7. H. Lock, R. A. Wind, G. E. Maciel, and N. Zumbulyadis, *Solid State Commun.*, 1987, **64**, 41.

8. H. Lock, G. E. Maciel, and C. E. Johnson, *J. Mater. Res.*, 1992, **7**, 2791.

9. H. Lock, R. A. Wind, G. E. Maciel, and C. E. Johnson, *J. Chem. Phys.*, 1993, **99**, 3363.

10. M. Afeworki, R. A. McKay, and J. Schaefer, *Macromolecules*, 1992, **25**, 4084.

11. Q. Z. Ni, E. Daviso, T. V. Can, E. Markhasin, S. K. Jawla, T. M. Swager, R. J. Temkin, J. Herzfeld, and R. G. Griffin, *Acc. Chem. Res.*, 2013, **46**, 1933.

12. A. Lesage, M. Lelli, D. Gajan, M. A. Caporini, V. Vitzthum, P. Mieville, J. Alauzun, A. Roussey, C. Thieuleux, A. Mehdi, G. Bodenhausen, C. Coperet, and L. Emsley, *J. Am. Chem. Soc.*, 2010, **132**, 15459.

13. A. J. Rossini, A. Zagdoun, M. Lelli, A. Lesage, C. Copéret, and L. Emsley, *Acc. Chem. Res.*, 2013, **46**, 1942.

14. T. Kobayashi, F. A. Perras, I. I. Slowing, A. D. Sadow, and M. Pruski, *ACS Catal.*, 2015, **5**, 7055.

15. A. S. L. Thankamony, J. J. Wittmann, M. Kaushik, and B. Corzilius, *Prog. Nucl. Magn. Reson. Spectrosc.*, 2017, **102**, 120.

16. P. C. A. van der Wel, K.-N. Hu, J. Lewandowski, and R. G. Griffin, *J. Am. Chem. Soc.*, 2006, **128**, 10840.

17. K. N. Hu, G. T. Debelouchina, A. A. Smith, and R. G. Griffin, *J. Chem. Phys.*, 2011, **134**, 125105.

18. A. J. Rossini, A. Zagdoun, F. Hegner, M. Schwarzwalder, D. Gajan, C. Coperet, A. Lesage, and L. Emsley, *J. Am. Chem. Soc.*, 2012, **134**, 16899.

19. A. C. Pinon, J. Schlagnitweit, P. Berruyer, A. J. Rossini, M. Lelli, E. Socie, M. Tang, T. Pham, A. Lesage, S. Schantz, and L. Emsley, *J. Phys. Chem. C*, 2017, **121**, 15993.

20. M. Lelli, D. Gajan, A. Lesage, M. A. Caporini, V. Vitzthum, P. Mieville, F. Heroguel, F. Rascon, A. Roussey, C. Thieuleux, M. Boualleg, L. Veyre, G. Bodenhausen, C. Coperet, and L. Emsley, *J. Am. Chem. Soc.*, 2011, **133**, 2104.

21. O. Lafon, A. S. L. Thankamony, T. Kobayashi, D. Carnevale, V. Vitzthum, I. I. Slowing, K. Kandel, H. Vezin, J.-P. Amoureux, G. Bodenhausen, and M. Pruski, *J. Phys. Chem. C*, 2013, **117**, 1375.

22. P. Berruyer, M. Lelli, M. P. Conley, D. L. Silverio, C. M. Widdifield, G. Siddiqi, D. Gajan, A. Lesage, C. Coperet, and L. Emsley, *J. Am. Chem. Soc.*, 2017, **139**, 849.

23. O. Lafon, M. Rosay, F. Aussenac, X. Lu, J. Trebosc, O. Cristini, C. Kinowski, N. Touati, H. Vezin, and J. P. Amoureux, *Angew. Chem. Int. Ed.*, 2011, **50**, 8367.

24. O. Lafon, A. S. L. Thankamony, M. Rosay, F. Aussenac, X. Y. Lu, J. Trebosc, V. Bout-Roumazeilles, H. Vezine, and J. P. Amoureux, *Chem. Commun.*, 2013, **49**, 2864.

25. V. K. Michaelis, B. Corzilius, A. A. Smith, and R. G. Griffin, *J. Phys. Chem. B*, 2013, **117**, 14894.

26. D. Lee, N. T. Duong, O. Lafon, and G. De Paepe, *J. Phys. Chem. C*, 2014, **118**, 25065.

27. A. Lund, M.-F. Hsieh, T.-A. Siaw, and S.-I. Han, *Phys. Chem. Chem. Phys.*, 2015, **17**, 25449.

28. A. S. L. Thankamony, S. Knoche, S. Bothe, A. Drochner, A. P. Jagtap, S. T. Sigurdsson, H. Vogel, B. J. M. Etzold, T. Gutmann, and G. Buntkowsky, *J. Phys. Chem. C*, 2017, **121**, 20857.

29. M. A. Hope, D. M. Halat, P. C. Magusin, S. Paul, L. Peng, and C. P. Grey, *Chem. Commun. (Camb.)*, 2017, **53**, 2142.

30. G. J. Gerfen, L. R. Becerra, D. A. Hall, R. G. Griffin, R. J. Temkin, and D. J. Singel, *J. Chem. Phys.*, 1995, **102**, 9494.

31. D. Gajan, A. Bornet, B. Vuichoud, J. Milani, R. Melzi, H. A. van Kalkeren, L. Veyre, C. Thieuleux, M. P. Conley, W. R. Gruning, M. Schwarzwalder, A. Lesage, C. Coperet, G. Bodenhausen, L. Emsley, and S. Jannin, *Proc. Natl. Acad. Sci. U. S. A.*, 2014, **111**, 14693.

32. K.-N. Hu, H.-H. Yu , T. M. Swager, and R. G. Griffin, *J. Am. Chem. Soc.*, 2004, **126**, 10844.

33. D. A. Hall, D. C. Maus, G. J. Gerfen, S. J. Inati, L. R. Becerra, F. W. Dahlquist, and R. G. Griffin, *Science*, 1997, **276**, 930.

34. C. Sauvee, M. Rosay, G. Casano, F. Aussenac, R. T. Weber, O. Ouari, and P. Tordo, *Angew. Chem. Int. Ed.*, 2013, **52**, 10858.

35. A. Zagdoun, A. J. Rossini, D. Gajan, A. Bourdolle, O. Ouari, M. Rosay, W. E. Maas, P. Tordo, M. Lelli, L. Emsley, A. Lesage, and C. Coperet, *Chem. Commun. (Camb.)*, 2012, **48**, 654.

36. T.-C. Ong, M. L. Mak-Jurkauskas, J. J. Walish, V. K. Michaelis, B. Corzilius, A. A. Smith, A. M. Clausen, J. C. Cheetham, T. M. Swager, and R. G. Griffin, *J. Phys. Chem. B*, 2013, **117**, 3040.

37. M. P. Conley, C. Copéret, and C. Thieuleux, *ACS Catal.*, 2014, **4**, 1458.

38. A. Zagdoun, A. J. Rossini, M. P. Conley, W. R. Gruning, M. Schwarzwalder, M. Lelli, W. T. Franks, H. Oschkinat, C. Coperet, L. Emsley, and A. Lesage, *Angew. Chem. Int. Ed.*, 2013, **52**, 1222.

39. A. J. Rossini, A. Zagdoun, M. Lelli, D. Gajan, F. Rascón, M. Rosay, W. E. Maas, C. Copéret, A. Lesage, and L. Emsley, *Chem. Sci.*, 2012, **3**, 108.

40. K. R. Thurber and R. Tycko, *J. Chem. Phys.*, 2014, **140**, 184201.

41. B. Corzilius, L. B. Andreas, A. A. Smith, Q. Z. Ni, and R. G. Griffin, *J. Magn. Reson.*, 2014, **240**, 113.

42. F. Mentink-Vigier, S. Paul, D. Lee, A. Feintuch, S. Hediger, S. Vega, and G. De Paepe, *Phys. Chem. Chem. Phys.*, 2015, **17**, 21824.

43. F. A. Perras, L.-L. Wang, J. S. Manzano, U. Chaudhary, N. N. Opembe, D. D. Johnson, I. I. Slowing, and M. Pruski, *Curr. Opin. Colloid Interface Sci.*, 2018, **33**, 9.

44. L. Piveteau, T.-C. Ong, A. J. Rossini, L. Emsley, C. Copéret, and M. V. Kovalenko, *J. Am. Chem. Soc.*, 2015, **137**, 13964.

45. J. Viger-Gravel, P. Berruyer, D. Gajan, J. M. Basset, A. Lesage, P. Tordo, O. Ouari, and L. Emsley, *Angew. Chem. Int. Ed.*, 2017, **56**, 8726.

46. J. R. Yarava, S. R. Chaudhari, A. J. Rossini, A. Lesage, and L. Emsley, *J. Magn. Reson.*, 2017, **277**, 149.

47. W. R. Grüning, A. J. Rossini, A. Zagdoun, D. Gajan, A. Lesage, L. Emsley, and C. Copéret, *Phys. Chem. Chem. Phys.*, 2013, **15**, 13270.

48. D. Lee, S. R. Chaudhari, and G. De Paëpe, *J. Magn. Reson.*, 2017, **278**, 60.

49. T. Gutmann, J. Liu, N. Rothermel, Y. Xu, E. Jaumann, M. Werner, H. Breitzke, S. T. Sigurdsson, and G. Buntkowsky, *Chem. Eur. J.*, 2015, **21**, 3798.

50. N. Eedugurala, Z. R. Wang, U. Chaudhary, N. Nelson, K. Kandel, T. Kobayashi, I. I. Slowing, M. Pruski, and A. D. Sadow, *ACS Catal.*, 2015, **5**, 7399.

51. M. K. Samantaray, J. Alauzun, D. Gajan, S. Kavitake, A. Mehdi, L. Veyre, M. Lelli, A. Lesage, L. Emsley, C. Coperet, and C. Thieuleux, *J. Am. Chem. Soc.*, 2013, **135**, 3193.

52. M. P. Conley, R. M. Drost, M. Baffert, D. Gajan, C. Elsevier, W. T. Franks, H. Oschkinat, L. Veyre, A. Zagdoun, A. Rossini, M. Lelli, A. Lesage, G. Casano, O. Ouari, P. Tordo, L. Emsley, C. Copéret, and C. Thieuleux, *Chem. Eur. J.*, 2013, **19**, 12234.

53. I. Romanenko, D. Gajan, R. Sayah, D. Crozet, E. Jeanneau, C. Lucas, L. Leroux, L. Veyre, A. Lesage, L. Emsley, E. Lacote, and C. Thieuleux, *Angew. Chem. Int. Ed.*, 2015, **54**, 12937.

54. T. Kobayashi, D. Singappuli-Arachchige, Z. Wang, I. I. Slowing, and M. Pruski, *Phys. Chem. Chem. Phys.*, 2017, **19**, 1781.

55. D. M. D'Alessandro, B. Smit, and J. R. Long, *Angew. Chem. Int. Ed.*, 2010, **49**, 6058.

56. A. J. Rossini, A. Zagdoun, M. Lelli, J. Canivet, S. Aguado, O. Ouari, P. Tordo, M. Rosay, W. E. Maas, C. Coperet, D. Farrusseng, L. Emsley, and A. Lesage, *Angew. Chem. Int. Ed.*, 2012, **51**, 123.

57. F. Pourpoint, A. S. Thankamony, C. Volkringer, T. Loiseau, J. Trebosc, F. Aussenac, D. Carnevale,

G. Bodenhausen, H. Vezin, O. Lafon, and J. P. Amoureux, *Chem. Commun. (Camb.)*, 2014, **50**, 933.

58. Z. Guo, T. Kobayashi, L.-L. Wang, T. W. Goh, C. Xiao, M. A. Caporini, M. Rosay, D. D. Johnson, M. Pruski, and W. Huang, *Chem. Eur. J.*, 2014, **20**, 16308.

59. K. J. Harris, A. Lupulescu, B. E. G. Lucier, L. Frydman, and R. W. Schurko, *J. Magn. Reson.*, 2012, **224**, 38.

60. T. Kobayashi, F. A. Perras, T. W. Goh, T. L. Metz, W. Huang, and M. Pruski, *J. Phys. Chem. Lett.*, 2016, **7**, 2322.

61. D. J. Kubicki, A. J. Rossini, A. Purea, A. Zagdoun, O. Ouari, P. Tordo, F. Engelke, A. Lesage, and L. Emsley, *J. Am. Chem. Soc.*, 2014, **136**, 15711.

62. F. Blanc, S. Y. Chong, T. O. McDonald, D. J. Adams, S. Pawsey, M. A. Caporini, and A. I. Cooper, *J. Am. Chem. Soc.*, 2013, **135**, 15290.

63. P. Wolf, M. Valla, A. J. Rossini, A. Comas-Vives, F. Núñez-Zarur, B. Malaman, A. Lesage, L. Emsley, C. Copéret, and I. Hermans, *Angew. Chem. Int. Ed.*, 2014, **53**, 10179.

64. W. R. Gunther, V. K. Michaelis, M. A. Caporini, R. G. Griffin, and Y. Roman-Leshkov, *J. Am. Chem. Soc.*, 2014, **136**, 6219.

65. Z. J. Berkson, R. J. Messinger, K. Na, Y. Seo, R. Ryoo, and B. F. Chmelka, *Angew. Chem. Int. Ed.*, 2017, **56**, 5164.

66. A. R. Mouat, C. George, T. Kobayashi, M. Pruski, R. P. van Duyne, T. J. Marks, and P. C. Stair, *Angew. Chem. Int. Ed.*, 2015, **54**, 13346.

67. A. R. Mouat, T. Kobayashi, M. Pruski, T. J. Marks, and P. C. Stair, *J. Phys. Chem. C*, 2017, **121**, 6060.

68. F. A. Perras, J. D. Padmos, R. L. Johnson, L. L. Wang, T. J. Schwartz, T. Kobayashi, J. H. Horton, J. A. Dumesic, B. H. Shanks, D. D. Johnson, and M. Pruski, *J. Am. Chem. Soc.*, 2017, **139**, 2702.

69. R. P. Sangodkar, B. J. Smith, D. Gajan, A. J. Rossini, L. R. Roberts, G. P. Funkhouser, A. Lesage, L. Emsley, and B. F. Chmelka, *J. Am. Chem. Soc.*, 2015, **137**, 8096.

70. A. Kumar, B. J. Walder, A. K. Mohamed, A. Hofstetter, B. Srinivasan, A. J. Rossini, K. Scrivener, L. Emsley, and P. Bowen, *J. Phys. Chem. C*, 2017, **121**, 17188.

71. Ü. Akbey, B. Altin, A. Linden, S. Özçelik, M. Gradzielski, and H. Oschkinat, *Phys. Chem. Chem. Phys.*, 2013, **15**, 20706.

72. D. Lee, G. Monin, N. T. Duong, I. Z. Lopez, M. Bardet, V. Mareau, L. Gonon, and G. De Paepe, *J. Am. Chem. Soc.*, 2014, **136**, 13781.

73. L. Protesescu, A. J. Rossini, D. Kriegner, M. Valla, A. de Kergommeaux, M. Walter, K. V. Kravchyk, M. Nachtegaal, J. Stangl, B. Malaman, P. Reiss, A. Lesage, L. Emsley, C. Coperet, and M. V. Kovalenko, *ACS Nano*, 2014, **8**, 2639.

74. T. C. Ong, W. C. Liao, V. Mougel, D. Gajan, A. Lesage, L. Emsley, and C. Coperet, *Angew. Chem. Int. Ed.*, 2016, **55**, 4743.

75. R. L. Johnson, F. A. Perras, T. Kobayashi, T. J. Schwartz, J. A. Dumesic, B. H. Shanks, and M. Pruski, *Chem. Commun. (Camb.)*, 2016, **52**, 1859.

76. M. Leskes, G. Kim, T. Liu, A. L. Michan, F. Aussenac, P. Dorffer, S. Paul, and C. P. Grey, *J. Phys. Chem. Lett.*, 2017, **8**, 1078.

77. Y. Jin, N. H. Kneusels, P. Magusin, G. Kim, E. Castillo-Martinez, L. E. Marbella, R. N. Kerber, D. J. Howe, S. Paul, T. Liu, and C. P. Grey, *J. Am. Chem. Soc.*, 2017, **139**, 14992.

78. V. K. Michaelis, E. Markhasin, E. Daviso, J. Herzfeld, and R. G. Griffin, *J. Phys. Chem. Lett.*, 2012, **3**, 2030.

79. F. Blanc, L. Sperrin, D. A. Jefferson, S. Pawsey, M. Rosay, and C. P. Grey, *J. Am. Chem. Soc.*, 2013, **135**, 2975.

80. F. A. Perras, T. Kobayashi, and M. Pruski, *J. Am. Chem. Soc.*, 2015, **137**, 8336.

81. F. A. Perras, Z. Wang, and P. Naik, *Angew. Chem. Int. Ed.*, 2017, **56**, 9165.

82. N. J. Brownbill, D. Gajan, A. Lesage, L. Emsley, and F. Blanc, *Chem. Commun. (Camb.)*, 2017, **53**, 2563.

83. F. Blanc, L. Sperrin, D. Lee, R. Dervisoglu, Y. Yamazaki, S. M. Haile, G. De Paepe, and C. P. Grey, *J. Phys. Chem. Lett.*, 2014, **5**, 2431.

84. D. Lee, C. Leroy, C. Crevant, L. Bonhomme-Coury, F. Babonneau, D. Laurencin, C. Bonhomme, and G. De Paepe, *Nat. Commun.*, 2017, **8**, 14104.

85. E. Pump, J. Viger-Gravel, E. Abou-Hamad, M. K. Samantaray, B. Hamzaoui, A. Gurinov, D. H. Anjum, D. Gajan, A. Lesage, A. Bendjeriou-Sedjerari, L. Emsley, and J. M. Basset, *Chem. Sci.*, 2017, **8**, 284.

86. W.-C. Liao, T.-C. Ong, D. Gajan, F. Bernada, C. Sauvée, M. Yulikov, M. Pucino, R. Schowner, M. Schwarzwälder, M. R. Buchmeiser, G. Jeschke, P. Tordo, O. Ouari, A. Lesage, L. Emsley, and C. Copéret, *Chem. Sci.*, 2017, **8**, 416.

Chapter 19

Growing Signals from the Noise: Challenging Nuclei in Materials DNP

Frédéric A. Perras[1], Takeshi Kobayashi[1], and Marek Pruski[1,2]

[1] *U.S. DOE Ames Laboratory, Ames, IA, USA*
[2] *Department of Chemistry, Iowa State University, Ames, IA, USA*

19.1 INTRODUCTION

Dynamic nuclear polarization (DNP) has revolutionized various areas of solid-state (SS) NMR spectroscopy. Materials science has been arguably one of the key beneficiaries of the recent remarkable advancement of the technique. Indeed, in many classes of materials, DNP offers sensitivity enhancements of two orders of magnitude, and often even larger savings of experimental time. Importantly, these gains in sensitivity are rarely accompanied

Handbook of High Field Dynamic Nuclear Polarization.
Edited by Vladimir K. Michaelis, Robert G. Griffin, Björn Corzilius and Shimon Vega.
© 2020 John Wiley & Sons, Ltd. This is a US Government work and is in the Public Domain in the United States of America
ISBN: 978-1-119-44164-9
Also published in eMagRes (online edition)
DOI: 10.1002/9780470034590.emrstm1556

by lower resolution since linewidths are typically controlled by the inhomogeneous effects due to material's heterogeneity and the effects of paramagnetic broadening are usually very minor in comparison. Finally, the recently developed dynamic nuclear polarization surface-enhanced NMR spectroscopy (DNP SENS) has endowed researchers with the capability of selectively sensitizing progressively smaller surface and interfacial regions of materials and eliciting responses from previously undetectable nuclei. As a result, DNP has already helped advance structural studies of catalysts, mesoporous solids, nanoparticles, glasses, polymers, and pharmaceuticals, and yet further progress is underway with new developments in ultralow-temperature, and fast, magic-angle spinning (MAS) technologies, microwave technology, the development of new polarizing agents, and theory.

In this chapter, we will review the most recent applications of DNP-enhanced SSNMR to materials, focusing specifically on two groups of measurements that pose insurmountable challenges to conventional SSNMR, at least without isotope enrichment: (i) the detection of important yet insensitive nuclei, such as ^{15}N, ^{17}O, ^{25}Mg, ^{35}Cl, ^{43}Ca, ^{79}Br, ^{89}Y, ^{119}Sn, and ^{195}Pt, which is mostly, but not exclusively, performed using one-dimensional (1D) schemes and (ii) the acquisition of two-dimensional (2D) correlation spectra involving dilute nuclei, such as ^{13}C and ^{29}Si. Our review includes primarily the applications reported following the resurgence of DNP in the past 20 years, subsequent to the development of low-temperature MAS,[1] modern gyrotrons,[2] and biradical agents for polarization transfer via the cross-effect.[3-6] Earlier applications of the technique, which relied primarily on the solid effect, were comprehensively reviewed by Wind.[7] Similarly, most of the experimental parameters have been omitted for the sake of brevity. We do, however, include information about the observed enhancements, as they define the technique's key impact on SSNMR spectroscopy. In most cases, the enhancements have been reported as the ratio of signal intensities observed with and without microwave irradiation ($\varepsilon_{on/off}$). These are not equivalent to the absolute sensitivity gains, which depend on a number of other parameters, including longitudinal relaxation, polarization buildup time, and quenching effects[8,9] and are rarely reported with high precision.

19.2 DETECTION OF CHALLENGING NUCLEI

19.2.1 ^{17}O

Oxygen is ubiquitous in nature and plays an important role in materials science, yet applied SSNMR studies of ^{17}O, the only NMR active isotope of oxygen, are very rare. This may appear surprising given the fact that ^{17}O has a very wide chemical shift range of approximately 1200 ppm, and, as a result, ^{17}O SSNMR spectra are generally very informative. Although ^{17}O is a quadrupolar nucleus ($I = 5/2$), it has one of the smallest quadrupole moments of any naturally occurring nuclide at -25.58 mb, and thus the linewidths are often quite manageable, allowing for the resolution of large numbers of resonances in solids, particularly at high magnetic fields.[10-13] This quadrupolar broadening is also valuable as it allows for the study of dynamics,[14] and additionally enables the determination of precise local and long-range structural information in a way that is not possible for spin-1/2 nuclei.[15,16] Notwithstanding these favorable features, the main hurdle that has limited the application of ^{17}O SSNMR is its extremely low natural abundance of 0.038%. Although natural abundance ^{17}O spectra can be, in principle, acquired by conventional SSNMR, using high magnetic fields, large sample quantities, and pulse sequences for enhancing the population difference across the central transition,[17] such experiments usually require prohibitively long data acquisition times. Isotopic enrichment, on the

other hand, has limited applications due to synthetic challenges and the lack of commercially available sources of ^{17}O, which are limited to costly ^{17}O$_2$ gas, H$_2$^{17}O, and a few small organics.

^{17}O is thus an exemplary nuclide for demonstrating the transformative force of DNP. The technique is theoretically capable of yielding sensitivity enhancements of up to $\gamma_e/\gamma_{^{17}O} \approx 4855$; if achieved, such enhancements would nearly double that obtainable by 100% ^{17}O enrichment. This opportunity was recognized early on, and the first ^{17}O SSNMR experiments performed on a noncubic solid, by Niebuhr *et al.*, were performed using DNP enhancement.[18] This seminal experiment was performed on a single crystal of ruby, using endogenous Cr^{3+} as the polarization source, at a magnetic field strength of 1 T and temperature of 1.9 K. No enhancement factors were quoted, but an ^{27}Al DNP enhancement factor of 360 was reported on the same material. Further advancements in ^{17}O DNP did not, however, take place until the development of high-field DNP instrumentation.

Michaelis *et al.*[19] were the first to apply modern high-field DNP to the ^{17}O nuclide. Using an indirect DNP approach, where electrons' polarization is transferred via a high-γ nucleus such as proton, they obtained a DNP enhancement, $\varepsilon_{on/off}$, of 80 on a ^{17}O-enriched water sample at a magnetic field strength of 5 T and a temperature of 85 K. Although this was not extensively discussed in their chapter, they also managed to detect the ^{17}O SSNMR signal in a natural abundance sample of 'DNP juice' (60/30/10% v/v d$_8$-glycerol/D$_2$O/H$_2$O with 20 mM TOTAPOL[20]) using the quadrupolar Carr-Purcell Meiboom-Gill (QCPMG) pulse sequence.[21] The earliest application of natural abundance ^{17}O DNP to materials was demonstrated by Blanc *et al.*[22] who again used indirect DNP, in combination with QCPMG, to acquire ^{17}O MAS spectra of Ca(OH)$_2$ and Mg(OH)$_2$ samples. DNP enabled them to perform these difficult experiments in a mere 10 min, thus facilitating the acquisition of a 2D ^{17}O{^{1}H} heteronuclear correlation (HETCOR) spectrum. In the same study, the application of direct ^{17}O DNP to ^{17}O-enriched and natural abundance MgO nanoparticles was reported. A sizeable DNP enhancement $\varepsilon_{on/off} = 43$ was achieved using a 20 mM solution of bTbK[4] in 1,1,2,2-tetrachloroethane, which could be further quadrupled with the use of a double-frequency sweep (DFS) pulse[23,24] to transfer hyperpolarization from the satellite transitions to the observed central transition of ^{17}O. Notably, this study enabled the enhancement and detection of an oxygen site at the MgO surface.

Perras *et al.*[25] later built on the work of Blanc and performed the first ^{17}O DNP SENS experiments on a mesoporous silica nanoparticle (MSN) sample. The success of this study rested on the use of PRESTO (phase-shifted recoupling effects a smooth transfer of polarization),[26] as opposed to CP, for the transfer of ^1H hyperpolarization to ^{17}O. Given that the application of a spin locking pulse can lead to large losses in magnetization for quadrupolar nuclei, the PRESTO sequence, which is considerably simpler

to apply at the ^{17}O frequency as it necessitates only the application of a spin echo, was shown to lead to ~5 times greater sensitivity.[25] Notwithstanding the detection of dilute silanol species on an MSN sample, PRESTO further enabled the acquisition of $^1H-^{17}O$ dipolar coupling data showing the presence of both hydrogen-bonded and mobile silanols. In a subsequent study, the same authors resolved these two silanol species and monitored the dehydroxylation of the hydrogen-bonded site upon thermal treatment (Figure 19.1a) by means of $^{17}O\{^1H\}$ HETCOR experiments.[27] Lastly, they used indirect DNP to

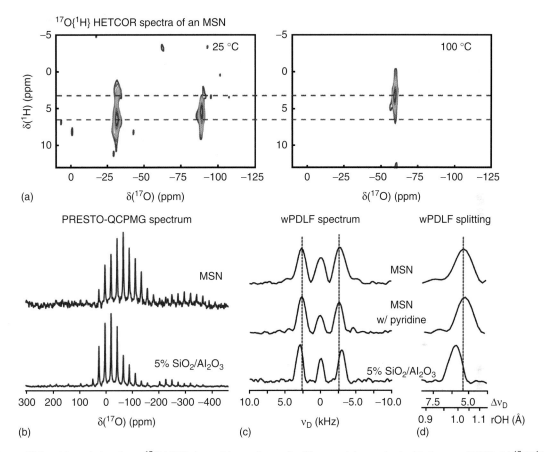

Figure 19.1.   Natural abundance ^{17}O NMR data of the surfaces of oxide materials acquired with the use of DNP. (a) $^{17}O\{^1H\}$ PRESTO-QCPMG HETCOR spectra of MSNs thermally treated at 25 and 100 °C showing the resolution of two types of silanols along the ^1H dimension, the hydrogen-bonded silanols being removed by thermal treatment . (b) 1D PRESTO-QCPMG spectra of hydroxyl moieties in silica and alumina materials. (c,d) $^{17}O\{^1H\}$ wPDLF spectra of silica and alumina materials showing the changes in bond lengths that are associated with Brønsted acidity. The addition of pyridine to the MSN also lengthens the O–H bond and provides evidence of the formation of intermolecular hydrogen bonding interactions at the surface. ((a) Reprinted with permission from F. A. Perras, U. Chaudhary, I.I. Slowing, and M. Pruski, J. Phys. Chem. C, 2016, 120, 11535. Copyright 2016 American Chemical Society. (b–d) Reproduced with permission from Ref. 29. © John Wiley and Sons, 2017)

perform $^{17}O\{^1H\}$ windowed proton-detected local field (wPDLF) experiments[28] on a series of different oxide materials.[29] The wPDLF experiment, which is insensitive to chemical shift anisotropy (CSA) and radio frequency (RF) field maladjustments and inhomogeneity, enabled the measurement of O–H bond lengths with subpicometer precision and provided confirmation of the longstanding assumption that the Brønsted acidity of surface sites is correlated to the hydroxyl group's O–H bond length (Figure 19.1b–d). Similar experiments performed on samples impregnated with pyridine were also able to conclusively show the formation of intermolecular hydrogen bonds at the silica surface, while the more Lewis-basic alumina surfaces form no such interactions with pyridine.[30]

Although significant insights into material surfaces could be obtained using indirect ^{17}O DNP, this approach is inherently limited to oxygens in the vicinity of hydrogen; in practice, only hydroxyls can be reliably detected. In order to tackle structural problems in a wider range of materials, direct ^{17}O DNP spectroscopy needed to be developed. Following the above-mentioned work by Blanc *et al.*,[22] Michaelis *et al.*[31] discovered that higher ^{17}O DNP enhancements could be obtained when using a narrow-line trityl radical as a polarization source. These authors reported enhancement factors ($\varepsilon_{on/off}$) of 115 for water, ≥ 80 for urea, and $\gg 100$ for phenol at a magnetic field of 5 T and a temperature of 85 K. Hope *et al.*[32] subsequently applied ^{17}O direct DNP to probe the surface structure of ceria nanoparticles. Importantly, they were able to show that ^{17}O direct DNP experiments of materials are surface selective, showing far faster signal buildup rates for the surface and subsurface sites than the bulk sites (Figure 19.2). The buildup rates and enhancement factors for the three first layers, however, did not correlate with their distance from the surface because the buildup rates were limited by the T_1 relaxation of ^{17}O, as is common for most DNP experiments. The studies by Hope *et al.* were performed at a magnetic field of 14.1 T, which showed strong promise for the application of ^{17}O DNP at even higher magnetic fields where the spectral resolution would be considerably improved. Indeed, Blanc *et al.* recently carried out ^{17}O indirect DNP experiments at an ultrahigh magnetic field of 18.8 T,[33,34] thereby demonstrating the possibility of using Overhauser effect-based DNP transfer for the measurement of highly resolved ^{17}O spectra in natural abundance materials.

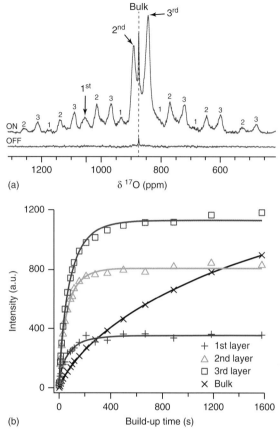

Figure 19.2. (a) ^{17}O direct DNP-enhanced MAS spectrum of a ^{17}O-enriched CeO_2 sample, with clearly identifiable resonances from the first three layers of the material. The spectrum acquired with microwave irradiation 'off' is shown at the bottom. (b) ^{17}O saturation recovery curves for the different oxygen environments of the material, which clearly demonstrate the faster buildup of the surface sites. (Hope *et al.*,[32] http://pubs.rsc.org/-/content/articlehtml/2017/cc/c6cc10145c. Licensed under CC BY 3.0)

19.2.2 ^{15}N

Nitrogen's prominence in many areas of materials science rivals that of oxygen, and so do the ways in which its NMR-active nuclei, ^{14}N and ^{15}N, pose spectroscopic challenges. Both these nuclei have low gyromagnetic ratios (γ), smaller than that of ^1H by factors of roughly 10 (^{15}N) and 14 (^{14}N), but they differ significantly in terms of natural abundance and their spin quantum number, I. The ^{14}N isotope is highly abundant (99.6%) but is an integer spin quadrupolar nuclide ($I = 1$). The resulting first-order quadrupolar broadening renders MAS ineffective and necessitates the use of complex wideline or indirect detection schemes, but even such methods can rarely provide practical site resolution in complex structures. For this reason, there have been concerted efforts to enable detection of the ^{15}N nuclide, for which $I = 1/2$. In contrast to ^{14}N, ^{15}N can thus yield highly resolved MAS spectra, but its very low natural abundance (0.4%) poses a sensitivity problem. To alleviate this problem, researchers have developed the indirectly detected heteronuclear correlation (idHETCOR) technique, which allows for the acquisition of $^1H\{^{15}N\}$ spectra of naturally abundant bulk materials, such as peptides and pharmaceuticals.[35] Nevertheless, natural abundance ^{15}N studies of surface and inter-facial regions have generally remained beyond the detection limits of SSNMR. Modern DNP-enhanced SSNMR has quickly emerged as a method of choice for such studies, enabling the acquisition of 1D and 2D natural abundance ^{15}N spectra of porous materials, such as mesoporous silica materials,[36–42] metal-organic frameworks (MOFs),[42–46] or organic porous polymers.[47] ^{15}N DNP has also been applied to studies of nonporous materials, for which providing access to the polarization source is an ongoing challenge;[48–51] in some cases in combination with isotopic enrichment.[52–55] In the subsequent paragraphs we will highlight a few of these studies.

The first application of modern ^{15}N DNP-enhanced SSNMR in materials science was reported in 2012 by Zagdoun *et al.*,[36] who tracked the stepwise postfunctionalization of a propyl azide moiety (PrN$_3$) attached to the surface of mesoporous silica, and the subsequent formation of imidazolium-containing materials (PrIm) using ^{15}N and ^{13}C NMR (see Figure 19.3). The ^{15}N NMR spectra unambiguously identified the chemical structures of the functional groups at each reaction step, which was not possible with the use of ^{13}C NMR spectroscopy alone. Gutmann *et al.*[38] later used a similar approach to study the interactions between surface amine functionalities and a dirhodium catalyst. They observed that the immobilization of Rh$_2$(CH$_3$COO)$_4$ led to a decrease in the intensity of the ^{15}N NMR signal assigned to amine functionalities and a concomitant appearance of new signals attributed to the interactions between the dirhodium catalyst and the amine moiety.

DNP has also been used to enhance the ^{15}N SS-NMR signals in natural abundance MOFs.[43,44] In particular, the technique was used to interrogate the host–guest interactions between Pt^{2+} and NH$_2$ groups in a UiO-66-NH$_2$ MOF (Figure 19.4).[44] In addition to the ^{15}N peaks at −315 and −242 ppm from the −NH$_2$ and −NH$_3^+$Cl moieties in UiO-66-NH$_2$, the Pt-containing sample showed a distinct peak at −388 ppm whose intensity increased linearly with the metal content. DNP-enhanced ^{15}N SSNMR thus provided credible evidence of chemical bonding between the Pt^{2+} and the −NH$_2$ moieties of the MOF. The use of a nonprotonated DNP solvent, DMSO-d$_6$, yielded a background-free 2D $^{15}N\{^1H\}$ HETCOR spectrum with well-resolved cross-peaks from −NH$_2$ and −NH$_2$···Pt^{2+} moieties in Pt/UiO-66-NH$_2$. This HETCOR experiment enabled the distinction of a ~1 ppm ^1H chemical shift increase upon the coordination of the amine group to Pt^{2+} (Figure 19.4b).[42]

As previously mentioned, DNP-enhanced ^{15}N SS-NMR spectroscopy has also been applied to nonporous materials. For instance, DNP-enhanced ^{15}N SSNMR of nitridated fibrous silica, which holds promise in applications for CO$_2$ capture and as a solid base catalyst, helped to determine that the catalytic deactivation of the material at higher nitridation temperatures was the result of a decrease in the number of Si-NH$_2$ sites, despite the concomitant increase of the overall nitrogen content.[53] In another study, ^{15}N DNP-enhanced SSNMR spectroscopy was used to monitor the dehydrogenation process of ammonia borane, a promising medium for hydrogen storage. The use of DNP, in combination with conventional ^{11}B multiple-quantum (MQ) MAS, $^1H\{^{15}N\}$ idHETCOR, and density functional theory (DFT) calculations of the magnetic shielding tensors, revealed that the oligomerization of ammonia borane proceeds in a 'head-to-tail' manner, ultimately leading to the formation of hexagonal boron nitride via a dehydrocyclization reaction, bypassing the formation of polyiminoborane.[48] At present, DNP investigations of nonporous materials or samples with very small pores, such as MOFs, rely on ^1H–^1H spin diffusion to transport the hyperpolarization into the bulk before being transferred to the nuclei of interest, ^{15}N in this case, by cross-polarization (CP).

19.2.3 ^{119}Sn

Given its high gyromagnetic ratio and I value of 1/2, ^{119}Sn may not seem a particularly challenging nucleus; however, the acquisition of ^{119}Sn SSNMR spectra in many materials of interest has been exceedingly difficult due to the large chemical shift distributions

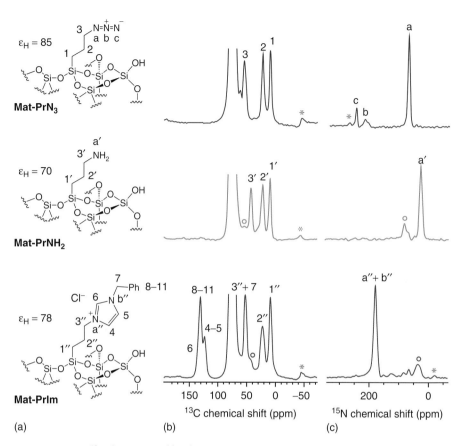

Figure 19.3. DNP-enhanced ^{13}C{^{1}H} (b) and ^{15}N{^{1}H} (c) CPMAS spectra of the functionalized hybrid organic–inorganic silica materials depicted in (a). (Reprinted with permission from A. Zagdoun, G. Casano, O. Ouari, G. Lapadula, A. J. Rossini, M. Lelli, M. Baffert, D. Gajan, L. Veyre, W. E. Maas, M. Rosay, R. T. Weber, C. Thieuleux, C. Copéret, A. Lesage, P. Tordo and L. Emsley, J. Am. Chem. Soc., 2012, 134, 2284. Copyright 2012 American Chemical Society)

and the low loadings typical of tin. As a result, ^{119}Sn SSNMR spectroscopy of materials has greatly benefitted from the development of DNP, which has enabled the acquisition of SSNMR spectra that were previously out of reach. The first application of DNP to ^{119}Sn SSNMR was in the elucidation of the core–shell structure of Sn/SnO$_x$ colloidal nanoparticles.[56] ^{119}Sn DNP SENS measurements, which feature solely signals from the particle surface due to a lack of ^{1}H spins in the bulk, were able to show that the outer shell of the nanoparticles was composed exclusively of SnO$_2$, while the SnO and Sn phases, detected by Mössbauer spectroscopy, must be located deeper inside the particle. ^{119}Sn DNP-enhanced SSNMR was also applied to the characterization of Sn-containing zeolites (Sn-BEA[57–59] and Sn-CHA[60]), which usually have tin

loadings of only a few percent; these low loadings typically prevent ^{119}Sn detection in the absence of DNP and isotope enrichment. Such ^{119}Sn DNP NMR spectra have been able to yield qualitative information regarding the coordination number of Sn and whether the site is coordinatively open or closed. The CSA parameters extracted using a magic-angle-turning (MAT) experiment[61] were also shown to aid in the assignment of certain resonances to specific sites within the crystal structure, potentially leading to site-specific explanations for the catalytic performance of different zeolites.[59] Importantly, it was revealed that typical nitroxide polarizing agents can coordinate to Sn sites within zeolites,[57] and thus larger, or dendritic,[62] polarizing agents that cannot access the Sn sites should be used. Lastly, ^{119}Sn DNP SENS has been performed

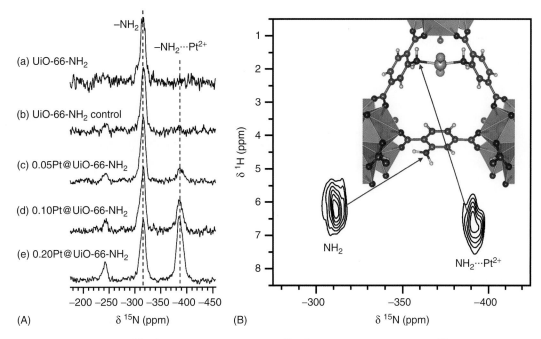

Figure 19.4. DNP-enhanced $^{15}N\{^{1}H\}$ CPMAS (A) and 2D $^{15}N\{^{1}H\}$ HETCOR spectra of Pt^{2+}/UiO-66-NH_2 (B). ((A) Reproduced with permission from Ref. 44. © John Wiley and Sons, 2014. (B) Reproduced with permission from Ref. 42. © Elsevier, 2017)

on silica-grafted organotin complexes in which the Sn loading was far too low to be detected without DNP.[63] These measurements were able to resolve the ^{119}Sn signals from mono- and bipodal Sn sites on silica and were used to suggest that bipodal complexes are formed by the reorganization of Q^2 sites, rather than vicinal Q^3 sites (where n in Q^n represents the number of next nearest neighbors that are Si).

19.2.4 Low-γ Nuclei

Given the fact that the theoretical gain in sensitivity from the application of DNP corresponds to the ratio of the γ value of the electron and the nucleus, one of the areas that shows the greatest promise for applications of DNP is SSNMR spectroscopy of low-γ nuclei. The working definition of what constitutes a low-γ nuclide is any nuclide that has a γ value which is smaller than that of ^{15}N (i.e., resonance frequency <40 MHz at a field of 9.4 T), as the NMR spectroscopy of these nuclides typically requires the use of additional specialized hardware.[64,65] These isotopes are very rarely studied not only due to hardware limitations but also due

to their unfavorable relaxation properties, low signal amplitudes, and poor CP dynamics, requiring the use of long, high-RF, spin-locking pulses. Nevertheless, the possibilities afforded by DNP have been recognized and, to date, four low-γ isotopes (^{14}N, ^{25}Mg, ^{43}Ca, and ^{89}Y) have been investigated with the use of DNP. Some of the results obtained for materials will be discussed here, while applications of DNP to ^{14}N have been limited to biomolecules[66–68] and are thus outside the scope of this chapter.

Although the theoretical gains in sensitivity for low-γ nuclei are massive, on the order of 10^4, obtaining these enhancements can be very difficult. As mentioned earlier, DNP can be performed using two separate approaches known as direct (i.e., $e^- \rightarrow X$) and indirect DNP (i.e., $e^- \rightarrow {}^{1}H \rightarrow X$). Direct DNP is clearly the more general approach, as it does not require the presence of an intermediate ^{1}H spin, but can be challenging for low-γ nuclei as much of the enhancement can be lost due to the overlap of the positive and negative enhancement maxima that are separated by only ν_0 for cross-effect DNP and $2\nu_0$ for solid-effect DNP.[69] The long T_1 relaxation times that are commonly observed for these nuclei can help improve the DNP enhancements, but at the cost of an increased recycle delay between scans. Despite this difficulty, direct DNP of ^{25}Mg via the solid effect has been performed at 4.2 K and 1.127 T on a Cr^{3+}-doped Mg$_2$SiO$_4$ crystal.[70] More recently, an 835-fold enhancement of an ^{89}Y signal was achieved by direct DNP in a nonspinning YDOTA sample at 1.4 K and 3.35 T.[71] All applications of MAS DNP to low-γ nuclei in materials, however, have opted instead for the use of the indirect DNP approach and have thus focused on the study of hydrogen-containing materials.

The first modern application of DNP for the characterization of materials via low-γ nuclei was performed by Blanc *et al.*,[72] who showed that the ^{89}Y CPMAS spectra of Y^{3+}-containing frozen solutions could be acquired in mere minutes; importantly noting that an added benefit of the cryogenic cooling used in DNP is the efficient heat dissipation during the long high-power spin locking pulses required to achieve the ^{1}H-^{89}Y CP transfer. They also studied ceramics made of hydrated Y-doped zirconate (BaZrO$_3$), a promising proton conductor with potential uses in solid oxide fuel cells. Only low enhancements of ~2–3 could be achieved in this case due to the low ^{1}H density of the material that impedes the ^{1}H–^{1}H spin diffusion. The improved sensitivity nevertheless enabled a ^{89}Y{^{1}H} HETCOR spectrum to be acquired in a reasonable amount of time (15 h), enabling the detection

of two different proton-trapping defect sites, which are important in mediating proton conduction.

Recently, ^{89}Y DNP SENS has also been applied to the characterization of silica-supported single-site Y catalysts. First, Eedugurala *et al.*[73] attempted to apply ^{89}Y DNP SENS to directly determine the podality of Y{N(SiHMe$_2$)tBu}$_3$ catalysts grafted onto MSNs that were thermally treated at 550 and 700 °C. Although the slight differences observed between the two catalysts were in agreement with the expected changes in podality, it was later discovered, using ^{15}N DNP SENS, that this catalyst had decomposed under the DNP conditions. More recently, Delley *et al.*[74] applied ^{89}Y DNP SENS on similar single-site catalysts (Y{N(SiMe$_3$)$_2$}$_3$ and Y{OSi(OtBu)$_3$}$_3${OHSi(OtBu)$_3$}), again with the goal of elucidating the coordination geometry of the surface-supported Y complexes. With the help of DFT calculations on model compounds, they were able to assign the ^{89}Y chemical shifts and show that tri-, tetra-, and pentacoordinated Y sites could all be found on the silica surface (Figure 19.5). The addition of nitrogen ligands was shown to slightly increase the ^{89}Y chemical shifts.

One of the most desirable NMR-active isotopes is ^{43}Ca, owing to the importance of Ca in both materials science and biology. Similar to ^{17}O, however, ^{43}Ca NMR is plagued by a combination of very low natural abundance (0.135%) and a low γ value (~15 times smaller than ^{1}H), which has limited the development

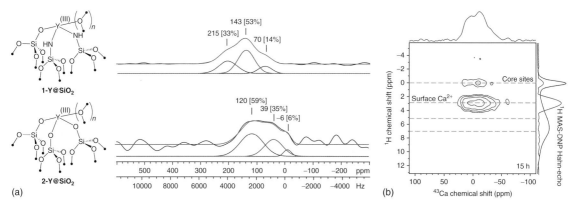

Figure 19.5. DNP-enhanced NMR data acquired for low-γ nuclei in materials. (a) DNP-enhanced ^{89}Y CP-CPMG MAS NMR spectra of the surface complexes depicted on the left. The spectra are deconvoluted to Y sites with varying coordination numbers. (b) ^{43}Ca{^{1}H} HETCOR spectrum acquired on a carbonated hydroxyapatite sample, with well-resolved resonances from surface and core ^{43}Ca sites. ((a) Reprinted with permission from M. F. Delley, G. Lapadula, F. Núñez-Zarur, A. Comas-Vives, V. Kalendra, G. Jeschke, D. Baabe, M. D. Walter, A. J. Rossini, A. Lesage, L. Emsley, O. Maury and C. Copéret, J. Am. Chem. Soc., 2017, 139, 8855. Copyright 2017 American Chemical Society. (b) Lee *et al.*,[75] https://www.nature.com/articles/ncomms14104. Licensed under CC BY 4.0)

of ^{43}Ca SSNMR.[76] Nevertheless, due to its small quadrupole moment and high spin quantum number ($I = 7/2$), ^{43}Ca often yields relatively narrow resonances, making it an ideal candidate for study by DNP-enhanced SSNMR.

The earliest ^{43}Ca DNP experiment was performed in 1974 by Abragam *et al.*, who used an indirect ^{19}F–^{43}Ca DNP approach to hyperpolarize and detect the ^{43}Ca NMR signal from a CaF$_2$ crystal.[77] They first hyperpolarized the ^{19}F nuclei directly by DNP, using a Tm^{2+} dopant as a polarization source, and transferred this hyperpolarization to the ^{43}Ca spins by adiabatic demagnetization in the rotating frame. According to their estimations, they achieved a ^{43}Ca nuclear polarization of $80 \pm 15\%$ using this approach. Surprisingly, only one study has been published using ^{43}Ca DNP-enhanced SSNMR since the technique's recent renaissance.[75] In this study, Lee *et al.* applied indirect ^{1}H–^{43}Ca MAS DNP for the study of hydroxyapatite nanoparticles, an important mineral found in bones. They were able to improve the sensitivity of ^{43}Ca SSNMR by a factor of 35 compared to conventional SSNMR and used a ^{43}Ca{^{1}H} HETCOR experiment to distinguish surface and core ^{43}Ca sites in carbonated hydroxyapatite. Unfortunately, the spectrum was found to be dominated by a surface species that was difficult to identify, perhaps due to poor spin diffusion of hyperpolarization into the bulk. As many important calcium-containing materials, such as ceramics and cements, have either a poor or nonexistent ^{1}H content, further expansion of the capabilities of direct DNP is needed to enable spectroscopic elucidation of bulk sites.

19.2.5 Ultrawideline NMR

Many compounds of interest in materials science contain heavy spin-1/2 nuclei (e.g., ^{195}Pt and ^{207}Pb) or quadrupolar nuclei (e.g., ^{14}N, $^{35/37}$Cl, and $^{79/81}$Br), which yield very broad SSNMR spectra, ranging from several hundred kilohertz to several megahertz, owing to their large CSA or quadrupolar interactions. The resulting spectral dispersion has a detrimental effect on sensitivity, which is often exacerbated by the low γ values and/or low natural abundances of many such nuclei. To date, the vast majority of the applications of DNP to materials have used MAS to average the CSA and the first-order quadrupolar broadening, as well as reduce the second-order quadrupolar broadening of the central transition of half-integer quadrupolar nuclei by

a factor of $\sim$3. In studies involving highly disordered materials with inhomogeneously broadened spectra, or in cases where the second-order quadrupolar interaction is simply too large, however, MAS can easily become ineffective, only yielding large manifolds of unresolved spinning sidebands. In such cases, it is often advantageous to acquire the spectra on static samples with the use of modern ultrawideline SSNMR methods. These methods include those making use of frequency-swept wideband, uniform-rate, and smooth-truncation (WURST) pulses to overcome limited excitation bandwidths of square RF pulses.[78] Along these lines, an important development that enables indirect ultrawideline DNP experiments is the broadband adiabatic inversion cross-polarization (BRAIN-CP) experiment. This experiment incorporates a WURST adiabatic inversion pulse in lieu of the spin locking pulse seen in conventional CP to cross-polarize broad lines in static samples.[79] Even with the use of these techniques, however, uniform excitation of broad spectra can rarely be achieved at any single transmitter offset, and thus a piecewise acquisition of ultrawideline spectra using the variable offset cumulative spectrum (VOCS) acquisition method is often required.[80] The sensitivity of these approaches can be further enhanced by detecting the signal with a WURST-based CPMG train of echoes (BRAIN-CP-WCPMG, or BCP).[81]

Recently, DNP has been coupled with the aforementioned BCP technique to further increase the sensitivity of ultrawideline SSNMR.[82–84] For example, Kobayashi *et al.*[82] acquired ultrawideline ^{195}Pt SSNMR spectra of Pt^{2+}-loaded UiO-66-NH$_2$ MOFs by combining DNP with ^{195}Pt{^{1}H} BCP (Figure 19.6a). Although the measurement of ^{195}Pt{^{1}H} BCP spectra, spanning over 10 000 ppm, required piecing together 14 subspectra, the final coadded spectrum was obtained in a practicable experimental time of $\sim$7.5 h. The spectral lineshapes, in conjunction with theoretical calculations, revealed the formation of both *cis-* and *trans-*coordinated Pt complexes in the UiO-66-NH$_2$. Notably, this conformational information can only be extracted via the CSA and is unobtainable by MAS-based ^{195}Pt SSNMR spectroscopy, an added value of ultrawideline SSNMR. The $\varepsilon_{on/off}$ factors achieved in this study were on the order of 5, which is considerably less than the values commonly reported for MAS DNP. This is due to the difficulty to satisfy the cross-effect condition in static samples,[85] as well as the fact that UiO-66-NH$_2$ MOF poses diffusional constraints for

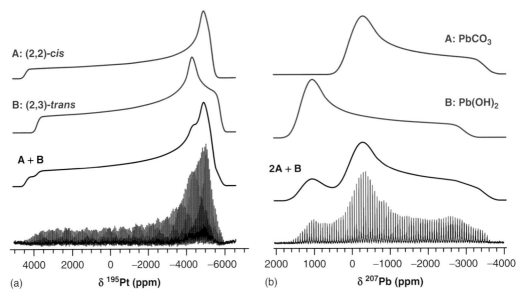

Figure 19.6. (a) DNP-enhanced ultrawideline $^{195}Pt\{^1H\}$ BCP spectrum of Pt^{2+}/UiO-66-NH$_2$ and the simulated powder patterns for the Pt^{2+} cations bound to $-NH_2$ in trans and cis configurations. (b) DNP-enhanced ultrawideline $^{207}Pb\{^1H\}$ BCP spectrum of basic lead white and its deconvolution into sites A and B in lead carbonate and lead hydroxide layers, respectively. ((a) Reprinted with permission from T. Kobayashi, F. A. Perras, T. W. Goh, T. L. Metz, W. Huang and M. Pruski, J. Phys. Chem. Lett., 2016, 7, 2322. Copyright 2016 American Chemical Society. (b) Reproduced from T. Kobayashi, F. A. Perras, A. Murphy, Y. Yao, J. Catalano, S. A. Centeno, C. Dybowski, N. Zumbulyadis and M. Pruski, Dalton Trans., 2017, 46, 3535 with permission from The Royal Society of Chemistry)

the radicals. Interestingly, Kubicki *et al.*[86] observed that the DNP enhancement can be increased by the addition of dielectric particles to the samples. Indeed, mixing the abovementioned Pt^{2+}/UiO-66-NH$_2$ with NaCl particles (75% w/w) increased the DNP enhancement by an additional factor of 5, which resulted in a higher overall sensitivity than a rotor containing pure Pt^{2+}/UiO-66-NH$_2$, despite the decreased sample amount.[82]

DNP-enhanced ultrawideline ^{207}Pb SSNMR spectroscopy has also been applied to characterize lead compounds relevant to cultural heritage science.[83] Specifically, DNP-enhanced ^{207}Pb SSNMR measurements enabled, for the first time, the observation of the basic lead carbonate phase in the lead white pigment (Figure 19.6b). The measurements also detected the formation of a lead soap in an aged paint film, thus demonstrating that this new methodology holds promise for the application of SSNMR to dilute and severely mass-limited archeomaterials and other objects of cultural significance.

Hirsh *et al.* have published the only current example of a DNP-enhanced BCP experiment on a quadrupolar nucleus. Notably, they demonstrated that the low signal enhancement in static samples can, at least in part, be alleviated using a technique dubbed 'spinning-on spinning-off' (SOSO), where the sample is spun to enhance the cross-effect condition during the recycle delay and is stopped for the data acquisition.[84] In favorable cases, using histidine HCl and ambroxol HCl, with sufficiently long ^1H T_1 relaxation times to allow for the spinning-down operation, the SOSO approach improved the DNP enhancement by a factor of 2 (Figure 19.7), but still fell short of the values expected under MAS.

Unlike the case for $I = 1/2$ nuclei, however, an improved sensitivity is not a guarantee for DNP-enhanced BCP spectra of quadrupolar nuclei. The rapid relaxation of quadrupolar nuclei may lead to an improved time performance for direct, non-DNP-enhanced, excitation. For example, Figure 19.7(a) and (b) show the ultrawideline ^{79}Br SSNMR spectra of 4-bromopyridine hydrobromide acquired using DNP-enhanced $^{79}Br\{^1H\}$ BCP and

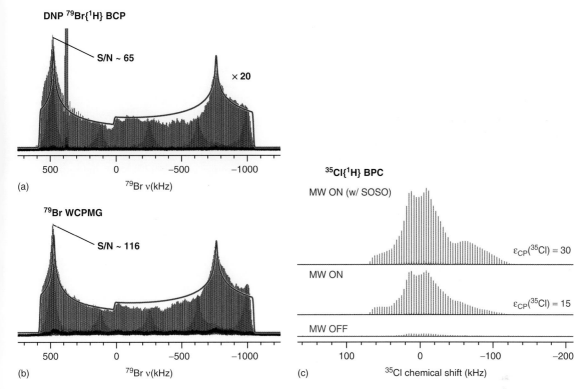

Figure 19.7. (a,b) ^{79}Br NMR spectra of 4-bromopyridine hydrobromide, obtained using ^{79}Br{^{1}H} DNP-enhanced BCP (a) and ^{79}Br WCPMG without microwave irradiation (b). The two spectra were acquired using equal experiment times and were both normalized to the number of scans. The spectra are fit to $\delta_{iso} = 150 \pm 100$ ppm, $C_Q = 33.2 \pm 0.2$ MHz, and $\eta = 0.15 \pm 0.01$. (c) ^{35}Cl{^{1}H} BPC spectra of ambroxol HCl showing the effect of the SOSO procedure

conventional ^{79}Br WCPMG, respectively. While the DNP experiment (Figure 19.7a, $\varepsilon_{on/off} \sim 4$) offered a higher signal intensity per scan than the ^{79}Br WCPMG experiment acquired without DNP (Figure 19.7b), by a factor of ~5, the time sensitivity of the conventional ^{79}Br WCPMG experiment was double that of the DNP experiment. This result was due to the long recycle delay required for DNP buildup (40 s, vs 0.5 s for WCPMG). Note that a significantly greater sensitivity could have been obtained at an ultrahigh magnetic field (~7.5 at 21.1 T) due to the $B_0^{5/2}$ dependence. Further advances in instrumentation, pulse sequences, and sensitizers that exploit DNP mechanisms other than the cross-effect are expected to expand the applications of ultrawideline DNP-enhanced SS-NMR, as seen, for instance, with the improved performance of trityl in static direct ^{17}O DNP experiments.[31]

19.3 CORRELATION SPECTROSCOPY INVOLVING UNRECEPTIVE AND/OR RARE NUCLEI

19.3.1 Acquisition of Challenging Homonuclear Correlation Experiments

Perhaps no experiment type better exemplifies DNP's capacity to 'grow signals from the noise' than the acquisition of 2D correlation spectra between rare nuclei, such as ^{13}C or ^{29}Si. Homonuclear and heteronuclear correlation experiments have unrivaled capacity for providing atomic-scale structural insight by determining intra- and intermolecular spin topology. Typically, such experiments involve highly sensitive ^{1}H nuclei, at least in one spectral dimension. Two-dimensional ^{1}H–^{1}H homonuclear correlation experiments have a particular advantage of high sensitivity[87]; however, their applications are often limited by low resolution, arising from the narrow chemical shift range and significant homonuclear dipolar broadening, present even under fast MAS and/or ^{1}H–^{1}H homonuclear decoupling. Heteronuclear ^{1}H–X spectroscopy takes advantage of higher resolution due to the wider chemical shift range of heteronuclei (X) at the expense of reduced sensitivity due to their lower γ and low natural abundance. In addition, the inefficiency of long-range polarization transfers and dipolar truncation effects strongly favor short-range intramolecular ^{1}H–X correlations. Homonuclear X–X spectroscopy, on the other hand, offers the highest resolution but has, until recently, been prohibitively insensitive under natural abundance. Note that the probability of finding an interacting pair of nuclei depends on the product of their natural abundances. The natural abundance ^{29}Si–^{29}Si double-quantum/single-quantum (DQ/SQ) correlation spectra obtained on 'NMR-friendly' crystalline zeolites are among the few examples of such studies reported by conventional NMR.[88–90]

Unprecedented signal enhancements offered by DNP have created new opportunities for the measurement of through-space and through-bond 2D homonuclear correlation spectra between unreceptive and low natural abundance nuclei within a practical experimental time.[49,91–101] In fact, some of these studies raised the sensitivity bar even higher by measuring correlation spectra of species that constituted only a fraction of the surface.[49,95,97,99,101] For example, both through-space and through-bond 2D DQ/SQ ^{29}Si–^{29}Si correlation spectra of a self-condensed organosiloxane layer formed on silica nanoparticles could be observed within a reasonable experimental time (Figure 19.8b).[95] The correlation spectra provided structural insights into the polymerization of organosiloxanes grafted onto silica nanoparticles; the modifier produced laterally self-condensed functionalizing groups without yielding significant core–shell oligomers above the surface.

We recently applied similar DNP-enhanced ^{29}Si–^{29}Si homonuclear correlation experiments to shed light on the long-standing conundrum regarding the spatial distribution of organic functional groups attached to the surface of MSNs via cocondensation and postsynthesis grafting of organosilane precursors.[97] The study established that T^3–T^3 site correlations could be observed in the DQ/SQ spectrum of cocondensed samples (Figure 19.9a), but not in the spectra of grafted samples (Figure 19.9b), unambiguously demonstrating that, contrary to some earlier reports, the latter synthetic method leads to a more homogeneous distribution of surface groups. Evidently, the organosilane precursors do not self-condense during the anhydrous grafting process, and are unlikely to bond to the silica surface in close proximity (<4 Å) due to the scarcity of suitably arranged hydroxyl groups.

We subsequently applied DNP-enhanced ^{13}C–^{13}C homonuclear correlation spectroscopy to detect intermolecular interactions between two types of MSN-bound organic functional groups: phenyl (Ph) and mercaptopropyl (MP).[100] Note that with the natural ^{13}C abundance of 1.1%, only 1 out of 8300 C–C pairs can produce correlation signals. The detection of ^{13}C–^{13}C correlations between species dispersed on the surface further exacerbates the sensitivity challenge. Nevertheless, the SQ/SQ ^{13}C–^{13}C homonuclear correlation experiment known as CP3 or CHHC, which relies on X-edited ^{1}H spin diffusion,[102,103] successfully detected the intermolecular Ph–MP correlations (Figure 19.9c) on MSN surfaces. This approach can also be applied for other combinations of nuclei, such as CHHN.

The ability to measure homonuclear correlation spectra under natural abundance adds the benefit of circumventing dipolar truncation effects, thereby allowing the measurement of long-range interactions and distances. For example, ^{13}C–^{13}C distances of up to 0.7 nm were measured in a self-assembled cyclic diphenylaniline peptide using DNP-enhanced ^{13}C–^{13}C DQ buildup curves.[13] In addition, such long-range correlations do not suffer from the broadening caused by one-bond scalar couplings, which improves the spectral resolution. Indeed, the DNP-enhanced ^{13}C–^{13}C correlation techniques can be used to access structural constraints, which are unavailable in isotopically-labeled systems owing to dipolar truncation effects. These structural parameters can in turn be used to elucidate the intermolecular order in organic compounds that cannot be prepared in a crystalline form suitable for X-ray diffraction, opening the door to *de novo* NMR crystallography.[49,95,101]

19.3.2 Acquisition of Challenging Heteronuclear Correlation Experiments

DNP has also been applied in a number of cases for the acquisition of very challenging heteronuclear correlation experiments involving rare nuclear

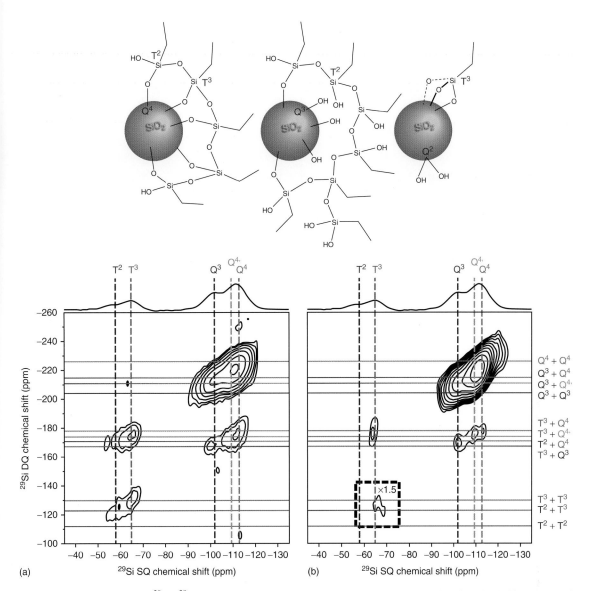

Figure 19.8. DNP-enhanced $^{29}Si-^{29}Si$ DQ/SQ correlation spectra of the polyethylsiloxane-functionalized silica nanoparticles via through-space dipolar coupling with the POST-C7 sequence (a), and through-bond scalar coupling with the INADEQUATE sequence (b). Each 2D experiment was recorded in ~5.5 h. (Reprinted with permission from D. Lee, G. Monin, N. T. Duong, I. Z. Lopez, M. Bardet, V. Mareau, L. Gonon and G. De Paëpe, *J. Am. Chem. Soc.*, 2014, 136, 13781. Copyright 2014 American Chemical Society, https://pubs.acs.org/doi/abs/10.1021/ja506688m)

isotopes in materials. As in the homonuclear case, the sensitivity of many of these experiments can be remarkably low given that the intensity depends on the multiplication of the natural abundances of the two isotopes and is further reduced by a factor of $(2I)^{-1}$ for half-integer quadrupolar

nuclei. The first such DNP-facilitated heteronuclear correlation experiment in materials, excluding ^1H-correlated experiments, was the $^{13}C\{^{27}Al\}$ dipolar–heteronuclear multiple-quantum correlation (D-HMQC)[104] spectrum acquired for MIL-100(Al) MOF.[105] In this case, >90% of the polarization was

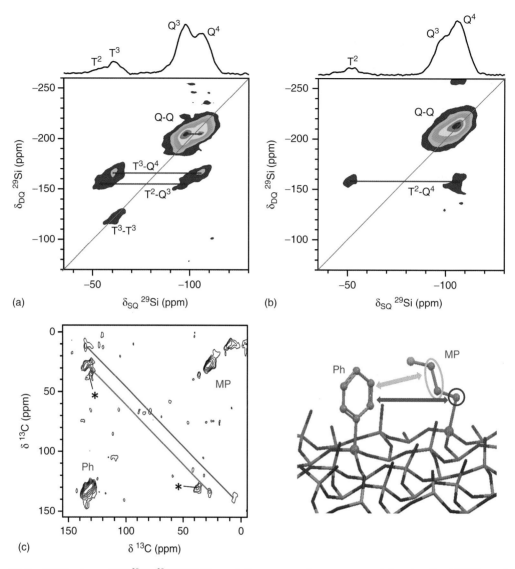

Figure 19.9. DNP-enhanced 2D ^{29}Si–^{29}Si DQ/SQ correlation spectra of mercaptopropyl-functionalized MSNs synthesized via cocondensation (a) and grafting (b), obtained using 5.3 ms of SPC-5 recoupling. (c) DNP-enhanced 2D ^{13}C–^{13}C SQ/SQ correlation spectrum of MP/Ph-bifunctionalized MSNs obtained using 200 μs of mixing time in the CHHC scheme. ((a,b) Reproduced from T. Kobayashi, D. Singappuli-Arachchige, Z. Wang, I. I. Slowing and M. Pruski, Phys. Chem. Chem. Phys., 2017, 19, 1781 with permission from The Royal Society of Chemistry. (c) Reprinted with permission from T. Kobayashi, I.I. Slowing, and M. Pruski, J. Phys. Chem. C, 2017, 121, 24687. Copyright 2017 American Chemical Society)

lost due to recoupling inefficiencies and satellite transition magnetization.

Piveteau *et al.*[106] have used DNP to enhance the ^{125}Te, $^{111/113}$Cd, ^{77}Se, and ^{31}P SSNMR signals from InP, CdSe, PbSe, CdTe, and PbTe quantum dots. In the

case of InP quantum dots, they were able to observe the signals from an oxidized surface, whereas separate core and shell ^{113}Cd resonances were observed for the CdSe sample. In order to assign these resonances, they performed a ^{13}C{^{111}Cd} D-HMQC experiment, showing that the carboxylate carbon of the oleate capping surfactant was in close proximity to one of the Cd sites, which also featured a far greater CSA. They thus concluded that this quantum dot had a core–shell structure with crystalline CdSe in the core and a surface that is capped by oleate ligands. The key to the success of this study was the dispersion of the quantum dots into mesoporous silica, which prevented their agglomeration and facilitated close contact between the polarizing agent and the quantum dot surface.

Silica-aluminas are important solid Brønsted acid catalysts of use in industry; however, the lack of crystallinity of these materials has prevented their characterization by crystallographic methods. As a result, much of their structure is unknown, and the nature of their Brønsted acidic sites has been hotly debated.[107] Using DNP, Valla et al.,[108] as well as Rankin et al.[109] in a later publication, were able to acquire challenging through-bond (refocused-INEPT) as well as through-space (D-INEPT) $^{27}Al\{^{29}Si\}$ correlation spectra on silicated alumina (Si/Al$_2$O$_3$) and aluminated silica (Al/SiO$_2$) (Figure 19.10). They found that, in general, the through-bond Si–O–Al linkages were predominantly to tetrahedral aluminum sites, while correlations to higher coordination numbers were observed in the through-space correlation spectra, particularly as the aluminum content increased. This led to the conclusion that the Brønsted acidic sites are associated with four-coordinate aluminum.

Aside from chemical shift correlation experiments, DNP has also enabled difficult heteronuclear dipolar recoupling experiments that have yielded unprecedented insights into the three-dimensional arrangements of atoms on surfaces. The potential of DNP to enable such difficult experiments was demonstrated by Pourpoint et al., who used it to considerably accelerate, and improve the quality of, $^{13}C\{^{27}Al\}$ rotational-echo saturation pulse double-resonance (RESPDOR)[110] experiments performed on MIL-100(Al) MOF.[105] Perras et al.[111] later built on this work and managed to obtain high-quality $^{13}C\{^{27}Al\}$ RESPDOR curves for the organic species situated at the surfaces of alumina and alumina-supported Pd catalysts. A novel simulation strategy was developed to accurately calculate the RESPDOR dephasing curves as a function of the distance between individual carbon atoms and the γ-Al$_2$O$_3$ surface. As a result, they were able to measure the distances between individual sites in methionine and a polyvinyl alcohol (PVA) coating to the catalyst support surface and determine their 3D conformations. This study showed that methionine,

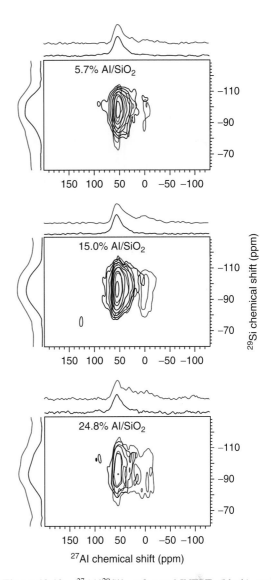

Figure 19.10. $^{27}Al\{^{29}Si\}$ refocused-INEPT (black) and D-INEPT (red) HETCOR spectra of aluminated silicas, with the aluminum content indicated on the spectra. The four-coordinate Al sites bind to the silica, while larger quantities of six-coordinate Al are situated near the silica as the aluminum concentration is increased. (Reprinted with permission from M. Valla, A. J. Rossini, M. Caillot, C. Chizallet, P. Raybaud, M. Digne, A. Chaumonnot, A. Lesage, L. Emsley, J. A. van Bokhoven and C. Copéret, J. Am. Chem. Soc., 2015, 137, 10710. Copyright 2014 American Chemical Society, https://pubs.acs.org/doi/abs/10.1021%2Fjacs. 5b06134)

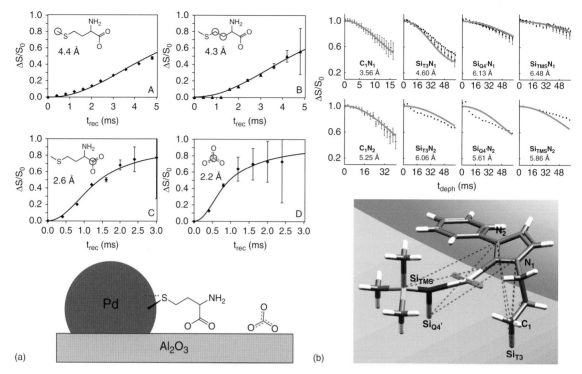

Figure 19.11. Dipolar recoupling measurements performed to determine the conformations of surface species. In (a) $^{13}C\{^{27}Al\}$ RESPDOR measurements were performed to extract the distance of individual carbon sites from the surface of the alumina support. Methionine was found to coordinate to both the alumina support and the Pd nanoparticle surface and to lay prone on the surface. In (b) $^{13}C\{^{15}N\}$ and $^{29}Si\{^{15}N\}$ REDOR measurements were performed to determine the 3D structure of a Pt complex; the best-fit structure is depicted on the bottom. ((a) Reprinted with permission from F. A. Perras, J. D. Padmos, R. L. Johnson, L.-L. Wang, T. J. Schwartz, T. Kobayashi, J. H. Horton, J. A. Dumesic, B. H. Shanks, D. D. Johnson and M. Pruski, J. Am. Chem. Soc., 2017, 139, 2702. Copyright 2017 American Chemical Society. (b) Reprinted with permission from P. Berruyer, M. Lelli, M. P. Conley, D. L. Silverio, C. M. Widdifield, G. Siddiqi, D. Gajan, A. Lesage, C. Copéret and L. Emsley, J. Am. Chem. Soc., 2017, 139, 849. Copyright 2017 American Chemical Society)

which is a catalyst poison, coordinates to an aluminum site at the surface and lays prone on the surface. Most notably, in the presence of Pd, methionine coordinates with the support, whereas the sulfur group bonds to the nanoparticle; this bimodal coordination to the catalyst likely exacerbates the poisoning effect of the compound (Figure 19.11). By contrast, the PVA chains were shown to interact with the surface through hydrogen-bonding interactions.

Sangodkar *et al.*[112] have used a similar strategy to establish the interactions between sucrose and tricalcium silicate, which is the primary constituent of Portland cement. Using $^{29}Si\{^{13}C\}$ rotational echo double-resonance (REDOR) experiments,[113] they were able to show that sucrose preferentially adsorbs onto Q^1 and Q^2 sites and slows the growth of calcium

silicate hydrate chains that are responsible for the strength of concrete.

Finally, REDOR experiments were applied to determine the high-resolution 3D structures of single-site catalysts on silica surfaces. For example, DNP-enhanced $^{13}C\{^{119}Sn\}$ REDOR measurements were carried out on butyltin organometallic complexes grafted onto a silica support to confirm that the butyl ligands were indeed bound to the Sn site.[63] Impressively, these experiments were performed at natural isotopic abundance and correspond to the detection of only 0.09% of all carbon-tin pairs in a surface-supported complex. More recently, Berruyer *et al.*[55] have used extensive $^{13}C\{^{15}N\}$ and $^{29}Si\{^{15}N\}$ REDOR measurements, in conjunction with extended X-ray absorption fine structure (EXAFS), to extract a large number of nontrivial structural constraints on a silica-supported organometallic Pt complex. The three-dimensional structure of this complex was then fit directly to the REDOR data (Figure 19.11). It was then discovered that all the conformations that were in good agreement with experiment were very similar, which led to the conclusion that this complex may feature a well-defined three-dimensional structure on the surface.

19.4 CONCLUDING REMARKS

DNP has revolutionized the SSNMR spectroscopy of materials and is pushing back the limits of what was once thought to be impossible. In just a few years, we have seen a number of previously unthinkable SS-NMR experiments performed with great success, such as the acquisition of natural abundance ^{15}N, ^{17}O, and ^{43}Ca spectra from surfaces. Similarly, ^{29}Si–^{29}Si and even ^{13}C–^{13}C correlations have been observed from the surfaces of mesoporous materials. Such experiments open the door to far greater insights into material structures and functions than was possible only a decade ago. Undoubtedly, the list of experiments and nuclides that still remain beyond our grasp will continue to shrink in the coming decade with the continued development of ultralow-temperature and pulsed DNP, which promise to further 'grow SSNMR signals from the noise'.

ACKNOWLEDGMENTS

This research is supported by the U.S. Department of Energy, Office of Basic Energy Sciences, Division of Chemical Sciences, Geosciences, and Biosciences through the Ames Laboratory. The Ames Laboratory is operated for the U.S. Department of Energy by Iowa State University under Contract No. DE-AC02-07CH11358. F.P. thanks the Natural Sciences and Engineering Research Council of Canada and the Government of Canada for a Banting Fellowship.

RELATED ARTICLES IN EMAGRES

Dynamic Nuclear Polarization and High-Resolution NMR of Solids

REFERENCES

1. A. B. Barnes, M. L. Mak-Jurkauskas, Y. Matsuki, V. S. Bajaj, P. C. A. van der Wel, R. DeRocher, J. Bryant, J. R. Sirigiri, R. J. Temkin, J. Lugtenburg, J. Herzfeld, and R. G. Griffin, *J. Magn. Reson.*, 2009, **198**, 261.

2. L. R. Becerra, G. J. Gerfen, B. F. Bellew, J. A. Bryant, D. A. Hall, S. J. Inati, R. T. Weber, S. Un, T. F. Prisner, A. E. McDermott, K. W. Fishbein, K. E. Kreischer, R. J. Temkin, D. J. Singel, and R. G. Griffin, *J. Magn. Reson. Ser. A*, 1995, **117**, 28.

3. K. N. Hu, H.-h. Yu, T. M. Swager, and R. G. Griffin, *J. Am. Chem. Soc.*, 2004, **126**, 10844.

4. Y. Matsuki, T. Maly, O. Ouari, H. Karoui, F. Le Moigne, E. Rizzato, S. Lyubenova, J. Herzfeld, T. Prisner, P. Tordo, and R. G. Griffin, *Angew. Chem. Int. Ed.*, 2009, **48**, 4996.

5. C. Sauvée, M. Rosay, G. Casano, F. Aussenac, R. T. Weber, O. Ouari, and P. Tordo, *Angew. Chem. Int. Ed.*, 2013, **52**, 10858.

6. A. Zagdoun, G. Casano, O. Ouari, M. Schwarzwälder, A. J. Rossini, F. Aussenac, M. Yulikov, G. Jeschke, C. Copéret, A. Lesage, P. Tordo, and L. Emsley, *J. Am. Chem. Soc.*, 2013, **135**, 12790.

7. R. A. Wind, eMagRes, 2007. doi:10.1002/9780470034590.emrstm0140.

8. T. Kobayashi, O. Lafon, A. S. L. Thankamony, I. I. Slowing, K. Kandel, D. Carnevale, V. Vitzthum, H. Vezin, J.-P. Amoureux, G. Bodenhausen, and M. Pruski, *Phys. Chem. Chem. Phys.*, 2013, **15**, 5553.

9. F. Mentink-Vigier, S. Paul, D. Lee, A. Feintuch, S. Hediger, S. Vega, and G. De Paepe, *Phys. Chem. Chem. Phys.*, 2015, **17**, 21824.

10. S. E. Ashbrook and M. E. Smith, eMagres, 2011. doi:10.1002/9780470034590.emrstm1213.

11. G. Wu, *Prog. Nucl. Magn. Reson. Spectrosc.*, 2008, **52**, 118.

12. I. P. Gerothanassis, *Prog. Nucl. Magn. Reson. Spectrosc.*, 2010, **57**, 1.

13. I. P. Gerothanassis, *Prog. Nucl. Magn. Reson. Spectrosc.*, 2010, **56**, 95.

14. I. Hung, G. Wu, and Z. Gan, *Solid State Nucl. Magn. Reson.*, 2017, **84**, 14.

15. F. A. Perras and D. L. Bryce, *J. Phys. Chem. C*, 2012, **116**, 19472.

16. C. P. Romao, F. A. Perras, U. Werner-Zwanziger, J. A. Lussier, K. J. Miller, C. M. Calahoo, J. W. Zwanziger, M. Bieringer, B. A. Marinkovic,

D. L. Bryce, and M. A. White, *Chem. Mater.*, 2015, **27**, 2633.

17. S. E. Ashbrook and I. Farnan, *Solid State Nucl. Magn. Reson.*, 2004, **26**, 105.

18. E. Brun, B. Derighetti, E. E. Hundt, and H. H. Niebuhr, *Phys. Lett. A*, 1970, **31**, 416.

19. V. K. Michaelis, E. Markhasin, E. Daviso, J. Herzfeld, and R. G. Griffin, *J. Phys. Chem. Lett.*, 2012, **3**, 2030.

20. C. Song, K.-N. Hu, C.-G. Joo, T. M. Swager, and R. G. Griffin, *J. Am. Chem. Soc.*, 2006, **128**, 11385.

21. F. H. Larsen, H. J. Jakobsen, P. D. Ellis, and N. C. Nielsen, *J. Phys. Chem. A*, 1997, **101**, 8597.

22. F. Blanc, L. Sperrin, D. A. Jefferson, S. Pawsey, M. Rosay, and C. P. Grey, *J. Am. Chem. Soc.*, 2013, **135**, 2975.

23. A. P. M. Kentgens and R. Verhagen, *Chem. Phys. Lett.*, 1999, **300**, 435.

24. F. A. Perras, J. Viger-Gravel, K. M. N. Burgess, and D. L. Bryce, *Solid State Nucl. Magn. Reson.*, 2013, **51-52**, 1.

25. F. A. Perras, T. Kobayashi, and M. Pruski, *J. Am. Chem. Soc.*, 2015, **137**, 8336.

26. X. Zhao, W. Hoffbauer, J. S. auf der Günne, and M. H. Levitt, *Solid State Nucl. Magn. Reson.*, 2004, **26**, 57.

27. F. A. Perras, U. Chaudhary, I. I. Slowing, and M. Pruski, *J. Phys. Chem. C*, 2016, **120**, 11535.

28. A. Gansmüller, J. P. Simorre, and S. Hediger, *J. Magn. Reson.*, 2013, **234**, 154.

29. F. A. Perras, Z. Wang, P. Naik, I. I. Slowing, and M. Pruski, *Angew. Chem. Int. Ed.*, 2017, **56**, 9165.

30. E. P. Parry, *J. Catal.*, 1963, **2**, 371.

31. V. K. Michaelis, B. Corzilius, A. A. Smith, and R. G. Griffin, *J. Phys. Chem. B*, 2013, **117**, 14894.

32. M. A. Hope, D. M. Halat, P. C. M. M. Magusin, S. Paul, L. Peng, and C. P. Grey, *Chem. Commun.*, 2017, **53**, 2142.

33. N. J. Brownbill, D. Gajan, A. Lesage, L. Emsley, and F. Blanc, *Chem. Commun.*, 2017, **53**, 2563.

34. T. V. Can, M. A. Caporini, F. Mentink-Vigier, B. Corzilius, J. J. Walish, M. Rosay, W. E. Maas, M. Baldus, S. Vega, T. M. Swager, and R. G. Griffin, *J. Chem. Phys.*, 2014, **141**, 064202.

35. T. Kobayashi, Y. Nishiyama, and M. Pruski, in Modern Methods in Solid-State NMR, ed. P. Hodgkinson, RSC Publishing: London, 2018.

36. A. Zagdoun, G. Casano, O. Ouari, G. Lapadula, A. J. Rossini, M. Lelli, M. Baffert, D. Gajan, L. Veyre, W. E. Maas, M. Rosay, R. T. Weber, C. Thieuleux, C. Copéret, A. Lesage, P. Tordo, and L. Emsley, *J. Am. Chem. Soc.*, 2012, **134**, 2284.

37. N. Eedugurala, Z. Wang, U. Chaundhary, N. Nelson, K. Kandel, T. Kobayashi, I. I. Slowing, M. Pruski, and A. D. Sadow, *ACS Catal.*, 2015, **5**, 7399.

38. T. Gutmann, J. Liu, N. Rothermel, Y. Xu, E. Jaumann, M. Werner, H. Breitzke, S. T. Sigurdsson, and G. Buntkowsky, *Chem. Eur. J.*, 2015, **21**, 3798.

39. B. Hamzaoui, A. Bendjeriou-Sedjerari, E. Pump, E. Abou-Hamad, R. Sougrat, A. Gurinov, K.-W. Huang, D. Gajan, A. Lesage, L. Emsley, and J.-M. Basset, *Chem. Sci.*, 2016, **7**, 6099.

40. G. S. Foo, J. J. Lee, C.-H. Chen, S. E. Hayes, C. Sievers, and C. W. Jones, *ChemSusChem*, 2017, **10**, 266.

41. J. Liu, P. B. Groszewicz, Q. Wen, A. S. L. Thankamony, B. Zhang, U. Kunz, G. Sauer, Y. Xu, T. Gutmann, and G. Buntkowsky, *J. Phys. Chem. C*, 2017, **121**, 17409.

42. T. Kobayashi, F. A. Perras, U. Chaudhary, I. I. Slowing, W. Huang, A. D. Sadow, and M. Pruski, *Solid State Nucl. Magn. Reson.*, 2017, **87**, 38.

43. A. J. Rossini, A. Zagdoun, M. Lelli, J. Canivet, S. Aguado, O. Ouari, P. Tordo, M. Rosay, W. E. Maas, C. Copéret, D. Farrusseng, L. Emsley, and A. Lesage, *Angew. Chem. Int. Ed.*, 2012, **51**, 123.

44. Z. Guo, T. Kobayashi, L.-L. Wang, T. W. Goh, C. Xiao, M. A. Caporini, M. Rosay, D. D. Johnson, M. Pruski, and W. Huang, *Chem. Eur. J.*, 2014, **20**, 16308.

45. W. R. Grüning, A. J. Rossini, A. Zagdoun, D. Gajan, A. Lesage, L. Emsley, and C. Copéret, *Phys. Chem. Chem. Phys.*, 2013, **15**, 13270.

46. T. K. Todorova, X. Rozanska, C. Gervais, A. Legrand, L. N. Ho, P. Berruyer, A. Lesage, L. Emsley, D. Farrusseng, J. Canivet, and C. Mellot-Draznieks, *Chem. Eur. J.*, 2016, **22**, 16531.

47. F. Blanc, S. Y. Chong, T. O. McDonald, D. J. Adams, S. Pawsey, M. A. Caporini, and A. I. Cooper, *J. Am. Chem. Soc.*, 2013, **135**, 15290.

48. T. Kobayashi, S. Gupta, M. A. Caporini, V. K. Pecharsky, and M. Pruski, *J. Phys. Chem. C*, 2014, **118**, 19548.

49. K. Marker, M. Pingret, J.-M. Mouesca, D. Gasparutto, S. Hediger, and G. De Paëpe, *J. Am. Chem. Soc.*, 2015, **137**, 13796.

50. L. Zhao, I. Smolarkiewicz, H.-H. Limbach, H. Breitzke, K. Pogorzelec-Glaser, R. Pankiewicz, J. Tritt-Goc, T. Gutmann, and G. Buntkowsky, *J. Phys. Chem. C*, 2016, **120**, 19574.

51. J. C. Mohandas, E. Abou-Hamad, E. Callens, M. K. Samantaray, D. Gajan, A. Gurinov, T. Ma, S. Ould-Chikh, A. S. Hoffman, B. C. Gates, and J.-M. Basset, *Chem. Sci.*, 2017, **8**, 5650.

52. M. Werner, A. Heil, N. Rothermel, H. Breitzke, P. B. Groszewicz, A. S. Thankamony, T. Gutmann, and G. Buntkowsky, *Solid State Nucl. Magn. Reson.*, 2015, **72**, 73.

53. A. S. L. Thankamony, C. Lion, F. Pourpoint, B. Singh, A. J. P. Linde, D. Carnevale, G. Bodenhausen, H. Vezin, O. Lafon, and V. Polshettiwar, *Angew. Chem. Int. Ed.*, 2015, **54**, 2190.

54. J. Leclaire, G. Poisson, F. Ziarelli, G. Pepe, F. Fotiadu, F. M. Paruzzo, A. J. Rossini, J.-N. Dumez, B. Elena-Herrmann, and L. Emsley, *Chem. Sci.*, 2016, **7**, 4379.

55. P. Berruyer, M. Lelli, M. P. Conley, D. L. Silverio, C. M. Widdifield, G. Siddiqi, D. Gajan, A. Lesage, C. Copéret, and L. Emsley, *J. Am. Chem. Soc.*, 2017, **139**, 849.

56. L. Protesescu, A. J. Rossini, D. Kriegner, M. Valla, A. de Kergommeaux, M. Walter, K. V. Kravchyk, M. Nachtegaal, J. Stangl, B. Malaman, P. Reiss, A. Lesage, L. Emsley, C. Copéret, and M. V. Kovalenko, *ACS Nano*, 2014, **8**, 2639.

57. W. R. Gunther, V. K. Michaelis, M. A. Caporini, R. G. Griffin, and Y. Roman-Leshkov, *J. Am. Chem. Soc.*, 2014, **136**, 6219.

58. P. Wolf, M. Valla, A. J. Rossini, A. Comas-Vives, F. Núñez-Zarur, B. Malaman, A. Lesage, L. Emsley, C. Copéret, and I. Hermans, *Angew. Chem. Int. Ed.*, 2014, **53**, 10179.

59. P. Wolf, M. Valla, F. Núñez-Zarur, A. Comas-Vives, A. J. Rossini, C. Firth, H. Kallas, A. Lesage, L. Emsley, C. Copéret, and I. Hermans, *ACS Catal.*, 2016, **6**, 4047.

60. J. W. Harris, W.-C. Liao, J. R. Di Iorio, A. M. Henry, T.-C. Ong, A. Comas-Vives, C. Copéret, and R. Gounder, *Chem. Mater.*, 2017, **29**, 8824.

61. Z. Gan, *J. Am. Chem. Soc.*, 1992, **114**, 8307.

62. W.-C. Liao, T.-C. Ong, D. Gajan, F. Bernada, C. Sauvée, M. Yulikov, M. Pucino, R. Schowner, M. Schwarzwälder, M. R. Buchmeiser, G. Jeschke, P. Tordo, O. Ouari, A. Lesage, L. Emsley, and C. Copéret, *Chem. Sci.*, 2017, **8**, 416.

63. M. P. Conley, A. J. Rossini, A. Comas-Vives, M. Valla, G. Casano, O. Ouari, P. Tordo, A. Lesage, L. Emsley, and C. Copéret, *Phys. Chem. Chem. Phys.*, 2014, **16**, 17822.

64. M. E. Smith, *Annu. Rep. NMR Spectrosc.*, 2001, **43**, 121.

65. N. G. Dowell, S. E. Ashbrook, and S. Wimperis, *J. Phys. Chem. B*, 2004, **108**, 13292.

66. V. Vitzthum, M. A. Caporini, and G. Bodenhausen, *J. Magn. Reson.*, 2010, **205**, 177.

67. A. J. Rossini, L. Emsley, and L. A. O'Dell, *Phys. Chem. Chem. Phys.*, 2014, **16**, 12890.

68. J. A. Jarvis, I. Haies, M. Lelli, A. J. Rossini, I. Kuprov, M. Carravetta, and P. T. F. Williamson, *Chem. Commun.*, 2017, **53**, 12116.

69. T. Maly, G. T. Debelouchina, V. S. Bajaj, K.-N. Hu, C.-G. Joo, M. L. Mak-Jurkauskas, J. R. Sirigiri, P. C. A. van der Wel, J. Herzfeld, R. J. Temkin, and R. G. Griffin, *J. Chem. Phys.*, 2008, **128**, 052211.

70. B. Derighetti, S. Hafner, H. Marxer, and H. Rager, *Phys. Lett. A*, 1978, **66**, 150.

71. L. Lumata, A. K. Jindal, M. E. Merritt, C. R. Malloy, A. D. Sherry, and Z. Kovacs, *J. Am. Chem. Soc.*, 2011, **133**, 8673.

72. F. Blanc, L. Sperrin, D. Lee, R. Dervişoğlu, Y. Yamazaki, S. M. Haile, G. De Paëpe, and C. P. Grey, *J. Phys. Chem. Lett.*, 2014, **5**, 2431.

73. N. Eedugurala, Z. Wang, K. Yan, K. C. Boteju, U. Chaudhary, T. Kobayashi, A. Ellern, I. I. Slowing, M. Pruski, and A. D. Sadow, *Organometallics*, 2017, **36**, 1142.

74. M. F. Delley, G. Lapadula, F. Núñez-Zarur, A. Comas-Vives, V. Kalendra, G. Jeschke, D. Baabe, M. D. Walter, A. J. Rossini, A. Lesage, L. Emsley, O. Maury, and C. Copéret, *J. Am. Chem. Soc.*, 2017, **139**, 8855.

75. D. Lee, C. Leroy, C. Crevant, L. Bonhomme-Coury, F. Babonneau, D. Laurencin, C. Bonhomme, and G. De Paëpe, *Nat. Commun.*, 2017, **8**, 14104.

76. D. L. Bryce, *Dalton Trans.*, 2010, **39**, 8593.

77. J. F. Jacquinot, W. T. Wenckebach, M. Goldman, and A. Abragam, *Phys. Rev. Lett.*, 1974, **32**, 1096.

78. R. Bhattacharyya and L. Frydman, *J. Chem. Phys.*, 2007, **127**, 194503.

79. K. J. Harris, A. Lupulescu, B. E. G. Lucier, L. Frydman, and R. W. Schurko, *J. Magn. Reson.*, 2012, **224**, 38.

80. D. Massiot, I. Farnan, N. Gautier, D. Trumeau, A. Trokiner, and J. P. Coutures, *Solid State Nucl. Magn. Reson.*, 1995, **4**, 241.

81. L. A. O'Dell and R. W. Schurko, *Chem. Phys. Lett.*, 2008, **464**, 97.

82. T. Kobayashi, F. A. Perras, T. W. Goh, T. L. Metz, W. Huang, and M. Pruski, *J. Phys. Chem. Lett.*, 2016, **7**, 2322.

83. T. Kobayashi, F. A. Perras, A. Murphy, Y. Yao, J. Catalano, S. A. Centeno, C. Dybowski, N. Zumbulyadis, and M. Pruski, *Dalton Trans.*, 2017, **46**, 3535.

84. D. A. Hirsh, A. J. Rossini, L. Emsley, and R. W. Schurko, *Phys. Chem. Chem. Phys.*, 2016, **18**, 25893.

85. K. R. Thurber and R. Tycko, *J. Chem. Phys.*, 2012, **137**, 084508.

86. D. J. Kubicki, A. J. Rossini, A. Purea, A. Zagdoun, O. Ouari, P. Tordo, F. Engelke, A. Lesage, and L. Emsley, *J. Am. Chem. Soc.*, 2014, **136**, 15711.

87. S. P. Brown, *Solid State Nucl. Magn. Reson.*, 2012, **41**, 1.

88. C. A. Fyfe, Y. Feng, H. Gies, H. Grondey, and G. T. Kokotailo, *J. Am. Chem. Soc.*, 1990, **112**, 3264.

89. D. H. Brouwer, R. J. Darton, R. E. Morris, and M. H. Levitt, *J. Am. Chem. Soc.*, 2005, **127**, 10365.

90. C. A. Fyfe, H. Grondey, Y. Feng, and G. T. Kokotailo, *J. Am. Chem. Soc.*, 1990, **112**, 8812.

91. A. J. Rossini, A. Zagdoun, F. Hegner, M. Schwarzwälder, D. Gajan, C. Copéret, A. Lesage, and L. Emsley, *J. Am. Chem. Soc.*, 2012, **134**, 16899.

92. H. Takahashi, D. Lee, L. Dubois, M. Bardet, S. Hediger, and G. De Paëpe, *Angew. Chem. Int. Ed.*, 2012, **51**, 11766.

93. A. J. Rossini, C. M. Widdifield, A. Zagdoun, M. Lelli, M. Schwarzwälder, C. Copéret, A. Lesage, and L. Emsley, *J. Am. Chem. Soc.*, 2014, **136**, 2324.

94. D. Lee, G. Monin, N. T. Duong, I. Z. Lopez, M. Bardet, V. Mareau, L. Gonon, and G. De Paëpe, *J. Am. Chem. Soc.*, 2014, **136**, 13781.

95. G. Mollica, M. Dekhil, F. Ziarelli, P. Thureau, and S. Viel, *Angew. Chem. Int. Ed.*, 2015, **54**, 6028.

96. Y. Matsuki, T. Idehara, J. Fukazawa, and T. Fujiwara, *J. Magn. Reson.*, 2016, **264**, 107.

97. T. Kobayashi, D. Singappuli-Arachchige, Z. Wang, I. I. Slowing, and M. Pruski, *Phys. Chem. Chem. Phys.*, 2017, **19**, 1781.

98. F. A. Perras, H. Luo, X. Zhang, N. S. Mosier, M. Pruski, and M. M. Abu-Omar, *J. Phys. Chem. A*, 2017, **121**, 623.

99. A. R. Mouat, T. Kobayashi, M. Pruski, T. J. Marks, and P. C. Stair, *J. Phys. Chem. C*, 2017, **121**, 6060.

100. T. Kobayashi, I. I. Slowing, and M. Pruski, *J. Phys. Chem. C*, 2017, **121**, 24687.

101. K. Märker, S. Paul, C. Fernández-de-Alba, D. Lee, J.-M. Mouesca, S. Hediger, and G. De Paepe, *Chem. Sci.*, 2017, **8**, 974.

102. M. Wilhelm, H. Feng, U. Tracht, and H. W. Spiess, *J. Magn. Reson.*, 1998, **134**, 255.

103. I. de Boer, L. Bosman, J. Raap, H. Oschkinat, and H. J. M. de Groot, *J. Magn. Reson.*, 2002, **157**, 286.

104. G. Tricot, J. Trebosc, F. Pourpoint, R. Gauvin, and L. Delevoye, *Ann. Rep. NMR Spectrosc.*, 2014, **81**, 145.

105. F. Pourpoint, A. S. L. Thankamony, C. Volkringer, T. Loiseau, J. Trébosc, F. Aussenac, D. Carnevale, G. Bodenhausen, H. Vezin, O. Lafon, and J.-P. Amoureux, *Chem. Commun.*, 2014, **50**, 933.

106. L. Piveteau, T.-C. Ong, A. J. Rossini, L. Emsley, C. Copéret, and M. V. Kovalenko, *J. Am. Chem. Soc.*, 2015, **137**, 13964.

107. Z. Wang, Y. Jiang, O. Lafon, J. Trébosc, K. D. Kim, C. Stampfl, A. Baiker, J.-P. Amoureux, and J. Huang, *Nat. Commun.*, 2016, **7**, 13820.

108. M. Valla, A. J. Rossini, M. Caillot, C. Chizallet, P. Raybaud, M. Digne, A. Chaumonnot, A. Lesage, L. Emsley, J. A. van Bokhoven, and C. Copéret, *J. Am. Chem. Soc.*, 2015, **137**, 10710.

109. A. G. M. Rankin, P. B. Webb, D. M. Dawson, J. Viger-Gravel, B. J. Walder, L. Emsley, and S. E. Ashbrook, *J. Phys. Chem. C*, 2017, **121**, 22977.

110. Z. Gan, *Chem. Commun.*, 2006, 4712.

111. F. A. Perras, J. D. Padmos, R. L. Johnson, L.-L. Wang, T. J. Schwartz, T. Kobayashi, J. H. Horton, J. A. Dumesic, B. H. Shanks, D. D. Johnson, and M. Pruski, *J. Am. Chem. Soc.*, 2017, **139**, 2702.

112. R. P. Sangodkar, B. J. Smith, D. Gajan, A. J. Rossini, L. R. Roberts, G. P. Funkhouser, A. Lesage, L. Emsley, and B. F. Chmelka, *J. Am. Chem. Soc.*, 2015, **137**, 8096.

113. T. Gullion and J. Schaefer, *J. Magn. Reson.*, 1989, **81**, 196.

Chapter 20

DNP-enhanced Solid-state NMR Spectroscopy of Active Pharmaceutical Ingredients

Li Zhao[1,2], Arthur C. Pinon[3], Lyndon Emsley[3], and
Aaron J. Rossini[1,2]

[1]Iowa State University, Ames, IA, USA
[2]US DOE Ames Laboratory, Ames, IA, USA
[3]Institut des Sciences et Ingénierie Chimiques, Ecole Polytechnique Fédérale de Lausanne (EPFL), Lausanne, Switzerland

20.1 INTRODUCTION

Approximately two-thirds of active pharmaceutical ingredients (APIs) are administered as solid dosage forms. The solid-state structure of APIs strongly affects the stability, solubility, and ultimately the efficacy of APIs within dosage forms.[1-4] Consequently, APIs are often prepared in different solid forms in order to identify the forms that are most suitable for formulation. For example, by varying recrystallization conditions, it is possible to isolate polymorphs of APIs that possess unique crystal structures and offer different stabilities and solubility.[5-9] APIs are often reacted with crystal coformers to obtain multicomponent solids (i.e., cocrystals or salts) that display enhanced solubility and/or stability.[10-15] Alternatively, some APIs are deployed as amorphous forms in order to enhance dissolution rates APIs must be stabilized to prevent crystallization, and this is usually accomplished by mixing the amorphous API with polymers to form amorphous solid dispersions (ASDs).[16-20] These examples demonstrate that the development and deployment of APIs in formulations require identification and characterization of the different possible solid forms of a particular API. Typically, the solid-state structure of APIs or multicomponent solids are determined with single-crystal X-ray diffraction; however, for many solids, it is challenging to isolate diffraction quality single crystals, and in the case of amorphous APIs, diffraction methods will likely

Handbook of High Field Dynamic Nuclear Polarization.
Edited by Vladimir K. Michaelis, Robert G. Griffin, Björn Corzilius and Shimon Vega
ISBN: 978-1-119-44164-9
Original Publication: Li Zhao, Arthur C. Pinon, Lyndon Emsley and Aaron J. Rossini, *Magnetic Resonance in Chemistry*, 2018, **56**, 583–609, DOI: 10.1002/mrc.4688, Reproduced by permission of John Wiley & Sons, Ltd.

not be applicable. Characterization of formulated APIs (i.e., tablets) is especially challenging because interference from the highly abundant excipients and the low concentration of the API hinders diffraction and other analytical techniques.

Solid-state NMR spectroscopy is now an established method for the characterization of pure, multicomponent (i.e., cocrystals and ASDs), and formulated APIs. Many excellent reviews and books describe the established applications of solid-state NMR spectroscopy to APIs.[21-37] One key application of solid-state NMR spectroscopy to APIs is the identification and differentiation of polymorphs. Early in the development of high-resolution solid-state NMR, it was shown that ^{13}C solid-state NMR spectroscopy could differentiate polymorphs of APIs on the basis of slight differences in the observed isotropic ^{13}C chemical shifts.[26,32,38-43] Consequently, ^{13}C solid-state NMR has been widely applied for identification (fingerprinting) and quantification of polymorphs of APIs in mixtures and/or dosage forms. Measurements of ^{1}H longitudinal relaxation times (T_1) and/or rotating frame longitudinal relaxation times ($T_{1\rho}$) are often used to obtain information about the domain or grain size of the APIs and determine the proximity of the API to excipients or coformers or stabilizers.[30,44,45] Advances in NMR technology and methods have also improved the analysis of APIs by solid-state NMR spectroscopy. For example, fast magic angle spinning (MAS)[46-50] and/or homonuclear ^{1}H decoupling (i.e., combined rotation and multiple-pulse spectroscopy)[51] are now routinely employed to obtain high-resolution ^{1}H solid-state NMR spectra of APIs and organic solids.[46,50,52,53] Solid-state ^{1}H chemical shifts are distinct for different polymorphic forms of APIs,[53-60] and ^{1}H solid-state NMR offers much higher sensitivity than ^{13}C solid-state NMR.

In favorable cases, crystal structures may be experimentally determined with solid-state NMR spectroscopy, possibly in conjunction with quantum chemical calculations. For example, ^{1}H–^{1}H distances measured by solid-state NMR were used to determine the crystal structure of β-L-aspartyl-L-alanine.[52,61] ^{1}H–^{1}H, ^{1}H–^{13}C, and ^{13}C–^{13}C internuclear distances were used to determine the molecular structure of ^{13}C-labeled L-tyrosine ethyl ester in the solid-state.[62] More recently, NMR crystallography structure determination protocols that combine experimental NMR data with density functional theory (DFT) calculations have been developed. Structure determination by NMR crystallography is usually accomplished by using crystal structure prediction (CSP)[63-69] to predict a large set of trial crystal structures. Plane-wave DFT calculations[54,70-72] are then used to calculate the ^{1}H and ^{13}C chemical shifts for the lowest energy trial structures.[54-56] Comparison of experimental and calculated chemical shifts allows identification of the trial crystal structure(s) that are in best agreement with the experimental structure.[54-56,58,73-75] Recently, modified NMR crystallography protocols that use genetic algorithms to generate trial structures and semiempirical calculations of ^{1}H chemical shifts to test the trial structures were described.[59] Plane-wave DFT and solid-state NMR spectroscopy are also often applied to validate and/or refine structures determined by X-ray diffraction.[69,76-82]

The availability of high-field NMR systems and specialized pulse sequences has enabled solid-state NMR experiments on APIs with quadrupolar nuclei such as ^{7}Li,[83] ^{14}N,[84,85] ^{17}O,[86-90] ^{23}Na,[91,92] and ^{35}Cl.[93-96] For example, Schurko and coworkers have demonstrated that ^{35}Cl solid-state NMR spectroscopy can be applied to hydrochloride salts of APIs to differentiate polymorphs and detect APIs within dosage forms.[93-95] Wu and coworkers have previously applied ^{17}O solid-state NMR spectroscopy to study ^{17}O-labeled pharmaceuticals such as acetylsalicylic acid (aspirin)[87] and warfarin.[90] Vogt used ^{17}O solid-state NMR, including two-dimensional (2D) ^{1}H–^{17}O heteronuclear correlation (HETCOR) experiments, to probe hydrogen bonding in several different solid forms of diflusinal.[88] Brown and coworkers have used ^{1}H–^{14}N double-resonance solid-state NMR experiments to classify multicomponent solid APIs as salts or cocrystals and probe for interactions between APIs and polymers in amorphous dispersions.[84]

Solid-state NMR experiments on many APIs generally suffer from poor sensitivity due to a combination of effects such as long ^{1}H T_1, unfavorable nuclear properties (Table 20.1), peak splitting or broadening, and the intrinsically low sensitivity of NMR spectroscopy. Consequently, solid-state NMR experiments on APIs are often restricted to basic 1D experiments on highly concentrated samples and receptive NMR-active nuclei (Table 20.1). This chapter describes how high-field dynamic nuclear polarization (DNP) can be applied to enhance the sensitivity of solid-state NMR experiments on pure and formulated APIs by several orders of magnitude. The sensitivity gains provided by DNP enable advanced multidimensional NMR experiments to be performed with unreceptive NMR nuclei on both

Table 20.1. Properties of NMR-active nuclei commonly found in active pharmaceutical ingredients[97]

Nucleus	Nuclear spin (I)	γ (10^7 rad T^{-1} S^{-1})	Natural isotopic abundance (%)
^{1}H	1/2	26.8	99.99
^{2}H	1	4.1	0.015
^{13}C	1/2	6.7	1.10
^{14}N	1	1.9	99.63
^{15}N	1/2	−2.7	0.37
^{17}O	5/2	−3.6	0.04
^{19}F	1/2	25.2	100.00
^{23}Na	3/2	7.1	100.00
^{29}Si	1/2	−5.3	4.67
^{31}P	1/2	10.8	100.00
^{33}S	3/2	2.1	0.76
^{35}Cl	3/2	6.7	75.53
^{43}Ca	7/2	−1.8	0.135

pure and formulated APIs, providing access to novel structural information about the molecular level and macroscopic structure of APIs.

20.1.1 Basic Aspects of High-field Dynamic Nuclear Polarization

Here, we give a brief overview of modern high-field DNP-enhanced solid-state NMR spectroscopy and instrumentation. The reader is referred to more thorough and comprehensive reviews and books on DNP.[98–101]

DNP was proposed by Overhauser in 1953 as a way to enhance the nuclear spin polarization of NMR-active nuclei within metals.[102] DNP was experimentally demonstrated shortly thereafter by Carver and Slichter with ^{7}Li DNP solid-state NMR experiments on Li metal.[103] In a modern DNP experiment, the large spin polarization of unpaired electrons is transferred to NMR-active nuclei, usually ^{1}H. Polarization transfer from the electrons to the NMR-active nuclei is typically achieved by saturating the electron paramagnetic resonance (EPR) at a specific frequency that depends upon the EPR properties of the polarizing agent (PA) and the DNP mechanism.[98–101] DNP-enhanced ^{1}H polarization can then be transferred to lower γ heteronuclei such as ^{13}C or ^{15}N by using cross-polarization (CP) or other techniques. DNP can potentially enhance the polarization and sensitivity of NMR experiments that use ^{1}H nuclei by a factor of up to 658. Historically, DNP at high magnetic fields (5 T and above) was challenging due to the lack of

high-power, high-frequency microwave sources. Due to the efforts of Griffin and coworkers[98,99,104,105] and with the introduction of commercial DNP instrumentation (Figure 20.1),[106] it is now possible to routinely perform DNP-enhanced solid-state NMR experiments at magnetic fields of up to 18.8 T. The commercial Bruker 9.4-T 263-GHz/400-MHz DNP solid-state NMR spectrometer[106] installed at the Department of Energy Ames Laboratory is shown in Figure 20.1, and several of the system components are labeled.

There are three key concepts developed by Griffin and coworkers that have enabled large DNP signal enhancements to be routinely obtained at high magnetic fields.[98,99,104,105] First, a gyrotron provides the high-power microwaves that are required to saturate the EPR of unpaired electrons and facilitate efficient polarization transfer to NMR-active nuclei.[104,106] A gyrotron is one of the few devices capable of producing high-power microwaves at frequencies above 200 GHz.[104–106] The microwaves are delivered into the low-temperature MAS probe and up to the stator by the waveguide. Second, DNP-enhanced solid-state NMR experiments are performed at low temperatures by using cooled nitrogen gas for both MAS and sample cooling.[107] The electron relaxation times of radicals significantly increase at low temperature, enabling more complete saturation of the EPR and higher DNP enhancements.[108] The nuclear relaxation times often increase as the sample temperature is reduced, which also improves DNP enhancements. The cryogenic sample temperatures also bring other benefits such as improved performance of the probe electronics and increased nuclear and electron thermal polarization.

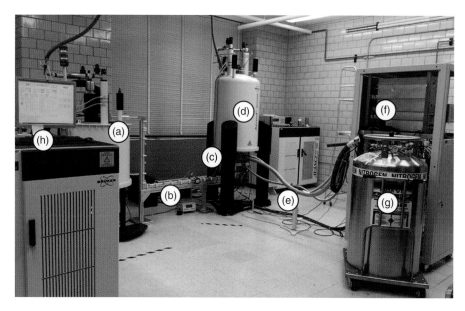

Figure 20.1. The commercial Bruker 9.4-T or 263-GHz dynamic nuclear polarization solid-state NMR spectrometer[106] setup at the Ames Laboratory. Continuous-wave 263-GHz microwaves are generated by a gyrotron (a) and transmitted via a waveguide (b) to the magic angle spinning probe (c) housed in the 9.4-T wide-bore NMR magnet (d). Sample temperatures of ~100 K are achieved by using cold nitrogen gas for sample spinning and sample cooling. The cold nitrogen gas is delivered to the magic angle spinning probe in an insulated transfer line (e). The nitrogen gas is cooled inside of a pressurized heat exchanger (f), which is fed with liquid nitrogen (g). The gyrotron control computer (h) can be used to turn the continuous-wave 263-GHz microwaves on or off and adjust the microwave power

Several research groups are currently exploring the possibility of using helium cooling to further reduce sample temperatures and obtain even larger DNP enhancements and higher absolute sensitivity.[109-112] Third, specially designed stable exogenous PAs are introduced into the sample to provide a source of unpaired electrons for DNP.[113,114] Currently, the most efficient PA for MAS DNP are typically based upon biradicals (e.g., TOTAPOL,[114] AMUPol,[115] TEKPol,[116] and TEMTriPol[117]). The structures of the radical PA that currently provide the largest ^{1}H DNP enhancements are shown in Figure 20.2. At fields of 9.4 T and below, the optimized nitroxide biradical PAs AMUPol[115] and TEKPol[116] can provide ^{1}H cross-effect (CE) DNP signal enhancements above 200. At magnetic fields above 9.4 T, the highest ^{1}H DNP signal enhancements have been obtained to date with a nitroxide-trityl biradical CE PA TEMTriPol[117] or with narrow-line carbon-centered monoradicals such as α,γ-Bisdiphenylene-β-phenylallyl (BDPA) that exploit the Overhauser effect (OE).[119,120] The design and optimization of PA for high magnetic

fields above 16.4 T is currently an active field of research. AMUPol, TEKPol, and BDPA are all commercially available. Below, we provide a practical description of how samples of organic solids and APIs are normally prepared for DNP solid-state NMR experiments.

20.2 PREPARING SAMPLES FOR DNP NMR EXPERIMENTS

A crucial aspect of DNP-enhanced solid-state NMR experiments is determining how to introduce (dope) the exogenous PA into the sample of interest. The preparation of the sample has a large impact on the DNP enhancements, nuclear relaxation rates, spectral resolution and absolute sensitivity of the NMR experiments. Modern DNP solid-state NMR experiments on biomolecules or isotopically labeled small molecules are typically performed by dissolving or suspending the analyte in an aqueous radical solution (Figure 20.3a).

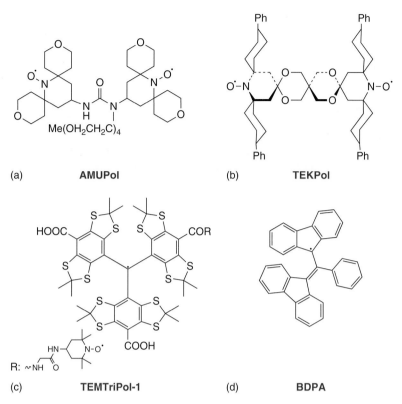

(a)   **AMUPol**   (b)   **TEKPol**

(c)   **TEMTriPol-1**   (d)   **BDPA**

Figure 20.2. The molecular structures of exogenous dynamic nuclear polarization (DNP) polarizing agents. (a) AMUPol,[115] (b) TEKPol,[116] (c) TEMTriPol-1,[117] and (d) BDPA.[118] The cross-effect polarizing agents AMUPol and TEKPol typically provide the largest ^{1}H DNP signal enhancements at fields of 9.4 T or lower. AMUPol is used with aqueous solvent mixtures, whereas TEKPol is used with organic solvents. TEMTriPol-1 and BDPA have provided the largest ^{1}H DNP enhancements at magnetic fields of 16.4 T and higher

Typically, partially deuterated water-glycerol solutions (glycerol-d_8:D$_2$O:H$_2$O, 60 : 30 : 10, vol%), sometimes referred to as *DNP Juice*, or partially deuterated dimethyl sulfoxide-water solutions (dimethyl sulfoxide:D$_2$O:H$_2$O, 60 : 36 : 4, vol%) are used for DNP experiments on biomolecules.[114,121] Dimethyl sulfoxide or glycerol acts as cryoprotectants that prevent crystallization of the water and ensure that an amorphous glass is obtained upon freezing. Glass formation is critical to maintain a homogenous distribution of the PA and analytes and obtain a high DNP enhancement. However, solid-state NMR experiments on APIs are normally performed on microcrystalline solids, with the goal being to probe or determine the solid-state structure. Therefore, nondestructive methods of doping the sample with PA that avoid perturbing the solid-state structure of APIs are required.

20.2.1 Relayed DNP for Microparticulate APIs and Organic Solids

Relayed DNP experiments may be applied to polarize the ^{1}H nuclei inside of nanoparticulate or microparticulate solids such as crystalline organic solids and APIs.[122,123] In a relayed DNP experiment, a powdered solid is impregnated[124] with just enough radical solution to coat the outside of the particles (Figure 20.3b).[122,123] Upon microwave irradiation, DNP-enhanced ^{1}H polarization builds up at the surface of the particles and is continuously transported into the interior of the particle by ^{1}H spin diffusion.[122,123] In this way, nanoparticulate or microparticulate solids, such as APIs, can be remotely polarized. The DNP-enhanced ^{1}H polarization can

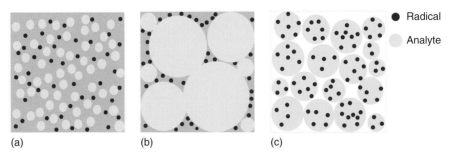

Figure 20.3. Cartoons of dynamic nuclear polarization sample preparations illustrating the distribution of analyte (yellow spheres) and radical polarizing agent (PA; red spheres). (a) The analyte is dissolved and homogenously distributed in the radical-containing solution; (b) an inhomogeneous nanoparticulate or microparticulate analyte is impregnated with a radical solution and the PA is restricted to the surface of the analyte domains; and (c) direct doping of the PA into the analyte particles

then be transferred via CP[125] or other methods to the heteronuclei inside the particles.

The solvent for the radical solution that is used in the impregnation step must be carefully chosen to be a nonsolvent for the solid material and should not perturb the solid-state structure of the analyte. Emsley and coworkers have previously identified several organic solvents that provide high DNP enhancements and are potentially compatible with relayed DNP.[126] Typically, 1,1,2,2-tetrachloroethane (TCE) is the best organic solvent for relayed DNP experiments; however, TCE may dissolve or cause polymorphic phase transitions in some solids.[127] In such cases, alternative solvents such as partially deuterated *ortho*-terphenyl,[120,127–130] 1,3-dibromobutane, or 1,1,2,2-tetrabromoethane may be used as solvents.[126] Notably, *ortho*-terphenyl has also allowed DNP enhancements of ~80 to be obtained with a high sample temperature of 240 K using TEKPol as the PA.[129] One other criterion for solvent selection is that the ^{13}C NMR signals of the solvent should not overlap with the analyte NMR signals, if possible. In cases where overlap with the solvent NMR signal is problematic, different solvent signal suppression methods based upon relaxation and/or dipolar couplings can be applied.[128,131] For example, pulse sequences with added spin echoes or spin-lock elements usually suppress the solvent signal.[128] Relayed DNP has been applied to many APIs and solids relevant to the pharmaceutical industry, and the observed DNP signal enhancements are summarized in Table 20.2.[122,133,140] Jannin and coworkers have also demonstrated that relayed DNP can be used to polarize organic solids for dissolution DNP experiments.[141]

The DNP-enhanced ^{1}H, ^{13}C, and ^{15}N solid-state NMR spectra of histidine hydrochloride monohydrate (histidine$\cdot$HCl$\cdot$H$_2$O) illustrate the sensitivity gains

that can be achieved with relayed DNP (Figure 20.4). The primary factors that determine the magnitude of the DNP enhancements obtained in a relayed DNP experiment are the size of the particles and the ^{1}H T_1 of the ^{1}H nuclei inside of the particles, with smaller particles and longer ^{1}H T_1 leading to larger DNP enhancements (see below for a discussion of modeling the propagation of DNP-enhanced ^{1}H polarization by spin diffusion).[122,123] The sample of histidine$\cdot$HCl$\cdot$H$_2$O was prepared by finely grinding it by hand with a mortar and pestle to minimize the particle size and reduce the distance over which the DNP-enhanced ^{1}H polarization must be transported by ^{1}H spin diffusion. The finely ground powder was then impregnated with a 16-mM TEKPol TCE solution. Histidine$\cdot$HCl$\cdot$H$_2$O is an ideal setup compound for relayed DNP experiments because it has a very long ^{1}H T_1 at low temperatures, estimated to be greater than 200 s, and it is unaffected by sample grinding or impregnation with TCE. Histidine$\cdot$HCl$\cdot$H$_2$O also has favorable dielectric properties that increase the magnitude of the microwave fields across the rotor volume, resulting in large DNP enhancements.[140]

A large ^{1}H DNP enhancement (ε_H) of ~350 was measured with ^{1}H spin echo solid-state NMR spectra for the TCE at the surface of the histidine$\cdot$HCl$\cdot$H$_2$O particles (Figure 20.4). Measurement of the ^{13}C CP-MAS DNP enhancement ($\varepsilon_{C\,CP}$) is performed by comparing the intensity of the ^{13}C solid-state NMR spectra acquired with and without microwave irradiation. For histidine$\cdot$HCl$\cdot$H$_2$O, an $\varepsilon_{C\,CP}$ of 180 was measured with a recycle delay of 10.0 s. The $\varepsilon_{C\,CP}$ of 180 indicates that relayed DNP has increased the spin polarization of the ^{1}H nuclei inside of the histidine particles by more than two orders of magnitude. In relayed DNP experiments on microcrystalline

Table 20.2.   Summary of APIs and select organic solids studied by DNP-enhanced solid-state NMR

API or solid	^{13}C CP DNP enhancement $(\varepsilon_{C\,CP})^a$	Polarizing agent	Solvent or matrix	Reference
Relayed DNP				
Acetylsalicylic acid	7	TEKPol	1,3-DBB	Hanrahan *et al.*[132]
Ambroxol HCl	92	TEKPol	TCE	Hirsh *et al.*[94]
Ambroxol HCl	54	TEKPol	OTP	Lelli *et al.*[129]
Cetirizine · 2HCl	20	TEKPol	1,3-DBB	Hirsh *et al.*[94]
Cetirizine · 2HCl	31	TEKPol	TCE	Rossini *et al.*[133]
Cetirizine · 2HCl (amorphous)	64	TEKPol	TCE	Rossini *et al.*[133]
Dicoumarol	10	AMUPol	Glycerol-d_8:D$_2$O:H$_2$O	Hanrahan *et al.*[132]
Diphenhydramine HCl	25	TEKPol	TCE	Hirsh *et al.*[94]
Furosemide	6	TEKPol	1,3-DBB	Kemp *et al.*[134]
Glucose	68	bCTbK	TBE	Rossini *et al.*[122]
Histidine · HCl · H$_2$O	225	TEKPol	TCE	Rossini *et al.*[135]
Histidine · HCl · H$_2$O	260	TEKPol	TCE	Hirsh *et al.*[94]
Ibuprofen	10	TEKPol	OTP	Lelli *et al.*[129]
Paracetamol	5	bCTbK	1,3-DBB	Rossini *et al.*[122]
Progesterone	—^{b}	AMUPol	Glycerol-d_8:D$_2$O:H$_2$O	Lee *et al.*[131]
Salicylic acid	75	TEKPol	TCE	Hanrahan *et al.*[132]
Sulfathiazole	51	TEKPol	1,3-DBB	Rossini *et al.*[122]
Theophylline Form II	11	TEKPol	TCE	Pinon *et al.*[127]
Theophylline Form I	4	TEKPol	Toluene-d_8:decalin	Pinon *et al.*[127]
Theophylline Form IV	2	TEKPol	TCE	Pinon *et al.*[127]
Theophylline Form M	2	AMUPol	Glycerol-d_8:D$_2$O:H$_2$O	Pinon *et al.*[127]
Theophylline Form II	17	TEKPol	TCE	Rossini *et al.*[136]
Theophylline Form II	12	TEKPol	EtBr$_4$	Mollica *et al.*[137]
Direct doping				
Indomethacin	14	bTtereph	—	Ong *et al.*[130]
Clotrimazole (1%, wt%)	17	AMUPol	—	Ni *et al.*[138]
Posaconazole (20%, wt%)	32	AMUPol	—	Ni *et al.*[138]
Cellulose	20	TOTAPOL	—	Takahashi *et al.*[139]

Note: API, active pharmaceutical ingredient; bTtereph, bis-TEMPO terephthalate; CP, cross-polarization; DNP, dynamic nuclear polarization; MAS, magic angle spinning; 1,3-DBB, 1,3-dibromobutane; OTP, *ortho*-terphenyl; TBE, 1,1,2,2-tetrabromoethane; TCE, 1,1,2,2-tetrachloroethane.
^{a}The ^{13}C CP-MAS DNP signal enhancement factor $(\varepsilon_{C\,CP})$ is determined by comparing the intensity of the NMR spectra acquired with and without with microwave irradiation and the same experimental conditions. See Figure 20.4.
^{b}The enhancement was not specified.

solids, the DNP enhancement for the ^{1}H nuclei in the interior of the particles is normally lower than that observed for the ^{1}H nuclei of the solvent at the surface of the particles because some of the DNP-enhanced ^{1}H spin polarization is lost to longitudinal relaxation during the transport process.[122,123] With the gain in sensitivity provided by DNP, it is possible to obtain natural isotopic abundance ^{13}C and ^{15}N solid-state NMR spectra with high signal-to-noise ratios in short experiment times; the ^{13}C solid-state NMR spectrum has a signal-to-noise ratio of ~700

after an 80-s experiment, and the ^{15}N solid-state NMR spectrum has a signal-to-noise ratio of ~90 after an 8-min experiment. The signal-to-noise ratios of the DNP-enhanced solid-state NMR spectra obtained with natural isotopic abundance are similar to those obtained with conventional solid-state NMR experiments on isotopically enriched samples. It is also important to note that the DNP-enhanced ^{13}C and ^{15}N solid-state NMR spectra of histidine · HCl · H$_2$O show good resolution with peak widths on the order of 0.7 ppm or less. This is a significantly better resolution

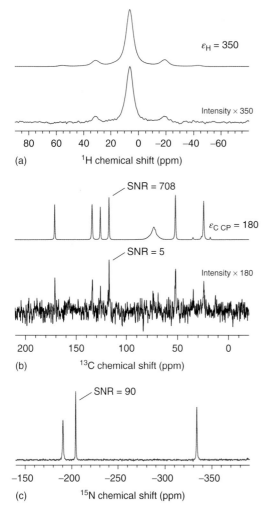

Figure 20.4. Examples of relayed dynamic nuclear polarization experiments on histidine · HCl · H₂O. (a) ¹H spin echo, (b) ¹³C cross-polarization magic angle spinning, and (c) ¹⁵N cross-polarization magic angle spinning solid-state NMR spectra of finely ground histidine · HCl · H₂O impregnated with a 16-mM TEKPol 1,1,2,2-tetrachloroethane solution. The sample temperature was ~110 K. The ¹H and ¹³C solid-state NMR spectra acquired with (blue traces) and without (black traces) microwave irradiation are compared to illustrate the signal enhancement provided by dynamic nuclear polarization. The signal-to-noise ratio (SNR) of the ¹³C and ¹⁵N solid-state NMR spectra are indicated. The ¹H, ¹³C, and ¹⁵N solid-state NMR spectra were obtained with coaddition of eight scans and a 3.0 s polarization delay, eight scans and a 10.0 s polarization delay, and eight scans and a 60.0 s polarization delay, respectively

than is typically observed for DNP solid-state NMR spectra of analytes dissolved in glass-forming radical solutions where peak widths may be on the order of 1 to 4 ppm. Relayed DNP experiments often result in high-resolution solid-state NMR spectra because the crystalline nature of the sample is unaffected by the sample preparation.[122,142]

One important issue to keep in mind when preparing samples for relayed DNP is that processes such as grinding, drying, impregnation, and tablet compression may cause some organic solids and APIs to convert to different polymorphs or result in amorphous phases.[10,26] This is an important issue for the development of formulated pharmaceuticals because undesired API forms may reduce the efficacy of the formulation.[143] Recently, Emsley and coworkers demonstrated that both room temperature grinding and impregnation with nonsolvents could cause phase transitions in some polymorphs and solvates of theophylline (Figure 20.5).[127] Therefore, care should be taken to ensure that the sample preparation steps of grinding and impregnation do not cause unwanted phase transitions when preparing samples for relayed DNP experiments. Polymorphic phase transitions can be easily monitored by comparing the ¹³C solid-state NMR spectra of the pristine material to those of the materials subjected to grinding and/or impregnation with various solvents (Figure 20.5).[127] For Form I of theophylline, it was determined that phase transitions to other polymorphs could be prevented by grinding at reduced temperatures and using PAs dissolved in a mixture of toluene and decalin for the impregnation step.[127] For some materials, it may not be possible to use relayed DNP because the particle sizes are too large, the ¹H T_1 of the solid is too short (i.e., 2 s or less), or it is not possible to identify compatible DNP solvents that avoid phase transitions or dissolution. In such cases, direct doping may be used to introduce the PA into the API.

20.2.2   Direct Doping with PA

DNP-enhanced solid-state NMR spectroscopy has previously been performed on amorphous APIs, polymers, API-polymer mixtures (ASDs), and crystalline organic solids that were prepared by directly doping the analyte with the PA (Figure 20.3c). Direct doping can be achieved in a couple of different ways. First, the PA may be introduced into the sample during the synthesis or precipitation of the material. Griffin and

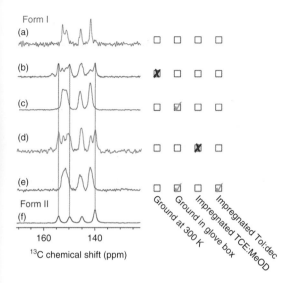

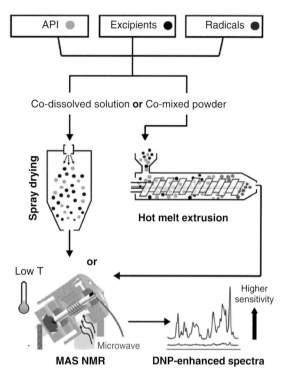

Figure 20.5. Dynamic nuclear polarization-enhanced ^{13}C crosspolarization magic angle spinning NMR spectra recorded at 105 K of theophylline Form I subjected to the different sample preparation steps associated with relayed dynamic nuclear polarization. (a) Pristine recrystallized Form I theophylline. (b) Theophylline Form I manually ground at room temperature, resulting in a mixture of Forms I and II. (c) Form I manually ground at a temperature of ~220 K in a glovebox. (d) Impregnation of ground Form I with a 1,1,2,2-tetrachloroethane (TCE)/d_4-methanol TEKPol solution results in a mixture of Forms I and II. (e) Ground Form I impregnated with a toluene-d_8/decalin TEKPol solution shows no conversion to Form II. (f) The reference spectrum of pure theophylline Form II. (Reprinted with permission from Pinon *et al.*[127] Copyright 2015 American Chemical Society)

Figure 20.6. Sample preparation protocols used to directly dope amorphous solid dispersions prepared by spray drying or hot-melt extrusion. API, active pharmaceutical ingredient; DNP, dynamic nuclear polarization; MAS, magic angle spinning. (Reprinted with permission from Ni *et al.*[138] Copyright 2017 American Chemical Society)

coworkers performed DNP-enhanced solid-state NMR on amorphous indomethacin obtained from hot-melt extrusion.[130] The PA was introduced by directly dissolving it in the molten indomethacin during the melt extrusion process.[130] De Paëpe and coworkers performed DNP-enhanced solid-state NMR experiments on diphenylalanine nanotubes that were directly doped by adding the PA into the solution used to crystallize and grow the nanotubes.[144] Recently, Su and coworkers have incorporated PA into ASDs that were prepared by either spray drying or hot-melt extrusion (Figure 20.6).[138] One obvious disadvantage of direct doping is that the PA must be incorporated during synthesis or preparation of the sample.

Alternatively, it may be possible to postsynthetically directly dope the sample with PA by using solvents that swell or dissolve the solid material. For example, De Paepe and coworkers performed DNP-enhanced solid-state NMR on cellulose, which was impregnated or swollen with aqueous TOTAPOL solutions.[139] Viel and coworkers have performed DNP-enhanced solid-state NMR experiments on polymers that were either dissolved in glass-forming PA solutions or swollen with PA solution.[145] In both of these experiments, the sample partially dissolves or is swollen with the radical PA solution. A potential disadvantage of using a solvent that swells or dissolves the solid analyte is that it may perturb the solid-state structure. In all of these direct-doping methods, it is also possible that there is some degree of phase segregation between the analyte and the PA, that is, a case intermediate to that of Figure 20.3(b,c). In such cases, ^1H spin diffusion likely assists with the distribution of DNP-enhanced ^1H polarization. Finally, it is also important to keep in mind that in directly doped samples, most of the nuclei in the sample are in close proximity to PA molecules and the doping process may alter the structure of the solid. Therefore, mechanisms such as signal broadening due to structural disorder, paramagnetic broadening or relaxation, and MAS-induced depolarization[146,147] may reduce the absolute sensitivity gains provided by DNP experiments on directly doped samples.[122,139,148] On the other hand, direct doping usually results in shorter, more favorable signal buildup times (T_B) because the radicals will be homogeneously distributed throughout the sample.

20.3 MODELING ^{1}H SPIN DIFFUSION IN RELAYED DNP EXPERIMENTS

Models of ^{1}H spin diffusion have been widely applied in solid-state NMR spectroscopy to understand diverse phenomena such as enhanced longitudinal relaxation and mixing and segregation of solid phases and to estimate the sizes of domains or particles.[44,122,123,149–156] Numerical and analytical models of ^{1}H spin diffusion can be used to obtain a fundamental understanding of the factors that determine the magnitude of the DNP enhancements in relayed DNP experiments.[122,123,155] As we describe in the applications section below, measurements of DNP enhancements can be combined with models of ^{1}H spin diffusion to estimate the diameter of API particles or domains within formulated pharmaceuticals.

The transport of DNP-enhanced polarization by ^{1}H spin diffusion in a heterogeneous system such as nanoparticles or microparticles that are remotely polarized (Figure 20.3b) can be modeled with Fickian diffusion and described by the following differential equation[123]:

$$\frac{\partial P}{\partial t} = D(x) \cdot \Delta P - \frac{P(x, t) - P_0(x)}{T_1(x)} \quad (20.1)$$

where x is the position from the border (crystal surface) between the radical solution and the analyte, t the time, P the instantaneous polarization at position x and time t, P_0 the local equilibrium polarization level, D the diffusion rate at position x, T_1 the longitudinal relaxation time of the analyte at position x, and ΔP the Laplacian of the polarization. The form of ΔP depends on the symmetry of the system. P_0 corresponds to the polarization that would be reached in the absence of spin diffusion. In the radical solution, the local equilibrium polarization without microwaves is the

Boltzmann polarization, which is arbitrarily assigned a value of 1. With microwaves, $P_0 = \varepsilon_0$, where ε_0 is the DNP enhancement of the solvent, which is typically between 100 and 200. Inside of the analyte domains, P_0 is always 1.

Griffin and coworkers used equation (20.1) to model relayed DNP in polypeptide nanocrystals suspended in a mixed water–glycerol solution with PA.[123] Assuming 1D symmetry and steady-state conditions (i.e., $\partial P/\partial t = 0$, realized at long polarization times) and by integrating the DNP signal enhancement over the 1D region, they obtained the following analytical expression for relayed DNP enhancement:

$$\varepsilon = \varepsilon_0 \frac{2\sqrt{DT_1}}{w} \tanh\left(\frac{w}{2\sqrt{DT_1}}\right) \quad (20.2)$$

In this equation, ε is the integrated DNP enhancement inside of the remotely polarized 1D region, ε_0 is the DNP enhancement outside of the remotely polarized region (i.e., at the surface of a crystal), T_1 is the nuclear longitudinal relaxation time inside the polarized region, w is the width of the remotely polarized region, and D is the spin diffusion rate, which is typically between 1×10^4 and 1×10^5 Å^2 s^{-1} in protonated organic solids. Equation (20.2) is very useful for quickly estimating the expected magnitude of relayed DNP enhancements for different particle sizes and T_1 values. For example, equation (20.2) predicts that for a 2-μm-wide 1D crystal, a ^{1}H T_1 of 100 s, a surface DNP enhancement (ε_0) of 100, and a spin diffusion rate of 1×10^5 Å^2 s^{-1} the integrated DNP enhancement observed inside the crystal would be a factor of 32 (i.e., 32% of the surface DNP enhancement). For the same conditions, but with $T_1 = 10$ s and $T_1 = 1000$ s, equation (20.2) predicts the DNP enhancements inside the 1D crystal are 10 and 76, respectively. These examples illustrate that larger relayed DNP enhancements are obtained for solids with longer proton T_1. Proton T_1s on the order of hundreds of seconds are commonly observed for many organic solids and APIs,[122] especially at the cryogenic sample temperatures used for DNP experiments. All of the solids and APIs where relayed DNP has provided high DNP enhancements (Table 20.2) likely have long intrinsic proton T_1s that are greater than 30 s at 105 K.

Emsley and coworkers numerically solved equation (20.1) under assumptions of spherical symmetry in order to obtain a more detailed model of relayed DNP in microcrystalline solids and predict how relayed DNP affects relaxation and buildup of the NMR signal.[122] Equation (20.1) must be numerically solved

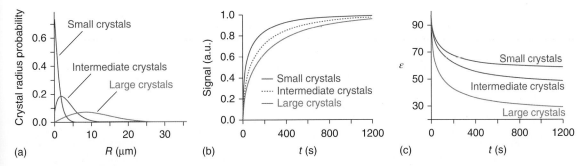

Figure 20.7. Dynamic nuclear polarization (DNP) enhancements and signal buildup rates for different spherical particle size distributions calculated by numerical solution of the diffusion equation and weighting the calculated enhancement by position, particle volume, and crystal probability. (a) Different model crystal radius Weibull distributions. (b) The predicted magnitude of the signal as a function of the polarization time. (c) Predicted DNP enhancements as a function of polarization time. In all cases, the proton T_1 inside the crystals was assumed to be 700 s and the DNP-enhanced proton polarization at the surface of the crystals was assumed to build up with a time constant of 1 ms, to a polarization 100 times that of the equilibrium level ($\varepsilon_0 = 100$). (Adapted with permission from Rossini *et al.*[122] Copyright 2012 American Chemical Society)

to determine the instantaneous polarization [$P(x, t)$] as a function of position and time in a spherically symmetric system because there is no general analytical solution. Figure 20.7 illustrates the signal buildup rates and magnitude of the DNP enhancement predicted by numerically solving equation (20.1), assuming spherical particles and averaging the calculated polarization over position, particle volume, and radius probability for several different particle radius distributions.[122] The spin diffusion model predicts that particles with smaller radii exhibit faster polarization buildup (Figure 20.7b) and larger DNP enhancements (Figure 20.7c); consequently, larger relayed DNP enhancements can be obtained for microparticulate powders by grinding the sample to reduce the average particle size. The magnetization builds up more quickly for smaller particles because the signal buildup is primarily driven by diffusion of the DNP-enhanced magnetization into the particles. The rate of the signal buildup also depends upon the gradient of the polarization inside and outside the crystal, so that larger surface DNP enhancements should translate to faster signal buildups and larger DNP enhancements inside the crystal. The spin diffusion model also predicts that the magnetization buildup can be fit with stretched exponential functions and that the apparent magnitude of the DNP enhancement will decrease as the polarization time is increased (Figure 20.7c). Both of these predictions have been confirmed by experiments.[122,133,155] Larger enhancements are obtained at shorter polarization times because the buildup of DNP-enhanced polarization at

the surface of the particles accelerates the diffusion of polarization into the particles, resulting in shorter signal buildup time constants in the DNP-enhanced NMR experiments as compared to the thermally polarized NMR experiments. Consequently, larger DNP enhancements are usually measured at short polarization delays, and DNP enhancement usually decreases as the polarization time is increased in relayed DNP NMR experiments. Note that in glassy frozen solutions where the radical PA is homogeneously distributed, the CE DNP enhancement usually shows no variation with the polarization delay.[155]

Figure 20.8 shows the steady-state DNP enhancements (ε_∞, the enhancement observed at long polarization times) predicted using numerical diffusion models for spherical particles with different radii and for several different proton T_1s inside the crystals. This plot is useful to estimate the magnitude of the relayed DNP enhancement if the particle size and proton T_1 of the solid are known. Recently, Pinon *et al.* have described phenomenological relations between ε_∞ and the radius of a spherical analyte (R)[155]:

$$\varepsilon_\infty = 1 + (\varepsilon_0 - 1)\frac{3\sqrt{DT_1}}{R}$$
$$\times \left[\coth\left(\frac{R}{\sqrt{DT_1}}\right) - \frac{\sqrt{DT_1}}{R}\right] \quad (20.3)$$

The definitions of the other terms in equation (20.3) are the same as those in equation (20.2). Phenomenological relationships between steady-state enhancement and the size of remotely polarized analytes were

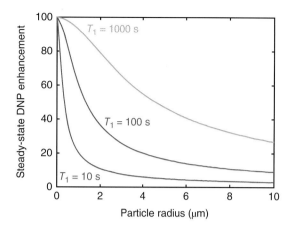

Figure 20.8. Simulated steady-state enhancement (ε_∞) of a spherical particle as a function of the particle radius. Curves are shown for proton T_1s inside the particle of 10, 100, and 1000 s. In the simulation, $D = 1.0 \times 10^5$ Å^2 s^{-1} inside of the particles, and the PA solution at the surface of the particles has a buildup time of 3.0 s and $\varepsilon_0 = 100$. DNP, dynamic nuclear polarization

also described for linear and cylindrical objects.[155] Equation (20.3) and/or Figure 20.8 can be used to estimate the particle radius for samples where the proton T_1 of the analyte, surface or solvent DNP enhancement, and steady-state DNP enhancement for the analyte have been measured. The application of relayed DNP to determine particle or domain sizes is described in more detail in the applications section below.

20.4 APPLICATIONS OF DNP-ENHANCED SOLID-STATE NMR SPECTROSCOPY TO PHARMACEUTICALS

In this section, we describe some of the advanced NMR experiments that have been performed with DNP on both pure and formulated APIs and emphasize the novel structural information provided by these experiments. Conventional solid-state NMR experiments on pure and formulated APIs are often hindered by poor sensitivity. Poor sensitivity typically limits NMR experiments to the acquisition of basic 1D ^{1}H and ^{13}C NMR experiments. DNP-enhanced solid-state NMR provides access to novel information about the molecular and macroscopic structure of pure and formulated APIs by enabling multidimensional NMR experiments that would normally be infeasible with conventional solid-state NMR spectroscopy. NMR experiments with unreceptive nuclei such as ^{14}N, ^{15}N, and ^{35}Cl also become much more amenable with DNP.

20.4.1 Multidimensional solid-state NMR Experiments on Pure APIs

Natural isotopic abundance double-quantum single quantum (DQ–SQ) ^{13}C–^{13}C homonuclear NMR-experiments are frequently accelerated or enabled by DNP.[122,139] DQ homonuclear correlation NMR experiments provide valuable structural information because they allow resonances to be assigned, the chemical bonding or connectivity of atoms can be directly observed, internuclear distances can be measured, and the relative orientations of the different NMR interaction tensors and local axes of functional groups can potentially be determined.[158–160] For example, ^{29}Si–^{29}Si and ^{31}P–^{31}P DQ–SQ correlation NMR experiments have previously been applied for structure determination of inorganic phosphates,[161] zeolites,[162] and silica materials.[163] ^{13}C DQ NMR experiments on ^{13}C-labeled samples have also been applied to determine torsion angles in organic solids and biomolecules.[164–166]

However, natural isotopic abundance ^{13}C homonuclear DQ NMR experiments typically require extremely long signal-averaging periods of days to weeks with conventional solid-state NMR spectroscopy.[58,160,167] The poor sensitivity arises because only 1 in 10 000 carbon atoms exists as ^{13}C–^{13}C spin pairs that give rise to observable DQ NMR signals. However, with DNP, both scalar and dipolar 2D ^{13}C–^{13}C homonuclear correlation solid-state NMR experiments can be easily obtained with experiment times that are typically on the order of 12 h or less.[122,139] For example, Emsley and coworkers applied relayed DNP to accelerate scalar and dipolar ^{13}C–^{13}C DQ–SQ homonuclear correlation spectra of organic solids such as glucose and sulfathiazole (Figure 20.9a).[122] In parallel, De Paëpe and coworkers applied DNP to cellulose swollen with TOTAPOL to accelerate dipolar ^{13}C–^{13}C DQ–SQ correlation experiments.[139]

Dipolar DQ-SQ ^{13}C homonuclear solid-state NMR experiments can also be applied to estimate carbon-carbon internuclear distance and probe crystal packing. For example, Mollica *et al.* applied relayed DNP to theophylline to measure dipolar ^{13}C–^{13}C DQ–SQ

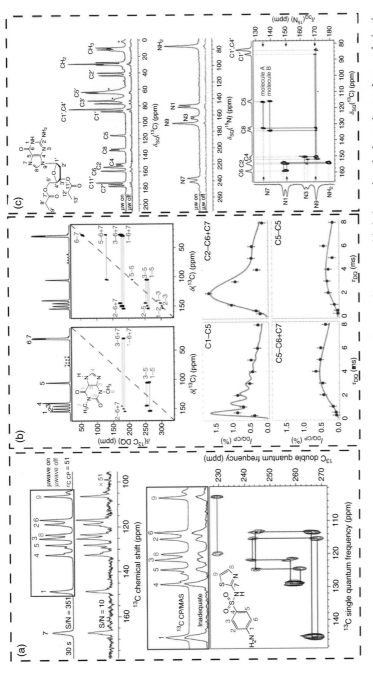

Figure 20.9. Examples of challenging multidimensional solid-state NMR experiments enabled by relayed dynamic nuclear polarization (DNP). (a) DNP-enhanced two-dimensional (2D) scalar $^{13}C-^{13}C$ double-quantum single-quantum (DQ-SQ) homonuclear correlation spectrum of sulfathiazole obtained with the refocused INADEQUATE pulse sequence in a total experiment time of 16 h. MAS, magic angle spinning. (Reprinted with permission from Rossini *et al.*[122] Copyright 2012 American Chemical Society). (b) DNP-enhanced 2D dipolar $^{13}C-^{13}C$ DQ-SQ homonuclear correlation spectra of theophylline Form II and the DQ signal buildup curves extracted from 2D spectra obtained with different mixing times. The DQ signal buildup curves corresponding to ideal buildup curves generated from the carbon–carbon distances and dipolar couplings observed in the crystal structure of theophylline. (Reprinted with permission from Mollica *et al.*[137] Copyright 2015 WILEY–VCH Verlag GmbH & Co. KGaA, Weinheim). (c) DNP-enhanced 2D $^{13}C-^{15}N$ cross-polarization (CP)– heteronuclear correlation spectrum of a self-assembled 2′-deoxyguanosine derivative. (Reprinted with permission from Märker *et al.*[157] Copyright 2015 American Chemical Society)

correlation NMR spectra and measure DQ coherence signal buildup curves (Figure 20.9b).[137] The carbon-carbon internuclear distances from CSP trial structures of theophylline were then used to generate simulated analytical DQ buildup curves. Comparison of the predicted and experimental DQ buildup curves showed that the true crystal structure gave the best agreement with the observed DQ buildup curves, whereas the predicted DQ buildup curves for other polymorphs showed significantly worse agreement with the experimental curve. These findings suggested that ^{13}C–^{13}C DQ buildup curves are sensitive to molecular conformation and crystal packing and may be used to distinguish polymorphs and identify the correct structure among CSP trial structures.

De Paëpe and coworkers applied DNP-enhanced solid-state NMR to obtain dipolar ^{13}C–^{13}C DQ–SQ correlation NMR spectra and DQ buildup curves for diphenylalanine nanotubes that were directly doped with PA.[144,168] With the carbon-carbon distances observed in the established crystal structure and with DFT-calculated chemical shift tensor orientations, it was possible to reproduce the experimentally determined ^{13}C–^{13}C DQ buildup curves.[168] It is worth noting that the DQ buildup curves could only be accurately simulated if carbon–carbon distances up to ~7 Å (dipolar couplings of ~20 Hz) were considered.[168] These results again highlight the sensitivity of dipolar DQ buildup curves to crystal packing and suggested that DQ–SQ ^{13}C–^{13}C NMR experiments can potentially be used to test the validity of proposed structural models. However, as noted by the authors, it is challenging to easily extract carbon–carbon distances from the observed DQ buildup curves because accurate simulation of the curve for a set of carbon resonances required knowledge of the chemical shift tensors for each carbon and the relative orientations with respect to the other carbon atoms within the lattice. Additionally, multiple carbon–carbon dipolar couplings must be included in the simulations because there are multiple sets of carbon spin pairs due to the presence of adjacent, symmetry-related molecules within the crystal lattice.

DNP has also been applied to accelerate 2D HETCOR solid-state NMR experiments. Two-dimensional ^{1}H–^{13}C and ^{1}H–^{15}N HETCOR experiments in conjunction with ^{1}H homonuclear decoupling sequences are widely employed to obtain high-resolution ^{1}H solid-state NMR spectra, measure heteronuclear dipolar couplings, and perform resonance assignments in organic solids and APIs.[55,57,58,169–172] Conventional 2D ^{1}H–^{13}C HETCOR NMR experiments typically require experiment times of hours to days depending upon the characteristics of the sample, whereas 2D ^{1}H–^{15}N HETCOR NMR experiments are often infeasible with natural isotopic abundance. Relayed DNP has previously been used to rapidly acquire 2D ^{1}H–^{13}C and ^{1}H–^{15}N HETCOR solid-state NMR spectra of the API cetirizine hydrochloride[133] and of a self-assembled, oligomeric organic solid formed by the reaction of CO_2 with amines.[173] With DNP, 2D ^{1}H–^{13}C or ^{1}H–^{15}N HETCOR NMR spectra can typically be obtained in experiment times of minutes to a few hours, even in tablets that contain less than 10 wt.% API loading (see below).

Very recently, it has been demonstrated that conventional and DNP-enhanced 2D ^{1}H–^{13}C HETCOR NMR spectra can be used to significantly enhance the resolution of solid-state NMR spectra by reducing signal broadening from anisotropic bulk magnetic susceptibility (ABMS).[132] ABMS broadening frequently occurs in solids with aromatic groups and results in large inhomogeneous ^{1}H and ^{13}C linewidths on the order of 1–2 ppm.[174,175] The low resolution caused by ABMS impedes the analysis of solid-state NMR spectra of APIs and organic solids. Two-dimensional HETCOR NMR spectra can reduce or eliminate inhomogeneous ABMS broadening because heteronuclear spin pairs experience the same local field, resulting in elongated, elliptical cross-peaks due to correlated inhomogeneous broadening.[176] NMR spectra with improved resolution can then be extracted from the individual rows or columns of a 2D HETCOR spectrum.[176] For example, the ^{13}C solid-state NMR spectra of salicylic acid extracted from the rows of a DNP-enhanced 2D ^{1}H–^{13}C HETCOR spectrum obtained with ^{1}H homonuclear decoupling showed a factor 2.7 direct improvement in resolution as compared to the 1D ^{13}C CP–MAS spectrum.[132] The direct gain in resolution depends upon the ratio of the ^{1}H and ^{13}C inhomogeneous linewidths to the ^{1}H homogeneous linewidth, the latter determined by the efficiency of the homonuclear decoupling.[132] Conventional and DNP-enhanced 2D ^{1}H–^{13}C HETCOR NMR spectra were used to reduce ABMS broadening and resolve the overlapping NMR signals associated with the Form 1 and Form 2 polymorphs of aspirin within a mixture of both forms.[132] Therefore, in addition to the indirect gains in resolution achieved by adding a second spectral dimension, HETCOR experiments may provide further direct gains in resolution by reducing inhomogeneous ABMS broadening.

In addition to HETCOR experiments involving protons, DNP has also enabled acquisition of 2D HETCOR spectra that correlate moderate-γ and low-γ nuclei (e.g., ^{13}C–^{15}N, ^{13}C–^{14}N, and ^{13}C–^{35}Cl). For instance, De Paëepe and coworkers recently applied relayed DNP to acquire natural isotopic abundance 2D ^{13}C–^{15}N dipolar HETCOR solid-state NMR spectra of a self-assembled 2′-deoxyguanosine derivative (Figure 20.9c).[157] This experiment is impossible without DNP because with the natural isotopic abundance of ^{13}C and ^{15}N, only 1 in 27 000 carbon–nitrogen atom pairs will correspond to the ^{13}C–^{15}N isotopomer. Two-dimensional ^{13}C–^{15}N HETCOR solid-state NMR spectra could potentially be very informative probes of structure in APIs because they could be used for resonance assignment, resolution enhancement, carbon–nitrogen distance measurements, and mapping of the connectivity or bonding of carbon and nitrogen atoms. In addition to correlations involving two spin-1/2 nuclei, DNP has also been used to correlate ^{13}C to quadrupolar nuclei. For example, Schurko and coworkers have recently demonstrated acquisition of a 2D ^{13}C{^{35}Cl} dipolar-heteronuclear multiple-quantum coherence (D-HMQC) correlation NMR spectrum of histidine ·HCl·H$_2$O.[94] O'Dell and coworkers also applied relayed DNP to acquire 2D ^{13}C{^{14}N^{OT}} correlation spectra.[135] Both of these results are discussed in more detail in the section describing DNP solid-state NMR experiments with quadrupolar nuclei.

20.4.2 DNP-enhanced Solid-state NMR Experiments with Quadrupolar Nuclei

Another application of DNP is to enable and/or accelerate NMR experiments with quadrupolar nuclei such as ^{2}H, ^{14}N, and ^{35}Cl. Approximately 73% of the spin active NMR isotopes in the periodic table are quadrupolar nuclei (nuclei with spin > 1/2). Some of the quadrupolar nuclei commonly encountered in APIs are listed in Table 20.1. Quadrupolar nuclei typically give rise to broad NMR signals because of broadening by the quadrupolar interaction (QI).[177–180] Useful structural information may be obtained from the solid-state NMR spectra of quadrupolar nuclei because the broadening and appearance of the spectrum depend upon the spherical and rotational symmetry at the nuclear site.[177–180] However, signal broadening

from the QI often leads to reduced sensitivity and poor resolution. DNP can address the limitation of poor sensitivity by enhancing the nuclear spin polarization. The resolution of solid-state NMR spectra of quadrupolar nuclei can be improved by adding a second spectral dimension that can resolve overlapping signals.

^{14}N is a spin-1 quadrupolar nucleus that is appealing for NMR experiments due to its high natural isotopic abundance of 99.6%. However, ^{14}N often gives rise to extremely broad (>1-MHz) solid-state NMR spectra due to broadening by the first-order QI. Consequently, direct acquisition of ^{14}N solid-state NMR spectra is typically performed on stationary samples with special Carr-Purcell Meiboom-Gill (CPMG)-based wideline solid-state NMR techniques.[181] The ^{14}N electric field gradient (EFG) tensor parameters determined in these experiments are useful for distinguishing polymorphs and probing the coordination environment or protonation state of the nitrogen atom.[85,182,183] However, the static wideline ^{14}N solid-state NMR experiments are often limited by poor sensitivity and resolution. As an alternative to static wideline experiments, fast MAS, proton detection, and rotor synchronized HMQC pulse sequences may be used to indirectly detect ^{14}N solid-state NMR spectra indirectly.[184–186] Fast MAS ^{1}H{^{14}N} HMQC experiments have previously been applied to classify multicomponent solid APIs as salts or cocrystals and probe for interactions between APIs and polymers in an ASD.[84] Unfortunately, the efficiency of proton-detected HMQC experiments is usually low when the nitrogen sites possess large quadru- polar coupling constants and/or when the ^{14}N nucleus is weakly coupled to ^{1}H. Bodenhausen and coworkers have previously applied DNP to enhance the sensitivity of 2D ^{13}C–^{14}N D-HMQC experiments on ^{13}C-labeled proline dissolved in glass-forming PA solution.[187]

Overtone ^{14}N (^{14}N^{OT}) NMR has been demonstrated as a method to improve the resolution of ^{14}N solid-state NMR spectra.[188,189] In a ^{14}N^{OT} NMR experiment, the ^{14}N spins are irradiated and observed at twice the fundamental Larmor frequency of the ^{14}N nucleus. The MAS ^{14}N^{OT} solid-state NMR signals typically have widths of less than 10 kHz because the ^{14}N^{OT} NMR spectra are unaffected by the first-order QI.[190,191] Another advantage of ^{14}N^{OT} NMR is that MAS experiments are straightforward because the ^{14}N^{OT} spectrum is tolerant to deviations from the magic angle. Unfortunately, the ^{14}N^{OT} NMR experiments suffer from poor sensitivity because

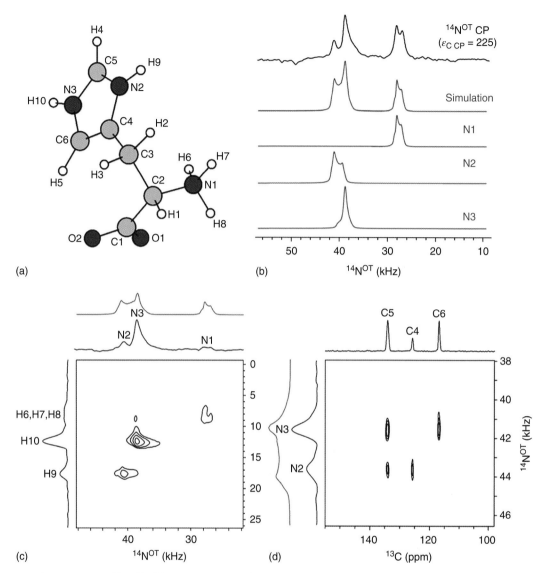

Figure 20.10. Overtone ^{14}N (^{14}N^{OT}) solid-state NMR spectra of histidine•HCl•H$_2$O obtained with relayed dynamic nuclear polarization. (a) Molecular structure of histidine•HCl•H$_2$O and atom labeling scheme. (b) Experimental ^{1}H–^{14}N^{OT} cross-polarization (CP) magic angle spinning solid-state NMR spectrum (black trace) and simulations of the ^{14}N^{OT} spectra of the three different nitrogen atoms. (c) Twodimensional ^{1}H–^{14}N^{OT} CP heteronuclear correlation spectrum obtained with ^{1}H homonuclear decoupling applied during the indirect dimension evolution period. The total experiment time was approximately 8 h. (d) Two-dimensional ^{13}C{^{14}N^{OT}} HMQC solid-state NMR spectrum. (Reproduced from Rossini *et al.*,[135] with permission from the Physical Chemistry Chemical Physics Owner Societies)

they require excitation of formally forbidden DQ transitions.

In order to address these limitations, O'Dell and coworkers applied relayed DNP to improve the

sensitivity of ^{14}N^{OT} solid-state NMR experiments on microcrystalline amino acids (Figure 20.10).[135] They demonstrated that under MAS, CP with short contact times could transfer DNP-enhanced ^{1}H polarization to the $^{14}N^{OT}$ transition. High-quality 1D $^1H-^{14}N^{OT}$ CP–MAS solid-state NMR spectra were typically obtained in about an hour with DNP. Notably, CP was found to provide excitation bandwidths similar to or better than direct $^{14}N^{OT}$ excitation pulses. The observed $^{14}N^{OT}$ second-order quadrupolar MAS powder patterns showed good agreement with numerically exact simulations[191] and allowed the ^{14}N quadrupolar parameters to be accurately determined from fits of the spectra (Figure 20.10). However, comparison of DNP-enhanced 1D $^{14}N^{OT}$ and ^{15}N solid-state NMR spectra showed showed that ^{15}N NMR provides significantly better sensitivity than $^{14}N^{OT}$ NMR, despite the 270-fold lower isotopic abundance of ^{15}N in comparison to ^{14}N.[135] Therefore, the primary motivation for performing $^{14}N^{OT}$ NMR experiments is to obtain high-resolution measurements of ^{14}N EFG tensor parameters, rather than to improve sensitivity.

With DNP, it was also possible to perform 2D $^{14}N^{OT}$ solid-state NMR experiments.[135] For example, a 2D $^1H-^{14}N^{OT}$ CP-HETCOR spectrum with ^1H homonuclear decoupling applied in the indirect dimension allowed the overlapping $^{14}N^{OT}$ NMR signals of histidine $\cdot$HCl$\cdot$H$_2$O to be resolved on the basis of correlations to different ^1H NMR signals in the indirect dimension (Figure 20.10c). A DNP-enhanced 2D $^{13}C\{^{14}N^{OT}\}$ HMQC spectrum was obtained in a total experiment time of about 3.8 h (Figure 20.10). The HMQC spectrum allowed overlapping $^{14}N^{OT}$ powder patterns to be resolved, and the connectivity or bonding of the carbon and nitrogen atoms was determined. Note that 2D $^{13}C\{^{14}N^{OT}\}$ HMQC experiments had an efficiency of ~2.8% in comparison to a standard ^{13}C CP-MAS spectrum. In comparison, natural abundance $^{13}C-^{15}N$ CP–HETCOR experiments can provide at most an efficiency of 0.37% (see above). Therefore, 2D $^{13}C\{^{14}N^{OT}\}$ or $^{13}C\{^{14}N\}$ HMQC is likely the highest sensitivity method to observe heteronuclear carbon and nitrogen correlations within natural isotopic abundance samples. However, 2D $^{13}C-^{15}N$ CP–HETCOR spectra will certainly provide better resolution and are more straightforward to interpret.

Hydrochloride salts in both pure and dosage forms can be studied with ^{35}Cl solid-state NMR spectroscopy.[93,95,96] ^{35}Cl solid-state NMR has previously been applied to differentiate the polymorphs of APIs and detect APIs within formulations.[93,95,96] ^{35}Cl is a highly abundant (natural abundance (NA) = 75.5%) half-integer quadrupolar nucleus ($I = 3/2$). Due to broadening by the second-order QI, the

^{35}Cl solid-state NMR spectra of hydrochloride salts are normally quite broad and acquisition is usually performed with static wideline CPMG techniques such as wideband, uniform rate, smooth truncation-CPMG (WCPMG) or broadband adiabatic inversion (BRAIN)–CP/WCPMG.[181] However, it may be challenging to obtain 1D ^{35}Cl solid-state NMR spectra of solids with unfavorable ^1H and ^{35}Cl relaxation times (long T_1s and short T_2 or $T_{1\rho}$) and/or in formulations where the loading of the API is low.

Schurko and coworkers applied relayed DNP to enhance the sensitivity of static $^1H-^{35}Cl$ BRAIN–CP/ WCPMG experiments on both pure and formulated APIs (Figure 20.11).[94] They obtained ^{35}Cl CP DNP enhancements ($\varepsilon_{Cl\,CP}$) on the order of 7–100 from stationary samples. The $\varepsilon_{Cl\,CP}$ obtained on static samples was between 3 and 8 times lower than those measured (with ^{13}C CP–MAS experiments) on samples undergoing MAS. It was previously shown that DNP enhancements obtained from nitroxide biradical PA are maximized with MAS frequencies between 2 and 8 kHz and that there was a large reduction in DNP enhancement for static samples.[106,116]

The reduced enhancements for static samples likely occurred because there are fewer radicals in the proper orientation to participate in CE DNP. The magnitude of the DNP enhancements in the static solid-state NMR experiments was increased substantially by slowly spinning the sample (~2-kHz MAS frequency) during the polarization delay; then sample spinning was halted prior to applications of pulses and acquisition of the static NMR spectra.[94] This procedure was termed *spinning-on/spinning-off (SOSO)*. However, this procedure was only viable for samples with long proton T_1 where there was enough time to start and stop MAS during the polarization delays.

With the large sensitivity gains from DNP and BRAIN-CP/WCPMG, wideline ^{35}Cl solid-state NMR spectra of pure APIs were obtained in total experiment times of a few minutes.[94] Given the short experiment times on the pure APIs, it was also possible to obtain DNP-enhanced ^{35}Cl solid-state NMR spectra of tablets with low API loading (see below). Finally, a 2D $^{13}C\{^{35}Cl\}$ D-HMQC spectrum of histidine $\cdot$HCl$\cdot$H$_2$O was also acquired. This spectrum allowed the proximity of the ^{13}C and ^{35}Cl spins to be probed. Two-dimensional $^{13}C\{^{35}Cl\}$ D-HMQC experiments could be useful for samples that give rise to multiple overlapping ^{35}Cl NMR signals. The high-resolution ^{13}C dimension in a 2D D-HMQC spectrum could

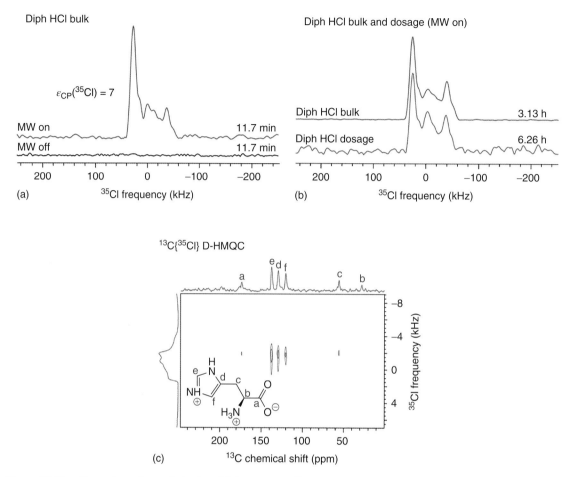

Figure 20.11. Dynamic nuclear polarization (DNP)-enhanced ^{35}Cl solid-state NMR spectra. (a) Static broadband adiabatic inversion–cross-polarization (CP)/wideband, uniform rate, smooth truncation–Carr–Purcell–Meiboom–Gill ^{35}Cl solid-state NMR spectrum of diphenhydramine HCl acquired with and without microwaves (MWs). The ^{35}Cl CP DNP enhancement was seven, and the total experiment time was 11.7 min. (b) Comparison of DNP- enhanced ^{35}Cl solid-state NMR spectra of bulk diphenhydramine HCl and a commercial formulation of diphenhydramine HCl, which was only ~6 wt.% active pharmaceutical ingredient (experiment time of 6.3 h). (c) DNP-enhanced two-dimensional ^{13}C{^{35}Cl} dipolar HMQC (D-HMQC) spectrum of histidine ·HC1·H$_2$O obtained with rotary resonance recoupling applied to ^{13}C. The total experiment time was 14 h. Reproduced from Hirsh *et al.*,[94] with permission from the Physical Chemistry Chemical Physics Owner Societies signals.

be used to resolve the overlapping ^{35}Cl NMR signals. However, extension of this method to sites with larger ^{35}Cl quadrupolar coupling constants likely requires faster MAS frequencies exceeding 40 kHz.[94]

^{2}H solid-state NMR is widely employed to obtain high-resolution hydrogen solid-state NMR spectra and probe dynamics by measurement of ^{2}H EFG tensors. ^{2}H NMR experiments are normally performed on ^{2}H-labeled materials given its

low 0.015% natural isotopic abundance. Relayed DNP was applied to theophylline and amino acids to obtain natural abundance MAS ^{2}H solid-state NMR spectra in experiment times on the order of 1 h (Figure 20.12).[136] CP was used to transfer DNP-enhanced ^{1}H polarization to ^{2}H, and the entire ^{2}H spinning sideband manifolds were uniformly excited. The ^{2}H EFG tensor parameters (c_Q and η_Q) were determined by fitting the MAS sideband manifolds. With knowledge of c_Q and η_Q, it was possible to correct

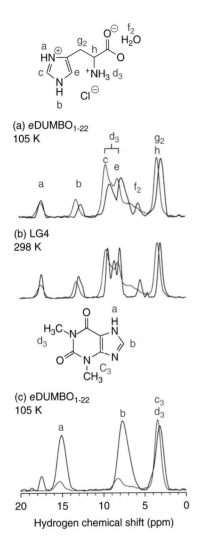

Figure 20.12. Comparison of dynamic nuclear polarization- enhanced natural isotopic abundance ^{2}H solid-state NMR spectra (red) and homonuclear decoupled ^{1}H solid-state NMR spectra (black) for histidine·HCl·H$_2$O (a and b) and theophylline (c). (Reproduced from Rossini *et al.*[136])

good agreement. Therefore, natural abundance ^{2}H solid-state NMR spectroscopy was suggested as an alternative method to ^{1}H homonuclear decoupling for acquisition of high-resolution hydrogen solid-state NMR spectra.

20.4.3 Characterization of Formulated APIs

Solid-state NMR spectroscopy is widely applied to determine and quantify the form of APIs within formulations.[143,194–196] However, the characterization of formulated APIs by solid-state NMR spectroscopy is often very challenging because of the dual problems of dilution and signal overlap. The API loading is often below 10 wt.% in formulations, which normally necessitates the use of long experiment times due to signal averaging.[143,194–196] Signal overlap occurs because the abundant excipients give rise to intense ^{1}H and ^{13}C NMR signals that will either overwhelm or overlap with the signals from dilute API. One way to overcome these limitations is to use ^{13}C isotopically labeled APIs. For example, ^{13}C-labeled steroid molecules could be detected at loadings as low as 0.5 wt.% in model formulations.[197] However, this approach requires synthesis of isotopically labeled materials and cannot be applied to analyze commercial products. DNP can directly address the problems of dilution and signal overlap, enabling low drug load formulations to be routinely studied by NMR with natural isotopic abundance. The large sensitivity enhancements provided by DNP can compensate any reductions in sensitivity due to dilution. Signal overlap can be addressed by performing 2D NMR experiments and/or observing the NMR spectra of elements such as nitrogen or chlorine that are likely only to be found within the API. Furthermore, with numerical models of spin diffusion, it is possible to use the measured DNP enhancements and signal buildup rates to probe the particle sizes of APIs within formulations.

20.4.4 Characterization of Commercial Tablets by DNP-enhanced Solid-state NMR Spectroscopy

Emsley and coworkers applied relayed DNP to several different commercial tablets of the antihistamine drug cetirizine dihydrochloride.[133] The tablets had API

the ^{2}H peak position for the quadrupole-induced shift and determine the isotropic hydrogen chemical shift. Comparison of MAS ^{1}H solid-state NMR spectra obtained with the eDUMBO$_{1-22}$[192] and LG4[193] homonuclear decoupling sequences and MAS ^{2}H solid-state NMR spectra showed that both methods provided similar resolution. The hydrogen chemical shifts determined by both methods were also in

loadings between 4.8 and 8.7 wt.%. The formulations were prepared for DNP experiments by breaking the tablet apart and then impregnating the powdered solid with a TEKPol TCE solution. Grinding of the tablet was not performed to avoid perturbing the particle size or phase of the different components. DNP-enhanced solid-state NMR experiments were also performed on the pure API and the individual excipients to measure the DNP enhancements and ^{13}C chemical shifts for each component of the tablet independently (Figure 20.13). Comparison of the ^{13}C solid-state NMR spectra of the excipients, pure APIs, and the formulation allowed the different ^{13}C NMR signals of the tablet to be assigned. The ^{13}C solid-state NMR spectrum of the tablet indicated that the API was likely present in an amorphous form. Within the formulation, the different components possessed different relayed DNP enhancements. The DNP enhancement was likely determined by the particle size and relaxation properties of each component. The largest DNP enhancements were obtained for the API and the excipient povidone. The similarity of the DNP enhancements of these two components suggested that they may be spatially proximate to one another and share a common bath of proton polarization. The high enhancement for the API was also beneficial because it helped to suppress the signals from the excipients in the DNP-enhanced NMR spectrum. The ^{13}C solid-state NMR spectrum of the tablet illustrates the problem of signal overlap; the chemical shifts of the different components coincide, which prevented resolution of the different signals.

DNP enables 2D NMR experiments and/or ^{15}N solid-state NMR experiments to be performed on tablets, solving the problem of signal overlap. For example, it was possible to obtain 2D ^{1}H–^{13}C CP–HETCOR spectra of the cetirizine tablets in experiment times of ~1 h. The added dispersion provided by the ^{1}H dimension assisted in resolving the NMR signals from the different components. With the large sensitivity enhancements provided by DNP, it was also possible to obtain 1D ^{15}N and 2D ^{1}H–^{15}N HETCOR solid-state NMR spectra (Figure 20.14). The 2D ^{1}H–^{15}N HETCOR NMR spectra were obtained in experiment times of less than 6 h, despite the API loadings of less than 9 wt.%. Nitrogen NMR is advantageous because only the API and one of the excipients (povidone) contain nitrogen, eliminating the problem of overlap between the NMR signals from the API and excipients. Figure 20.14 compares the DNP-enhanced 2D ^{1}H–^{15}N CP–HETCOR spectra

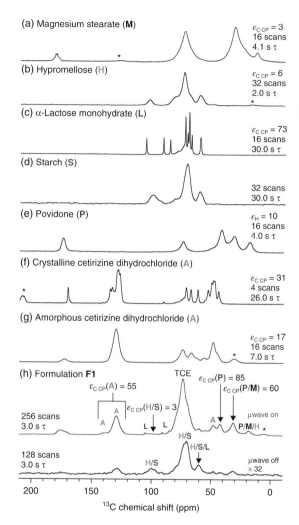

Figure 20.13. Dynamic nuclear polarization-enhanced ^{13}C cross-polarization magic angle spinning NMR spectra of (a) magnesium stearate, (b) hypromellose, (c) α-lactose monohydrate, (d) starch, (e) povidone, (f) crystalline cetirizine dihydrochloride, (g) amorphous cetirizine dihydrochloride, and (h) a commercial formulation of cetirizine dihydrochloride with 8.7 wt.% active pharmaceutical ingredient loading. Dynamic nuclear polarization enhancements for each compound and the different components of the formulation are indicated. All solids were ground and impregnated with 1,1,2,2-tetrachloroethanesolutions of TEKPol except for starch where a spectrum was acquired from the pure solid without addition of 1,1,2,2-tetrachloroethane. (Reprinted with permission from Rossini *et al.*[133] Copyright 2014 American Chemical Society)

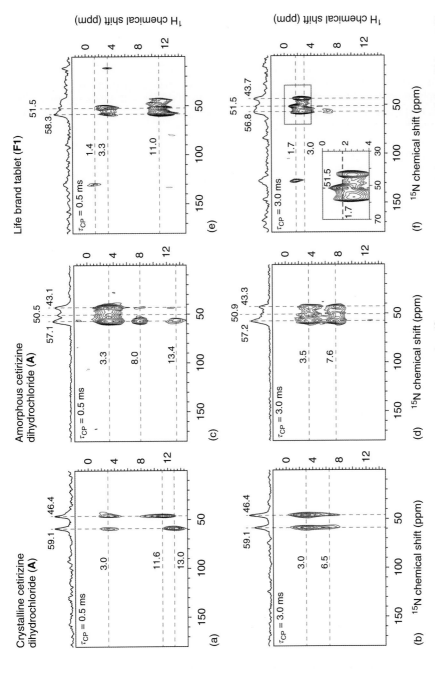

Figure 20.14. Dynamic nuclear polarization-enhanced two-dimensional ^{1}H–^{15}N dipolar heteronuclear correlation spectra of (a and b) crystalline cetirizine dihydrochloride, (c and d) amorphous cetirizine dihydrochloride, and (e and f) an 8.7 wt.% API tablet. The spectra were acquired with contact times (τ_{CP}) of 0.5 ms (top spectra) and 3.0 ms (lower spectra) to probe for short- and long-range ^{1}H–^{15}N distances, respectively. The total experiment times were less than 6 h. (Reprinted with permission from Rossini *et al.*[133] Copyright 2014 American chemical Society)

of crystalline API, amorphous API, and a tablet that contains 8.7 wt.% API. This comparison confirms that the API was present in an amorphous form within the formulation. The 2D HETCOR spectra of the formulation showed that the ^{15}N NMR signal at ~51.5 ppm correlated to the povidone ^{1}H NMR signal at a chemical shift of 1.4 ppm, which provided direct evidence for a molecular-level interaction between povidone and the API. The interaction between povidone and the amorphous API is consistent with the fact that povidone is frequently used to form ASDs to stabilize amorphous APIs.

Schurko and coworkers have used DNP to obtain ^{35}Cl solid-state NMR spectra of APIs in commercial tablets with low API loadings.[94] Usually, the API will be the only component within a formulation that contains chloride anions that can give rise to observable ^{35}Cl solid-state NMR signals. Therefore, ^{35}Cl solid-state NMR spectra of the tablets are free of interfering excipient signals and allow APIs to be directly interrogated.[94,95] For example, the DNP-enhanced static ^{35}Cl solid-state NMR spectra of pure crystalline diphenhydramine HCl and of a commercial tablet were identical in appearance, which confirmed that the form of the API was identical in both pure and dosage forms (Figure 20.11).[94] With the sensitivity enhancement provided by DNP and the BRAIN-CP/WCPMG pulse sequence, it was generally possible to obtain ^{35}Cl solid-state NMR spectra in less than 12 h from tablets with API loadings of less than 7 wt.%.

20.4.5 Determination of API Domain Sizes

Relaxation and spin diffusion solid-state NMR experiments have been broadly applied to probe the macroscopic structure and ordering or mixing within solid materials such as polymers.[154,156,198–200] In the context of pharmaceutical formulations, analysis of T_1 relaxation time constants, rotating-frame longitudinal relaxation times ($T_{1\rho}$), and spin diffusion rates are commonly used to assess the degree of mixing between excipients, polymers, coformers, and APIs.[28,30,138,201,202] Here, we describe how experimental measurements of relayed DNP enhancements and signal buildup rates can be fit to numerical models of spin diffusion (described above) to determine the particle or domain size of APIs within formulations. This method can potentially be applied to any kind of pharmaceutical formulation, with the only requirement being that there be phase segregation between the radical and the analyte and that the NMR signals from the analyte be resolved.

Rossini *et al.* analyzed the ^{13}C signal buildup rates and DNP enhancements with spin diffusion models in order to determine the particle or domain sizes of cetirizine dihydrochloride within commercial tablets.[133] Figure 20.15(a) shows the ^{13}C CP–MAS signal buildup for the aromatic signals of the API obtained from saturation recovery experiments performed with and without microwaves to drive DNP (see Figure 20.13 for the corresponding ^{13}C solid-state

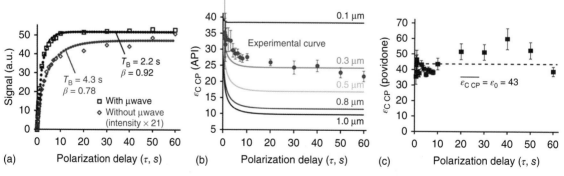

Figure 20.15. (a) Signal buildup observed for the active pharmaceutical ingredient (API) ^{13}C peak with a saturation recovery cross polarization (CP) pulse sequenced with (black) and without (red) microwave irradiation. Filled circles are fits to stretched exponential functions of the form. (b) Comparison between experimental and simulated $\varepsilon_{C\,CP}$s of the API as a function of the polarization delay using the numerical spherical model. Simulations are shown for different API particle radii. In the simulations, $\varepsilon_0 = 43$, $T_{1,API} = 5.3$ s, $T_{B,Source} = 2.3$ s (the T_1 at the surface of the particles), and $D = 1 \times 10^5$ Å^2 s^{-1}. (c) Experimental enhancement of the povidone ^{13}C peak as a function of time. (Reprinted with permission from Rossini *et al.*[133] Copyright 2014 American Chemical Society)

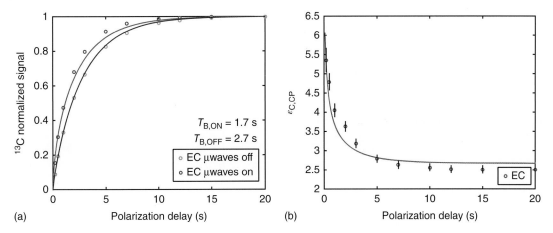

Figure 20.16. (a) Normalized ^{13}C cross-polarization (CP) magic angle spinning signal buildup of the ethyl cellulose (EC) peak in an EC-HPC film without and with microwave irradiation to drive dynamic nuclear polarization. (b) ^{13}C CP dynamic nuclear polarization enhancement of the EC peak as a function of the polarization delay. The blue line corresponds to numerical simulations of a one-dimensional diffusion model with the following parameters: $L = 200$ nm, $T_{B,Source} = 5$ ms, $T_{1,EC} = 3.5$ s, and $D = 2 \times 10^4$ Å^2 s^{-1}. (Reprinted with permission from Pinon *et al.*[155] Copyright 2017 American Chemical Society)

NMR spectra of the tablet). The ^{1}H saturation recovery experiments were performed with ^{13}C signal detection in order to use the high-resolution ^{13}C dimension to resolve and differentiate the longitudinal ^{1}H relaxation rates associated with different components of the formulation. The ^{13}C signal buildup curves for the API could be fit to stretched exponential functions. Note that the signal buildup rate was observed to be accelerated with DNP. Both of these observations were consistent with the predictions of numerical spin diffusion simulations of relayed DNP. The DNP enhancement was determined for each polarization delay by dividing the signal intensity obtained with microwaves by the signal intensity without microwaves (Figure 20.15b). Figure 20.15(c) illustrates that povidone, which was hypothesized to coat the outside of the API particles, shows no clear variation in DNP enhancement with the polarization delay. The observed povidone DNP enhancement and measured relaxation times were incorporated into a numerical spin diffusion model that was used to predict the DNP enhancements as a function of the polarization delay for particles with different radii. Comparison of the observed DNP enhancements and the predicted enhancements for different particle sizes suggested that the API particles or domains had a radius of approximately 0.3 μm. Using the

phenomenological equation (20.3) with $\varepsilon_0 = 43$, $\varepsilon_{analyte,\infty} = 23$, $T_{1,API} = 5.3$ s, and $D = 10^5$ Å^2 s^{-1} gives a particle radius of 0.34 μm, in very good agreement with the value that was found with numerical simulations.

Relayed DNP NMR was also applied to determine the size of the ethyl cellulose (EC) and hydroxypropyl cellulose (HPC) domains within blended films that are used in controlled release formulations.[155] The EC or HPC films were impregnated with a glycerol-d_8:D$_2$O: H$_2$O (60:30:10 by vol%) solution with a 16-mM TOTAPOL concentration[155] On the basis of observed enhancements and relaxation properties, it was hypothesized that the PA solution penetrates and swells the amorphous HPC domains, whereas the PA does not enter the insoluble EC domains. The buildup of the polarization occurred in the HPC domains and was relayed by spin diffusion into the EC domains. Therefore, with the analysis of the DNP enhancements and signal buildup times, it was possible to measure the size of the EC domains (Figure 20.16). Simulation of the variation in the relayed DNP enhancements with polarization time indicated that the length of the EC domains was 0.2 μm, in good agreement with measurements of the EC domain size made using paramagnetic relaxation enhancement.[154]

20.4.6 Characterization of Directly Doped Amorphous Solid Dispersions

An ASD consists of an amorphous API dissolved or suspended within a polymer matrix.[16–19] ASDs typically show superior drug solubility and bioavailability as compared to crystalline APIs.[16–19] Solid-state NMR is a powerful probe of structure in ASDs because it is possible to monitor phase segregation (crystallization of the API), measure API dynamics, and probe for molecular-level interactions between amorphous API and the polymer matrix.[30,143,201,203,204] Recently, Su and coworkers have performed DNP-enhanced solid-state NMR experiments on ASDs that were directly doped by incorporating the PA during the spray drying or hot-melt extrusion procedures that are used to prepare ASDs (Figure 20.6).[138] The advantage of the direct-doping method is that the PA is directly incorporated during synthesis; postsynthetic introduction of the PA into ASDs by swelling of the polymer matrix could potentially perturb the structure.

Samples of the amorphous API clotrimazole, the polymer copovidone, and a clotrimazole–copovidone ASD were prepared for DNP experiments by directly adding TOTAPOL during hot-melt extrusion or into the methanol solution used for spray drying.[138] The final concentration of the TOTAPOL PA in the solid materials was between 0.5 and 2 wt.%. DNP enhancements of 12, 4, and 5 were obtained for samples of clotrimazole, copovidone, and the ASD, respectively, all of which were prepared by spray drying. Samples prepared by hot-melt extrusion exhibited similar DNP enhancements. For both spray drying and hot-melt extrusion, the largest DNP enhancements were typically obtained with final PA concentrations of ~1 wt.%. The proton T_1 of the spray-dried or extruded solids was also observed to substantially decrease with increasing radical concentration due to paramagnetic relaxation enhancement. DNP-enhanced NMR experiments were also performed on a posaconazole–^{2}H vinyl acetate ASD (~20 wt.% API), which was prepared by spray drying and contained 1 wt.% of AMUPOL (Figure 20.17). The use of deuterated vinyl acetate and AMUPOL resulted in a DNP enhancement of ~32 for the ASD, which was three to four times higher than the DNP enhancement measured for an ASD prepared with protonated vinyl acetate. With the large DNP enhancement, it was possible to obtain 2D ^{1}H–^{13}C and ^{1}H–^{15}N HETCOR NMR spectra in experiment times of 7 h (Figure 20.17). In summary,

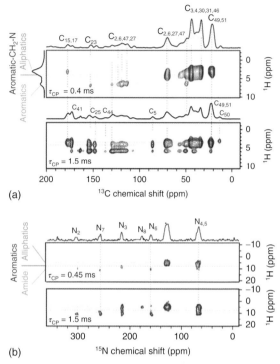

Figure 20.17. Dynamic nuclear polarization-enhanced (a) ^{1}H–^{13}C and (b) ^{1}H–^{15}N heteronuclear correlation solid-state NMR spectra of a posaconazole–^{2}H vinyl acetate amorphous solid dispersion directly doped with ~1 wt.% AMUPol. The amorphous solid dispersion was prepared by spray drying. The two-dimensional heteronuclear correlation solid-state NMR spectra were obtained with total experiment times of 7 h each. (Reprinted with permission from Ni *et al.*[138] Copyright 2017 American Chemical Society)

these results illustrate that direct doping of an ASD can be achieved by incorporating the PA during spray drying or hot-melt extrusion. Significant DNP enhancements can be obtained with this approach, enabling acquisition of 2D solid-state NMR spectra of ASDs.

20.5 CONCLUSIONS AND FUTURE PERSPECTIVES

This review has described how DNP-enhanced solid-state NMR spectroscopy may be applied for the characterization of APIs. Relayed DNP experiments can be applied to externally polarize microcrystalline solids. Numerical spin diffusion models describe the transport of DNP-enhanced polarization in samples with heterogeneous distributions of the PA. As an alternative to relayed DNP, the PA may be directly doped into APIs during synthesis or crystallization. With modern DNP instrumentation and state-of-the-art designer PA, it is possible to routinely obtain DNP signal enhancements of one to two orders of magnitude in both pure and formulated APIs. The large gains in sensitivity provided by DNP enable solid-state NMR experiments that were previously very challenging or impossible. For example, with DNP, it is possible to rapidly obtain homonuclear $^{13}C-^{13}C$ DQ–SQ correlation solid-state NMR spectra of pure organic solids and APIs. DNP has also permitted acquisition of ^{15}N and ^{35}Cl solid-state NMR spectra of formulated APIs with low drug loads. The ^{15}N and ^{35}Cl solid-state NMR spectra are free of interfering signals from excipients, allowing the API to be directly probed. Measurement of relayed DNP signal enhancements combined with numerical models of spin diffusion allowed the size of the API particles within tablets or formulations to be estimated. These examples demonstrate the unique insight into the molecular level and macroscopic structure of APIs that can potentially be obtained with DNP-enhanced solid-state NMR spectroscopy.

There are several interesting future directions for DNP-enhanced solid-state NMR experiments on APIs. First, high-field DNP is an emerging technique, with commercial instrumentation introduced only 7 years ago.[106] Currently, many research groups are developing hardware and polarization schemes to expand the capabilities of DNP and obtain further gains in absolute sensitivity. For example, helium-cooled systems allow MAS DNP experiments to be performed with sample temperatures of 50 K or less.[109–112] DNP at helium sample temperatures could likely provide a gain of three to four orders of magnitude in absolute sensitivity because DNP enhancements and Boltzmann polarization of the nuclear spins are both improved at lower temperatures. The development of next-generation microwave sources[134,205,206] could also bring about further substantial gains in sensitivity by enabling pulsed DNP experiments[108,206–209] and electron decoupling.[210] The development of PA designed for OE DNP or biradical PA containing radicals with an isotropic EPR spectrum has allowed DNP enhancements greater than 50 to be obtained at magnetic fields above 16.4 T.[117,119] Probes capable of MAS at frequencies above 40 kHz have also been demonstrated.[120,211] Notably, fast MAS has allowed OE DNP enhancements of over 100 to be obtained at a high magnetic field of 18.8 T.[120] With fast MAS, it should be possible to combine DNP and proton detection to obtain further improvements in resolution and absolute sensitivity.

Second, novel structural information obtained from DNP-enhanced solid-state NMR experiments will be implemented into NMR crystallography protocols to perform crystal structure determination or validation for both pure and formulated APIs. For example, the previously described DNP-enhanced homonuclear and HETCOR solid-state NMR experiments will provide access to novel structural constraints such as carbon–carbon and carbon–nitrogen interatomic distances.[122,139,144,157,168,173] Finally, we note that DNP has also been applied to obtain solid-state NMR spectra of unreceptive nuclei such as ^{17}O and ^{43}Ca found in inorganic materials.[212–214] We therefore expect that DNP should provide access to the natural isotopic abundance solid-state NMR spectra of unreceptive nuclei such as ^{17}O, ^{33}S, and ^{43}Ca that are also commonplace in APIs. We speculate that with further gains in sensitivity provided by new DNP schemes and hardware, the types of experiments described here could one day be applied to perform in situ NMR crystallography on APIs in low-drug-load formulations.

ACKNOWLEDGMENTS

This material is based upon work supported by the National Science Foundation under Grant 1709972 to A. J. R. A. J. R. is also grateful for support from Genentech Inc., a subsidiary of Roche. A. J. R. thanks Iowa State University and the Ames Laboratory (Royalty Account) for additional support. The Ames Laboratory is operated for the U.S. DOE by Iowa State University under contract DE-AC02-07CH11358. L. E. acknowledges ERC Advanced Grant 320860 and Swiss National Science Foundation Grant 200021_160112 for supporting this work. The authors thank their numerous coauthors and collaborators who have worked on the development of DNP NMR methods over the years.

REFERENCES

1. D. Hörter and J. Dressman, *Adv. Drug Delivery Rev*, 1997, **25**, 3.

2. S. R. Chemburkar, J. Bauer, K. Deming, H. Spiwek, K. Patel, J. Morris, R. Henry, S. Spanton, W. Dziki, and W. Porter, *Org. Process Res. Dev*, 2000, **4**, 413.

3. L.-F. Huang and W.-Q. T. Tong, *Adv. Drug Delivery Rev.*, 2004, **56**, 321.

4. J. Bauer, J. Morley, S. Spanton, F. Leusen, R. Henry, S. Hollis, W. Heitmann, A. Mannino, J. Quick, and W. Dziki, *J. Pharm. Sci.*, 2006, **95**, 917.

5. R. Hilfiker, Polymorphism: In the Pharmaceutical Industry, John Wiley & Sons: Weinheim, 2006.

6. D. Braga, F. Grepioni, L. Maini, and M. Polito, in 'Molecular Networks', eds: M. W. Hosseini, Springer: Berlin, 2009, p 87.

7. S. Aitipamula, P. S. Chow, and R. B. Tan, *CrystEngComm*, 2014, **16**, 3451.

8. H. G. Brittain, Polymorphism in Pharmaceutical Solids, CRC Press: New York, 2016.

9. E. H. Lee, *Asian. Phar. Sci.* 2014, **9**, 163.

10. Ö. Almarsson and M. J. Zaworotko, *Chem. Commun.*, 2004, 1889.

11. C. B. Aakeröy, M. E. Fasulo and J. Desper, *Mol. Pharm.*, 2007, **4**, 317.

12. S. L. Childs, G. P. Stahly and A. Park, *Mol. Pharm.*, 2007, **4**, 323.

13. G. P. Stahly, *Cryst. Growth Des.*, 2007, **7**, 1007.

14. T. Friščić and W. Jones, *J. Pharm. Pharmacol.*, 2010, **62**, 1547.

15. W. Jones, W. D. S. Motherwell and A. V. Trask, *MRS Bull.*, 2011, **31**, 875.

16. W. L. Chiou and S. Riegelman, *J. Pharm. Sci.* 1971, **60**, 1281.

17. F. Qian, J. Huang, and M. A. Hussain, *J. Pharm. Sci.*, 2010, **99**, 2941.

18. A. Newman, G. Knipp, and G. Zografi, *J. Pharm. Sci.*, 2012, **101**, 1355.

19. S. Baghel, H. Cathcart, and N. J. O'Reilly, *J. Pharm. Sci.*, 2016, **105**, 2527.

20. A. Singh and G. den Van Mooter, *Adv. Drug Delivery Rev.*, 2016, **100**, 27.

21. D. E. Bugay, *Pharm. Res.*, 1993, **10**, 317.

22. D. E. Bugay, *Adv. Drug Delivery Rev.*, 2001, **48**, 43.

23. G. A. Stephenson, R. A. Forbes, and S. M. Reutzel-Edens, *Adv. Drug Delivery Rev.*, 2001, **48**, 67.

24. B. C. Hancock, E. Y. Shalaev, and S. L. Shamblin, *J. Pharm. Pharmacol.* 2002, **54**, 1151.

25. S. Reutzel-Edens and J. Bush, *Am. Pharmaceut. Rev.*, 2002, **5**, 112.

26. P. A. Tishmack, D. E. Bugay, and S. R. Byrn, *J. Pharm. Sci.*, 2003, **92**, 441.

27. A. W. Newman and S. R. Byrn, *Drug Discov. Today*, 2003, **8**, 898.

28. J. W. Lubach and E. J. Munson, Solid-State Nuclear Magnetic Resonance of Pharmaceutical Formulations. Encyclopedia of Analytical Chemistry, John Wiley & Sons, Ltd: Chichester, UK, 2013.

29. R. K. Harris, *J. Pharm. Pharmacol.*, 2007, **59**, 225.

30. M. Geppi, G. Mollica, S. Borsacchi, and C. A. Veracini, *Appl. Spectrosc. Rev.*, 2008, **43**, 202.

31. R. T. Berendt, D. M. Sperger, E. J. Munson, and P. K. Isbester, *TrAC Trends Anal. Chem.*, 2006, **25**, 977.

32. R. K. Harris, *Analyst*, 2006, **131**, 351.

33. F. G. Vogt, *Futures*, 2010, **2**, 915.

34. K. A. Bakeev, Process Analytical Technology: Spectroscopic Tools and Implementation Strategies for the Chemical and Pharmaceutical Industries, John Wiley & Sons: Oxford, UK, 2010.

35. U. Holzgrabe, NMR Spectroscopy in Pharmaceutical Analysis, Elsevier: Oxford, UK, 2011.

36. F. G. Vogt, Solid-State NMR in Drug Discovery and Development in New Applications of NMR in Drug Discovery and Development, Royal Society of Chemistry: Cambridge, UK, 2013, **43**.

37. E. Pindelska, A. Sokal, and W. Kolodziejski, *Adv. Drug Delivery Rev.*, 2017, **117**, 111.

38. J. A. Ripmeester, *Chem. Phys. Lett.*, 1980, **74**, 536.

39. R. A. Fletton, R. K. Harris, A. M. Kenwright, R. W. Lancaster, K. J. Packer, and N. Sheppard, *Spectrochim. Acta, Part A*, 1987, **43**, 1111.

40. S. R. Byrn, P. A. Sutton, B. Tobias, J. Frye, and P. Main, *J. Am. Chem. Soc.*, 1988, **110**, 1609.

41. H. G. Brittain, K. R. Morris, D. E. Bugay, A. B. Thakur, and A. T. Serajuddin, *J. Pharm. Biomed. Anal.*, 1993, **11**, 1063.

42. D. C. Apperley, R. A. Fletton, R. K. Harris, R. W. Lancaster, S. Tavener, and T. L. Threlfall, *J. Pharm. Sci.*, 1999, **88**, 1275.

43. R. K. Harris, *Solid State Sci.*, 2004, **6**, 1025.

44. J. W. Lubach, D. Xu, B. E. Segmuller, and E. J. Munson, *J. Pharm. Sci.*, 2007, **96**, 777.

45. S. Schantz, P. Hoppu, and A. Juppo, *J. Pharm. Sci.*, 2009, **98**, 1862.

46. D. H. Zhou and C. M. Rienstra, *Angew. Chem. Int. Ed.*, 2008, **47**, 7328.

47. H. W. Spiess, *eMag. Res.*, 2012.

48. M. K. Pandey, H. Kato, Y. Ishii, and Y. Nishiyama, *Phys. Chem. Chem. Phys.*, 2016, **18**, 6209.

49. R. Zhang, K. H. Mroue, and A. Ramamoorthy, *Acc. Chem. Res.*, 2017, **50**, 1105.

50. E. Salager, R. S. Stein, S. Steuernagel, A. Lesage, B. Elena, and L. Emsley, *Chem. Phys. Lett.*, 2009, **469**, 336.

51. L. Ryan, R. Taylor, A. Paff, and B. Gerstein, *J. Chem. Phys.*, 1980, **72**, 508.

52. B. Elena and L. Emsley, *J. Am. Chem. Soc.*, 2005, **127**, 9140.

53. J. M. Griffin, D. R. Martin, and S. P. Brown, *Angew. Chem. Int. Ed.*, 2007, **46**, 8036.

54. C. J. Pickard, E. Salager, G. Pintacuda, B. Elena, and L. Emsley, *J. Am. Chem. Soc.* 2007, **129**, 8932.

55. E. Salager, R. S. Stein, C. J. Pickard, B. Elena, and L. Emsley, *Phys. Chem. Chem. Phys.*, 2009, **11**, 2610.

56. E. Salager, G. M. Day, R. S. Stein, C. J. Pickard, B. Elena, and L. Emsley, *J. Am. Chem. Soc.*, 2010, **132**, 2564.

57. L. Mafra, S. M. Santos, R. Siegel, I. Alves, F. A. Almeida Paz, D. Dudenko, and H. W. Spiess, *J. Am. Chem. Soc.*, 2012, **134**, 71.

58. M. Baias, C. M. Widdifield, J.-N. Dumez, H. P. G. Thompson, T. G. Cooper, E. Salager, S. Bassil, R. S. Stein, A. Lesage, G. M. Day, and L. Emsley, *Phys. Chem. Chem. Phys.*, 2013, **15**, 8069.

59. S. M. Santos, J. Rocha, and L. Mafra, *Cryst. Growth Des.*, 2013, **13**, 2390.

60. K. Maruyoshi, D. Iuga, A. E. Watts, C. E. Hughes, K. D. M. Harris, S. P. Brown, *J. Pharm. Sci.* 2017, **106**, 3372. DOI: 10.1016/j.xphs.2017.07.014.

61. B. Elena, G. Pintacuda, N. Mifsud, and L. Emsley, *J. Am. Chem. Soc.*, 2006, **128**, 9555.

62. K. Seidel, M. Etzkorn, L. Sonnenberg, C. Griesinger, A. Sebald, and M. Baldus, *J. Phys. Chem. A*, 2005, **109**, 2436.

63. J. P. M. Lommerse, W. D. S. Motherwell, H. L. Ammon, J. D. Dunitz, A. Gavezzotti, D. W. M. Hofmann, F. J. J. Leusen, W. T. M. Mooij, S. L. Price, B. Schweizer, M. U. Schmidt, B. P. van Eijck, P. Verwer, and D. E. Williams, *Acta Crystallogr. B*, 2000, **56**, 697.

64. G. M. Day, W. D. S. Motherwell, H. L. Ammon, S. X. M. Boerrigter, R. G. Della Valle, E. Venuti, A. Dzyabchenko, J. D. Dunitz, B. Schweizer, B. P. van Eijck, P. Erk, J. C. Facelli, V. E. Bazterra, M. B. Ferraro, D. W. M. Hofmann, F. J. J. Leusen, C. Liang, C. C. Pantelides, P. G. Karamertzanis, S. L. Price, T. C. Lewis, H. Nowell, A. Torrisi, H. A. Scheraga, Y. A. Arnautova, M. U. Schmidt, and P. Verwer, *Acta Crystallogr. B*, 2005, **61**, 511.

65. W. D. S. Motherwell, H. L. Ammon, J. D. Dunitz, A. Dzyabchenko, P. Erk, A. Gavezzotti, D. W. M. Hofmann, F. J. J. Leusen, J. P. M. Lommerse, W. T. M. Mooij, S. L. Price, H. Scheraga, B. Schweizer, M. U. Schmidt, B. P. van Eijck, P. Verwer, and D. E. Williams, *Acta Crystallogr. B*, 2002, **58**, 647.

66. G. M. Day, T. G. Cooper, A. J. Cruz-Cabeza, K. E. Hejczyk, H. L. Ammon, S. X. M. Boerrigter, J. S. Tan, R. G. Della Valle, E. Venuti, J. Jose, S. R. Gadre, G. R. Desiraju, T. S. Thakur, B. P. van Eijck, J. C. Facelli, V. E. Bazterra, M. B. Ferraro, D. W. M. Hofmann, M. A. Neumann, F. J. J. Leusen, J. Kendrick, S. L. Price, A. J. Misquitta, P. G. Karamertzanis, G. W. A. Welch, H. A. Scheraga, Y. A. Arnautova, M. U. Schmidt, J. van de Streek, A. K. Wolf, and B. Schweizer, *Acta Crystallogr. B*, 2009, **65**, 107.

67. A. M. Reilly, R. I. Cooper, C. S. Adjiman, S. Bhattacharya, A. D. Boese, J. G. Brandenburg, P. J. Bygrave, R. Bylsma, J. E. Campbell, R. Car, D. H. Case, R. Chadha, J. C. Cole, K. Cosburn, H. M. Cuppen, F. Curtis, G. M. Day, R. A. DiStasio Jr.,, A. Dzyabchenko, B. P. van Eijck, D. M. Elking, J. A. van den Ende, J. C. Facelli, M. B. Ferraro, L. Fusti-Molnar, C.-A. Gatsiou, T. S. Gee, R. de Gelder, L. M. Ghiringhelli, H. Goto, S. Grimme, R. Guo, D. W. M. Hofmann, J. Hoja, R. K. Hylton, L. Iuzzolino, W. Jankiewicz, D. T. de Jong, J. Kendrick, N. J. J. de Klerk, H.-Y. Ko, L. N. Kuleshova, X. Li, S. Lohani, F. J. J. Leusen, A. M. Lund, J. Lv, Y. Ma, N. Marom, A. E. Masunov, P. McCabe, D. P. McMahon, H. Meekes, M. P. Metz, A. J. Misquitta, S. Mohamed, B. Monserrat, R. J. Needs, M. A. Neumann, J. Nyman, S. Obata, H. Oberhofer, A. R. Oganov, A. M. Orendt, G. I. Pagola, C. C. Pantelides, C. J. Pickard, R. Podeszwa, L. S. Price, S. L. Price, A. Pulido, M. G. Read, K. Reuter, E. Schneider, C. Schober, G. P. Shields, P. Singh, I. J. Sugden, K. Szalewicz, C. R. Taylor, A. Tkatchenko, M. E. Tuckerman,

F. Vacarro, M. Vasileiadis, A. Vazquez- Mayagoitia, L. Vogt, Y. Wang, R. E. Watson, G. A. de Wijs, J. Yang, Q. Zhu, and C. R. Groom, *Acta Crystallogr. B*, 2016, **72**, 439.

68. J. D. Hartman, G. M. Day, and G. J. Beran, *Cryst. Growth Des.*, 2016, **16**, 6479.

69. C. M. Widdifield, H. Robson, and P. Hodgkinson, *Chem. Commun.*, 2016, **52**, 6685.

70. P. Hohenberg and W. Kohn, *Phys. Ther. Rev.*, 1964, **136**, B864.

71. T. Charpentier, *Solid State Nucl. Magn. Reson.*, 2011, **40**, 1.

72. C. Bonhomme, C. Gervais, F. Babonneau, C. Coelho, F. Pourpoint, T. Azaïs, S. E. Ashbrook, J. M. Griffin, J. R. Yates, and F. Mauri, *Chem. Rev.*, 2012, **112**, 5733.

73. M. Sardo, R. Siegel, S. M. Santos, J. Rocha, J. R. B. Gomes, and L. Mafra, *J. Phys. Chem. A*, 2012, **116**, 6711.

74. K. Kalakewich, R. Iuliucci, and J. K. Harper, *Cryst. Growth Des.*, 2013, **13**, 5391.

75. S. E. Ashbrook and D. McKay, *Chem. Commun.*, 2016, **52**, 7186.

76. J. K. Harper, R. Iuliucci, M. Gruber, and K. Kalakewich, *CrystEngComm*, 2013, **15**, 8693.

77. T. Pawlak and M. J. Potrzebowski, *J. Phys. Chem. B*, 2014, **118**, 3298.

78. K. Kalakewich, R. Iuliucci, K. T. Mueller, H. Eloranta, and J. K. Harper, *J. Chem. Phys.*, 2015, **143**, 194702.

79. H. E. Kerr, H. E. Mason, H. A. Sparkes, and P. Hodgkinson, *CrystEngComm*, 2016, **18**, 6700.

80. M. K. Dudek, G. Bujacz, and M. J. Potrzebowski, *CrystEngComm*, 2017, **19**, 2903.

81. A. Hofstetter and L. Emsley, *J. Am. Chem. Soc.*, 2017, **139**, 2573.

82. C. M. Widdifield, S. O. Nilsson Lill, A. Broo, M. Lindkvist, A. Pettersen, A. Svensk Ankarberg, P. Aldred, S. Schantz, and L. Emsley, *Phys., Chem. Chem. Phys.* 2017, **19**, 16650.

83. A. Haimovich, U. Eliav, and A. Goldbourt, *J. Am. Chem. Soc.*, 2012, **134**, 5647.

84. A. S. Tatton, T. N. Pham, F. G. Vogt, D. Iuga, A. J. Edwards, and S. P. Brown, *Mol. Pharm.*, 2013, **10**, 999.

85. S. L. Veinberg, K. E. Johnston, M. J. Jaroszewicz, B. M. Kispal, C. R. Mireault, T. Kobayashi, M. Pruski, and R. W. Schurko, *Phys. Chem. Chem. Phys.*, 2016, **18**, 17713.

86. K. J. Pike, V. Lemaitre, A. Kukol, T. Anupold, A. Samoson, A. P. Howes, A. Watts, M. E. Smith, and R. Dupree, *J. Phys. Chem. B*, 2004, **108**, 9256.

87. X. Kong, M. Shan, V. Terskikh, I. Hung, Z. Gan, and G. Wu, *J. Phys. Chem. B*, 2013, **117**, 9643.

88. F. G. Vogt, H. Yin, R. G. Forcino, and L. Wu, *Mol. Pharm.*, 2013, **10**, 3433.

89. G. Wu, *Solid State Nucl. Magn. Reson.*, 2016, **73**, 1.

90. X. Kong, Y. Dai, and G. Wu, *Solid State Nucl. Magn. Reson.*, 2017, **84**, 59.

91. K. Burgess, F. A. Perras, A. Lebrun, E. Messner-Henning, I. Korobkov, and D. L. Bryce, *J. Pharm. Sci.*, 2012, **101**, 2930.

92. N. M. Dicaire, F. A. Perras, and D. L. Bryce, *Can. J. Chem.*, 2013, **92**, 9.

93. H. Hamaed, J. M. Pawlowski, B. F. Cooper, R. Fu, S. H. Eichhorn, R. W. Schurko, *J. Am. Chem. Soc.*, 2008, **130**, 11056.

94. D. A. Hirsh, A. J. Rossini, L. Emsley, and R. W. Schurko, *Phys. Chem. Chem. Phys.*, 2016, **18**, 25893.

95. A. M. Namespetra, D. A. Hirsh, M. P. Hildebrand, A. R. Sandre, H. Hamaed, J. M. Rawson, and R. W. Schurko, *CrystEngComm*, 2016, **18**, 6213.

96. M. Hildebrand, H. Hamaed, A. M. Namespetra, J. M. Donohue, R. Fu, I. Hung, Z. Gan, and R. W. Schurko, *CrystEngComm*, 2014, **16**, 7334.

97. R. K. Harris, E. D. Becker, S. M. Cabral de Menezes, R. Goodfellow, and P. Granger, *Magn. Reson. Chem.*, 2002, **40**, 489.

98. A. B. Barnes, G. D. Paepe, P. C. A. van der Wel, K. N. Hu, C. G. Joo, V. S. Bajaj, M. L. Mak-Jurkauskas, J. R. Sirigiri, J. Herzfeld, R. J. Temkin, and R. G. Griffin, *Appl. Magn. Reson.*, 2008, **34**, 237.

99. Q. Z. Ni, E. Daviso, T. V. Can, E. Markhasin, S. K. Jawla, T. M. Swager, R. J. Temkin, J. Herzfeld, and R. G. Griffin, *Acc. Chem. Res.*, 2013, **46**, 1933.

100. Ü. Akbey, W. T. Franks, A. Linden, M. Orwick-Rydmark, S. Lange, and H. Oschkinat, 'Dynamic Nuclear Polarization Enhanced NMR in the Solid-State in Hyperpolarization Methods in NMR Spectroscopy', Springer: Heidelberg, 2013, Vol. **338**.

101. A. S. L. Thankamony, J. J. Wittmann, M. Kaushik, and B. Corzilius, *Prog. Nucl. Magn. Reson. Spectrosc.*, 2017, 102-103, **120**.

102. A. W. Overhauser, *Phys. Ther. Rev.*, 1953, **92**, 411.

103. T. R. Carver and C. P. Slichter, *Phys. Ther. Rev.*, 1953, **92**, 212.

104. L. R. Becerra, G. J. Gerfen, R. J. Temkin, D. J. Singel, and R. G. Griffin, *Phys. Rev. Lett.*, 1993, **71**, 3561.

105. T. Maly, G. T. Debelouchina, V. S. Bajaj, K.-N. Hu, C.-G. Joo, M. L. Mak-Jurkauskas, J. R. Sirigiri, P. C. van der Wel, J. Herzfeld, and R. J. Temkin, *J. Chem. Phys.*, 2008, **128**, 02B611.

106. M. Rosay, L. Tometich, S. Pawsey, R. Bader, R. Schauwecker, M. Blank, P. M. Borchard, S. R. Cauffman, K. L. Felch, and R. T. Weber, *Phys. Chem. Chem. Phys.*, 2010, **12**, 5850.

107. D. A. Hall, D. C. Maus, G. J. Gerfen, S. J. Inati, L. R. Becerra, F. W. Dahlquist, and R. G. Griffin, *Science*, 1997, **276**, 930.

108. T. V. Can, Q. Z. Ni, and R. G. Griffin, *J. Magn. Reson.*, 2015, **253**, 23.

109. K. R. Thurber, W.-M. Yau, and R. Tycko, *J. Magn. Reson.*, 2010, **204**, 303.

110. Y. Matsuki, K. Ueda, T. Idehara, R. Ikeda, I. Ogawa, S. Nakamura, M. Toda, T. Anai, and T. Fujiwara, *J. Magn. Reson.*, 2012, **225**, 1.

111. E. Bouleau, P. Saint-Bonnet, F. Mentink-Vigier, H. Takahashi, J. F. Jacquot, M. Bardet, F. Aussenac, A. Purea, F. Engelke, S. Hediger, D. Lee, and G. De Paepe, *Chem. Sci.*, 2015, **6**, 6806.

112. K. Thurber and R. Tycko, *J. Magn. Reson.*, 2016, **264**, 99.

113. K.-N. Hu, H.-h. Yu, T. M. Swager, and R. G. Griffin, *J. Am. Chem. Soc.*, 2004, **126**, 10844.

114. C. Song, K.-N. Hu, C.-G. Joo, T. M. Swager, and R. G. Griffin, *J. Am. Chem. Soc.*, 2006, **128**, 11385.

115. C. Sauvée, M. Rosay, G. Casano, F. Aussenac, R. T. Weber, O. Ouari, and P. Tordo, *Angew. Chem. Int. Ed.*, 2013, **125**, 11058.

116. A. Zagdoun, G. Casano, O. Ouari, M. Schwarzwälder, A. J. Rossini, F. Aussenac, M. Yulikov, G. Jeschke, C. Copéret, A. Lesage, P. Tordo, and L. Emsley, *J. Am. Chem. Soc.*, 2013, **135**, 12790.

117. G. Mathies, M. A. Caporini, V. K. Michaelis, Y. Liu, K.-N. Hu, D. Mance, J. L. Zweier, M. Rosay, M. Baldus, and R. G. Griffin, *Angew. Chem. Int. Ed.*, 2015, **54**, 11770.

118. C. F. Koelsch, *J. Am. Chem. Soc.*, 1957, **79**, 4439.

119. T. V. Can, M. A. Caporini, F. Mentink-Vigier, B. Corzilius, J. J. Walish, M. Rosay, W. E. Maas, M. Baldus, S. Vega, T. M. Swager, and R. G. Griffin, *J. Chem. Phys.*, 2014, **141**, 064202.

120. S. R. Chaudhari, D. Wisser, A. C. Pinon, P. Berruyer, D. Gajan, P. Tordo, O. Ouari, C. Reiter, F. Engelke, C. Copéret, M. Lelli, A. Lesage, and L. Emsley, *J. Am. Chem. Soc.*, 2017, **139**, 10609.

121. Y. Matsuki, T. Maly, O. Ouari, H. Karoui, F. Le Moigne, E. Rizzato, S. Lyubenova, J. Herzfeld, T. Prisner, P. Tordo, and R. G. Griffin, *Angew. Chem. Int. Ed.*, 2009, **48**, 4996.

122. A. J. Rossini, A. Zagdoun, F. Hegner, M. Schwarzwalder, D. Gajan, C. Coperet, A. Lesage, and L. Emsley, *J. Am. Chem. Soc.*, 2012, **134**, 16899.

123. P. C. van der Wel, K.-N. Hu, J. Lewandowski, and R. G. Griffin, *J. Am. Chem. Soc.*, 2006, **128**, 10840.

124. A. Lesage, M. Lelli, D. Gajan, M. A. Caporini, V. Vitzthum, P. Mieville, J. Alauzun, A. Roussey, C. Thieuleux, A. Mehdi, G. Bodenhausen, C. Coperet, and L. Emsley, *J. Am. Chem. Soc.*, 2010, **132**, 15459.

125. A. Pines, M. Gibby, and J. Waugh, *Chem. Phys. Lett.*, 1972, **15**, 373.

126. A. Zagdoun, A. J. Rossini, D. Gajan, A. Bourdolle, O. Ouari, M. Rosay, W. E. Maas, P. Tordo, M. Lelli, L. Emsley, A. Lesage, and C. Coperet, *Chem. Commun.*, 2012, **48**, 654.

127. A. C. Pinon, A. J. Rossini, C. M. Widdifield, D. Gajan, and L. Emsley, *Mol. Pharm.*, 2015, **12**, 4146.

128. J. R. Yarava, S. R. Chaudhari, A. J. Rossini, A. Lesage, and L. Emsley, *J. Magn. Reson.*, 2017, **277**, 149.

129. M. Lelli, S. R. Chaudhari, D. Gajan, G. Casano, A. J. Rossini, O. Ouari, P. Tordo, A. Lesage, and L. Emsley, *J. Am. Chem. Soc.*, 2015, **137**, 14558.

130. T.-C. Ong, M. L. Mak-Jurkauskas, J. J. Walish, V. K. Michaelis, B. Corzilius, A. A. Smith, A. M. Clausen, J. C. Cheetham, T. M. Swager, and R. G. Griffin, *J. Phys. Chem., B* 2013, **117**, 3040.

131. D. Lee, S. R. Chaudhari, and G. De Paëpe, *J. Magn. Reson.*, 2017, **278**, 60.

132. M. P. Hanrahan, A. Venkatesh, S. L. Carnahan, J. L. Calahan, J. W. Lubach, E. J. Munson, and A. J. Rossini, *Phys. Chem. Chem. Phys.*, 2017, **19**, 28153.

133. A. J. Rossini, C. M. Widdifield, A. Zagdoun, M. Lelli, M. Schwarzwalder, C. Coperet, A. Lesage, and L. Emsley, *J. Am. Chem. Soc.*, 2014, **136**, 2324.

134. T. F. Kemp, H. R. W. Dannatt, N. S. Barrow, A. Watts, S. P. Brown, M. E. Newton, and R. Dupree, *J. Magn. Reson.*, 2016, **265**, 77.

135. A. J. Rossini, L. Emsley, and L. A. O'Dell, *Phys. Chem. Chem. Phys.*, 2014, **16**, 12890.

136. A. J. Rossini, J. Schlagnitweit, A. Lesage, and L. Emsley, *J. Magn. Reson.*, 2015, **259**, 192.

137. G. Mollica, M. Dekhil, F. Ziarelli, P. Thureau, and S. Viel, *Angew. Chem. Int. Ed.*, 2015, **54**, 6028.

138. Q. Z. Ni, F. Yang, T. V. Can, I. V. Sergeyev, S. M. D'Addio, S. K. Jawla, Y. Li, M. P. Lipert, W. Xu, R. T. Williamson, A. Leone, R. G. Griffin, and Y. Su, *J. Phys. Chem., B* 2017, **121**, 8132.

139. H. Takahashi, D. Lee, L. Dubois, M. Bardet, S. Hediger, and G. De Paepe, *Angew. Chem. Int. Ed.*, 2012, **51**, 11766.

140. D. J. Kubicki, A. J. Rossini, A. Purea, A. Zagdoun, O. Ouari, P. Tordo, F. Engelke, A. Lesage, and L. Emsley, *J. Am. Chem. Soc.*, 2014, **136**, 15711.

141. X. Ji, A. Bornet, B. Vuichoud, J. Milani, D. Gajan, A. J. Rossini, L. Emsley, G. Bodenhausen, and S. Jannin, *Nat. Commun.* 2017, **8**, 13975.

142. A. B. Barnes, B. Corzilius, M. L. Mak-Jurkauskas, L. B. Andreas, V. S. Bajaj, Y. Matsuki, M. L. Belenky, J. Lugtenburg, J. R. Sirigiri, and R. J. Temkin, *Phys. Chem. Chem. Phys.*, 2010, **12**, 5861.

143. J. Lubach, B. Padden, S. Winslow, J. Salsbury, D. Masters, E. Topp, and E. Munson, *Anal. Bioanal. Chem.*, 2004, **378**, 1504.

144. H. Takahashi, B. Viverge, D. Lee, P. Rannou, and G. De Paepe, *Angew. Chem. Int. Ed.*, 2013, **52**, 6979.

145. D. Le, G. Casano, T. N. T. Phan, F. Ziarelli, O. Ouari, F. Aussenac, P. Thureau, G. Mollica, D. Gigmes, P. Tordo, and S. Viel, *Macromolecules*, 2014, **47**, 3909.

146. K. R. Thurber and R. Tycko, *J. Chem. Phys.*, 2014, **140**, 184201.

147. F. Mentink-Vigier, Ü. Akbey, Y. Hovav, S. Vega, H. Oschkinat, and A. Feintuch, *J. Magn. Reson.*, 2012, **224**, 13.

148. A. J. Rossini, A. Zagdoun, M. Lelli, D. Gajan, F. Rascón, M. Rosay, W. E. Maas, C. Coperet, A. Lesage, and L. Emsley, *Chem. Sci.*, 2012, **3**, 108.

149. N. Bloembergen, *Phys. Ther.*, 1949, **15**, 386.

150. H. Rorschach, *Phys. Ther.*, 1964, **30**, 38.

151. D. Tse and I. Lowe, *Phys. Ther. Rev.*, 1968, **166**, 292.

152. K. R. Brownstein and C. E. Tarr, *Phys. Rev. A*, 1979, **19**, 2446.

153. I. Bertini, C. Luchinat, and G. Parigi, 'Solution NMR of Paramagnetic Molecules: Applications to Metal-lobiomolecules and Models', Elsevier: Amsterdam, 2001, Vol. **2**.

154. J. Schlagnitweit, M. Tang, M. Baias, S. Richardson, S. Schantz, and L. Emsley, *J. Am. Chem. Soc.*, 2015, **137**, 12482.

155. A. C. Pinon, J. Schlagnitweit, P. Berruyer, A. J. Rossini, M. Lelli, E. Socie, M. Tang, T. Pham, A. Lesage, and S. Schantz, *J. Phys. Chem. C*, 2017, **121**, 15993.

156. K. Schmidt-Rohr and H. W. Spiess, Multidimensional Solid-State NMR and Polymers, Academic Press: San Diego, CA, USA, 1994.

157. K. Marker, M. Pingret, J.-M. Mouesca, D. Gasparutto, S. Hediger, and G. De Paepe, *J. Am. Chem. Soc.*, 2015, **137**, 13796.

158. X. Feng, P. J. E. Verdegem, M. Eden, D. Sandstrom, Y. K. Lee, P. H. M. Bovee-Geurts, W. J. de Grip, J. Lugtenburg, H. J. M. de Groot, and M. H. Levitt, *J. Biomol. NMR*, 2000, **16**,1.

159. A. Lesage, M. Bardet, and L. Emsley, *J. Am. Chem. Soc.*, 1999, **121**, 10987.

160. G. De Paëpe, A. Lesage, S. Steuernagel, and L. Emsley, *ChemPhysChem*, 2004, **5**, 869.

161. W. A. Dollase, M. Feike, H. Förster, T. Schaller, I. Schnell, A. Sebald, and S. Steuernagel, *J. Am. Chem. Soc.*, 1997, **119**, 3807.

162. D. H. Brouwer, R. J. Darton, R. E. Morris, and M. H. Levitt, *J. Am. Chem. Soc.* 2005, **127**, 10365.

163. D. H. Brouwer, S. Cadars, J. Eckert, Z. Liu, O. Terasaki, and B. F. Chmelka, *J. Am. Chem. Soc.*, 2013, **135**, 5641.

164. X. Feng, Y. K. Lee, D. Sandstrom, M. Eden, H. Maisel, A. Sebald, and M. Levitt, *Chem. Phys. Lett.*, 1996, **257**, 314.

165. K. Schmidt-Rohr, *J. Am. Chem. Soc.*, 1996, **118**, 7601.

166. X. Feng, P. Verdegem, Y. Lee, D. Sandstrom, M. Eden, P. Bovee-Geurts, W. De Grip, J. Lugtenburg, H. De Groot, and M. Levitt, *J. Am. Chem. Soc.*, 1997, **119**, 6853.

167. R. K. Harris, S. A. Joyce, C. J. Pickard, S. Cadars, and L. Emsley, *Phys. Chem. Chem. Phys.*, 2006, **8**, 137.

168. K. Marker, S. Paul, C. Fernandez-de-Alba, D. Lee, J. M. Mouesca, S. Hediger, and G. De Paepe, *Chem. Sci.*, 2017, **8**, 974.

169. B. J. van Rossum, C. P. de Groot, V. Ladizhansky, S. Vega, and H. J. M. de Groot, *J. Am. Chem. Soc.*, 2000, **122**, 3465.

170. F. G. Vogt, J. S. Clawson, M. Strohmeier, A. J. Edwards, T. N. Pham, and S. A. Watson, *Cryst. Growth Des.*, 2009, **9**, 921.

171. W. Liu, W. D. Wang, W. Wang, S. Bai, and C. Dybowski, *J. Phys. Chem., B* 2010, **114**, 16641.

172. A. Abraham, D. C. Apperley, S. J. Byard, A. J. Ilott, A. J. Robbins, V. Zorin, R. K. Harris, and P. Hodgkinson, *CrystEngComm*, 2016, **18**, 1054.

173. J. Leclaire, G. Poisson, F. Ziarelli, G. Pepe, F. Fotiadu, F. M. Paruzzo, A. J. Rossini, J.-N. Dumez, B. Elena-Herrmann, and L. Emsley, *Chem. Sci.*, 2016, **7**, 4379.

174. D. L. Vanderhart, W. L. Earl, and A. N. Garroway, *J. Magn. Reson.*, 1981, **44**, 361.

175. D. H. Barich, J. M. Davis, L. J. Schieber, M. T. Zell, and E. J. Munson, *J. Pharm. Sci.*, 2006, **95**, 1586.

176. G. Kervern, G. Pintacuda, Y. Zhang, E. Oldfield, C. Roukoss, E. Kuntz, E. Herdtweck, J. M. Basset, S. Cadars, A. Lesage, C. Coperet, and L. Emsley, *J. Am. Chem. Soc.*, 2006, **128**, 13545.

177. A. P. M. Kentgens, *Geoderma*, 1997, **80**, 271.

178. S. E. Ashbrook and M. J. Duer, *Concepts Magn. Reson..A*, 2006, **28A**, 183.

179. S. E. Ashbrook, *Phys. Chem. Chem. Phys.*, 2009, **11**, 6892.

180. R. E. Wasylishen, S. E. Ashbrook, and S. Wimperis, NMR of Quadrupolar Nuclei in Solid Materials, John Wiley & Sons: Chichester, UK, 2012.

181. R. W. Schurko, *Acc. Chem. Res.*, 2013, **46**, 1985.

182. L. A. O'Dell, R. W. Schurko, K. J. Harris, J. Autschbach, and C. I. Ratcliffe, *J. Am. Chem. Soc.*, 2010, **133**, 527.

183. S. L. Veinberg, Z. W. Friedl, K. J. Harris, L. A. O'Dell, and R. W. Schurko, *CrystEngComm*, 2015, **17**, 5225.

184. Z. Gan, *J. Am. Chem. Soc.*, 2006, **128**, 6040.

185. S. Cavadini, A. Lupulescu, S. Antonijevic, and G. Bodenhausen, *J. Am. Chem. Soc.*, 2006, **128**, 7706.

186. S. Cavadini, A. Abraham, and G. Bodenhausen, *Chem. Phys. Lett.*, 2007, **445**, 1.

187. V. Vitzthum, M. A. Caporini, and G. Bodenhausen, *J. Magn. Reson.*, 2010, **205**, 177.

188. M. Bloom and M. A. LeGros, *Can. J. Phys.*, 1986, **64**, 1522.

189. R. Tycko and S. Opella, *J. Chem. Phys.*, 1987, **86**, 1761.

190. L. A. O'Dell and C. I. Ratcliffe, *Chem. Phys. Lett.*, 2011, **514**, 168.

191. L. A. O'Dell and A. Brinkmann, *J. Chem. Phys.*, 2013, **138**, 064201.

192. B. Elena, G. de Paëpe, and L. Emsley, *Chem. Phys. Lett.*, 2004, **398**, 532.

193. M. E. Halse and L. Emsley, *J. Phys. Chem. A*, 2013, **117**, 5280.

194. P. J. Saindon, N. S. Cauchon, P. A. Sutton, C.-J. Chang, G. E. Peck, and S. R. Byrn, *Pharm. Res.*, 1993, **10**, 197.

195. L. M. Katrincic, Y. T. Sun, R. A. Carlton, A. M. Diederich, R. L. Mueller, and F. G. Vogt, *Int. J. Pharm.*, 2009, **366**, 1.

196. R. K. Harris, P. Hodgkinson, T. Larsson, and A. Muruganantham, *J. Pharm. Biomed. Anal.*, 2005, **38**, 858.

197. K. J. Booy, P. Wiegerinck, J. Vader, F. Kaspersen, D. Lambregts, H. Vromans, and E. Kellenbach, *J. Pharm. Sci.*, 2005, **94**, 458.

198. J. Clauss, K. Schmidt-Rohr, and H. W. Spiess, *Acta Polym.*, 1993, **44**, 1.

199. D. E. Demco, A. Johansson, and J. Tegenfeldt, *Solid State Nucl. Magn. Reson.*, 1995, **4**, 13.

200. D. L. VanderHart and G. B. McFadden, *Solid State Nucl. Magn. Reson.*, 1996, **7**, 45.

201. F. G. Vogt and G. R. Williams, *Pharm. Res.*, 2012, **29**, 1866.

202. T. N. Pham, S. A. Watson, A. J. Edwards, M. Chavda, J. S. Clawson, M. Strohmeier, and F. G. Vogt, *Mol. Pharm.*, 2010, **7**, 1667.

203. A. Paudel, M. Geppi, and G. V. d. Mooter, *J. Pharm. Sci.*, 2014, **103**, 2635.

204. Y. Song, X. Yang, X. Chen, H. Nie, S. Byrn, and J. W. Lubach, *Mol. Pharm.*, 2015, **12**, 857.

205. H. J. Kim, E. A. Nanni, M. A. Shapiro, J. R. Sirigiri, P. P. Woskov, and R. J. Temkin, *Phys. Rev. Lett.*, 2010, **105**, 135101.

206. D. E. M. Hoff, B. J. Albert, E. P. Saliba, F. J. Scott, E. J. Choi, M. Mardini, and A. B. Barnes, *Solid State Nucl. Magn. Reson.*, 2015, **72**, 79.

207. A. Henstra, P. Dirksen, J. Schmidt, and W. T. Wenckebach, *Magn. Reson.* (1969), 1988, **77**, 389.

208. G. Mathies, S. Jain, M. Reese, and R. G. Griffin, *Phys. Chem. Lett.*, 2016, **7**, 111.

209. T. V. Can, R. T. Weber, J. J. Walish, T. M. Swager, and R. G. Griffin, *Angew. Chem. Int. Ed.*, 2017, **56**, 6744.

210. E. P. Saliba, E. L. Sesti, F. J. Scott, B. J. Albert, E. J. Choi, N. Alaniva, C. Gao, and A. B. Barnes, *J. Am. Chem. Soc.*, 2017, **139**, 6310.

211. S. R. Chaudhari, P. Berruyer, D. Gajan, C. Reiter, F. Engelke, D. L. Silverio, C. Coperet, M. Lelli, A. Lesage, and L. Emsley, *Phys. Chem. Chem. Phys.*, 2016, **18**, 10616.

212. F. Blanc, L. Sperrin, D. A. Jefferson, S. Pawsey, M. Rosay, and C. P. Grey, *J. Am. Chem. Soc.*, 2013, **135**, 2975.

213. F. A. Perras, T. Kobayashi, and M. Pruski, *J. Am. Chem. Soc.* 2015, **137**, 8336.

214. D. Lee, C. Leroy, C. Crevant, L. Bonhomme-Coury, F. Babonneau, D. Laurencin, C. Bonhomme, and G. De Paepe, *Nat. Commun.*, 2017, **8**, 14104.

Chapter 21

In Vivo Hyperpolarized ^{13}C MRS and MRI Applications

Irene Marco-Rius[1] and Arnaud Comment[2]

[1]*Cancer Research UK Cambridge Institute, University of Cambridge, Cambridge, UK*
[2]*General Electric Healthcare, Chalfont St Giles, UK*

21.1 INTRODUCTION

Magnetic resonance (MR) is a powerful imaging modality providing high temporal and spatial resolution of morphological details. The modality is widespread, well accepted, and contrast agents are used extensively for improved imaging or perfusion examination. MR is also a unique technique to obtain in vivo metabolic maps using the spectroscopic information that can be extracted from time-resolved acquisitions. In particular, it is possible to monitor the biochemical transformations of specific metabolic substrates that are delivered to subjects. Because it gives access to the kinetics of the conversion of substrates into metabolites, magnetic resonance spectroscopy (MRS) on the carbon nuclide (^{13}C) is one of the most compelling techniques to investigate intermediary metabolism. ^{13}C MR has notably brought essential information on the use of energy substrates in the brain, the liver, and the heart.[1-6] However, as a direct consequence of the inherently low ^{13}C MR sensitivity, conventional in vivo ^{13}C MR only gives access to the most concentrated metabolites; many metabolites cannot be detected. The poor sensitivity, which stems from the weak ^{13}C spin polarization, also limits the temporal resolution of conventional measurements to several minutes when many biochemical transformations occur within seconds.

So-called hyperpolarization techniques were recently developed to greatly increase the sensitivity of ^{13}C MR. Among these techniques, dissolution dynamic nuclear polarization (dissolution DNP) is by far the most versatile.[7] Dissolution DNP was developed within Amersham Health Research and Development AB, a company that was later purchased by General Electric (GE) Healthcare.[8] The idea behind DNP is to transfer to nuclear spins of the substrate the large low-temperature ($\sim$1 K) and high-field electron spin polarization of polarizing agents (free radicals) that are incorporated into a frozen solution containing

Handbook of High Field Dynamic Nuclear Polarization.
Edited by Vladimir K. Michaelis, Robert G. Griffin, Björn Corzilius and Shimon Vega
© 2020 John Wiley & Sons, Ltd. ISBN: 978-1-119-44164-9
Also published in eMagRes (online edition)
DOI: 10.1002/9780470034590.emrstm1592

the substrate. The polarization transfer is typically achieved in a magnetic field of 3–7 T by irradiating the frozen solution with microwaves at a frequency close to the difference between the electron spin resonance and the nuclear MR frequencies. A body-temperature solution containing hyperpolarized molecules can be obtained by rapid melting of the frozen solution in superheated water, and, once injected into cells, a tissue or an animal, the recorded MR signal is transiently enhanced by up to five orders of magnitude in standard MR fields (1–14 T).

A prerequisite for applying this technique to metabolic studies is that the substrate should contain at least one nuclear spin with a relatively long longitudinal relaxation time (T_1) in order to maintain the polarization from the low-temperature environment

to the biological target. This is, for instance, the case for molecules containing a carboxyl functional group isotopically enriched in ^{13}C, which typically exhibits a room-temperature T_1 on the order of one minute even at low magnetic field. Although the signal enhancement provided by dissolution DNP is only available for a limited time, it offers the opportunity to perform in vivo MR with a typical time resolution of one second, allowing metabolic imaging in vivo to be essentially acquired in real time.[9] Numerous preclinical applications have demonstrated the enormous potential of this technology for in vivo metabolic imaging.[10]

Because changes in metabolism are often potent indicators of pathological conditions, hyperpolarized substrates prepared via dissolution DNP have been

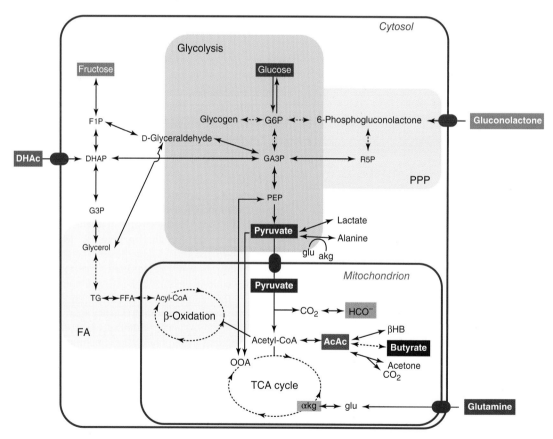

Figure 21.1. Scheme of the metabolic pathways probed with dissolution DNP mentioned in the text. Abbreviations: FA, fatty acid metabolism; PPP, pentose phosphate pathway; FFA, free fatty acids; TG, triglycerides; F1P, fructose 1-phosphate; DHAc, dihydroxyacetone; DHAP, dihydroxyacetone phosphate; G3P, glycerol 3-phosphate; G6P, glucose 6-phosphate; R5P, ribose 5-phosphate; OOA, oxaloacetate; AcAc, acetoacetate; βHB, β-hydroxybutyrate; glu, glutamate; αkg, α-ketoglutarate; PEP, phosphoenolpyruvate; GA3P, glyceraldehyde 3-phosphate

proposed as MR imaging contrast agents for cancer or cardiac failure diagnosis and therapy monitoring through the detection of metabolic impairments in vivo.[11-14] The idea behind hyperpolarized ^{13}C metabolic contrast agents is to directly image the rate of formation of compounds in healthy and disease tissue to obtain a comparison between both. This technology has recently been translated to humans, and several research hospitals are currently performing studies on patients. Motivated by the results obtained in preclinical models, the main area of applications being explored are cancer and cardiac metabolism. Specific metabolic pathways known to be altered in cancer cells can be probed using glucose and glutamine as substrates.[15] Likewise, it is known that heart failure is associated with change in cardiac metabolism,[16] and by probing the utilization of glucose, fatty acids, and ketone bodies in vivo, clues on the state of the myocardium can be extracted from the hyperpolarized ^{13}C MR measurements.

Pyruvate can be considered a surrogate for glucose since it is the end product of glycolysis (Figure 21.1). It is only one metabolic step away from lactate, the production of which is generally upregulated in cancer through the so-called Warburg effect. Pyruvate is also only one step away from intramitochondrial acetyl-CoA, which is at the entry of the tricarboxylic acid (TCA) cycle and at the crossroad of carbohydrate, fatty acid, and ketone body metabolism that yields the energy necessary to maintain proper heart function. Pyruvate can therefore provide crucial information for diagnosing patient health, and it has been selected as the first hyperpolarized ^{13}C-labelled substrate for in vivo metabolic studies in humans.

21.2 EARLY RESULTS IN HUMANS

About 10 years after Ardenkjaer-Larsen et al.[8] introduced the dissolution technology, hyperpolarized [1-^{13}C]pyruvate solutions were prepared in a large, clean room containing a prototype DNP polarizer and injected for the first time into humans.[17] The study demonstrated that pyruvate is safe to use at the doses required to obtain quantitative data. Following this pioneering work performed at the University of California San Francisco (UCSF), a dedicated DNP polarizer allowing the production of sterile doses of hyperpolarized ^{13}C-substrates in a standard room was designed and implemented.[18] This instrument was built around the concept of sealed 'fluid paths' (also referred to as

'pharmacy kits' in the context of clinical studies) to avoid contamination of the substrate and all fluids required for the preparation of the injectable solutions. This DNP polarizer commercialized by GE Healthcare is named SPINlab™, and it led to the dissemination of this technology for human use to other research institutions. To date, seven sites worldwide have started in-human trials to evaluate the clinical potential of hyperpolarized ^{13}C MRI in various organs and diseases: UCSF, Memorial Sloan Kettering Cancer Center (MSKCC, USA), University of Cambridge (UK), University of Oxford (UK), University of Toronto (Canada), University of Texas Southwestern (USA), and University College London (UK).

21.2.1 Protocol for Clinical Applications

The protocol for clinical applications of hyperpolarized ^{13}C MR consists of four main steps: sample preparation, dissolution DNP, quality control (QC) and injection, and finally MR data acquisition (Figure 21.2).

21.2.1.1 Sample Preparation

A critical part of the preparation of an injectable solution of hyperpolarized ^{13}C-molecule is the incorporation of the doped substrate (e.g., ^{13}C-pyruvic acid mixed with polarizing agent), dissolution medium (water), as well as neutralization medium (NaOH solution mixed with tris buffer) into the pharmacy kits inside a controlled environment (pharmacy clean room). This procedure can be referred to as filling process. Various methods have been designed at different sites, but all of them can be separated in two different concepts: the first one is to consider that the content of the pharmacy kits should be sterile at the end of the filling process, i.e., when the kits are released from the pharmacy clean room; the second one is to rely on terminal sterilization through a 0.2 μm filter at the time of dissolution, i.e., when each injectable solution is actually produced. The latter is based on the approach taken by radiopharmacists and was initially proposed by the team at MSKCC.[19] One of the main issues with the former is that all initial products including the substrate, the polarizing agent, the dissolution medium, the neutralization medium, and the plastic fluid path itself have to be sterile or at least have a low bioburden. The terminal sterilization approach is therefore more flexible, and it seems that in the future it will be the preferred protocol for most sites.

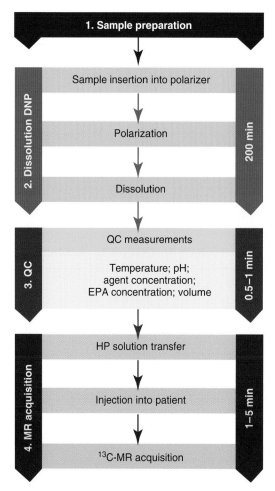

Figure 21.2. Timescale of dissolution DNP experiments

21.2.1.2 Dissolution DNP

Once the pharmacy kits are filled, they can be loaded inside the DNP polarizer. The current implementation of SPINlab™ operates at 5 T and at a temperature below 1 K (typically around 0.8 K). The substrate/polarizing agent mixture contained inside the cryogenic vial of the filled pharmacy kit is irradiated by microwaves generated around 140 GHz and the ^{13}C spins are typically polarized in a few hours. At the end of the DNP procedure, once the patient has been catheterized, placed inside the MRI, and a set of anatomical ^{1}H MR images have been recorded, the superheated dissolution medium is released from the dissolution syringe of the pharmacy kit and directed into the cryogenic vial to dissolve the frozen mixture.

The resulting acidic solution runs through a filter that collects most of the precipitated polarizing agent prior to be neutralized and buffered to physiological pH by mixing with the neutralization medium contained in the receiver vessel of the pharmacy kit.

21.2.1.3 Quality Control (QC)

Following dissolution, filtration, and neutralization, the following parameters are checked prior to injection: temperature, volume, pH, pyruvate concentration, residual polarizing agent concentration, and optionally ^{13}C polarization level. The exact accepted range depends on the release criteria of the local regulatory authorities and qualified people involved in the procedure. However, there is a general agreement on the gross range for hyperpolarized [1-^{13}C]pyruvate solutions: the target pyruvate concentration should be 250 ± 30 mM, polarizing agent concentration less than $5 \, \mu$M, and a pH between 6.8 and 8.4. These checks are performed on a small aliquot (a few ml) using optical measurements, which only take a fraction of a second to provide the results. Temperature and minimum volume are dictated by the current implementation of the QC hardware, which is set to provide at least 40 ml of solution at around body temperature (37 °C). Finally, the ^{13}C polarization level can be determined by comparing the ^{13}C NMR measurement of the aliquot with the signal obtained from a standard sample, and it is expected to be at least 5%. From these QC results, a pharmacist or qualified person will reject or release the dose for injection into a subject.

21.2.1.4 MR Data Acquisition

MR data acquisition can be synchronized with the injection with a potential delay to catch the bolus as it reaches the area of interest. Since high polarization of the ^{13}C-substrates only lasts for about a minute, the use of rapid and efficient ^{13}C-MR acquisition techniques is imperative. The challenge set for metabolic imaging by hyperpolarized ^{13}C MRI is to record dynamic magnetic resonance spectroscopic imaging (MRSI), which consists of five-dimensional (three spatial, one chemical, and one temporal) acquisitions. Signal-to-noise ratio (SNR) and resolution can be dramatically improved by prioritizing the dimensions that provide the most valuable information for each scan.

When several radiofrequency (RF) pulses are required to follow the spectral dynamics of the substrate and its products (chemical and temporal dimensions), variable flip angle schemes may be used to equalize the usage of hyperpolarized ^{13}C magnetization at every acquisition.[20] Dynamic hyperpolarized ^{13}C MRS acquisitions can also benefit from the use of spectral-spatial RF pulses, which deliver metabolite-specific flip angles to the spatial location of interest. In this case, a larger flip angle may be applied on the metabolic products compared to the substrate in order to preserve the hyperpolarized ^{13}C magnetization of the substrate while maximizing the overall SNR.[21] Rapid MRSI that reduces the scan time and RF power deposition can be achieved with several k-space trajectories, such as echo-planar spectroscopic imaging (EPSI), radial, spiral, and concentric rings.[22–24] For an even faster MRI acquisition, single-metabolite imaging can be performed applying a spectrally selective RF on a specific metabolite, then using another spectrally selective RF pulse designed to acquire another metabolite in the next image.[25,26] Alternatively, spatiotemporal encoding allows for rapid chemical-shift imaging in a single shot.[27] In order to shorten the acquisition time further, acceleration strategies such as compressed sensing[28] and parallel imaging[29] have also been used for hyperpolarized ^{13}C MRSI. All these methods have their benefits and drawbacks, and prior knowledge of the information sought would help carefully select the pulse sequence best suited for the MRI and MRS acquisition. More details on pulse sequences for hyperpolarized ^{13}C MRS can be found in Ref. 30.

21.2.2 Prostate Cancer

The translation of dissolution DNP into the clinic started with a Phase I clinical study to demonstrate the safety and feasibility of hyperpolarized [1-^{13}C]pyruvic acid.[17] Thirty-one human subjects with biopsy-proven prostate cancer were injected with three dose levels of a solution containing 230 mM pyruvate (0.13, 0.28, and 0.43 ml kg^{-1} actual body weight), and spectroscopic images were subsequently acquired using a 3 T MRI scanner. Signals of both hyperpolarized pyruvate and its metabolic product lactate were detected in the prostate 20 s after the beginning of the injection, showing that prostate cancerous tissue has a higher lactate-to-pyruvate ratio than healthy prostate tissue. The study concluded that 0.43 ml kg^{-1} of 230 mM hyperpolarized ^{13}C-pyruvate solution (equivalent to a dose of 0.1 mmol kg^{-1}) gives the highest SNR without causing any negative side effect to the subject. Further work on prostate patients carried out at UCSF demonstrated the value of hyperpolarized ^{13}C-pyruvate as a metabolic marker of treatment response[31]: after a 6-week androgen ablation treatment, the tumor size observed in T_2-weighted images and the apparent diffusion coefficient (ADC) only changed modestly, while the hyperpolarized lactate-to-pyruvate ratio dropped.

21.2.3 Brain Cancer

In a study performed at UCSF, eight patients diagnosed with glioma were injected intravenously with a hyperpolarized [1-^{13}C]pyruvate solution (0.1 mmol kg^{-1}) and signal acquired in a 3 T MRI scanner to develop acquisition methods for patients with brain tumors.[32] The study revealed that ^{13}C-pyruvate is transported across the blood–brain barrier in humans by detecting signal of the injected ^{13}C-pyruvate and ^{13}C-pyruvate hydrate, as well as its downstream products ^{13}C-lactate and ^{13}C-bicarbonate.

Recently, another study carried out at MSKCC looked further into the altered cancer metabolism in brain by injecting hyperpolarized [1-^{13}C]pyruvate into four patients with untreated or recurrent brain tumors.[19] As an example, Figure 21.3 shows the high contrast relative to the background achieved with ^{13}C MRI of hyperpolarized ^{13}C-pyruvate and its metabolic product, ^{13}C-lactate, in a patient with untreated melanoma metastasis.

21.2.4 Other Ongoing Clinical Studies in Cancer

Clinical studies involving the use of dissolution DNP to detect impaired metabolism are rapidly increasing. For example, there are ongoing studies of breast cancer metabolism in women,[33] and another study is focusing on the hepatic conversion of pyruvate to lactate through lactate dehydrogenase (LDH) in patients with liver metastasis.[34]

21.2.5 Cardiac Metabolism

The group led by Dr. Cunningham in Toronto presented the first hyperpolarized ^{13}C MR images in the human heart.[35] Four healthy subjects were

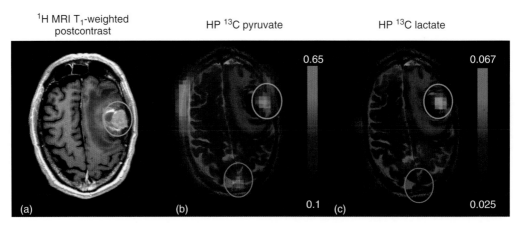

¹H MRI T₁-weighted postcontrast   HP ¹³C pyruvate   HP ¹³C lactate

Figure 21.3. Hyperpolarized pyruvate imaging of a metastasis. (a) The left-middle frontal gyrus melanoma metastasis (green circle) has solid components laterally and hemorrhagic components medially. (b) The hyperpolarized pyruvate map shows high signal corresponding to the metastasis, likely due to perfusion. The volume normalized pyruvate signal in the lesion is 2.8-fold higher than the entire brain. (c) The hyperpolarized lactate map shows significantly high intensity corresponding to the solid component of the metastasis, absent from the medial perilesional hemorrhagic component. The volume normalized lactate signal in the lesion is 6.7-fold higher than the entire brain. Importantly, lactate signal in the superior sagittal sinus is low, consistent with the detected lactate being made locally in the brain, not due to delivery from outside the brain. (Courtesy of Kayvan R. Keshari)

placed in a 3 T MRI scanner, injected hyperpolarized [1-¹³C]pyruvate solution (0.1 mmol kg⁻¹). Signal arising from the injected substrate appeared within the heart chambers but not the muscle. As expected from preclinical studies, the downstream metabolic products detected were [1-¹³C]lactate and ¹³C-bicarbonate. The former was detected both in the chambers and the myocardium, while the latter appeared in the left ventricular myocardium. This reveals that hyperpolarized ¹³C MR can be used to obtain a direct measurement of the flux through pyruvate dehydrogenase (PDH) in the human myocardium.

21.3 PROBLEMS AND CHALLENGES

To date, hyperpolarized ¹³C-pyruvate is the only molecule that has been injected into humans. The main reason for this limitation is the lack of pharmacology and toxicity data required by regulatory bodies (e.g. U.S. Food and Drug Administration (FDA) and European Medicines Agency (EMA)) for authorizing the use of other hyperpolarized ¹³C-compounds. Several other molecules are however currently being evaluated for injection into patients, and the commercially available DNP polarizer designed for clinical applications will be equipped with the appropriate

hardware to perform the QC on these compounds. It can therefore be expected that within the coming few years, urea, fumarate, lactate, and possibly acetate and α-ketoisocaproate will be used in clinical trials.

Nevertheless, important obstacles still exist for hyperpolarization technology to unlock its enormous potential. The principal drawback of dissolution DNP is that the production of hyperpolarized ¹³C-molecules has to be thoroughly synchronized with their injection, which means that the hyperpolarized ¹³C-substrates cannot be stored. As a consequence, each MRI suite has to be equipped with a DNP polarizer placed in proximity to their scanner, and the hyperpolarized ¹³C-labeled compounds have to be produced promptly prior injection. This severe limitation is due to the relatively short lifetime of the hyperpolarized state characterized by the ¹³C longitudinal relaxation time constant, T_1. In the context of hyperpolarized ¹³C MR, $1/T_1$ is the rate at which the longitudinal component of the ¹³C magnetization exponentially decays toward its thermal equilibrium value. Using a dedicated transfer/injection device, the delay between production and injection into animals can be minimized to a few seconds,[37] although this method would be difficult to translate to human applications because of potential safety issues and the necessary QC checks prior to injection. Some developments have been done to exploit longer-lived spin states to preserve the ^{13}C polarization in biocompatible solutions, but the relaxation time constant of those states remained comparable to T_1.[38,39] Another issue is that the dissolution step restricts the production rate as well as the concentration of hyperpolarized solutions, therefore forcing large bolus injections to reach sufficient sensitivity in vivo. A higher production rate with increased concentration would allow nearly continuous injection of hyperpolarized ^{13}C-substrates during an extended time window and hence provide the ability to average the in vivo ^{13}C signal beyond the inherent lifetime of the hyperpolarized state.

The physical reason underlying the necessity to include an in situ dissolution step following DNP is the high relaxivity of unpaired electron spins in the frozen solution of ^{13}C-molecules doped with polarizing agents, preventing the extraction of the substrates from the DNP polarizer without destroying the enhanced ^{13}C polarization. If the electron spins could be annihilated inside the frozen solution once the DNP process has taken place, it would be possible to obtain long-lived hyperpolarized ^{13}C-molecules in the solid state, which could be extracted in a solid form, avoiding the dissolution process. The presence of free radicals in the final solution is also a health concern, and it needs to be filtered out prior to injection into humans. This process adds an additional potential failure point to the whole process. One of the consequences of the dissolution step is therefore that it restricts the number and type of ^{13}C MR scans that can be performed and preclude the hospitals that are not equipped with the full equipment from using this technology.

21.4 NEW APPROACHES

In the authors' opinion, hyperpolarized ^{13}C MR can only be rapidly and widely disseminated and, in a way, eventually democratized if it becomes possible to decouple the preparation of the hyperpolarized ^{13}C molecules from the injection. In other words, it is crucial to develop a method to store and transport polarized ^{13}C-molecules to hospitals MRI suites in a fashion similar to ^{18}F-fludeoxyglucose (^{18}F-FDG) for positron emission tomography (PET) scans. An alternative to dissolution DNP that would make this approach feasible has been recently proposed. It is based on radicals that are photoinduced at liquid nitrogen temperature and persist long enough to be used for DNP at subliquid

helium temperature.[40] Using a rapid thermalization process, these radicals can be annihilated while maintaining the enhanced ^{13}C polarization inside the frozen solution, resulting in a lifetime about 20-fold longer at 4.2 K, from 53 ± 3 min to 967 ± 17 min in [1-^{13}C]pyruvic acid (Figure 21.4).[36] This means that hyperpolarized ^{13}C-substrates could be produced in high concentration, stored, transported, and delivered in a quasi-continuous manner, with the added benefit that there is no need to use potentially toxic persistent radicals.[41] The main benefit of such solution would be that a large number of doses could be prepared in a dedicated facility and delivered to clinics ahead of the MR scans, thus reducing the complexity and alleviating the need for synchronizing the preparation and the injection of hyperpolarized ^{13}C-substrates. It would also relax the need for highly qualified technical staff to maintain and run the DNP polarizer inside the MRI suite. An added benefit would be that the number of scans per day could be greatly increased.

No clinically compatible solution for storing or transporting hyperpolarized ^{13}C-subtrates is available today, and only a restricted number of sites worldwide are equipped with a DNP polarizer enabling clinical studies. Although the current SPINlab™ operates at 5 T, it has been recently shown that increasing the field to 7 T can lead to larger polarization (up to 80%) at least at 1 K for the most widely used radicals, namely trityls.[42] A more recent study demonstrated that at 6.7 T and 1.1 K, photo-induced radicals in pyruvic acid can yield a ^{13}C polarization of 30%.[43] It is therefore becoming increasingly clear that photoinduced radicals are competitive at higher field. In addition to the development of a clinical DNP polarizer compatible with this technology, a dedicated transport device capable of maintaining the ^{13}C polarization for an extended time could be implemented. If the polarized [1-^{13}C]pyruvate is kept in a field of 1 T at a temperature of 77 K (liquid nitrogen temperature) or below, the ^{13}C T_1 is expected to be at least around 1 h.[44] Eventually, this technology could be available for routine clinical applications.

21.5 PRECLINICAL AND CELL WORK AS A TESTBED FOR DNP INNOVATIONS

Prior to clinical translation, hyperpolarized ^{13}C substrates and ground-breaking technological approaches are tested in vivo in animals and cells.

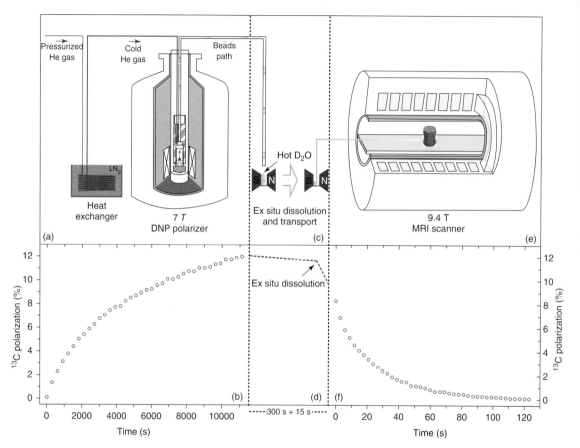

Figure 21.4. Ex situ dissolution. (a) Solid sample extraction procedure. (b) ^{13}C polarization time evolution measured by ^{13}C NMR at 7 T and 1 K in 10 ultraviolet-irradiated frozen beads; once the ^{13}C polarization had reached $12 \pm 0.5\%$, the ultraviolet-induced radicals were annihilated using the thermalization process described in the text (c). Schematic representation of the ex situ dissolution and transfer of the hyperpolarized [1-^{13}C]PA aqueous solution from the side of the polarizer to the custom-designed injection pump placed inside the MRI scanner. (d) Estimated ^{13}C polarization behavior during all intermediate operations: the cyan dotted line represents the decay of the ^{13}C polarization at 7 T and 4.2 K during the 300 s required to cool down the extraction line; the fuchsia dotted line sketches the liquid-state ^{13}C longitudinal relaxation after ex situ dissolution at 100 mT in hot D_2O during the 15 s needed to transport the hyperpolarized solution in proximity of the 9.4 T MRI scanner. (e) Sketch of the injection pump equipped with a ^{13}C Alderman–Grant NMR coil and placed at the isocenter of the 9.4 T MRI scanner. (f) Liquid-state ^{13}C NMR longitudinal relaxation measurement performed at 9.4 T and room temperature inside the injection pump. (Source: Capozzi *et al.*,[36] https://www.nature.com/articles/ncomms15757. Licensed under CC BY 4.0)

21.5.1   Recent In Vivo Applications with Alternative Substrates

While the preparation and applications of most hyperpolarized ^{13}C substrates introduced to date have been extensively reported elsewhere,[45–48] this section aims at discussing some of the alternative compounds to pyruvate that have been proposed to assess cancer and

cardiac metabolism (Table 21.1). Even for preclinical studies, most hyperpolarized ^{13}C MR scans were performed using [1-^{13}C]pyruvate. Although this substrate allows probing LDH and PDH activities, therefore providing a measurement related to upregulated glycolysis through the lactate signal as well as a measurement of the formation of glucose-derived acetyl-CoA feeding the TCA cycle, it does not give a direct reading of

Table 21.1. List of hyperpolarized ^{13}C-probes used in vivo and discussed in this chapter

DNP probe	Polarizer	Concentration (mol l⁻¹)	Solvent/Glassing agent	P (%)	T_1 (s)	B_0 (T)	References
[U-^{2}H, U-^{13}C]Glucose	HS	3.0	H₂O/DMSO	13	10–13	9.4	49
	7 T	3.0	H₂O	19–22			49
[2,3,4,6,6-^{2}H$_5$, 3,4-^{13}C$_2$]Glucose	HS	3.6	D₂O or D₂O/d-DMSO	15	9	7.0	50
	HS	3.0	H₂O/DMSO	30	11–14	9.4	49
[2-^{13}C]fructose	HS	4.0	H₂O	12	16	3.0	51
					14	11.7	
[2-^{13}C]Dihydroxyacetone	HS	8.0	H₂O/DMSO	15	40	3.0	21,52,53
					32	9.4	
[1-^{13}C]Glycerate (sodium salt)	SL	3.0	H₂O/glycerol	20	60	3.0	54
[1-^{13}C]Gluconolactone	HS	4.0	H₂O/glycerol	6	18–20	9.4	55
[1-^{13}C]Glutamine	SL	4.0	NaOH/H₂O/glycerol	35	31	1.0	56
					25	9.4	
[5-^{13}C]Glutamine	HS	3.5	DMSO	34	20	14.1	57
[1-^{13}C]α-ketoglutaric acid	HS	5.9	H₂O/glycerol	16	52	3.0	58
					19	11.7	
[1-^{13}C]Butyric acid	HS	8	H₂O/DMSO	28	56	3.0	59
	HS	8.4	H₂O/DMSO	7	20	11.7	60
	7 T	5.5	Pyruvic acid (no trityl radicals)	3		9.4	61
[1-^{13}C]Octanoic acid	7 T	4.0	DMSO	11	29	9.4	62
[1-^{13}C]Acetoacetate (sodium)	HS	6.5	H₂O/DMSO	8	28	11.7	63
[1,3-^{13}C$_2$]Acetoacetate (sodium)	HS	1.0	DMSO	7	52–58	3.0	64
[1,5-^{13}C$_2$]Zymonic acid	HS	4.0	DMSO	22	43–51	3.0	65

Key features of the predissolution process are detailed, such as what type of DNP polarizer was used, at what concentration, and with which glassing agent. Polarization level in the liquid state after dissolution and T_1 in the specified magnetic field are also indicated. The preparations on this table use trityl radical unless stated otherwise. Abbreviations: HS, Hypersense; SL, SPINlab^{TM}; 5 T/7 T, custom-built polarizer; P, polarization level; T_1, longitudinal relaxation time constant; B_0, magnetic field.

glycolysis or other pathways related to cellular uptake of other essential substrates such as fatty acids, ketone bodies, or amino acids. Since these other pathways are also known to be altered in cancer or cardiac diseases, they can bring additional information that is complementary to pyruvate metabolism. Figure 21.1 displays the metabolic network in which these ^{13}C substrates (shaded background) are involved.

21.5.1.1 Probing Intermediate Steps of Glycolysis and Pentose Phosphate Pathway

Hyperpolarized ^{13}C-glucose is the natural substrate for probing glycolysis but the T_1 of protonated ^{13}C-glucose is too short ($\sim 1-2$ s) to retain the large solid-state ^{13}C polarization that can be reached by DNP all the way through the dissolution process and the injection into an animal. Fortunately, deuterated ^{13}C-glucose has an extended ^{13}C T_1 of ~ 10 s,[66] and it has been used to measure cerebral metabolism in healthy mice,[49] as well as tumor metabolism in the EL4 xenograft mouse model.[50] Hyperpolarized [2-^{13}C]fructose, with a T_1 relaxation time constant of 16 s, has also been proposed as a glucose analogue to probe glycolysis in a transgenic mouse model of prostate cancer.[51] Alternative hyperpolarized ^{13}C-substrates with longer T_1 for probing glycolysis have been proposed. One of them is [2-^{13}C]dihydroxyacetone, which has a T_1 of ~ 40 s at 3 T and enters the Embden–Meyerhof pathway at the intermediate triose phosphate level, with the potential to report on both gluconeogenesis and glycolysis as well as on the glycerol synthesis pathway. This substrate has been hyperpolarized and injected in perfused mouse liver[52] and in rat kidney and liver in vivo.[21,53] Similarly, [1-^{13}C]glycerate (T_1 of ~ 60 s at 3 T) rapidly entered the glycolytic pathway and [1-^{13}C]pyruvate and [1-^{13}C]lactate peaks were observed in vivo, allowing the detection of differences between the fasted and fed states of the liver in healthy rats.[54]

Hyperpolarized [U-^{2}H, U-^{13}C]glucose was also used to probe the pentose phosphate pathway (PPP), which is upregulated in cancer and trauma.[50,67] Alternatively, hyperpolarized [1-^{13}C]gluconolactone has been suggested as a tracer to specifically target PPP, reporting its metabolic conversion into gluconate and bicarbonate in a perfused mouse liver.[55]

21.5.1.2 Amino Acid Metabolism in Cancer

For cancer and other proliferating cells, glutamine is an important substrate that can be used as a source of carbon and nitrogen. However, the metabolic uptake of amino acids is usually slow compared to the timescale of hyperpolarized ^{13}C MR experiments. Despite this limitation, it has been shown that hyperpolarized [5-^{13}C]glutamine can be used to probe metabolism in vivo, in particular in rat or mouse cancer liver models where its conversion into glutamate was detected.[57,68] More recently, it was also shown that 2-hydroxyglutarate can be rapidly generated from [1-^{13}C]glutamine in IDH1/2 mutant tumors in vivo.[56] Similarly, [1-^{13}C]α-ketoglutarate had previously been injected into glioma-bearing mice and its metabolic conversion into 2-hydroxyglutarate was observed due to the mutant IDH1 gene expressed in such tumor cells.[58]

21.5.1.3 Fatty Acid and Ketone Bodies Metabolism in the Heart

Fatty acids, ketone bodies, and carbohydrates are the main fuel sources to meet the heart's energetic requirements. Hyperpolarized ^{13}C-substrates that can provide insights into myocardium energetic sources and how they shift in disease and after treatment have been investigated. The metabolic conversion of [1-^{13}C]butyrate, a short-chain fatty acid, into acetylcarnitine, glutamate, butyrylcarnitine, and acetoacetate has been followed in the rat heart in vivo in fed and fasted states.[59–61] The medium-chain fatty acid [1-^{13}C]octanoate has also been hyperpolarized and its metabolic product after β-oxidation, acetylcarnitine, has been detected in healthy rats in vivo.[62] Ketone bodies have also been explored as hyperpolarized ^{13}C substrates. Kennedy et al.[69] developed hyperpolarized ^{13}C-probes based on acetoacetate and β-hydroxybutyrate derivatives and explored their potential to probe cardiac metabolism in mice in vivo. However, it was not until recently that their value as metabolic substrates was proven. The injection of [1-^{13}C]acetoacetate showed a decrease in acetylcarnitine and bicarbonate in the rat heart in vivo after an overnight fasting.[63] The metabolic conversion of [1,3-^{13}C$_2$]acetoacetate into β-hydroxybutyrate has also been observed in the rat kidneys in vivo, where renal mitochondrial redox state was modulated in rats treated with metformin.[64]

21.5.1.4   Hyperpolarized ^{13}C-pH Probes

Pathological processes, such as ischemia, infection, inflammation, or cancer cause localized pH changes within the affected tissue. In 2008, [^{13}C]bicarbonate was the first hyperpolarized ^{13}C substrate used to probe pH in mice in vivo. pH was calculated by measuring the ^{13}C signal intensity of $H^{13}CO_3^-$ and $^{13}CO_2$ and using a modified Henderson–Hasselbalch equation.[70] However, this ratiometric approach yields low spatial resolution, mainly due to the low SNR of $^{13}CO_2$. Recently, an alternative set of ^{13}C-pH probes that rely on chemical shift changes induced by pH has been proposed.[65,71] Zymonic acid is the only hyperpolarized ^{13}C probe of this class to show MR pH maps in vivo, showing that a tumor was 0.1–0.2 pH units more acidic than the surrounding tissue in a tumor-bearing mouse.[65] The chemical shift difference between the two ^{13}C of hyperpolarized [1,5-$^{13}C_2$]zymonic acid is a sensitive MR extracellular pH sensor, with a pH resolution of 0.1 pH units and independent of concentration, temperature, ionic strength, and protein concentration. Although to date they have not been tested in vivo, other promising pH sensors such as N-(2-acetoamido)-2-aminoethanesulfonic acid (ACES)[72] and dicarboxylic acids like [2-^{13}C, D10]diethylmalonic acid[71] have also been successfully hyperpolarized.

21.5.2   Additional Technological Considerations for Hyperpolarized ^{13}C MR

Metabolic and molecular imaging is a very dynamic field, and the list of applications of hyperpolarized ^{13}C MR is constantly growing. In order to enable advanced applications of this rapidly expanding field, new techniques and hardware are constantly being developed. This section provides a brief overview of some of the latest technological advancements that can be used in conjunction with hyperpolarized ^{13}C MR.

21.5.2.1   Co-injection of Hyperpolarized ^{13}C-substrates and PET Tracers

While hyperpolarized ^{13}C MR provides metabolic information with high temporal resolution albeit relatively low spatial resolution, PET reports on substrate uptake with higher spatial resolution but gives no information on its metabolic conversion. These two modalities have been compared by sequentially acquiring hyperpolarized ^{13}C MR and PET in the same animal.[73,74] More recently, both hyperpolarized ^{13}C-substrates (^{13}C-fumarate and ^{13}C-pyruvate) and a PET tracer (^{18}F-2-fluoro-2-deoxy-D-glucose, FDG) were simultaneously injected into a rat model of necrosis, and ^{13}C MR and PET data were acquired in a MR/PET scanner.[75] As expected, uptake of ^{13}C-pyruvate and ^{18}F-FDG was observed from the same tissue. The conversion from fumarate to malate, as well as the high lactate-to-pyruvate ratio, confirmed necrotic tissue. The combination of these two imaging modalities can provide complementary information and could allow for better evaluation of the disease aggressiveness or treatment response.

21.5.2.2   Low-field vs High-field Hyperpolarized ^{13}C MR

In general, clinical MR scanners operate at magnetic field strengths between 1 and 3 T, while preclinical scanners operate at higher magnetic fields, up to 14 T. For nonhyperpolarized MR, higher magnetic fields are advantageous because they provide better chemical shift resolution and higher sensitivity. For in vivo hyperpolarized ^{13}C MR, the size of the RF coil is usually such that the data is acquired in a sample-dominated noise regime,[76] which means that the SNR only depends on the polarization level and is therefore the same at 1 T as it is at 14 T. Since ^{13}C has an inherent large chemical shift dispersion, there is essentially no advantage in using a higher magnetic field for hyperpolarized ^{13}C MR (there are obviously some specific cases in which an increased chemical shift resolution is necessary to distinguish MR signals with very similar resonance frequencies, like those of [1-^{13}C]acetate and [1-^{13}C]α-ketoglutarate in Ref. 77). However, since T_1 is often field dependent, the choice of field strength can affect the SNR of the hyperpolarized ^{13}C MR scans.

To maximize the chances of a successful clinical translation, it might therefore be advantageous to design and perform animal hyperpolarized ^{13}C MR experiments on a preclinical MR scanner operating at standard clinical fields. Keshari et al. used a benchtop 1 T permanent magnet for MR experiments to characterize the hyperpolarized ^{13}C-substrate and perform initial cell extract studies, and a 1 T MR/PET system to perform in vivo studies in mice.[78] Also using a benchtop 1 T permanent magnet, they designed a setup that allowed them to perform hyperpolarized ^{13}C metabolic experiments in cells with 10^3 more sensitivity than with conventional high-field MR equipment.[79] They overcame the generally low temporal and spatial resolution associated with permanent magnets by implementing a microcoil that resonates at both the ^{1}H and ^{13}C Larmor frequency and using the ^{1}H resonance mode to correct for first-order field inhomogeneity prior to ^{13}C-MR acquisition.

21.5.2.3 *RF Coils*

MR sensitivity directly depends on the RF coil design used for acquiring data, and over the years ^{1}H MR has benefited from tremendous improvements in coil technology.[80] Until the advent of hyperpolarization, only few research groups had developed dedicated ^{13}C coils, and their focus was mostly limited to surface coils designed for localized spectroscopy rather than imaging.[81,82] The implementation of optimized ^{13}C coils for clinical 1.5 and 3 T scanners is therefore another important technological development that is expected to further improve the quality of ^{13}C MR acquisitions. To obtain homogeneous B_1 excitation across a large field of view, it would be interesting to have a dual-tuned ^{1}H/^{13}C volume coil, the ^{1}H coil being necessary for anatomical images and localization. Such coils exist for the head,[83] but not yet for the whole human body. It would also be of interest to develop improved receive surface coils that are very close to the patient's body such as those developed for pig heart experiments.[84]

21.6   CONCLUSIONS

Numerous preclinical applications based on dissolution DNP have demonstrated the ability of hyperpolarized ^{13}C MR to assess metabolism in vivo and that it can be used to provide information on the health status of tissue in various organs. It also seems quite clear by now that hyperpolarized ^{13}C MR has an enormous potential for becoming a clinically relevant modality. Many technical challenges still need to be tackled to make this technology available for routine examinations, but the fundamental issues have been resolved and the initial concerns related to safety and sensitivity were rapidly overcome. The nonradiative nature of MR and the fact that hyperpolarized ^{13}C agents are endogenous should help convincing the medical community as well as the various national health institution

systems that this imaging modality will be beneficial to patients, notably those undergoing cancer treatment.

ACKNOWLEDGMENTS

This work is part of a project that has received funding from the European Union's Horizon 2020 European Research Council (ERC Consolidator Grant) under grant agreement No 682574 (ASSIMILES).

RELATED ARTICLES IN EMAGRES

Radiofrequency Systems and Coils for MRI and MRS

Hyperpolarized Carbon-13 MRI and MRS Studies

Pulse Sequences for Hyperpolarized MRS

^{13}C MRS in Human Tissue

REFERENCES

1. R. Gruetter, E. J. Novotny, S. D. Boulware, G. F. Mason, D. L. Rothman, G. I. Shulman, J. W. Prichard, and R. G. Shulman, *J. Neurochem.*, 1994, **63**, 1377.

2. R. Gruetter, E. R. Seaquist, S. Kim, and K. Ugurbil, *Dev. Neurosci.*, 1998, **20**, 380.

3. E. A. Maher, I. Marin-Valencia, R. M. Bachoo, T. Mashimo, J. Raisanen, K. J. Hatanpaa, A. Jindal, F. M. Jeffrey, C. Choi, C. Madden, *et al.*, *NMR Biomed.*, 2012, **25**, 1234.

4. R. G. Shulman and D. L. Rothman, *Annu. Rev. Physiol.*, 2001, **63**, 15.

5. K. C. C. Van De Ven, B. E. De Galan, M. Van Der Graaf, A. A. Shestov, P.-G. Henry, C. J. J. Tack, and A. Heerschap, *Diabetes*, 2011, **60**, 1467.

6. M. Krššák, in 'Encyclopedia of Magnetic Resonance', eds R. Harris and R. Wasylishen, John Wiley & Sons, Ltd: Chichester, 2016, p 1027.

7. W. Köckenberger, in 'Encyclopedia of Magnetic Resonance', eds R. Harris and R. Wasylishen, John Wiley & Sons, Ltd: Chichester, 2014, p 161.

8. J. H. Ardenkjaer-Larsen, B. Fridlund, A. Gram, G. Hansson, L. Hansson, M. H. Lerche, R. Servin, M. Thaning, and K. Golman, *Proc. Natl. Acad. Sci. U. S. A.*, 2003, **100**, 10158.

9. K. Golman, R. in't Zandt, and M. Thaning, *Proc. Natl. Acad. Sci. U. S. A.*, 2006, **103**, 11270.

10. A. Comment, *J. Magn. Reson.*, 2016, **264**, 39.

11. S. E. Day, M. I. Kettunen, F. A. Gallagher, D.-E. Hu, M. Lerche, J. Wolber, K. Golman, J. H. Ardenkjaer-Larsen, and K. M. Brindle, *Nat. Med.*, 2007, **13**, 1382.

12. J. Kurhanewicz, D. B. Vigneron, K. Brindle, E. Y. Chekmenev, A. Comment, C. H. Cunningham, R. J. DeBerardinis, G. G. Green, M. O. Leach, S. S. Rajan, *et al.*, *Neoplasia*, 2011, **13**, 81.

13. R. Sriram, J. Kurhanewicz, and D. B. Vigneron, in 'Encyclopedia of Magnetic Resonance', eds R. Harris and R. Wasylishen, John Wiley & Sons, Ltd: Chichester, 2014, Vol. **3**, p 311.

14. C. R. Malloy, M. E. Merritt, and A. D. Sherry, *NMR Biomed.*, 2011, **24**, 973.

15. W. Kaelin and C. Thompson, *Nature*, 2010, **465**, 563.

16. W. C. Stanley, F. A. Recchia, and G. D. Lopaschuk, *Physiol. Rev.*, 2005, **85**, 1093.

17. S. J. Nelson, J. Kurhanewicz, D. B. Vigneron, P. E. Z. Larson, A. L. Harzstark, M. Ferrone, M. V. Criekinge, J. W. Chang, I. Park, G. Reed, *et al.*, *Sci. Transl. Med.*, 2014, **5**, 198.

18. J. H. Ardenkjaer-Larsen, A. M. Leach, N. Clarke, J. Urbahn, D. Anderson, and T. W. Skloss, *NMR Biomed.*, 2011, **24**, 927.

19. V. Z. Miloushev, K. L. Granlund, R. Boltyanskiy, S. K. Lyashchenko, L. M. DeAngelis, E. Sosa, Y. W. Guo, A. P. Chen, J. Tropp, F. Robb, *et al.*, *Cancer Res.*, 2018, **78**, 3755.

20. J. Maidens, S. Member, J. W. Gordon, M. Arcak, and P. E. Z. Larson, *IEEE Trans. Med. Imaging*, 2016, **1**, 1.

21. I. Marco-Rius, P. Cao, C. von Morze, M. Merritt, K. X. Moreno, G.-Y. Chang, M. A. Ohliger, D. Pearce, J. Kurhanewicz, P. E. Larson, and D. B. Vigneron, *Magn. Reson. Med.*, 2017, **77**, 1419.

22. D. Mayer, Y. S. Levin, R. E. Hurd, G. H. Glover, and D. M. Spielman, *Magn. Reson. Med.*, 2006, **56**, 932.

23. M. S. Ramirez, J. Lee, C. M. Walker, V. C. Sandulache, F. Hennel, S. Y. Lai, and J. A. Bankson, *Magn. Reson. Med.*, 2014, **72**, 986.

24. W. Jiang, M. Lustig, and P. E. Z. Larson, *Magn. Reson. Med.*, 2016, **75**, 19.

25. H. Shang, S. Sukumar, C. von Morze, R. A. Bok, I. Marco-Rius, A. Kerr, G. D. Reed, E. Milshteyn, M. A. Ohliger, J. Kurhanewicz, *et al.*, *Magn. Reson. Med.*, 2017, **78**, 963.

26. A. Z. Lau, J. Miller, and D. J. Tyler, *Magn. Reson. Med.*, 2017, **77**, 1810.

27. R. Schmidt, C. Laustsen, J. N. Dumez, M. I. Kettunen, E. M. Serrao, I. Marco-Rius, K. M. Brindle, J. H. Ardenkjaer-Larsen, and L. Frydman, *J. Magn. Reson.*, 2014, **240**, 8.

28. M. Lustig, D. L. Donoho, and J. M. Pauly, *Magn. Reson. Med.*, 2007, **58**, 1182.

29. A. Arunachalam, D. Whitt, K. Fish, R. Giaquinto, J. Piel, R. Watkins, and I. Hancu, *NMR Biomed.*, 2009, **22**, 867.

30. J. W. Gordon and P. E. Z. Larson, in 'Encyclopedia of Magnetic Resonance', eds R. Harris, and R. Wasylishen, John Wiley & Sons, Ltd: Chichester, 2016, Vol. **5**, p 1229.

31. R. Aggarwal, D. B. Vigneron, and J. Kurhanewicz, *Eur Urol*, 2017, **72**, 1028.

32. I. Park, P. E. Z. Larson, J. W. Gordon, L. Carvajal, H. Chen, R. Bok, M. V. Criekinge, M. Ferrone, J. B. Slater, D. Xu, *et al.*, *Magn. Reson. Med.*, 2018, **80**, 864.

33. K. L. Granlund, E. A. Morris, H. A. Vargas, S. K. Lyashchenko, P. J. Denoble, V. A. Sacchini, R. A. Sosa, M. A. Kennedy, Y. W. Guo, A. P. Chen, J. Tropp, H. Hrica, and K. A. Keshari, In *Proceedings of the International Society for Magnetic Resononance in Medicine*, Singapore, 2016, 3690.

34. Z. Zhu, I. Marco-Rius, M. A. Ohliger, L. Carvajal, J. W. Gordon, H.-Y. Chen, I. Park, P. Cao, P. J. Shin, E. Milshteyn, C. von Morze, M. Ferrone, J. Slater, Z. Wang, P. Larson, R. Aggarwal, R. Bok, J. Kurhanewicz, P. Munster, and D. Vigneron, In *Proceedings of the International Society for Magnetic Resononance in Medicine*, 2017, 1115.

35. C. H. Cunningham, J. Y. C. Lau, A. P. Chen, B. J. Geraghty, W. J. Perks, I. Roifman, G. A. Wright, and K. A. Connelly, *Circ. Res.*, 2016, **119**, 1177.

36. A. Capozzi, T. Cheng, G. Boero, C. Roussel, and A. Comment, *Nat. Commun.*, 2017, **8**, 1.

37. T. Cheng, M. Mishkovsky, J. A. Bastiaansen, O. Ouari, P. Hautle, P. Tordo, B. van den Brandt, and A. Comment, *NMR Biomed.*, 2013, **26**, 1582.

38. I. Marco-Rius, M. C. D. Tayler, M. I. Kettunen, T. J. Larkin, K. N. Timm, E. M. Serrao, T. B. Rodrigues, G. Pileio, J. H. Ardenkjaer-Larsen, M. H. Levitt, *et al.*, *NMR Biomed.*, 2013, **26**, 1696.

39. M. C. D. Tayler, I. Marco-Rius, M. I. Kettunen, K. M. Brindle, M. H. Levitt, and G. Pileio, *J. Am. Chem. Soc.*, 2012, **134**, 7668.

40. T. R. Eichhorn, Y. Takado, N. Salameh, A. Capozzi, T. Cheng, J. N. Hyacinthe, M. Mishkovsky, C. Roussel, and A. Comment, *Proc. Natl. Acad. Sci. U. S. A.*, 2013, **110**, 18064.

41. A. Comment, *Imaging Med.*, 2014, **6**, 1.

42. H. A. I. Yoshihara, E. Can, M. Karlsson, M. H. Lerche, J. Schwitter, and A. Comment, *Phys. Chem. Chem. Phys.*, 2016, **18**, 12409.

43. A. Capozzi, M. Karlsson, J. R. Petersen, M. H. Lerche, and J. H. Ardenkjaer-Larsen, *J. Phys. Chem. C*, 2018, **122**, 7432.

44. M. L. Hirsch, B. A. Smith, M. Mattingly, A. G. Goloshevsky, M. Rosay, and J. G. Kempf, *J. Magn. Reson.*, 2015, **261**, 87.

45. M. Karlsson, P. R. Jensen, J. Duus, S. Meier, and M. H. Lerche, *Appl. Magn. Reson.*, 2012, **43**, 223.

46. M. M. Chaumeil, C. Najac, and S. M. Ronen, *Methods Enzymol.*, 2015, **561**, 1.

47. K. R. Keshari and D. M. Wilson, *Chem. Soc. Rev.*, 2014, **43**, 1627.

48. A. Comment and M. E. Merritt, *Biochemistry*, 2014, **53**, 7333.

49. M. Mishkovsky, B. Anderson, M. Karlsson, M. H. Lerche, A. D. Sherry, R. Gruetter, Z. Kovacs, and A. Comment, *Sci. Rep.*, 2017, **7**, 4.

50. T. B. Rodrigues, E. M. Serrao, B. W. C. Kennedy, D.-E. Hu, M. I. Kettunen, and K. M. Brindle, *Nat. Med.*, 2014, **20**, 93.

51. K. R. Keshari, D. M. Wilson, A. P. Chen, R. Bok, P. E. Larson, S. Hu, M. Van Criekinge, J. M. Macdonald, D. B. Vigneron, and J. Kurhanewicz, *J. Am. Chem. Soc.*, 2009, **131**, 17591.

52. K. X. Moreno, S. Satapati, R. J. DeBerardinis, S. C. Burgess, C. R. Malloy, and M. E. Merritt, *J. Biol. Chem.*, 2014, **289**, 35859.

53. I. Marco-Rius, C. von Morze, R. Sriram, P. Cao, G.-Y. Chang, E. Milshteyn, R. A. Bok, M. A. Ohliger, D. Pearce, J. Kurhanewicz, *et al.*, *Magn. Reson. Med.*, 2017, **77**, 65.

54. J. M. Park, M. Wu, K. Datta, S.-C. Liu, A. Castillo, H. Lough, D. M. Spielman, and K. L. Billingsley, *J. Am. Chem. Soc.*, 2017, **139**, 6629.

55. K. X. Moreno, C. E. Harrison, M. E. Merritt, Z. Kovacs, C. R. Malloy, and A. Dean Sherry, *NMR Biomed.*, 2017, **30**, 1.

56. L. Salamanca-Cardona, H. Shah, A. J. Poot, F. M. Correa, V. Di Gialleonardo, H. Lui, V. Z. Miloushev,

57. C. Cabella, M. Karlsson, C. Canapè, G. Catanzaro, S. Colombo Serra, L. Miragoli, L. Poggi, F. Uggeri, L. Venturi, P. R. Jensen, *et al.*, *J. Magn. Reson.*, 2013, **232**, 45.

58. M. M. Chaumeil, P. E. Z. Larson, H. A. I. Yoshihara, O. M. Danforth, D. B. Vigneron, S. J. Nelson, R. O. Pieper, J. J. Phillips, and S. M. Ronen, *Nat. Commun.*, 2013, **4**, 2429.

59. A. Flori, G. Giovannetti, M. F. Santarelli, G. D. Aquaro, S. De Marchi, S. Burchielli, F. Frijia, V. Positano, L. Landini, and L. Menichetti, *Spectrochim. Acta A Mol. Biomol. Spectrosc.*, 2018, **199**, 153.

60. D. R. Ball, B. Rowlands, M. S. Dodd, L. Le Page, V. Ball, C. A. Carr, K. Clarke, and D. J. Tyler, *Magn. Reson. Med.*, 2014, **71**, 1663.

61. J. A. M. Bastiaansen, H. A. I. Yoshihara, A. Capozzi, J. Schwitter, R. Gruetter, M. E. Merritt, and A. Comment, *Magn. Reson. Med.*, 2018, **79**, 2451.

62. H. Yoshihara, J. A. Bastiaansen, M. Karlsson, M. H. Lerche, A. Comment, and J. Schwitter, *J. Cardiovasc. Magn. Reson.*, 2015, **17**, O101.

63. J. J. Miller, D. R. Ball, A. Z. Lau, and D. J. Tyler, *NMR Biomed.*, 2018, **31**, e3912.

64. C. von Morze, M. A. Ohliger, I. Marco-Rius, D. M. Wilson, R. R. Flavell, D. Pearce, D. B. Vigneron, J. Kurhanewicz, and Z. J. Wang, *Magn. Reson. Med.*, 2018, **79**, 1862.

65. S. Düwel, C. Hundshammer, M. Gersch, B. Feuerecker, K. Steiger, A. Buck, A. Walch, A. Haase, S. J. Glaser, A. Schwaiger, and J. Schilling, *Nat. Commun.*, 2017, **8**, 15126.

66. T. Harris, H. Degani, and L. Frydman, *NMR Biomed.*, 2013, **26**, 1831.

67. K. N. Timm, J. Hartl, M. A. Keller, D. E. Hu, M. I. Kettunen, T. B. Rodrigues, M. Ralser, and K. M. Brindle, *Magn. Reson. Med.*, 2015, **74**, 1543.

68. P. Xu, M. H. Oosterveer, S. Stein, H. Demagny, D. Ryu, N. Moullan, X. Wang, E. Can, N. Zamboni, A. Comment, *et al.*, *Genes Dev.*, 2016, **30**, 1255.

69. B. W. C. Kennedy, M. I. Kettunen, D.-E. Hu, S. E. Bohndiek, and K. M. Brindle, *Proc. Int. Soc. Mag. Reson. Med.*, 2012, **20**, 4326.

70. F. A. Gallagher, M. I. Kettunen, S. E. Day, D.-E. Hu, J. H. Ardenkjær-Larsen, R. in't Zandt, P. R. Jensen,

K. L. Granlund, S. S. Tee, J. R. Cross, *et al.*, *Cell Metab.*, 2017, **26**, 830.

M. Karlsson, K. Golman, M. H. Lerche, *et al.*, *Nature*, 2008, **453**, 940.

71. D. E. Korenchan, C. Taglang, C. von Morze, J. E. Blecha, J. W. Gordon, R. Sriram, P. E. Z. Larson, D. B. Vigneron, H. F. VanBrocklin, J. Kurhanewicz, *et al.*, *Analyst*, 2017, **142**, 1429.

72. R. R. Flavell, C. von Morze, J. E. Blecha, D. E. Korenchan, M. Van Criekinge, R. Sriram, J. W. Gordon, H.-Y. Chen, S. Subramaniam, R. A. Bok, *et al.*, *Chem. Commun.*, 2015, **51**, 14119.

73. M. I. Menzel, E. V. Farrell, M. A. Janich, O. Khegai, F. Wiesinger, S. Nekolla, A. M. Otto, A. Haase, R. F. Schulte, and M. Schwaiger, *J. Nucl. Med.*, 2013, **54**, 1113.

74. K. R. Keshari, V. Sai, Z. J. Wang, H. F. Vanbrocklin, J. Kurhanewicz, and D. M. Wilson, *J. Nucl. Med.*, 2013, **54**, 922.

75. A. Eldirdiri, A. Clemmensen, S. Bowen, A. Kjær, and J. H. Ardenkjær-Larsen, *NMR Biomed.*, 2017, **30**, 1.

76. L. Darrasse and J.-C. Ginefri, *Biochimie*, 2003, **85**, 915.

77. M. Mishkovsky, A. Comment, and R. Gruetter, *J. Cereb. Blood Flow Metab.*, 2012, **32**, 2108.

78. S. S. Tee, V. DiGialleonardo, R. Eskandari, S. Jeong, K. L. Granlund, V. Miloushev, A. J. Poot, S. Truong, J. A. Alvarez, H. N. Aldeborgh, *et al.*, *Sci. Rep.*, 2016, **6**, 1.

79. S. Jeong, R. Eskandar, S. M. Park, J. Alvarez, S. S. Tee, R. Weissleder, M. G. Kharas, H. Lee, and K. R. Keshari, *Sci. Adv.*, 2017, **3**, e1700341.

80. W. A. Edelstein, in 'Encyclopedia of Magnetic Resonance', eds R. K. Harris and R. L. Wasylishen, John Wiley & Sons, Ltd: Chichester, 2007.

81. G. Adriany and R. Gruetter, *J. Magn. Reson.*, 1997, **125**, 178.

82. J. den Hollander, K. Behar, and R. Shulman, *J. Magn. Reson.*, 1984, **57**, 311.

83. J. T. Grist, M. A. McLean, S. S. Deen, F. Riemer, C. J. Daniels, A. B. Gill, F. Zaccagna, R. F. Schulte, S. F. Hilborne, J. P. Mason, J. W. McKay, A. Comment, A. Chhabra, V. Fernandes, H. Loveday, M.-C. Laurent, I. Patterson, R. Hernandez, R. A. Slough, T. Matys, I. B. Wilkinson, B. Basu, C. Trumper, D. J. Tyler, D. J. Lomas, M. J. Graves, A. J. Coles, K. M. Brindle, and F. A. Gallagher, *Proc. Int. Soc. Mag. Reson. Med.*, 2018, **26**, 0277.

84. F. Frijia, M. F. Santarelli, U. Koellisch, G. Giovannetti, T. Lanz, A. Flori, M. Durst, G. D. Aquaro, R. F. Schulte, D. De Marchi, *et al.*, *J. Med. Biol. Eng.*, 2016, **36**, 53.

Chapter 22

Dissolution Dynamic Nuclear Polarization

Walter Köckenberger

University of Nottingham, Nottingham, UK

22.1 INTRODUCTION

Dissolution DNP (dissDNP) was originally proposed and implemented as a strategy to generate solutions containing highly polarized ^{13}C spin systems for use in in vivo molecular MRI applications.[1,2] The strategy was developed in Klaes Golman's research laboratory in Malmö, Sweden, which was initially part of Nycomed and then changed hands from Amersham to GE Healthcare. The successful demonstration of this experimental technique by Golman *et al.* generated substantial interest in the magnetic resonance community, and a steadily increasing number of research laboratories worldwide are currently involved in optimizing and extending the strategy for both medical applications and in vitro NMR spectroscopy. The key idea underpinning dissDNP involves the use of DNP, in conjunction with a cryogenic sample temperature, to increase the nuclear spin polarization in a solid-state sample. Subsequently, in a second step, the sample is dissolved in a hot solvent to obtain a liquid-state sample with highly polarized nuclear spins that can be used in NMR imaging or spectroscopy experiments. The generation of high nonthermal nuclear spin polarization, which can exceed by orders of magnitude the thermal polarization arising from the Zeeman effect, is frequently referred to as *spin hyperpolarization*.

22.2 PRINCIPLE

In dissDNP experiments, samples are prepared by homogeneously mixing the chemical compounds that are to be polarized with stable radicals carrying unpaired electrons. Usually, the radical concentration is between 5 and 50 mM, depending on the chemical nature of the radical. A glass-forming agent is also added, to avoid the formation of crystal structures when cooling the sample to cryogenic temperatures, as ice crystal formation has a detrimental effect on the homogeneous distribution of the radicals. It has also been reported that addition of a low concentration ($\sim$0.5 mM) of the lanthanide ion Gd^{3+} has a positive effect on the final nuclear spin polarization.[3] The use of a cryogenic temperature in combination with an appropriate magnetic field ensures that a high thermal polarization of the unpaired electron spins is established, and relaxation processes that cause loss of nuclear polarization are kept at a minimum. For instance, at a medium field strength of 3.4 T and temperatures around 1 K, electron spins are almost fully polarized (Figure 22.1).

Handbook of High Field Dynamic Nuclear Polarization.
Edited by Vladimir K. Michaelis, Robert G. Griffin, Björn Corzilius and Shimon Vega
© 2020 John Wiley & Sons, Ltd. ISBN: 978-1-119-44164-9
Also published in eMagRes (online edition)
DOI: 10.1002/9780470034590.emrstm1311

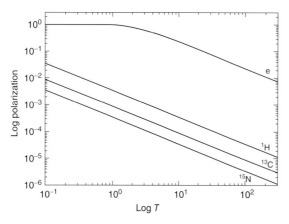

Figure 22.1. Electron and nuclear spin polarization as a function of sample temperature, calculated for a magnetic field $B_0 = 3.4$ T. The spin polarization is shown for a temperature range from 0.1 to 300 K

The high electron spin polarization is subsequently transferred by DNP from the unpaired electrons to the surrounding nuclear spin system that interacts with the electron spins through the hyperfine interaction. The sample is irradiated with a microwave field at an appropriate frequency in the electron spin resonance spectrum to mediate the transfer. DNP is a complex spin dynamics process in which the number of coupled electrons that interact with the nuclear spins plays a decisive role in determining which pathway dominates in the transfer of the polarization from the electrons to nuclear spins.[4-11] After the nuclear polarization built up has reached a steady-state value, the sample temperature is rapidly raised by injection of a hot solvent in which the sample dissolves to yield a solution that contains nuclear spin systems with high-spin polarization. The NMR signal enhancement factor $\varepsilon^\dagger$, which compares the potential NMR signal intensity I_{DNP} after dissDNP and the signal intensity $I_{thermal}$, arising from a thermally polarized sample kept at ambient temperature, can be estimated by

$$\varepsilon^\dagger = \frac{I_{DNP}}{I_{thermal}} = \varepsilon_{DNP} \frac{T_{detect}}{T_{DNP}} \frac{B_{DNP}}{B_{detect}} \qquad (22.1)$$

where ε_{DNP} is the enhancement of the NMR signal due to DNP carried out at cryogenic temperature T_{DNP}; T_{detect}, the temperature at which the NMR signal is detected after dissolution; B_{DNP}, the magnetic field strength at which DNP was carried out; and B_{detect}, the field strength at which the NMR signal is acquired.[12] Theoretically, the maximal enhancement that can be

generated by DNP alone is given by the ratio of the gyromagnetic constants of the unpaired electrons and the interacting nuclear spins, $\varepsilon_{DNP} = \left|\frac{\gamma_e}{\gamma_n}\right|$, where $\gamma_e = -\frac{g\mu_B}{\hbar}$ with g, the Landé factor; μ_B, the Bohr magneton; and $\hbar$, the Planck's constant divided by 2π. For instance, for hydrogen nuclei, ε_{DNP} has a theoretical maximal value close to 660, and for ^{13}C nuclei, a theoretical limit of 2600. The actual enhancement that can be obtained by DNP in an experiment depends in a complex manner on multiple factors including the solubility and chemical structure of the radical, the relaxation time constants of the unpaired electrons, the linewidth of the electron spectrum (which itself depends on relaxation), g-anisotropy, and hyperfine structure. Furthermore, the type of nuclear spins, their relaxation parameters, and the nuclear spin density also play a role in determining how much polarization from the unpaired electrons can be transferred to the nuclear spin system.[13] It should be noted that equation (22.1) provides only a crude estimate for the potential enhancement because it does not take into account any losses caused by relaxation during dissolution and transfer of the sample to another magnet. However, the use of equation (22.1) is instructive in demonstrating that the signal enhancement can exceed four orders of magnitude even if only modest DNP enhancement is achieved at cryogenic temperatures. In a typical dissDNP experiment, DNP is carried out at a temperature $T_{DNP} = 1.5$ K and the signal is detected at $T_{detect} = 300$ K after dissolution. Taking into account the change of magnetic field strength, if, for example, the sample is transferred from a 3.4 T magnetic field to 9.4 T, an enhancement $\varepsilon_{DNP} = 150$ arising purely from DNP would result in a final overall enhancement factor $\varepsilon^\dagger$ of more than four orders of magnitude, provided that relaxation during dissolution and shuttling only weakly affect the final polarization level:

$$\varepsilon^\dagger = 150\frac{300 \text{ K}}{1.5 \text{ K}}\frac{3.4 \text{ T}}{9.4 \text{ T}} \approx 1.1 \times 10^4 \qquad (22.2)$$

Another factor which is neglected in this basic calculation is the relatively long time (30 min to several hours) that is necessary to generate the high polarization of the nuclear spins. A more meaningful enhancement factor would be the signal enhancement achieved per unit time.

The nonthermal spin polarization relaxes back to its thermal equilibrium after dissolution, with a longitudinal relaxation time constant T_1 that depends on the motional correlation times of the compounds, the temperature of the liquid-state sample, the field strength, and the concentration of paramagnetic ions in the sample.[14,15] The loss of polarization through relaxation provides a time window lasting only a few T_1 in which the signal can be acquired with an appreciable enhancement. After a time longer than 5 T_1 has elapsed, the spin polarization has returned back to thermal polarization. In applications of dissDNP to medical diagnostics, the time taken between dissolution and detection is usually appreciable on the NMR relaxation time scale. Therefore, it is crucial in these applications that the T_1 values of the nuclei of interest are relatively large. A particularly well-suited compound in this respect is [1-^{13}C] pyruvate, which has a rather long T_1 of 55 s at 1.5 T[16] for the ^{13}C label. For in vitro spectroscopic applications, the requirement for a long T_1 time constant of the polarization-carrying nuclei is less strict because the delay between dissolution and signal detection can be made relatively short compared to the longer delay required for in vivo applications. However, for the observation of an appreciable signal enhancement, the longitudinal relaxation time constant T_1 of the nuclei should not be less than a third of the time required to transfer the solution after dissolution to the site of detection. DissDNP applied to in vitro NMR spectroscopy always introduces a weighting of the resonance line intensities according to their corresponding T_1 values.

Owing to the time required for the generation of the high nuclear spin polarization and the unavoidable dilution of the sample during dissolution, it is usually not possible to repeat the dissDNP experiment immediately, and data acquisition is limited to the signal arising from a single excitation or to a set of time data obtained by a few consecutive low flip angle excitation pulses.

22.3 HARDWARE

The dedicated experimental setup for dissDNP that involves a superconductive magnet, a cryostat, and a microwave source, as well as accessories for temperature control and sample handling and transfer (Figure 22.2), has been termed a *DNP polarizer* in the NMR and MRI communities.

22.3.1 Cryostat

To cool the sample to cryogenic temperatures, a cryostat is used that makes it possible to immerse the

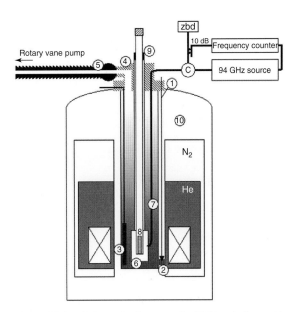

Figure 22.2. Schematic diagram of a DNP polarizer with stabilized microwave source. 1: needle valve handle, 2: needle valve, 3: He level gauge, 4: variable temperature insert, 5: vacuum line with cutoff valve, 6: microwave shield, 7: waveguide, 8: sample cup with sample holder, 9: vacuum-tight seal, 10: 3.35 T magnet. C, circulator; zbd, zero-bias diode for microwave detection. (R. Panek, J. Leggett, N.S. Andersen, J. Granwehr, A.J. Perez Linde, W. Köckenberger: Magnetic Resonance Microscopy, eds S.L. Codd, J.D. Seymour. 2009. Copyright Wiley-VCH Verlag GmbH & Co. KGaA. Reproduced with permission)

sample in a bath of liquid helium. For the generation of sample temperatures below 4.2 K, the helium bath is pumped with a strong rotary vane pump. Using this principle, temperatures of approximately 1 K can be achieved.[17,18] Two different strategies have been described that can be used to keep a sample space at cryogenic temperature within a superconductive magnet. In the first implementation, the sample cryostat is integrated with the magnet and the helium reservoir of the main solenoid of the superconductive magnet is used as a source of liquid helium.[1,19,20] In this case, the liquid helium is allowed to flow from the reservoir via a thin capillary, which is controlled by a needle valve, into the sample space located within the cold bore of the magnet (Figure 22.2). In the second variant, a separate sample cryostat is inserted into the warm bore of a superconductive magnet.[21] The first setup does not require the use of an external liquid helium dewar for the experiments, but has the disadvantage that the inner bore of the magnet can only be thoroughly cleaned by warming up the whole magnet. As in the second setup the cryostat is a separate piece of equipment, it can be warmed up to ambient temperatures and cleaned more easily. The liquid helium consumption is comparable for the two setups.

22.3.2 Microwave Irradiation

In dissDNP experiments, the sample is irradiated with a continuous wave field at 95 GHz or higher. Microwave sources operating at this frequency (W-band) are relatively cheap while still providing a sufficiently high power output of 200–400 mW. Because the volume of a cavity resonating at this frequency is small (the wavelength λ at 95 GHz is about 3 mm) and because usually a sample volume of several 10–100 μL is desirable, the microwaves are directed onto the sample without using any resonating structure. As a consequence, the strength of the microwave field generated is rather low (in the order of 5–50 kHz). A shield is used to confine the microwaves to the sample space. As the buildup time for ^{13}C polarization is quite long, with experimental times of 30 min to several hours, an important extension of the basic polarizer setup, in particular for applications in medical MRI, is the addition of multiple sample handling capabilities and simultaneous irradiation of multiple samples with microwaves.[22,23] The storage of polarized samples is possible because of the long longitudinal relaxation time constants (in the order of hours) of ^{13}C nuclei kept at cryogenic temperatures around 1 K.

22.3.3 Sample Dissolution and Shuttling

Dissolution of the sample is performed by guiding a small amount (a few milliliters) of boiling solvent onto the frozen sample. The boiling solvent is prepared outside of the cryogenic environment, using a resistive heater wound around a small reservoir. On triggering of the dissolution process, the hot solvent is guided through a capillary into the cryogenic sample space and injected onto the cold sample. The sample is dissolved in the continuous flow of the solvent and flushed through an outlet capillary to the outside of the cryogenic sample space, where it is collected for use in medical imaging applications or guided through an

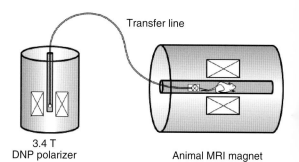

Figure 22.3. Experimental setup for automated injection of a solution containing highly polarized molecules into an animal kept in a horizontal bore magnet. After dissolution, the solution is guided to the horizontal magnet and into the reservoir of an infusion pump. On arrival, the perfusion pump is automatically triggered and injects the solution with the highly polarized molecules into the model animal at a preset injection rate

extended transfer line to a separate high-field magnet for NMR spectroscopy experiments.

A specially designed infusion pump has been described for the automated administration of solutions containing highly polarized spin systems to animal models in a controlled manner, using a phase separator to avoid any gas bubbles in the infusion flow[21] (Figure 22.3). It is crucial that none of the materials with which the solution gets into contact during transfer are paramagnetic, so as to avoid the loss of polarization by paramagnetic relaxation.

For spectroscopy experiments, the sample is shuttled pneumatically between the polarizer and the high-field magnet, followed by injection into an NMR sample tube (Figure 22.4a). The injection into the NMR tube generates a complex sample motion which takes a few seconds to decay. In addition, because of the rapid drop in pressure after injection, this procedure leads to the generation of small gas bubbles which have a detrimental effect on the line width of the NMR spectra.[24] One strategy to avoid out-gassing of the liquid is based on maintaining a high gas pressure in the NMR tube during NMR signal acquisition. This approach was developed by Hilty's group,[25] using a pneumatically controlled multiport valve which first directs the sample arriving from the polarizer into a sample loop and then subsequently injects the sample into the NMR tube while maintaining a high nitrogen gas pressure of 18 bar during the NMR experiment. The advantage of this strategy is that out-gassing of

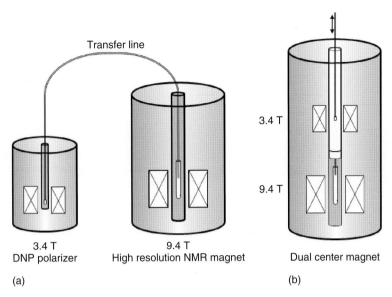

Figure 22.4.   Two different strategies currently used for linking dissDNP to high-resolution spectroscopy. (a) After dissolution the liquid sample is pneumatically shuttled through a transfer line to a high-field magnet and injected into an NMR sample tube. (b) A purpose-designed magnet with two isocenters makes it possible to minimize the distance over which the sample is shuttled between the cryogenic environment (at the 3.4 T isocenter) and the high-resolution isocenter (here 9.4 T). In this setup, the sample is shuttled to the high-field center while still solid, and only dissolved close to the high-field center

the transferred sample is suppressed and only a short delay is required after injection.

During pneumatic shuttling from the polarizer to the high-field NMR magnet, the liquid sample is moved through the spatially varying magnetic field that arises from the stray fields of the polarizer and the main magnet. As relaxation of the nonthermal sample polarization in the liquid state shows a strong field dependence, with low field strengths accelerating the loss of the high-spin polarization, particularly for larger molecules, it is advantageous to keep the field strength as high as possible during the transfer of the sample from the polarizer to the main magnet.

Polarization loss in liquid state by paramagnetic relaxation due to the free electrons of the radical compounds in solution contributes to the loss of nuclear spin polarization during the sample transfer. An improvement in preserving the polarization can be obtained by rapidly converting the paramagnetic radicals into diamagnetic molecules. If nitroxide radicals are used as sources for unpaired electrons in the solid sample, these radicals can be converted after sample dissolution by adding frozen ascorbic acid beads to the sample.[26]

An alternative strategy for improving the preservation of polarization during transfer involves the use of two immiscible solvents for sample dissolution. This has the effect that the concentration of hydrophilic molecules carrying the high-spin polarization can be increased using only a small volume of aqueous solvent and a larger portion of organic solvent.[27] In addition, the two different solvent phases can be used to extract a nonpolar radical from the aqueous phase during transfer, thus decreasing the effect of paramagnetic relaxation on the molecules dissolved in the aqueous phase.[28]

Another improvement for preserving the nonthermal spin polarization in a dissDNP experiment is an experimental setup which makes it possible to transfer the sample in solid form from the cryogenic environment to a higher magnetic field center, retaining a cryogenic sample temperature.[29] Longitudinal relaxation processes are usually several orders of magnitudes slower in the solid state at cryogenic temperatures than they are in the liquid state. After transfer as a solid, the sample is subsequently dissolved close to the high magnetic field isocenter. A prototype system was developed based on a dual isocenter magnet with a cryogenic sample space at 3.4 T and a 9.4 T isocenter kept at ambient temperature (Figure 22.4b). The distance between the two isocenters was 0.8 m.

The sample was transferred in the solid state by a computer-controlled actuator, followed by rapid dissolution using hot water at a location very close to the 9.4 T isocenter. As in this experimental setup the time for which the sample is in the liquid state is cut down to 700 ms, high spin polarization can be detected even for ^{13}C nuclei with rather short T_1 time constants.

22.3.4   Commercial Instruments

Two types of commercial instruments for dissDNP experiments have been developed. The HyperSense, manufactured by Oxford Instruments Ltd,[30] is an automated version of the original experimental setup of Golman's laboratory. GE HealthCare offers the Diamond Spin Lab DNP polarizer, for use in a clinical environment.[31] This polarizer is based on a sorption pump that can be regenerated overnight, and can work at temperatures just below 1 K. The polarizer can also handle multiple samples, which can be polarized and dissolved to give sterile solutions.

22.3.5   Radicals

Two different classes of radicals have been predominately used in dissDNP experiments. Various trityl (triphenylmethyl) derivates show excellent performance in generating very high polarization of ^{13}C spin systems (up to 65%). These trityl compounds have very narrow lines, with only weak *g*-anisotropy and spurious signals arising from hyperfine coupling.[32] These EPR properties arise from the symmetric molecular structure and the high localization of the unpaired electron on the center carbon. The center carbon is well shielded from solvent access by bulky aromatic groups.[33]

Nitroxides such as various TEMPO (2,2,6,6-tetramethylpiperidine-1-oxyl) derivates have been used to generate both relatively high heteronuclear polarization[34] and high polarization of hydrogen spins.[35] TEMPO derivatives possess EPR resonance lines with strong *g*-anisotropy, and hyperfine structure due to the electron coupling to the nitrogen nuclei.

22.4   STRATEGIES AND VARIANTS

22.4.1   Cross Polarization

Under the low microwave power irradiation that is commonly used in dissDNP and without a resonating cavity, the buildup time constant τ_{DNP} is in the same order as the nuclear longitudinal relaxation time constant T_1. The buildup time constants for hydrogen nuclei when TEMPO derivatives are used for DNP are in the order of 10 min, while buildup of polarization of the ^{13}C spin system is in the order of hours. Polarization transfer from the hydrogen spin system to the ^{13}C spin system by CP can be used to shorten polarization buildup times for heteronuclear spins.[36,37] This strategy was initially demonstrated for MAS DNP NMR.[38] However, in the cryogenic environment of dissDNP, helium gas, which has a relatively high electric polarizability, can prevent the use of high-power radiofrequency pulses for CP. Adiabatic half-passage radiofrequency sweeps can be used to overcome this problem.[39,40] As the polarization of the hydrogen spins can be quickly regenerated, a set of consecutive CP experiments can be carried out to achieve polarization buildup time constants for heteronuclei which are comparable to the buildup time constants for the hydrogen nuclei.

22.4.2   Co-polarizing Agents

It has been demonstrated that a higher level of nonthermal ^{13}C polarization on molecules with natural ^{13}C abundance can be generated by adding to the sample mixture molecules with chemical groups that carry ^{13}C nuclei with long relaxation time constants.[41] Possible explanations for the observed increase in the polarization in liquid state after dissolution include a higher level of polarization generated by DNP, due to a higher concentration of ^{13}C nuclei in the solid-state sample, or a transfer of ^{13}C polarization after dissolution from the nuclei with long T_1 time constants to nuclei with shorter T_1 time constants through intermolecular cross relaxation in a heteronuclear Overhauser effect. In a related dissDNP experiment, the ^{13}C concentration in the sample matrix was increased using ^{13}C dimethylsulfoxide, but only a decrease in the polarization buildup time constant τ_{DNP} was found and no increase in the final polarization was observed.[42]

22.4.3 Melting

An interesting strategy proposed by Griffin and group[12] for raising the temperature in samples after enhancing the nuclear polarization by DNP is based on the use of an infrared laser to melt the sample. The advantages are twofold: (i) the sample is not diluted further by solvent and (ii) the experiment can be repeated after signal acquisition in the liquid state by refreezing the sample and irradiating again with microwaves at a suitable frequency. Using this procedure in conjunction with repetitively polarized samples, two-dimensional NMR spectroscopy experiments could be performed.[43] This melting strategy was implemented together with DNP carried out at about 100 K.

22.4.4 Higher Magnetic Field

Several groups have reported an increase in achievable nuclear spin polarization by moving to higher field strengths for the applied external magnetic field. Using TEMPO derivatives as radicals for the dissDNP experiment, 50% more polarization was obtained by moving from 3.4 to 5 T magnetic field strength.[44] For trityl radicals, it was shown that increasing the field strength from 3.4 to 4.7 T resulted in an approximate doubling of the nuclear spin polarization.[45,46]

22.4.5 Xenon Polarization

As an alternative to the established route for polarizing xenon by spin exchange optical pumping, an experimental protocol was developed that makes it possible to generate highly polarized xenon gas using a modified version of dissDNP. Xenon gas was condensed together with a solvent that contained radicals to yield a glassy solid sample. After DNP at cryogenic temperature, xenon was separated from the radicals by sublimation.[47] A subsequent rise in temperature released the xenon as gas with an NMR signal that was enhanced by three to four orders of magnitude compared to the ambient temperature thermal equilibrium signal at 7 T. The transport of polarization by spin diffusion and the formation of xenon clusters were described for solid-state xenon DNP at both 1.5 and 5 T field strengths.[48]

22.4.6 Long-Lived Spin States and Hyperpolarization

The high nonthermal nuclear spin polarization arising from dissDNP has a limited lifetime, and can only be detected in a time window with a duration of a few T_1 time constants. However, it has been demonstrated that it is possible to preserve the polarization in long-lived spin states of spin pairs.[49] These states can be sustained either by suppressing the chemical shift difference of the spin pair using an appropriate radiofrequency pulse sequence, or by keeping the sample at very low magnetic field.[50–55] Interestingly, during the dissolution process and transfer of the sample from the relatively high magnetic field inside of the polarizer magnet to the low magnetic field outside, a portion of the high polarization is preserved in the long-lived singlet state.[56] An important step toward the use of long-lived spin states as a reservoir for preserving the high-spin polarization generated by dissDNP in medical diagnostic imaging was made by the demonstration that the high polarization of ^{13}C spins could be transferred into, and retrieved from, singlet spin order in a small animal MRI scanner.[57] Other applications that will benefit from the combination of long-lived spin states and the generation of high-spin polarization using dissDNP include measurements of slow molecular dynamics and of slow diffusion processes.

22.4.7 Nonlinear NMR Effects

Nonlinear effects have been reported in conjunction with the high-spin polarization produced by dissDNP. Spontaneous emission of signal was observed from ^1H spin systems, which were hyperpolarized to negative spin temperature. The multiple signal beats arose without any radiofrequency excitation. A model that combines high-spin polarization within the detected region inside the radiofrequency coil, radiation damping, and an additional supply of polarization outside of the detected region was able to give a phenomenological explanation of the observation.[58] Long-distance intermolecular MQCs have been used in highly polarized partly deuterated methanol (CD_3OH) to generate an enhancement of the signal arising from these coherences, with an enhancement factor of more than 10^6.[59]

22.5 APPLICATION

22.5.1 Medical Diagnostic Imaging

Medical diagnostic imaging is the predominating area where dissDNP is currently making strong impact. Only a few representative examples of the many active research initiatives in this field are included in this section. The key compound used in dissDNP experiments for medical applications is [1-^{13}C] pyruvate. Pyruvate is an endogenous molecule that is one of the intermediates of the breakdown of glucose in glycolysis. The spins of the ^{13}C label can be highly polarized in free pyruvic acid together with a trityl derivate as the radical. With a dissDNP polarizer operating at 4.6 T magnetic field strength, a polarization of $64 \pm 5\%$ was reached.[60] Pyruvate is either metabolized to lactate under hypoxic conditions, converted into the amino acid alanine, or oxidized further to carbon dioxide under normal aerobic conditions.[61] As the oxygenation state of tissue determines the direction of the metabolic conversion, it has been found that the conversion of lactate from highly polarized pyruvate in cancer tissue can be used as a diagnostic indicator for the metabolic activity of tumor cells.[62]

The conversion of [1-^{13}C] pyruvate carrying high spin polarization after a dissDNP experiment into lactate can be monitored by the appearance of a new resonance line in the NMR spectrum, as this metabolic step requires only few seconds (Figure 22.5). Using fast spectroscopic imaging, it was possible to generate metabolic maps of lactate distribution that indicate cancerous tissue. This strategy shows great potential

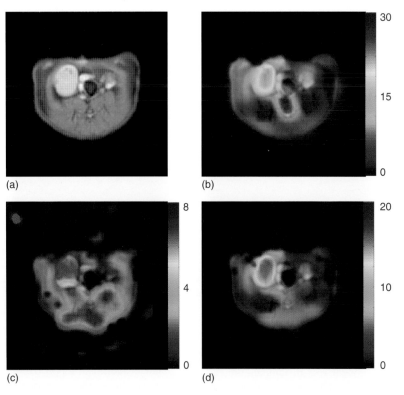

(a)   (b)   (c)   (d)

Figure 22.5. Application of ^{13}C-labeled pyruvate in conjunction with dissDNP to visualize a tumor in a mouse model. (a) The ^{1}H image shows in bright contrast the tumor on the ventral side of the mouse. The metabolic maps were acquired with directly detected ^{13}C-spectroscopic imaging. (b) After injection of the highly polarized [1-^{13}C] pyruvate there is high signal intensity within the tumor, indicating a high concentration of the metabolite in this tissue. (c) In the healthy tissue, pyruvate is converted into ^{13}C alanine, whereas in the anaerobic cancer tissue the pyruvate is predominately metabolized into ^{13}C lactate (d). (The images are courtesy of J.-H. Ardenkjaer-Larsen, Technical University Denmark)

for monitoring fast changes in lactate level that could be used as an indicator for response to chemical treatment of the cancer condition.[63] [1-^{13}C] Pyruvate has also been used to monitor metabolism in heart muscle and to visualize infarcted regions.[64,65] As the nonthermal polarization of the ^{13}C label in pyruvate and lactate can be used only for a relatively short time, and as it is useful to acquire several metabolic maps during this short time window, it is important for this purpose to develop reliable and robust spectroscopic imaging protocols. In particular, sparse sampling techniques of k-space, also known as compressed sensing, can be used to shorten the acquisition time per image while retaining the important spectral and spatial information.[66,67]

Several alternative compounds have been suggested as probes to be used in conjunction with dissDNP for measuring in vivo metabolic activities and for serving as indicators for medical diagnosis.[68] This included [1,4-^{13}C(2)] fumarate[69] and ^{15}N-labeled choline.[70] Another interesting application of dissDNP is its use for enhancing the polarization of ^{13}C-labeled bicarbonate ion.[71] As the ratio between bicarbonate and carbon dioxide in tissue depends on the local pH, it was possible using the intensities of the two different ^{13}C NMR resonance lines arising from bicarbonate and carbon dioxide, respectively, to generate pH maps that reveal low pH in tumors in vivo.

Apart from ^{13}C and ^{15}N nuclear spins, several other nuclei have been polarized and suggested for medical diagnostics. Yttrium-89, which can be used in a complex as a pH indicator, has been polarized by dissDNP,[72,73] to yield an enhancement $\varepsilon^{\ddagger} = 60 \times 10^3$, and its complexation was monitored exploiting the high signal intensity.[74] Lithium-6 has a long longitudinal relaxation time constant T_1 and has been detected after dissDNP in rat brain by localized spectroscopy.[75]

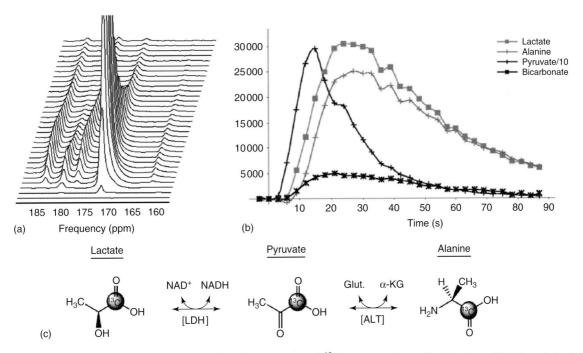

Figure 22.6. (a) In vivo spectroscopy showing the appearance of ^{13}C resonance lines after injection of highly polarized [1-^{13}C] pyruvate in a rat. Spectra were acquired every 3 s from a 90 mm thick section of the rat. Lactate, pyruvate hydrate, alanine, pyruvate, and bicarbonate resonance lines (from the left to the right) are all well resolved. The pyruvate resonance lines of the early spectra are truncated. (b) Plots of metabolite signal intensities vs time. The signal intensities have been corrected for the loss of magnetization caused by the use of the 10° flip angles at each excitation. Note that the signal intensities of the pyruvate peak were scaled down by a factor of 10 for plotting purposes. (c) The two possible pathways for the metabolic conversion of pyruvate. (a,b: Adapted from Ref. 81. © John Wiley & Sons, Ltd)

22.5.2 In Vitro Spectroscopy

A number of attempts have been made to exploit the high signal enhancement provided by dissDNP in in vitro spectroscopy experiments. The relatively long time required in dissDNP to build up the nuclear spin polarization, and more importantly the sample dilution by the dissolution process, prevents an immediate repetition of the experiment using the same sample. Conventional multidimensional NMR spectroscopy requires repetitive excitation and signal acquisition while simultaneously incrementing evolution delays in the pulse sequence to obtain multidimensional data sets. A technique, based on the use of fast frequency-swept radiofrequency pulses in combination with oscillating magnetic field gradients and the creation of multiple echoes, has been proposed by Frydman *et al.*[76] that makes it possible to acquire the data required for the generation of two-dimensional spectra using a single sample excitation. This strategy, termed *ultrafast spectroscopy*, can be combined with dissDNP to acquire two-dimensional correlation spectra with an acquisition time in the order of 100 ms (Figure 22.7).[77,78] Alternative strategies for the acquisition of two-dimensional spectra based on the use of small flip angle excitation pulses that use only a fraction of the available spin polarization have been suggested,[41,79] and the use of alternating switching coherence selection gradients in a pulse sequence is termed *single-point amplitude-separated multi-dimensional NMR*.[80]

A number of studies have demonstrated that the use of dissDNP can provide novel insights into in vitro metabolic turnover rates. The NMR signal arising from cell samples or organs that were perfused with a solution containing the highly polarized spins was acquired using a number of small flip angle excitation pulses. Quantitative turnover rates of various metabolites could be extracted from the time-resolved spectra (Figure 22.6).[81,82–84]

Another important research area where novel dissDNP NMR experiments have started to make an impact involves studies of the dynamics of proteins and ligands. First reports describing the observation of binding of highly polarized low-concentrated ligands to proteins have been published.[85] Using two competitively binding ligands for the protein kinase A and increasing the spin polarization on one of the ligands by dissDNP, the high spin polarization could be transferred first onto spins within the binding pocket of the enzyme, and subsequently onto the

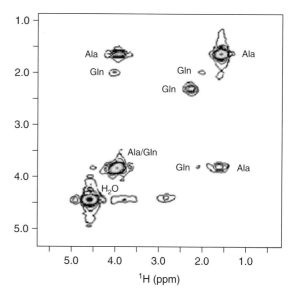

Figure 22.7. COSY spectrum of the dipeptide Ala-Gln acquired in a single scan using ultrafast 2D spectroscopy in conjunction with dissDNP. A sample containing the dipeptide and the radical TEMPO was irradiated with microwaves for 30 min at 1.3 K. After dissolution, the sample was shuttled to a high-field magnet (9.4 T) and the data for the spectrum were acquired in 44 ms. The final concentration of the dipeptide was 1.5 mM and the liquid-state enhancement $\varepsilon^\dagger$ was 150. (Adapted from Ref. 82 with permission from the PCCP Owner Societies)

second ligand, using a hyperpolarized binding pocket NOE.[86] Another promising strategy is to create high spin polarization of fluorine labels on ligands to determine the dissociation constants of ligand–protein complexes.[87] The important advantage arising from the use of dissDNP is the very low ligand concentration that can be used in these experiments, due to the increase in detection sensitivity afforded by the high spin polarization.

22.6 OUTLOOK

DissDNP is a complex experimental strategy that provides an unprecedented increase in nuclear spin polarization and concomitant increase in detection sensitivity. Applications of this technique are currently predominant in the area of in vivo detection of ^{13}C-labeled metabolic compounds for molecular imaging by MRI. In particular, the strategy makes an impact in cancer diagnostics and treatment, by offering access to novel information such as in vivo turnover rates of endogenous metabolites. However, there is clear evidence that this technique also makes it possible to gain novel data from in vitro spectroscopy and NMR spectroscopy of ligand–protein complexes. There is still a need for further hardware development to be carried out to optimize the complex experimental strategy and make it more robust and easier to control.

RELATED ARTICLES IN EMAGRES

Dynamic Nuclear Polarization: Applications to Liquid-State NMR Spectroscopy

Dynamic Nuclear Polarization and High-Resolution NMR of Solids

Cryogenic NMR Probes: Applications

Low Spin Temperature NMR

High-Frequency Dynamic Nuclear Polarization

Ultrafast Multidimensional NMR: Principles and Practice of Single-Scan Methods

Polarization of Noble Gas Nuclei with Optically Pumped Alkali Metal Vapors

Long-Lived States and Coherences for Line Narrowing, DNP, and Study of Interactions

Cross Polarization in Solids

Molecular Motions: T_1 Frequency Dispersion in Biological Systems

REFERENCES

1. J. H. Ardenkjaer-Larsen, B. Fridlund, A. Gram, G. Hansson, L. Hansson, M. H. Lerche, R. Servin, M. Thaning, and K. Golman, *Proc. Natl. Acad. Sci. U. S. A.*, 2003, **100**, 10158.

2. K. Golman, R. In't Zandt, and M. Thaning, *Proc. Natl. Acad. Sci. U. S. A.*, 2003, **100**, 11270.

3. J. H. Ardenjaer-Larsen, S. Macholl, and H. Jóhannesson, *Appl. Magn. Reson.*, 2008, **34**, 509.

4. Y. Hovav, A. Feintuch, and S. Vega, *J. Magn. Reson.*, 2010, **207**, 176.

5. Y. Hovav, A. Feintuch, and S. Vega, *J. Magn. Reson.*, 2012, **214**, 29.

6. D. Shimon, Y. Hovav, A. Feintuch, D. Goldfarb, and S. Vega, *Phys. Chem. Chem. Phys.*, 2012, **14**, 5729.

7. Y. Hovav, O. Levikron, A. Feintuch, and S. Vega, *Appl. Magn. Reson.*, 2012, **43**, 21.

8. A. Karabanov, A. van der Drift, L. J. Edwards, I. Kuprov, and W. Köckenberger, *Phys. Chem. Chem. Phys.*, 2012, **41**, 2658.

9. A. Karabanov, G. Kwiatkowski, and W. Köckenberger, *Appl. Magn. Reson.*, 2012, **43**, 43.

10. A. A. Smith, B. Corzilius, A. B. Barnes, T. Maly, and R. G. Griffin, *J. Chem. Phys.*, 2012, **136**, 015101.

11. T. Maly, G. T. Debelouchina, V. S. Bajaj, K. N. Hu, C. G. Joo, M. L. Mak-Jurkauskas, J. R. Sirigiri, P. C. A. van der Wel, J. Herzfeld, R. J. Temkin, and R. G. Griffin, *J. Chem. Phys.*, 2008, **128**, 052211.

12. C.-G. Joo, K.-N. Hu, J. A. Bryant, and R. G. Griffin, *J. Am. Chem. Soc.*, 2006, **128**, 9428.

13. F. Kurdzesau, B. van den Brandt, A. Comment, P. Hautle, S. Jannin, J. J. van der Klink, and J. A. Konter, *J. Phys. D*, 2008, **41**, 155506.

14. C. Luchinat and G. Parigi, *Appl. Magn. Reson.*, 2008, **34**, 379.

15. P. Miville, S. Jannin, and G. Bodenhausen, *J. Magn. Reson.*, 2011, **210**, 137.

16. K. Golman, R. In't Zandt, and M. Thaning, *Proc. Natl. Acad. Sci.*, 2006, **100**, 11270.

17. L. E. DeLong, O. G. Symko, and J. C. Wheatley, *Rev. Sci. Instr.*, 1971, **42**, 147.

18. E. T. Swartz, *Rev. Sci. Instr.*, 1986, **57**, 2848.

19. J. Wolber, F. Ellner, B. Fridlund, A. Gram, H. Johannesson, G. Hansson, L. H. Hansson, M. H. Lerche, S. Mansson, R. Servin, M. Thaning, K. Golman, and J. H. Ardenkjaer-Larsen, *Nucl. Instrum. Methods Phys. Res. Sect. A*, 2004, **526**, 173.

20. R. Panek, J. Leggett, N. S. Andersen, J. Granwehr, A. J. Perez Linde, and W. Köckenberger, in Magnetic Resonance Microscopy, eds S. L. Codd and J. D. Seymour, Wiley-VCH Verlag GmbH: Weinheim, 2009.

21. A. Comment, B. van der Brandt, K. Uffmann, F. Kurdzesau, S. Jannin, J. A. Konter, P. Hautle, W. T. Wenckebach, R. Gruetter, and J. J. van der Klink, *Concepts. Magn. Reson.*, 2007, **31B**, 255.

22. M. Batel, M. Krajewski, K. Weiss, O. With, A. Dapp, A. Hunkeler, M. Gimersky, K. P. Pruessmann,

P. Boesiger, B. H. Meier, S. Kozerke, and M. Ernst, *J. Magn. Reson.*, 2012, **214**, 166.

23. Y. Crémillieux, F. Goutailler, B. Moncel, D. Grant, G. Vermeulen, and P.-E. Wolf, *Appl. Magn. Reson.*, 2012, **43**, 167.

24. J. Granwehr, R. Panek, J. Leggett, and W. Köckenberger, *J. Chem. Phys.*, 2010, **132**, 244507.

25. S. Bowen and C. Hilty, *Phys. Chem. Chem. Phys.*, 2010, **12**, 5766.

26. P. Miéville, P. Ahuja, R. Sarkar, S. Jannin, P. R. Vasos, S. Gerber-Lemaire, M. Mishkovsky, A. Comment, R. Gruetter, O. Ouari, P. Tardo, and G. Bodenhausen, *Angew. Chem. Int. Ed.*, 2010, **49**, 6182.

27. E. T. Peterson, J. W. Gordon, M. G. Erickson, S. B. Fain, and I. J. Rowland, *J. Magn. Reson. Imaging*, 2011, **33**, 1003.

28. T. Harris, C. Bretschneider, and L. Frydman, *J. Magn. Reson.*, 2011, **211**, 96.

29. J. Leggett, R. Hunter, J. Granwehr, R. Panek, A.J. Perez-Linde, A.J. Horsewill, J. McMaster, G. Smith, and W. Köckenberger, *Phys. Chem. Chem. Phys.*, 2010, **12**, 5883.

30. www.oxford-instruments.com/products/spectrometers/ nuclear-magnetic-resonance-(nmr)/hypersense (accessed Dec 2012)

31. J. H. Ardenkjaer-Larsen, A. M. Leach, N. Clarke, J. Urbahn, D. Anderson, and T. W. Skloss, *NMR Biomed.*, 2011, **24**, 927.

32. J. H. Ardenkjaer-Larsen, I. Laursen, I. Leunbach, G. Elnholm, L.-G. Wistrand, J. S. Petersson, and K. Golman, *J. Magn. Reson.*, 1998, **133**, 1.

33. M. K. Bowman, C. Mailer, and H. J. Halpern, *J. Magn. Reson.*, 2005, **172**, 254.

34. A. Comment, J. Rentsch, F. Kurdzesau, S. Jannin, K. Uffmann, R. B. van Heeswijk, P. Hautle, J. A. Konter, B. van Brandt, and J. J. van der Klink, *J. Magn. Reson.*, 2008, **194**, 152.

35. P. Ahuja, R. Sarkar, S. Jannin, P. R. Vasos, and G. Bodenhausen, *Chem. Commun.*, 2010, **46**, 8192.

36. S. Jannin, A. Bornet, S. Colombo, and G. Bodenhausen, *Chem. Phys. Lett.*, 2011, **517**, 234.

37. A. Bornet, R. Melzi, S. Jannin, and G. Bodenhausen, *Appl. Magn. Reson.*, 2012, **43**, 107.

38. D. A. Hall, D. C. Maus, G. J. Gerfen, S. J. Inati, L. R. Becerra, F. W. Dahlquist, and R. G. Griffin, *Science*, 1997, **276**, 930.

39. A.J. Linde-Perez, Application of cross polarization techniques to dynamic nuclear polarization dissolution experiments, PhD ethesis, University of Nottingham ethesis, 2010.

40. M. Batel, M. Krajewski, A. Däpp, A. Hunkeler, B. H. Meier, S. Kozerke, and M. Ernst, *Chem. Phys. Lett.*, 2012, **554**, 72.

41. C. Ludwig, I. Marin-Montesinos, M. G. Saunders, and U. L. Günther, *J. Am. Chem. Soc.*, 2009, **132**, 2508.

42. L. Lumata, Z. Kovacs, C. Malloy, A. D. Sherry, and M. Merritt, *Phys. Med. Biol.*, 2011, **56**, N85.

43. C.-G. Joo, A. Casey, C. J. Turner, and R. G. Griffin, *J. Am. Chem. Soc.*, 2009, **131**, 12.

44. S. Jannin, A. Comment, F. Kurdzesau, J. A. Konter, P. Hautle, B. van den Brandt, and J. J. van der Klink, *J. Chem. Phys.*, 2008, **128**, 241102.

45. H. Johanneson, S. Macholl, and J. H. Ardenkjaer-Larsen, *J. Magn. Reson.*, 2009, **197**, 167.

46. S. Macholl, H. Johannesson, and J. H. Ardenkjaer-Larsen, *Phys. Chem. Chem. Phys.*, 2010, **12**, 5804.

47. A. Comment, S. Jannin, J.-H. Hyacinthe, P. Miéville, R. Sarkar, P. Ahuja, P. R. Vasos, X. Montet, F. Lazeyras, J.-P. Vallée, P. Hautle, J. A. Konter, B. van den Brandt, J.-Ph. Ansermet, R. Grütter, and G. Bodenhausen, *Phys. Rev. Lett.*, 2010, **105**, 018104.

48. N. N. Kuzma, M. Pourfathi, H. Kara, P. Manasseh, R. K. Gosh, J. H. Ardenkjaer, S. J. Kadlecek, and R. R. Rizi, *J. Chem. Phys.*, 2012, **137**, 104508.

49. P. R. Vasos, A. Comment, R. Sarkar, P. Ahuja, S. Jannin, J.-P. Ansermet, J. A. Konter, P. Hautle, B. van den Brandt, and G. Bodenhausen, *Proc. Natl. Acad. Sci. U. S. A.*, 2009, **106**, 18469.

50. M. Carravetta, O. G. Johannessen, and M. H. Levitt, *Phys. Rev. Lett.*, 2004, **92**, 153003.

51. M. Carravetta and M. H. Levitt, *J. Am. Chem. Soc.*, 2004, **126**, 6228.

52. M. Carravetta and M. H. Levitt, *J. Chem. Phys.*, 2005, **122**, 214505.

53. G. Pileio and M. H. Levitt, *J. Chem. Phys.*, 2009, **130**, 214501.

54. E. Vinogradov and A. K. Grant, *J. Magn. Reson.*, 2007, **188**, 176.

55. M. H. Levitt, *Ann. Rev. Phys. Chem.*, 2012, **63**, 89.

56. M. C. D. Tayler, I. Marco-Rius, M. I. Kettunen, K. M. Brindle, M. H. Levitt, and G. Pileio, *J. Am. Chem. Soc.*, 2012, **134**, 7668.

57. C. Laustsen, G. Pileio, M. C. D. Tayler, L. J. Brown, R. C. D. Brown, M. H. Levitt, and J. H. Ardenkjaer-Larsen, *Magn. Reson. Med.*, 2012, **68**, 1262.

58. H. Y. Chen, Y. Lee, S. Bowen, and C. Hilty, *J. Magn. Reson.*, 2011, **208**, 204.

59. M. Mishkovsky, U. Eliav, G. Navon, and L. Frydman, *J. Magn. Reson.*, 2009, **200**, 142.

60. H. Johanneson, S. Macholl, and J. H. Ardenkjaer-Larsen, *J. Magn. Reson.*, 2009, **197**, 167.

61. K. Golman, R. In't Zandt, M. Lerche, R. Pehrson, and J. H. Ardenkjaer-Larsen, *Cancer Res.*, 2006, **66**, 10855.

62. M. J. Albers, R. Bok, A. P. Chen, C. H. Cunningham, M. L. Zierhut, V. Y. Zhang, S. J. Kohler, J. Tropp, R. E. Hurd, Y. F. Yen, S. J. Nelson, D. B. Vigneron, and J. Kurhanewicz, *Cancer Res.*, 2008, **68**, 8607.

63. S. E. Day, M. I. Kettunen, F. A. Gallagher, D.-E. Hu, M. Lerche, J. Wolber, K. Golman, J. H. Ardenkjaer-Larsen, and K. M. Brindle, *Nat. Med.*, 2007, **13**, 1382.

64. K. Golman, J. S. Petersson, P. Magnusson, E. Johansson, P. Akeson, C. M. Chai, G. Hansson, and S. Mansson, *Magn. Reson. Med.*, 2008, **59**, 1005.

65. K. Weiss, E. Mariotti, D.K. Hill, M.R. Orton, J.T. Dunn, R.A. Medina, R. Southworth, S. Kozerke, and T.R. Eykyn. *Appl. Magn. Reson.*, 2012, **43**, 275.

66. S. Hu, M. Lustig, A. Balakrishnan, P. E. Z. Larson, R. Bok, J. Kurhanewicz, S. J. Nelson, A. Goga, J. M. Pauly, and D. B. Vigneron, *Magn. Reson. Med.*, 2010, **63**, 312.

67. P. E. Z. Larson, S. Hu, M. Lustig, A. B. Kerr, S. J. Nelson, J. Kurhanewcz, J. M. Pauly, and D. B. Vigneron, *Magn. Reson. Med.*, 2011, **65**, 610.

68. M. Karlsson, P. R. Jensen, J. O. Duus, S. Meier, and M. H. Lerche, *Appl. Magn. Reson.*, 2012, **43**, 223.

69. F. A. Gallagher, M. I. Kettunen, D. E. Hu, P. R. Jensen, R. In't Zandt, M. Karlsson, A. Gisselsson, S. K. Nelson, T. H. Witney, S. E. Bohndiek, G. Hansson, T. Peitersen, M. H. Lerche, and K. M. Brindle, *Proc. Natl. Acad. Sci. U. S. A.*, 2009, **106**, 19801.

70. R. Sarkar, A. Comment, P. R. Vasos, S. Jannin, R. Gruetter, G. Bodenhausen, H. Hall, D. Kirik, and V. P. Denisov, *J. Am. Chem. Soc.*, 2009, **131**, 16014.

71. F. A. Gallagher, M. I. Kettunen, S. E. Day, D.-E. Hu, J. H. Ardenkjaer-Larsen, R. In't Zandt, P. R. Jensen,

M. Karlsson, K. Golman, M. H. Lerche, and K. M. Brindle, *Nature*, 2008, **453**, 940.

72. L. Lumata, A. K. Jindal, M. E. Merritt, C. R. Malloy, A. D. Sherry, and Z. Kovscs, *J. Am. Chem. Soc.*, 2011, **133**, 8673.

73. A. K. Jindal, M. E. Merritt, E. H. Suh, C. R. Malloy, A. D. Sherry, and Z. Kovacs, *J. Am. Chem. Soc.*, 2010, **132**, 1784.

74. P. Mieville, S. Jannin, L. Helm, and G. Bodenhausen, *J. Am. Chem. Soc.*, 2010, **132**, 5006.

75. R. B. van Heeswijk, K. Uffmann, A. Comment, F. Kurdzesau, C. Perazzolo, C. Cudalbu, S. Jannin, J. A. Konter, P. Hautle, B. van den Brandt, G. Navon, J. J. van der Klink, and R. Gruette, *Magn. Reson. Med.*, 2009, **61**, 1489.

76. L. Frydman, T. Scherf, and A. Lupulescu, *Proc. Natl. Acad. Sci. U. S. A.*, 2002, **99**, 15858.

77. L. Frydman and D. Blazina, *Nat. Phys.*, 2007, **13**, 415.

78. M. Mishkovsky and L. Frydman, *Chem. Phys. Chem.*, 2008, **9**, 2340.

79. H. Zeng, S. Bowen, and C. Hilty, *J. Magn. Reson.*, 2009, **199**, 159.

80. K. J. Donovan and L. Frydman, *J. Magn. Reson.*, 2012, **225**, 115.

81. S. J. Kohler, Y. Yen, J. Wolber, A. P. Chen, M. J. Albers, R. Bok, V. Zhang, J. Tropp, S. Nelson, D. B. Vigneron, J. Kurhanewicz, and R. E. Hurd, *Magn. Reson. Med.*, 2007, **58**, 65.

82. R. Panek, J. Granwehr, J. Leggett, and W. Köckenberger, *Phys. Chem. Chem. Phys.*, 2010, **12**, 5771.

83. T. Harris, G. Eliyahu, L. Frydman, and H. Degani, *Proc. Natl. Acad. Sci. U. S. A.*, 2009, **106**, 18131.

84. M. E. Merritt, C. Harrison, C. Storey, F. M. Jeffrey, A. D. Sherry, and C. R. Malloy, *Proc. Natl. Acad. Sci. U. S. A.*, 2007, **104**, 19773.

85. M. H. Lerche, S. Meier, P. R. Jensen, H. Baumann, B. O. Petersen, M. Karlsson, J. O. Duus, and J. H. Ardenkjaer-Larsen, *J. Magn. Reson.*, 2010, **203**, 52.

86. Y. Lee, H. Zeng, A. Mazur, M. Wegstroth, T. Carlomagno, M. Reese, D. Lee, S. Becker, C. Griesinger, and C. Hilty, *Angew. Chem. Int. Ed.*, 2012, **51**, 5179.

87. Y. Lee, H. Zeng, S. Ruedisser, A. D. Gossert, and C. Hilty, *J. Am. Chem. Soc.*, 2012, **134**, 17448.

Index